Deformation Mechanisms, Rheology and Tectonics

Geological Society Special Publications
Series Editor J. BROOKS

THIS VOLUME IS DEDICATED TO THE
WORK OF HENDRIK JAN ZWART

GEOLOGICAL SOCIETY SPECIAL PUBLICATION NO. 54

Deformation Mechanisms, Rheology and Tectonics

EDITED BY

R. J. KNIPE
Department of Earth Sciences
Leeds University
UK

&

E. H. RUTTER
Department of Geology
Manchester University
UK

ASSISTED BY

S. M. AGAR
Department of Earth Sciences
Leeds University
UK

R. D. LAW
Department of Geological Sciences
Virginia University
USA

D. J. PRIOR
Department of Earth Sciences
Liverpool University
UK

R. L. M. VISSERS
Institute of Earth Sciences
University of Utrceht
Netherlands

1990

Published by

The Geological Society

London

THE GEOLOGICAL SOCIETY

The Geological Society of London was founded in 1807 for the purposes of 'investigating the mineral structures of the earth'. It received its Royal Charter in 1825. The Society promotes all aspects of geological science by means of meetings, special lectures and courses, discussions, specialist groups, publications and library services.

It is expected that candidates for Fellowship will be graduates in geology or another earth science, or have equivalent qualifications or experience. All Fellows are entitled to receive for their subscription one of the Society's three journals: *The Quarterly Journal of Engineering Geology*, the *Journal of the Geological Society* or *Marine and Petroleum Geology*. On payment of an additional sum on the annual subscription, members may obtain copies of another journal.

Membership of the specialist groups is open to all Fellows without additional charge. Enquiries concerning Fellowship of the Society and membership of the specialist groups should be directed to the Executive Secretary, The Geological Society, Burlington House, Piccadilly, London W1V 0JU.

Published by the Geological Society from:

The Geological Society Publishing House
Unit 7
Brassmill Enterprise Centre
Brassmill Lane
Bath
Avon BA1 3JN
UK
(*Orders*: Tel. 0225 445046)

Distributor USA:
AAPG Bookstore
PO Box 979
Tulsa
Oklahoma 74101-0979
USA
(*Orders*: Tel: (918) 584-2555)

First published 1990

British Library Cataloguing in Publication Data
Deformation mechanisms, rheology and tectonics.
1. Rocks. Mechanics
I. Knipe, R. J. (Robert John) *1952–* II. Rutter, E. H. (Ernest Henry) *1946–* III. Series 624.15132

ISBN 0–903317–58–3

Printed in Great Britain at the Alden Press, Oxford

Contents

Flow mechanisms and flow laws

Rock fabrics

Deformation of weak sediments

Experimental modelling using analogue materials

Preface

It is my great pleasure to introduce this volume dedicated to my long-standing colleague and friend, Henk Zwart; the more so since I missed the opportunity of addressing him on the occasion of his retirement from the chair of structural and applied geology in the State University at Utrecht in 1988.

Hendrik Jan Zwart enrolled as a student of geology in the State University at Leiden in 1946. In my welcome speech as chairman of the student's geology club at that time I included him among the 'rolling stones of the post-war basal conglomerate' of enrollment, and I must admit that he has never gathered any moss!

Although initially absorbed by Quaternary geology, Henk opted for structural geology at a later stage taking his MSc and PhD with L. U. de Sitter as supervisor. At that time the stamping ground of de Sitter was located in the soft-rock sector of the central Pyrenees, but Henk managed to persuade de Sitter (no mean task anyway!) to let him study the crystalline basement of the Saint Barthélemy Massif for his dissertation.

Armed with the structural and petrographic experience gained from this study of first honours degree quality, he began to apply microstructural techniques to the study of dynamothermal rock sequences in the central Pyrenees. With great insight, enthusiasm and perseverance he established himself as a pioneer in the meso- and microstructural fields, proposing clear criteria and a useful reference matrix (popularly known as Zwart's chicken pen) from which the thermal and dynamic history of a metamorphic orogenic belt can be deduced.

After three years of tenure of the chair of geology in the University of Aarhus, where the Caledonides of Sweden came within his compass, Henk was invited to return to Leiden in 1969 to succeed de Sitter in the chair of structural and applied geology. With his collaborators and students he set out to compare the structural and metamorphic histories of the Pyrenees, the Caledonides and the Alps leading to his concept of the dualism of orogenic belts. Close collaboration with British and Australian workers in the microstructural field, some of whom were persuaded to join his staff, led to the establishment of a sort of Anglo-Dutch school. Moreover, much credit is to be given him for building up a first-rate structural laboratory, first at Leiden and later in Utrecht as well.

Enterprising as he is, Henk took the initiative to organize an international seminar on 'Fabrics, microstructures and microtectonics' that was held near Leiden in 1976, thus paving the way for biennial meetings on similar topics which have created a close fraternity and a fruitful integration of a broad gamut of structural disciplines. The Leeds Meeting on Deformation Mechanisms, Rheology and Tectonics is the latest in this series.

His international reputation gained even more from the almost Sisyphean task of editorship of the Metamorphic Map of the World, carried out at the request of the IUGS between 1966 and 1974. It was not surprising that he was then asked to occupy a number of important offices in the framework of IUGS: as secretary and later chairman of the Commission on Tectonics, as member of several working groups of the Interunion Commission on Geodynamics, and lately in the Interunion Commission on the Lithosphere of which he is at present the Secretary-General.

These merits of Henk Zwart formed the basis for his election as Member of the Royal Netherlands Academy of Sciences and Arts, and for the award of the prestigious van Waterschoot van der Gracht Medal by the Royal Netherlands Geological and Mining Society.

In conclusion I should like to mention that Henk's personality is greatly appreciated by his colleagues, co-workers and students alike. I wish Henk every conceivable pleasure and satisfaction in his years of retirement.

Emile den Tex

Foreword

This volume is a collection of 47 original research and review papers on the theme of Deformation Mechanisms, Rheology and Tectonics. It is dedicated to Prof. Henk Zwart, on the occasion of his 65th birthday, in recognition of his own personal contributions to this subject area and of the stimulus he has provided to its development, particularly in Europe. An important part of this stimulus was the first international conference on tectonics and microstructures which he organized in Leiden in 1976, and which was the forerunner of a highly successful series held at various European centres since then. Most of the papers collected here were presented at the latest of such meetings, held at Leeds University in March 1989.

The papers are gathered into groups which are aimed at reflecting current research themes, ranging from geologically-oriented rock mechanics, through structural and microstructural studies of naturally-deformed rock masses to large-scale tectonics. In some cases thematic groups contain a 'keynote' paper, containing a substantial review component. The thematic grouping adopted for the present volume has, of course, depended on the nature of the papers submitted, so that not all of the sections contain review papers. Such papers will present an introductory framework for those new to the subject but the volume is dominated by original research papers.

To some extent the grouping of papers into thematic sets is arbitrary, because common threads often link the various sections. One of these is the role of water in rock deformation. The first section is headed 'Role of fluids in rock deformation', but various aspects of the role of water recur in later sections. **Carter *et al.*** provide a concise yet comprehensive review of the topic. The following paper by **Sibson** discusses the idea that a fault may periodically allow the drainage of a volume of high-pressure water which it intersects, following a seismic event: the 'fault-valve' hypothesis. Finally, **Evans** describes the role of water in the chemical alteration of granitic rocks, in turn modifying their deformability.

The following section deals with aspects of rock fracture and faulting. Contributions range from observational (**Agar; Hippler & Knipe; Main *et al.*; Zulauf *et al.*; Stewart & Hancock**), through mechanical experiments (**Cox & Wong**) to attempts at modelling aspects of brittle rock failure in compression (**Casey & Wust; Cox & Paterson; Wong**). **Agar** describes the temporal evolution of fracturing, hydrothermal alteration and mineralization of upper oceanic crustal rocks from a DSDP borehole. **Hippler & Knipe** describe the microstructural modification of a pre-existing plastic mylonite during a superposed event causing cataclastic granulation. **Main *et al.*** review theoretical and experimental studies of the distribution of flaw sizes in rock bodies, from microcracks to lithospheric faults. These commonly exhibit scale invariance over a range of length scales, each characterized by a particular fractal dimension. The geometric configuration of flaws is expected to be important in the failure behaviour, and the authors develop a fracture mechanics model for the temporal variations of the fractal dimension. **Zulauf *et al.*** describe the microstructures of cataclastically deformed rocks from the KTB-VB pilot research borehole, which is close to the site of the projected German superdeep hole. Finally, **Stewart & Hancock** describe the characteristics of brecciation and damage in the near-surface region during the upwards penetration of extensional fault tips in the Aegean region.

Until recently, experimental studies of rock friction have been limited by attainable shear displacements of only a few millimetres. Rotary shear experiments are important way of overcoming this problem, but another approach is described by **Cox** in this volume. He uses a large direct-shear configuration in which large effective displacements are accumulated through repeated reversals of the direction of motion. He uses the state-variable friction-law approach to describe the results of velocity-stepping experiments, and finds that a friction law with two state variables is required to describe the experimental results.

Increasing amounts of effort are being applied to the problem of the theoretical description and modelling of the accumulation of damage (in the form of oriented arrays of extensional cracks) during loading of a brittle rock through failure. Related to this is the problem of instability, and whether the failure remains distributed (macroscopically ductile) or becomes localized into bands or faults. Three papers in this volume address the former question. **Casey & Wust** use a finite element approach to estimate the anisotropic elastic properties of rocks containing a regular array of oriented cracks. They show that a concentration of cracks in an inclined planar zone could result in an instability leading to shear faulting. **Cox & Paterson** take a similar approach to examine the distribution of microcrack damage in a material modelled as a collection of deformable elements with a specified distribution of strength and elasticity. Finally, **Wong** describes experimental results on pore collapse in sandstones of a range of porosities, and uses these to help constrain fracture mechanics modelling of the process. He shows how the critical pressure for pore collapse depends on porosity and grain-size, and how volume reduction through pore collapse counteracts the destabilizing effects of dilatancy in the brittle deformation of low porosity rocks, favouring a degree of stable, cataclastic flow in porous rocks.

The mechanics of the development of instabilities during the deformation rocks, manifesting themselves as localized faults or shear zones, is receiving increasing amounts of attention, and it is appropriate that a section be devoted to a group of papers addressing this problem. A keynote contribution from **Hobbs *et al.*** attempts to apply a criterion of instability developed by materials scientists to rocks. Clarifying misconceptions which have crept into the geological literature, they show how zones of localized flow can develop even when the material within the zone does not suffer any change in rheological characteristics. Developing the same approach, **Ord** presents numerical models which display periodically-spaced localization of flow into shear bands, and discusses their significance for the focussing of fluid flow when the bands are dilatant.

There is a wide range of processes which modify rock properties and which can therefore favour localization of deformation. Some of these are illustrated by other contributions within the section. **Olgaard** shows how second-phase particles, through their inhibiting effect on grain-growth in rocks otherwise sufficiently fine to permit grain-size sensitive flow, can lead to *petrographic* controls on flow localization. **Klaper** describes an example of how softening associated with a *metamorphic reaction* appears to have led to flow localization. On a larger scale, using numerical modelling, **Davies & Fletcher** address the problem of localization of periodically spaced ramps in a duplex. **Green & Burnley** describe new high pressure/temperature experiments on Mg-germanate olivine which suggest that the instability which leads to deep-focus earthquakes might arise from the weakness of transiently ultrafine-grained reaction products of the olivine-spinel transformation. Important new microstructural observations suggest that the instability may depend in some way on the coalescence into a fault plane of ellipsoidal 'anticracks', oriented normal to the compression direction. Although they draw analogy with the coalescence of axially-oriented cracks into a low-temperature brittle fault, further studies will be required to discover exactly how a discrete fault zone develops in this case. There is no doubt, however, that this will be a fruitful new line of research.

The formation of stylolites represents a particular kind of localization phenomenon in the flow of rocks by pressure solution. The paper by **Carrio-Schaffhauser *et al.*** describes a new and perhaps exotic way to examine them. They used X-ray tomography to construct a section through a stylolite, revealing the porosity structure ahead and on either side of the stylolite. Compared to the bulk rock remote from the stylolite the advance of the stylolite tip causes porosity increase, followed by porosity decrease as the stylolite tip passes through the 'process' zone.

Papers on intracrystalline-plastic and diffusional flow processes and the constitutive laws used to describe them are assembled in the following section. The rock deformation group at Utrecht have in recent years examined many aspects of the flow of halite rocks, and the first two papers (**Franssen & Spiers; Spiers *et al.***) deal with this material. The former considers the implications of mechanical behaviour for the rate controlling process in intracrystalline plasticity. The latter attempts to draw together and synthesize the considerable body of data they have assembled on flow and compaction by pressure solution of halite. Recognizing that in nature halite shows evidence of flow both by pressure solution and intracrystalline plasticity, they attempt to assess the natural conditions under which the mechanism transition occurs.

The concept of 'superplasticity' applied to rock deformation has long been contentious. Whilst there is an argument for a purely geometric definition (the capacity for an 'extreme' degree of ductile strain, but without the deformation mechanism being specified), there is compelling reason in this case for associating the term with diffusion- or plasticity-accommodated grain-boundary sliding. In their paper, **Gilotti & Hull** express the former view. However, there was strong dissent amongst the referees of this paper, all three of whom raised the point that during rock deformation under entirely compressive stresses several mechanisms are capable of accommodating extreme strains. Readers are left to form their own opinions, or perhaps to doubt the advisability of using the term at all in the Earth sciences.

The following two papers are on themes related to the superplasticity issue. **Burkhard** describes the microstructural characteristics of highly deformed micritic limestones from the Alps and interprets them in terms of flow by grain-boundary sliding. **Walker *et al.*** describe a sequence of high temperature experiments on synthetic, hot-pressed calcite rocks of controlled fine grain-sizes. These demonstrate grain-size sensitive flow of calcite rocks by grain-boundary sliding and the transition to relatively grain-size insensitive plastic flow with increasing grain-size.

Retaining the theme of calcite deformation, **de Bresser & Spiers** describe high-temperature experiments on calcite, in which they find that r^+ and f^+ slip in a single crystal is characterized by much more non-linear flow than in a polycrystal at the same temperature.

Understanding the apparent weakening role of water in the plastic flow of quartz has proved remarkably difficult, despite a quarter-century of effort. **Paterson & Luan** review the 'state of the art', to which **Paterson** and his coworkers have made seminal contributions. They argue that except for vacuum dried sample materials, the flow law parameters, stress exponent and activation enthalpy, are not significantly different in most of the experimental programs which have been run, and that water activity affects mainly the pre-exponential 'constant' in the flow law. They suggest a 'representative' quartz rheology and argue that apparent stress exponents of less than 4 in the flow law may be 'unnatural' owing to undesirable contributions (in experiments) from cracking and partial melting. In a second paper dealing with quartz deformation, **Prior *et al.*** use variations in recrystallized grain-size around rigid inclusions, together with stratigraphic constraints on strain-rates, to infer rates of microstructural equilibration in mylonites. The final two papers in the section are concerned with transmission electron microscopic studies of dislocation microstructures in hornblende (**Skrotzki**) and albite (**White**) in mylonitic rocks.

Five papers follow, dealing with aspects of measurement and interpretation of crystallographic fabrics in deformed rocks. **Law** provides a review of some of the recent work in this field, emphasising the ways in which petrofabric studies can be used to help solve problems in tectonics (an aim which this conference and proceedings was intended to stimulate). By means of numerical simulations, **Jessel & Lister** attempt to predict quartz fabrics that might arise from the combined operation of intracrystalline slip and 'selective' dynamic recrystallization, recognizing that the fabric types might change with temperature, owing to changes in relative critical resolved shear stresses for slip on different systems and the facilitation of recrystallization with increasing temperature. **Ree** effectively deals with the same problem, but by means of high homologous temperature deformation experiments using octachloropropane as a mineral analogue. In these experiments dynamic grain-growth makes a major contribution to the final microstructure. **Olesen & Schmidt** provide an example of the utility of electron channelling in the scanning microscope for petrofabric studies. They show that Dauphiné twinning in quartz can be detected by electron channelling but that Brazil twins cannot. Finally, **Schaeben *et al.*** provide an illustrated account of a method of calculating the orientation distribution function from pole figure data, subject to the maximization of an entropy-like function.

Going hand-in-hand with the exploration of submarine accretionary complexes by drilling and seismic profiling is a growing appreciation of the rock types and their mechanical properties in such complexes. The following section, entitled 'Deformation of Weak Sediments' commences with an assessment (**Karig**) of the behaviour of the zone of active accretion at prism toes, taking into account their geometric characteristics and the growing body of experimental data on the mechanics of weak, water-saturated rocks. **Yassir** describes some new experimental data of this type on the undrained behaviour of a group of mud-volcano derived silty clays.

The structural and microstructural study of eroded ancient accretionary and related rock types also provides information essential to the evolution of a model for this environment of natural rock deformation. The final two papers of this section provide examples of this type of study. **Nell** infers deformation mechanisms from the structures and microstructures of melange belts on Alexander Island, Antarctica, and **Pickering *et al.*** examine the relationships between veins and pore fluids during the deformation of Miocene volcaniclastic rocks of SE Japan.

In the following section, on analogue modelling of the development of large-scale structures, **Liu & Dixon** describe centrifuge models which show how duplex structures throughout a wedge-shaped 'thrust-belt' can be continuously activated in order to maintain the equilibrium geometry of the wedge. **McClay** describes a suite of scaled models which demonstrate stages in the development of extensional fault systems and the way in which the underlying detachment configuration controls the cover fault template development. In a third paper using analogue modelling methods, **Wilson & Will** demonstrate the formation of slickenline features on a fault surface when the wall-rocks are themselves non-rigid.

The volume is concluded with a section which we have entitled 'Deformation mechanisms and tectonics'. In this we have grouped contributions which attempt to use structural and microstructural observations to help constrain interpretations of large-scale tectonic processes. **Gratier & Gamond** discuss the range of mid-crustal processes which can occur under conditions commonly equated with the 'brittle-plastic' transition, and which complicate the oversimplified interpretations commonly made of the depth limit of shallow seismicity. **Coli & Sani** and **Sani** describe occurrences from two different parts of the northern Apennines (Italy) of relationships between vein formation and overthrusting. **Holdsworth & Grant** attempt to relate 'anomalous' shear sense observations in the Moine nappe of N Scotland to dynamic spreading of the orogenic wedge. **Iliffe & Lerche** describe numerical modelling experiments which attempt to predict zones of sedimentary compaction associated with rotation of fault blocks. Finally, **Urai *et al.*** describe the microstructures produced during Alpine deformation and metamorphism on Naxos (Greece), and show how shear sense indicators most consistently indicate extensional unroofing of the higher-grade metamorphic complex through movement of the upper plate towards the north.

In conclusion, we would like to express our appreciation to all those who have been instrumental in bringing to fruition the conference and the present volume of proceedings. These are the members of the organizing committee, the referees of the 65 submitted papers (listed below), staff and graduate students of Leeds University Earth Science department, staff of the Geological Society Publishing House and the meeting sponsors (listed below) and exhibitors whose financial contributions helped defray costs.

E. H. Rutter & R. J. Knipe
March 1990

Referees

S. Agar
A. Barker
J. Behrmann
M. Blanpied
T. Blenkinsop
J. Boland
E. Bombalakis
K. Brodie
S. Burley
M. Casey
P. Cobbold
D. Cowan
M. Coward
S. Cox
D. Davis
D. Dietrich
M. Drury
C. Ellis
B. Evans
J. Evans
D. Fisher
W. Fitches
R. Fletcher
J. Gammond
A. Gibbs
J. P. Gratier
H. Green II
P. Hancock
S. Hencher
S. Hippler
B. Hobbs
R. Holdsworth
P. Huddleson
M. Jessell
P. Jordan
D. Karig
E. Klaper
R. Knipe
S. Laubach
R. Law
G. Lloyd
J. Logan
D. Mainprice
A. Maltman
A. McCaig
W. Means
J. C. Moore
P. Nell
G. Norrell
M. Norton
N. Oleson
D. Olgaard
A. Ord
M. Paterson
K. Pickering
J. Platt
W. Power
G. Price
D. Prior
S. Reddy
J. Ridley
D. Rubie
E. Rutter
S. Schmid
R. Sibson
B. Smith
C. Spiers
D. Spratt
C. Talbot
J. Urai
I. van der Molen
H. van Roermund
R. Vissers
J. Watterson
J. Wheeler
J. White
P. Williams
T-F. Wong
N. Yassir

Meeting sponsors

Department of Earth Sciences, University of Leeds
The Geological Society of London
Cambridge Scanning Co Ltd.
Gatan Ltd.
Wild Leitz
Link Systems Ltd.
Logitech Ltd.
Philips (Pye Unicam Ltd.)
Blackwell Scientific Publications
Pergamon Press plc.
J. Wiley & Sons Ltd.

Control of fluids on deformation of rocks

N. L. CARTER, A. K. KRONENBERG, J. V. ROSS[1] & D. V. WILTSCHKO

Center for Tectonophysics, Texas A & M University, College Station, TX 77843

[1] *On leave from: Deparment of Geological Sciences, University of British Columbia, Vancouver, BC. V6T 2B4 Canada.*

Abstract: Fluids of many compositions, concentrations and pressures, are ubiquitous throughout the continental lithosphere, exerting strong control on the deformation properties and processes of rocks both by mechanical means and by complex chemical rock-fluid interactions. Fluids of meteoric and juvenile origin, released by compaction, dehydration reactions, melting, and degassing, commonly during large-scale tectonic events, flow by means of thermal convection, advection (infiltration), and surface and intracrystalline diffusion. These fluids transport mass for distances ranging from the grain scale to hundreds of kilometres; fracture zones provide favourable conduits for flow. Abnormal pore pressures, recorded at all metamorphic grades, develop intermittently during syntectonic deformation, enhancing fluid infiltration by promoting increased porosity and permeability, hydraulic fracturing and severe grain size reductions. The infiltrating fluids enhance hydrolytic weakening, several grain boundary mechanisms, and reaction kinetics in a feedback manner so that strain is commonly localized into semibrittle and ductile shear zones. Large-scale detachments may take place along these shear zones at virtually any depth below the uppermost few kilometres, which, when combined with softening resulting from depth-dependent petrological and geochemical segregations, form a rheological stratigraphy. The rheology of the lithosphere through time has been governed by a combination of bulk rock flow and localized deformation is shear zones, both of which have been aided or controlled by pervasive dynamic rock-fluid interactions.

The nearly ubiquitous presence of fluids of various types, compositions, concentrations and pressures throughout the lithosphere exert a very strong control on the nature and extent of fracturing, faulting, shear zone development and bulk flow of rocks, from the scale of crystal defects (10^{-10} m) to that of major global plates (10^{7} m). On the lithospheric scale, Fyfe *et al.* (1978) argue that fluids are generally available at depth, and Fyfe & Kerrich (1985) maintain that massive fluid transport must occur at abnormal fluid pressures in regions of large-scale thrusting, such as sites of subduction, collision and thin-skinned tectonics. Extensive stable isotope studies have shown that shear zones can provide conduits for massive fluid movement from various reservoirs (including those at the surface) to depths as great as 25 km (e.g. Labato *et al.* 1983; Kerrich *et al.* 1984; Kerrich 1986; Burkhard & Kerrich 1988; McCaig 1988). Shimamoto (1985) argues for abundant fluid release and restricted regions of abnormal pore pressure during progressive metamorphism to 25 km based upon low seismicity, ductile deformation and inter-plate decoupling in shallow subducting plates. Etheridge *et al.* (1984) suggest that during moderate- to high-grade metamorphism, pore fluid pressures may exceed minimum principal compressive stresses leading to high porosities and permeabilities. Such conditions also lead to natural syntectonic hydrofracturing with compelling evidence recorded at all metamorphic grades (e.g., Ross & Lewis 1989).

Oliver (1986) suggests that continental margins buried beneath thrust sheets expel fluids that are then transported into foreland basins and continental interiors giving rise to faulting, magma generation, metamorphism and migration of hydrocarbons and mineral-bearing fluids. In accord with this postulate, recent studies of Mississippi Valley-type ore deposits as well as the thermal maturation of coal and petroleum have led many workers to conclude that hot fluids have moved hundreds of kilometres from orogenic belts into the craton (e.g., Leach & Rowan 1986; Jackson *et al.* 1985). Thus, there appears to be ample evidence for the widespread availability of fluids during rock deformation on the lithospheric scale; its importance in governing the mechanical response of rocks as well as the nature and occurrence of natural resources cannot be over-emphasized. In the following sections, we attempt to summarize very briefly mechanisms of fluid transport and major effects of fluids on rock

From Knipe, R. J. & Rutter, E. H. (eds), 1990, *Deformation Mechanisms, Rheology and Tectonics*, Geological Society Special Publication No. 54, pp. 1–13.

deformation processes, providing a few examples; space limitations prohibit a more comprehensive review.

Modes of fluid transport

Fluids released by such processes as compaction, dehydration reactions and melting must be transported in order to account for the phenomena outlined above. Palciaukas & Domenico (1989) give a general relation for conservation of fluid mass in a deforming medium:

$$\frac{1}{V}\left(\frac{DV}{Dt}\right) + \frac{1}{\rho_f}\nabla \bullet (\rho_f \mathbf{q}) + \frac{\phi}{\rho_f}\left(\frac{D_f}{Dt}\right) + \frac{1-\phi}{\rho_s}\left(\frac{D\rho_s}{Dt}\right) = 0 \qquad (1)$$

where ϕ is porosity, t is time, ρ is density, $\mathbf{q}$ is the volume flux of fluid, V is the volume of an infinitesimal element and subscripts f and s refer to fluid and solid, respectively. The first term reflects the volume strain rate of the material as changes in stress, fluid pressure, chemistry and temperature are applied. The second term describes the volume fluid flux, $\mathbf{q}$, relative to the solid matrix and is defined in terms of velocities, υ, as

$$\mathbf{q} = \phi(\upsilon_f - \upsilon_s) \qquad (2)$$

The last two terms in (1) reflect changes in fluid and solid density and are regarded as second order in importance. An explicit expression of (1) depends on the micromechanical processes governing the deformation, a discussion of which forms the main focus of this paper.

Thermal convection is an efficient means of fluid transport but for most lithospheric deformation, this mechanism may be restricted to the uppermost few kilometres, (Fyfe *et al.* 1978; McCaig 1988). Another important mechanism of rapid fluid migration, which appears to be required, is advection or infiltration, emphasized by Etheridge *et al.* (1984). Solute transport by this mechanism, enhanced at low effective pressures and hence high porosities and permeabilities, is given by Fletcher & Hoffman (1974):

$$X_A = \frac{\kappa \bullet \overrightarrow{VP}}{\nu \bullet \phi} t \qquad (3)$$

where κ is the permeability, $\overrightarrow{VP}$ is the net pressure gradient, ν is the fluid viscosity, and X_A is distance. As pointed out both by Fletcher & Hoffman (1974) and Etheridge (1984), regional scale solute transport must occur by this mechanism as opposed to solute diffusion along chemical potential gradients within a stationary aqueous solvent (which can transport solutes by distances of < 1 m). Solute diffusion is, however, an important process involved in wide variety of pressure solution mechanisms in polyphase rocks throughout the lithosphere (e.g., Elliott 1973; Fletcher & Hoffman 1974; Fletcher 1982; Rutter 1983; Knipe & Wintsch 1985; Hickman & Evans 1988; Ross & Lewis 1989). Diffusional solute transport may be expressed by:

$$X_D = (2D_f\phi t)^{1/2} \qquad (4)$$

where D_f is the diffusivity in the fluid and ϕ is porosity; this process dominates advective transport when $vl/D_f<1$ where v is velocity and l is length (Dewers & Ortoleva pers. comm.).

Seismic or dilatancy pumping provides still another transport mechanism for fluid movement in fault zones at high levels (Sibson, 1977) and at deeper levels within ductile shear zones (Etheridge *et al.* 1984; Reynolds & Lister 1987; McCaig 1988). This mechanism requires that in situ stress levels be close to the strength of the rock so that significant dilatancy can be introduced prior to failure, drawing in surrounding fluids. Following failure, stress levels drop rapidly, forcing the fluids out as fracture networks close. Generally, the fluids migrate upward but McCaig (1988) has suggested a model whereby they may move downward (Fig. 1), as is required in some areas from $\delta^{18}O$ determinations.

Important questions to be addressed are the volumes and sources of fluids migrating along structural discontinuities and the temporal relation between the discontinuities and major deformation events (Suppe & Witke 1977). Wiltschko *et al.* (1985) and Kilsdonk & Wiltschko (1987) established that, in ramp regions of the Pine Mountain thrust sheet (PMTS), joints, veins and tectonic stylolites have a characteristic orthogonal relationship that may be placed in temporal order. For the northwest ramp of the PMTS, the sequence is as follows: (1) early strike stylolites (bedding and transport normal); (2) transport veins and joints (bedding normal, transport parallel); (3) strike joints; and (4) late faults cross-cutting all other fabrics. They also identified a third set of bedding parallel veins.

Budai & Wiltschko (1987) sought to determine the time of fluid migration relative to the temporal sequence of development of these structures as well as the extent and significance of rock–fluid interactions during vein formation. They found, in the Haystack Peak area within the Absoroka thrust sheet of the Idaho–

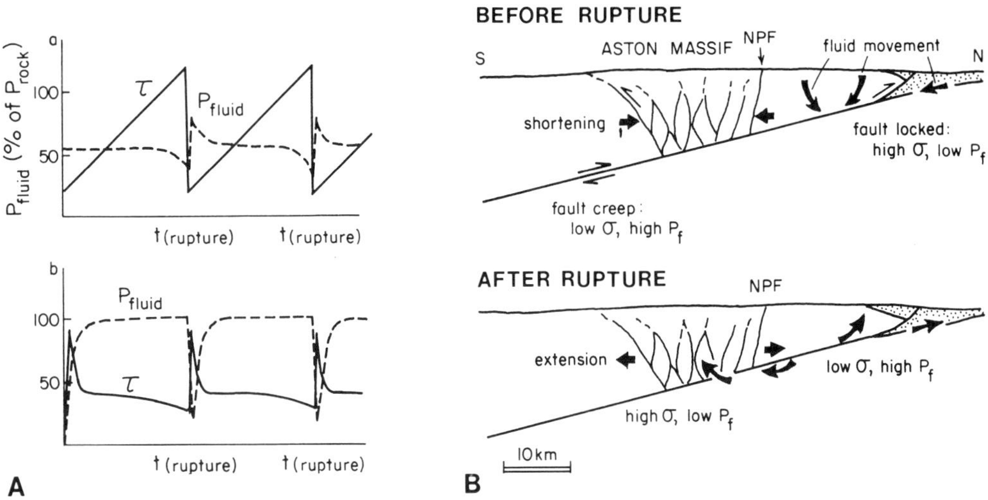

Fig. 1. Seismic or dilatancy pumping model (after McCaig 1988, figs 2 and 3). (**A**) Schematic variation of shear stress (τ) and fluid pressure with time (t) over two rupture cycles. (**B**) Application of model to Pyrenees. Before rupture, high stress leads to low fluid pressures in seismogenic zone; creep on low angle fault in semi-brittle regime leads to tip strains with shortening across shear zone. During and after rupture, fractures propagate downward leading to fluid flow downward then up into shear-zone network, now in tension.

Wyoming–Utah thrust belt, the following temporal sequence: (1) strike stylolites (termed Group I); (2) transport and dip joints (the latter termed Group II); (3) strike, transport and bed-parallel veins (the latter termed Group III); (4) Groups II and III fault surfaces; and (5) Group I fold axes. One of the most significant results of their studies of $\delta^{18}O$ and $\delta^{13}C$ compositions in the carbonate veins and adjacent host rock is shown in Fig. 2. Transport (Group II) veins (cluster of poles in SE quadrant of Fig. 2A) are best developed and show a moderate deviation of $\delta^{13}C$ values from the host rock and a very large difference in $\delta^{18}O$ composition (Fig. 2B). Strike (Group I) veins show a similar pattern. By contrast, bed-parallel veins (Group III in Fig. 2C) show very little difference in isotopic compositions from the adjacent host rock (Fig. 2D). Based on these observations and other lines of evidence, Budai and Wiltschko (1987) conclude that Group I and II veins record the presence of isotopically light and perhaps hot fluids migrating during the early stages of thrusting. Group III veins, believed to have formed later, may reflect either cooling of vein fluids and equilibration with the Madison reservoir near the end of thrusting or increased incorporation of granulated host rock during bedding-parallel slip. Other workers have demonstrated mixing of meteoric and basin brines during thrusting (e.g., Rye & Bradbury 1988).

Effects of fluids on rock deformation

Purely mechanical effects of fluids on rock deformation, through variations in effective pressure with depth and tectonic regime, are reasonably well-understood (e.g., Hubbert & Rubey 1959; Handin *et al.* 1963). Strongly coupled and equally important are chemical effects on deformation, effects that are generally not understood so well because of their manifold and complex nature. These coupled effects are in some instances cooperative and in others competitive in governing brittle versus ductile mechanical responses. In general, however, the overall behaviour of the lithosphere may be regarded as transitional or semibrittle (Carter & Kirby 1978; Carter & Tsenn 1987; Ross & Lewis 1989). As has been emphasized by many authors (e.g., Hobbs 1981, 1984), metamorphic reaction kinetics and thermally-activated deformation rate processes are tightly interwoven.

Fractures, faults and shear zones

Atkinson (1984) has provided a thorough treatise on subcritical crack growth in geological materials, a most important process involving aqueous solutions. For a two-dimensional crack in any mode of loading, the stress intensity factor is given by:

$$K = Y\sigma(\pi c)^{1/2} \qquad (5)$$

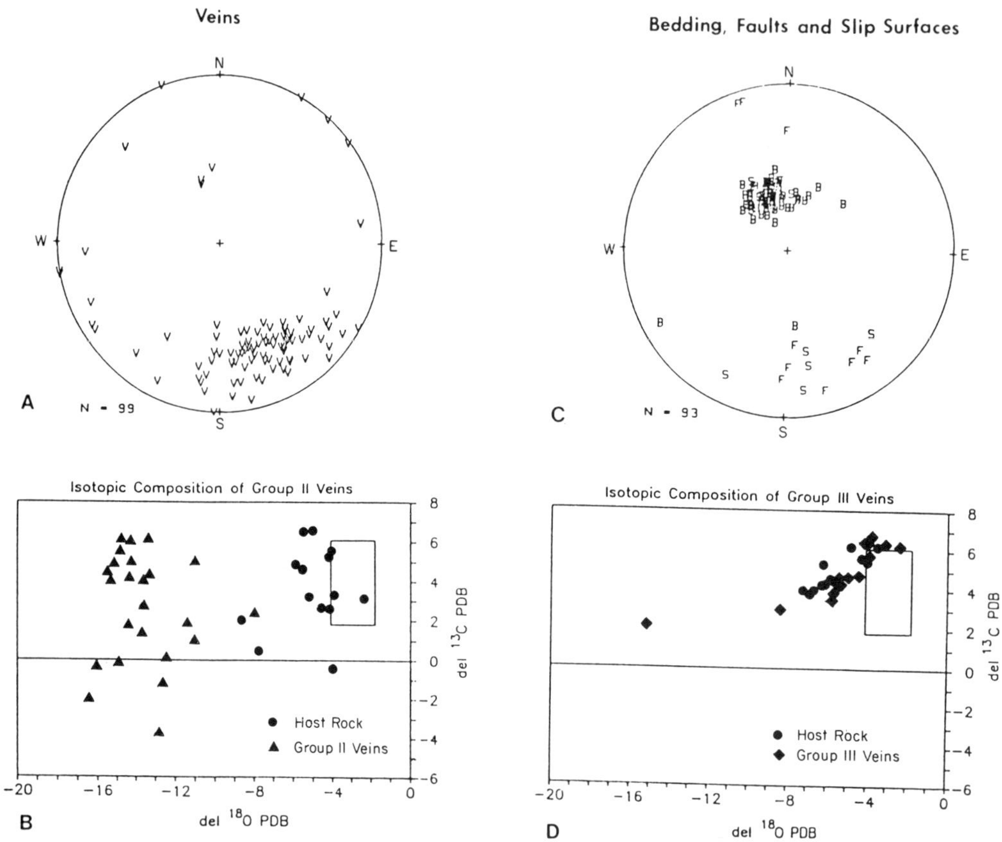

Fig. 2. Orientations and isotopic compositions of some calcite-filled veins and adjacent rocks in the Haystack Peak region, Wyoming (after Budai & Wiltschko 1987, figs 3 and 8). (**A**) Orientations of poles to veins. (**B**) Isotopic compositions of Group II (transport-parallel) veins (SE-quadrant cluster in A) and adjacent host rock;rectangle encloses composition of typical Madison Group carbonate. (**C**) Orientations of poles to other planar surfaces. (**D**) Compositions of Group III (bedding-parallel) veins and adjacent host rocks.

where σ is a far-field applied stress, Y is a geometrical factor and c is the half length of a penny-shaped internal crack. The growth velocity, v, may be expressed as:

$$v = \upsilon_0 \exp(-H/RT)K^n \qquad (6)$$

where H is the activation enthalpy, R is the gas constant, T is in K, and υ_0 and n are constants. The value of n, known as the subcritical crack growth index, apparently depends on the mechanisms of growth which include stress corrosion, dissolution, ion exchange, diffusion and microplasticity. All of these mechanisms are expected to be influenced by chemical effects and some of them result in crack-tip blunting, retarding the growth velocity. Atkinson (1984) points out that at low velocities, crack paths tend toward grain boundaries so that corrosive properties will then be controlled by details of the surface chemistry and nature of intergranular boundaries.

Competing with subcritical crack growth, and also aided by aqueous solutions, in crack healing (Smith & Evans 1984; Wanamaker & Evans 1985; Hickman & Evans 1988). For quartz, cracks are healed, leaving an array of fluid inclusions, as silica is transported by diffusion along the crack surface and/or through the fluid. The activation parameters estimated suggest that at temperatures above 200°C, cracks in naturally deformed quartz heal rapidly (Smith & Evans 1984). The healing processes vary depending on the physical conditions and chemical environment; at sufficiently high temperature, solid diffusive healing may occur (Wanamaker & Evans 1985). Capillary forces play an important role in solution-aided processes and crack healing rates may set lower bounds for sub-critical crack propagation velocities (Atkinson 1984; Smith & Evans 1984).

Paterson (1978), Brace & Kohlstedt (1980) and Kirby (1980, 1983, 1985) have summarized effects of several physical variables on the strengths of rocks deforming in the brittle field. While there appears to be no acceptable theory which predicts macroscopic fracture strength under general states of stress, a lower bound is placed by the constraint that the stress difference required can not fall below that necessary for frictional sliding along existing, optimally-oriented faults (Kirby 1983). Estimates of frictional resistance to sliding are commonly based on Byerlee's (1968) relation developed from extensive experiments on dry rock-on-rock surfaces. However, this relation yields an upper limit to stress differences required as even trace quantities of water on the surfaces result in a marked lowering of the friction coefficient (Dieterich & Conrad 1984). Fracture surfaces readily adsorb chemical species derived from aqueous solutions, reducing surface energies and hence the critical stress for sliding (Dunning *et al.* 1984). Large departures from Byerlee's relation (and in the mode of slip) are also observed when gouges of various types and compositions are placed between the rock surfaces. The complex mechanical and chemical effects of aqueous fluids and various gouge types on frictional resistance at depth are currently being researched vigorously (e.g., reviews by Kirby & Kronenberg 1987; Carter & Tsenn 1987). Results to date lead to the conclusion that rock strengths in the brittle field are much lower than have been generally supposed.

Localized deformation in shear zones throughout the continental crust and extending into the upper mantle may be responsible for large fractions of the displacements at the lithospheric scale and has received considerable attention for more than a decade (e.g., Sibson 1977, 1983, Kirby 1985; Kirby & Kronenberg 1987; Rutter & Brodie 1988; Rubie 1990). Regardless of depth, tectonic regime or specific nucleation mechanism, a large grain-size reduction is the key factor in providing the enhanced creep strains and strain rates associated with these zones. This reduction, whether by: (1) comminution (e.g., Tullis & Yund 1987); (2) dynamic recrystallization (e.g., Behrmann & Mainprice 1987) or (3) by metamorphic reactions and phase changes (Rubie 1990) permits extensive fluid infiltration and acceleration of the several softening processes. Mass-balance calculations have shown, for example, that some mylonites have reacted with many times their volume of hydrous fluid (e.g., McCaig 1988). While most studies have emphasized softening aspects of shear zones, Hobbs *et al.* (this volume) has pointed out that Coulomb materials in the upper crust may localize these zones under strain-hardening conditions and Ord (this volume) finds that migrating fluid may be focused along such structurally-controlled zones.

Kirby & Kronenberg (1987, and references therein) summarize both mechanisms of nucleation of semibrittle ductile shear zones and processes associated with continued strain localization along them. They suggest the following nucleation mechanisms: (1) localization in zones slightly softer than the bulk rock mass deforming by non-linear processes such as dislocation creep (e.g., Kirby 1985); (2) strain localization along pre-existing zones of weakness resulting from stress concentrations or from varying material properties, including grain size. Pre-existing fractures may also serve as conduits for fluid infiltration and a large number of water-weakening processes (e.g., Segall & Simpson 1986); (3) distributed microcracking permitting fluid infiltration leading to cyclic hydrothermal alteration and ductile deformation (e.g., White & White 1983). Once nucleated, localization and softening, enhanced by shear heating, may continue by: (a) increase of dislocation mobility, enhancing dislocation creep; (b) grain boundary migration, dissolution, diffusional transport, precipitation and sliding processes are all facilitated; and (c) reaction kinetics are accelerated. In addition, external crystallographic rotations resulting from intracrystalline slip may facilitate continued deformation (geometric softening). Strain rates may be further accelerated by shear heating, metamorphic transformations which release latent heat, change in volume or growth of softer phases. Reciprocally, many of these deformation processes may enhance metamorphic reaction kinetics leading to strong coupling between deformation and metamorphism.

Mechanisms of fluid-assisted weakening and penetrative deformation have been suggested by Kronenberg *et al.* (1990) based on observations on a small ductile shear zone in granitic rocks of the central Sierra Nevada. Tensile fracturing of granodiorite host and an aplite dyke preceded a ductile shearing event and the subsequent shear strains localized on a pre-existing sealed fracture. The original fracture is shown stippled in Fig. 3C and foliation development of the adjacent granodiorite is indicated. Fluids were present at the time of tensile fracturing, precipitating fracture-filling quartz and hydrous minerals. An aplite dyke (blank) with a sharp

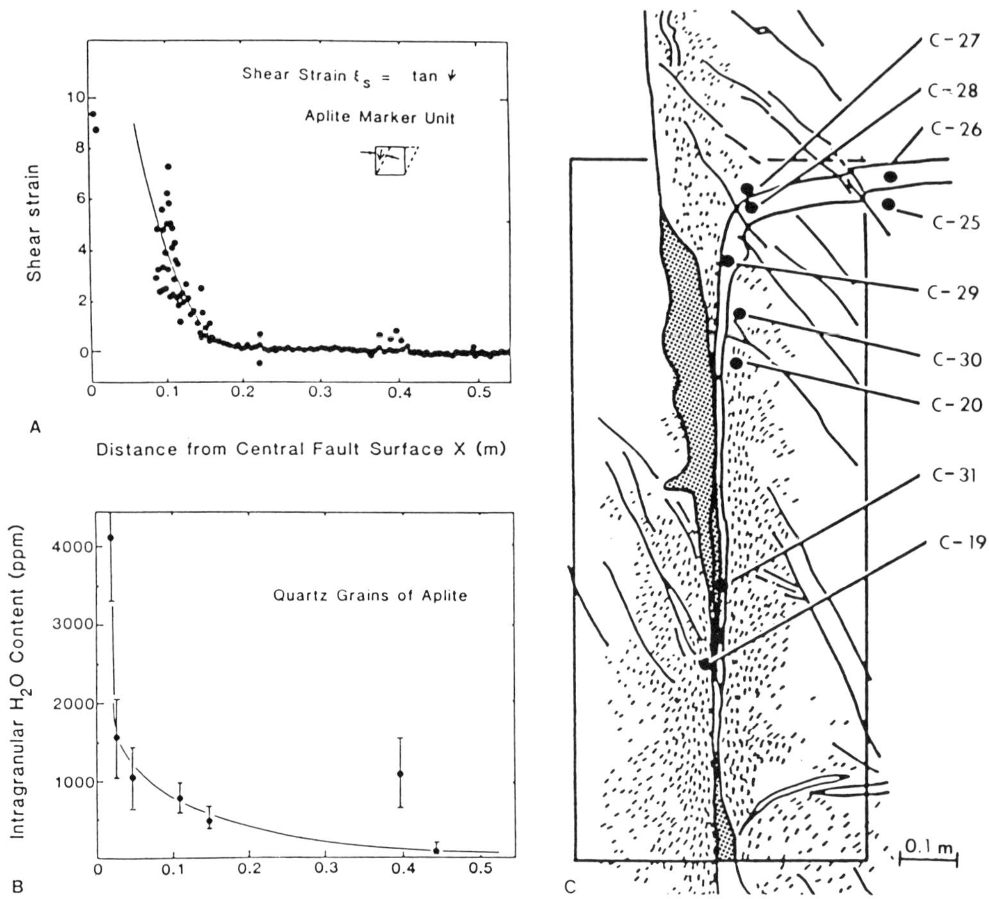

Fig. 3. Ductile shear zone in aplite–granodiorite in the central Sierra Nevada (after Kronenberg *et al.* 1990, figs 1 and 7). (**A**) Shear strain in aplite dyke as a function of distance from zone centre. (**B**). IR-determined H_2O concentration in quartz corresponding to (A). (**C**) Sketch of aplite dyke (blank area with sample localities) in granodiorite with dashed marks indicating trace of foliation; pre-existing fracture (water reservoir) is stippled.

bend from its original orientation (upper right of Fig. 3C) to its sheared orientation near the shear zone center served as a shear strain marker. Figure 3A shows the shearing strain as a function of distance from the central fault surface as determined by the deflection of the aplite dyke and Fig. 3B shows the corresponding water content in quartz crystals as determined by infrared spectroscopy. IR signatures at 298 and 77K show that most of the intragranular water resides in fluid inclusions: (a) in planar arrays along the pre-existing tensile fractures; and (b) as extremely small (20–140 nm) inclusions along dislocation nodes and free dislocations associated with the later ductile shearing. These combined field, IR and TEM observations suggest that water-related defects gained access to grain interiors and dislocation cores by fluid infiltration along open microcracks, followed by pipe diffusion along mobile dislocations.

Bulk rock creep, dilatancy and water weakening

Apart from solution transfer creep (pressure solution) phenomena which are now known to be pervasive throughout the lithosphere and the many grainsize-sensitive mechanisms operative in shear zones, coarse-grained rock rheologies at depth are dominated by processes giving rise to nonlinear relationships between stress and strain rate. Polymineralic crystalline rocks, comprising most of the lithosphere, have been accorded relatively little attention by experimentalists and theoreticians because of complexities associated with strain partitioning, but promising progress is being made (e.g., Handy 1990). Most extensive experimental work on common rocks, aimed at determining the flow properties and processes in the steady-state regime, has concentrated on single crystals and monomineralic aggregates of calcite, halite, quartz and olivine. Of these minerals and rocks, chemical water weakening of some type is most pronounced in the last three and our brief discussion is confined to them alone.

Quartz

Since the discovery of hydrolytic weakening in quartz (Griggs & Blacic 1965) and its possible extension to other silicates (Griggs 1967), the nature of intracrystalline water-weakening of quartz has been studied actively (see digest and issue edited by Kirby 1984 and references in Kirby & Kronenberg 1987). Current questions concerning the nature of hydrolytic weakening processes center around the manner in which H_2O accesses the interiors of quartz grains, the chemical species responsible for weakening and specific mechanisms by which dislocation mobility is achieved.

Most recently Paterson (1990) has provided an excellent summary of this complex topic as well as an assessment of the various water-related species that affect the mechanical behavior of quartz. Water may be involved in three ways: (1) molecular water aggregates, including adsorbed species at quartz–water interfaces, serving as reservoirs; (2) species in solid solution which promote diffusion; and (3) species in dislocations involved in promoting their glide mobility. Paterson (1990) points out that internal molecular water, whether introduced during growth or later along microcracks (Kronenberg *et al.* 1986, 1990), does not reside in the quartz crystal structure but occurs in caged clusters (non-freezable aggregates) and as fluid inclusions (freezable).

Controversy continues, however, regarding the means by which water-related defects are incorporated within quartz grain interiors. Kronenberg *et al.* (1986) show that hydrogen diffuses rapidly as H^+ (proton) interstitials with diffusion coefficients many orders of magnitude greater than those of oxygen (Dennis 1984; Giletti & Yund 1984) at comparable laboratory conditions. A marked discrepancy between the hydrothermal oxygen diffusion data of Dennis (1984) and Giletti & Yund (1984), regarding the influence of $f(H_2O)$, $f(H_2)$ and $f(O_2)$, has not yet been resolved. A solution to this difficulty may lie in distinguishing between the proton activities which were set by the different experimental techniques. Farver & Yund (1989) find that for alkali feldspar, hydrothermal oxygen diffusion is strongly correlated with $f(H_2O)$ but not $f(H_2)$ or $f(O_2)$, implying that oxygen diffuses as H_2O. While Kronenberg *et al.* (1986) have argued that diffusive transport of oxygen (in addition to hydrogen) is required to incorporate the defects necessary for hydrolytic weakening within quartz grain interiors, Rovetta *et al.* (1989) have suggested that hydrogen diffusion alone and reaction with lattice oxygen may introduce hydroxyl defects which contribute to hydrolytic weakening.

Paterson (1990) concludes that the applicable creep flow law for quartz-dominated rocks will depend fundamentally on the fugacity of the water with which they are in equilibrium. Regardless of the specific mechanism, water facilitaties the nucleation and propagation of kinks along dislocations at relatively low temperature, so that hydrolytic dislocation glide controls the creep rate; core diffusion may be important in this regime. At high temperatures, where dislocation climb controls the creep rate, the role of water is to assist self-diffusion of silicon and/or oxygen so that dislocations may climb over obstacles.

Olivine

Blacic (1972) first confirmed earlier indications (Carter & Ave' Lallemant 1970; Post 1970) of hydrolytic weakening in olivine by demonstrating that dry olivine crystals are stronger by a factor of 2 than those deformed plastically in the presence of water. This strength difference for single crystals has since been confirmed by the carefully controlled experiments of Mackwell *et al.* (1985) who showed that the water-related reduction in strength is due to enhancement of the rate of dislocation climb. Chopra & Paterson (1984) made the significant discovery that olivine in polycrystals containing < 100 ppm H_2O exhibit mechanical behavior identical to that of wet (>0.1 wt % H_2O) aggregates.

Kohlstedt & Hornack (1981) have, in addition, shown that the creep rate of olivine increases with $a_{O_2}^{1/6}$ and Hobbs (1983) found the creep rate to be about ten-fold higher for olivine annealed in the presence of enstatite rather than with magnesiowustite. Hobbs (1983) interpreted these combined observations as indicating that propagation of positively charged kinks along dislocation lines provide the rate-controlling creep mechanism in dry olivine.

Most recently, Mackwell & Kohlstedt (1990) have investigated the kinetics of diffusion of hydrogen in olivine single crystals under hydrothermal conditions and have shown the diffusivities to be rapid and anisotropic, with fastest transport along the [100] crystal axis. They observed that: (1) the diffusivity of water-derived species is independent of the concentration of the diffusing species and of oxygen fugacity; (2) the mobile defect does not include an oxygen ion; (3) the diffusion coefficient for H^+ diffusion along [100] is similar to that for quartz (Kronenberg *et al.* 1986) under comparable conditions. From this information, Mackwell & Kohlstedt (1990) also suggest that the mobile defects in olivine are protons, possibly moving as interstitials. Following considerations of various hydrogen defects, they conclude that hydrogen ions associated with silicon vacancies should have the most direct effect on creep of olivine. They conclude also that low hydrogen concentrations in the structure and fluid inclusions of olivine in mantle-derived xenoliths may result from the rapid decrease in pressure, and hence hydrogen fugacity, experienced by the ascending hot xenoliths, permitting dehydration of the olivine. Thus, the Earth's upper mantle may not be anhydrous.

Rocksalt

Hydrolytic weakening has not yet been shown to operate in halite. Observations on naturally deformed Salina rocksalt suggest that dislocation mobility is enhanced in the vicinity of brine inclusions (Carter *et al.* 1982). However, strengths of natural NaCl crystals from Grand Saline and Avery Island at elevated temperature (Handin 1966), which are virtually identical to Harshaw crystals at comparable conditions, may argue against hydrolytic weakening.

Some of the H_2O-related defects that have been identified in NaCl, primarily utilizing IR to UV optical absorption, Raman scattering, and magnetic resonance techniques, include: (1) defects of both hydrogen atoms (H^0) and hydride ions (H^-) which may occur either as substitutions for Cl^- or as interstitials (i). U, U_1, and U_2 centres are H^-_s, H^-_i and H^0_i, respectively (e.g. Schaffer 1960; Goodvaert *et al.* 1983); (2) O^-_2 ions may also occupy Cl^- sites (Känzig & Cohen 1951; Rolfe *et al.* 1961) and may react with Ca^{2+} in Na^+ sites to reduce the formation of vacancies, thus depressing the extrinsic electrical conductivity (Witham & Calderwood 1975); and (3) OH^- substitutional defects which, in concentrations as low as 10 ppm, depress the thermal conductivity by 10^2 (Wedding & Klein 1969). The ionic conductivity may be similarly depressed by the formation of $Mg^{2++}-OH^-$ complexes which also remove vacancies (Staebe 1967). Other defects may occur, including gamma-irradiation induced F-centres (anion vacancies with a trapped electron) and H-centres (Cl_2^- molecular anions); these defects may coalesce into clusters of colloidal sodium and trapped molecular chlorine (e.g., Celma *et al.* 1988). All these defects may interact, forming pairs or complexes and their concentrations may be interdependent. For example, U-centres form with a concentration proportional to the OH^- content which also affects the growth of sodium colloid clusters (Klein *et al.* 1968).

The influence of these defects on various physical properties of NaCl, especially at low temperatures (<23°) has been reasonably well-documented but little is known of their effects on the mechanical properties. Alybakov *et al.* (1970; translated ref. Zhim 1971) found by testing microhardness that both x-ray irradiation and introduction of anion impurities (including OH^-) strengthened all alkali halides; they attributed the strengthening to point defect-dislocation reactions. The removal of vacancies noted above may also inhibit dislocation mobility. Celma *et al.* (1988) suggest that gamma-irradiated NaCl containing sodium colloids may recrystallize more rapidly than that devoid of colloids. The introduction of >1000 ppm foreign cations strengthens rocksalt significantly (Heard & Ryerson 1986) but few natural rocksalts contain the required concentrations. The presence of H_2O in intragranular brine inclusions in natural and reagent-grade salt may soften the material. The observation of Friedman *et al.* (1984) that optically visible fluid inclusions disappear during deformation only to reappear upon annealing may suggest either that the inclusions disperse to form submicroscopic inclusions and/or that some species of H_2O diffuses to dislocation cores and lattice vacancies, enhancing dislocation mobility.

Figure 4 shows infrared absorption spectra normalized to 10 mm thick cleavage slabs, of a melt-grown Harshaw crystal, two Harshaw crystals deformed to 10% strain at 25°C, 200 MPa confining pressure at a strain rate of $10^{-4}s^{-1}$ (406 and 505; Carter & Heard 1970) and from Avery Island (AI) domal salt. On the basis of integrated absorbance values, Δ, in the

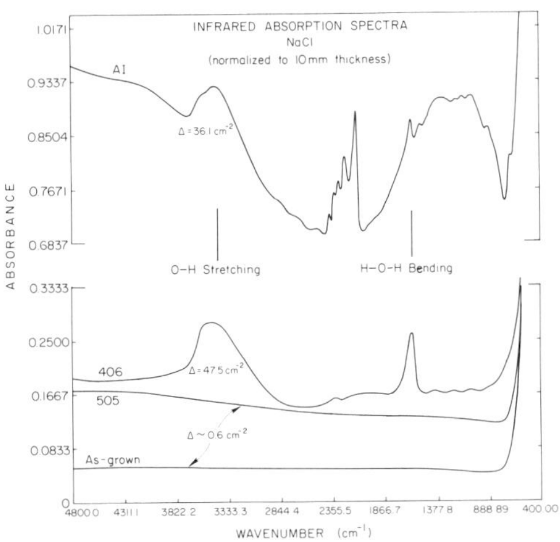

Fig. 4. Infrared absorption spectra of {100} cleavage plates of Avery Island (AI) rocksalt and synthetic single crystals 406 and 505, both deformed at 200 MPa pressure, 25°C at a strain rate of $10^{-4}s^{-1}$. Δ is the integrated absorbance. H–O–H bending and O–H stretching bands are as indicated. Large peak near 2300 cm^{-1} in AI may reflect organic contaminants.

O–H stretch region near 3400 cm^{-1}, and the calibration of Wedding & Klein (1969), experimentally deformed crystal 405 contains 4700 ppm H, AI contains 3570 ppm H and both experiment 505 and the as-grown crystal contain <60 ppm H (below the resolution measuring spectra at room temperature). However, the calibration of Wedding & Klein (1969) appears to differ from the molecular absorption coefficient for liquid water by a factor of 300 so that H_2O concentrations may be much lower. An additional difficulty appears in the spectrum of crystal 406, which shows an appreciable concentration of H_2O relative to the as-grown crystal, crystal 505 and several other experimentally deformed crystals examined; this crystal appears to be anomalous.

Urai *et al.* (1986) and Spiers *et al.* (1988) have shown that the presence of water, inherent or added, weakens both natural Asse rocksalt (provided the confining pressure is sufficiently high to suppress dilatancy) and fine-grained synthetic aggregates. Spiers *et al.* (1988) separate the mechanical response of rocksalt according to whether or not dilatant behaviour is observed, independent of H_2O content and of the grain-size of the starting material. In the dilatant field, the creep rate is controlled by dislocation processes. In the non-dilatant field, the creep rate is controlled by fluid-assisted diffusional transport (Spiers *et al.* this volume), where the strain rate ε_f is given by

$$\dot{\varepsilon}_f = 4.7 \times 10^{-4} \exp(-24.7/RT)\ \sigma/Td^3 \quad (7)$$

where Q_c is in kJ mol^{-1}, T is in K and d is the grain diameter in mm.

In addition to these studies, extensive creep and constant strain rate (including incremental and decremental) tests have been carried out by other groups on several natural rocksalts in the last decade (e.g., Carter & Hansen 1983; Wawersik & Zeuch 1986; Handin *et al.* 1986; Senseny & Hansen 1987; Horseman & Handin 1990; Russell *et al.* 1990). The largest and most comprehensive data base at present is that for Avery Island domal salt, a nearly pure, equigranular, relatively fine grained (7.5 mm average grain diameter), nominally dry (16 ppm H_2O) domal rocksalt. These experiments have been conducted on dry cylinders to *c.* 10% strain at temperatures from 25°C to 200°C, strain rates from 10^{-4} to $10^{-9}s^{-1}$, confining pressures from 0.1–30 MPa and differential stresses in the range 2–30 MPa. Flow rates at temperatures ≤100°C and strain rates $\geq 10^{-6}s^{-1}$ are controlled by the cross-slip of screw dislocations. In the temperatures range 100°C to 200°C at strain rates $\leq 10^{-6}s^{-1}$ steady state flow of AI rocksalt follows a power creep law (Carter & Hansen 1983):

$$\dot{\varepsilon}_p = 7.6 \times 10^{-4} \exp(-66.5/RT)\ \sigma^{4.5} \quad (8)$$

with the creep activation energy, Q_c, expressed in kJ $mole^{-1}$ and σ, in MPa. Excellent subgrain formation occurs under these conditions and the activation energy obtained is near that for Na^+ pipe diffusion (76 kJ/mole; Heard 1972), which could be rate-controlling.

Wawersik & Zeuch (1986), Wawersik (1988) and Skrotski & Haasen (1988) contend, however, that the creep rate to *c.* 200°C is probably controlled by the cross-slip of screw dislocations. Their contention is based primarily on an observed stress dependence on Q_c and temperature dependence on the stress exponent, n. No such dependencies were observed under conditions upon which the power law creep relation above (eqn. 8) was based and this difficulty remains to be resolved; Poirier (1985) reports that such dependencies are not reliable criteria for cross-slip.

Assuming that equations (7) and (8) are representative of fluid-assisted diffusional transport, and of dislocation-controlled creep, we have plotted in Fig. 5, in temperature-log strain rate space, curves of equal differential stresses (from 0.1 to 10 MPa) using equation (8; bold curves) and for two grainsizes (1 mm, dotted

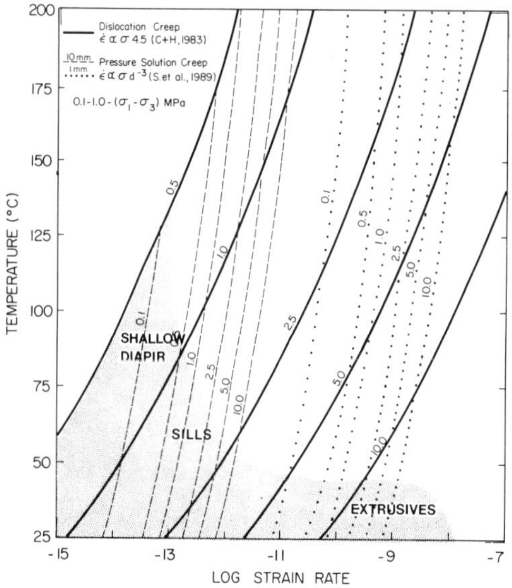

Fig. 5. Temperature versus log strain rate plot of equal stress differences ranging from 0.1–10 MPa calculated from power law creep equation (8; bold curves) and from fluid-assisted diffusion creep equation (7) for grainsizes 10 mm (dashed) and 1.0 mm (dotted). Superimposed on the diagram are conditions representative of gulf-coast-type rocksalt extrusives and shallow intrusives (after Jackson & Talbot 1986).

and 10 mm, dashed) using equation (7). Inasmuch as the micromechanical mechanisms represented by equations (7) and (8) are independent, that giving rise to the highest creep rate will dominate the creep strain. Thus, for example, at 2.5 MPa stress difference and a grainsize of 10 mm, ε_p is higher at temperatures above 30°C and would be expected to dominate salt deformation in the shallow regime (stippled region from Jackson & Talbot 1986). The same generalization for pressure solution creep may be made, for all reasonable stresses, if the grainsize is as small as 1 mm.

The conclusion derived from Fig. 5 is that fluid-assisted diffusional creep and creep limited by dislocation motion compete over most physical conditions appropriate to the formation of shallow rocksalt structures. However, this conclusion is contrary to experience with the micromechanical behavior of coarse-grained aggregates in general and specifically to observations on eleven bedded, domal and anticlinal salts (Carter & Hansen 1983) whose substructures are identical to those in epxeriments upon which equation (8) is based. In this regard, it is important to note that Wawersik & Hannum (1979) observed dilatancy well within the non-dilatant field of Spiers *et al.* (1988), in three experiments on Salado bedded salt. Clearly, this topic requires further investigation which may require flow-through experiments on coarse-grained natural rocksalt as functions of the various physical variables (T, P_f, P_e, $\dot{\varepsilon}$, σ) to determine the relative contributions to the creep strain of fluid-assisted diffusion and dislocation mechanisms including an evaluation of the effects of H_2O-related defects.

The most significant observation of dilatancy during long-term steady-state creep of Salado salt at the extreme conditions 21 MPa confining pressure, 200°C at a low constant stress difference of 7 MPa (Wawersik & Hannum 1979) has considerably greater fundamental importance than interpretations of rocksalt structural evolution alone. Rocksalt is the most ductile of common rock-forming materials and, under steady-state conditions, possesses the largest number of independent slip systems. Inasmuch as dilatancy occurs during high temperature, steady-state creep of this material at high pressure, we suspect that silicate aggregates also dilate under these same homologous conditions. Dilatancy has been observed in calcite under such conditions (M. S. Paterson, pers. comm. 1989) but not at the high temperatures and pressures required for steady-state flow of silicates; currently such measurements are technologically infeasible. If we are correct, then the empirically determined steady-state flow parameters include microcrack opening and healing, perhaps at a constant rate, and the micromechanical behavior is semi-brittle. We suspect, too, that rocks composing the lithosphere and flowing naturally in the steady-state exhibit similar microstructural behavior and that the crack opening and healing process is yet another grain-scale mechanism for fluid-assisted mass transfer.

We are grateful to E. H. Rutter for reviewing the manuscript and to both the University of Leeds and Geological Society of London for hosting a stimulating conference. N.L.C. and A.K.K. acknowledge support from DOE/BES Grant DE-FGUS-87ER13711 and D.R.W. acknowledges support by NSF Grant EAR-8417136.

References

ATKINSON, B. K. 1984. Subcritical crack growth in geological materials. *Journal of Geophysical Research*, **89**, 4077–4114.

BEHRMANN, J. H. & MAINPRICE, D. 1987. Deformation mechanisms in a high-temperature quartz–feldspar mylonite–evidence for superplastic flow in the lower crust. *Tectonophysics*, **140**, 297–305.

BLACIC, J. D. 1972. Effect of water on the experimental deformation of olivine. *American Geophysical Union Monograph*, **16**, 109–115.

BRACE, W. & KOHLSTEDT, D. L. 1980. Limits on lithospheric stress imposed by laboratory experiment. *Journal of Geophysical Research*, **85**, 6248–6252.

BUDAI, J. M. & WILTSCHKO, D. V. 1987. Structural controls on syntectonic diagenesis within Haystack Peak region of the Absaroka thrust sheet, Idaho–Wyoming–Utah thrust belt. *In*: MILLER, W. R. (ed.) *The thrust belt revisited*. Wyoming Geological Association 38th Annual Guidebook, Casper, 55–68.

BURKHARD, M. & Kerrich, R. 1988. Fluid regimes in the deformation of the Helvetic nappes, Switzerland, as inferred from stable isotope data. *Contributions to Mineralogy and Petrology*, **99**, 416–429.

BYERLEE, J. D. 1968. Brittle–ductile transition in rocks. *Journal of Geophysical Research*, **73**, 4741–4750.

CARTER, N. L. & AVE'LALLEMANT, H. G. 1970. High-temperature flow of dunite and peridotite. *Geological Society of America Bulletin*, **81**, 2181–2202.

—— & HANSEN, F. D. 1983. Creep of rocksalt. *Tectonophysics*, **92**, 275–333.

—— & HEARD, H. C. 1970. Temperature and rate dependent deformation of halite, *American Journal of Science*, **269**, 193–249.

—— & KIRBY, S. H. 1978. Transient creep and semibrittle behavior of crystalline rocks. *Pure and Applied Geophysics*, **116**, 807–839.

—— & TSENN, M. C. 1987. Flow properties of continental lithosphere. *Tectonophysics*, **136**, 27–63.

——, HANSEN, F. D. & SENSENY, P. E. 1982. Stress magnitudes in natural rock salt. *Journal of Geophysical Research*, **87**, 9289–9300.

CELMA, A. G., URAI, J. L. & SPIERS, C. J. 1988. A laboratory investigation into the interaction of recrystallization and radiation damage effects in polycrystalline salt rocks. *Commission of European Communities report*. EUR 11849EN.

CHOPRA, P. N. & PATERSON, M. S. 1984. The role of water in the deformation of dunite. *Journal of Geophysical Research*, **89**, 7861–7876.

DENNIS, P. F. 1984. Oxygen self-diffusion in quartz under hydrothermal conditions. *Journal of Geophysical Research*, **89**, 4047–4057.

DIETERICH, J. H. & CONRAD, G. 1984. Effect of humidity on time- and velocity-dependent friction in rocks. *Journal of Geophysical Research*, **89**, 4196–4202.

DUNNING, J. D., PETROVSKI, D., SCHMYLER, J. & OWENS, A. 1984. The effects of aqueous chemical environments on crack propagation in quartz. *Journal of Geophysical Research*, **89**, 4115–4123.

ELLIOT, D. 1973. Diffusion flow laws in metamorphic rocks. *Geological Society of America, Bulletin*, **84**, 2645–2664.

ETHERIDGE, M. A., WALL, V. J. & COX, S. F. 1984. High fluid pressures during regional metamorphism and deformation: implications for mass transport and deformation mechanism. *Journal of Geophysical Research*, **89**, 4344–4358.

FARVER, J. R. & YUND, R. A. 1989. The effect of H_2, O_2 and H_2O fugacities on oxygen diffusion in alkali feldspar. *EOS, Transactions of American Geophysical Union*, **70**, 501.

FLETCHER, R. C. 1982. Coupling of diffusional mass transport and deformation in a tight rock. *Tectonophysics*, **83**, 275–291.

—— & HOFFMAN, A. W. 1974. Simple models of diffusion and combined diffusion-infiltration metasomatism. *In*: HOFFMAN *et al.* (eds) *Geochemical, Transport and Kinetics*, Publication 634. Carnegie Institute, Washington, Washington, DC, 242–262.

FRIEDMAN, M., DULA, W. F., Gangi, A. F. & AZONAS, G. A. 1984. Structural petrology of experimentally deformed synthetic rocksalt, *In*: *Proceedings of the 1st Conference on Mechanical Behavior of Salt*. Transactions of Technical Publications, Hanover, Germany, 19–36.

FYFE, W. S. & KERRICH, R. 1985. Fluids and thrusting. *Chemical Geology*, **49**, 353–362.

——, PRICE, N. J. & THOMPSON, A. B. 1978. *Fluids in the Earth's Crust*. Elsevier, Amsterdam.

GILETTI, B. J. & Yund, R. A. 1984. Oxygen diffusion in quartz. *Journal of Geophysical Research*, **89**, 4039–4046.

GOODVAERT, E., ZHOU, Z.-F. & SCHOEMAKER, D. 1983. Raman scattering of pure singly- and doubly-perturbed interstitial hydrogen atom centers in alkali halides. *Radiation Effects*, **72**, 81–87.

GRIGGS, D. T. 1967. Hydrolytic weakening of quartz and other silicates, *Geophysical Journal*, **14**, 19–31.

—— & BLACIC, J. D. 1965. Quartz: Anomalous weakness of synthetic crystals. *Science*, **147**, 292–295.

HANDIN, J. 1966. Strength and ductility. *In*: CLARK, S. P. (ed.) *Handbook of Physical Constant–Revised Edition*. Geological Society of America Memoir. **97**, 223–289.

——, HAGER, R. V., Jr., FRIEDMAN, M. & FEATHER, J. N. 1963. Experimental deformation of sedimentary rocks under confining pressure: pore pressure tests. *American Association of Petroleum Geology Bulletin*, **47**, 717–755.

——, RUSSELL, J. E. & CARTER, N. L. 1986. Experimental deformation of rocksalt. *In*: HOBBS, B. E. & HEARD, H. C. (eds) *Mineral and Rock Deformation: Laboratory Studies*. American Geophysical Union Monograph, **36**, 117–160.

HANDY, M. R. 1990. The solid-state flow of polymineralic rocks. *Journal of Geophysical Research*, **95**, 8647–8661.

HEARD, H. C. 1972. Steady-state flow in polycrystalline halite at pressures of 2 kilobars. *American Geophysical Union Monograph*, **16**, 191–210.

—— & RYERSON, F. J. 1986. Rheology of polycrystalline halite containing cation impurities. *In*: HOBBS, B. E. & HEARD, H. C. (eds) *Mineral and Rock Deformation: Laboratory Studies*, American Geophysical Union Monograph. **36**, 99–115.

HICKMAN, S. & EVANS, B. 1988. Experimental pressure solution in halite bicrystals. *EOS Transactions of American Geophysical Union*, **69**, 481.

HOBBS, B. E. 1981. The influence of metamorphic environment upon the deformation of minerals. *Tectonophysics*, **78**, 335–383.

—— 1983. Constraints on the mechanism of deformation of olivine imposed by defect chemistry. *Tectonophysics*, **92**, 35–69.

—— 1984. Point defect chemistry of minerals under a hydrothermal environment. *Journal of Geophysical Research*, **89**, 4026–4038.

——, MÜHLAUS, H-B. & ORD, A. 1990. Instability, softening and localization of deformation. *This volume*.

HORSEMAN, S. T. & HANDIN, J. 1990. Triaxial-compression tests on rocksalt at temperatures from 50° to 200°C and strain rates from 10^{-4} to 10^{-9}/s. *American Geophysical Union Monograph Series*, (in press).

HUBBERT, M. K. & RUBEY, W. W. 1959. Role of fluid pressure in mechanics of overthrust faulting, 1, Mechanics of fluid-filled porous solids and its application to overthrust faulting. *Geological Society of America Bulletin*, **70**, 115–166.

JACKSON, M. P. A. & TALBOT, C. J. 1986. External shapes, strain rates and dynamics of salt structures. *Geological Society of America Bulletin*, **97**, 305–323.

JACKSON, M., MCCAIG, C., BULLARD, M. M. & VAN DER VOO, R. 1985. Magnetite authigenesis and diagenetic paleotemperatures across the northern Appalachian. *Geology*, **16**, 592–595.

KÄNZIG, W. & COHEN, M. H. 1951. Paramagnetic resonance of oxygen in alkali halides. *Physical Review Letters*, **3**, 509–510.

KERRICH, R. 1986. Fluid infiltration into fault zones: chemical, isotopic, and mechanical effects. *Pure and Applied Geophysical*, **124**, 225–268.

——, LATOUR, T. E. & WILLMORE, L. 1984. Fluid participants in deep fault zones: evidence from geological, geochemical and $^{18}O/^{16}O$ relations. *Journal of Geophysical Research*, **89**, 4337–4343.

KILSDONK, M. W. & WILTSCHKO, D. V. 1987. Deformation mechanisms in the southeastern ramp region of the Pine Mountain block. Tennessee. *Geological Society of America Bulletin*, **100**, 653–664.

KIRBY, S. H. 1980. Tectonic stresses in the lithosphere: constraints provided by the experimental deformation of rocks. *Journal of Geophysical Research*, **85**, 6353–6363.

—— 1983. Rheology of the lithosphere. *Reviews of Geophysics and Space Physics*, **21**, 1458–1487.

—— 1984. Introduction and digest to the special issue on chemical effects of water on the deformation and strengths of rocks. *Journal of Geophysical Research*, **89**, 3991–3995.

—— 1985. Rock mechanics observations pertinent to the rheology of the continental lithosphere and the localization of strain along shear zones. *Tectonophysics*, **119**, 1–27.

—— & KRONENBERG, A. K. 1987. Rheology of the lithosphere: Selected topics. *Reviews in Geophysics*, **25**, 1219–1244.

KLEIN, M. V., KENNEDY, S. O., Gie, T. I. & Wedding, B. 1968. The hydroxylion in alkali halids crystals—crystal growth and characterization. *Materials Research Bulletin*, **3**, 677–686.

KNIPE, R. J. & WINTSCH, R. P. 1985. Heterogeneous deformation, foliation development, and metamorphic processes in a polyphase mylonite. *In*: THOMPSON, A. B. & RUBIE, D. C. (eds) *Advances in Physical Chemistry*, **4**, Springer-Verlag, New York, 180–210.

KOHLSTEDT, D. L. & HORNACK, P. 1981. Effect of oxygen partial pressure on the creep of olivine. *In*: STACEY, F. D. *et al.* (eds) *Anelasticity in the earth. American Geophysical Union*, **4**, 101–107.

KRONENBERG, A. K., KIRBY, S. H., AINES, R. D. & ROSSMAN, G. R. 1986. Solubility and diffusional uptake of hydrogen in quartz at high water pressures: implications for hydrolytic weakening. *Journal of Geophysical Research*, **91**, 12,723–12,744.

——, SEGALL, P. & WOLF, G. H. 1990. Hydrolytic weakening and penetrative deformation within a natural shear zone. *American Geophysical Union, Monograph Series*, (in press).

LABATO, L. M., FORMAN, J. M. A., FAZIKAWA, K., FYFE, W. S. & KERRICH, R. 1983. Uranium in overthrust Archean basement, Bahia, Brazil. *Canadian Mineralogist*, **21**, 647–654.

LEACH, D. L. & ROWAN, E. L. 1986. Genetic link between Ouachita foldbelt tectonism and Mississippi Valley-type lead–zinc deposits of the Ozarks. *Geology*, **14**, 931–935.

MACKWELL, S. J. & KOHLSTEDT, D. L. 1990. Diffusion of hydrogen in olivine: implications for water in the mantle. *Journal of Geophysical Research*, (in press).

——, KOHLSTEDT, D. L. & PATERSON, M. S. 1985. The role of water in the deformation of olivine single crystals. *Journal of Geophysical Research*, **90**, 11,319–11,333.

MCCAIG, A. M. 1988. Deep circulation in fault zones. *Geology*, **16**, 867–870.

OLIVER, J. 1986. Fluids expelled tectonically from orogenic belts: their role in hydrocarbon migration and other geologic phenomena. *Geology*, **14**, 99–102.

ORD, A. 1990. Mechanical controls on dilatant shear zones. *This volume*.

PALCIAUSKAS, V. V. & DOMENICO, P. A. 1989. Fluid pressures in deforming porous rocks. *Journal of Geophysical Research*, **25**, 203–213.

PATERSON, M. S. 1978. *Experimental Rock Deformation—The Brittle Field*. Springer Verlag, Berlin.

—— 1990. The interaction of water with quartz and its influence in dislocation flow—an overview. *In*: KARATO, S. & TORIUMI, M. (eds) *Rheology of solids and of the Earth*. Oxford Univ. Press, London, (in press).

POIRIER, J. P. 1985. Creep of crystals. *Cambridge University Press*.

POST, R. L., Jr. 1970. The flow laws of Mt. Burnette dunite at 750°–1150°C. *EOS, Transactions of American Geophysical Union*, **51**, 424.

REYNOLDS, S. J. & LISTER, G. S. 1987. Structural aspects of rock-fluid interactions in detachment zones. *Geology*, **15**, 362–366.

ROLFE, J., LIPSETT, F. R. & KING, W. J. 1961. Optical absorption and fluorescence of oxygen in alkali halide crystals. *Physical Reviews*, **123**, 447–454.

ROSS, J. V. & LEWIS, P. D. 1989. Brittle-ductile transition: Semi-brittle behavior. *Tectonophysics*, **167**, 75–79.

ROVETTA, M. R., BLACIC, J. D., HERVIG, R. L. & HOLLOWAY, J. R. 1989. An experimental study of hydroxyl in quartz using infrared spectroscopy and ion microprobe techniques. *Journal of Geophysical Research*, **94**, 5840–5850.

RUBIE, D. C. 1990. Mechanisms of reaction-enhanced deformability in minerals and rocks. *In*: BARBER, D. J. & MEREDITH, P. G. (eds) *Deformation Processes in Minerals, Ceramics and Rocks*. Unwin Hyman, London, 262–295.

RUSSELL, J. E., CARTER, N. L. & WALKER, S. C. 1990. A material model for Avery Island Rocksalt. *American Geophysical Union Monograph Series*, (in press).

RUTTER, E. H. 1983. Pressure solution in nature, theory and experiment. *Journal of Geological Society, London*, **140**, 725–740.

—— & BRODIE, K. H. 1988. The role of tectonic grain size reduction in the rheological stratification of the lithosphere. *Geologische Rundschau*, **77**, 295–308.

RYE, D. M. & BRADBURY, H. J. 1988. Fluid flow in the crust: an example from a Pyrenean thrust ramp. *American Journal of Science*, **288**, 197–235.

SCHAEFFER, G. 1960. Das ultrarote spektrum des U-zentrums. *Journal of Physics and Chemistry of Solids*, **12**, 233–244.

SEGALL, P. & SIMPSON, C. 1986. Nucleation of ductile shear zones on dilatant fractures. *Geology*, **14**, 56–59.

SENSENY, P. E. & HANSEN, F. D. 1987. *Potential constitutive models for salt: survey of phenomonology, micromechanisms and equations*, RE/SPEC Inc. Topical Rept. RSI-0284.

SHIMAMOTO, T. 1985. The origin of large or great thrust-type earthquakes along subducting plate boundaries. *Tectonophysics*, **119**, 37–65.

SIBSON, R. H. 1977. Fault rocks and fault mechanisms. *Journal of Geological Society of London*, **133**, 191–213.

—— 1983. Continental fault structure and the shallow earthquake source. *Journal of Geological Society of London*, **140**, 741–767.

SKROTSKI, W. & HAASEN, P. 1988. The role of cross slip in the steady state creep of salt, *In*: HARDY, H. R., Jr. & LANGER, M. (eds) *Proceedings of Second Conference on Mechanical Behavior of Salt*. Transactions of Technical Publications, Hanover, W. Germany; 69–82.

SMITH, D. L. & EVANS, B. 1984. Diffusional crack healing in quartz. *Journal of Geophysical Research*, **89**, 4125–4135.

SPIERS, C. J., PEACH, C. J., BRZESOWSKY, R. H., SCHUTJENS, P. M. T. M., LIEZENBERG, J. L. & ZWART, J. J. 1988. *Long-term rheological and transport properties of dry and wet salt rocks*. Final Report, Commission of the European Communities.

——, SCHUTJENS, P. M. T. M., BRZASOWSKY, R. H., PEACH, C. J., LIEZENBERG, J. L. & ZWART, H. J. 1990. Experimental determination of constitutive parameters governing creep of rocksalt by pressure solution. *This volume*.

STAEBE, T. G. 1967. Influence of OH^- ions on infrared absorption and ionic conductivity in lithium fluoride crystals. *Journal of Physics and Chemistry of Solids*, **28**, 1375–1382.

SUPPE, J. & WITKE, J. H. 1977. Abnormal pore-fluid pressures in relation to stratigraphy and structure in the active fold-and-thrust belt of northwestern Taiwan. *Petroleum Geology, Taiwan*, **14**, 11–24.

TULLIS, J. & YUND, R. H. 1987. Transition from cataclastic flow to dislocation creep of feldspar: mechanisms and microstructures. *Geology*, **15**, 606–609.

URAI, J. L., SPIERS, C. J., HENDRIK, H. J., ZWART, H. J. & LISTER, G. S. 1986. Weakening of rock-salt by water during long-term creep. *Nature*, **324**, 554–557.

WANAMAKER, B. J. & EVANS, B. 1985. Experimental diffusional crack healing in olivine. *Geophysical Monograph* **31**. *Mineral Physics*, 194–210.

WAWERSIK, W. R. 1988. Alternatives to a power-law creep model for rock salt at temperatures below 160°C. *In*: HARDY, H. R., Jr. & LANGES, M. (eds) *Proceedings of Second Conference on Mechanical Behavior of Salt*. Transactions of Technical Publications, Hanover, W. Germany, 103–128.

—— & HANNUM, D. W. 1979. *Interim summary of Sandia creep experiments on rocksalt from the WIPP study area, southeastern New Mexico*. DOE Contract DE-AC04–76DP0079.

—— & ZEUCH, D. H. 1986. Modelling and mechanistic interpretation of creep of rocksalt below 200°C. *Tectonophysics*, **121**, 125–152.

WEDDING, B. & KLEIN, M. V. 1969. Infrared absorption of the hydroxyl ion in alkali halide crystals. *Physical Reviews*, **177**, 1274–1288.

WHITHAM, W. & CALDERWOOD, J. H. 1975. Oxygen impurity complexes in sodium chloride crystals. *Journal of Physics D: Applied Physics*, **8**, 1305–1310.

WHITE, J. C. & WHITE S. H. 1983. Semi-brittle deformation within the Alpine fault zone, New Zealand. *Journal of Structural Geology*, **5**, 579–589.

WILTSCHKO, D. V., MEDWEDEFF, D. M. & MILLSON, H. E. 1985. Distribution and mechanism of strain within rocks on the northwest ramp of the Pine Mountain block, southern Appalachian foreland: a field test of theory. *Geological Society of America Bulletin*, **96**, 426–435.

ZHIM, K. 1972. Optical–mechanical properties of alkali halides with singly charged anion impurities. *Chemical Abstracts*, **76**, 387.

Conditions for fault-valve behaviour

RICHARD H. SIBSON

Institute for Crustal Studies and Department of Geological Sciences, University of California, Santa Barbara California 93106, USA

Present address: Geology Department, University of Otago, PO Box 56, Dunedin, New Zealand

Abstract: Evidence for fluid pressures well in excess of hydrostatic during crustal deformation is provided by direct measurements of pressure levels in active tectonic regions and by hydrothermal vein systems in ancient orogenic belts. Faults that act as impermeable seals except immediately postfailure, when they become highly permeable channelways for fluid discharge, may behave as fluid-pressure-activated *valves* wherever they transect a suprahydrostatic fluid pressure gradient. Such *fault-valve* behaviour, causing abrupt fluctuations in fluid pressure linked to the earthquake cycle, is particularly likely for faults that remain active while unfavourably oriented for frictional reactivation in prevailing stress fields. The most extreme fault-valve action is likely to be associated with high-angle reverse faults. Valve activity may be especially prevalent near the base of the seismogenic zone when unstable frictional faulting gives way to aseismic shearing with increasing depth. At such structural levels, cyclic variations in the style of deformation may accompany fluid pressure cycling.

Fault-valve activity plays an important role in the development of much fault-hosted mineralization, and may also assist the migration of hydrocarbon fluids in some structural settings.

Fluid pressure (P_f) is a key parameter affecting the mechanical response of rock within the Earth's crust. In a fluid-saturated rock mass, the fluid pressure counteracts all normal stresses (σ_n) in accordance with the principle of effective stress (Hubbert & Rubey 1959), so that the *effective* normal stresses are given by:

$$\sigma_n' = \sigma_n - P_f \tag{1}$$

At a depth, z, in the earth's crust the level of fluid pressure is conveniently defined in terms of the pore-fluid factor,

$$\lambda_v = P_f/\sigma_v = P_f/(\rho_r g z) \tag{2}$$

where σ_v is the vertical stress or overburden pressure, ρ_r is the average rock density and g is the gravitational acceleration. The effective overburden pressure may then be written,

$$\sigma_v' = \sigma_v - P_f = \rho_r g z(1 - \lambda_v) \tag{3}$$

so that the effect of confining pressure on rock strength and ductility may be counteracted by increases in fluid pressure.

At shallow depths, where fractures are interconnected through to the surface, fluid pressure is simply given by:

$$P_f = \rho_f g z \tag{4}$$

where ρ_f is the fluid density, and the gradient is said to be hydrostatic. In such circumstances, $\lambda_v = \rho_f/\rho_r \approx 0.4$. However a transition towards suprahydrostatic fluid pressures at depths of a few kilometres has been noted in many sedimentary basins (Fertl *et al.* 1976). Of particular interest here is the transition towards lithostatic fluid pressure levels noted in deformed sedimentary sequences adjacent to major active faults within the San Andreas system of California (Berry 1973; Yeats 1983; Yerkes *et al.* 1985). Over extensive areas, measured fluid pressure levels increase from initial hydrostatic values towards λ_v values of *c*.0.9 over the depth range 2–5 km (Fig. 1). Bearing in mind the general expectation that fluid pressures may approach lithostatic values ($\lambda_v \rightarrow 1$) in regions undergoing prograde metamorphism at depth (Etheridge *et al.* 1983), this raises a number of interesting questions. Do suprahydrostatic fluid pressures also occur in basement rocks adjacent to major faults? If so, are the high fluid pressure levels concentrated in aseismic shear zones at subseismogenic depths, or are they uniformly distributed? Does the transition from hydrostatic to lithostatic pressure remain constant with time or does it fluctuate?

Active faults are of particular relevance to this last question because, when cutting across a suprahydrostatic gradient, fault rupture may breach a permeability barrier causing an episode of fluid discharge and a local reversion towards a hydrostatic fluid pressure gradient. Here we attempt to define the structural conditions and tectonic settings under which such *fault-valve* behaviour is most likely to occur.

From Knipe, R. J. & Rutter, E. H. (eds), 1990, *Deformation Mechanisms, Rheology and Tectonics*, Geological Society Special Publication No. 54, pp. 15–28.

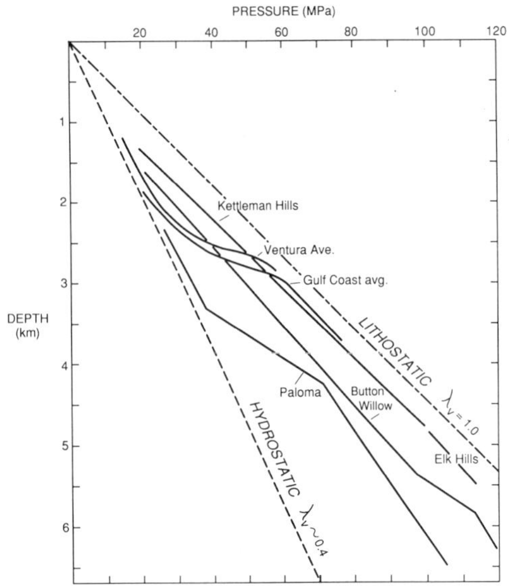

Fig. 1. Measurements of fluid pressure vs depth in sedimentary basins within the San Andreas fault system, California (after Yerkes *et al.* 1985). Button Willow, Elk Hills, Kettleman Hills and Paloma are oil fields lying in a swathe along the southwestern margin of the San Joaquin Valley ≤50 km from the San Andreas fault; Ventura Avenue field is flanked by seismically active faults within the Western Transverse Ranges *c.*70 km southwest of the San Andreas fault; Gulf Coast average trend presented for reference.

Fluid pressure and fault stability within seismogenic crust

The seismogenic regime occupies the upper part of deforming continental crust and appears to represent the zone of unstable frictional sliding. Background microearthquake activity extends to depths of 10–15 km in regions of moderate to high heat flow (60–100 mW m^{-2}), but deepens significantly in regions of rapid thrust convergence (> 10 mm a^{-1}) such as the thrust-front of the Himalaya. Larger shocks (**M** > 5.5) tend to nucleate towards the bottom of this seismogenic zone, with their ruptures expanding mostly laterally and upwards over existing fault surfaces. There is however, some uncertainty as to how far large ruptures may propagate downward beneath the background microseismic zone (Strehlau 1986; Scholz 1988). In quartzo-feldspathic crust, the base of the seismogenic zone is inferred to coincide with the onset of significant quartz plasticity under greenschist facies metamorphic conditions at temperatures greater than *c.*350°C (Sibson 1977, 1983). It appears to represent the gradual transition from unstable frictional (FR) faulting, where shear resistance is dominantly pressure-dependent and increases with depth, to quasi-plastic (QP) aseismic shearing in mylonitic belts where flow shear resistance is temperature-dependent and decreases with depth (Fig. 2). Peak shear resistance is therefore inferred to occur in the vicinity of the FR/QP transition, the zone of large rupture nucleation, where a mixture of continuous and discontinuous deformation must occur over a large range of strain rates.

Fault stability

Within the seismogenic zone the shear resistance and stability of fault surfaces is likely to be governed approximately by an empirical relationship of Coulomb form:

$$\tau = C + \mu_s(\sigma_n - P_f) \qquad (5)$$

where τ and σ_n are respectively the shear and normal stresses on the fault (Fig. 3), C is its cohesive or cementation strength (which in the case of an existing fault may be rather low), and μ_s is the coefficient of static friction. Experimental studies have shown that $0.6 < \mu_s < 0.85$, virtually independent of rock type (apart from a few clay minerals) (Byerlee 1978), a good representative value over a broad range of normal stress being $\mu_s = 0.75$. It is clear from this relationship that fault instability may be triggered either by an increase in τ due to elastic strain accumulation, by a decrease in σ_n, or by an increase in P_f causing a reduction in effective normal stress across potential slip planes. Clear evidence for this last instability mechanism within at least the top few kilometres of the crust is provided by cases of reservoir-induced seismicity, and by earthquakes triggered by fluid injection during mining, forced oil recovery and waste disposal (Healy *et al.* 1968; Raleigh *et al.* 1976; Talwani & Acree 1985). This is not to deny the possible role of other water-related effects such as stress corrosion contributing to time-dependent fault failure (Das & Scholz 1981). Additional evidence of the interplay between faulting and fluid flow is provided by fluctuations in gas/oil/water-well pressures accompanying earthquakes, by instances of post-seismic fluid discharge from well-consolidated rocks, and by fault-hosted hydrothermal mineralization with textures recording incremental precipitation (Phillips 1972; Sibson 1981). In this regard, it is also interesting to note the global correlation between active seismic belts and the distribution of CO_2-rich

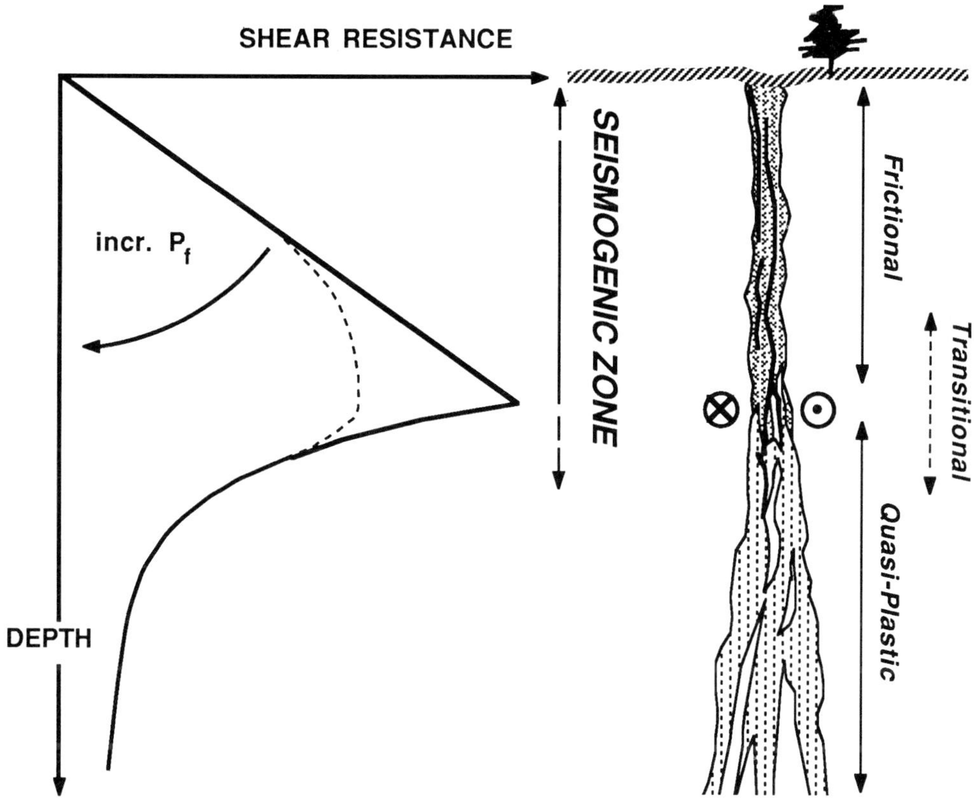

Fig. 2. Shear resistance profile for a transcrustal fault zone, showing the smoothed-out peak in shear resistance around the frictional/quasi-plastic transition, and the effect of increasing fluid pressure within the frictional regime (after Sibson 1984). Note that increasing fluid pressure will tend to deepen the FR/QP transition.

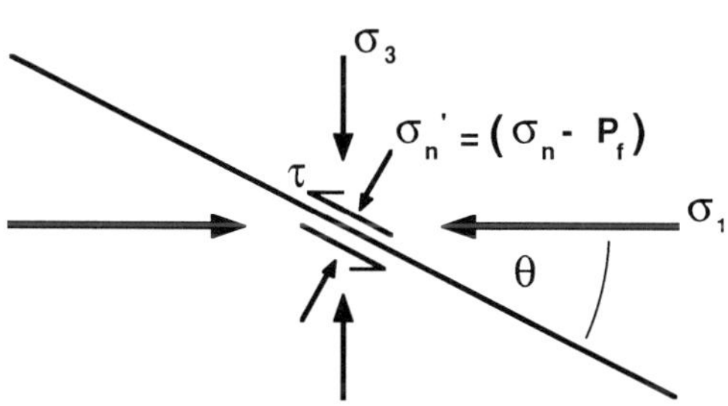

Fig. 3. Resolved components of shear stress (τ) and effective normal stress (σ_n') affecting the stability of a fault whose pole lies in the σ_1/σ_3 plane, shown in relation to the principal compressive stresses.

springs (>1000 ppm HCO_3^-) (Barnes *et al.* 1984).

Favourable versus unfavourable fault reactivation

In homogeneous isotropic rock under triaxial stress (principal compressive stresses, $\sigma_1 > \sigma_2 > \sigma_3$), brittle faults appear to form in accordance with the Coulomb criterion for shear failure of intact material (Anderson 1905). Because the coefficient of internal friction for most rocks lies in the range $0.5 < \mu_i < 1.0$ (Jaeger & Cook 1979), first-formed faults therefore tend to lie in planes which contain the σ_2 axis and lie at $\theta = \pm 0.5\tan^{-1}(1/\mu_i) = \pm 22-32°$ to the maximum compressive stress, σ_1 (Fig. 3).

The ease of reactivation of *existing* faults varies with their orientation in the prevailing stress field. This is simply illustrated in an analysis of the conditions for frictional reactivation of cohesionless faults obeying the simplified criterion:

$$\tau = \mu_s(\sigma_n - P_f) \qquad (6)$$

(Sibson 1985). We restrict ourselves here to 2-D stress analysis because it allows us to neglect the effect of σ_2, whose relative value to the other principal stresses is only rarely constrained. The ratio of effective principal stresses, $\sigma_1'/\sigma_3' = (\sigma_1 - P_f)/(\sigma_3 - P_f)$, needed for frictional reactivation of an existing fault with $\mu_s = 0.75$, whose pole lies in the σ_1/σ_3 plane, is plotted against the angle of reactivation, θ_r, in Fig. 4. Clearly for Byerlee-type friction, the optimal orientation for fault reactivation, θ_r^*, lies close to the fault's original orientation with respect to the stress field. Faults that remain close to *Andersonian* attitudes (thrusts dipping at 22–32° in compressional regimes where $\sigma_v = \sigma_3$, normal faults dipping at 58–68° in extensional regimes where $\sigma_v = \sigma_1$, and vertical strike-slip faults lying at 22–32° to σ_1 where $\sigma_v = \sigma_2$) are therefore *favourably* oriented for reactivation.

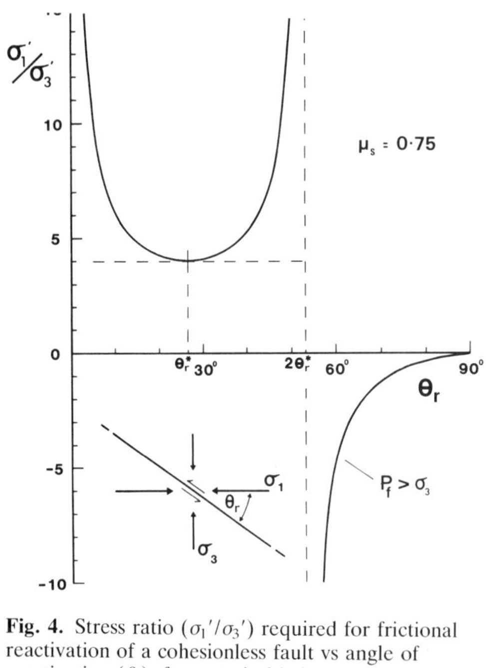

Fig. 4. Stress ratio (σ_1'/σ_3') required for frictional reactivation of a cohesionless fault vs angle of reactivation (θ_r), for a static friction coefficient, $\mu_s = 0.75$ (after Sibson 1985).

However, as the reactivation angle departs further from the optimal angle, by more than ±15°, say, the stress ratios required for reshear increase markedly and the faults become *unfavourably* oriented for reactivation. Under such circumstances, reactivation is only likely to continue under elevated fluid pressures with $\sigma_3' \to 0$; otherwise the differential stresses required for reactivation exceed those required for the formation of a new through-going fault (Sibson 1989). For reshear at reactivation angles, $\theta_r > 2\theta_r^*$, a *necessary* condition for reactivation is that $\sigma_3' < 0$ or $P_f > \sigma_3$. This special requirement for faults that are *severely misoriented* plays an important role in *extreme* fault-valve behaviour (see below). Unless it is met, faults become frictionally locked-up as $\theta_r \to 2\theta_r^*$.

In this regard it is interesting to note the dip range of 70–30° observed for normal-slip faults that are seismically active today (Jackson 1987). The range is broadly consistent with the faults initiating as steep *Andersonian* normal faults with σ_1 vertical, 'dominoing' to lower dips with increasing regional extension, and locking-up as the reactivation angle approaches 60°. These observations also lend support to the application of Byerlee-type friction coefficients to natural fault systems.

Shear resistance profiles

On the basis of Byerlee's (1978) friction laws and laboratory-derived flow laws for dislocation creep in quartz and/or feldspar rich rocks, it is possible to construct shear resistance profiles defining the likely depth of frictional interaction in transcrustal fault zones for a given geotherm and lithology (Fig. 2). Such profiles account fairly satisfactorily for observed depth distributions of seismic activity (Sibson 1983, 1984) but for a variety of reasons seem likely to provide only an upper bound to actual strength profiles (for an alternative view, see Scholz 1988). Among their many shortcomings is the neglect of diffusional flow mechanisms such as pressure solution, which are probably important for fine-grained quartz-bearing fault rocks in the temperature interval 200–400°C (McClay 1977), and which may notably round off the shear resistance peak in the vicinity of the FR/QP transition.

Of particular concern here, however, is that the profiles are generally constructed on the *assumption* of hydrostatic fluid pressures within the frictional regime. A trend towards suprahydrostatic fluid pressures with depth would curve the frictional shear resistance line towards the depth axis and would reduce the amplitude of the intercept with the plastic flow curve (Fig. 2).

Fault-valve behaviour

Fault-valve activity depends on the ability of faults to behave as impermeable seals in the interseismic period, but to form highly permeable channelways for fluid flow immediately postfailure as a consequence of the inherent roughness of natural rupture surfaces. The sealing properties of inactive faults are well known in the oil industry and may arise through the presence of clay-rich gouge or through hydrothermal cementation, perhaps assisted by processes of solution transfer in fine-grained fault material (Angevine *et al*. 1982). Also pertinent is the observation that the highest permeability zones in geothermal steam fields tend to be found in the vicinity of the most recently active faults, older fault breccia zones having become choked by hydrothermal deposition (Grindley & Browne 1976).

For fault-valve behavior to occur, a fault must cut across a vertical fluid pressure gradient that exceeds the hydrostatic gradient of *c*. 10 MPa km^{-1} (Fig. 5). Suprahydrostatic gradients may occur over broad regions, or may be confined to the vicinity of the fault itself when, for example, a fault zone penetrates an anhydrous

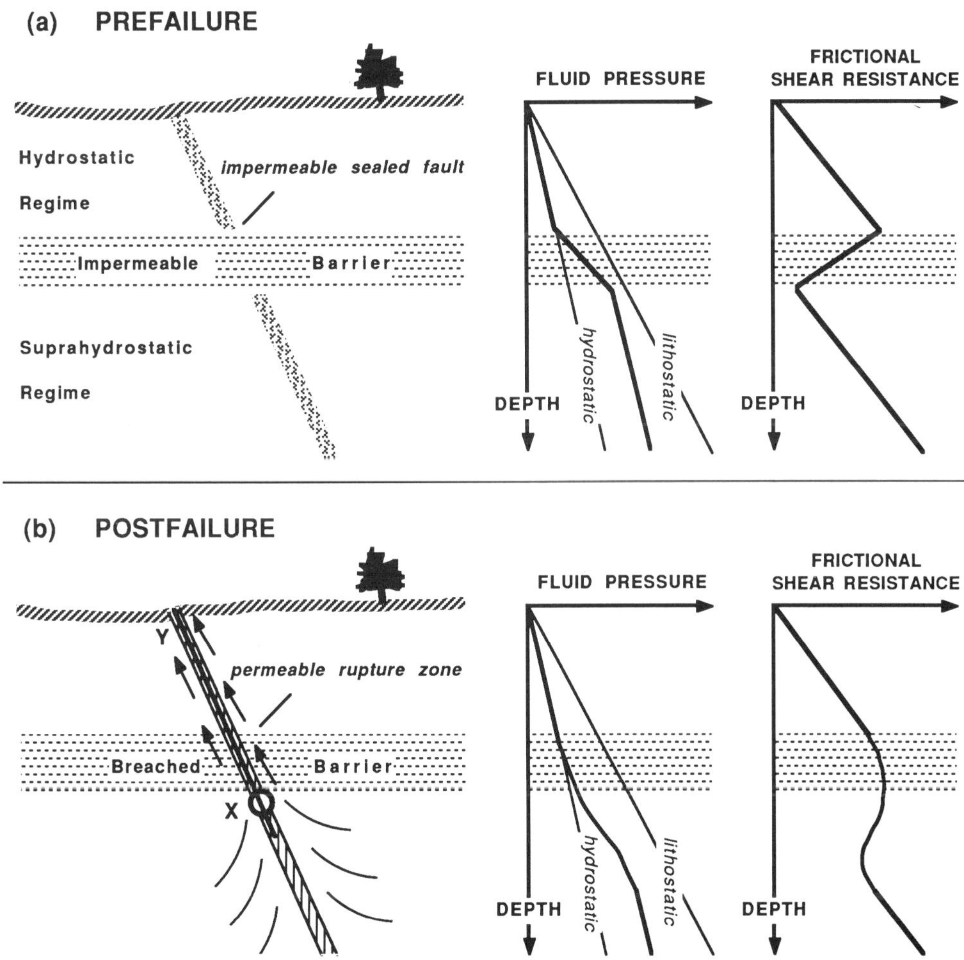

Fig. 5. Potential for fault-valve behaviour (**a**) Impermeable barrier separating hydrostatic and suprahydrostatic fluid pressure regimes. (**b**) Breaching of barrier by fault rupture X-Y, leading to an upwards discharge of fluids.

assemblage of impermeable crystalline rocks. Development of such gradients requires the existence of localized or regionally extensive low permeability barriers. These may take various forms such as laterally extensive sequences of shales or evaporites, a crystalline rock assemblage, or a horizon of hydrothermal cementation and sealing that may be widespread or localized within the fault zone itself (Etheridge *et al.* 1983).

Suprahydrostatic fluid pressures may develop beneath such barriers from a variety of causes including aquathermal pressuring, sediment compaction from overburden or tectonic loading, and progressive metamorphic dehydration (the latter being enhanced by any accompanying plutonism), or through permeable connections to some remote high fluid pressure source at greater depth (Fyfe *et al.* 1978). Valve action initiates when the shear failure condition (eqn. 5) is met through the accumulation of shear stress and/or fluid pressure within the overpressured zone (e.g. point X in Fig. 5b) and the permeability barrier is breached by the ensuing fault rupture (X–Y). The resulting upwards discharge of fluids along the fault from the overpressured zone goes on until the entire hydraulic gradient reverts to hydrostatic, or the fault reseals. The abrupt drops in fluid pressure accompanying discharge seem likely to be responsible for much hydrothermal deposition in lenticular 'ride-overs' along the deeper portions of crustal fault zones (Sibson *et al.* 1988).

Upper bounds to λ_v. The maximum values of λ_v likely to occur in the three fundamental tectonic environments, where σ_v is equal to one or other of the three principal compressive stresses (Anderson 1905), may be explored on the assumption that the upper bound to fluid pressure levels is limited by the formation of hydraulic extension fractures which would allow fluids to escape from the overpressured zones. Under conditions of low differential stress ($(\sigma_1 - \sigma_3) < 4T$, where T is the tensile rock strength), the criterion for hydraulic fracturing is:

$$P_f = \sigma_3 + T \qquad (7)$$

Thus, in the stress regimes associated with extensional normal faulting ($\sigma_v = \sigma_1$) and strike-slip faulting ($\sigma_v = \sigma_2$), where σ_3 is horizontal and less than the overburden pressure, vertical extension fractures may form at $\lambda_v < 1$, the precise value depending on the relative values of differential stress and overburden pressure (Sibson 1981). The upper bound to λ_v in an extensional regime will in general be somewhat less than in a strike-slip stress regime. However, in a thrust stress regime where $\sigma_v = \sigma_3$, the condition $\lambda_v > 1$ may prevail so long as the rocks retain some finite tensile strength. Any hydraulic extension fractures formed in such a setting would tend to be subhorizontal, and would provide less suitable drainage channels for the relief of fluid overpressures. Under such circumstances, *metastable water sills* may form beneath impermeable layers, perhaps persisting for extended time periods (Price 1975; Fyfe *et al.* 1978).

Importance of unfavourably oriented faults. Faults that remain active though unfavourably oriented for reactivation in the prevailing stress field are especially prone to fault-valve behaviour because, as previously noted, their continued reactivation is likely to depend on the presence of elevated pore fluid pressures. For activity to be maintained, an additional requirement is the absence of any throughgoing faults that are more favourably oriented for reactivation. Note especially that the presence of cohesionless favourably oriented faults prevents the condition $P_f > \sigma_3$ from being achieved, since shear failure leading to discharge is then always induced before σ_3' becomes less than zero.

In general, therefore, the likelihood of valve action increases with the degree of active fault misorientation in the stress field. However, severe misorientation at $\theta_r > 2\theta_r^*$ may lead to especially marked fault-valve behaviour because of the prefailure requirement, $P_f \geq \sigma_3$. As a result of this requirement, arrays of hydraulic extension fractures oriented perpendicular to σ_3 may open up in the overpressured zone prior to fault failure, providing an extensive fluid reservoir for rapid postfailure discharge. If Byerlee friction prevails, severely misoriented faults liable to this enhanced form of fault-valve activity would include low-angle normal faults with dips $<c.35°$, vertical strike-slip faults oriented at high angles ($>c.55°$) to σ_1, and high-angle reverse faults with dips $>c.55°$.

Optimal conditions for fault-valve behaviour

Under what circumstances is fault-valve action likely to be optimized with a rapid, high-volume discharge following fault rupture? First, it is clear that a regime of horizontal compression with $\sigma_v = \sigma_3$ would allow the highest degree of overpressuring prior to failure ($\lambda_v \geq 1$). Any hydraulic extension fractures that formed in such a regime would tend to be subhorizontal and would therefore not breach low-permeability barriers. Second, rapid upwards discharge along the steepest fluid pressure gradient would be favoured by a steeply inclined fault (Fig. 5). Third, for a major discharge to occur postfailure, a large volume of fluid should be stored within the overpressured zone with ready access to the fault, perhaps in a high-permeability fracture mesh. Given the preceding discussion on the fluid pressure requirements for reactivation of high-angle reverse faults, it becomes clear that these structures are capable of the most extreme fault-valve behaviour, with fluid pressures potentially fluctuating from prefailure lithostatic to post-failure hydrostatic levels. The depth of the lithostatic-hydrostatic interface is thus likely to be time-dependent in the vicinity of such structures.

Examples of fault-valve activity

To illustrate the changes in stress and fluid pressure accompanying fault-valve behaviour, consider first some examples of postseismic fluid discharge that may result from present-day valve action and then the mechanics of two fault-hosted vein systems in compressional regimes, one favourably for reactivation, the other unfavourably oriented.

Surface discharges from modern valve action. One of the best modern examples of a postseismic discharge possibly arising from fault-valve action is the outpouring which followed the M7.5 Kern County, California, earthquake of 1952. Rupturing involved left-reverse slip on the steeply dipping (*c*.65°) White Wolf fault at the southeastern end of the San Joaquin Valley. The rupture could be traced at the surface for *c*.35 km, but probably extended along strike for three times that distance at depth. It was followed by the surface effusion of *c*.$10^7 m^3$ of water over a period of two months during the aftershock phase, much of the discharge coming out of the crystalline batholith rocks of the hanging wall (Briggs & Troxell 1955, Sibson 1981). Casing pressures in oil wells situated within the hanging wall block, some 12 km south of the surface trace, also rose dramatically in the days after the event and then declined over a period of about a fortnight (Johnston 1955). Surface discharges of hydrocarbon fluids have also been recorded accompanying major historical earthquakes in the western Transverse Ranges of California, where fluid pressures in the actively deforming regions are known to approach lithostatic values (see Ventura Ave trend, Fig. 1) and the faulting is likewise dominantly of high-angle reverse or reverse-oblique character (Hamilton *et al.* 1969; Yerkes & Lee 1987). Postseismic discharges are also known from steep reverse-oblique faults in other arid seismically active terrains such as Iran (e.g. Tchalenko & Berberian 1974), but are generally unquantifiable.

Valve action on favourably oriented faults. Regions of elevated fluid pressure are likely to be loci for the formation of new faults because of the marked local reduction in crustal strength. In such circumstances, fault-valve action may follow fault inception. As an example, consider the mesothermal gold-quartz veins of Grass Valley which occupy a conjugate set of thrust faults cutting a greenstone-dominant assemblage intruded by an early Cretaceous granodiorite (Johnston 1940; Bohlke & Kistler 1986). The faults intersect each other at 50–60° (Fig. 6a) and are zones of intense fracturing, generally of the order of a metre or so in thickness, with varying proportions of gouge, breccia, and hydrothermal quartz and carbonate. Carbonate alteration of the wallrocks is widespread. Although total shear displacements across the fault-veins appear to be only of the order of a few metres, the vein textures record multiple episodes of quartz deposition alternating with slip increments. It seems probable that the fault system hosting the veins developed under suprahydrostatic fluid pressures, with the thrusts acting as self-sealing fault-valves through the hydrothermal precipitation accompanying each discharge episode and drop in fluid pressure.

Effective stress conditions at failure and immediately postfailure for an *Andersonian* thrust fault acting as a fault-valve are illustrated in Fig. 6a. Note that while suprahydrostatic fluid pressures at failure are inferred to give rise to postfailure discharge, there is here no direct control on the extent to which hydrostatic pressure was exceeded, nor on the amplitude of the ensuing fluid pressure fluctuation. From the Mohr diagram, it is clear that shear failure under suprahydrostatic fluid pressures, leading to fault-valve discharge, causes reductions in both the differential stress and the fluid pressure. For failure to recur, either differential stress must build up to a much greater level under hydrostatic fluid pressures, or suprahydrostatic fluid pressures must reaccumulate beneath sealed portions of the faults.

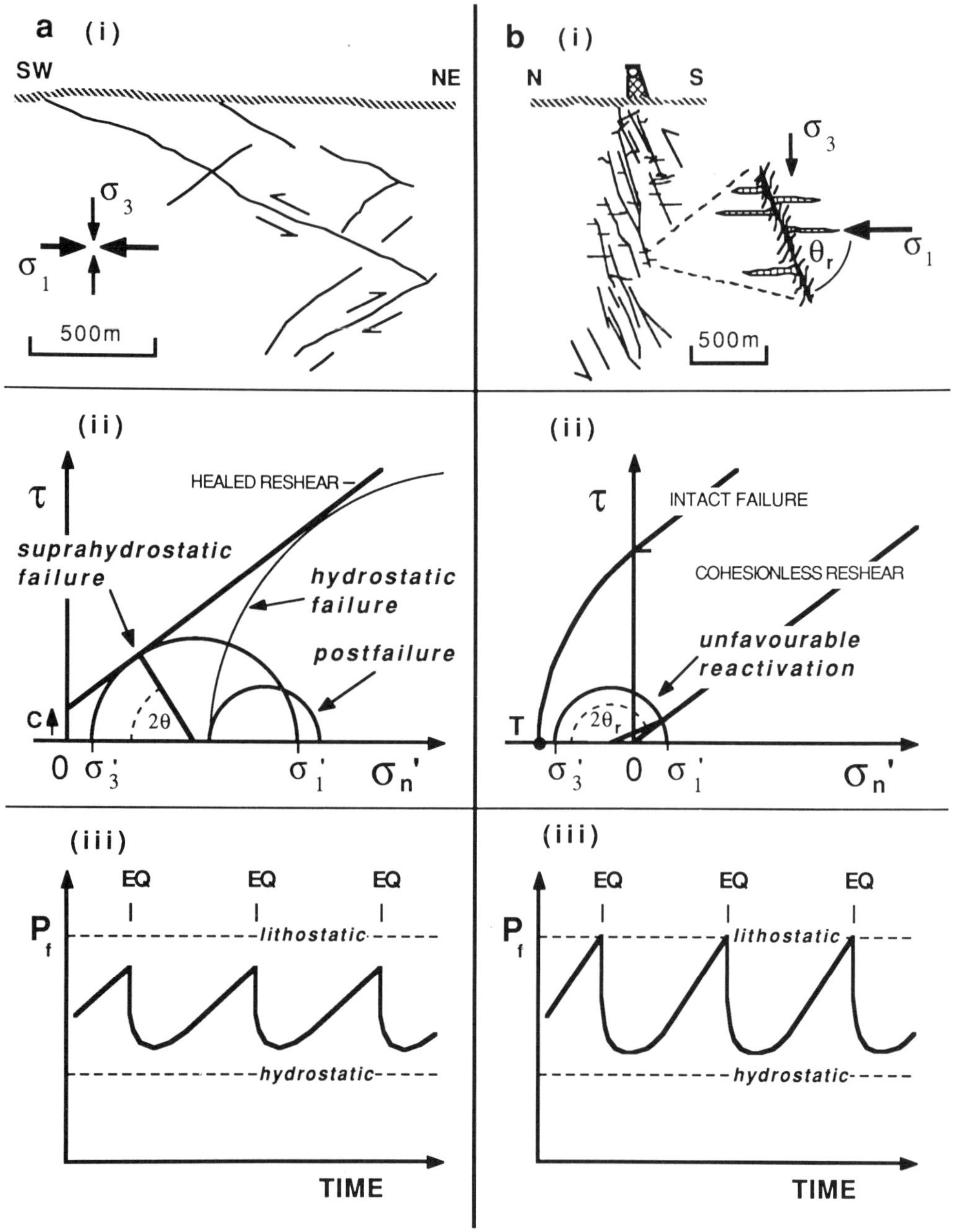

Fig. 6. Examples of vein systems attributable to fault-valve behaviour. (**a**) *Favourably oriented faults*: (**i**) section through North Star and subsidiary Au–quartz lodes occupying a set of conjugate thrust faults, Grass Valley, California (after Johnston 1940); (**ii**) Mohr diagram illustrating stress conditions at failure and postfailure; (**iii**) fluid pressure cycles inferred to accompany successive earthquake (EQ) ruptures.
(**b**) *Unfavourably oriented faults*: (**i**) section through Au–quartz lodes hosted by a system of high-angle reverse faults and associated extension fractures ($\theta_r \approx 70°$), Sigma Mine, Val d'Or, Quebec (after Robert & Brown 1986); (**ii**) Mohr diagram illustrating stress condition which allows reactivation of a severely misoriented fault without causing failure of surrounding intact rock; (**iii**) resultant fluid pressure cycling accompanying successive earthquake (EQ) ruptures.

Valve action on unfavourably oriented faults. Many mesothermal gold-quartz lodes are hosted within granite-greenstone terrains on faults that were undergoing high-angle reverse or reverse-oblique motion at the time of mineralization. Examples include the Cretaceous Mother Lode vein system of California (Knopf 1929), and many of the Archean lode gold deposits in Canada such as those at Yellowknife (Henderson & Brown 1965), or those associated with the Porcupine–Destor and Kirkland Lake–Cadillac breaks within the Abitibi belt (Roberts 1987).

A representative example of these last is the gold-bearing quartz–tourmaline vein system at the Sigma Mine in Val d'Or, Quebec, formed during almost pure reverse motion on steeply-dipping (*c*.70°) shears (Robert & Brown 1986) (Fig. 6b). As with many deposits of this kind the veins have tremendous vertical extent, in this instance approaching 2 km. They are hosted in a mesh of mixed continuous–discontinuous ('brittle–ductile') shear zones in which L–S tectonite fabrics were syntectonically disrupted by discrete shears and vein fractures. Fluid inclusion studies show that the veins were precipitated from a low salinity, mixed H_2O-CO_2 fluid at temperatures of *c*.300–400°C (Robert & Kelly 1987), consistent with the greenschist assemblage of the shear zones. As at Grass Valley, carbonate alteration of the wallrocks, in this case andesitic metavolcanic rocks intruded by a porphyritic diorite, is widespread. Two main vein-sets occur (Fig. 6b); lenticular *fault-veins* lying subparallel to schistosity within the steeply dipping shear zones, and *flats* (subhorizontal extension veins). Both types of vein range up to a metre or so in thickness. At Sigma, the flats are known to extend tens of metres from the shears, but in other mines they have been traced laterally for several hundred metres. Within each vein-set, composite vein textures record histories of incremental deposition, the purely extensional character of the flats being revealed by relict crack-seal microstructure (Robert & Brown 1986). No consistent cross-cutting relationship exists between the different vein-sets, with the implication that the two sets developed incrementally at different stages of a repeating cycle.

The vein-sets at Sigma Mine are interpreted as resulting from fault-valve action on severely misoriented high-angle reverse faults, with the flats opening up whenever the prefailure condition of supralithostatic fluid pressures ($P_f > \sigma_3$) was met, and the fault-veins developing during episodes of postfailure discharge upwards through the shear zones (Sibson *et al.* 1988). Potentially, such valve action could result in a sawtooth oscillation between slightly supralithostatic and hydrostatic levels of fluid pressure, representing the *extreme* form of fault-valve behaviour (Fig. 6b). Phase separation of CO_2 and intense hydrothermal precipitation are inferred to have accompanied the abrupt pressure drops accompanying discharge. Stress conditions for this extreme fault-valve behaviour at large reactivation angles are illustrated in the Mohr diagram in Fig. 6b. Note that fault reactivation of this kind necessarily occurs under low differential stress, otherwise more favourably oriented faults may form. Steep fault-veins in several of these deposits are, in fact, quite commonly disrupted by late shallow-dipping *Andersonian* thrusts, implying that reactivation of these unfavourably oriented structures persisted only while fluid pressures could be restored to approximately lithostatic levels before differential stress built up to a level sufficient to induce shear failure in the surrounding intact rock.

Tectonic settings for fault-valve behaviour

It is apparent from the foregoing that the suprahydrostatic fluid pressure gradients needed for fault-valve behaviour are most likely to develop in compressional regimes. In such regions, crustal thickening through accretion and thrust stacking leads to progressive dewatering of the thickened crust and the maintenance of suprahydrostatic fluid pressure gradients over long periods. While fault-valve behaviour may set in with the inception of favourably oriented faults in overpressured regions (as appears to have been the case at Grass Valley), valve activity is likely to be most intense where fluid pressures at depth approach lithostatic values and the only through-going faults are very unfavourably oriented for reactivation. As previously discussed, the most extreme fault-valve behaviour is likely to be associated with high-angle reverse faults, which form the focus for our attention from here on. However, it should be borne in mind that less intense fault-valve activity may also occur on other severely misoriented faults, such as vertical strike-slip faults lying at a high angle to σ_1 and very low-angle normal faults (dips $<c.35°$), provided suprahydrostatic fluid pressure gradients can be maintained in their vicinity.

High-angle reverse faults generally appear to have developed through the reactivation of earlier-formed faults, or through progressive misorientation of faults in the prevailing stress

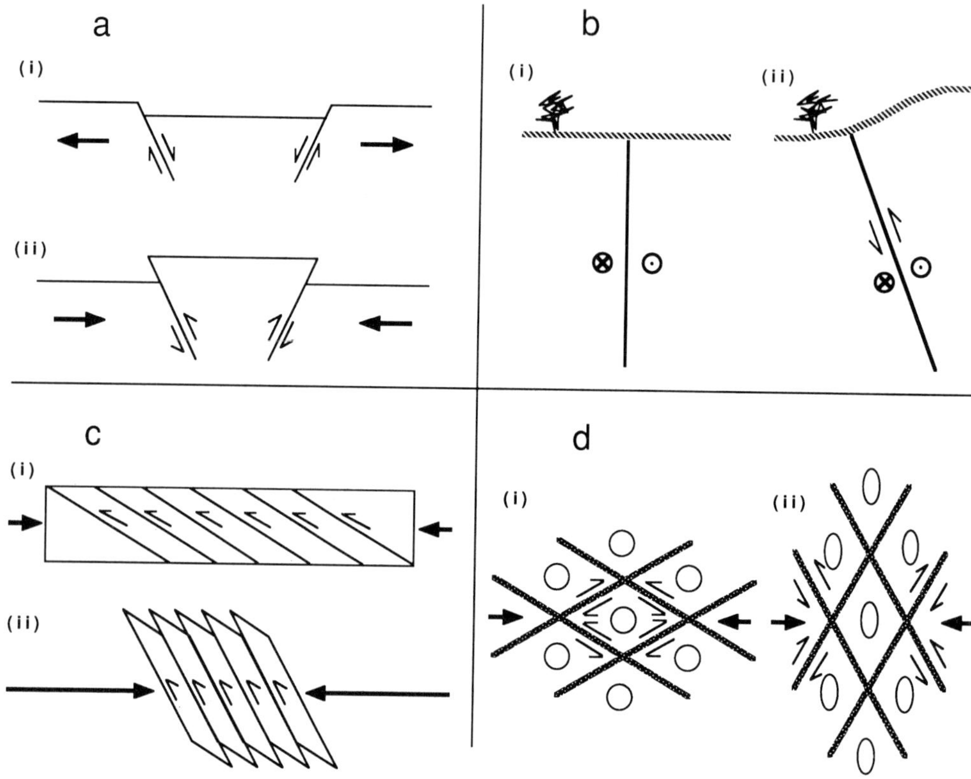

Fig. 7. Possible origins for high-angle reverse faults: (**a**) shortening of formerly extended and rifted crust; (**b**) transpression across originally vertical strike-slip faults; (**c**) dominoing of thrust stacks; (**d**) bulk shortening of a mesh of conjugate shears.

field. It is appropriate, therefore, to explore the tectonic settings under which such faults may evolve and remain active, giving rise to the most extreme forms of fault-valve activity (Fig. 7).

Shortening of formerly extended and rifted crust

Wherever formerly extended crust comes under compression, inherited steep normal faults may become reactivated as high-angle reverse faults (Fig. 7a). This situation is particularly likely to arise through the Wilson Cycle of ocean opening and closure, when formerly rifted continental margins collide or back-arc basins close (e.g. Winslow 1981). A present-day example is the high-angle reverse faulting currently active beneath the Zagros fold-belt (Jackson 1980).

Transpression across originally vertical strike-slip faults

Transpression involving high-angle reverse or reverse-oblique faulting may develop locally at antidilational jogs along major strike-slip faults (Sibson 1986), or may affect an entire transform fault system on a regional scale when the inter-plate slip vector has changed with time (Fig. 7b). Such seems to have been the case with the Alpine fault system of New Zealand where steep dextral-reverse slip now prevails along the Alpine fault itself (Sibson *et al.* 1979), and subsidiary faults bounding adjacent trans-tensional basins formed in the mid-Tertiary have likewise, since the late Miocene, been re-activated with large components of high-angle reverse slip (Norris *et al.* 1978). Some of the steeply dipping reverse faults within the western

Transverse Ranges of California seem likely to have had a similar origin (Yerkes & Lee 1987).

Dominoing of thrust stacks

Sets of initially low-dipping thrusts may steepen progressively by 'dominoing' to lock up as an imbricate stack of steep reverse faults (Fig. 7c). Such imbricate stacks are well known within the Moine Thrust and other thrust belts (e.g. McClay & Coward 1981), but may also develop on larger scales. Subduction-accretion complexes form one setting where imbrication and steepening of early-formed thrusts as a consequence of underthrust stacking may lead to an environment especially suitable for fault-valve behaviour. Dewatering of the underthrust material provides a mechanism for the maintenance of high fluid pressures, allowing continuing activity on the reverse faults as they progressively steepen (Fig. 8). Steep reverse faults at the rear of accretionary complexes appear to form particularly favourable hosting sites for mesothermal gold–quartz lodes (e.g. Mother Lode of California, Bohlke & Kistler 1986; Juneau gold belt, Alaska, Goldfarb *et al.* 1988), with fault-valve activity presumably acting as the precipitating mechanism.

Bulk shortening of conjugate shear zone meshes

A conjugate mesh of initially shallow-dipping shears may steepen progressively through bulk inhomogeneous shortening of the lozenge-shaped areas between the shears (Bell 1982). For example, under plane strain conditions, a bulk shortening of *c*.42% will steepen an initial conjugate set of $\pm 30°$ dipping Andersonian thrusts to a mesh of high-angle reverse faults dipping at $\pm 60°$ (Fig. 7d). This mechanism, combined with some dominoing of originally low-angle thrust stacks, may be responsible for the dominance of steep reverse faults within terrains such as the Palaeozoic Lachlan Fold Belt of southeastern Australia (Gray 1988), where gold–quartz lodes associated with the reverse faults provide much evidence for fault-valve activity (Cox *et al.* 1990).

Favoured level for valve activity

Some form of fault-valve behaviour may be expected at any level within overpressured portions of seismically active crust, wherever an active fault cuts an impermeable barrier. However, several lines of evidence suggest that valve activity may be particularly prevalent near the base of the seismogenic zone. Though absolute depth control is poor, the P–T conditions generally inferred for fault-hosted mesothermal gold-quartz lodes ($300° < T < 400°C$; $0.2 < P = \sigma_v < 0.4$ GPa. Kerrich 1986; Robert & Kelly 1987) are broadly consistent with the low- to mid-greenschist metamorphic environment believed to delimit the base of seismic activity in the mid-crust. Many, though not all, of the hosting shear zones themselves contain greenschist tectonite-vein assemblages which exhibit the mixed continuous-discontinuous ('brittle–ductile') style of fault deformation inferred for the FR/QP transition. On these grounds, and because formation of the veins appears linked to the pre- and postfailure fluid pressure conditions for severely misoriented faults, the vein systems are inferred to represent *fossil* nucleation sites for moderate to large earthquake ruptures (Sibson *et al.* 1988).

Why should fault-valve activity be especially prevalent at this crustal level? It may be, in accordance with the ideas of Etheridge *et al.* (1983), that the base of the seismogenic zone acts as a precipitation cap to regions of prograde metamorphism at depths where processes of solution transfer are operative over broad areas, and the actively deforming shear zones form the principal conduits for fluid discharge (see also Cox *et al.* 1986).

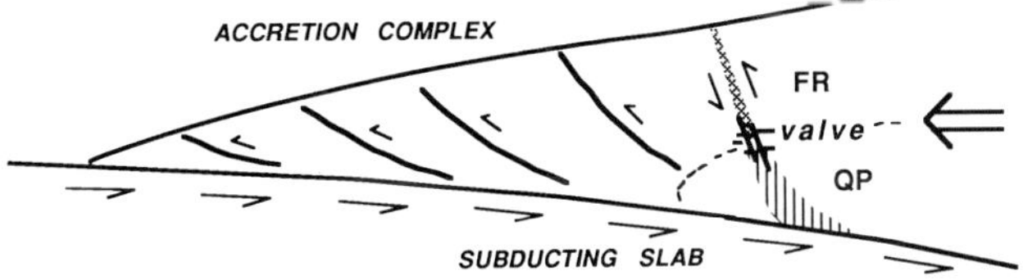

Fig. 8. Schematic representation of fault-valve behaviour at the rear of a subduction-accretion complex (after Goldfarb *et al.* 1988).

Discussion

A case has been presented that fault-valve behaviour, leading to episodic postfailure discharge upwards along fault zones, is probably common in regions of crustal accretion, shortening and thickening where fluid overpressuring may be widespread. Valve action is likely to be especially pronounced in terrains where the only through-going faults are unfavourably oriented for reactivation, the most extreme activity being associated with high-angle reverse faults. Although the reactivation analysis has not been extended to 3-D, it is apparent that valve action is also likely on steep reverse-oblique faults, and perhaps also on vertical strike-slip faults oriented at high angles to σ_1. Valve activity may occur at any level within the seismogenic zone wherever faults transect permeability barriers, but appears to be especially prevalent near its base where unstable frictional faulting gives way to aseismic shearing with increasing depth.

Structural and tectonic implications

The structural implications of fault-valve behaviour are many. It is clear first that, because of the fluid pressure cycling, shear resistance profiles for faults exhibiting valve behaviour must be *time-dependent* with marked strength reductions occurring beneath impermeable barriers prior to fault failure (Fig. 5). Moreover, the depth to the FR/QP transition within fault zones exhibiting valve behaviour, and the style of shearing at a particular crustal level, may cycle with the changing fluid pressure. Evidence for these effects comes from the mixed continuous-discontinuous style of deformation observed in may of the shear zones which host mesothermal vein systems attributable to fault-valve action. Observed field relations (e.g. Robert & Brown 1986; Roberts 1987) are consistent with the penetrative L-S tectonite fabrics having developed through aseismic shearing under comparatively low fluid pressures ($P_f \rightarrow$ hydrostatic?), with discrete seismic slip surfaces and associated vein-fractures forming at the same level during the intervals of high fluid pressure ($P_f \rightarrow$ lithostatic.) This is entirely in accord with expectations from experimental rock deformation, where increases or decreases in fluid pressure may cause reversions from ductile to brittle, and from brittle to ductile deformation, respectively, under a fixed confining pressure (Rutter 1974).

In compressional regimes, the concentrations of flat hydraulic extension fractures that may develop near the base of the seismogenic zone as the result of fault-valve activity on steep reverse faults also have broader tectonic significance. Arrays of subhorizontal fluid-filled cracks (water sills), which in at least some instances may extend laterally for hundreds of metres, must contribute to the formation of a weak decoupling horizon at this structural level. This may aid the detachment of upper crustal thrust flakes. As noted by Oxburgh (1972), such thrust flakes are typically of the order of 10–15 km in thickness, corresponding approximately to the thickness of the continental seismogenic zone.

Identifying fault-valve behaviour

Recognition of fault-valve activity in ancient fault zones is probably easiest for the extreme cases where mutual cross-cutting relationships can be established between prefailure extension veins lying in the σ_1/σ_2 plane and postfailure discharge veins lying within the fault zones (see Fig. 6b). However, less extreme forms of valve action (where $P_f < \sigma_3$) may occur without prefailure hydraulic extension fracturing. Postfailure *discharge* veins with textures recording incremental deposition alternating with episodes of fault slip then provide the main evidence for valve activity. Such veins may lie along brittle faults (as at Grass Valley), or may have developed syntectonically within mixed 'brittle-ductile' shear zones. Lenticular fault-veins of quartz, lying subparallel to mylonitic foliation are, in fact, fairly common within greenschist shear zones (e.g. White *et al.* 1982). More subtle microstructural evidence for fluid pressure cycling may also be preserved in some circumstances (White & White 1983; McCaig 1987; O'Hara 1988).

In the sense that they represent *transient discontinuities* within otherwise continuously deforming shear zones, hydrothermal fault-veins of this latter kind can be regarded as the *wet* equivalent of the smeared-out pseudotachylites believed to have formed under comparatively *dry* conditions at comparable crustal levels (Sibson 1980). The depth within fault zones to which such wet transient discontinuities extend has relevance to the vexed question of how deep large ruptures may propagate beneath the background microseismic zone (Strehlau 1986).

Special thanks are due to N. N. Brown, S. F. Cox, F. Robert, and K. H. Poulsen for the many overpressured discussions which led to this paper, to Giant Yellowknife Mines and Western Mining Corporation for providing access to their mines, and to the organisers for the opportunity to attend the Leeds Meeting. Work leading to this paper was supported by National Science Foundation grant EAR89-04571. This paper is contribution number 0036-03EQ-18TC from the Institute for Crustal Studies at the University of California, Santa Barbara.

References

ANDERSON, E. M. 1905. The dynamics of faulting. *Transactions of the Geological Society of Edinburgh*, **8**, 393-402.

ANGEVINE, C. L., TURCOTTE, D. L. & FURNISH, M. D. 1982. Pressure solution lithification as a mechanism for the stick-slip behavior of faults. *Tectonophysics*, **1**, 151-160.

BARNES, I., IRWIN, W. P. & WHITE, D. E. 1984. *Global distribution of carbon dioxide discharges and major zones of seismicity. US Geological Survey Miscellaneous Series Map*, **I-1528**.

BELL, T. H. 1982. Foliation development—the contribution, geometry, and significance of progressive, bulk, inhomogeneous shortening. *Tectonophysics*, **75**, 273-296.

BERRY, F. A. F. 1973. High fluid potentials in California Coast Ranges and their tectonic significance. *American Association of Petroleum Geologists Bulletin*, **57**, 1219-1249.

BOHLKE, J. K. & KISTLER, R. W. 1986. Rb-Sr, K-Ar, and stable isotope evidence for the ages and sources of fluid components of gold-bearing quartz veins in the northern Sierra Nevada foothills metamorphic belt, California. *Economic Geology*, **81**, 296-322.

BRIGGS, R. C. & TROXELL, H. C. 1955. Effect of Arvin-Tehachapi earthquake on spring and stream flow. *Californian Division of Mines Bulletin*, **171**, 81-98.

BYERLEE, J. D. 1978. Friction of rocks. *Pure and applied Geophysics*, **116**, 615-626.

COX, S. F., ETHERIDGE, M. A. & WALL, V. J. 1986. The role of fluids in syntectonic mass transport and the localization of metamorphic vein-type ore deposits. *Ore Geology Reviews*, **2**, 65-86.

——, WALL, V. J., ETHERIDGE, M. A. & POTTER, T. F. 1990. Deformational and metamorphic processes in the formation of mesothermal vein-hosted gold deposits – examples from the Lachlan Fold Belt in central Victoria, Australia. *Ore Geology Reviews*, (in press).

DAS, S. & SCHOLZ, C. H. 1981. Theory of time-dependent rupture in the earth. *Journal of Geophysical Research*, **86**, 6039-6051.

ETHERIDGE, M. A., WALL, V. J., & VERNON, R. H. 1983. The role of the fluid phase during regional metamorphism and deformation. *Journal of Metamorphic Geology*, **1**, 205-226.

FERTL, W. H., CHILINGARIAN, G. V. & RIEKE, H. H. 1976. *Abnormal Formation Pressures*. Elsevier, Amsterdam.

FYFE, W. S., PRICE, N. J. & THOMPSON, A. B. 1978. *Fluids in the Earth's Crust*. Elsevier, Amsterdam.

GOLDFARB, R. J., LEACH, D. L., PICKTHORN, W. J. & PATERSON, C. J. 1988. Origin of lode-gold deposits of the Juneau gold belt, southeastern Alaska. *Geology*, **16**, 440-443.

GRAY, D. 1988. Structure and tectonics. *In*: DOUGLAS, J. D. & FERGUSON, J. A. (eds) *Geology of Victoria*. Victorian Division, Geological Society of Australia, 1-36.

GRINDLEY, G. W. & BROWNE, P. R. L. 1976. Structural and hydrological factors controlling the permeabilities of some hot-water geothermal fields. *In: Proceedings of the 2nd United Nations Symposium on Development and Use of Geothermal Resources*. New York, 377-86.

HAMILTON, R. M., YERKES, R. F., BROWN, R. D., BURFORD, R. O. & DENOYER, J. M. 1969. Seismicity and associated effects, Santa Barbara region. *US Geological Survey Professional Paper*, **679**, 47-72.

HEALY, J. H., RUBEY, W. W., GRIGGS, D. T. & RALEIGH, C. B. 1968. The Denver earthquakes. *Science*, **161**, 1301-10.

HENDERSON J. F. & BROWN, J. C. 1965. Geology and structure of the Yellowknife greenstone belt, District of McKenzie. *Geological Survey of Canada Bulletin* **141**.

HUBBERT, M. K. & RUBEY, W. W. 1959. Role of fluid pressure in the mechanics of overthrust faulting. *Geological Society of America Bulletin*, **70**, 115-205.

JACKSON, J. A. 1980. Reactivation of basement faults and crustal shortening in orogenic belts. *Nature*, **283**, 343-346.

—— 1987. Active normal faulting and crustal extension. *In:* COWARD, M. P., DEWEY, J. F. & HANCOCK, P. L. (eds) *Continental Extensional Tectonics* Geological Society, London, Special *Publication*, **28**, 3-17.

JAEGER, J. G. & COOK, N. G. W. 1979. *Fundamentals of Rock Mechanics*. 3rd edn. Chapman & Hall, London.

JOHNSTON, R. L. 1955. Earthquake damage to oil fields and to the Paloma cycling plant in the San Joaquin Valley. *Californian Division of Mines Bulletin* **171**, 221-226.

JOHNSTON W. D. 1940. The gold quartz veins of Grass Valley, California. *US Geological Survey Professional Paper*, **194**.

KERRICH, R. 1986. Fluid infiltration into fault zones: chemical, isotopic and mechanical effects. *Pure and Applied Geophysics*, **124**, 225-268

KNOPF, A. 1929. The Mother Lode system of California. *US Geological Survey Professional Paper*, **157**.

MCCLAY, K. 1977. Pressure solution and Coble creep in rocks and minerals: a review. *Journal of the Geological Society London*, **134**, 57-70.

—— & COWARD, M. P. 1981. The Moine Thrust Zone: an overview. *In*: MCCLAY K. R. & PRICE, N. J. (eds) *Thrust and Nappe Tectonics*. Geological Society, London Special Publication, **9**, 241-260.

MCCAIG, A. M. 1987. Deformation and fluid-rock interaction in metasomatic dilatant shear bands.

Tectonophysics, **135**, 121–132.

Norris, R. J., Carter, R. M. & Turnbull, I. M. 1978. Cainozoic sedimentation in basins adjacent to a major continental transform in southern New Zealand. *Journal of the Geological Society, London*, **135**, 191–205.

O'Hara, K. 1988. Fluid flow and volume loss during mylonitization: an origin for phyllonite in an overthrust setting, North Carolina, U.S.A. *Tectonophysics*, **156**, 21–36.

Oxburgh, E. H. 1972. Flake tectonics and continental collision. *Nature*, **239**, 202–204.

Phillips, W. J. 1972. Hydraulic fracturing and mineralization. *Journal of Geological Society, London*, **128**, 337–359.

Price, N. J. 1975. Fluids in the crust of the earth. *Science Progress*, Oxford, *62*, 59–87.

Raleigh, C. B., Healy, J. H. & Bredehoeft, J. D. 1976. An experiment in earthquake control at Rangely, Colorado. *Science*, **191**, 1230–37.

Robert, F. & Brown, A. C. 1986. Archean gold-bearing quartz veins at the Sigma Mine, Abitibi greenstone belt, Quebec: Part I. Geologic relations and formation of the vein system. *Economic Geology*, **81**, 578–592.

—— & Kelly, W. C. 1987. Ore-forming fluids in Archean gold-bearing quartz veins at the Sigma Mine, Abitibi greenstone belt, Quebec, Canada. *Economic Geology*, **82**, 56–74.

Roberts, R. G. 1987. Ore deposit models #1 — Archean lode gold deposits. *Geoscience Canada*, **14**, 37–52.

Rutter, E. H. 1974. The influence of temperature, strain rate and interstitial water in the experimental deformation of calcite rocks. *Tectonophysics*, **22**, 311–34.

Scholz, C. H. 1988. The brittle-plastic transition and the depth of seismic faulting. *Geolische Rundschau*, **77**, 319–328.

Sibson, R. H. 1977. Fault rocks and fault mechanisms. *Journal of the Geological Society, London*, **133**, 191–214.

—— 1980. Transient discontinuities in ductile shear zones. *Journal of Structural Geology*, **2**, 165–171

—— 1981. Fluid flow accompanying faulting: field evidence and models. *In*: Simpson, D. W. & Richards, P. G. (eds) *Earthquake Prediction: an International review*. Maurice Ewing Series **4**, American Geophysical Union, Washington D. C. 593–603.

—— 1983. Continental fault structure and the shallow earthquake source. *Journal of the Geological Society, London*, **140**, 741–167.

—— 1984. Roughness at the base of the seismogenic zone. *Journal of Geophysical Research*, **89**, 5791–5799.

—— 1985. A note on fault reactivation. *Journal of Structural Geology*, **7**, 751–4.

—— 1986. Earthquakes and lineament infrastructure. *Philosophical Transactions of the Royal Society, London*, **A317**, 63–79.

—— 1989. High-angle reverse faulting in northern New Brunswick, Canada, and its implications for fluid pressure levels. *Journal of structural Geology*, **11**, 873–878.

——, Robert, F. & Poulsen, K. H. 1988. High-angle reverse faults, fluid pressure cycling and mesothermal gold–quartz deposits. *Geology*, **16**, 551–5.

——, White, S. H. & Atkinson, B. K. 1979. Fault rock distribution and structure within the Alpine Fault Zone: a preliminary account. *In*: Walcott, R. I. & Cresswell, M. M. (eds) *The Origin of the Southern Alps*. Royal Society of New Zealand Bulletin, **18**, 55–65.

Strehlau, J. 1986. A discussion of the depth extent of rupture in large continental earthquakes. *In*: Das, S., Boatwright, J. & Scholz, C. H. (eds) *Earthquake Source Mechanics*. American Geophysical Union Geophysical Monograph, **37** (Maurice Ewing Series **6**), 147–155.

Talwani, P. & Acree, S. 1985. Pore pressure diffusion and the mechanism of reservoir-induced seismicity. *Pure and Applied Geophysics*, **122**, 947–65.

Tchalenko, J. S. & Berberian, M. 1974. The Salmas earthquake of May 6th, 1930. *Annali di Geofisica*, **27**, 151–212.

White, J. C. & White, S. H. 1983. Semi-brittle deformation within the Alpine fault zone, New Zealand. *Journal of Structural Geology*, **5**, 579–589.

White, S. H., Evans, D. J. & Zhong, D. -L. 1982. Fault rocks of the Moine Thrust Zone: microstructures and textures of selected mylonites. *Textures & Microstructures*, **5**, 33–62.

Winslow, M. A. 1981. Mechanism for basement shortening in the Andean foreland fold belt of southern South America. *In*: McClay, K. R. & Price, N. J. (eds) *Thrust and Nappe Tectonics*. Geological Society, London, Special Publication, **9**, 513–528.

Yeats, R. S. 1983. Large-scale Quaternary detachments in Ventura basin, southern California. *Journal of Geophysical Research*, **88**, 569–83.

Yerkes, R. F. & Lee, W. H. K. 1987. Late Quaternary deformation in the western Transverse Ranges. *US Geological Survey Professional Paper*, **1339**, 71–82.

Yerkes, R. F., Levine, P. & Wentworth, C. M. 1985. Abnormally high fluid pressures in the region of the Coalinga earthquakes — a preliminary report. *US Geological Survey Open-File Report*, **85–44**, 344–375.

Textures, deformation mechanisms, and the role of fluids in the cataclastic deformation of granitic rocks

JAMES P. EVANS

Department of Geology, Utah State University, Logan, UT 84322–4505, USA

Abstract: Optical and scanning electron microscopy were used to examine the textures and infer deformation mechanisms in rocks deformed at upper crustal conditions in the Geesaman oblique-slip fault, southeastern Arizona, and the Washakie thrust fault, northwestern Wyoming. Feldspar in slightly and moderately deformed rocks in both faults deformed by formation of extension and shear fractures along cleavage planes, and quartz grains deformed by brittle fracture. Fracture density is greater in feldspar than in quartz. Intergranular fractures and faults developed by the linking of intragranular fractured feldspars, which in turn resulted in the development of foliated cataclasites in the Washakie fault and phyllonites in the Geesaman fault. Clays and mica developed by the syntectonic alteration of feldspar, and with increasing mica or clay content quartz grains deformed by fracture in a progressively softer matrix. Fracturing and healing of fractures in quartz resulted in irregularly-shaped, elongate pods of quartz in a mica matrix in the Geesaman fault and in a clay and feldspar fragment matrix in the Washakie fault. The syntectonic alteration of feldspars, the presence of iron oxides in the faults, and late-stage quartz veins attest to the flow of water during deformation. The fault zones developed by early fracture in feldspar which gave way to cataclastic flow in feldspars and subsequent development of clay-rich cataclasites at low temperature and of phyllonites at high temperatures. Most of the slip in mature cataclasites was localized by slip and cataclasis of clays. Slip in the phyllonites was localized by plastic deformation of micas. The brittle to ductile transition in the rocks was the result of fracture and subsequent cataclasitic flow in feldspars, followed by deformation of clay or mica, and for the most part excluded the plastic deformation of quartz.

Detailed field and microstructural studies of grain-scale mechanisms and textures in fault-related rocks which developed in the brittle and semi-brittle regime reveal much about the conditions under which faults nucleate and grow and about the mechanical behavior of the fault zones (Engelder 1974; Sibson 1977; White & White 1983; Mitra 1984; Blenkinsop & Rutter 1986; Rutter *et al.* 1986; Simpson 1986; Chester & Logan 1986; Janecke & Evans 1988; Kamineni *et al.* 1988). Examination of the deformation mechanisms, the distribution of slip within fault zones, and the role of fluid phases also help establish the long-term rheology and strength of fault zones (Zoback *et al.* 1987). The number of studies which combine field and microstructural studies of cataclastically deformed rocks is increasing, but pales in comparison to the number of studies of rocks which have deformed primarily by crystal plasticity. Relatively little work has examined the details of grain-scale deformation mechanisms and textures in cataclasites, and few studies have determined the relationship between experimentally and naturally deformed cataclastic rocks (cf. Rutter *et al.* 1986).

This paper presents the results of detailed examination of textures and deformation mechanisms of two faults which formed in granitic rocks at relatively shallow levels. The objectives of this work are to: (1) study the processes responsible for the nucleation and growth of the faults, (2) determine the chemical and mechanical effects of fluids present during fault development; (3) examine the correlation between mechanisms and textures produced in experimentally-deformed cataclasites and naturally-formed faults in order better to understand the micromechanisms of the fault development (cf. Blenkinsop & Rutter 1986).

Study areas and fault structures

The fault-related rocks examined in this paper are from the Washakie thrust system, north-central Wyoming, and the Geesaman fault, southeastern Arizona. The Washakie thrust system thrust Archean granites and granitic gneisses to the southwest during Early Eocene time (Winterfeld & Conard 1983). Faults in the granitic rocks were responsible for the emplacement and internal deformation of the thrust

From Knipe, R. J. & Rutter, E. H. (eds), 1990, *Deformation Mechanisms, Rheology and Tectonics*, Geological Society Special Publication No. 54, pp. 29–39.

sheets. Restorable cross sections of the Washakie system show that the fault-related rocks exposed at the surface formed at a depth of 6 to 10 km, and were thrusted over folded and faulted Paleozoic and Mesozoic sedimentary rocks. Microstructural observations of the fault-related rocks (Evans 1988) show that brittle deformation of feldspars was the dominant deformation mechanism. The syntectonic alteration of feldspars to clay minerals resulted in the formation of foliated cataclasites in which elongate pods of quartz grains remain as porphryoclasts which fractured and healed during the alteration and cataclasis of the feldspar.

The faults within the thrust sheet are narrow; frontal thrusts of the system are wide fault zones (Evans 1988). Narrow faults are 1 cm to 10 cm wide and are composed of well-foliated, indurated and fine-grained gouge. Contacts between the gouge and protolith are sharp along the narrow faults. Wide faults are 10 cm to 30–40 m wide which contain red, fine-grained gouge and display a crude to well-defined foliation defined by elongate, wavy lozenges and zones of less-deformed protolith (Fig. 1). The fault studied here consists of a central 3–4 m wide zone of highly indurated and foliated gouge which is bordered by a fractured and faulted zone up to 8 m wide. This fractured layer has diffuse, gradational contacts with the nearly undeformed protolith. Slightly deformed, late-stage quartz veins are parallel to the wide faults or cut the wide gouge zones at a shallow angle to the faults. Fault dips range between 20° and 70°, with most of the faults

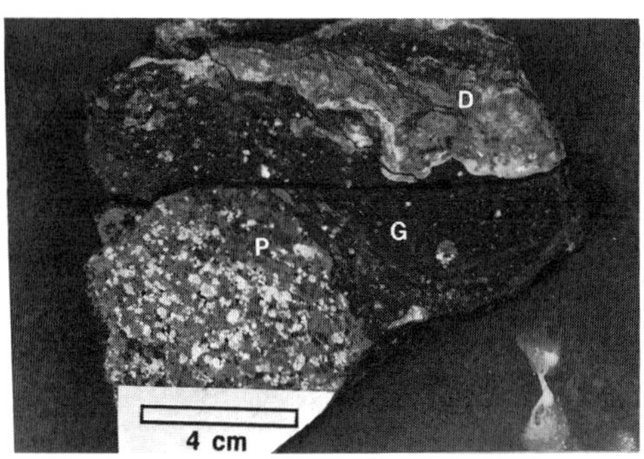

Fig. 1. Hand sample of a fault rock from the Washakie thrust. Sample contains a rounded porphyroclast of protolith granite (P), fine-grained, dark-red gouge (G), and an irregularly-shaped zone of highly deformed granite (D) which developed by grain-scale cataclasis. This sample is characteristic of the wide fault zones in the Washakie thrust system.

dipping 25–45°. Slip along the faults was primarily dip slip. Samples described in this paper were collected along a traverse perpendicular to the strike of the fault and include gouge, damaged zone and protolith layers.

Variably deformed Proterozoic Oracle Granite in the lower plate of the Santa Catalina metamorphic core complex in southeastern Arizona is exposed within and adjacent to the Geesaman fault zone (Janecke 1987). Early Tertiary east-directed shearing and regional metamorphism on the northeast flank of the Catalina complex (Bykerk-Kauffman 1986; Bykerk-Kauffman & Janecke 1987) were coeval with left-lateral strike-slip movement along the shallowly south-dipping Geesaman fault zone (Janecke 1987). The Geesaman fault was initiated as a normal fault in Late Jurassic and Early Cretaceous time and was reactivated as a strike-slip fault during the Early Tertiary event (Janecke 1987).

The Oracle Granite and the Proterozoic through Mesozoic sedimentary and metasedimentary cover rocks were intruded before, during, and after ductile deformation and regional metamorphism by Late Cretaceous through Early Tertiary plutons (Bykerk-Kauffman 1986; Bykerk-Kauffman & Janecke 1987; Keith *et al.* 1980). Andalusite in the contact aureoles of two of these plutons constrains the pressure during deformation to less than approximately 3.8 kbar, equivalent under lithostatic conditions to 13 km depth (Palais & Peacock 1987). The sub-greenschist to greenschist facies mineral assemblages in pelitic and carbonate rocks adjacent to the Geesaman fault zone indicate temperatures of approximately 250–400°C during deformation.

The fault rocks examined in this study developed in the Proterozoic Oracle Granite, and Janecke & Evans (1988) show that syntectonic feldspar alteration to micas resulted in a foliated phyllonite in which quartz remained relatively undeformed. The shear zone textures suggest that strength of fault zones may vary dramatically from the strengths predicted from experimentally derived flow laws (Janecke & Evans 1988). The undeformed granite consists of 32% quartz, 29% K-feldspar, 28% plagioclase, 9% biotite and 2% other minerals (averaged values reported in Banks 1980). There is no muscovite in the undeformed granite.

The Geesaman fault in the Oracle Granite is approximately 30–40 m wide, and the contact between protolith and shear zone is transitional over a 5–10 m zone. The gross structure within the fault is characterized by a quartz–mica phyllonite near the center of the fault giving way to quartz–feldspar mica cataclasites near the edges of the zone. In detail, phyllonite grades into cataclasite in the fault zone and the distribution of cataclasite and phyllonite is variable. At least one, and usually two foliations defined by narrow mica–chlorite rich faults or shear zones can be recognized in the field and in hand samples (see Janecke & Evans 1988, for photographs of macroscopic samples). Narrow shear zones anastomose throughout the outcrops and in highly deformed regions the fault-related rocks consist of quartz-mica phyllonite in which quartz form augen in a well-foliated mica matrix. The samples discussed in the following sections were collected on a traverse across the fault and are representative of the dominant domains of deformation within the fault.

Textures and microstructures

The optical microscopic observations of these rocks presented to date (Evans 1988; Janecke & Evans 1988) reveal the gross microstructures and textures of the rocks, but several important points concerning the development of these cataclasites have not been resolved. The previous work did not examine in any detail fracture distribution, densities, or history of fracturing, which can provide important clues to the generation of the fault-textures and the development of cataclasites (Blenkinsop & Rutter 1986; Mitra 1984). In addition, some of the textures examined at the optical scale could be the result of crystal plastic or diffusional mass transfer processes, and in general these mechanisms cannot be confirmed at the optical scale (Tullis & Yund 1987; Knipe 1989), although some authors (e.g., Mitra 1984; Chester & Logan 1986; Shimamoto 1989) suggest that diffusional mass transfer processes have been documented in cataclasites.

Textures of the fault-related rocks were examined with transmitted and reflected light optical microscopy and scanning electron microscopy. Crack lengths, grain sizes, fracture densities and fracture orientations were determined by measurements made along at least 10 traverses perpendicular to the macroscopic foliation of the fault-related rocks in each of two mutually perpendicular thin sections. The optical studies also examined the orientation of phyllosilicates which developed in the faults. Standard petrographic and 5 × 7.6 cm thin sections were used to examine microstructures and fracture patterns. Scanning Electron Microscopy (SEM) studies were also used to provide greater insight into deformation mechanisms and textures of the fine-grained portions of the fault-related rocks. While SEM analysis is limited in its ability to resolve the type of deformation mechanisms which were operative in deformed rocks when compared with TEM, SEM observations are extremely useful in determining the development of textures within brittle and semibrittle rocks (Rutter *et al.* 1986; Schedl *et al.* 1986). Samples from both study areas were sequentially etched and polished (see Schedl *et al.* 1986 for details of polishing and etching procedures) in order to reveal the microstructures responsible for deformation. X-ray diffraction studies were also conducted to study the composition and distribution of minerals in the fault rocks which formed at different stages of fault development.

The textures of the Geesaman fault are grouped into three deformation regimes for convenience in describing the rocks. These domains grade into one another in the fault zone. These groups are: (1) slightly deformed rocks from the margins of the fault and in regions within the fault, in which a moderately well developed foliation exists and fractured feldspar grains are recognizable in hand sample; (2) moderately deformed rocks in which a foliation is well-developed, defined by mica–chlorite shear zones, quartz grains are fractured and feldspars grains are highly fractured and altered; (3) a quartz–mica phyllonite, in which a well developed foliation in the micas wraps around quartz grains which exhibit fracture and some evidence for crystal plastic deformation. There is little pre-fault texture or structure to the protolith; quartz grains exhibit no irregular extinction and fractures are few in number.

The microstructures in the slightly deformed rocks are dominated by intragranular fractures along and at 45–60° to the albite cleavage in both microline and albite, with lesser amounts of intragranular fractures in quartz (Fig. 2a and Table 1). Some fractures nucleated along grain boundaries and at the edges of biotite grains. Intragranular fracture densities are greater in feldspar than in quartz, and the crack lengths are shorter in feldspar than in quartz at this stage of development, resulting in angular, fracture-bounded fragments of feldspar which are smaller than the adjacent quartz grains (Fig. 3). The orientation of microfractures indicate that early stages of fault development were dominated by cleavage-controlled intragranular fracture of feldspars and intragranular cracks in quartz, neither of which reflect the orientation of the macroscopic fractures and faults which define the macroscopic foliation (Fig. 4). At

Fig. 2. Photomicrographs of the Geesaman fault zone rocks under cross-polarized light. (**A**) Moderately deformed sample consisting of muscovite-lined fractures and small faults (F) which define a foliation that wraps around the stiffer quartz grain (Q). Fractures form along (001) cleavage and as fractures within the albite grain (A). Deformation in quartz in this view consists of grain-scale fracture, and possibly some intracrystalline slip as indicated by undulose extinction. (**B**) Highly deformed Geesaman fault rocks consisting of well foliated quartz-muscovite-chlorite phyllonite. Slightly deformed quartz grain (Q) lies in a matrix of intergrown quartz and muscovite (Q/M) and kinked muscovite (M). Fine-grained quartz-muscovite intergrowths form in pressure shadows of the quartz grain.

this first stage of deformation, fractures appear to have been paths for fluids to and through individual grains, as most of the fractures in feldspar contain muscovite or muscovite–chlorite intergrowths (Fig. 2a). Many fractures in quartz are partly healed and decorated by trails of fluid inclusions.

The moderately deformed fault rocks are near the boundary of the fault with the protolith and in irregular layers within the fault. These samples exhibit two distinct foliations; one is parallel to or within 20° of the boundary of the fault zone and direction of slip as determined by Janecke (1987), and the second foliation is approximately 30° to the first. Both foliations are

Table 1. *Linear crack density, in number of fractures per cm, in quartz and feldspar grains from 12 thin sections from the slightly and moderately deformed parts of the Geesaman fault*

Quartz	Feldspar
Slightly deformed samples	
4.5	16.6
5.4	25.0
5.0	9.7
5.0	22.2
6.8	13.6
9.7	25.0
Moderately deformed samples	
8.3	29.5
6.2	20.3
7.1	13.8
7.0	15.0
15.0	20.0
10.5	21.9
20.4	20.8

30 grains were examined in each thin section and the number of intragranular fractures were counted in traverses perpendicular and parallel to the dominant foliation observed in each thin section.

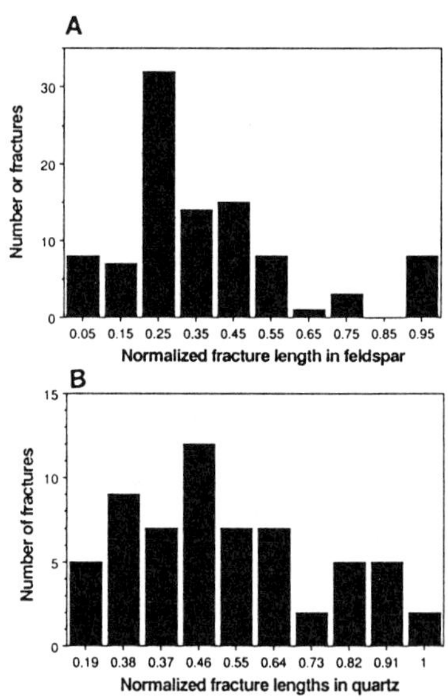

Fig. 3. Fracture lengths in slightly deformed samples from the Geesaman fault. Fracture lengths are normalized with respect to the mean grain size to eliminate a bias in the data towards the larger grains. Crack lengths are shorter in feldspar than in quartz and result in a first-stage gouge which has fracture-bounded feldspar fragments which are smaller than the quartz grains. (**A**) Fracture lengths in feldspar grains. (**B**) Fracture lengths in quartz.

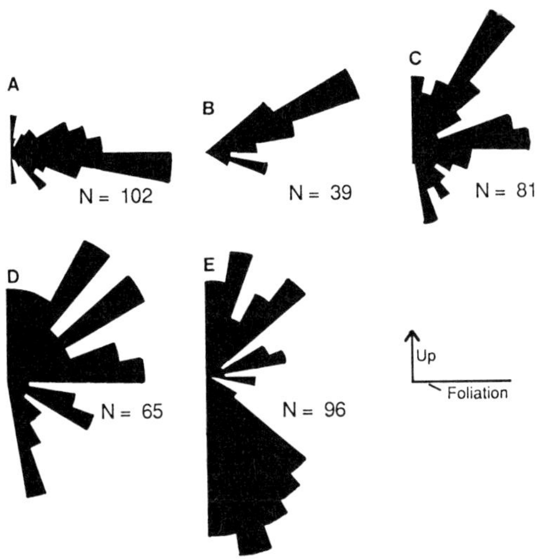

Fig. 4. Equal-area rose diagrams of the orientation of microfractures in slightly and moderately deformed Geesaman fault samples. Orientations are from this sections cut parallel to the slip direction and perpendicular to the dominant macroscopic foliation of the sample. Macroscopic foliation is defined by transgranular fractures and phyllosilicate-rich zones which weave through the sample. (**A**) Orientation of intergranular fractures in large (5 × 7.6 cm) thin sections define the macroscopic foliation. (**B**) Orientation of intragranular microfractures in feldspars in slightly deformed samples. Fractures lie primarily along the (001) cleavage in both albite and microcline, and are 20° to 30° from the macroscopic foliation. (**C**) Orientation of intragranular microfractures in quartz in slightly deformed samples. Two dominant sets of fractures developed, one parallel to the foliation and the second approximately 50° from the foliation. (**D**) Orientation of intragranular microfractures in feldspar from moderately deformed samples from the Geesaman fault. Microfracture orientations in feldspar follow cleavage planes, and are not parallel to the macroscopic foliation defined by mica-rich shear zones. (**E**) Microfracture orientations in quartz in moderately deformed samples formed a high angle to the dominant foliation.

defined by muscovite–biotite–chlorite intergrowths, and weave evenly through the rock (Fig. 2b). The fractures in feldspar are both intra- and intergranular, fracture lengths are longer in feldspar than in quartz, and two groups of microfractures are approximately 20° and 90° to the macroscopic foliations (Fig. 4). Feldspars are in most cases highly altered, with only ghosts of twin patterns and grain boundaries of the original feldspar grains visible in many places. The rock consists of at least 10% muscovite, and biotites have altered to chlorite. At the optical scale the fine-grained muscovite appears to be oriented with the (001) planes parallel to the boundaries of the fractures, and is more randomly oriented where it formed within individual feldspar grains.

The deformation mechanisms in quartz consists of intragranular fracture and possibly minor amounts of crystal plastic deformation which gives rise to patchy, undulose extinction and possible subgrains. The microfracture orientation in quartz is independent of the macroscopic foliation (Fig. 4e), and healed or partly healed microfractures in quartz are common. Intergranular fractures appear to have developed by the linking up of intragranular fractures in feldspar and form through-going shear zones along which muscovite developed and slip was localized. This texture indicates that the intergranular fractures were relatively efficient conduits for fluids through the fault which corroded the feldspars.

Highly deformed rocks consist of quartz augen in a fine-grained matrix of muscovite, chlorite and quartz. The phyllosilicates define a well-developed foliation, and feldspar has been entirely consumed in the conversion of feldspar to muscovite. Quartz grains exhibit intragranular fractures and undulose extinction and lie in a matrix of very fine-grained muscovite and quartz (Fig. 2b). The quartz porphyroclasts have quartz–muscovite intergrowths in pressure shadows and exhibit evidence for rotation within the fine-grained muscovite–quartz matrix. The muscovite-rich regions have deformed by kinking which in places appears to have nucleated at the more rigid quartz porphyroclasts (Fig. 2b).

SEM observations of the Geesaman fault rocks provide a view of the semi-brittle behavior and alteration processes which led to the development of the quartz–muscovite phyllonites. Secondary electron and backscatter electron images of highly polished samples were made and the samples were subsequently etched and examined again to reveal fractures in feldspar and quartz grains and textures in phyllosilicates. The intergranular fractures and faults nucleated as intragranular fractures in feldspar grains which coalesced to form small faults (Fig. 5a). Quartz grains exhibit far fewer fractures than the feldspar grains and the early stages of deformation resulted in fracture and fault zones which isolated the quartz grains from quartz–quartz grain contacts (Fig. 5a). The slightly and moderately deformed samples of the Geesaman fault rocks also show the distribution of early-formed muscovite in strain shadows of the

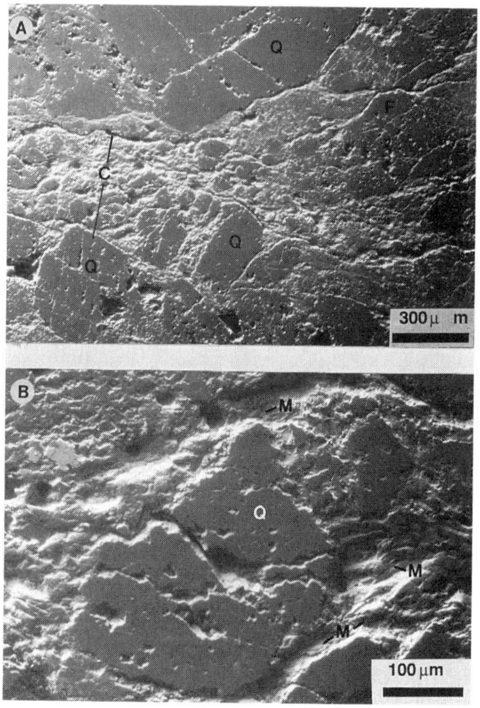

Fig. 5. (A) Secondary electron image of slightly-etched sample from the Geesaman fault showing the transition from a small zone of cataclasis (C) to intragranular fractures in feldspar (F). Smoother-textured grains (Q) are quartz along the edge of and within the zone of cataclasis. Sample was etched for 1 minute in concentrated HF − HCl solution. (**3**) Secondary electron image of etched sample from the Geesaman fault which shows the development of oriented muscovite (M) along intergranular cracks, behind a quartz grain (Q).

quartz grains and along intergranular fractures (Fig. 5b). All samples examined showed that the muscovite grains developed with the long axis parallel to fractures, which indicates that the well-developed orientation of the highly-deformed samples (Fig. 2b) is most likely a result of early alteration in the rocks rather than due to a later rotation during subsequent slip in the fault zones.

SEM and optical microscopic observations indicate that the reactions of feldspar to muscovite took place along feldspar grain boundaries, intragranular fractures, and more rarely as a pervasive intragranular alteration (e.g. sericitization). Given the protolith mineralogy and the inferred conditions at which the deformation took place, the development of the muscovite most likely proceeded according to the reactions:

$$\underset{\text{microcline}}{3KAlSi_3O_8} + 2H^+ = \underset{\text{muscovite}}{KAl_2(AlSi_3O_{10})(OH)_2} + \underset{\text{quartz (aqueous)}}{6SiO_2(aq)} + 2K^+ \quad (1)$$

$$\underset{\text{albite}}{3NaAlSi_3O_8} + 2H^+ = \underset{\text{Na-mica}}{NaAl_2(AlSi_3O_{10})(OH)_2} + 2Na^+ + \underset{\text{quartz (aqueous)}}{6SiO_2(aq)} \quad (2)$$

Hydrogen is added by the penetration of water into the rocks along intergranular fractures, and the development of mica and quartz result in the release of K^+ and Na^+ ions. Late-stage quartz veins which exhibit some cataclasis along their borders and are parallel to or cut the deformation fabrics in outcrop and the very fine-grained quartz which is intergrown with the muscovite (Fig. 2b) attest to the syntectonic development of quartz. None of the SEM observations recorded regions of concentrated K^+ or Na^+, which suggests these ions remained in solution and were deposited outside of the regions studied.

Microstructures of the Washakie rocks exhibit evidence for cataclastic flow and alteration at much lower temperatures than the conditions which were dominant during development of the Geesaman rocks. However, all of the trends in microstructures and textures documented in the Geesaman fault hold for the Washakie fault. Feldspar sustained most of the deformation, with alteration to kaolinite resulting in a foliated cataclasite with fractured quartz grains and aggregates of grains forming porphryoclasts in the cataclasites (Fig. 6a). In some cases an ultracataclasite (Blenkinsop & Rutter 1986) formed by the extreme brittle grain size reduction and cataclasis of quartz, feldspar, and clay minerals (Fig. 6b). In many of the fault-related rocks an irregular foliation is defined by concentration of opaque minerals separated by irregularly-shaped zones of quartz (Fig. 6c). Slightly deformed rocks have an approximately orthogonal macroscopic fracture pattern in the plane perpendicular to the main foliation (Fig. 7a), and the intragranular fractures in both quartz and feldspar roughly mimic the macroscopic fracture pattern (Fig. 7b and c). Thin sections cut perpendicular to both foliation and the slip direction show that the fracture pattern is blocky, with several intergranular fractures defining the foliation plane and a second set perpendicular to the foliation (Fig. 7d).

Fig. 6. Photomicrographs of deformed sample from the Washakie thrust system. (A) Moderately deformed sample under cross-polarized light composed of fine-grained kaolinite–chlorite mixtures (K/C) wrap around fractured and cataclastically deformed quartz (Q). (B) Ultrabreccia to protolith transition across a narrow fault under cross-polarized light. Ultrabreccia (U) is approximately 0.3 mm wide and X-ray diffraction and elemental analyses on SEM micrographs show that the ultrabreccia is composed of quartz, hematite, kaolinite, chlorite and small amounts of albite. Breccia (B) adjacent to the ultrabreccia consists of cataclastically deformed quartz, feldspar, and kaolinite. Protolith (P) is almost completely undeformed. (C) Plane polarized photomicrograph of breccia shows a wavy zone of cataclasis marked by dark regions of iron oxides. Two episodes of hydrothermal iron oxide deposition are marked by the lighter (1) hematite zone bordered by the darker (2) hematite zone. Note darker staining along cleavage fractures in feldspar (F).

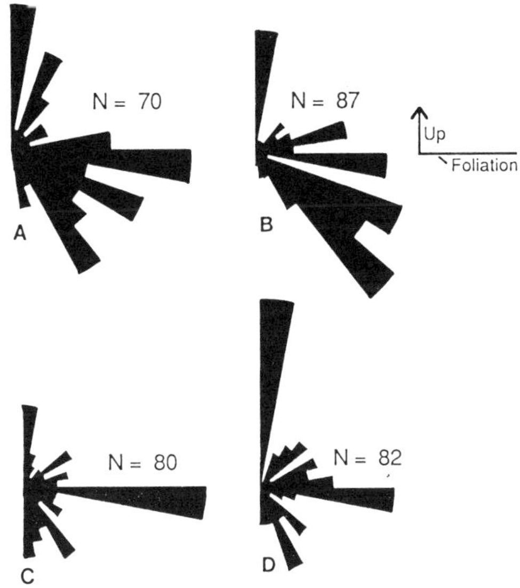

Fig. 7. Rose diagrams showing the orientation of fractures in moderately deformed rocks from the Washakie system. A, B, and C show oriented thin sections cut parallel to the slip direction and perpendicular to the dominant macroscopic foliation of the sample. D shows orientations from thin section cut perpendicular to foliation and slip direction. (A) Orientation of transgranular fractures from a large (5 × 7.6 cm) thin section define a horizontal foliation and a set of fractures perpendicular to the foliation. (B) Orientation of intragranular microfractures in feldspars. (C) Orientation of intragranular microfractures in quartz. (D) Orientation of microfractures in both feldspar and quartz in thin section cut perpendicular to foliation and slip direction.

SEM analysis of polished and etched samples from the Washakie fault system shows the overwhelming dominance of brittle deformation in both feldspar and quartz. Polished samples show the contrast between feldspar and quartz grains in the regions within and adjacent to slip horizons. Fractures in the feldspar appear to end at quartz grain boundaries (Fig. 8a and b) and the feldspar grains are subdivided into fracture-bounded particles. Quartz grains in the unetched samples exhibit few signs of deformation (Fig. 8a and b). The textures in the breccias and cataclasites are brought out in samples etched for 30 seconds to 1 minute. Many of the fractures lie along the (010) and (001) cleavages of the feldspars (Fig. 8b). Etching also shows that the microcataclasites are composed of very fine-grained kaolinite and quartz (Fig 8c and d) and kaolinite grains in the

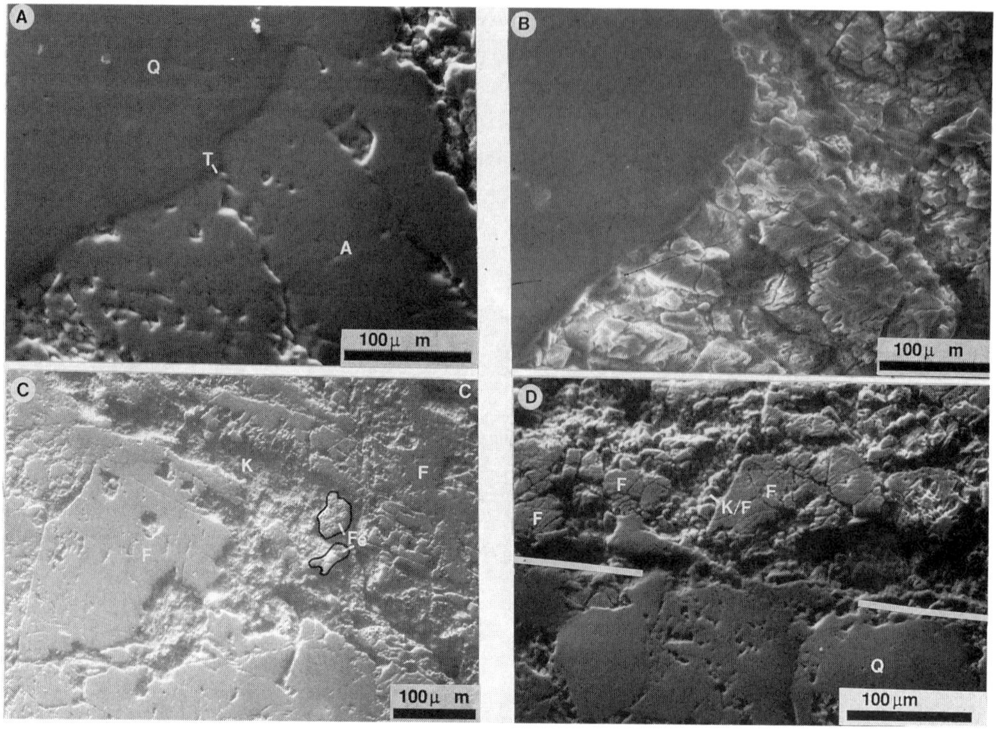

Fig. 8. Scanning Electron microscope (SEM) images of samples from the Washakie system. (**A**) Highly polished, unetched sample of part of a quartz grain (Q) and albite grain (A) from moderately deformed sample shows several fractures in feldspar truncated at grain boundary (T). (**B**) Approximately the same view as in (A) etched for 1 hour, shows no dislocation pile-ups in quartz at the fracture tips. Most fractures are deeply etched in the 1 hour treatment. (**C**) Moderately deformed sample etched for 30 seconds to 1 minute shows fracture patterns in feldspars (F), kaolinite-rich region along a small fault (F) and rounded pods of pure iron (Fe) within fault zone. Elemental analyses of the fault show virtually no iron along fault except in the rounded blebs. (**D**) Sample from the boundary (shown by white line) between the undeformed rock and a narrow cataclastic zone. Sample was etched for 30 seconds to 1 minute and shows smooth quartz grain (Q) in the undeformed region and feldspar (F) and fine-grained, unoriented kaolinite (K/F) in the fault zone.

breccias do not exhibit a strong preferred orientation (Fig. 8d). The early stages of the development of the fine-grained portions of the gouge appears to have developed by the alteration of feldspar to kaolinite according to the reaction:

$$\underset{\text{microcline}}{2KAlSi_3O_8} + 2H^+ + 9H_2O = \underset{\text{kaolinite}}{H_4Al_2Si_2O_9} + 2K^+ + \underset{\text{dissolved silica}}{4H_4SiO_4} \qquad (3)$$

This reaction resulted in the formation of very fine-grained breccias along which subsequent slip was localized. Water penetrated the faults to provide the H^+ ions, and the dissolved silica was most likely transported and deposited in late-stage veins which cut the faults.

The Washakie fault rocks exhibit a foliation which is in part defined by wavy bands of dark red and black iron oxides that follow or are subparallel to the highly comminuted and altered regions in the fault zones (Fig. 6c). Under reflected light and cathodoluminscence these seams appear as semi-continuous to discontinuous deposits of hematite. SEM observations show that the iron oxides formed as small, irregularly shaped to rounded zones within the faults, and in general are not localized along curviplanar seams (Fig. 8c). This morphology indicates that the Fe-oxides were deposited as a hydrothermal material with associated elements that include traces of silver. Both iron oxide and silver are undeformed or only slightly deformed, suggesting that the deposition of these elements occurred late in the development of the fault zone. Similar iron oxide rich zones have been interpreted by Mitra (1984) to be the insoluble residue of pressure solution in faults, but the morphology of these seams indicates a hydrothermal origin for the metals in this case. The SEM observations point towards a fault which evolved almost exclusively by brittle deformation. No evidence for grain truncations, apparent offsets, or dissolution-precipitation structures were found in the Washakie fault rocks, indicating that no pressure solution took place in these rocks.

Discussion and conclusions

It is useful to compare the results of field-based studies of fault zones with laboratory investigations of shear zones and intact samples to test how well experiments may predict the formation of natural fault zones. Comparison of the textures and deformation mechanisms recorded in this study with those of moderate and high temperature and high pressure experiments on granitic rocks (Tullis & Yund 1977; Carter *et al.* 1981) indicate that brittle deformation may be more dominant in nature than in experiments. The textures and inferred deformation mechanisms interpreted here resemble the results of experiments performed at low to moderate temperature (200–400°C) on intact granitic rocks (Tullis & Yund 1977, 1987, 1989). Fracture and cataclastic flow in feldspar largely controlled the mechanical behaviour of the rocks and mechanically isolated quartz in the rock. The morphology of the early, intragranular extension fractures and the evidence for crack nucleation at mica-rich regions agrees with the work of Kranz (1979) and Tapponnier & Brace (1976). The textures of the cataclasites and gouge rocks resemble those developed in experiments on polymineralic gouge (Logan *et al.* 1981; Blanpied *et al.* 1988). Deformation mechanisms in the quartz and feldspars are nearly identical, with the exception that little experimental work has reproduced the syntectonic alteration of feldspars. The deformation mechanisms in micas resemble those produced experimentally (Shea & Kronenberg 1989; Shea, pers. comm., 1989). However, slip is not localized along the boundary of faults as in the case of Logan *et al.* 1981; rather slip becomes concentrated along regions of high mica content or along regions of cataclasis within the fault, similar to textures described by Blanpied *et al.* (1988).

It is dangerous to infer temporal proximity of fault processes and textures based on the spatial proximity of structures within fault zones (Mitra 1990), however in the two faults examined here, the transition across a relatively small distance (5 to 20 m) from undeformed to highly-deformed rocks suggests that the textures may represent the evolution of the faults from early to late stages of development. The Washakie fault rocks are from a region in which undeformed rocks grade into a damaged zone which in turn grades into the indurated gouge along which most of the slip was accommodated. Thus the samples examined here very likely represent a growth sequence. It is more difficult to see such a gradation in the outcrop exposures of the Geesaman fault, as the three dominant textures described here are distributed throughout the fault and only a crude segregation of the macroscopic textures may be made. However at the handsample and larger scales the spatial transition from fractured feldspar–quartz cataclasites to quartz–mica–feldspar cataclasites to quartz–muscovite phyllonites is clearly seen, and the link between the three textures is clear.

The cataclastic processes responsible for the nucleation and growth of the faults were dominated by fracture and alteration of feldspar. Slightly deformed regions which represent the early stages of deformation are characterized by intragranular fractures in feldspar, the development of phyllosilicates, and the fracture of quartz. Fracture of feldspar grains resulted in reduced feldspar grain size whereas quartz grains fractured but fractures healed to preserve larger quartz grains. The early development of the cleavage-controlled fracture of the feldspar isolated quartz grains from intense cataclasis by forming soft zones around the more rigid quartz grains. Intragranular fracture orientations in feldspar and quartz do not reflect the orientation of the macroscopic foliation in the slightly deformed rocks.

Moderately deformed rocks represent the intermediate stages of fault development in which cataclasis gives way to phyllosilicate-dominated behaviour. In the Geesaman fault, muscovite developed an initial orientation in intra- and intergranular fractures which allowed slip to occur along mica basal planes, whereas in the lower temperature Washakie rocks the kaolinite formed as disordered grains which deformed by cataclasis and slip.

The highly-deformed zones represent the late stages of fault development in which phyllosilicate deformation dominated the behaviour of the rock. Slip along clays and micas effectively shielded the quartz grains from brittle or plastic deformation, and the quartz grains rolled in the very fine-grained matrix of muscovite and quartz in the Geesaman fault and in kaolinite and feldspar fragments in the Washakie fault.

The brittle fracture in these rocks was followed by the alteration to layer silicates. Intergranular fractures and faults which nucleated as intragranular fractures in feldspars allowed fluids to penetrate the fractured feldspars early in the deformation history of the faults. Through-going fractures permitted fluids to move through the fault, and formed softer clay or mica-rich regions around the quartz grains. Since the feldspar is volumetrically dominant in these rocks, feldspar deformation dominated the development of the faults. Additional slip took place along micas, and the end result is a strain-softened phyllonite or clay-rich cataclasite. The textures described here, or those resembling the ones described here, have been described in several other studies of faults and shear zones (Bruhn *et al.* 1987; Mitra & Frost 1981; O'Hara 1988; Parry *et al.* 1988; Schweitzer 1988; Stierman & Williams 1985; White 1983) and indicates that the phenomenon of feldspar and mica dominated fault rheology may be common in fault and shear zones in the upper crust. Many workers have ascribed textures similar to those documented here to be the result of rocks going through a brittle to ductile transition. This term has commonly been equated with the onset of crystal plasticity, but should perhaps be restricted to mean the onset of ductile deformation by whatever mechanism may be the cause (Rutter 1986). The rocks examined here may have behaved ductilely in the latter stages of fault development, but the transition from brittle to ductile behavior was the result of early fracture and subsequent cataclastic flow of feldspars accompanied by the reaction-softening to mica and clay. Crystal plasticity and diffusional mass transfer were relatively unimportant deformation mechanisms in the development of these faults.

Reviews by S. U. Janecke and two anonymous reviews greatly improved early versions of this paper. I would also like to thank R. Knipe for editorial efforts. Grateful acknowledgement is made to the Donors to the Petroleum Research Fund, administered by the American Chemical Society, for support of this research.

References

BANKS, N. G. 1980. Geology of the zone of metamorphic core complexes in southeastern Arizona. *In*: CRITTENDEN, M. D., JR., CONEY, P. J. & DAVIS, G. H. (eds) *Cordilleran metamorphic core complexes*. Geological Society of America Memoir **153**, 177–215.

BLANPIED, M. L., TULLIS, T. E. & WEEKS, J. D. 1988, Textural and mechanical evolution of granite gouge in high-displacement sliding experiments. *EOS, Transactions of the American Geophysical Union*, **69**, 1463.

BLENKINSOP, T. G. & RUTTER, E. H. 1986. Cataclastic deformation of quartzite in the Moine thrust zone. *Journal of Structural Geology*, **8**, 669–681.

BRUHN, R. L., YONKEE, W. A. & PARRY, W. T. 1987. Faulting processes in the brittle-ductile transition regime. *EOS, Transactions of the American Geophysical Union*, **68**, 1472.

BYKERK-KAUFFMAN, A. 1986. Multiple episodes of ductile deformation within the lower plate of the Santa Catalina metamorphic core complex, southeastern Arizona. *In*: BEATTY, B. & WILKINSON. P. A. K. (eds) Frontiers in geology and ore deposits of Arizona and the southwest. *Arizona Geological Society Digest*, **16**, 460–463.

—— & JANECKE, S. U. 1987. Late Cretaceous to Early Tertiary ductile deformation, Catalina-Rincon metamorphic core complex, southeastern Arizona. *Geology*, **15**, 462–465.

CARTER, N. L., ANDERSON, D. L., HANSEN, F. D., and KRANZ, R. L., 1981, Creep and creep rupture of granitic rocks. *In*: CARTER, N. L., FRIEDMAN, M., LOGAN, J. M. & STEARNS, D. W. (eds) *Mechanical Behavior of Crustal Rocks*. American Geophysical Union Monograph, **24**, 61–82.

CHESTER, F. M. & LOGAN, J. M. 1986. Implications for mechanical properties of brittle faults from observations of the Punchbowl fault zone, California. *Pure and Applied Geophysics*, **124**, 80–106.

ENGELDER, J. T. 1974. Cataclasis and the generation of fault gouge. *Geological Society of America Bulletin*, **85**, 1515–1522.

EVANS, J. P. 1988. Deformation mechanisms in granitic rocks at shallow crustal levels. *Journal of Structural Geology*, **10**, 437–443.

JANECKE, S. U. 1987. Mesozoic and early Tertiary structural history of the Geesaman Wash area northeastern Catalina-Rincon metamorphic core complex. *In*: DICKINSON, W. R. & KLUTE, M. A. (eds) 1987. Mesozoic rocks of southern Arizona and adjacent areas. *Arizona Geological Society Digest*, **18**, 189–201.

—— & EVANS, J. P. 1988. Feldspar-influenced rock rheologies. *Geology*, **16**, 1064–1067.

KAMINENI, D. C., THIVIERGE, R. H. & STONE, D. 1988. Development of a cataclasitic fault zone in an Archean granitic pluton of the Superior Province; Structural, Geochemical, and Geophysical Characteristics. *American Journal of Science*, **288**, 458–494.

KEITH, S. B., REYNOLDS, S. J., DAMON. P. E., SHAFIQULLAH, M., LIVINGSTON, D. E. & PUSHKAR, P. D. 1980. Evidence for multiple episodes of deformation within the Santa Catalina-Rincon-Tortolita crystalline complex, southeastern Arizona. *In*: CRITTENDEN, M. D., JR., CONEY, P. J. & DAVIS, G. H. (eds) *Cordilleran Metamorphic Core Complexes*. Geological Society of America Memoir, **153**, 217–257.

KNIPE, R. J. 1989. Deformation mechanisms—recognition from natural tectonites. *Journal of Struc-*

tural Geology, **11**, 127–146.

KRANZ, R. L. 1979. Crack growth and development during creep of Barre granite. *International Journal of Rock Mechanics, Mineral Sciences and Geomechanics Abstracts*, **16**, 23–35.

LOGAN, J. M., HIGGS, N. G. & FRIEDMAN, M. 1981. Laboratory studies on natural gouge from the U.S. Geological Survey Dry Lake Valley No. 1 well, San Andreas fault zone. *In*: CARTER, N. L., FRIEDMAN, M., LOGAN, J. M., STEARNS, D. W. (eds) *Mechanical Behavior of Crustal Rocks*. American Geophysical Union Monograph **24**, 121–134.

MITRA, G. 1984. Brittle to ductile transition due to large strains along the White Rock thrust, Wind River Mountains, Wyoming. *Journal of Structural Geology*, **6**, 51–61.

—— 1990. Deformation of granitic basement rocks along fault zones at shallow to intermediate crustal levels. *In*: MITRA, S. (ed.) *Structural Geology of Fold and Thrust Belts*. Johns Hopkins University Press, (in press).

MITRA, G. & FROST, R. B. 1981. Mechanisms of deformation within Laramide and Precambrian deformation zones in basement rocks of the Wind River Mountains. *Contributions to Geology*, **19**, 161–173.

O'HARA, K. 1988. Fluid flow and volume loss during mylonitization: an origin for phyllonite in an overthrust setting, North Carolina, U.S.A. *Tectonophysics*, **156**, 21–36.

PALAIS, D. G. & PEACOCK, S. M., 1987, P-T indicators in metamorphic rocks from the Santa Catalina metamorphic core complex, SE Arizona: Contact metamorphism at shallow crustal levels. *Geological Society of America Abstracts with Programs*, **19**, 438.

PARRY, W. T., WILSON, P. N. & BRUHN, R. L. 1988. Pore-fluid chemistry and chemical reactions on the Wasatch normal fault, Utah. *Geochimica et Cosmichimica Acta*, **52**, 2053–2063.

RUTTER, E. H. 1986. On the nomenclature of mode of failure transitions in rocks. *Tectonophysics*, **122**, 381–387.

—— MADDOCK, R. H., HALL, S. H. & WHITE, S. H. 1986. Comparative microstructures of natural and experimentally produced clay-bearing fault gouges. *Pure and Applied Geophysics*, **124**, 3–30.

SCHEDL, A., KRONENBERG, A. K. & TULLIS, J. 1986. Deformation microstructures of Barre granite: An optical, SEM and TEM study. *Tectonophysics*, **122**, 149–164.

SCHWEITZER, J. 1988. Interaction of physical and chemical processes during chlorite breccia development, Sacramento Mountains, SE California. *Geological Society of America Abstracts with Programs*, **20**, A 212.

SHEA, W. T., JR. & KRONENBERG, A. K. 1989. Experimental deformation of biotite schist. *EOS, Transactions of the American Geophysical Union*, **70**, 477.

SHIMAMOTO, T. 1989. The origin of S-C mylonite and a new fault-zone model. *Journal of Structural Geology*, **11**, 51–64.

SIMPSON, C. 1986. Fabric development in brittle-to-ductile shear zones. *Pure and Applied Geophysics*, **124**. 269–288.

SIBSON, R. H. 1977. Fault rocks and fault mechanisms. *Journal of the Geological Society, London*, **133**, 191–214.

STIERMAN, D. J. & WILLIAMS, A. E. 1985. Hydrologic and geochemical properties of the San Andreas Fault at the Stone Canyon well. *Pure and Applied Geophysics*, **122**, 402–424.

TAPPONNIER P. & BRACE, W. F. 1976. Development of stress-induced microcracks in Westerly granite. *International Journal of Rock Mechanics, Mineral Sciences and Geomechanics Abstracts*, **13**, 103–112.

TULLIS, J. & YUND, R. A. 1977. Experimental deformation of dry Westerly Granite. *Journal of Geophysical Research*, **82**, 5705–5718.

—— 1987. The brittle-ductile transition in feldspathic rocks. *Transactions American Geophysical Union*, **68**, 1464.

—— 1989. The brittle-ductile transition in granitic rocks. *Geological Society of America Abstracts with Programs*, **21**, A 176.

WHITE, J. C. & WHITE, S. H., 1983, Semi-brittle deformation with the Alpine fault zone, New Zealand. *Journal of Structural Geology*, **5**, 579–589.

WINTERFELD, G. F. & CONARD, J. B. 1983. Laramide tectonics and deposition, Washakie Range and northwestern Wind River Basin Wyoming. *In*: LOWELL, J. D. & GRIES, R. R. (eds) *Rocky Mountain Foreland Basins and Uplifts*. Rocky Mountain Association of Geologists, 1983 Symposium, Denver, Colorado, 137–148.

ZOBACK, M. D., ZOBACK, M. L., MOUNT, V. S., SUPPE, J., EATON, J. P., HEALY, J. H., OPPENHEIMER, D., REASENBERG, P, JONES, L., RALEIGH, C. B., WONG, I. B., SCOTTI, O. and WENTWORTH, C. 1987. New evidence on the state of stress of the San Andreas Fault system. *Science*, **238**, 1105–1111.

Fracture evolution in the upper ocean crust: evidence from DSDP hole 504B

SUSAN M. AGAR

Department of Earth Sciences, University of Leeds, Leeds LS2 9JT, UK
Present address: Geological Sciences Department, Northwestern University, Evanston, Illinois 60208, USA

Abstract: Detailed structural studies of ocean crust specimens from DSDP borehole 504B have been integrated with a well defined alteration history to yield fracture evolution paths during movement away from the mid-ocean ridge. Fracturing was the dominant deformation process in the upper ocean crust of 504B. A broad range of fracture types exists. The type and orientation of most early fractures were related to lithology. Late fracture orientations and intensities were less dependent on lithology and may have been more closely related to tectonic stress conditions. Prolonged deformation histories are evident in the transition zone where non-coaxial deformation overprints successive generations of dilational veins. Changes in the orientation of the local stress field between successive generations of fractures occurred between and within lithological units. This suggests the presence of numerous mechanical discontinuities in the upper ocean crust, particularly in the transition zone. Changes in fracture porosity between alteration stages were often large and, together with changes in the characteristics of fractures, would have influenced local thermal gradients. The upper ocean crust rheology would have been extremely time dependent as a result. Zones of weakness were created by clay alteration products in basalts and were located in the most permeable horizons. Movement on these zones was promoted by elevated pore-fluid pressures with major failure most likely during major upwelling associated with a hydrothermal system. Subsequent precipitation of high temperature minerals from circulating fluids may have strengthened permeable horizons causing migration of detachment surfaces to another level.

Despite widespread acceptance that the creation and motion of ocean crust is directly controlled by a balance of global tectonic forces, detailed structural studies of in situ ocean crust have been neglected. The Deep Sea Drilling Project (DSDP) and the Ocean Drilling Program (ODP) have directly sampled the upper kilometre of the ocean crust. Structural and microstructural studies on this material yield information on deformation mechanisms and histories and their relationship to mid-ocean ridge magmatism and hydrothermal circulation. These data are crucial to models of ocean crust rheology and tectonics and provide a comparison for structural studies in ophiolites where the distinction of ocean floor and obduction deformation may not be clear.

DSDP borehole 504B (Cann *et al.* 1983; Anderson *et al.* 1985; CRRUST1982; Becker *et al.* 1988) lies 200 kms south of the Costa Rica rise (Fig. 1a) and penetrates 1.5 km into 5.9 Ma year old ocean crust. The Costa Rica rise was spreading at a half rate of around 36 mm per year at the time the ocean crust at 504B was being generated (Klitgord *et al.* 1975). The drilled section penetrates geophysically defined layers 1, 2A, 2B & 2C to a depth of 1562.3 m below the sea floor (Fig. 1b). The geochemical and geophysical features have been extensively documented by previous DSDP and ODP research but the structure and microstructure of 504B core material has not, until now, been systematically investigated. Although borehole televiewer data provide an instantaneous picture of present day fractures in the borehole wall (Anderson & Zoback 1982), they cannot constrain the detailed deformation history, kinematics or mechanisms. This paper is intended to contribute to a database for the deformation processes which are active in in situ, young upper ocean crust. 504B is situated more than 70 km away from the nearest major fracture zones (Fig. 1a) and is unlikely to have been deformed as a result of slip along these. The deformation histories recorded in the core should therefore relate to a comparatively simple deformation path followed by ocean crust moving away from the ridge. Structural interpretations presented here were facilitated greatly by previous studies on core alteration which defined the timing of deposition of fracture fill phases (Honnorez *et al.* 1983; Alt *et al.* 1985) and the detailed lithostratigraphy constructed by Adamson (1985). This short paper focuses on the occurrence, nature and evolution of fractures in DSDP hole 504B. A few examples

From Knipe, R. J. & Rutter, E. H. (eds), 1990, *Deformation Mechanisms, Rheology and Tectonics*, Geological Society Special Publication No. 54, pp. 41–50.

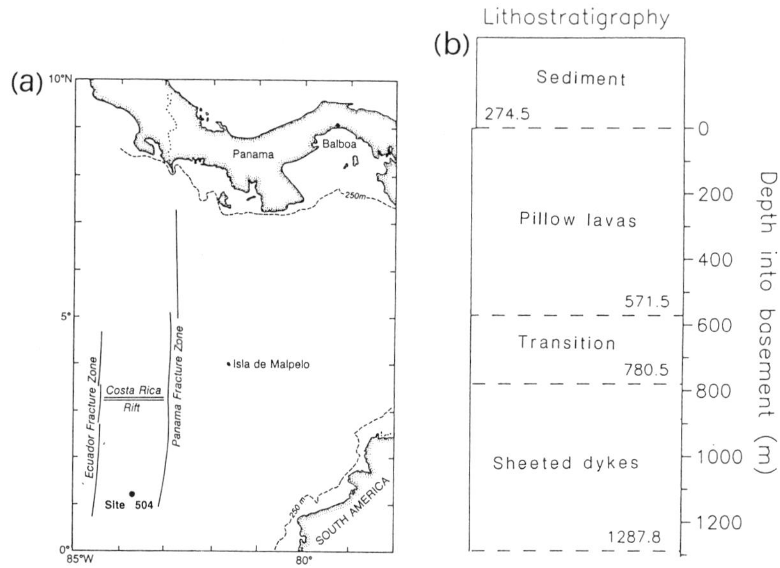

Fig. 1. (**a**) Location of DSDP borehole 504B (from Cann *et al.* 1983). (**b**) Summary of lithological and geophysical divisions in 504B (from Becker *et al.* 1988).

are discussed in detail to demonstrate the variation in structures and deformation histories at different depths within 504B.

Fracturing in the upper kilometre of the ocean crust

In the following examples references to the position in 504B core are given according to the DSDP/ODP core referencing system. The first number is the core number, the second represents the core section and the third number represents the piece number. If more specific intervals are required these are given as intervals in centimetres measured from the top of the core section. Backscattered (BSE) investigations were carried out on a Camscan Series 4 scanning electron microscope (SEM) using a solid state four quadrant detector, working distances of less than 8 mm, accelerating voltages of 20–30 kV and a beam current of 175 nA. Electron channelling patterns (Lloyd 1987) were used to observe changes in crystallographic orientation across some fractures. Quantification of fracture orientations, lengths and areas in two dimensional sections (from core and thin section micrographs) were carried out using digipad software (Prior, unpublished program).

Example 1: pillow basalts in the upper zone

Pillow basalts in the upper zone of 504B (274.5 to 846 mbsf: Adamson 1985) are heavily fractured throughout and contain several zones of intense brecciation. The recovery overall in 504B is low, averaging 20% in the upper part of the borehole and reducing to 12% in the dyke section (Cann *et al.* 1983; Anderson *et al.* 1985; Becker *et al.* 1988). As the unrecovered sections were probably the most fractured some structural features are likely to have been missed and the deformation observed represents less deformed portions of the upper ocean crust (Adamson 1985).

Mesoscopic fractures with filling related to two phases of alteration (Alt *et al.* 1985) are documented. Fracture fills are predominantly clays related to early low temperature (0–50°C) alteration (Alt *et al.* 1985). The earliest microcracks (up to 3 mm long) are located in phenocryst phases (olivine, clinopyroxene and plagioclase). They commonly follow cleavage planes and rarely penetrate the surrounding groundmass. The cracks probably nucleated during differential contraction of phases in the cooling basalt (Richter & Simmons 1974) with subsequent dilation as a result of nucleation and growth of alteration products (Fig. 2a). It

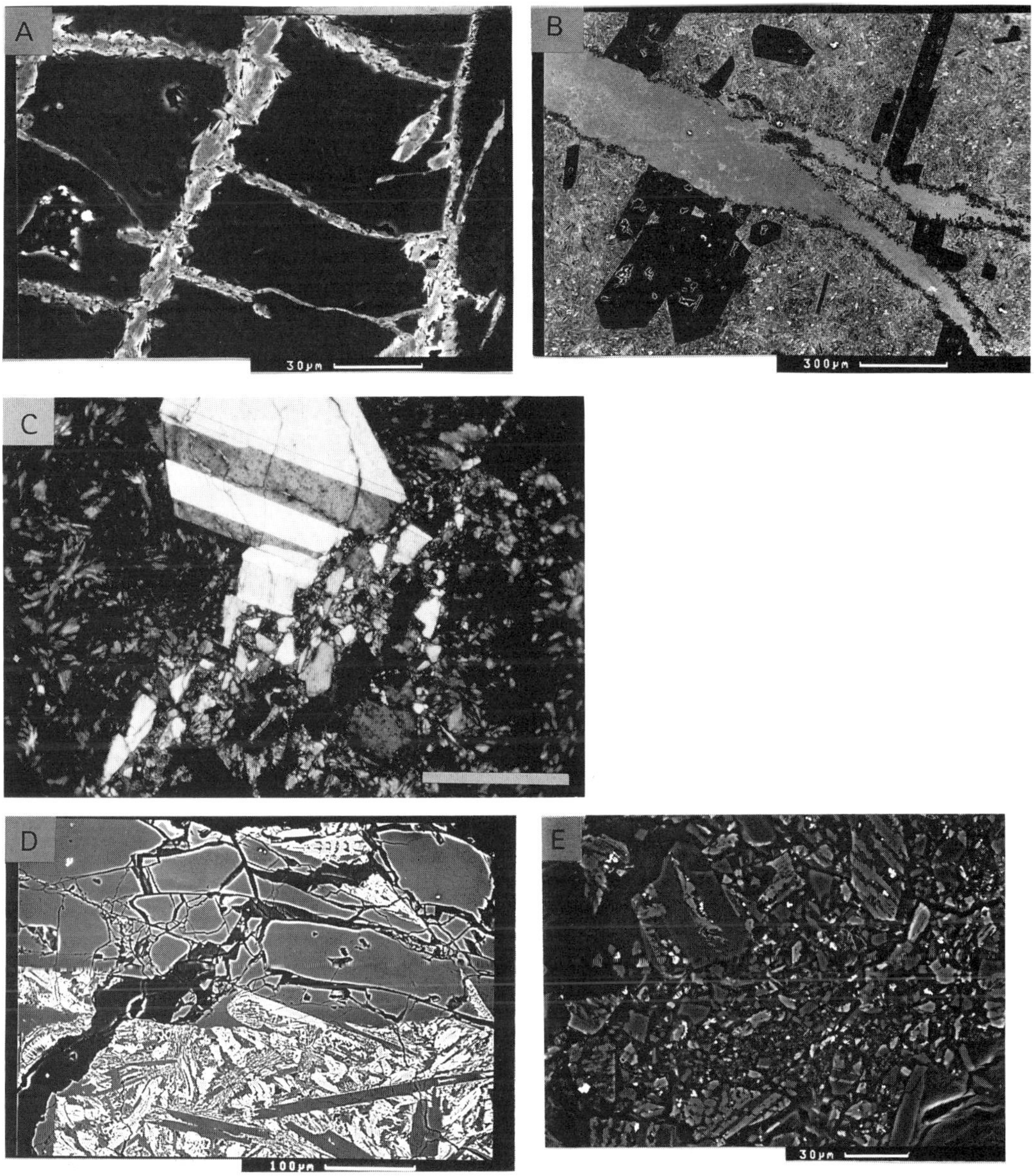

Fig. 2. Examples of fractures in pillow sequences in layers 2A and 2B of DSDP hole 504B. (**a**) Back-scattered SEM image showing fractured plagioclase (dark) in-filled with mixed-layer clays (grey) and chlorite (light). (4–5–347, 27–30 cm). (**b**) Back-scattered SEM image showing dilational fracture in pillow basalt in upper zone of 504B. Lack of fabric within the veins (clays and chlorite) and local growth of alteration phases perpendicular to the vein margin suggest model 1 cracking. The small offsets in the plagioclase phenocrysts (dark) are probably only apparent. (37–2–373, 70–73 cm). (**c**) Photomicrograph showing disaggregation of fractured plagioclase phenocryst with fragments being asymmetrically sheared along a narrow alteration/cataclastic zone. Scale bar is 1 mm. (49–2–945, 74–79 cm). (**d & e**) Back-scattered SEM images of 'hyaloclastite' unit in core 21. (d) shows glass seal around fragments of basalt in breccia associated with fault zone (see core description, Cann *et al.* 1983). (e) Intense, grainscale deformation of phenocryst phase occurs within some glassy breccias. (d = 21–2–1218, 145–146 cm, e = 21–2–1215, 125–128 cm).

is conceivable that some microcracks were induced solely by volume changes accompanying hydration reactions, for example, the transformation of olivine to saponite involves a volume increase of about 60%. Early fractures within partially devitrified glass commonly occur as polygonal sets. Branching networks of clays, unrelated to crystal cleavage dissect some phenocrysts but optical crystallography and electron channelling show that there is rarely any crystallographic mismatch, and hence no rotation, between fragments. These fractures are therefore likely to have been subcritical (Atkinson 1987), with dissolution and replacement processes controlling their geometry (Prior 1988).

Mesoscopic fractures filled with early alteration phases rarely exceed 2 mm in width and tend to occur as irregular branching networks. Oxidation zones around some fractures reach up to 10–20 mm in width. There are small zones (up to 100 mm) where mesoscopic fractures show a moderate to strong preferred orientation (see later). A later and less extensive phase of fracture filling in the upper pillows is dominated by carbonates and calcic zeolites and is thought to be related to alteration during the upwelling of a hydrothermal convection cell (Alt *et al.* 1985). These later fractures are also characterized by dominantly mode 1 (extensional), branching cracks which dissect both phenocrysts and groundmass (Fig. 2b). Dilation rarely exceeds 2 mm. Both generations of fracture fill occur as microcrystalline aggregates, randomly oriented flakes and fibres. Locally, calcite forms coarser grains (1–2 mm). These are usually untwinned or contain narrow twins, probably formed during crystallization. Fibres are usually straight, suggesting mode 1 (extensional) cracking, with no rotations of the local stress field (Ramsay 1980). Some fibrous fills record evidence of more than one opening event, with aligned fluid inclusions marking the boundaries of each fracturing event. Each opening event records the same extension direction. More than ten successive openings in one fracture have been observed. Evidence for mode 2 (shear) failure is indicated in a few core pieces. Close to the edges of some of the mode 2 fractures plagioclase and clinopyroxene phenocrysts are more intensely fractured. In contrast to the homogeneous, early microcracking, gradients in fracture densities both within phenocrysts and towards phenocryst boundaries (Fig. 2c) have developed and fragments of phenocryst phases have been asymmetrically distributed along fractures. These features contrast strongly with early thermal contraction features and suggest non-coaxial deformation.

Breccias within the upper zone are varied both in clast shape and size and the clast-matrix ratio. One end member is represented by pillow breccias where fragments of pillows have been partially or wholly replaced by alteration phases and the breccia is matrix supported. The resulting fragments are rounded, preserving little evidence of the original fracturing processes. Moderate preferred shape orientations of clasts are present in some pillow breccias. Clasts in these can often be matched with neighbouring clasts suggesting that little or no differential movement has occurred. The pillow breccias contrast with less common breccias of clast supported angular fragments also found in the upper zone. Some of these breccias clearly contain fragments of pillow breccias which have been reworked either as a result of near-surface instabilities shortly after extrusion or during subsequent faulting. The earliest breccias are cut by mesoscopic fractures containing early alteration phases. Breccia clast shape may have been controlled by pre-existing fractures generated during cooling.

Breccias also include units described as hyaloclastites which contain abundant angular basalt and glass fragments sealed in a partially altered glassy matrix. Many of the clasts have retained their original conchoidal fracture surfaces and probably formed by quenching of extruded lavas. SEM studies suggest however that the interpretation of some glassy breccias is not straightforward. Previously cooled units have been fractured and sealed by later glass injections (Fig. 2d) and strongly fragmented phenocryst phases are suspended within the glass (Fig. 2e). One of these is located in core 21 which is interpreted as a shallow fault zone because of the occurrence of small fragments of a variety of lithologies and the presence of slickensides (Cann *et al.* 1983, p. 113). The intense grainscale deformation in this breccia may be related to fault slip, with glass injected from migrating magmas during fault induced dilation. In this setting the relationship between faulting and dyke intrusion will be closely linked and distinction between these processes is unrealistic.

Example 2: fracturing in pillow basalts in the 'stockwork' alteration zone

A detailed lithostratigraphy constructed by Adamson (1985) shows the presence of a stockwork zone between 910 and 930 m in the upper section of a thick pillow sequence (lithological unit 63). The term 'stockwork' was adopted to represent the characteristic alteration which developed under local temperatures of 250–350°C (Alt *et al.* 1985). At the highest temperatures the geothermal gradient at the top of the transition zone is estimated to have been as high as 2.5°C m^{-1} (Alt *et al.* 1986a). The alteration characteristics and hydraulic disruption in the stockwork zone suggest that this was the location of a focused fluid flow associated with a discharge cycle (Alt *et al.* 1986*a*, *b*; Harper *et al.* 1988).

The highest volume percent of veining is found in the stockwork zone (see later). The pillows in the stockwork zone have been affected by major brecciation which postdates early breccias and fractures, similar to those described from the upper zone. The later breccias appear to be preferentially developed in pillow rims which are also sites of extensive early alteration and fracturing. Fragments are often isolated or entrained as suspended material in the fracture fill suggesting that hydraulic fracture is an important process in generation of these breccias (Phillips 1972). Hydraulic breccias are sealed mainly by quartz, epidote and sulphides. Rapid deposition of fracture fills inherent to the hydraulic fracture model (Phillips 1972) is supported by equant grains and a lack of preferred crystallographic orientation in fracture fills. Individual stockwork-related fractures in this zone locally exceed the core diameter (60 mm) and dilations of 5–10 mm are common. The majority of veins are dilational but, in contrast to the upper zone, their main orientation rarely follows that of earlier fractures and generally only one fracture opening event is preserved (Fig. 3). These characteristics may suggest that strain and fluid flow rates were higher than in the upper zone. Inconsistent cross-cutting relations between fractures containing the same fill but in different orientations suggests that fracture geometries were more strongly influenced by hydrostatic stresses than non-hydrostatic tectonic stresses. These relationships are a good indication of pore-fluid pressures in excess of lithostatic pressures.

Fractures were not completely sealed during stockwork alteration and calcite and zeolites have filled cavities which remained, usually in the centres of fractures. The calcite is commonly more densely twinned than similar grain sizes of calcite in the upper zone pillows but the degree of twinning is variable. Palaeostress estimates using the calcite twinning palaeopiezometry calibration of Rowe & Rutter (1990) suggest that differential stress magnitudes were locally high (100–150 MPa) (Agar, unpublished data.)

Fig. 3. Early fractures in a pillow basalt from the stockwork zone are cross-cut by later, more discrete fractures. Some reactivation of earlier fractures also occurs. (80–1–7A).

Example 3: massive Flow/Dyke units in the transition zone (846–1055 mbsf)

Massive flows and dykes are less altered than pillow basalts and thin flows. There are two phases of alteration one corresponding to temperatures of *c.* 100°C and a later phase at 100–250°C (Alt *et al.* 1985). Early fractures, although less pervasive in these units have similar characteristics to those in the upper zone.

Narrow (<0.5 mm) cataclastic zones, which are cut by dilational veins of the later alteration phases (Fig. 4a), are developed in many coarser grained units. These appear to have nucleated along zones of early alteration and are developed in isolation or in branching networks with no mesoscopic preferred orientation. These zones usually surround intact lenses of coarser, undeformed material. Shattering and rotation of fragments suggest that dynamic fragmentation is the dominant process causing grain-size reduction (Fig. 4b). The rounded grain

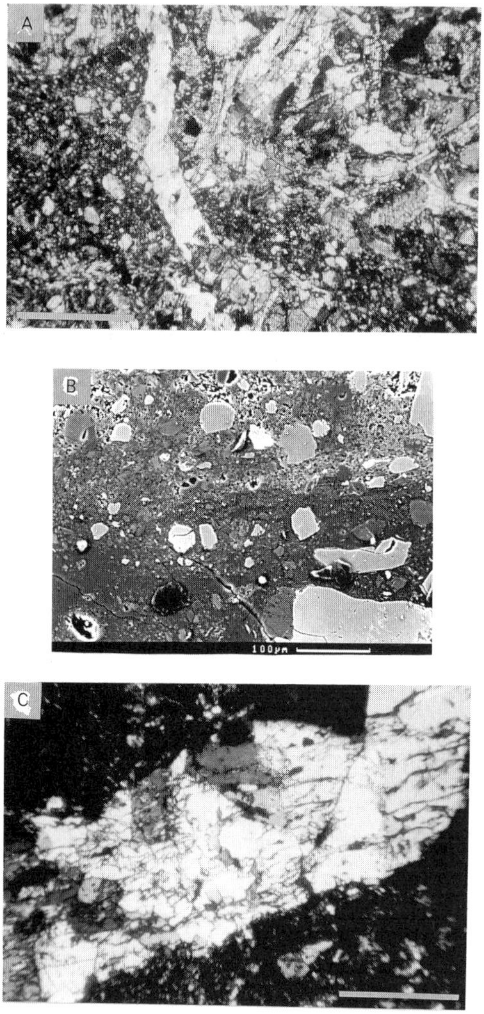

Fig. 4. Examples of cataclastic deformation. (**a**) Narrow cataclastic zone developed in thin dyke unit. Grain size reduction has been achieved through both dissolution and cataclasis. There is a rapid reduction in grainsize towards the zone. This deformation pre-dated quartz veining in the sample. (79–3–6, 88–91 cm). Scale bar is 1 mm. (**b**) BSEM of a cataclastic zone where grain size reduction within the zone clearly involves both cataclasis and dissolution. Few fragments within the zone can be matched with their neighbours. Incipient fragmentation of clasts is evident. Corroded grain margins show dissolution. (73–2–IC, 48–52 cm). (**c**) Quartz vein forms part of a branching network of veins sealing a breccia. Quartz grain margins have been corroded by laumontite which also fills fractures dissecting the vein. The late fractures are oriented subparallel to the quartz vein margins. Both fracture generations are dilational. Other examples show a rotation of the local stress field between alteration stages. (79–1–4, 30–33 cm). Scale bar is 1 mm.

margins suggest that grainsize reduction is achieved by dissolution as well as cataclasis.

There is extensive brecciation at the top and bottom of the transition zone. Despite the reduction in early alteration, brecciation in massive units is as significant as it is in pillow units. Typical breccias have angular fragments with brecciation often restricted to zones narrower than individual core pieces and intensifying towards a fault or fracture. Clear examples of tectonic breccias (Sibson 1977) exist in the central portions of dykes and massive flow units (Agar 1990). Slickensides, a rare feature in breccias of hydrothermal origin (Hulen & Nielson 1988) are present in some breccias. Recementation of pillow breccias is also evident. In contrast, brecciation at the margins of lithological units could be associated with intrusion or a rubbly flow top. Hybrid magmatic-tectonic origins are likely as strong mechanical contrasts exist at these locations. Observations of chilled margins illustrate that these contain some of the most complicated fracture histories.

The transition zone enjoyed a prolonged deformation history relative to the upper zone. Not all fractures are dilational. Some epidote and chlorite veins are filled wih fibres elongate parallel to the fracture edges and their slight obliquity may indicate shearing parallel to the fractures (Ramsay 1980). Shearing may also be indicated by deflection of phenocrysts. Noncoaxial deformation has overprinted many dilational fractures. This deformation includes crystal plasticity and fracturing. These processes are responsible for the boudinage and disaggregation of quartz veins, development of deformation bands, subgrains and fractures of laumontite (Agar 1990). Laumontite fills fractures locally oriented preferentially subparallel to the margins of earlier quartz fractures showing the same extension direction during later alteration (Fig. 4c). Not all laumontite is in discrete fractures, some replaces quartz resulting in broader, irregular deposits. Disaggregation and fracturing of epidote is also observed.

Fracture orientations

The current database of fracture orientations, areas, fill and cross-cutting relations contains several thousand individual fractures. A full account of fracture orientations in the core is impractical in this short paper. It is relevant at this point however to include some discussion of the orientations of different generations of fractures. Figures (5 and 6) show the orien-

Fig. 5. Large rotations between different stages of alteration are indicated locally. The veins in this photograph have been examined in thin sections and their fabrics suggest that they are mode 1 fractures. The numbers on the fractures show the sequence of generation. (note that parts of later fractures are slightly deflected by pre-existing veins, not offset by them). The rotation of the least principal stress indicated by this sequence was about 120°. (92–1–2).

tations of early and later veins in different lithological units from the different crustal zones discussed. The division into early and late fractures is oversimplified for many areas in the core but mesoscopic observations usually resolve only two main phases of fracturing. Three or four phases of fracturing are locally observed in core sections and routinely at thin section scale. In many of the units the earliest fractures can be related to the predicted cooling stresses patterns (Adamson 1985). Although some early fractures are reactivated by subsequent hydrothermal activity, later veins, with different fill phases are often oblique to early failure planes. Successive fractures record changes in the orientation of the local stress field, most notably, the minimum compressive stress. Although cross-cutting relations between veins can be ambiguous (P. F. Williams, pers. comm) one specimen (Fig. 5) apparently preserves a rotation of about 120° of the local minimum compressive stress direction. More typical minimum rotations range between 30° and 90°. Sometimes the fracture sets are bimodal or even polymodal in which case the rotation between generations cannot be easily assessed but a change in the local stress field is still evident. The variation in degrees of rotation between and within different units, combined with the evidence for faulting indicates the presence of numerous mechanical discontinuities in the upper ocean crust. Where it has been possible to measure fault surface orientations there has been a wide variation in dips. The restricted view provided by the core material makes it difficult to ascertain which fault surfaces are representative of the orientation of major detachments. Larger scale rotations in the upper ocean crust by listric normal faulting in 504B have been postulated by Kinoshita *et al.* (1989).

Fracture porosity

Quantification of fracture area in 2-dimensional sections gives a semi-quantitative estimate of fracture porosity. The fracture porosity data for different phases of fracture integrated with the deformation history enables a reconstruction of fluid flow paths. Examples of fracture porosity variations in different lithologies are shown in Table 1. Fracture densities for brecciated zones are not included. The data therefore represent minimum fracture densities, especially as the most fractured portions of the core were not recovered. Some overestimation for the 'later' generation of fractures however is likely as it was not generally possible to distinguish more detailed fracture history in core samples. In many units the total porosity changes little, although the variation in fracture geometry in some units between different generations would have strongly influenced the permeability and permeability anisotropy.

In the examples from the upper zone the early dyke fracture porosity is significantly less than the pillow or flow units. The porosity related to later fractures cutting through these units was more evenly distributed. All three units show a reduction in porosity with time, most notable in the pillow basalt.

At the top of the transition zone the porosity in the pillow unit shows little change with time but the porosity in the faulted dyke unit has

Fig. 6. Examples of changes in fracture orientation during deformation in the upper ocean crust. For each pair of rose diagrams early fractures are on the left hand column, later fractures are on the right. The divisions into early and later fractures are generalized and are based on the nature of the fracture fill and cross-cutting relations. More than two generations can be distinguished in thin section but only occasionally in hand specimens. The diagrams show the cumulative fracture length for a given degree of dip (10° intervals) summed for the range of fractures measured in a given core interval. This means that fractures with irregular traces can be included in the plots. Oriented (way-up) core pieces only have been used. As no azimuth is available for the core pieces only one quadrant of the rose diagram is used. Notice that the length scale is not linear. The diagrams have not been normalized for the number of fractures measured in each interval. 79–1 (12 fractures), 79–2 (33 fractures) and 79–3 (12 fractures) show a sequence through a dyke unit. 80–1 (52 fractures) and 80–2 (37 fractures) are in the uppermost stockwork zone. 85–1 to 85–2–2 (59 fractures) is at the base of a thick pillow sequence. 91–2–6A to 15A (35 fractures) is a thin flow. 91–3 (24 fractures) is a massive flow unit. 92–1–9 to 92–2–5B (30 fractures) is a dyke unit.

Table 1. *Examples of changes in fracture porosity during the evolution of the upper ocean crust in 504B*

Lithological Unit	Core	Thickness m	Fracture porosity %	
			Early	Late
Thin flow	6-2-414 – 6-2-426	1.5	2.6	1.1
Pillows	7-1-432 – 7-4-508	6.0	3.2	0.9
Dyke	8-1-1 – 8-5-567	10.0	1.8	1.6
Pillows	75-2-1 – 76-1-8	15.2	2.9	2.5
Dyke	78-2-1F – 78-2-7	2.5	1.72	4.9
Pillows	80-1-1 – 80-1-7D	4.0	3.2	6.0
Pillows	80-3-2 – 80-4-2	6.0	5.2	3.0
Thin flow	91-2-6A – 91-2-15B	2.9	3.5	14.0
Massive	91-3-1 – 92-1-8	4.3	4.3	7.0
Dyke	92-1-9 – 92-2-5B	3.2	5.8	1.1

The porosity data were calculated from fracture density data collected from digitized photographs. These data are a refinement of preliminary estimates of fracture intensity (Adamson 1984) calculated from the number of fractures per unit length and not making allowances for different generations of fractures. Although inaccuracies are included by this simple two-fold division of alteration generally only two clear phases of fracturing are evident in the core samples. Microstructural analysis demonstrates that there are often three or four phases in the transition zone, but the last phases are not volumetrically significant compared to the main high temperature phase of alteration. See discussion in text.

nearly doubled. Towards the base of the transition zone, in a tectonized zone, there is an increase in the porosity of all three lithological units shown with time. In the example from the stockwork the pillow basalts at the top nearly double their porosity during the stockwork alteration sequence but there is a decrease in porosity during progressive alteration lower down in the stockwork.

Discussion

Even in a relatively simple tectonic setting the fracture porosity of the upper ocean crust in 504B is extremely variable both with depth and with time. There does not appear to be a simple relation between lithology, unit thickness and fracture density. Changes in the distribution and characteristics of fractures have influenced the location and nature of fluid flow and local thermal gradients. In the top 1.5 km of ocean crust temperature ranges may not have affected the bulk rheology that much. However, failure would have occurred at the weakest, pervasively altered zones in the upper ocean crust whose location would have been influenced by fracture networks. Zones of weakness will be utilized during subsequent fluid flow but the changing temperatures and fluid composition will vary the nature of the alteration minerals and may strengthen these zones.

In 504B a change from a network of finescale fractures to a more focused flow at the top of the transition zone is recorded. This bears a close resemblance to the flow transition proposed by Goldfarb & Delaney (1988) but they use the transition from 'ductile' to 'brittle' deformation to explain the generation of such a transition. As 504B is dominantly in the 'brittle' field an alternative explanation is required. The spatial association of fault rocks with a zone of high temperature alteration minerals suggests that faulting and focused fluid flow were closely linked. Fluid may have been 'sucked' in during dilation associated with a fault. Such faulting events may have been localized by weak, alteration zones within the pillows and the mechanical contrast at the top of the transition zone. Continued slip with grainsize reduction and precipitation of minerals altering the fracture network on the fault would have led to transient high pore fluid pressures and steep pressure gradients forming the hydraulic breccias observed in this zone (Phillips 1972; Sibson 1987). Faulting may have been triggered by a discharge flow focusing at the top of the transition zone. The linked faulting and fluid flow processes may have resulted in a runaway situation enabling rapid massive mineralization of the stockwork zone. As Goldfarb & Delaney (1988) pointed out there is a closely linked feedback system between the evolution of fractures and hydrothermal circulation. In order to understand the rheological evolution of the ocean crust these feedback systems need to be assessed in detail. The sealing of fault horizons by focused fluid flow may have caused detachments to migrate to other weak horizons or the onset of more widespread deformation. The numerous detachments inferred in the upper ocean crust of 504B are likely to have developed in this way causing independent rotation of blocks and further complicating interpretation of the fracture network evolution (e.g. Verosub & Moores 1981; Varga & Moores 1985). The complex deformation histories observed in the 504B section show that the mechanical behaviour of the ocean crust changes during progressive alteration. The rheology is clearly time dependent and models employing a depth dependent rheology need to take this into account.

Summary

The earliest fractures preserved in the 504B core samples are thermal cracks associated with cooling and those generated during volume changes as a result of pervasive alteration of basalts. Early mesoscopic and microscopic cracks were extended by the growth of early alteration phases. Intense early fracturing was associated with pillow breccias, hyaloclastite formation and possibly shallow level faulting. Later fractures in the upper zone often reactivate early fractures but towards the transition zone different generations of fractures dissect each other suggesting changes in the local stress field. Although many of the later fractures are dilational many examples of subsequent noncoaxial deformation overprint the fracture fills. Tectonic breccias were generated during later alteration. The changing orientations of fractures and the presence of fault rocks indicate numerous discontinuities within the upper ocean crust. Early faulting in the section was probably located on the most altered, clay rich horizons but later faulting is observed in a variety of lithologies. Mesoscopic fracture porosity variations in the upper ocean crust are recorded by the quantity and location of fracture fill phases of different generations of alteration. Fluid flow during later deformation was transiently focused and appears to have been spatially associated with faulting.

This work was supported by a NERC special ODP Fellowship. G. Norrell is thanked for a careful and constructive review. A. Adamson, J. Cann, R. Knipe, D. Prior & H. Richards helped to initiate this work and have provided stimulating discussions during its progress. Staff at the ODP core repositories, particularly J. Bode, S. Prince and C. Mato went out of their way to smooth my path through all those cores.

References

ADAMSON, A. C. 1984. *Hydrothermal petrology in the Costa Rica Rift*. PhD thesis, University of Newcastle upon Tyne.

—— 1985. Basement lithostratigraphy, Deep Sea Drilling Project Hole 504B. *In*: ANDERSON, R. N., HONNOREZ, J., BECKER, K. *et al. Initial Reports of the Deep Sea Drilling Project*, **83**, US government printing office, Washington, 121–128.

AGAR, S. M. 1990. Microstructural evolution of a fault zone in the upper ocean crust. *In*: NICOLAS, A. (ed.) *Ophiolites and the ocean crust*. Special Volume of Journal of Geodynamics, (in press)

ALT, J. C., LAVERNE, C. & MUEHLENBACHS, K. 1985. Alteration of the upper oceanic crust: Mineralogy and processes in Deep Sea Drilling Project Hole 504B. *In*: ANDERSON, R. N., HONNOREZ, J., BECKER, K. *et al. Initial Reports of the Deep Sea Drilling Project*, **83**, US government printing office, Washington, 217–248.

——, HONNOREZ, J., LAVERNE, C. & EMMERMAN, R. 1986*a*. Alteration of a 1 km section through the upper oceanic crust, DSDP Hole 504B: The mineralogy, chemistry and evolution of basalt-seawater interactions. *Journal of Geophysical Research*, **91**, 10309–10335.

——, MUEHLENBACHS, K. & HONNOREZ, J. 1986*b*. An oxygen isotopic profile through the upper kilometre of the ocean crust, DSDP hole 504B. *Earth and Planetary Science Letters*, **80**, 217–229.

ANDERSON, R. N. & ZOBACK, M. D. 1982. Permeability, underpressures and convection in the oceanic crust near the Costa Rica Rist, eastern equatorial Pacific. *Journal of Geophysical Research*, **87**, 2860–2868.

ANDERSON, R. N., HONNOREZ, J., BECKER, K. *et al.* 1985. *Initial Reports of the Deep Sea Drilling Project*, **83**, US government printing office, Washington.

ATKINSON, B. K. 1987. *Fracture mechanics of rock*. Academic Press London.

BECKER, K., SAKAI, H. *et al.* 1988. *Proceedings of the Ocean Drilling Program*, **111**, Part A.

CANN, J. R., LANGSETH, M. G., HONNOREZ, J. *et al.* 1983. *Initial Reports of the Deep Sea Drilling Project*, **69**, US government printing office, Washington.

CRRUST (COSTA RICA RIFT UNITED SCIENTIFIC TEAM) 1982. Geothermal regimes of the Costa Rica Rift, east Pacific, investigated by drilling, DSDP-IPOD Legs 68, 69 and 70. *Geological Society of America Bulletin*, **93**, 862–875.

GOLDFARB, M. S. & DELANEY, J. R. 1988. Response of two phase fluids to fracture configurations within submarine hydrothermal systems. *Journal of Geophysical Research*, **93**, 4585–4594.

HARPER, G. D. BOWMAN, J. R. & KUHNS, R. 1988. A field, chemical and stable isotope study of subseafloor metamorphism of the Josephine Ophiolite, California–Oregon. *Journal of Geophysical Research*, **93**, 4625–4656.

HONNOREZ, J., LAVERNE, C., HUBBERTEN, H. W., EMMERMAN, R. & MUEHLENBACHS, K. 1983. Alteration processes in layer 2 basalts from Deep Sea Drilling Project Hole 504B, Costa Rica Rift. *In*: CANN, J. R., LANGSETH, M. G., HONNOREZ, J. *et al.* **69**, US government printing office, Washington, 509–546.

HULEN J. B. & NIELSON, D. L. 1988. Hydrothermal brecciation in the Jemez fault zone, Valles Caldera, New Mexico. Results from Continental Scientific Drilling Program Core Hole VC-1. *Journal of Geophysical Research*, **93**, 6077–6089.

KINOSHITA, H., FURUTA, T. & PARISO, J. 1989. Downhole magnetic field measurements and palaeomagnetism, hole 504B, Costa Rica Ridge. *In*: BECKER, K., SAKAI, H. *et al. Proceedings of the Ocean Drilling Program, Scientific Results*, **111**, 147–156.

KLITGORD, K. D., MUDIE, J. D., HUESTIS, S. P. & PARKER, R. L. 1975. An analysis of near-bottom magnetic anomalies: Sea floor spreading and the magnetized layer. *Geophysical Journal of the Royal Astronomical Society*, **43**, 387–424.

LLOYD, G. E. 1987. Atomic number and crystallographic contrast images with the SEM: a review of backscattered electron techniques. *Mineralogical Magazine*, **51**, 3–19.

PHILLIPS, W. J. 1972. Hydraulic fracture and mineralization. *Journal of the Geological Society, London*, **128**, 337–359.

PRIOR, D. J. 1988. Fractures and retrogression in garnets from the Alpine fault Mylonites, New Zealand. *Journal of the Geological Society, London*, **146**, 335.

RAMSAY, J. G. 1980. The crack-seal mechanism of rock deformation. *Nature*, **284**, 135–139.

RICHTER, D. & SIMMONS, D. 1974. Thermal expansion behaviour in rocks. *International Journal of Rock Mechanisms and Mining Science*, **11**, 403–411.

ROWE, K. J. & RUTTER, E. H. 1990. Palaeostress estimation using calcite twinning: experimental calibration and application to nature. *Journal of Structural Geology*, **12**, 1–17.

SIBSON, R. H. 1977. Fault rocks and fault mechanisms. *Journal of the Geological Society, London*, **133**, 191–213.

——. 1987. Earthquake rupturing as a hydrothermal mineralizing agent. *Geology*, **15**, 701–704.

VARGA, R. J. & MOORES, E. M. 1985. Spreading structure of the Troodos ophiolite, Cyprus. *Geology*, **13**, 846–850.

VEROSUB, K. L. & MOORES, E. M. 1981. Tectonic rotations in extensional regimes and their palaeomagnetic consequences for oceanic basalts. *Journal of Geophysical Research*, **86**, 6335–6349.

The calculation of bulk rheologies of structured materials and its application to brittle failure in shear

MARTIN CASEY & GILLES WUST

Geologisches Institut, ETH-Z, CH8092 Zürich, Switzerland

Abstract: A finite element method is presented which allows the calculation of the bulk rheological properties of a material consisting of a regular array of heterogeneous sub-regions. The method is used to calculate the elastic properties of rocks with aligned cracks containing air or fluid. It is found that the cracked material has anisotropic properties. A compressible fluid filling makes the bulk rock less stiff. The stiffness perpendicular to the long axis of the cracks is reduced more strongly than that parallel to the cracks. It is shown that a concentration of aligned cracks in an inclined planar zone could result in an instability leading to shear failure.

Continuum mechanics is concerned with the properties of materials idealized so that they can be treated as being homogeneous. Real materials are heterogeneous at all scales and the use of continuum mechanics requires that an appropriate scale be defined at which the continuum approximation is valid (Means 1976). Often it is of interest to study processes at one scale, e.g. the 0.1 mm scale, but to have an idealisation of the material properties at the 10 cm scale. An example of this could be the effect of microcracks on the elastic moduli (Brace 1965; Budiansky & O'Connell 1976). In considering the development of preferred crystallographic orientation in deforming crystalline aggregates, hypotheses about the micromechanics of dislocation motion are made and a means of deriving the macroscopic rheology of the rock is required. The latter problem has been approached by the Taylor, Bishop-Hill method (Taylor 1938; Bishop & Hill 1952) or by self-consistent averaging (Kröner 1961; Hill 1965; Canova 1988). In the self consistent method each grain of an aggregate of grains with various orientations is considered as being an ellipsoidal inclusion in a matrix of average properties, which are not known *a priori*, but for which it is possible to derive and solve non-linear equations, Hill (1965). A method is presented in this paper which tackles the problem in a different way: the material is considered to be made up of a regular array of sub-regions each of which contains enough heterogeneity to be representative of the material at the small scale. A derivative of the finite element method is used to calculate an effective stress/strain or stress/strain-rate relation for the material at the larger scale.

Bulk elastic properties of a heterogeneous continuum

For a homogeneous material:

$$U=\frac{1}{2}c_{ij}\epsilon_i\epsilon_j V \tag{1}$$

where U is the elastic internal energy, c_{ij} is the elastic stiffness, ϵ_i is the strain and V is the volume. The Einstein summation convention is used, i.e. repetition of a suffix indicates summation over that suffix. From equation 1 it can be seen that:

$$c_{ij} = \frac{\partial^2 U}{\partial\epsilon_i \partial\epsilon_j} V^{-1} \tag{2}$$

That is, the elastic stiffness can be obtained by differentiating an expression for the elastic energy twice with respect to imposed strain. It is postulated that this procedure can be used for a heterogeneous material, provided that an expression relating elastic energy and imposed deformation can be derived. This postulate may be justified with the following argument. The stress in an elastic continuum is defined as the first derivative of elastic energy with respect to strain (Truesdell & Toupin 1960). If this is done, then an expression for the stress as a function of strain results. When the elastic strain energy is a quadratic function of strain, stress will be a linear function of strain and the elastic stiffness can be derived by differentiation with respect to strain.

Linear elastic finite element analysis gives the following equation for the elastic energy:

$$U = \frac{1}{2} u_i \int_v k_{ij} \mathrm{d}v u_j \tag{3}$$

From Knipe, R. J. & Rutter, E. H. (eds), 1990, *Deformation Mechanisms, Rheology and Tectonics*, Geological Society Special Publication No. 54, pp. 51–55.

where the u_i are the nodal displacements and $\int_V k_{ij} dv = K_{ij}$ is the global stiffness matrix.

Finite element analysis also gives the following expression for the nodal displacements:

$$K_{ij} u_j = R_i \tag{4}$$

where R_i are the nodal forces. This equation can, in principle, be inverted to give:

$$u_i = K_{ij}^{-1} R_j \tag{5}$$

Figure 1 shows a finite element model in which the boundary conditions are all prescribed displacements or constraint relations among nodal displacements resulting from symmetry.

Consider a boundary node for which the displacements are prescribed. The value of the prescribed displacements are given by:

$$\begin{pmatrix} u_1^* \\ u_2^* \end{pmatrix} = \begin{pmatrix} X_1 & 0 & \frac{1}{2}(X_1 + X_2) \\ 0 & X_2 & \frac{1}{2}(X_1 + X_2) \end{pmatrix} \begin{bmatrix} \epsilon_1 \\ \epsilon_2 \\ \gamma_{12} \end{bmatrix} \tag{6}$$

Application of the boundary conditions is achieved in practice by eliminating the equations for the prescribed components in equation 4 and setting the displacement components to their prescribed values. To achieve the latter the stiffness terms in the corresponding columns of the matrix K_{ij} are multiplied by the prescribed displacement values and moved to the right hand sides of the equations where they contribute to the nodal force vector, R_i.

This gives, for one component of the nodal force vector:

$$R_i = - u_j^* K_{ij} \tag{7}$$

where the summation on j runs over all prescribed displacement components. Rewriting equation 6 as:

$$u_j^* = n_{jl} \epsilon_l \tag{8}$$

where n_{jl} is a function of nodal coordinates, gives:

$$\mathbf{R}_i = -n_{ml} \mathbf{K}_{im} \epsilon_l \tag{9}$$

Substituting in equation 5:

$$u_i = K_{ij}^{-1} \mathrm{n}_{\mathrm{ml}} K_{jm} \epsilon_l \tag{10}$$

or:

$$u_i = l_{il} \epsilon_l \tag{11}$$

with

$$l_{il} = K_{ij}^{-1} n_{ml} K_{jm} \tag{12}$$

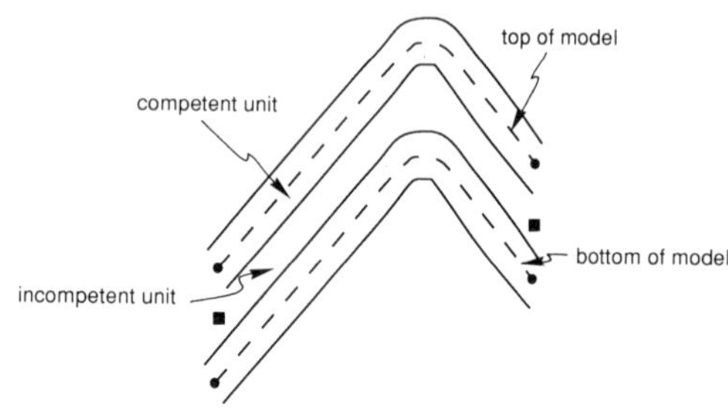

Fig. 1. An asymmetric similar fold in an infinitely repeating sequence of competent and incompetent units. Points marked have prescribed displacements and are centres of symmetry. The velocities of all other boundary points are related to symmetrically related points by constraints. The displacement at a point on the upper dashed line differs from that at the equivalent point on the lower dashed by an amount dependent only on the imposed strain and the relative positions of the points.

Substituting equation 11 into equation 3 gives:

$$U = \frac{1}{2} \epsilon_i K_{ij} l_{jm} \epsilon_m \tag{13}$$

Thus the elastic energy can be written as a quadratic function of imposed strain and the stiffness obtained by differentiating twice with respect to ϵ_i.

$$c_{im} = l_{il} K_{lj} l_{jm} \tag{14}$$

A finite element program has been adapted to perform the calculations needed to obtain the elastic stiffness. The calculation of l_{il} using equation 12 was performed without explicit inversion of the global stiffness matrix, but using a standard finite element technique with several right hand sides.

Calculation of elastic parameters of a material with aligned cracks

As an example and to provide results to be used in an attempt to explain brittle failure in shear, the technique outlined above was applied to the calculation of the elastic stiffness of a material containing aligned cracks. Previous theoretical calculations of the elastic moduli of cracked solids have treated random distributions of cracks (Budiansky & O'Connell 1976) giving isotropic behaviour. Microcracks which result from loading a material will be statistically aligned and give anisotropic elastic behaviour. A non-random orientation distribution of microcracks is idealized by an array of elliptical flaws as shown in Fig. 2. The unit cell of the array and boundary constraints appropriate to the imposition of a general deformation are shown in Fig. 3.

The elastic properties were calculated for various crack densities for both air and water filled cracks. To simulate air filled cracks the material in the elliptical flaw was given very low rigidity and reciprocal compressibility. To simulate water the material was given very low rigidity and compressibility appropriate to water. Typical values for the two dimensional stiffness are given in Table 1 and plots of selected terms of the stiffness matrix as a function of crack density are shown in Fig. 4.

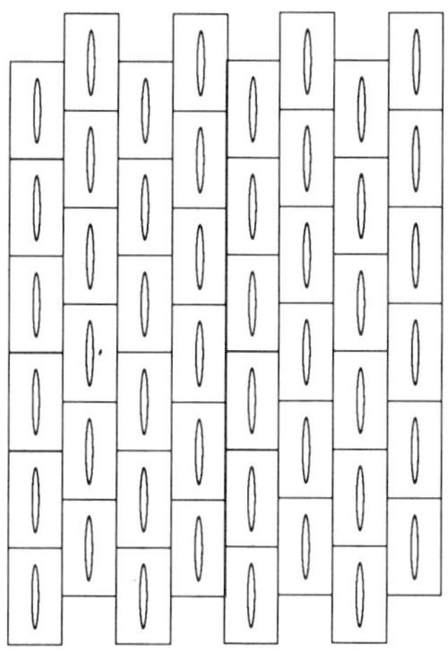

Fig. 2. A cracked material idealized as a material with a regular array of cracks.

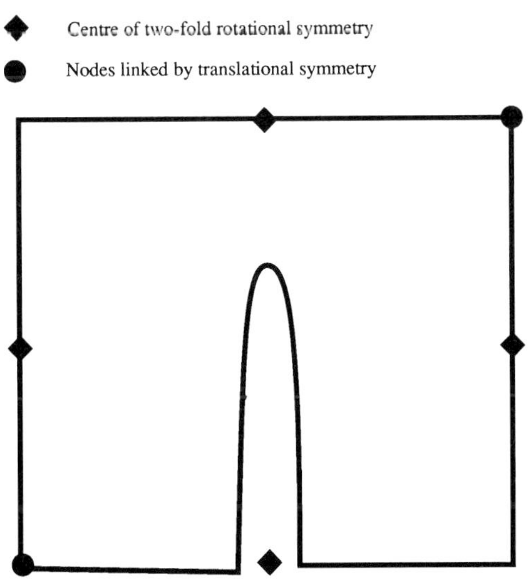

Fig. 3. The unit cell of the array in Fig. 2 with boundary conditions.

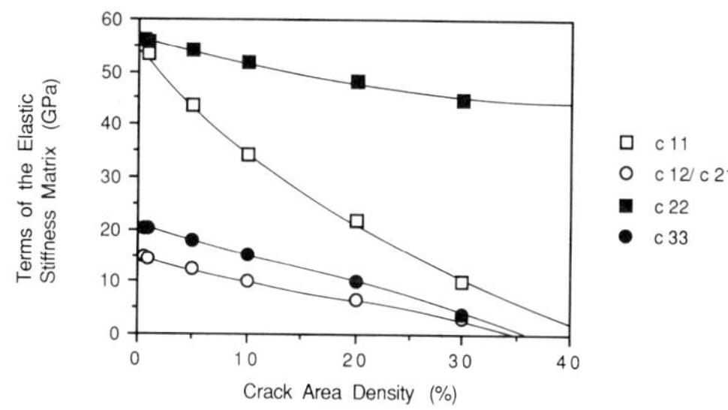

Fig. 4. The effect of increasing density of fluid filled cracks on the terms of the elastic stiffness matrix. The intact rock has a Young's modulus of 50.0 GPa and a Poisson's ratio of 0.21 giving a reciprocal compressibility of 28.7 GPa. The fluid has a reciprocal compressibility of 287.0 MPa

Table 1. *The stiffness matrix (terms are GPa/10) for 10% crack area density for air and water filled cracks in limestone.*

Intact rock		
5.628	1.496	0.0
1.496	5.228	0.0
0.0	0.0	2.0661
Air filled crack		
3.356	0.987	0.0
0.987	5.179	0.0
0.0	0.0	1.534
Fluid filled crack		
3.440	1.017	0.0
1.017	5.190	0.0
0.0	0.0	1.534

The intact rock has a Young's modulus of 50.0 GPa and a Poisson's ratio of 0.21 giving a reciprocal compressibility of 28.7 GPa. The fluid has a reciprocal compressibility of 287.0 MPa

Brittle failure in shear

The failure of a material under tensile stress conditions is well understood in terms of tensile crack propagation. When a critical stress is reached the most dangerous flaw propagates unstably through the material and it breaks.

Although it is possible in theory for cracks to propagate in shear-mode II and mode III failure, such cracks are not observed. When a material is stressed such that all principal stresses are compressive, microcracks occur but they are mostly mode I and their propagation is stable, a higher-differential stress being required to cause further extension. Shear failure appears to occur by the development of a zone of microcracking which develops as an instability until the material is so much weakened that a through-going failure plane forms (Hallbauer *et al.* 1973; Casey 1980; Li 1987). In order for such an instability to develop increased microcrack density should so modify the stress that cracks propagate further. In order to analyse such a process the following are needed.

(1) The ability to calculate the effective elastic stiffness of a cracked material.
(2) The calculation of stress in a zone of cracked material.
(3) A criterion for crack extension.

The first problem is solved above. The second requirement can be met by making some simplifying assumptions about the problem. The weak zone is taken as being planar, of infinite extent and oriented at an angle ϕ to the principal extension direction, see Fig. 5. This simple case certainly begs some of the questions related to the problem. A more realistic case would be an ellipsoidal zone which intensifies and propagates. Using the planar zone it is possible only to study the intensification of the instability, but this is a first condition for feasibility of the mechanism and as it allows of a simple solution it is useful to follow it up.

The boundary conditions between the matrix and weak zone are, with reference to Fig. 5:

$$\epsilon_1'^M = \epsilon_1'^Z \tag{15}$$

$$\sigma_2'^M = \sigma_2'^Z \tag{16}$$

$$\tau_{12}'^M = \tau_{12}'^Z \tag{17}$$

That is, strain parallel to the zone boundary is continuous as are the normal and shear stresses on the boundary.

In the weak zone the equation for $\epsilon_1'^Z$ in terms of stresses is:

$$\epsilon_1'^Z = s_{11}'^Z\,\sigma_1'^Z + s_{12}'^Z\sigma_2'^Z + s_{13}'^Z\,\tau_{12}'^Z \tag{18}$$

where $s_{ij}'^Z$ is the compliance of the weak zone. This equation may be solved for $\sigma_1'^Z$, the remaining stress component:

$$\sigma_1'^Z = \frac{\epsilon_1'^Z - s_{12}'^Z\sigma_2'^Z - s_{13}'^Z\tau_{12}'^Z}{s_{11}'^Z} \tag{19}$$

For the weak zone shown in Fig. 5 it is assumed that the cracks occur in an array with their long axes parallel to the shortening direction in the matrix. The finite element program can be used to calculate a stiffness matrix corresponding to a small but finite density of microcracks. A solid with a Young's modulus of 951 MPa and a bulk compressibility of 754 MPa, containing cracks with an aspect ratio of 15:1 and a density of 10% was used. The crack filling had very low rigidity and reciprocal compressibility. This stiffness matrix is anisotropic and can be inverted to give the corresponding compliance matrix. The compliance matrix in the primed coordinate system can be obtained by transformation as described in Casey & Huggenberger (1985), equations 13 to 16.

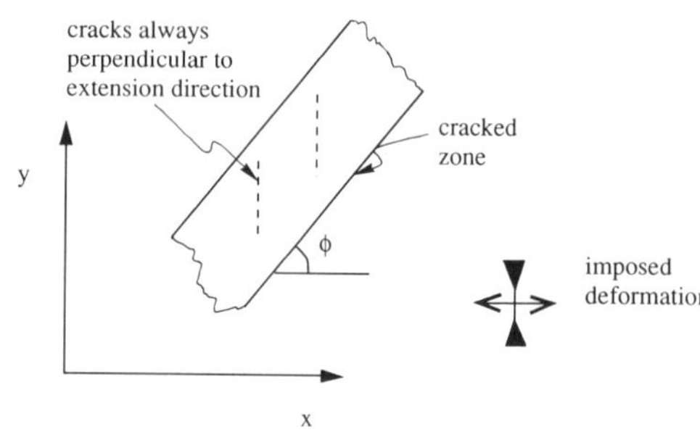

Fig. 5. Cracked zone model used in the analysis.

Modification of the stress in the weak zone as a function of the angle between the zone boundary and the extension direction is shown in Fig. 6. As yet a criterion of crack extension is lacking but from Fig. 6 it appears that when the weak zone is around 30° to the shortening direction the stresses are modified so as to be less compressive and this may be expected to favour extension. Thus it appears that a weak zone oriented at around 30° would intensify.

The modification of the stress in the present case is stronger than that reported by Casey (1980) for a cracked zone with isotropic elastic properties and an assumed 5% reduction of Young's modulus.

Discussion and conclusions

Microcrack localization is an example of a problem where it is useful to apply continuum mechanics approximations at two scales: that of the microcracks and that of the zone of increased microcrack density. The finite element method given here allows the considerations at these scales to be linked. Other situations where this approach could be applied are as follows.

(1) The study of two phase materials where interactions of grains of the weaker phase are important to the bulk rheological behaviour (Jordan 1987).
(2) The calculation of effective rheologies for rock masses with periodic structures such as folds. This could be important for calculating seismic velocity anisotropy for seismic waves with wave-length comparable to the structural wavelength.

The presence of aligned microcracks induces

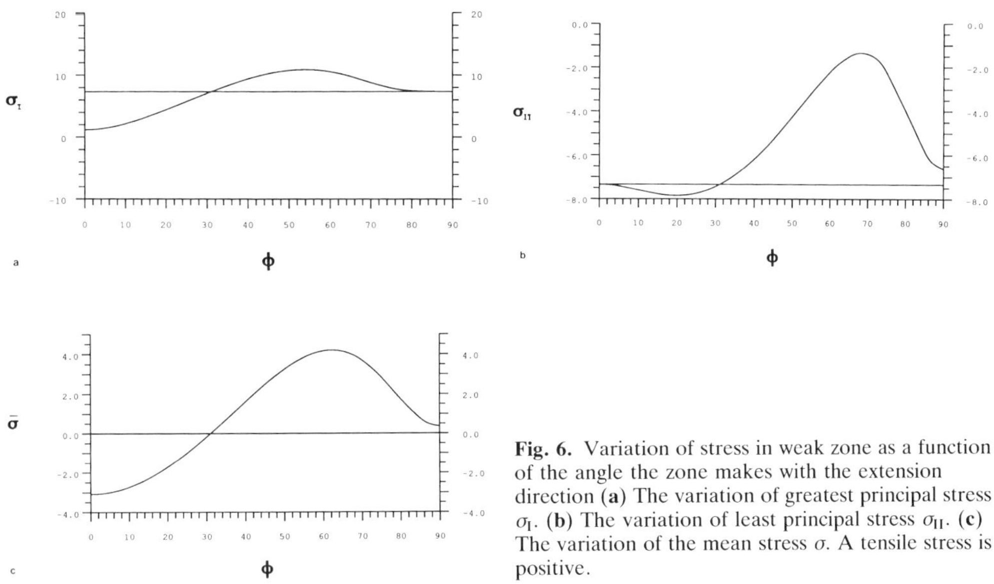

Fig. 6. Variation of stress in weak zone as a function of the angle the zone makes with the extension direction (**a**) The variation of greatest principal stress σ_I. (**b**) The variation of least principal stress σ_{II}. (**c**) The variation of the mean stress σ. A tensile stress is positive.

anisotropy in the material. If the crack filling is more compressible than the rock material, increasing crack density makes the bulk material less stiff.

Concentration of aligned cracks in a zone at an angle to the compression direction modifies the stresses in the weak zone. An angle of around 30° to the shortening direction modifies the stress in such a way that crack extension might be favoured. The modification of the stress using the calculated anisotropic stiffness is more marked than when a simple reduction of Young's modulus is used.

A public domain program is available from Martin Casey on any IBM or Macintosh diskette sent to him.

The authors are thankful for helpful reviews from A. Ord, G. Lloyd and comments on the manuscript by J. Ramsay. Financial support from Swiss Nationalfonds project no. 2.214–0.86 is gratefully acknowledged.

References

BISHOP, J. F. W. & HILL, R. 1952. A theory of the plastic distortion of a poly-crystalline aggregate under combined stresses, *Philosophical Magazine*, **42**, 414–427.

BRACE, W. F. 1965. Some new measurements on the linear compressibility of solids, *Journal of Geophysical Research*, **70**, 391–411.

BUDIANSKY, B. & O'CONNELL, R. J. 1976. Elastic moduli of a cracked solid, *International Journal of Solid Structures*, **12**, 81–97.

CANOVA, G. R. 1988. Effect of texture and grain shape on anisotropy, *Revue Physique Applique*, **23**, 533–548.

CASEY, M. 1980. Mechanics of shear zones in isotropic dilatant materials, *Journal of Structural Geology*, **2**, 143–147.

—— & HUGGENBERGER, P. 1985. Numerical modelling of finite amplitude similar folds developing in general deformation histories. *Journal of Structural Geology*, **7**, 103–114.

HALLBAUER, D. K., WAGNER, H. & COOK, N. G. W. 1973. Some observations concerning the microscopic and mechanical behaviour of quartzite specimens in stiff, triaxial compression tests, *International Journal of Rock Mechanics & Mining Science*, **10**, 713–726.

HILL, R. 1965. Continuum micro-mechanics of elastoplastic crystals, *Journal of Mechanics and Physics of Solids*, **13**, 89–101.

JORDAN, P. 1987. The deformational behaviour of bimineralic limestone–halite aggregates, *Tectonophysics*, **135**, 185–197.

KRÖNER, E. 1961. Zur plastischen verformung des vielkristalls, *Acta Metallurgican*, **9**, 155–161.

LI, V. 1987. Mechanics of shear rupture applied to earthquake zones, *In*: ATKINSON, B. K. (ed.) *Fracture Mechanics of Rock*. Academic Press, 351–428.

MEANS, W. D. 1976. *Stress and Strain*. Springer-Verlag.

TAYLOR, G. I. 1938. Plastic strain in metals. *Journal of the Institute of Metals*, **62**, 307–324.

TRUESDELL, C. & TOUPIN, R. 1960. The classical field theories. *In*: FLUGGE, S. (ed.) *Handbuch der Physik vol III/I. Principles of Classical Mechanics and Field Theory*. Springer Verlag, 226–733.

Damage development during rupture of heterogeneous brittle materials: a numerical study

S. J. D. COX & L. PATERSON

CSIRO Division of Geomechanics, PO Box 54, Mt Waverley, Vic 3149 Australia

Abstract: We have used a finite element method with an iterative algorithm to simulate rupture development in complex brittle materials such as rocks. A simple set of continuous distributions of properties has been used, and only tensile failure of elements allowed. Wider strength distribution leads to more damage and less abrupt mechanical breakdown characteristics. Variations in strength are found to have a greater effect on the geometry of breakdown than variations in elastic properties. The material behaviour simulated here gives a basic explanation for the extent to which natural fractures in the earth, such as joints and dykes, have a non-planar form.

The deformation of real materials is dominated by the influence of mechanical defects. In plastic deformation the role of lattice dislocations as strain concentrators has been widely discussed, while for brittle behaviour attention has focussed on stress concentrations caused by geometry and cracks.

If we wish to use the insights gained from the analysis of individual flaws, we must consider the combined effects of multiple defects to account for bulk macroscopic behaviour. Significant progress to this end has been made for plastic behaviour, for example in Taylor–Bishop–Hill theory which considers the effects of multiple slip systems in polycrystals [Lister *et al.* 1978]. In the brittle field a modified Griffith theory has been developed which predicts the shape of the failure envelope in normal-shear stress space from a consideration of the action of microcracks at a variety of azimuths (McLintock & Walsh 1962; Murrell & Digby 1970*a*, *b*; Horii & Nemat-Nasser, 1985). However, finer details of damage development, and therefore macroscopic stress-strain behaviour upon which stability analyses depend, have not been so comprehensively analysed.

The pertinent observation is that real materials, and especially rocks, are always to some degree heterogeneous. This includes grain-to-grain mismatch (in composition and orientation), the presence of grain boundaries, inclusions and porosity, and distributed fractures at various scales from sub grain-size microcracks up to joints and faults. The direct influence of these on brittle rupture, in particular, is shown by the fact that real tensile fracture surfaces are not usually as smooth as is deduced from a simple application of linear elastic fracture mechanics assuming a uniform material. Thus, models of joint formation such as that of Lachenbruch (1961), Pollard *et al.* (1982) Pollard (1987) and Olson & Pollard (1988) fail to predict the full complexity of fracture shapes and arrays observed in the field.

It is appealing to treat the defect structure as a stochastic process, with elements in a model subject to some random distribution of properties. In this case the success of the model will be measured in statistical terms. In attempting to analyze such a model, however, we are confronted with the possibility of a multiplicity of types of heterogeneity (anisotropy, strength, elastic properties) each with an unknown spatial and probability distribution.

Put simply, there is only one way to distribute properties in a uniform fashion, but there is an infinity of possible random distributions. We may then ask questions such as the following.

(1) Which type of property has most influence on the mechanical behaviour of the solid?
(2) What are the effects of various forms of distribution functions?

Answers to these may give us some insight into what microprocesses will dominate the behaviour of materials, and hence will allow us to focus attention on the important effects in analytical and numerical simulations of real material behaviour.

We have used a finite element method for performing such a sensitivity analysis, and have studied rupture in models of heterogeneous materials under tensile loading. We have moni-

From Knipe, R. J. & Rutter, E. H. (eds), 1990, *Deformation Mechanisms, Rheology and Tectonics*, Geological Society Special Publication No. 54, pp. 57–62.

tored the sequence of damage development and the macroscopic mechanical properties of the model.

Experiments

Numerical methods

Breakdown problems are found in various areas of physics, including fracture, electrical breakdown, percolation, fusion and crystallisation. Collectively, these may all be considered to be analogous to order-disorder phase transitions. The fracture problem has been studied using a number of models.

(1) The simplest has been using networks of fuses as an analogue (de Arcangelis *et al.* 1985; Duxbury *et al.* 1986; Kahng *et al.* 1988). In numerical treatments this only involves the solution of scalar field equations (volts, amps).

(2) The next level of sophistication is networks of rigid bars (Paterson 1988). This requires the determination of force vectors in the network, but full coupling between all force components (Poisson's ratio effects) is not handled conveniently.

Both of these types of models are most easily solved using *finite difference* techniques.

(3) Probably the most 'natural' representation of continuous media uses *finite elements*. In this method the model is discretised into elements, which are in themselves continuous, and therefore a solution of all components of the stress tensor is determined at all points of the model with no interpolation needed.

We have developed a finite element code to study the progress of damage development and rupture by elastic-brittle processes for models with heterogeneous property distributions. Details of the model and algorithm are given in Cox & Paterson (1989), so only a few important points will be repeated here.

The model

A very simple configuration of a 2D square mesh of square elements under tensile axial loading has been used. Linear, isotropic elasticity with a tensile cut-off was assumed, i.e. only one inelastic deformation process was allowed, tensile failure of individual elements.

The effects of variation in the strength and the elastic modulus were examined separately. We assigned properties to the individual elements randomly. Two simple probability distributions $P(s)$ for the property s were used,

(1) uniform distributions $P(s) = \frac{1}{2}\omega$ for s between $1-\omega$ and $1+\omega$ with ω varied between 0 and 1. These have a mean value of 1 and a variance of $\omega^2/2$

(2) a negative exponential distribution, $P(s) = e^{-s}$ which has a mean and variance of 1.

The spatial distribution of properties was assumed to be isotropic and uncorrelated.

Rupture was achieved by breaking single elements in turn until a broken strand connected across the model. We found that models with side lengths, L, of up to 40 could be ruptured within a reasonable computer run time (40 × 40 models took up to 2 days on a 1 MIPS machine with a floating point accelerator). Different random number seeds were used to generate at least ten starting meshes for each set of parameters.

Results

Fracture geometry and damage

Examples of the damage distribution at breakdown for different strength distributions are shown in Fig. 1.

Two parameters have been used to summarize this damage distribution, in order to characterize the geometric results:

<*N*> the total number of broken elements

<*C*> the number of broken elements in the largest connected cluster, i.e. the amount of the damage associated with the dominant rupture.

The notation <> indicates that the results represent means over the sets of meshes. In Fig. 2 we show how the damage parameters vary with ω and L. Rupture of the model takes a minimum of L broken elements so we normalise the damage parameters by this dimension.

We find that <*N*>/*L* increases with ω and L (Fig. 2a, b), and that the exponential strength distribution leads to the greatest amount of damage amongst the distributions examined. For the wider strength distributions, however, the additional damage was mostly disconnected from the dominant rupture, as indicated by the divergence between <*C*> and <*N*> in Fig. 2a. Furthermore, <*N*>/*L* only diverges significantly from 1 for $\omega > 0.4$.

Kahng *et al.* (1988) found that the relationship between <*N*> and L could be described using a power-law. For our data, although the relationship in log–log space is clearly non-linear for small values of L, for the larger models (which will be of most interest for practical application) a power law <*N*>/*L* = $\alpha L^{\beta-1}$ appears to fit the data quite well. The lines on Fig. 2b are

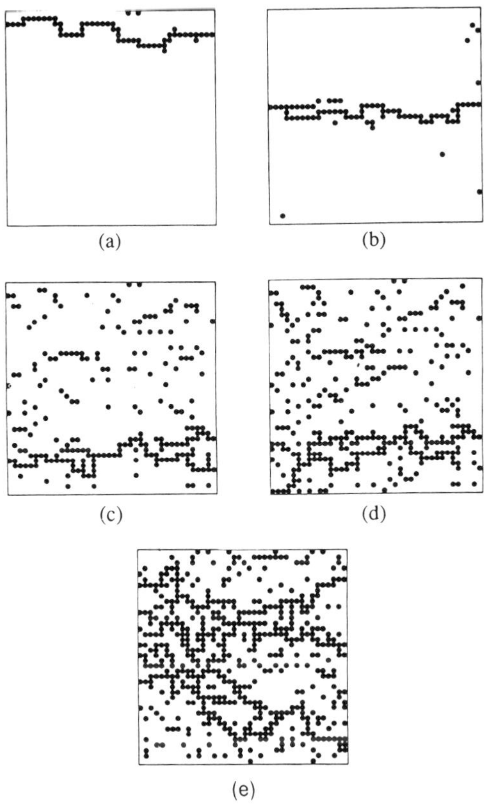

Fig. 1. Examples of damage distribution at breakdown for (**a**) $\omega = 0.5$; (**b**) $\omega = 0.6$; (**c**) $\omega = 0.8$; (**d**) $\omega = 1.0$; (**e**) exponential strength distribution.

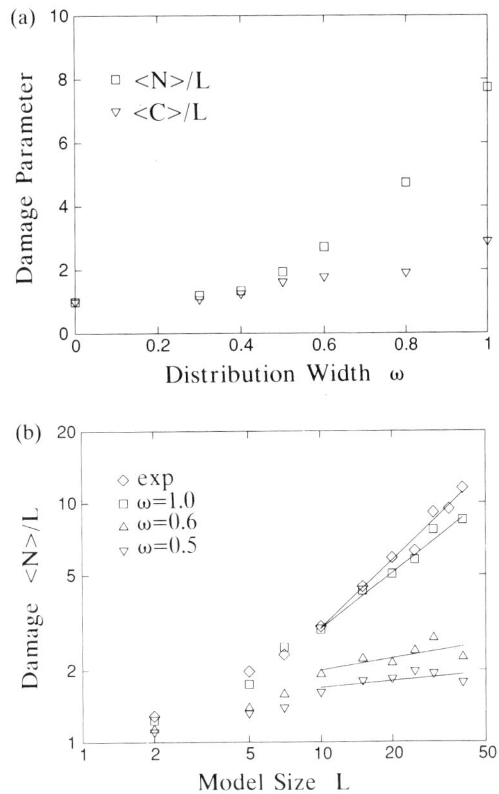

Fig. 2. (**a**) Damage parameters for 30 × 30 meshes with varying widths of strength distributions; (**b**) Total damage at breakdown for different sized models, with best-fit lines for the range $10 \leq L$.

linear regressions to the data for $10 \leq L \leq 40$ and the slopes give values for $\beta - 1$ of 0.94 for the exponential case and 0.77, 0.17 and 0.09 for $\omega = 1.0$, 0.6 and 0.5 respectively. It is trivial to see, for large L, $0 \leq \beta - 1 \leq 1$.

We found that the distribution of elastic modulus has a smaller effect than the distribution of strengths: for an exponential distribution of moduli $\beta - 1 = 0.54$.

Mechanical data

Stress-strain histories may be derived for the models. Those presented here (Fig. 3a) assume a monotonically increasing strain, so may be compared with tensile fracture tests on rock in a stiff testing machine. The simulations show that wider strength distributions yield an earlier onset of non-linear behaviour and lower peak stresses, but a more gradual loss of strength after the peak.

The importance of sample size for strength tests is well illustrated in Fig. 3b. For a continuous elemental strength distribution, larger volumes are likely to contain larger flaws, so the mean strength would continuously decrease with sample size. However, these strength distributions do not produce the power-law relationship predicted elsewhere (Weibull 1939) and we find that reductions in strength become minimal for large L for some strength distributions here.

Discussion

Experimental data on rocks

Peng (1975) gives complete tensile load-displacement curves for four different rock types tested under displacement control in a servo-controlled machine. Contrasting pre-peak and post-peak behaviour is most pronounced between experiments on Berea sandstone and Tennessee marble. The former is a highly porous

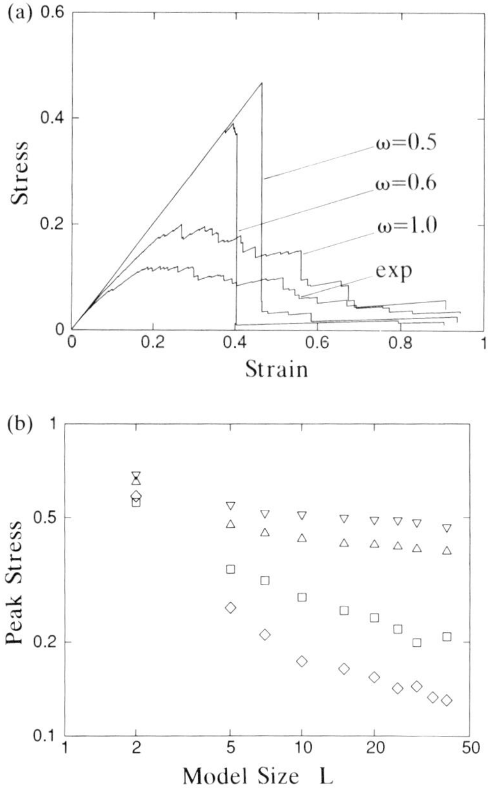

Fig. 3. (**a**) Examples of stress-strain curves for various strength distributions; (**b**) Peak stress for different sized models. Symbols as for Fig. 2b.

material with a relatively large proportion of clays, whereas the latter is a fairly pure, massive, low porosity rock. Thus, the gentler behaviour which is observed for the sandstone is as expected for the material with a wider initial strength distribution.

The observations of crack propagation accompanying those experiments, however, do not match the pattern we would expect from our results here, since the sandstone appears to have a simpler damage structure. We suspect the explanation is that the microscopic characterization is incomplete: evidence for this is that almost no damage is observed prior to the peak stress, although non-linear behaviour commenced substantially earlier and in this material this must be almost entirely due to cracking.

In fracture mechanics experiments described by Labuz *et al.* (1985) a steeper post peak load-displacement curve is observed for the finer grained and more homogeneous of the two different granites studied (Charcoal Black). Although their study was designed to measure fracture lengths directly, Labuz *et al.* found that it was not possible to monitor the crack tip extensions in the larger grain-sized material (Rockville) due in part to a 'discontinuous main crack and multiple cracks' so in this case the more distributed damage correlates with a slower loss of strength, as observed in the simulations.

A correlation between material homogeneity and more 'brittle' behaviour has also been observed experimentally for rocks in compressive and shear rupture (Paterson 1978, p. 156; Cox & Scholz 1988).

Fracture mechanics

Since the focus of this investigation was on the development of damage distributed throughout a rupturing body, rather than on the extension of a single crack, we have not attempted to analyse these data in terms of energy dissipation rates and fracture mechanics. It is useful, however, to make some general comparisons with experimental data from fracture mechanics in view of the attention paid to the relationship between micromechanics and rupture in these experiments.

With a narrow distribution of strengths the results follow the predictions of linear elastic fracture mechanics (LEFM) and a single rupture strand develops due to the dominance of the stress concentration. Crack branching and distributed damage will only be seen if the strength variation is greater than the stress concentration. In this discretized case we found that with a single broken element the stress in the adjacent elements was *c.* 1.46 times the mean, so a strength variation of at least this much is needed to cause deviation in the crack path. This correlates well with the observation from Fig. 2a that divergence in the damage parameters only occurs for $\omega > 0.4$.

For physical experiments on rocks several studies have used a modified 'effective' crack length in fracture toughness calculations to account for the effect of the damaged and softened material apart from the main rupture strand (Labuz *et al.* 1985, 1987; Peck *et al.* 1985*a*). Furthermore, the fracture energy increases during rupture extension (the so-called R-curve behaviour), and a steady value for fracture energy may only be attained after a finite crack length has been grown. Peck *et al.* (1985*b*) found that this length was large for rocks with 'a preexisting network of interconnected microcracks or when the texture of the rock provides multiple incipient fracture surfaces.' Thus, reproducible values for strength, which truly represent a material property may only be measured for samples larger than some critical size, which depends on the initial microstructure. This may be compared with our data in Fig. 3b here.

These non-linear elastic extensions to fracture mechanics are mainly 'small sample' devices to allow for the fact that reasonable, and useful, sized samples fail to satisfy the LEFM requirement that the breakdown zone dimension is small enough to be unaffected by the boundary conditions. Li & Liang's (1986) treatment explicitly considers these methods to be transitional between rupture criteria based strictly on crack extension, and those based on more general bulk yielding, and examined the range of validity of the non-linear models. The constitutive behaviour of our material elements is, however, much simpler than assumed in the continuum models considered by Li & Liang. We have found here that realistic softening behaviour may be produced by a distribution of small, perfectly brittle events.

Damage models

We find that the damage exponent, $\beta-1$, approaches the theoretical maximum value of 1 for the exponential strength distribution. This result might be expected since this distribution can be shown to have the highest possible entropy of any continuous distribution (Chan *et al.* 1986, 1988). At this value the amount of damage at breakthrough is proportional to the area of the model.

Weibull (1939) made the first thorough statistical analysis of the effects of elemental strength distributions on macroscopic strengths for different sample sizes. He found that many natural materials could be described with a similar class of elemental strength distributions. However, the solution is very non-unique, and we now have a method for a more thorough examination of the possibilities. Here, we have described a first step using a very simple set of continuous distributions of properties, with a single deformation mechanism, and yet we still will find complex behaviour. This is in contrast to a model by Okubo & Nishimatsu (1986) where a complex set of material properties was assumed and failure was allowed by a mixture of tensile and shear processes, with a fixed strength ratio between these.

Implications and applications

A *critical size* has been proposed as a component in translating laboratory strength measurements for geomechanical analysis of full scale mining excavations (Bieniawski 1984) with the understanding that the intrinsic flaw structure of various rocks will have an upper scale limit which correlates with this critical size. Our results are for a material whose only intrinsic scale is the size of the elements, and furthermore we explicitly chose a poisson distribution of properties (no spatial correlation) for these experiments, so we do not expect 'big' flaws to be constructed regularly from clusters of small ones. Hence, we conclude from the data shown in Fig. 3b that the critical size is actually a large multiple of the flaw dimension, if we take weak elements to represent intrinsic flaws.

There has been some interest in applying a fracture mechanics type analysis of stress concentrations directly to the phenomenon of joint and dyke formation if the earth (Lachenbruch 1961; Anderson & Grew 1977; Rudnicki 1980; Pollard *et al.* 1982; Pollard 1987; Olson & Pollard 1988). The analyses typically assume that the rock is homogeneous except for a small number of prescribed initiating flaws. Thus although they make reasonable predictions of the large scale form of natural fractures, there is no account taken of secondary damage. This omission may be significant in the consideration of some problems: for example extensive veining in hydrothermal systems indicates the importance of the overall damage array in coupled deformation and fluid transport systems.

Conclusions

From simulations of brittle rupture in solids with some simple distributions of material properties we found the following.

(1) Variation in strengths has more effect on the geometry of brittle rupture than variation of elastic modulus. This may be contrasted with the fundamental assumption of linear elastic fracture mechanics in which stress concentrations due to elastic heterogeneities dominate.

(2) The most disordered elemental strength distributions lead to damage proportional to the amount of material present (area). However, a major fraction of that damage is disconnected from the main rupture strand.

(3) For a particular distribution, the mean strength (peak stress supported) of a sample decreases as the sample dimension increases.

(4) Wider distributions of strengths lead to more gradual mechanical breakdown behaviour.

J. Boland provided a helpful review. Comments from T.-F. Wong helped us improve the discussion relating this work to the results of physical experiments and geological applications. CSIRO Division of Geomechanics Research Paper 545.

References

Anderson, O. L. & Grew, P. C. 1977. Stress corrosion theory of crack propagation with applications to geophysics *Reviews of Geophysics and Space Physics*, **15**, 77–104.

Bieniawski, Z. T. 1984. *Rock mechanics design in mining and tunnelling*. Balkema, Rotterdam.

Chan, D. Y. C., Hughes, B. D. & Paterson, L. 1986. Fluctuations, viscous fingering and diffusion limited aggregation. *Physical Review, A*, **34**, 4079–4082.

——, ——, —— & Sirakoff, C. 1988. Simulating flow in porous media. *Physical Review A*, **38**, 4106–4120.

Cox, S. J. D. & Paterson, L. 1989. Tensile fracture of heterogeneous solids with distributed breaking strengths. *Physical Review B*, **40**, 4690–4695.

—— & Scholz, C. H. 1988. On the formation and growth of faults: an experimental study. *Journal of Structural Geology*, **10, 413–430.**

De Arcangelis, L., Redner, S. & Herrmann, H. J. 1985. A random fuse model for breaking processes. *Journal de Physique Lettres*, **46**, L585–590.

Duxbury, P. M., Beale, P. D. & Leath, P. L. 1986. Size effects of electrical breakdown in quenched random media. *Physical Review Letters*, **57**, 1052–1055.

Horii, H., & Nemat-Nasser, S. 1985. Compression-Induced Microcrack Growth in Brittle Solids: Axial Splitting and Shear Failure. *Journal of Geophysical Research*, **90**, 3105–3125.

Kahng, B., Batrouni, G. G., Redner, S., De Arcangelis, L. & Herrmann, H. J. 1988. Electrical breakdown in a fuse network with random, continuously distributed breaking strengths. *Physical Review B*, **37**, 7625–7637.

Labuz, J. F., Shah, S. P. & Dowding, C. H. 1985. Experimental analysis of crack propagation in granite. *International Journal of Rock Mechanics and Mining Science and Geomechanical Abstracts*, **22**, 85–98.

——, —— & —— 1987. The fracture process zone in granite: evidence and effect. *International Journal of Rock Mechanics and Mining Science and Geomechanics Abstracts*, **24**, 235–246.

Lachenbruch, A. H. 1961. Depth and spacing of tension cracks. *Journal of Geophysical Research*, **66**, 4273–4292.

Li, V. C. & Liang, E. 1986. Fracture processes in concrete and fibre reinforced cementitious composites. *Journal of Engineering Mechanics*, **112**, 566–586.

Lister, G. S., Paterson, M. S. & Hobbs, B. E. 1978. The simulation of fabric development in plastic deformation and its application to quartzite: the model. *Tectonophysics*, **45**, 107–158.

McLintock, F. A. & Walsh, J. B. 1962. Friction on Griffith cracks in rocks under pressure. *Proceedings of the 4th US National Congress on Applied Mechanics, Vol. II* American Society of Mechanical Engineers, New York, 1015–1021.

Murrell, S. A. F. & Digby, P. J. 1970*a*. The theory of brittle fracture initiation under triaxial stress conditions I. *Geophysical Journal of the Royal Astronomical Society*, **19**, 309–334.

—— & —— 1970*b*. The theory of brittle fracture initiation under triaxial stress conditions II. *Geophysical Journal of the Royal Astronomical Society*, **19**, 499–512.

Okubo, S. & Nishimatsu, Y. 1986. Computer modelling of stochastic rock failure during uniaxial loading. *International Journal of Rock Mechanics and Mining Science*, **23**, 363–370.

Olson, J. & Pollard, D. D. 1988. Inferring stress states from detailed joint geometry. *In*: Cundall, P. A., Sterling, R. L. & Starfield, A. M. (eds) *Key questions in rock mechanics: Proceedings of the 29th US symposium on Rock Mechanics* Balkema, Rotterdam, 159–167.

Paterson, L. 1988. Serrated fracture growth with branching. *In*: Cundall, P. A., Sterling, R. L. & Starfield, A. M. (eds) *Key questions in rock mechanics: Proceedings of the 29th U.S. symposium on Rock Mechanics*. Balkema, Rotterdam, 351–358.

Paterson, M. S. 1978. *Experimental rock deformation – the brittle field*. Springer, New York.

Peck, L., Nolen-Hoeksema, R. C., Barton, C. C. & Gordon, R. B. 1985*a*. Measurement of the resistance of imperfectly elastic rock to the propagation of tensile cracks. *Journal of Geophysical Research*, **90**, 7827–7836.

——, Barton, C. C. & Gordon, R. B. 1985b. Microstructure and the resistance of rock to tensile fracture. *Journal of Geophysical Research*, **90**, 11533–11546.

Peng, S. S. 1975. A note on the fracture propagation and time-dependent behavior of rocks in uniaxial tension. *International Journal of Rock Mechanics and Mining Science*, **12**, 125–127.

Pollard, D. D. 1987. Elementary fracture mechanics applied to the structural interpretation of dykes. *In*: Hall, H. C. & Fahrig, W. F. (eds) *Mafic Dyke Swarms. Geological Society of Canada Special Paper*, **34**, 5–24.

——, Segall, P. & Delaney, P. T. 1982. Formation and interpretation of dilatant echelon cracks. *Geological Society of American Bulletin*, **93**, 1291–1303.

Rudnicki, J. W. 1980. Fracture mechanics applied to the earth's crust. *Annual Reviews of Earth and Planetary Science*, **8**, 489–525.

Weibull, W. 1939. A statistical theory of the strength of materials. *Ingeniorsvetenskapsakademiens*, **151**, Handlingar, Stockholm.

Velocity-dependent friction in a large direct shear experiment on gabbro

S. J. D. COX

CSIRO Division of Geomechanics, PO Box 54, Mt Waverley, Vic 3149 Australia

Abstract: Velocity-stepping sliding experiments on a 0.1 m^2 interface between two gabbro blocks under 0.5–2.0 MPa normal stress have shown consistent velocity-weakening behaviour. The parameter values are in a similar range to those measured on other materials in other laboratories. However, simulation of the complete transient behaviour requires a second state-variable which has a characteristic decay distance *c*. 100 × longer than the first, and with the opposite sign (velocity-strengthening). The second delayed effect is not sufficient to cancel out the short period effect, and is not expected to control the onset of unstable stick-slip sliding. These experiments extend the observed realm of velocity-weakening behaviour to larger interfaces and lower normal stresses than has been previously reported.

Since the suggestions by Bridgman (1936) and Brace & Byerlee (1966) that 'stick-slip' friction on sliding interfaces provided an explanation for earthquakes, much experimental effort has gone towards examination of the detailed frictional behaviour of rock. The principal feature of the model is that the first order Amontons' friction law (that shear resistance is proportional to normal load) is overlaid by second order effects, yielding more than one frictional strength during the stick-slip cycle. Furthermore, the familiar simple description of this, using two values of the coefficient of friction for the static and dynamic strengths, has been found to be just one simplified case of a more general class of constitutive frictional relations involving time-, displacement- and velocity-dependent effects (Dieterich 1979; Rice & Ruina 1983; Ruina 1983; Tullis 1988).

Experimental observations of velocity dependent behaviour show that this effect is generally small: of the order of 1% changes in shear resistance for 10× changes in steady sliding velocity. Techniques to maximize this signal have therefore become standardized with the velocity-stepping procedure being most commonly used (Fig. 1). The essential features observed from this include (i) an instantaneous effect, which may be characterized simply by an amplitude, and (ii) a delayed effect, which needs at least two parameters – an amplitude and a characteristic wavelength.

The most commonly used formalisation of this laboratory behaviour is the so-called Dieterich–Ruina constitutive law (Tullis 1988). This employs a notation derived from non-linear dynamics to follow the transient behaviour following a perturbation in terms of state-variables, which evolve asymptotically over the characteristic displacements to steady-state values for a steady sliding velocity. In the generalized theory a key requirement, which allows the possibility of repeated unstable behaviour, is that the sliding surface shows (at least short-term) velocity weakening behaviour [Gu *et al.* 1984]. Within the framework of such velocity-dependent friction laws, also taking some depth and temperature dependent effects into account, a very complex model of earthquake cycles in an elastic lithosphere has been shown to be possible (e.g. Tse & Rice 1986; Marone & Scholz 1988; Hobbs 1988).

In the particular case of homogeneous slip seen on small (laboratory sized) samples, stability is controlled by interaction between the specimen and the machine stiffness (Gu *et al.* 1984). The stability envelope is fixed by the size of the velocity perturbation together with the parameters in the friction law. Lower machine stiffness, higher amplitude weakening and shorter characteristic displacements cause less stable behaviour. In this study I have used the version of the Dieterich–Ruina law expressed in Tullis (1988), with normalised amplitude and length parameters as given in Table 1.

Experimental determinations of the parameter values have been made in several configurations, including direct shear, sandwich direct shear, obliquely split cylinders in triaxial loading, obliquely split biaxial blocks and annular rotary shear. Each of these has its own advantages in allowing access to a particular regime of normal stress, sliding velocity and displacement, but none is comprehensive. In

From Knipe, R. J. & Rutter, E. H. (eds), 1990, *Deformation Mechanisms, Rheology and Tectonics*, Geological Society Special Publication No. 54, pp. 63–70.

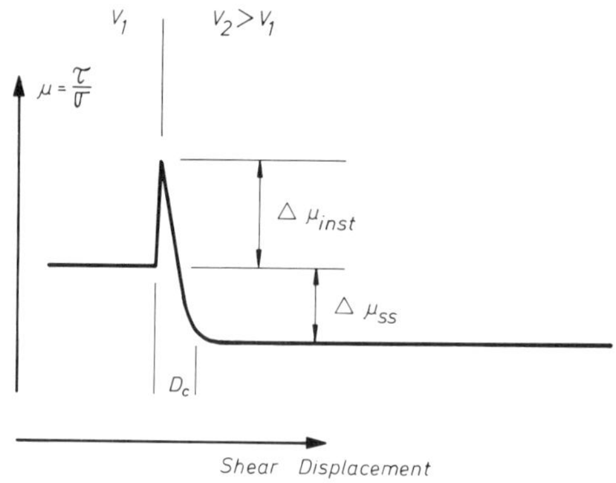

Fig. 1. Changes in shear resistance during a step increase in steady-sliding velocity (idealized).

Table 1. *Constitutive equations and parameters*

Shear stress τ is normalised by normal stress σ to give

$$\mu = \tau / \sigma$$

which has transient dependence

$$\mu = \mu_* + a \ln(V/V_*) + b_i \Psi_i \ldots$$

where Ψ are state-variables which evolve as

$$\frac{d\Psi_i}{dt} = -\frac{V}{D_{ci}} \cdot (\Psi_i + \ln (V/V_*)).$$

a is the amplitude of the immediate effect,
b_i are the amplitudes of the delayed effects, where the subscripts i refer to different state-variables,
V is the instantaneous velocity, normalized by
D_{ci}, the characteristic decay distance for the particular state-variable, and
V_* & μ_* are arbitrary reference values.
The steady-state shear resistance is therefore

$$\mu_{ss} = \mu_* + (a - \Sigma b) \cdot \ln(V/V_*).$$

The stability criterion for small velocity perturbations for a single state variable law is

$$k > -\frac{(a-b)}{D_c}$$

where k is the unloading stiffness.

particular, it seems that the behaviour evolves, with steady-state conditions only being established after large accumulated displacements (or *damage*) (Biegel *et al.*, 1987; M. L. Blanpied pers. comm.). Furthermore, when considering the scaling of laboratory data to the earth, we are hampered by the limited scale covered by experimental sample sizes.

In the experiments described here, I have been using a new apparatus to address some of these issues using a large sample of a medium grained Australian gabbro. Short-displacement velocity-weakening effects have been verified for this new material. I have also found that description of longer displacement behaviour consistently requires the use of a second state-variable, whose effect on frictional resistance has the opposite sense to the primary state-variable.

Experimental procedure

A direct shear box was used to slide two rectangular blocks of gabbro (Imperial Gabbro, from South Australia) under servo-controlled shear and normal loads (Fig. 2). The bottom block has low friction roller bearings underneath. Its upper surface is undercut so that the frictional sliding interface has a constant dimension of 250 × 400 mm. The apparatus used has a maximum normal force of 1 MN so mean normal stresses up to 10 MPa are accessible, though in the experiments reported here normal stresses were only up to 2 MPa. The shear actuator has a stroke of 150 mm, which is therefore the maximum displacement in a single cycle.

A pair of displacement transducers mounted on either side of the yoke of the bottom sample box and gauging against the end of the top block were used for control of the shear motion. A second pair of transducers was used to correct the recorded signal for the relative displacements between the yoke and the bottom block. The static shear stiffness of this configuration was measured from pre-sliding loading cycles to be 0.01–0.02 MPa μm^{-1}

Additional transducers were mounted directly on the sides of the bottom block perpendicular to and 10 mm from the upper surface at the four corners of the active part of the surface. These have hardened 2 mm radius spherical probe tips which slid with a very low gauging force on precision machined strips on the overhanging edge of the top block. These were used to measure directly the changes in normal separation (dilation) of the interface during shear.

The sliding surfaces were used initially bare, as supplied by the mason, with a coarsely ground finish with an rms roughness of approximately 20 μm. These were conditioned before the experiments reported here by sliding the system back and forth under a low normal stress (0.5 MPa) until a total displacement of approximately 7 m had been accumulated and a fine wear debris (gouge) of thickness 100–300 μ had been generated in the interface. Changes in sliding direction due to this reciprocating tech-

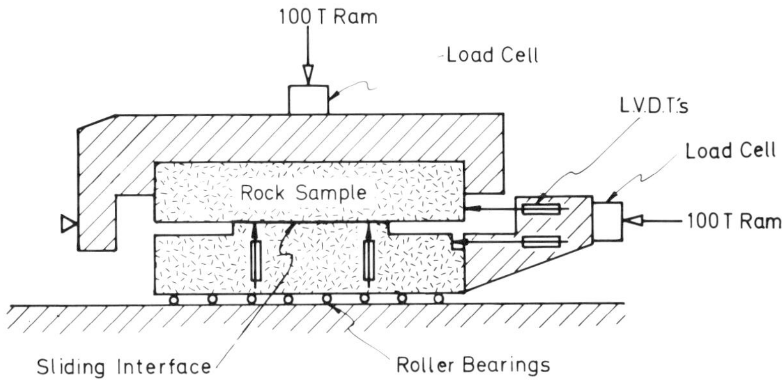

Fig. 2. Geometry of shear-box used in the experiments, indicating the sample and the positions of the displacement and load transducers.

nique do not appear to have a significant effect on the properties of the interface, as assessed by the mechanical data; transient effects correlating with the turn-around have never been observed past about 2 mm sliding distance.

A sequence of experiments stepping stable sliding-velocities in the range 0.1 μm s^{-1} to 3000 μm s^{-1} under constant normal stresses of 0.5–2.0 MPa are the basis of this report; at higher normal stresses stable sliding could not be controlled (due to the reduced normalized stiffness of the machine; see discussion of k below). Data from all the displacement and load transducers was recorded digitally at rates of 2–100 samples per second.

Results

Very consistent velocity-weakening was observed over repeated cycles of 10× velocity changes (e.g. Fig. 3a). A simple determination of the average velocity-weakening parameter $(a-b)$ may be derived simply from the separation between the portions of the curve for each sliding velocity, which also allows the secular changes in the friction coefficient to be removed. From sequences where stable sliding intervals between velocity steps were relatively short (< 1000 μm) such as those in Fig. 3a, I found $(a-b)$ to fall in the range −0.001 to −0.003 with a preferred value around −0.002 (Fig. 3b). No dependence on either normal stress or sliding velocity was observed within the range investigated.

The measurements of dilation of the interface, however, showed no signal directly associated with changes in sliding velocity, within the precision available (better than 0.1 μm, Fig. 4).

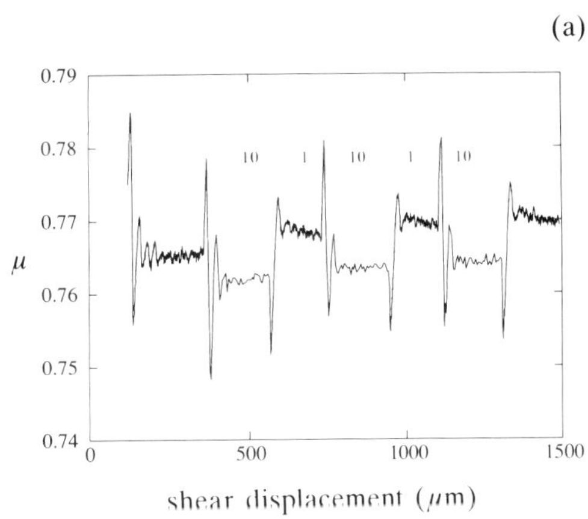

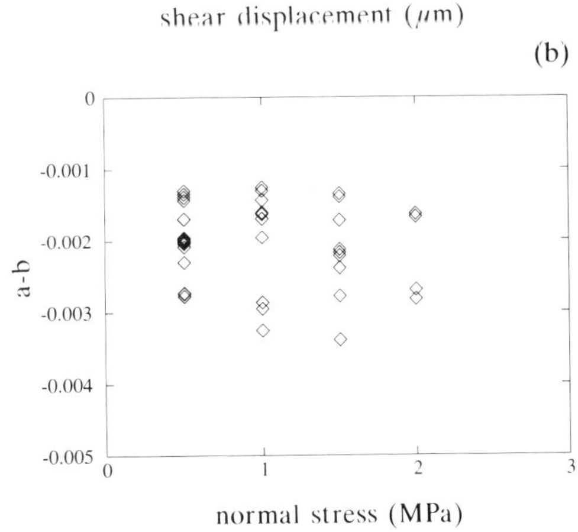

Fig. 3. (**a**) Normalised shear stress during a sequence of changes in sliding velocity, between 1 and 10 μm/s^{-1} as indicated above the curve. (**b**) Summary of measurements of the 'slip-weakening parameter' from short displacement experiments.

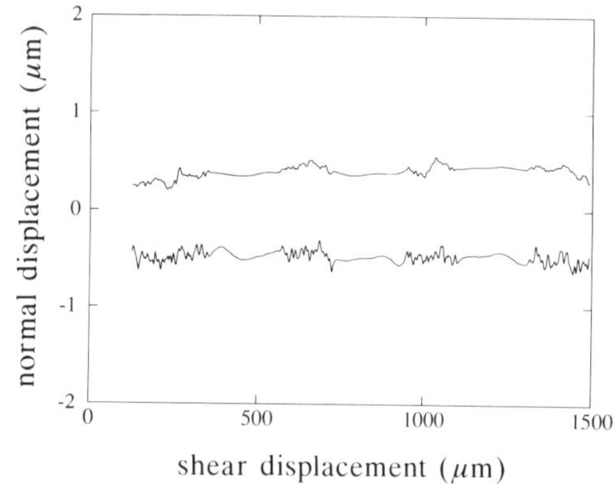

Fig. 4. Dilatation of the interface measured directly by transducers at opposite corners of the sliding interface during the velocity sequence shown in Fig. 3(a).

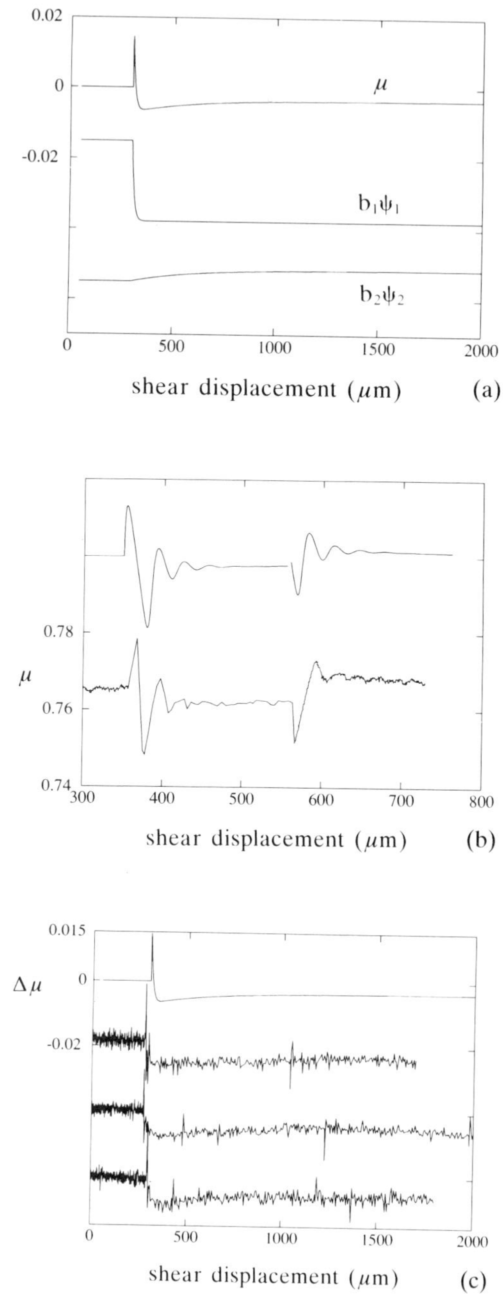

Fig. 5. (**a**) Transient changes in shear resistance (top) and the two state-variables (middle and bottom) for a velocity step with the mixed model discussed here. (**b**) Simulations matching short period data for a velocity increase and decrease: a = 0.012, D_{c1} = 3 μm, b_1 = 0.0135, D_{c2} = 300 μm, b_2 = −0.001 (**c**) Long period simulation with three matching data portions from a single experiment a = 0.007, D_{c1} = 10 μm, b_1 = 0.010, D_{c2} = 300 μm, b_2 = −0.0015.

More rigorous determination of the values of a and b and particularly the characteristic lengths (D_c), may be made by matching the observations to curves generated theoretically from a model including the coupled effects of the loading machine stiffness (k) and the constitutive properties of the sample. The procedures have been discussed at some length by Gu *et al.* (1984), Rice & Tse (1986), Blanpied & Tullis (1986) and Horowitz (1988) amongst others, so will not be rehearsed here. The simulations shown (Fig. 5) were derived numerically using double precision arithmetic and a fourth order Runge–Kutta integration scheme with adaptive step size control [Press *et al.* 1986].

Attempting to match the observations with theoretical curves, however, showed that it was not possible to get adequate solutions using a single state-variable law. A second state-variable with the opposite sign to the primary one (i.e. causing some delayed strengthening, rather than weakening, Fig. 5a) allowed very good fits to be made. The amplitude of this second effect, though not completely uniform in this series of experiments, was always much less than the short displacement effect, so an overall steady-state velocity-weakening effect wasy found in all cases. The mixed mode effect was observed with similar parameters for velocity steps in both directions (Fig. 5b). Furthermore, the two delayed effects had decay distances separated by up to two orders of magnitude, so were largely decoupled as far as the simulations, and therefore probably the physical mechanism, were concerned. In order to make good measurements of the second delayed effect it was necessary to allow steady sliding over distances greater than 1000 μm after each velocity step (Fig. 5c).

The disadvantage of using more complex friction laws is that for each extra state-variable, two extra parameters (one amplitude and one wavelength) must be extracted from the same data. This increases the uncertainty in each one. Going through the parameters in turn, the confidence in each of the determinations may be assessed.

k (machine stiffness). This should be independent of the other parameters, since it can be measured from the initial loading curve in the experiment. However, that value is the long-term large-signal stiffness, and in a real system servo-controlled on displacement, the small-amplitude or high frequency response will be less than optimal. The effective machine stiffness is therefore reduced, and furthermore is non-linear with respect to velocity and offset amplitude. The slope of the stress-displacement curve achieving the 'instantaneous' effect is the main constraint on the appropriate value of k to use in the simulations, which were found to be as much as two orders of magnitude less than the ideal value. Also, at normal stresses above c. 2 MPa the machine stiffness, normalized by the normal stress (i.e. in the same non-dimensional form as μ), was reduced such that the stability envelope [Gu *et al.*, 1984] was very close to the steady-state point. Stable sliding could not be controlled and velocity stepping experiments were not useful at higher stresses.

D_{c1} (characteristic distance). The shape of the oscillations initially after a velocity step is governed by the coupling between D_{c1} and k, so the uncertainty in k also affects D_{c1}. In all cases for this well-conditioned thin gouge, however, D_{c1} is constrained to fall in the range 3–15 μm, which is in good agreement with other recent observations (Biegel *et al.* 1987; M. L. Blanpied pers. comm.).

D_{c2} (characteristic distance). Because of the large difference between the two characteristic distances, and also since D_{c2} is large enough to be decoupled from k, the value of 300–500 μm can be determined with a high confidence level.

a (amplitude of instantaneous effect). The uncertainty in k also affects the determination of a, and this varies considerably between simulations matching different velocity steps, in the range 0.004–0.015.

b_1 (amplitude of delayed effect). Because of the decoupling between the two state-variables, the amplitude of short displacement velocity-dependence, $(a-b_1)$ is well constrained, in the range −0.002 to −0.006, with a preferred value around −0.003. The uncertainty in a is reflected in the determination of an absolute value for b_1.

b_2 (amplitude of delayed effect). This is independent of all the parameters above, due to the large difference between D_{c1} and D_{c2}, but over the larger displacements the effects of longer term changes in the value of the background friction coefficient are noticed. I have no explanation for these drifts. The measurements varied in the range −0.001 to −0.004, but the value was always less than $(a-b_1)$, as overall velocity weakening was always seen. The overall velocity weakening parameter $(a-\Sigma_i b_i)$, was in the range 0 to −0.002.

Micromechanical models

A conceptual model of frictional shear resistance is based on the understanding that surfaces are always rough at some scale, so interface contact and deformation is limited to discrete points (Bowden & Tabor 1954; Archard 1957; Rabinowicz 1965; Byerlee 1967). Scholz & Engelder (1976), Dieterich (1979) and Dieterich & Conrad (1984) introduced dynamic effects to the model by considering time and environment dependent indentation creep or contact growth, and time dependent contact shear strength.

In this micromechanical model, the characteristic displacement parameter is associated with some sliding distance over which all the contacts will be renewed. This view is supported by the observed correlation between the initial value of D_c and the roughness of the bare surface, or the particle dimension in an artificial gouge layer [Dieterich 1981; Biegel *et al.* 1987].

Biegel *et al.* [1989] have attempted to generalize the simple model for the characteristic distance to include interfaces containing debris, but were not able to correlate D_c with any observed dimension within a mature gouge. They found that the size distribution of particles within the debris could be characterised at any stage by a (band limited) power law, and although they concluded that particle size reduction continues over large displacements, D_c achieves a relatively stable value after initial conditioning of the interface, which is independent of the initial surface preparation or gouge particle size distribution.

However, in view of the correlation between the initial value of D_c and the surface preparation noted above, identifying D_c with some feature of the microgeometry of the interface seems unavoidable. The question here must therefore centre on what would lead to two distinct characteristic displacements, with up to 100× difference between them.

Microscopic examinations have shown that (after initial compaction) deformation is typically concentrated in bands within and on the margins of both natural and artificial gouge (Logan *et al.* 1981; Rutter *et al.* 1986). In laboratory tests, on samples whose surfaces have only short wavelength roughness, shear zones close to the gouge-rock interface are particularly important. Two particular orientations of shear bands seem to be dominant for gouges both rich and poor in phyllosilicates: the Y or boundary parallel shears, and the R_1 (Riedel) shears which dip 10–20° beneath the displacement vector (Rutter *et al.* 1986; Moore (*et al.* 1988; Marone *et al.* 1989; M. L. Blanpied, T. E. Tullis & J. D. Weeks pers. comm.). Even in the large total displacement experiments of Blanpied *et al.* (pers. comm.), R_1 shears are almost always present, although the Y shears appear to be the main displacement-accommodating features (which is not surprising since oblique shears are not kinematically stable for large strains). Since deformation is heterogeneous, at the scale of the whole interface layer it may be appropriate to consider the gouge to behave mechanically as a *structure*, with parallelogram-prism shaped domains of essentially non-deforming material bounded by high strain rate zones.

Analysis of shear zone formation within granular materials under simple shear indicates that, within material of a uniform grain-size, the zones will have a characteristic thicknesses of between 10 and 20 grain diameters (Muhlhaus & Vardoulakis 1987). Multiple shear zones in a single specimen will need separations of several times this, thus the minimum domain dimension will be around two orders of magnitude larger than the characteristic particle dimension. Thus, purely on the basis of the relative magnitudes of the characteristic displacements, I suggest that if the first state-variable is associated with some critical element dimension within the interface (comprising the whole surface-gouge-surface system), then the second state-variable may be due to the re-equilibration of a structure of interlinked zones of localized deformation within the gouge layer. Alternatively, of course, long wavelength features on the surface of the substrate may lead to a second decay distance. Locating the source of the critical dimensions remains a major challenge, since in this system the grain size is neither uniform nor-constant (Marone & Scholz 1989; Biegel *et al.* 1989; Blanpied *et al.* pers. comm.). This is particularly important when we consider scaling laboratory data to the dimensions of crustal faults (e.g. Scholz 1988).

Why the longer displacement effect should be velocity-strengthening, however, is not clear. In a study of the early stages of deformation of a thick artificial gouge layer, Marone *et al.* (1989) consistently observed velocity-strengthening, and proposed a model in which an 'intrinsic' velocity-weakening was overridden by strengthening due to a change in dilatancy rate. The model was based on careful measurements of pore-volume changes, which were combined with the observed localized zone orientations to yield net resolved displacements normal to the gouge layer, and which did work against a constant normal stress. Those experiments were over quite small total displacements (< 13 mm); rapid grain size reduction and was still proceeding and the overall porosity level was still evolving, so it is not surprising that there is a contrast with the overall velocity-weakening seen in 'natural' wear products here. Although we do not have such detailed observations for these experiments, the null result found when monitoring normal displacements during velocity steps is consistent with the idea that volume changes are not a dominant short term effect in the more mature gouge here. Furthermore, dilatancy-strengthening is not appealing as a long displacement effect, since it is due to a positive rate of change of volume with displacement, which cannot continue indefinitely. In this case the other mechanisms must be appealed to, and more detailed microstructural observations will be needed to show how these may cause strengthening.

Discussion

Though there is variation in experimental observations of velocity-dependent friction, some common features of behaviour in the strictly brittle field are now emerging. Two phases of deformation may be distinguished, as follows. (1) Prior to the accumulation of high strains, behaviour is dominated by the geometry of the starting material. Velocity-weakening is seen for bare surfaces (which may be considered to be an 'intrinsic' effect), with a characteristic decay distance directly correlated with the initial roughness. When artificial gouge layers are introduced the transients have a characteristic decay distance which correlates with the particle size. In thick artificial gouge layers, initial velocity-strengthening, probably due to dilatancy effects, overrides the velocity weakening. (2) After sufficient damage has accumulated, typically after 5–15 mm sliding for laboratory samples, a quasi-equilibrium structure of localized zones of deformation is established in the deformed 'gouge' or wear debris layer. Background friction reaches a constant value and overall velocity-weakening is seen, with a short and constant characteristic distance of 5–10 μm, independent of the starting state (for granite at least) (Biegel *et al.* 1987; Blanpied *et al.* pers. comm.).

The primary observation in the experiments reported here, of velocity-weakening in a mature, thin gouge layer, fits into phase 2 of this scheme. The characteristic distance determined for the short distance weakening is comparable to that reported for large displacement experiments on other materials, so this now appears to be a fairly general parameter.

Two state-variable laws have been discussed before, typically to effect a second order correction to the decay curve (e.g. Blanpied & Tullis 1986; Tullis & Weeks 1986), but also including the so-called mixed effect of state-variables with opposite signs (Weeks & Tullis 1985). Horowitz (1988) analysed an (unpublished) observation by Ruina, similar to that here, where the state-variables have such differing characteristic displacements that they may be considered to be essentially decoupled. In this case the system stability will be controlled simply by the short wavelength effect. Wong & Zhao (1989) have also observed mixed mode behaviour using particular mixtures of artificial gouge in short displacement triaxial tests, but the effect was only seen for velocity increases and not for velocity decreases.

Cases not fitting the two phase scheme can generally be understood as due to a modification of the deformation mechanisms. Blanpied *et al.* (pers. comm.) found that above a critical normal stress and sliding velocity, velocity strengthening behaviour may be explained by shear heating which they suggested causes accelerated contact adhesion. In Shimamoto & Logan's (1986) observations of salt, the experiments were all carried out at a much higher homologous temperature than most of the silicate rock experiments, and plastic processes were probably involved. Of course, the simple brittle micromechanical model is limited in its applicability to the earth because of these considerations. Over the substantially larger time scales involved in the earthquake cycle, even very mature faults will be subject to modification during the interseismic periods, with strength recovery through different mechanisms, particularly involving fluid assisted mass transfer, which has not been considered here.

Conclusions

I have used a new direct shear machine to study large rock interfaces sliding very large total displacements under relatively low normal stresses.

For experiments on a mature, thin gouge layer derived by wear from an initially bare gabbro: (i) short-displacement velocity-weakening is always seen, with an amplitude of *c.* −0.003 and a characteristic decay distance of *c.* 5–10 μm; however (ii) a second state-variable is needed to fit the long displacement experiments, with the opposite sign to the first (i.e. transient velocity-strengthening), but the decay distance of 300–500 μm makes this largely decoupled from the first. The second state-variable must correlate with structures in the gouge or longer wavelength features in the topography of the substrate.

These data confirm the presence of velocity weakening behaviour for a new material with a large sample size and at low normal stresses which had not been examined previously.

Discussions with and pre-prints from M. Blanpied, C. Marone, T. Tullis and T.-F. Wong have been invaluable in putting together the discussion part of this paper. P. Torok and G. Crawford gave assistance with figure preparation, and A. Swallow, A. White and R. Cox in the experimental program. The efforts of P. Muir of Layton's Granite in preparing large parallel sided samples were appreciated. Helpful reviews were provided by C. Marone and W. Power. The project was partially supported by USGS grant 14–08–0001–G1191 to B. E. Hobbs. CSIRO Division of Geomechanics Research Paper 544.

References

ARCHARD, J. F. 1957. Elastic deformation and the laws of friction. *Proceedings of the Royal Society of London*, **A243**. 190–205.

BIEGEL, R. L., SAMMIS, C. G. & DIETERICH, J. H. 1987. Frictional properties of a simulated fractal gouge layer in Westerly granite. *EOS, Transactions of the American Geophysical Union*, **68**, 1478.

——, —— & —— 1989. Frictional properties of a simulated gouge layer with a fractal particle distribution. *Journal of Structural Geology*, **11**, 827–846.

BLANPIED, M. L. & TULLIS, T. E. 1986. The stability and behaviour of a frictional system with a two state variable constitutive law. *Pure & Applied Geophysics*, **124**, 415–444.

BOWDEN, F. P. & TABOR, D. 1954. *The Friction and Lubrication of Solids*. Oxford University Press.

BRACE, W. F. & BYERLEE, J. D. 1966. Stick-slip as a mechanism for earthquakes. *Science*, **153**, 990–992.

BRIDGMAN, P. W. 1936. Shearing phenomena at high pressure of possible importance for geology. *Journal of Geology*, **44**, 653–669.

BYERLEE, J. D. 1967. Theory of friction based on brittle fracture. *Journal of Applied Physics*, **38**, 2928–2934.

DIETERICH, J. H. 1979. Modeling of rock friction: 1. Experimental results and constitutive equations. *Journal of Geophysical Research*, **84**, 2161–2168.

—— 1981. Constitutive properties of faults with simulated gouge. *In*: CARTER, N. L., FRIEDMAN, M., LOGAN, J. M. & STEARNS, D. W. (eds) *Mechanical behaviour of crustal rocks*. Geophysical Monograph, **24**. American Geophysical Union, Washington DC, 103–120.

—— & CONRAD, G. 1984. Effect of humidity on time- and velocity-dependent friction in rocks. *Journal of Geophysical Research*, **89**, 4196–4202.

GU, J. C., RICE, J. R., RUINA, A. L. & TSE, S. T. 1984. Slip motion and stability of a single degree of freedom elastic system with rate and state dependent friction. *Journal of the Mechanics and Physics of Solids*, **32**, 167–196.

HOBBS, B. E. 1988. Chaotic behavior of frictional shear instabilities. *2nd International Symposium on Rockbursts and Seismicity in Mines*. University of Minnesota, Minneapolis, Minn., 93–103.

HOROWITZ, F. G. 1988. Mixed state variable friction laws: some implications for experiments and a stability analysis. *Geophysical Research Letters*, **15**, 1243–1246.

LOGAN, J. M., HIGGS, N. G. & FRIEDMAN, M. 1981. Laboratory studies on natural gouge from the U.S.G.S. Dry Lake valley No. 1 well, San Andreas Fault Zone. *In*: CARTER, N. L., FRIEDMAN, M., LOGAN, J. M. & STEARNS, D. W. (eds) *Mechanical behaviour of crustal rocks*. Geophysical Monograph, **24**, American Geophysical Union, Washington DC, 121–131.

MARONE, C. & SCHOLZ, C. H. 1988. The depth of seismic faulting and the upper transition from stable to unstable slip regimes. *Geophysical Research Letters*, **15**, 621–624.

—— & —— 1989. Particle-size distribution and microstructures within simulated fault gouge. *Journal of Structural Geology*, **11**, 799–814.

——, RALEIGH, C. B. & SCHOLZ, C. H. 1990. Frictional behavior and constitutive modeling of simulated fault gouge. *Journal of Geophysical Research*, (in press).

MOORE, D. E., SUMMERS, R. & BYERLEE, J. D. 1988. Relationship between textures and sliding motion of experimentally deformed fault gouge: Application to fault zone behavior. *In*: CUNDALL, P. A., STERLING, R. L. & STARFIELD, A. M. (eds) *Key Questions in Rock Mechanics*. Proceedings of the 29th US Symposium on Rock Mechanics, Balkema, Rotterdam, 103–110.

MUHLHAUS, H. B. & VARDOULAKIS, I. 1987. The thickness of shear bands in granular materials. *Geotechnique*, **37**, 271–283.

PRESS, W. H., FLANNERY, B. P., TEUKOLSKY, S. A. & VETTERLING, W. T. 1986. *Numerical recipes: the art of scientific computing*. Cambridge University Press, New York.

RABINOWICZ, E. 1965. *Friction and wear of materials*. John Wiley & Sons, New York.

RICE, J. R. & RUINA, A. L. 1983. Stability of steady frictional slipping. *Transactions of ASME. Journal of Applied Mechanics*, **50**, 343–349.

—— & TSE, S. T. 1986. Dynamic motion of a single degree of freedom system following a rate and state dependent friction law. *Journal of Geophysical Research*, **91**, 521–530.

RUINA, A. 1983. Slip instability and state variable friction laws. *Journal of Geophysical Research*, **88**, 10359–10370.

RUTTER, E. H., MADDOCK, R. H., HALL, S. H. & WHITE, S. H. 1986. Comparative microstructures of natural and experimentally produced clay-bearing fault gouges. *Pure & Applied Geophysics*, **124**, 3–30.

SCHOLZ, C. H. 1988. The critical slip distance for seismic faulting. *Nature*, **336**, 761–763.

—— & ENGELDER, J. T. 1976. The role of asperity indentation and ploughing in rock friction. *international Journal of Rock Mechanics and Mining Science and Geomechanics Abstracts*, **13**, 149–154.

SHIMAMOTO, T. & LOGAN, J. M. 1986. Velocity-dependent behavior of simulated halite shear zones: an analogue for silicates. *In*: DAS, S., BOATWRIGHT, J. & SCHOLZ, C. H. (eds) *Earthquake Source Mechanics*, Geophysical Monograph, **37**, American Geophysical Union, Washington DC 49–64.

TSE, S. T. & RICE, J. R. 1986. Crustal earthquake instability in relation to the depth variation of frictional slip properties. *Journal of Geophysical Research*, **91**, 9452–9472.

TULLIS, T. E. 1988. Rock friction constitutive behavior from laboratory experiments and its implications for an earthquake prediction field monitoring program. *Pure and Applied Geophysics*, **126**, 555–588.

—— & WEEKS, J. D. 1986. Constitutive behaviour and stability of frictional sliding of granite. *Pure and Applied Geophysics*, **124**, 383–414.

WEEKS, J. D. & TULLIS, T. E. 1985. Frictional sliding of dolomite: a variation in constitutive behaviour. *Journal of Geophysical Research*, **90**, 7821–7826.

WONG, T. F. & ZHAO, Y. 1990. Effects of load point velocity on frictional instability behaviour. *Tectonophysics*, **175**, 177–195.

The evolution of cataclastic fault rocks from a pre-existing mylonite

SUSAN J. HIPPLER [1,2] & ROBERT J. KNIPE[1]

[1] *Department of Earth Sciences University of Leeds, Leeds LS2 9JT, UK*

[2] *Present address: Exxon Production Research Co., PO Box 2189, Houston, Texas, 77252–2189, USA*

Abstract: A microstructural investigation of cataclastic fault rock evolution from a quartzite with an original mylonitic microstructure is reported. The fault rocks produced range from clast dominated microbreccias to matrix dominated ultracataclasites. The recrystallized grain size and the sub-grain size in the original mylonite appear to control the development of the fine-grained matrix in the microbreccias and cataclasites by focusing fracture along sub-grain and grain boundaries. The ultracataclasite generation involves further grain size reduction which is dominated by transgranular fracturing. The host rock clasts present in the fault zones show a significant increase in dislocation density indicating that a component of low temperature crystal plasticity is associated with the faulting. In addition the fault rocks show evidence of partial cementation by the growth of quartz and carbonate cements. The evolution of the fault rocks studied in terms of the clast size and the clast/fine-grained matrix ratios are not a simple function of the displacement magnitude.

The identification of fracture mechanisms in rocks is an important element in the understanding of faulting processes. Although a large amount of progress has been made in the analysis of faulting processes (Sibson 1977, 1986, 1989), and in the application of fracture mechanics to rocks (Rudnicki 1980; Atkinson 1987; Pollard & Aydin 1988) there are still a number of problems associated with the recognition of fracture mechanics in natural fault zones (Knipe 1989*a*). Very few studies of the grain-scale microstructural evolution of natural fault rocks have been reported (Rutter *et al.* 1986; Knipe 1989*b*; Lloyd & Knipe 1990).

This contribution presents field and microstructural data on the development of a range of cataclastic fault rocks associated with the evolution of a normal fault array. Detailed microstructural observations from optical and transmission electron microscopy (TEM) of quartzite fault rock products from the array are described with the aim of: (1) assessing the detailed microstructural changes resulting from cataclasis of quartzite with an original mylonitic texture; (2) describing the grain size distribution and its evolution in the fault zone; (3) inferring the deformation mechanisms contributing to cataclastic evolution and grain size reduction during faulting. The regional setting and range of cataclastic fault rocks observed in the field are outlined before the detailed microstructural features in the cataclasites are discussed. The discussion will consider the deformation mechanisms contributing to the cataclastic grain size reduction, and outline their involvement during the fault rock evolution.

Regional setting

The fault rocks studied developed from Cambrian quartz mylonites during the evolution of a normal fault array exposed near Durness, NW Scotland, (Fig. 1). The NE–SW/NW–SE trending extensional fault array in NW Sutherland forms the margin of a series of basins located to the north of the Scottish mainland (see Smythe *et al.*, 1982; Enfield & Coward 1987; Kirton & Hitchen 1987). The deep seismic reflection data from the Moine and Outer Isles Seismic Traverse (MOIST) obtained from this region by the British Institutions Reflection Profiling Syndicate (BIRPS), show a series of easterly-dipping faults to the west of the Orkney Islands, which are thought to represent half-graben on which the West Orkney Basin formed (Smythe *et al.*, 1982; Brewer & Smythe 1984, 1986). Enfield & Coward (1987) suggested that initial extension in the West Orkney Basin occurred in Devonian times and was controlled by reactivation on low-angle thrust faults indicated on the MOIST profile. The steeply dipping NE–SW trending structures observed in the NW Sutherland region and along the northern coast in Durness (Fig. 1) dip primarily to the

From Knipe, R. J. & Rutter, E. H. (eds), 1990, *Deformation Mechanisms, Rheology and Tectonics*, Geological Society Special Publication No. 54, pp. 71–79.

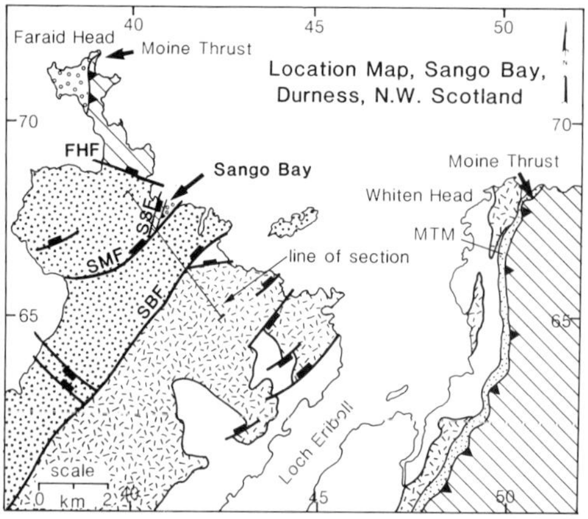

Fig. 1. Location map for Sango Bay, Durness, NW Scotland. FHF, Faraid Head Fault: SSF, Sango Sands Fault; SMF, Sangomore Fault: SBF, Sangobeg Fault. Lewisian basement rocks shown by random hatched ornament; Cambrian quartzite, no ornament; Durness Limestone Sequence, dot ornament; and Moine rocks, line ornament. Moine Thrust mylonites (MTM) are derived from the foreland succession, and are shown by the small dot ornament. Lewisian mylonites shown by open dot ornament. Line of section shown in Fig. 2. For details of Sango Bay, see Fig. 3.

NW. The age of the NW-dipping fault system is unclear, as there is an absence of syn-rift sediments exposed onshore, and Coward & Enfield (1987) state that the NW-dipping faults cannot be seen on the seismic reflection profiles offshore. However, Coward & Enfield (1987) suggest that the NW-dipping faults appear to be confined to the eastern margin of the Minch Basin, a large half graben to the west of the Scottish mainland, and that the NW-dipping faults most probably represent antithetic faults associated with extension in this basin. The Minch Basin is considered to have initiated in the Mesozoic (Steel & Wilson, 1975), or earlier, in the Torridonian (Kilenyi & Standley, 1985), or Devonian (Enfield & Coward 1987). Laubach & Marshak (1987) studied the fault system onshore, and concluded that the fault orientation had been controlled by basement-fabric geometry, and that the principal extension direction was NW–SE.

The rocks exposed in the study area at Sango Bay, Durness (Fig. 1) are mylonites derived from Cambrian quartzite and Lewisian gneiss, deformed during the Caledonian Orogeny. The mylonites lie structurally below a thrust which carries Moine rocks in its hangingwall (Fig. 2). The thrust package consisting of the Moine rocks and underlying mylonites are thrust over the younger and unfoliated Cambro-Ordovician rocks of the Durness Limestone Sequence (Figs 2 and 3). Lewisian mylonites are also observed to be thrust over the Moine rocks in Sango Bay (Figs 2 and 3), indicating that the Moine Thrust has been breached at this locality. A cross-section through the area (Fig. 2) shows that just over 2 km of vertical offset down to the NW has occurred along the NE–SW trending faults near Durness. Figure 3 shows the dominant NE-SW orientation of the steep extensional faults in Sango Bay which lie in the hangingwall of the Sangomore Fault. Offsets of the Caledonian thrust sheets in Sango Bay can be used to estimate displacements along the extensional faults cutting through the area. Figure 3 shows the range of vertical displacements on these faults and their spacing. Minor faults within the block west of the Sango Bay fault commonly show dip-slip movement indicators (slickensides) and their vertical displacements range from 2 m to 10 m.

This paper focuses upon cataclastic fault rock evolution which has affected a Caledonian mylonitic quartzite thrust sheet approximately five meters in thickness. These mylonites provide an opportunity to study the development

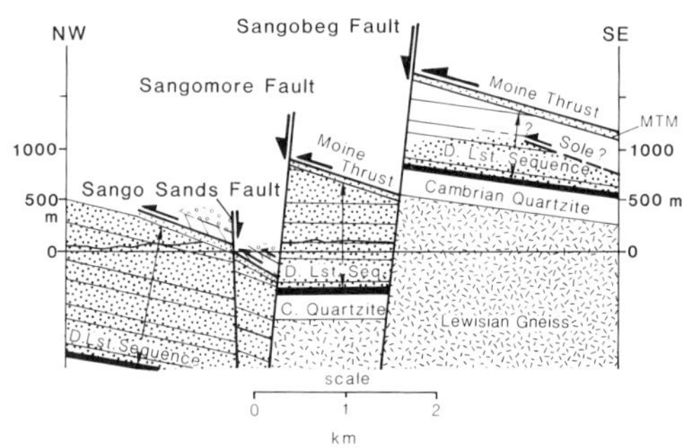

Fig. 2. Cross section through the Durness area showing the Moine Thrust plane (projected from the thrust zone to the east of Loch Eriboll) carrying Moine rocks (line ornament) to the WNW over Lewisian basement (random hatched ornament), Cambrian quartzite (no ornament), Fucoid beds and Serpulite grit (solid black) and the Durness Limestone Sequence (dot ornament). The Moine Thrust and underlying mylonites are displaced vertically by steeply-dipping normal faults near Durness and are exposed at Sango Bay, in the hangingwall of the Sangomore Fault. A thrust carrying the Moine Thrust mylonites (see Fig. 1) is shown just below the Moine Thrust. Sole? is Sole Thrust Plane. Location of section shown in Fig. 1.

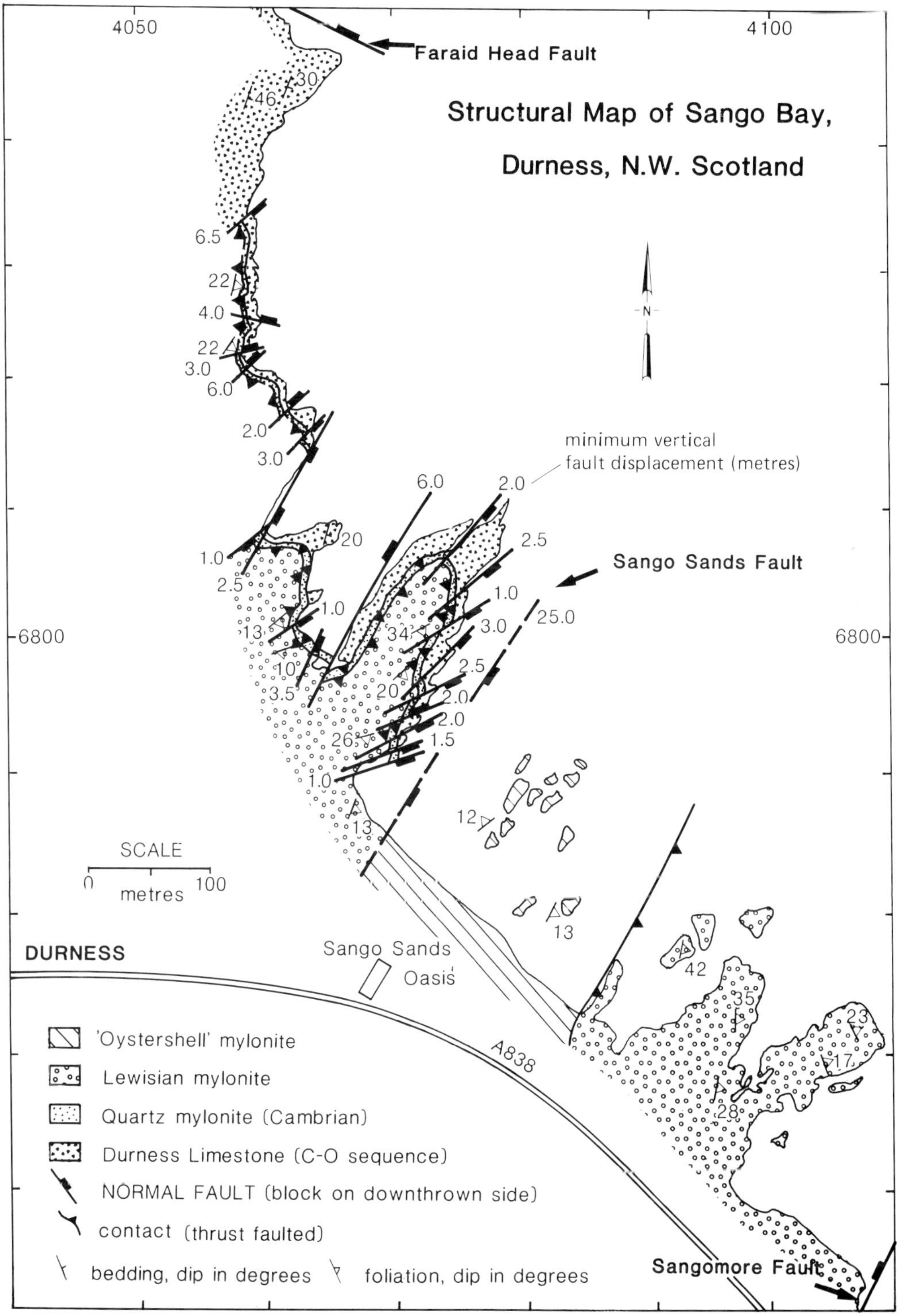

Fig. 3. Structural map of Sango Bay, Durness, showing location of major normal faults (≥ 25 m vertical displacement) (Faraid Head, Sango Sands and Sangomore Faults), and minor faults with 1–6.5 m vertical displacement. Note that in this diagram, the Moine Thrust mylonites are divided into those derived from the Lewisian basement (open dot ornament), and the Cambrian quartzite (small dot ornament). The thrust which carries Lewisian mylonite over the Moine 'Oystershell' mylonite (line ornament) is a probable late thrust which breaches the Moine Thrust. Limit of exposure shown by the extent of ornament.

of cataclastic fault rocks associated with varied fault displacements and to assess the influence of the initial mylonite microstructure on the fracture processes. The fracturing and cataclasis involved in the development of the extensional fault rocks studied in Sango Bay suggests that the rocks deformed at depths less than approximately 10 km. At normal geothermal gradients of *c*. 25°C km_{-1}, this corresponds to temperatures of $\leqslant$ 250°C.

Initial quartz mylonite

The quartz mylonites unaffected by the extensional faulting in Sango Bay contain microstructures associated with the evolution of the Caledonian Moine Thrust system, and are similar to those described by Weathers *et al.* (1979), White (1979*a*, *b*), Ord & Christie (1984), Law *et al.* (1986), and Knipe & Law (1987). The quartzite contains domains of elongate quartz ribbons 100–500 μm in width and up to 3 mm in length, and totally recrystallized zones with grain size ranging from 10–20 μm (Fig. 4a). The mean recrystallized grain size is 16.1 ± 4.0 μm (Fig. 5a). (The mean grain size quoted is the mean of the diameter of a circle equivalent to the area of each grain measured. This method was used in determining all mean grain sizes which are quoted in this paper.) On the thin-section scale, the recrystallized domains occur in 10–30 mm wide zones.

TEM of the quartz ribbons has identified a well-developed sub-grain structure (Fig. 4b). Measurements of 375 subgrains on the TEM has revealed that the mean subgrain size within the quartz ribbons is 2.6 $\mp$ 1.2/−1.0 μm (Fig. 5b). Dislocation densities in the undeformed samples range from 1.7–5.0 × 10^8 cm cm^{-3}.

Cataclastic rocks

There is a range of cataclastic fault rocks derived from the quartz mylonite thrust sheet in terms of the width of the deformation zone (the cross-sectional dimension perpendicular to the strike and dip of the zone), the clast size and shape, the clast sorting, the clast/matrix ratio, and the estimated displacement associated with each zone. The width of the deformation zone is used as the main criteria to distinguish two types of cataclastic zone present in Sango Bay: i.e., (a) narrow cataclastic zones 1–10 cm wide and (b) broad breccia/cataclastic zones 10–100 cm wide.

The narrow cataclastic zones are spaced 1–5 m apart, are 1–5 m in length, and show displacements of several centimetres up to 1 m. These zones form a 3-dimensional micro-fault network where one set is sub-parallel to the major fault orientation. The displacement on each of the cataclastic zones is not simply related to the amount of grain size reduction within the zone, nor the finite width of the zone. For example, the amount of matrix ranges from 25% to 75% and the clast size ranges from < 1 cm to > 20 mm in zones with identical displacements. These zones are similar in character to the attrition breccias (related to progressive frictional wear along slip surfaces) described by Sibson (1986).

The broad breccia/cataclastic zones (10–100 cm wide) contain angular to blocky clasts ranging from < 1–20 cm in length, in which matrix accounts for < 25% of the zone volume. These broad zones are often discontinuous in that they occur in patches along faults or within the fractured quartz mylonite.

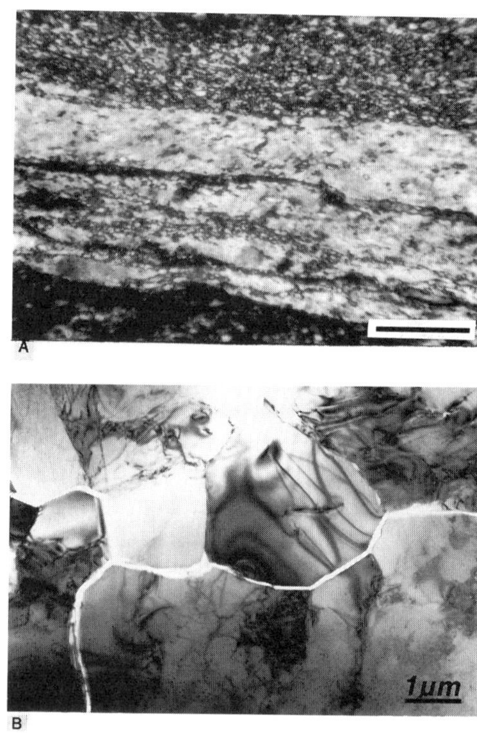

Fig. 4. (**a**) Optical micrograph of quartz mylonite from Sango Bay not affected by later extensional faulting, showing domains of elongate grains and recrystallised grains. Scale bar 0.5 mm. (**b**) TEM micrograph of subgrain structure from elongate grain in quartz mylonite not affected by later extensional faulting in Sango Bay. Scale bar 1 μm.

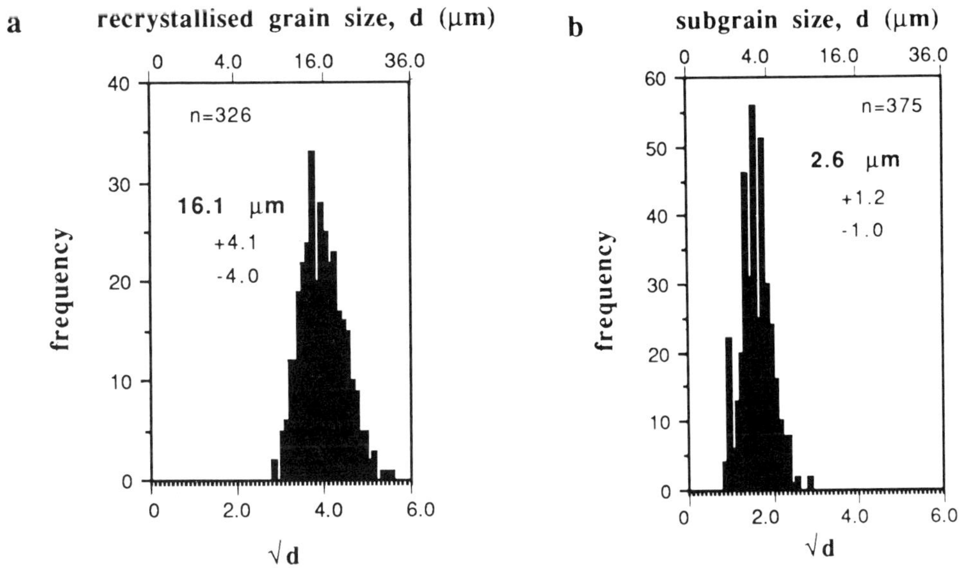

Fig. 5. (**a**) Frequency distribution for the recrystallised grain size (**d**) calculated from the grain area from optical micrographs. Mean grain size and standard deviation (shown inside of graph) calculated assuming a normal distribution of $\sqrt{d}$. Number of grains measured, n = 326. (**b**) Frequency distribution for subgrain size (**d**) calculated from the grain area from TEM micrographs. Mean subgrain size and standard deviation calculated as in (a). Number of grains measured, n = 375.

They are very similar to the implosion breccias described by Sibson (1986), which form by the linking of extension fracture systems during rapid dilation events.

On the thin-section scale, several cataclastic features can be identified within the fault rocks. These range from microfractures (with displacements < 500 μm) to deformation zones (> 50 μm wide) which contain the complete suite of cataclastic rocks from micro-breccias (< 10% fine-grained matrix) to ultracataclasites (> 75% fine-grained matrix). Each of these different cataclastic features is described separately below.

The microfractures present are descrete, semi-planar features which are generally 2–3 cm in length, < 10 μm in width, and usually have displacements of < < 5500 μm (Fig. 6a). Discrete microfractures with > 500 μm displacement have lengths greater than a thin-section (> 5 cm). A second type of microfracture is present and is characterized by a dentate morphology and has effective fracture widths of 10–30 mm. The wavelengths and amplitude of the dentate fracture morphology is irregular but is always similar to the adjacent subgrain or recrystallized grain size. That is, the microfracture morphology appears to be controlled by the subgrain size or recrystallised grain size. A similar control on fracture morphology was noted by Lloyd & Knipe (1990).

The fine-grained matrix present in the microbreccias, the cataclasites and the ultracataclastites is usually composed of an aggregate of angular clasts (Fig. 6b). The clast sizes and shapes in the deformation zones are varied on the thin-section scale. In addition, the clast edges in the microbreccias exhibit a range in morphology. Many clasts present in the fault rocks contain features characteristic of the more intact quartzite (i.e. microfractures and cataclastic zones). For example, some clast edges are similar to the microfractures described above and are semi-planar (e.g., Fig. 6b). However a large number of the clast edges are dentate. The irregular dentate boundaries appear to be following the subgrain or recrystallized grain boundaries in the clasts (Fig. 6c). In the ultracataclasite zones, there is a predominance of sub-rounded to rounded clasts, but the

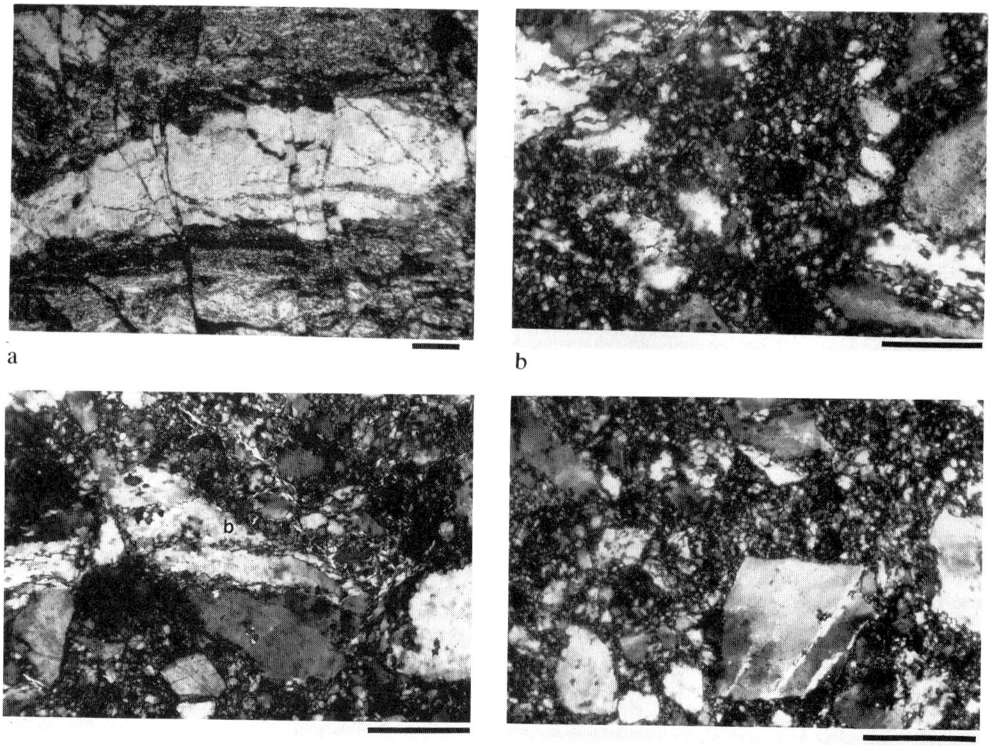

Fig. 6. (**a**) Optical micrograph of discrete, planar microfractures associated with late extensional faulting, offsetting foliation in quartz mylonite. Displacements 20 to 100 μm; many individual fractures have effective fracture widths of < 10 μm. Scale bar 0.5 mm. (**b**) Optical micrograph of microbreccia zone with grain sizes ≤ 25 μm between larger clasts (≤ 1000 μm) with semi-planar boundaries. Scale bar 0.5 mm. (**c**) Optical micrograph of microbreccia zone adjacent to a largest clast (≤ 1000 μm) with irregular, dentate boundary b. Scale bar 0.5 mm. (**d**) Optical micrograph of cataclastic zone containing clasts ≤ 1.5 mm in size floating in fine-grained matrix with clasts ≤ 25 μm in diameter. Sclae bar 1 mm.

clast size can vary, such that large clasts, ≥ 2 mm, are observed (Fig. 6d). Note that even the largest clasts incorporated into the matrix zones have sub-rounded to rounded shapes. Measurements of optical thin sections indicate that the clast size is 6.8 +2.4/−2.3 μm in the microbreccias and cataclasites (Fig. 7a).

TEM shows the angular nature of the clasts in the matrix zones of the ultracataclasite, and the grain impringement (grain-to-grain) relationships (Fig. 8a). In the ultracataclasites the mean matrix clast size is 1.0 +0.9/−0.7 μm (Fig. 7b) (based on TEM measurements of 461 microclasts). Patches of the matrix are characterized by clasts possessing straight grain boundaries, often with 120° junctions (Fig. 8b). The 120° junctions represent overgrowths of quartz on the fine clasts, and emphasize the growth of cement from fluids which were present within the fault zone. The cement does not completely fill the pore space between quartz grains and some porosity (5% − 10%) is preserved within these patches (Fig. 8c). Occasionally a carbonate cement occurs and indicates fluid infiltration from the underlying limestone thrust sheet.

TEM analysis of the dislocation densities within the clasts which are ≤ 2 mm in size reveals that they are higher (4.1×10^8 to 1.2×10^9 cm cm^{-3}) than that of the background range in the 'undeformed' quartzite, (< 5.0×10^8 cm cm^{-3}). The increase in dislocation densities within the clasts suggests that low temperature crystal plasticity contributed to the deformation of the clasts during the cataclastic deformation. In addition, the TEM has shown that there is a significant increase in the development of voids along subgrain and grain boundaries as well as healed fractures in clasts within the cataclastic zones (Fig. 8d). The significance of both these features is discussed below.

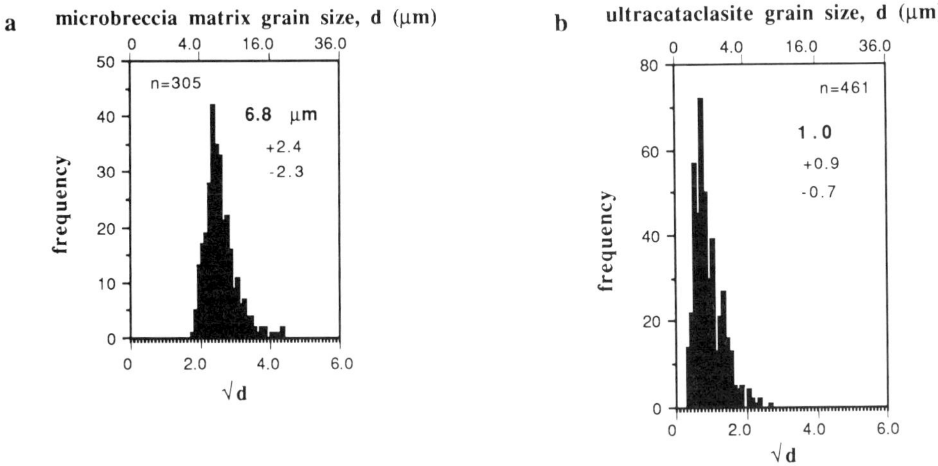

Fig. 7. (**a**) Frequency distribution for the grain size (**d**) in microbreccia matrix zone, calculated from the grain area from optical micrographs. Mean grain size and standard deviation calculated as in Fig. 5. Number of grains measured, $n = 305$. (**b**) Frequency distribution for the grain size (**d**) in ultracataclasite, calculated from the grain area from TEM micrographs. Mean grain size and standard deviation calculated as in Fig. 5. Number of grains measured, $n = 461$.

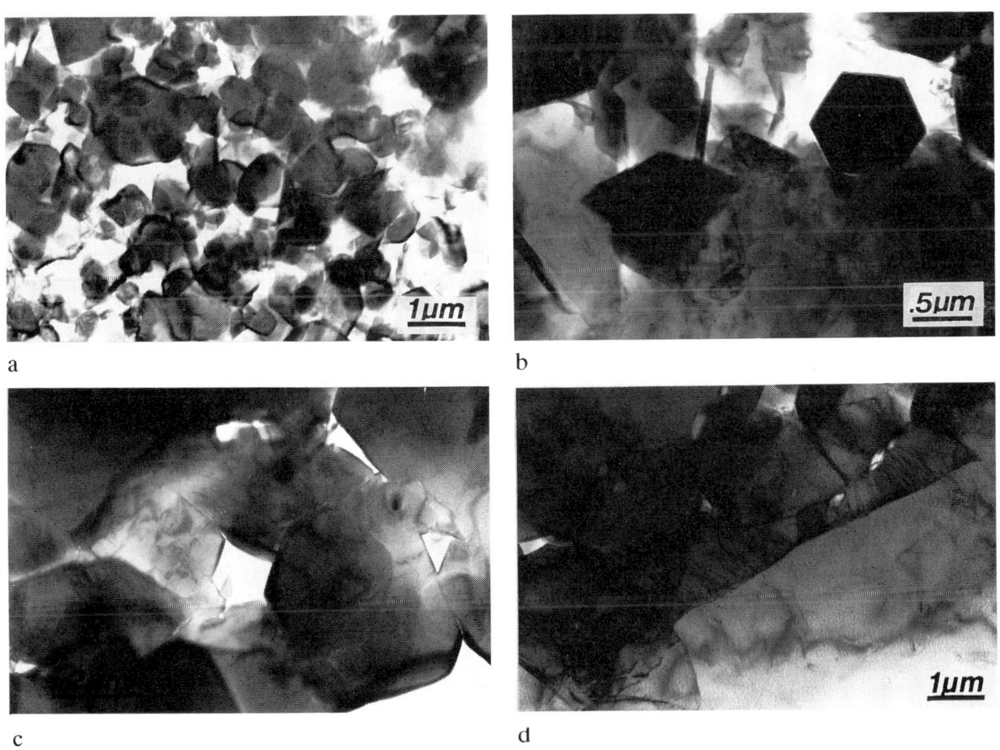

Fig. 8. (**a**) TEM micrograph of fine-grained ultracataclasite zone showing point contacts between clasts. (**b**) TEM micrograph of 1 μm microclast adjacent to larger grains in microbreccia zone. Note straight boundaries and 120° junctions on the microclast. (**c**) TEM micrograph of matrix clasts in microbreccia zone showing straight boundaries on microclasts and preserved porosity. (**d**) TEM micrograph of voids along subgrain boundaries associated with a healed fracture in a clast within a cataclastic zone.

Discussion

This section outlines the deformation mechanisms contributing to the cataclastic grain size reduction by reviewing and interpreting the microstructural observations presented above. The role each mechanism contributes to the grain size evolution in the fault zones studied and in producing the different types of cataclastic fault rocks described is also discussed.

In the relatively intact quartzite the microfractures present are dominated by planar/semi-planar morphologies and represent transgranular fractures (Fig. 9a). These microfractures are important to the initial fracturing of the quartz mylonite and the presence of clast margins with similar morphologies suggests that these are used during the isolation of clasts which accompanies the fault rock evolution.

The increased dislocation density and the number of voids along boundaries within the clasts of the microbreccias, cataclasites and ultracataclasites suggests that crystal plastic deformation processes, possibly accommodated by grain boundary sliding, operated during the faulting. Both these features represent damage which has accumulated in the clasts after their initial formation and isolation from the intact rock. It is likely that they developed during the rotation and internal straining during displacement events. This is also indicated by the rounded edges of many of the clasts in the microbreccias, which suggests wear of the clasts during deformation (e.g., Fig. 6d).

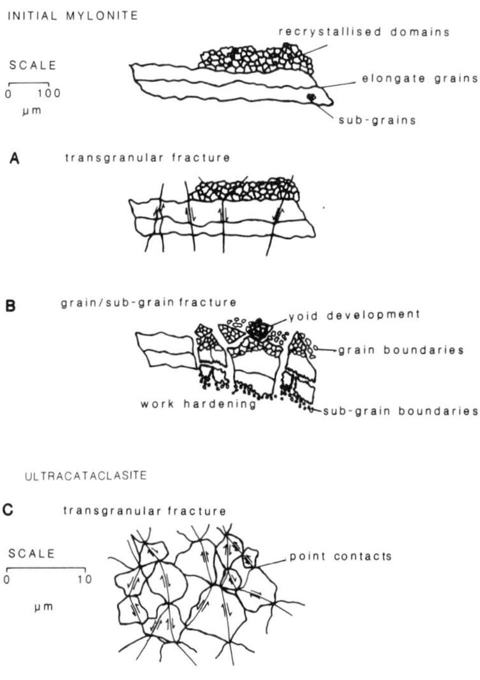

Fig. 9. Schematic diagram illustrating the mechanisms involved in cataclastic grain size reduction in a pre-existing mylonite. (**a**) Transgranular fractures cross-cutting the initial mylonitic foliation defined by the elongate grains and domains of recrystallised grains. The fractures act to disaggregate the mylonite during the early stages of faulting. (**b**) Fracturing along the grain and sub-grain boundaries during continued straining. (**c**) Transgranular fracturing of microclasts in the ultracataclasites due to stress concentrations at point contacts between clasts.

The development of the voids along the sub-grain and grain boundaries within the clasts has important implications for the progressive grain size reduction and fracturing of the clasts during continued deformation. Although the initial formation of the clast may have been dominated by the generation and linking of planar transgranular microfractures, the void growth during the later straining of the clast may cause a change in the fracturing process to one focussed along sub-grain boundaries or grain boundaries (Fig. 9b). Lloyd & Knipe (1990) have also noted the role of sub-grain boundaries and recrystallized grain boundaries in localizing fracturing.

The clast size in the matrix of the ultracataclasites is significantly smaller than in the cataclasites indicating that continued fracturing of the disaggregated mylonite takes place during their development. Point contacts between clasts in the cataclastites and ultracataclastites are common (e.g., Fig. 8a), and the continued clast size reduction may arise from fractures initiated by the stress concentrations developed at these contacts in the manner described by Gallagher (1974) (Fig. 9c).

The precipitation of cements in the fault zones also has implications for the evolution of the fault rocks. The presence of a cement in the pore space between clasts results in the transfer of stress concentrations at point contacts to a stress which is distributed along the length of the boundaries between the clasts and cement. The reduced number of point contacts between clasts may inhibit further fracture in these zones and promote preservation of the early fault rocks as clasts. In addition, the distribution of such cements may also influence the location of subsequent fracture events by generating stronger domains.

The above observations suggest that as the fault rocks evolve several fracture mechanisms contribute to the grain size reduction process. Initially, transgranular microfracturing dominates the deformation. The generation of the cataclasites appears to be controlled by fracture along existing sub-grain boundaries or grain boundaries weakened by the growth of voids. The production of the ultracataclasites involves further clast size reduction by transgranular fracture processes which appear to be controlled by stress concentrations at point contacts. Work hardening of clasts by dislocation activity may also contribute to the fracturing during the development of both the cataclasites and the ultracataclasites.

Two anonymous reviewers are thanked for improving an earlier version of this paper. S.J.H. gratefully acknowledges a British Overseas Research Award, and a Leeds University Tetley & Lupton Research Scholarship. Robertson Group, North Wales, provided funding for field work. The people in the village of Durness are thanked for their generous hospitality.

References

ATKINSON, B. K. 1987. *Fracture mechanics of rock*. Academic Press, London

BREWER, J. A. & SMYTHE, D. K. J. 1984. MOIST and the continuity of crustal reflector geometry along the Caledonian-Appalachian orogen. *Journal of the Geological Society, London*, **141**, 105–120.

—— & —— 1986. Deep structure of the foreland to the Caledonian Orogen, NW Scotland: Results of the BIRPS WINCH profile. *Tectonics*, **5**, 171–194.

COWARD, M. P. & ENFIELD, M. E. 1987. The structure of the West Orkney and adjacent basins: *In*: BROOKS, J. & GLENNIE, K. (eds) *Petroleum Geology of North West Europe*, 687–696.

ENFIELD, M. E., & Coward, M. P. 1987. The structure of the West Orkney Basin, northern Scotland. *Journal of the Geological Society, London*, **144**, 871–84.

GALLAGHER, J. J., FRIEDMAN, M., HANDIN, J. & SOWERS, G. M. 1974. Experimental studies relating to microfracture in sandstone. *Tectonophysics*, **21**, 203–247.

KILENYI, T. & STANLEY, R. 1985. Petroleum prospects in the northwestern seaboard of Scotland. *Oil and Gas Journal*, **7–10**, 100–108.

KIRTON, S. R. & HITCHEN, K. 1987. Timing and style of crustal extension north of the Scottish Mainland. In: COWARD, M. P., DEWEY, J. F. & HANCOCK, P. L., (eds) *Continental Extensional Tectonics* Geological Society, London, Special Publication, 501–510.

KNIPE, R. J. 1989*a*. Deformation mechanisms — recognition from natural tectonites. *Journal of Structural Geology*, **11**, 127–146.

—— 1989*b*. Microstructural analysis and tectonic evolution in thrust systems. *In*: BARBER, D. J., MERIDITH, P. G. (eds) *Deformation Processes in Minerals, Ceramics and Rocks*. The Mineralogical Society, Series 1, 228–261.

—— & LAW, R. D. 1987. The influence of crystallographic orientation and grain boundary migration in microstructural and textural evolution in an S.C. mylonite. *Tectonophysics* **135**, 153–169.

LAUBACH, S. E. & Marshak, S. 1987. Fault patterns generated during extensional deformation of crystalline basement, NW Scotland. *In*: COWARD, M. P., DEWEY, J. F. & HANCOCK, P. L., (ed) *Continental Extensional Tectonics*. Geological Society of London, Special Publication, **26**, 495–99.

LAW, R. D., CASEY, M. & KNIPE, R. J. 1986. Kinematic and tectonic significance of microstructures and crystallographic fabrics within quartz mylonites from Assynt and Eriboll regions of the Moine Thrust Zone. *Transactions of the Royal Society of Edinburgh, Earth Sciences*, **77**, 99–125.

LLOYD, G. E. & KNIPE, R. J. 1990. Deformation mechanisms accommodating faulting of quartzite under upper crustal conditions. (Journal of Structural Geology), (in press).

ORD, A. & CHRISTIE, J. M. 1984. Flow stresses from microstructures in mylonitic quartzites from the Moine Thrust Zone, Assynt area, Scotland. *Journal of Structural Geology*, **6**, 639–654.

POLLARD, D. D. & AYDIN, A. 1988. Progress in understanding jointing over the past century. *Geological Society of America Bulletin*, **100**, 1181–1204.

RUDNICKI, J. W. 1980. Fracture mechanics applied to the Earth's crust. *Annual Review of Earth and Planetary Sciences*, **8**, 489–525.

RUTTER, E. H., MADDOCK, R. H., HALL, S. H. & WHITE, S. H. 1986. Comparative microstructures of natural and experimentally produced clay bearing fault gouges. *Pure and Applied Geophysics*, **124**, 3–30.

SIBSON, R. H. 1977. Fault rocks and fault mechanisms. *Journal of the Geological Society, London*, **133**, 191–213.

—— 1986. Brecciation processes in fault zones. Inferences from earthquake rupturing *Pure and Applied Geophysics*, **124**, 159–175.

—— 1989. Earthquake faulting as a structural process. *Journal of Structural Geology*, **11**, 1–4.

SMYTHE, D. K., DOBINSON, A., MCQUILLIN, R., BREWER, J. A., MATTHEWS, D. H., BLUNDELL, D. J. & KELK, B. 1982. Deep structure of the Scottish Caledonides revealed by the MOIST reflection profile. *Nature*, **299**, 338–40.

STEEL, R. J. & WILSON, A. C. 1975. Sedimentation and tectonism (?Permo-Triassic) on the margin of the North Minch Basin, Lewis. *Journal of the Geological Society, London*, **131**, 183–202.

WEATHERS, M. S., BIRD, J. M., COOPER, R. F. & KOHLSTEDT, D. C. 1979. Differential stress determined from deformation induced microfractures of the Moine Thrust Zone. *Journal of Geophysical Research*, **84**, 7495–7509.

WHITE, S. H. 1979*a*. Difficulties associated with palaeostress estimates. *Bulletin of Mineralogy*, **102**, 210–215.

—— 1979*b*. Grain and sub-grain size variations across a mylonite zone. *Contributions to Mineralogy and Petrology*, **70**, 193–202.

Influence of fractal flaw distributions on rock deformation in the brittle field

IAN G. MAIN,[1] PHILIP G. MEREDITH,[2] PETER R. SAMMONDS[2] & COLIN JONES[3]

[1] *Department of Geology & Geophysics, University of Edinburgh, James Clerk Maxwell Building, Mayfield Road, Edinburgh EH9 3JZ, UK*

[2] *Rock Physics Laboratory, Department of Geological Sciences, University College London, Gower Street, London WC1E 6BT, UK*

[3] *Rock and Fluid Physics Group, Schlumberger Cambridge Research, Madingley Road, Cambridge CB2 3BE, UK*

Abstract: The geometrical distribution of flaws plays a crucial role in the physical behaviour of geological materials under stress. Flaws are present in the earth on all scales, from microcracks to plate-rupturing faults. They may be distributed on one characteristic length scale (e.g. joints, 'characteristic' earthquakes), or more commonly exhibit scale-invariance over a specified range of sizes. Scale-invariance implies that the discrete length distribution in a finite range is a power law of negative exponent D, where $1 \leq D < 3$. Fault systems where motion is concentrated on a dominant fault (e.g. San Andreas) have $D \approx 1$, but more diffuse fault systems have D near 2. D is one of the fractal dimensions of the fracture system. The length distribution of faults or microcracks may be inferred from the slope b of the log-linear frequency–magnitude distribution of earthquakes, or laboratory-scale acoustic emissions, since it can be shown that $D = 3b/c$. The scaling factor c depends on the relative time constants of the seismic event and the recording instrument, and is usually equal to 3/2. b is found experimentally to be negatively correlated with the stress intensity on the dominant flaw, which depends in turn on the applied stress and the flaw length. Thus a fracture mechanics model of rock failure which includes a range of flaw sizes can be tested by seismic monitoring.

We describe a fracture mechanics model of rock failure for a variety of styles of deformation, ranging from elastic failure to quasi-static cataclastic flow, and predict the time-dependence of D and the seismic b-value at different times up to and including failure. Critical coalescence of microcracks during dynamic failure (e.g. earthquake foreshocks) occurs when $D = 1$ ($b = 0.5$); random processes (e.g. cataclastic flow, background seismicity) are associated with $D = 2$ ($b = 1$); positive feedback in the concentration of stress on the dominant flaw (e.g. during strain softening and shear localisation) occurs when $D < 2$ ($b < 1$); negative feedback in stress concentration (e.g. during the early stages of dilatancy), and where a highly diffuse fracture system is produced, occurs at low stress intensities and is associated with $D > 2$ ($b > 1$).

It has long been a goal of structural geologists to measure stress on rocks, since most geometrical signatures of deformation are strain-related. We show that stress is not usually as significant in rock fracture as stress intensity, and furthermore that the geometric signature of the length distribution of microcracks is well-correlated with the stress intensity.

The deformation of rocks and of the Earth's crust in the brittle field is known to be strongly dependent on the presence and geometry of pre-existing flaws in the material of interest. These may take the form of distributed microcracks, joints or shear faults on a scale of fractions of a mm up to several hundred km. Microcracks themselves nucleate from pre-existing flaws or stress concentrators such as grain boundaries, pores or lattice defects, and compressive failure by shear faulting in the laboratory is preceded by a phase of tensile microcrack nucleation, growth and coalescence. Thus the distribution and coalescence of microcracks exerts a controlling influence on the macroscopic style of deformation of stressed rock. In this paper we present an overview of a theoretical model which predicts changes in

From Knipe, R. J. & Rutter, E. H. (eds), 1990, *Deformation Mechanisms, Rheology and Tectonics*, Geological Society Special Publication No. 54, pp. 71–79.

the relative distribution of microcrack sizes accompanying a wide variety of styles of deformation, and show that the results of laboratory monitoring of acoustic emissions (microseismicity) are consistent with the predictions of the model. Although the model was developed from observations of tensile microcracking in the laboratory, we show that it may also be applicable to the seismicity preceding earthquakes. For example, both laboratory results and field observations support the prediction of intermediate-term and short-term precursors to failure.

The present paper reviews a wide body of recent theoretical and experimental work (Main 1988, 1990; Main *et al.* 1989; Main & Meredith 1989; Meredith *et al.* 1990), but also presents a new and extended discussion of the scale-invariant or fractal nature of the distribution of flaws in geological materials. It is well known that the only distribution of features with a characteristic length scale which is scale-invariant is a power law (Mandelbrot 1982). This takes the form

$$N(l) = Cl^{-D}, \qquad (1)$$

where N, in our case, is the number of flaws (faults or microcracks) in the discrete length interval $l - \delta l/2 \leq l < l + \delta l/2$. C is a constant, and the power-law exponent D is one of the fractal dimensions of the system (Mandelbrot 1977, 1982; Turcotte 1989). A 'cumulative', or time-integrated value of D can be found by examination of the final structure of a rock which has undergone fracture, and an 'instantaneous' value of D can be found either by examining the sample for evidence of new, recent fracture, or more easily by recording acoustic emissions or earthquakes in a discrete time interval. However it is important to recognise that these seismic determinations of D are of necessity indirect and rest on assumptions to be discussed later.

A new idea presented in the present work is that changes in the 'instantaneous' value of seismically-inferred D can be associated with positive or negative feedback in local stress concentration during deformation, and in some cases these can be identified directly with mechanisms of strain hardening and strain softening. For example, dilatancy hardening inhibits further crack growth at a particular microcrack after it has opened, and is associated with $D > 2$. In contrast, random seismicity and cataclastic flow is associated with $D = 2$; and positive feedback during the accelerated growth of a dominant shear fault in the final stages of quasi-static failure is associated with $D < 2$.

Scale-invariance and the fractal geometry of fracture systems

It is well known that geological fabrics are often scale-invariant (Tchalenko 1970) otherwise it would not be necessary to include scale bars or the ubiquitous geological hammer on photographs of rock outcrop. Exceptions to this general picture are the existence of characteristic-sized features, for example joint-spacing in intrusive sills, or characteristic earthquakes in seismically active zones (e.g. Schwartz & Coppersmith 1984). In the former the characteristic spacing may be related to the size of convection cells and hence to the width of the intrusive sill, whereas in the latter the finite depth of the seismogenic zone of the earth's crust may strongly control the size of characteristic earthquakes (Main & Burton 1986; Turcotte 1989). The existence of characteristic features such as these implies that even where scale-invariance does hold, it is restricted to length scales between some upper and lower limits. For example, in rock deformation experiments involving microcracking, the limits may be set by the grain size and the sample size (e.g. Brown & Scholz 1985). Having qualified its general validity, we now proceed to discuss the evidence for scale-invariance in the distribution and geometry of the great majority of fracture systems, within the context of Mandelbrot's (1977, 1982) concept of the fractal geometry of nature.

Shear fracture systems and dynamic earthquake faults

Figure 1, after Shaw & Gartner (1986), shows convincing evidence for the scale-invariance of shear fault systems with one dominant throughgoing fault, over a range of scales from fractions of a millimetre to hundreds of kilometres. Without the scale bars and annotation it would be virtually impossible to distinguish laboratory clay-box experiments from plate-rupturing faults. Figure 2 superposes the size distribution of the four fault systems shown in Fig. 1, with each normalized to the length of the maximum throughgoing fault. Two aspects emerge instantly. The first is that all the fault systems have the same relative size distribution of subsidiary faults, and second that this distribution is a power-law of negative exponent $D = 1$. According to Mandelbrot's (1977, 1982) concept of the fractal geometry of many natural systems (e.g. rivers, trees, topography, coastlines), this power-law exponent is one of the fractal dimensions of the system (Turcotte 1989). Note

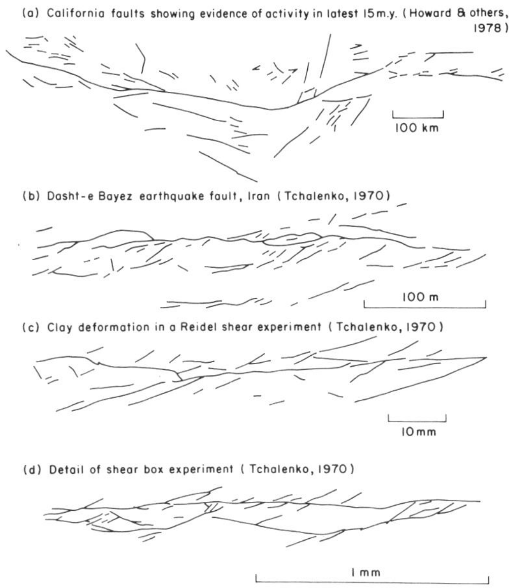

Fig. 1. Geometry of shear faulting over a range of scales, after Shaw & Gartner (1986). On the scales shown, each example is dominated by one major throughgoing fault.

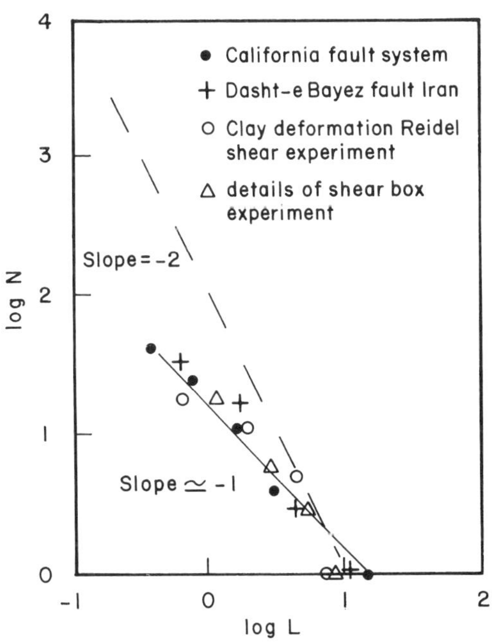

Fig. 2. Normalized discrete frequency–length distribution of the faults shown in Fig. 1. All four data sets are consistent with a power law of negative slope $D = 1$ (solid line). A dashed line of slope -2 is shown for reference.

that in this case D results from the cumulative production of fracture over one complete cycle of deformation, leading eventually to one dominant throughgoing fault.

The most appropriate, currently-available, mathematical description of the fault distribution mapped on Fig. 1 is perhaps the Cantor set (Mandelbrot 1982; Turcotte 1989), which describes a dominant straight line drawn on a two-dimensional plane, with more and more identical replicas on a cascade of smaller scales, and with a power-law frequency distribution in length. An extension to three dimensions would require a dominant fault plane with a cascade of smaller fault planes.

A second aspect of the scale-invariance of faults is the systematic increase in measured fault length with decreasing size of the measuring element used. This increase in length is due to the roughness of natural topography, the significance of which was first recognised by Mandelbrot (1977). We shall call this the 'ruler' fractal dimension, D_R. For example, Aviles *et al.* (1987) have estimated a 'ruler' fractal dimension for the San Andreas fault in the range $1.0008 < D_R < 1.0191$, compared to $D_R = 1$ for a perfectly smooth ideal fault. Thus the dynamic failure of faults in shear mode, where one dominant throughgoing fault is produced, results in fault systems with both fractal dimensions D and D_R near unity. However, in general, these two different fractal dimensions need not be identical.

In contrast to Fig. 2, Fig. 3 shows the frequency-length distribution of all of the mapped faults showing evidence of Holocene activity in the contiguous US, after Shaw & Gartner (1986). The distribution is still a power-law over the scale range 10–200 km, but has a higher 'cumulative' fractal dimension $D = 1.76$. This higher value of D implies an overall fault system where no single throughgoing dominant fault exists, and the deformation is taken up by a more distributed system of unconnected, relatively minor faults. For unbounded scale-invariance, the maximum permissible value for D in a Euclidean volume is 3, when the whole volume is eventually filled by a cascade of ever smaller fractures. With this physical limitation, we would always expect $1 \leq D < 3$, where lower D represents a greater concentration of deformation on a few relatively large faults, and higher D represents a more diffuse system dominated by a greater proportion of relatively small faults. In practice the existence of a finite lower limit to the length scale, or the possibility of repeated fracture on numerous small flaws, allows a greater number of smaller events than

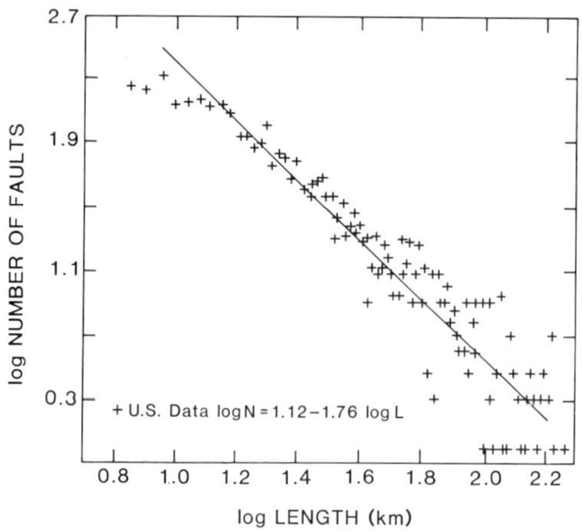

Fig. 3. Discrete frequency–length distribution of active surface faults in the US mapped by Howard *et al.* (1978), after Shaw & Gartner (1986). The data are consistent with a power law distribution of fault lengths with negative exponent $D = 1.76$, equivalent to the slope of the best-fitting line.

predicted by unbounded scale-invariance. Thus the upper bound to D of 3 should be treated as only approximate, especially if scale-invariance is restricted to a narrow range of length scales.

The fractal interpretation has also been widely validated by seismological studies of faults. For example, the log-linear frequency-magnitude distribution

$$\log N = a - bm, \tag{2}$$

where N is the number of earthquakes in the magnitude range $m - \delta m/2 \leq m < m + \delta m/2$, and a, b are empirical constants (Richter 1958), can be shown to be a direct consequence of the relative constancy of earthquake stress drops and a power-law distribution of fault lengths (Caputo 1976; Aki 1981; King 1983; Main & Burton 1984).

The seismic moment is defined by

$$M_0 = \mu A s, \tag{3}$$

and is related to the fault length l and stress drop $\Delta\sigma$ by

$$M_0 = C_0 \Delta \sigma l^3 \tag{4}$$

where μ is the rigidity modulus, A is the fault area, and s is the average fault slip. Expressions for the dimensionless constant C_0 are given in Kanamori & Anderson (1975). Seismic moment is related to magnitude by an expression of the form

$$\log M_0 = c\,m + d, \tag{5}$$

(Kanamori 1978), where d depends on the stress drop averaged over the fault area and c depends on the relative time constants of the seismic event and the recording instrument (Kanamori & Anderson 1975).

Thus, if the following assumptions hold: (i) moment scales as length cubed (eq. 4), (ii) stress drop is constant, and (iii) moment scales exponentially with magnitude (eq. 5), then it follows from equations (1) and (2) that the seismological b-value is related to the power-law exponent D by

$$D = 3\,b/c. \tag{6}$$

For events with very long durations compared with the natural period of the recording instrument $c = 3$, and for very small events with short durations $c = 1$. However the most common case for intermediate earthquakes is $c = 3/2$ (Kanamori & Anderson 1975). Indeed, the new moment-magnitude scale, M_W, which compensates for the effect of band-limited saturation of the magnitude scale, has been defined with $c = 3/2$ (Kanamori 1978). For this common value of c, $D = 2b$.

Seismological investigations with a high degree of statistical significance have established a typical 'background' value of $b \approx 1$ ($D \approx 2$), with b fluctuating in general between $0.5 \leq b < 1.5$, ($1 \leq D < 3$), and earthquake foreshocks occurring when $b \approx 0.5$ ($D \approx 1$) (von Seggern 1980; Main 1988). This is in very good agreement with the observations and predictions of the range of D derived from direct fault mapping (Figs 2 and 3), but with the advantage of better resolution with depth for the smaller faults which do not break the surface. The disadvantage is that some measure of inference is inherent in the seismological approach, albeit founded on the three assumptions outlined above, the validity of which has been generally established in the seismological literature (e.g. Kanamori 1978).

Tensile microcrack systems and quasi-static subcritical crack growth

Earthquakes are an example of a dynamic fracture mechanism where crack growth occurs at sub-sonic velocities due to the limitations imposed by inertia. In classic linear elastic fracture mechanics, this occurs when the rate of energy supply G equals or exceeds a critical rate G_c, or equivalently when the stress intensity factor K exceeds the fracture toughness K_c of the material (Atkinson & Meredith 1987). However even when G or K are lower than their critical values quasi-static crack growth can occur in the presence of an active chemical environment (Anderson & Grew 1977). The most significant mechanism for such subcritical grack growth in geological materials in the upper crust is considered to be stress corrosion (see Atkinson 1984, for an extensive review). Even under these subcritical conditions crack advance occurs by discrete, individually dynamic events, but with very slow *average* rupture velocities. The average velocity of crack propagation is found experimentally to be correlated to the stress intensity, i.e. the level of stress concentration at the growing crack tip, as well as the chemical activity, pressure and temperature of the pore fluid. In particular water is a very effective reagent, even in trace amounts (Atkinson 1984; Atkinson & Meredith 1987). The fracture surfaces produced by low rupture velocities tend to be rougher than those produced during dynamic crack propagation at stress intensities comparable to the fracture toughness of the material (Fig. 4), because subcritical crack growth exploits natural weaknesses such as grain boundaries. Figure 4 is taken from samples used in the experiments of Meredith & Atkinson (1983).

Subcritical crack propagation also causes more distributed damage around the propagating crack compared to dynamic failure (Fig. 5, also taken from samples analysed in Meredith & Atkinson 1983). The observation of more distributed damage is qualitatively consistent with the high fractal dimensions D found for distributed shear faults on Fig. 3. Although D can, in principle, be measured directly on tensile test samples, it is much simpler to use acoustic emission statistics to infer D via the seismic b-value, since a proper volume count can be obtained by this latter method. Figure 6 shows how b and the inferred instantaneous value of D are correlated to the stress intensity K. In this diagram, K is normalized to the fracture toughness K_c, and data are plotted for three different rock types broken under the same double-torsion loading configuration in both air-dry and water-saturated environments at ambient laboratory temperatures and pressures. The experimental arrangement is described in Meredith & Atkinson (1983), and for these conditions it can be inferred that $c = 3$, or $b = D$ (Main *et al.* 1989). Figure 6 illustrates one of the most fundamental aspects of this paper. The x axis on the diagram represents the physics of the process (stress intensity, fracture toughness, crack propagation velocity) and the y axis is the resulting fractal geometry. The link between these variables

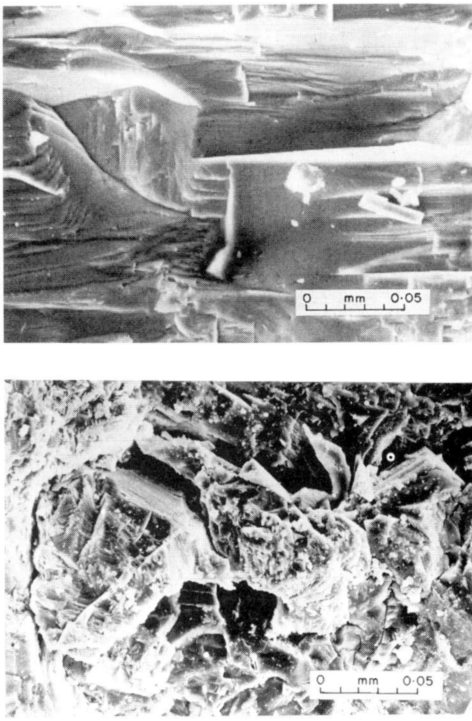

Fig. 4. Scanning electron micrographs of the fracture surfaces of Whin Sill dolerite double torsion specimens. **(a)** Rapid mechanical fracture at high crack velocity (> 10 ms^{-1}) during a fracture toughness test carried out at ambient pressure in air at 20°C. Note the dominance of smooth transgranular fracture, cleavage steps and hackle marks. **(b)** Subcritical crack growth at low velocity (*c.* 10^{-7} ms^{-1}) in water at ambient pressure at 20°C. The crack surface is dominated by rough, grain-boundary microcracking even though some transgranular fracture is still evident.

depends on the chemical activity of the fluid-rock interaction. A range of conclusions can be drawn from Fig. 6, including the following:

(i) b and the inferred value of D are negatively correlated to the normalized stress intensity K/K_c;

(ii) the inferred value of D is in the range $1 \leq D < 3$, similar both to that observed in field exposure and that inferred seismologically for earthquake faults;

(iii) $D \approx 1$ at critical stress intensities corresponding to the fracture toughness $K = K_c$, similar to the inferred value of D from earthquake foreshock sequences;

(iv) for the three crystalline rock types tested so far, all the available data fall on the

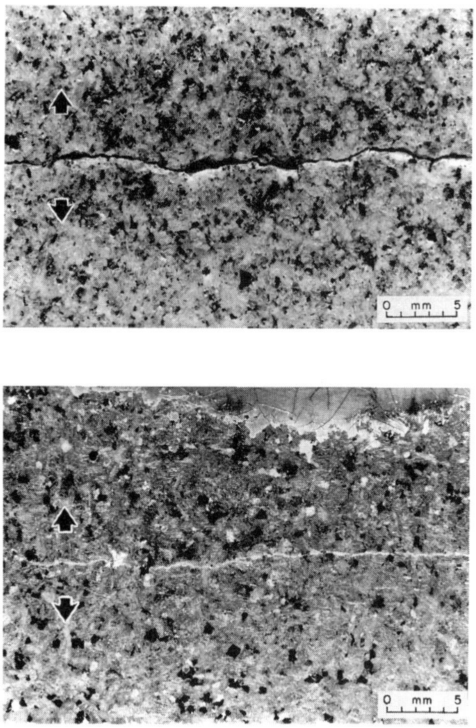

Fig. 5. Optical reflection micrographs of samples of Westerly granite fractured in double torsion tests. Tensile load applied in the direction of the arrows. (**a**) Rapid mechanical fracture at high crack velocity ($> 10\ \mathrm{ms}^{-1}$) during a test carried out in air at 20°C. Note the presence of a small number of aligned tensile microcracks. (**b**) Subcritical crack growth at a low velocity (about $10^{-7}\ \mathrm{ms}^{-1}$) in water at ambient pressure at 20°C. Pervasive microcracking aligned perpendicular to the applied tensile load is shown throughout the sample (a thin section) in this case.

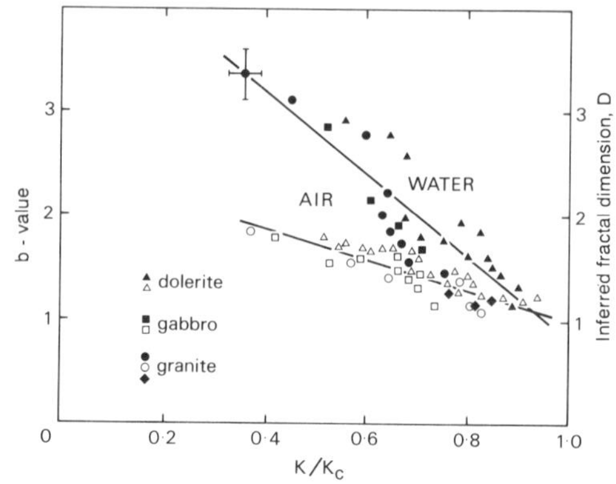

Fig. 6. Synoptic diagram of the variation in seismic *b*-value and the inferred fractal dimension *D* with stress intensity *K* and crack tip humidity during tensile crack propagation for a variety of crystalline rocks. The data were derived from tensile crack growth experiments performed in air or water at ambient pressure and temperature using the double torsion technique, and are normalised relative to the fracture toughness K_c. Solid lines are least squares fits to the data points and converge approximately at the point ($K/K_c = 1$, $b = 1$, $D = 1$), and typical error bars are shown. *b* depends on the time constants of the event and the recording instrument, so *D* is the more fundamental parameter. For these tests the relevant relation between *b* and *D* is $b = D$, but for most earthquake studies $b = D/2$ (Main *et al.* 1989).

same curve for a given humidity within the resolution of the experiments;

(v) humidity exerts a strong control on the crack distribution during subcritical crack growth, and in particular $D > 2$ requires water to be present above the levels found in ambient laboratory air;

(vi) dynamic crack growth (when $K = K_c$) produces $D = 1$ irrespective of humidity.

Main (1988) showed that $D = 1$ corresponds to unstable critical coalescence of tensile microcracks, by considering the effect of varying *D* on the average crack length $<l>$ in the range l_{min} to l_{max}, since it can be shown that

$$<l> = \frac{l_{min}^{1-D} - l_{max}^{1-D}}{l_{min}^{-D} - l_{max}^{-D}} \frac{D}{D-1} \qquad (3)$$

for a power-law distribution of negative exponent *D*. For unbounded scale-invariance ($l_{max}/l_{min} \rightarrow \infty$), $<l>$ remains finite as long as $D > 1$. However, as $D \rightarrow 1$, $<l> \rightarrow \infty$ analogous to a critical mean free path criterion, this represents a critical instability in the fracture process. This theoretical lower limit to *D* should also be treated as approximate, particularly when scale-invariance is restricted to a narrow range of length scales.

More recently Main (1990) has shown that $D = 1$ also corresponds to the maximum concentration of overall seismic stress relief on the largest fault or fracture, whereas $D = 2$ corresponds to an even distribution of stress relief throughout the magnitude range, with smaller events contributing equally to the overall stress relief because of their greater numbers (see the Appendix for a short summary). For $D > 2$ the stress relief is dominated by even larger relative numbers of small events, corresponding to distributed damage without crack coalescence. Thus $D < 2$ corresponds to a situation of stress concentation and hence positive feedback in the fracture process at high stress intensities. $D = 2$ corresponds to a memoryless process with stress relief distributed evenly through the magnitude range, and $D > 2$ corresponds to negative feedback at low stress intensities, resulting in relatively large numbers of distributed, isolated smaller cracks or faults. For $D = 1$ the probability of fracture recurring at a particular site is increased to a maximum because of the resulting critical stress concentration; for $D = 2$ the probability is independent of previous events and for $D > 2$ the probability of immediate recurrence is reduced locally in a system of distributed, isolated flaws. This is consistent with the seismological observation that random background seismicity is associated with $b \approx 1$ and by inference $D \approx 2$, whereas foreshocks ($b \approx 0.5$) imply $D \approx 1$.

The above discussion has highlighted many of the similarities between laboratory-scale tensile fracture experiments and large, natural shear fracture systems, in particular the observation $1 \leq D < 3$, with $D \approx 1$ corresponding to dynamic failure and the production of a through-going fault. In the next section we shall show that similar parallels occur in the time-dependence of D under a wide variety of loading conditions.

A fracture mechanics model of temporal variations in *D* accompanying rock deformation

Figure 7 schematically shows predictions of the time dependence of D for a range of loading histories under microscopically brittle conditions and constant remote strain rate. The model was first presented in Meredith *et al.* (1990), and refers to instantaneous rather than cumulative values of D. Model A corresponds to ideal macroscopically brittle behaviour, with instantaneous crack acceleration from zero to near-sonic velocity at failure. The stress intensity is assumed proportional to the applied effective stress and square root of the dominant flaw length (Lawn & Wilshaw 1975), and D is negatively correlated to K. Model B represents failure at peak stress after a period of strain hardening, and predicts a more gradual, concave-downward decrease in D to a critical value $D_c = 1$ at failure. Model C represents failure after periods of strain hardening and strain softening, the most realistic case for modelling natural systems (e.g. Paterson 1978). In this case the behaviour of D depends on the mechanism of strain softening. If slip behind the accelerating crack front is responsible for the stress decrease (dashed lines on Fig. 7), then we would expect an inflexion or hiatus in K at peak stress, because local stress release by this mechanism results in greater concentration of stress at the growing crack tip. A similar inflexion in D is predicted, after an initial convex-downward decrease. However, strain softening by stress release off the dominant crack, or by an increase in pore pressure, followed eventually by accelerating crack growth, results in two maxima in the stress intensity and two minima in D (solid lines on Fig. 7). Main & Meredith (1989) have suggested the occurrence of two peaks in K as a possible explanation for the existence of intermediate-term and short-term earthquake precursors. In this interpretation the first broad peak in stress intensity is responsible for intermediate-term precursors whose duration scales with magnitude, and the second sharp peak corresponds to short-term precursors whose duration is independent of magnitude (Rikitake 1987). Finally, model D involves no stress drop and the mode of failure is quasi-static cataclastic flow. In this case there is no stress concentration and K is therefore undefined. Here we have simply assumed D to be negatively correlated to the stress, as suggested by Scholz (1968). D simply mirrors the stress and tends to a steady state value above the critical level $D_c = 1$.

For brevity we have not reproduced the mathematical basis for Fig. 7 and the interested reader is referred to Meredith *et al.* (1990) for a complete derivation. However, its empirical basis is firmly rooted in experimental observations of tensile subcritical crack growth, even though there are promising signs that Model C could also be applied to earthquakes; in particular the occurrence of intermediate-term and short-term earthquake precursors.

Figure 7 shows how the instantaneous values of b and D vary with time in the build-up to a single dynamic failure event, and not the integrated value of D which could be measured if earth materials were transparent. In those variants of the model, where b and D are monotonically decreasing with time, we would expect the integrated value of D to be greater than that obtained in a single, discrete time interval. The shape of the cumulative anomaly in D is thus less responsive to sudden changes in the fracture distribution. However, it will nevertheless be important in future work to establish a correlation between the instantaneous and cumulative values of D for comparison with field observations. At this stage it is also important to note that an integrated value of D during cataclasic flow is impossible to define because of continuous overprinting, and a change of deformation mechanism from fracturing to grain crushing and rotation. In the next sections we

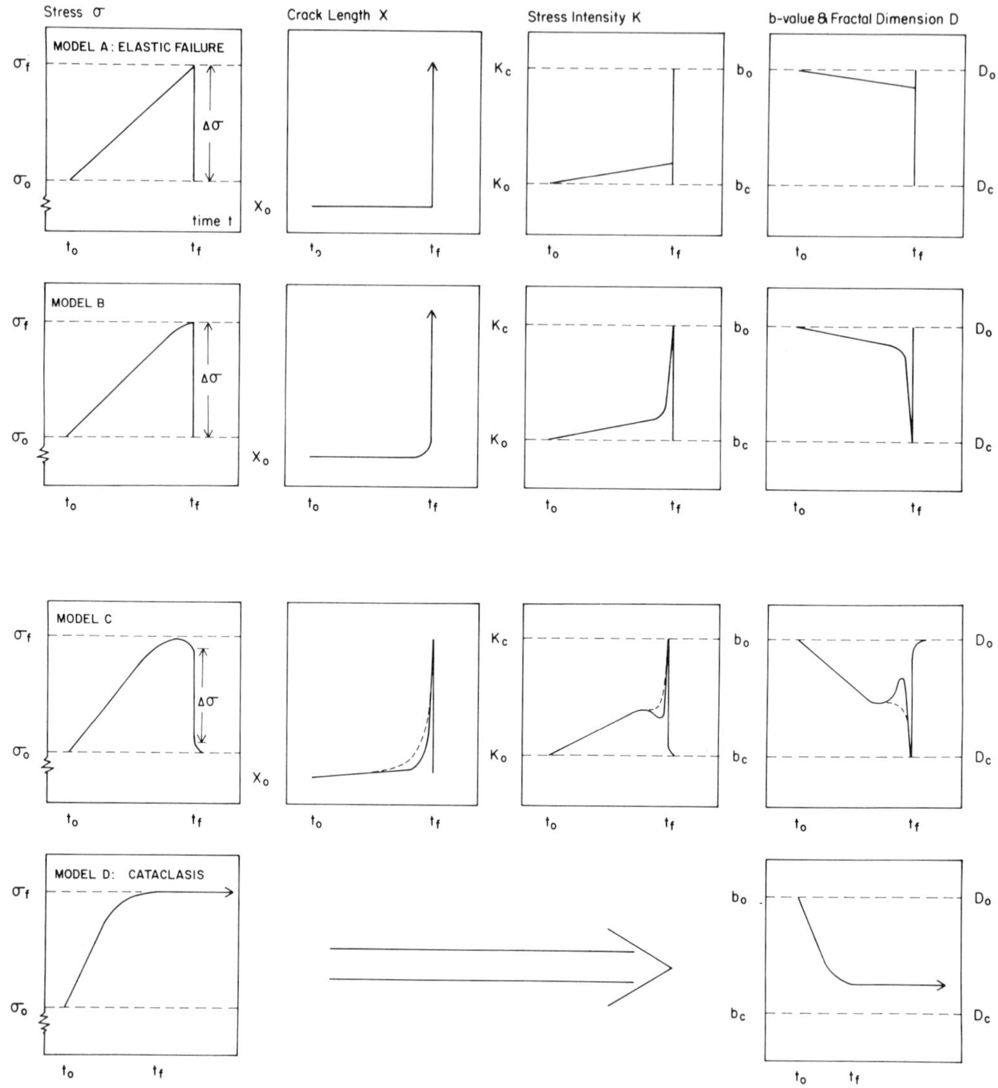

Fig. 7. Four variants of the proposed *b*-value anomaly model associated with different stress histories and levels of stress concentration. The stress, crack or fault length, stress intensity and *b*-value are plotted in columns, with the rows corresponding to the four model variants, viz: (A) elastic–brittle failure; (B) strain hardening prior to dynamic rupture; (C) strain softening prior to dynamic rupture; (D) quasi-static deformation by cataclastic flow. All diagrams are normalised to the start time t_0 and the failure time t_f, and to the peak stress $\Delta\sigma_f$. X_0, $\Delta\sigma_0$, K_0 and b_0 are respectively initial values of the crack length, stress, stress intensity and *b*-value at time t_0. K_c and b_c are critical values of the stress intensity and *b*-value, and *D* is the power-law exponent of the crack length distribution during a discrete time interval. $D_c = 1$ and $D < 3$, so $b_c = 0.5$ and $b < 1.5$ for intermediate-size earthquakes, since $D = 2b$.

Models A to D represent a gradation from purely elastic-brittle behaviour to macroscopically more ductile behaviour, with the dashed line in model C corresponding to precursory stress drop dominated by slip behind an accelerating crack tip (which results in a decrease in stress but increase in stress intensity). These gradations can be modelled by gradations in the stress corrosion index, *n*, with higher values corresponding to more 'brittle' behaviour for models A, B and C. For model A, $n \rightarrow \infty$. For model D there is no stress concentration.

discuss the experimental evidence for the general validity of the model proposed in Fig. 7, and to what extent it may be applied to precursory seismicity statistics.

Acoustic emission monitoring of compressive rock deformation

Results from a series of laboratory-scale, triaxial compression experiments on various rocks are shown in Figs 8 to 10. These experiments were designed to test as many variants as possible of the proposed model of Fig. 7, by contemporaneous monitoring of both deformation behaviour and acoustic emission (AE) signals. All data are plotted against time for ease of comparison with both field data and the model predictions. Details of the experimental arrangement and methodology, and of the test materials used, are given in Meredith *et al.* (1990), where the results were first presented, and so are not repeated here. For all of these experiments it can be inferred that $c = 3/2$ in equation (6), (Main *et al.* 1989), and hence $D = 2b$. All samples were deformed at a nominally constant applied strain rate of $10^{-5}\ s^{-1}$.

Figure 8 shows data from a test on an air-dry sample of Westerly granite deformed under a confining pressure of 100 MPa. Note that the stress/time curve (upper diagram) is quasi-linear up to about 85% of the peak stress, and then exhibits a dilatant, strain hardening phase leading to dynamic failure at peak stress. Such mechanical behaviour corresponds to our previously-defined model B. No significant level of AE activity (as defined by the AE event rate, middle diagram) was recorded during the linear elastic phase, but when the stress/time curve deviates from linearity, indicating the growth of new microcracks, there is a corresponding increase in AE activity up to the point of failure.

Temporal variations in *b*-value derived from the peak amplitudes of individual AE events according to the maximum likelihood method of Aki (1965) are shown on the lower diagram. Since very few emissions were recorded during linear elastic loading, it has not been possible to determine *b*-values for most of the loading cycle. However, where a *b*-value has been calculated, it decreases monotonically with increasing stress, and is concave downward as predicted by model B.

Figure 9 shows the same set of data as Fig. 8, but for a test on a water-saturated sample of Darley Dale sandstone deformed at 50 MPa confining pressure. The stress/time curve for this material is markedly non-linear, with peak stress followed by a significant precursory de-

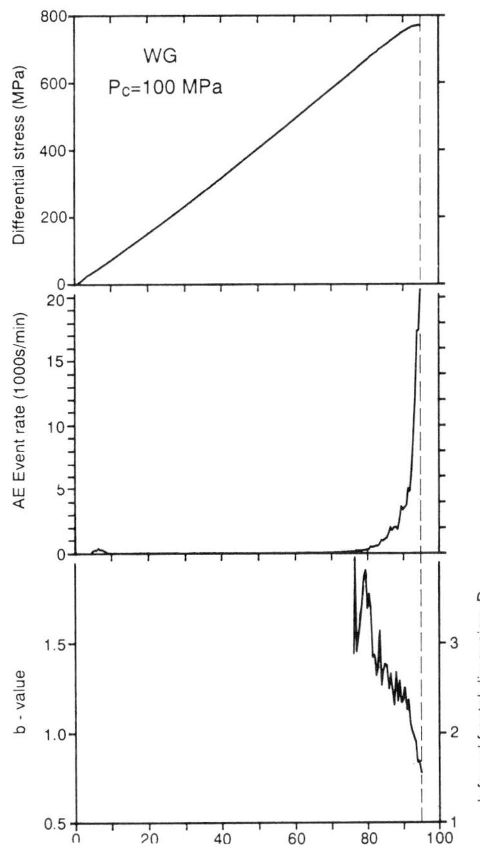

Fig. 8. Contemporaneous measurements of differential stress, acoustic emission (AE) event rate, and *b*-value as functions of time for an air dry sample of Westerly granite (WG) deformed at a nominally constant strain rate of $10^{-5}\ s^{-1}$ under a confining pressure (P_c) of 100 MPa. The stress/time curve is quasi-linear over most of the loading cycle, and dynamic failure (dashed line) occurs at peak stress, after a short strain hardening phase. AE activity is essentially absent during the quasi-linear phase, but increases exponentially during strain hardening. *b*-values and *D* both decrease monotonically during the strain hardening phase, and the curve is concave downwards as predicted for model B behaviour. Here $D = 2b$.

crease in stress prior to dynamic failure on a well-defined fault plane and, finally, stable sliding along the fault (model C mechanical behaviour). Macroscopic dynamic failure occurs when the negative slope of the stress/time curve is a maximum (i.e. when the rate of stress drop is a maximum).

Again, the AE rate increases rapidly with the onset of dilatancy, which in this case occurs at

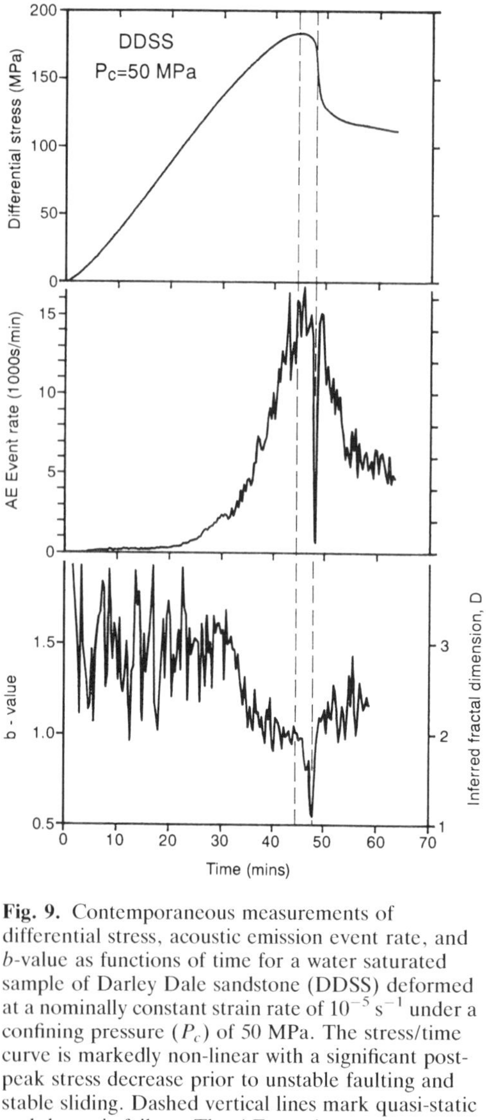

Fig. 9. Contemporaneous measurements of differential stress, acoustic emission event rate, and b-value as functions of time for a water saturated sample of Darley Dale sandstone (DDSS) deformed at a nominally constant strain rate of $10^{-5}\,s^{-1}$ under a confining pressure (P_c) of 50 MPa. The stress/time curve is markedly non-linear with a significant post-peak stress decrease prior to unstable faulting and stable sliding. Dashed vertical lines mark quasi-static and dynamic failure. The AE rate increases exponentially during the strain hardening phase leading to peak stress, but falls dramatically close to dynamic failure. Post-failure recovery and subsequent decay may be associated with aftershocks. A decreasing b-value associated with increasing stress leads to a broad minimum at peak stress followed by a short-term low b-value anomaly at failure. The critical b-value is close to the predicted value (model C) of 0.5 ($D = 1$ for $D = 2b$). Post-failure the b-value recovers.

little more than 50% of the peak stress. We can usefully separate dilatancy into two phases; the first dominated by the growth of new microcracks in the period up to peak stress, and the second dominated by the interaction and coalescence of these microcracks to form a throughgoing fault in the period of post-peak strain softening. Around peak stress, the AE rate at first flattens out and then falls dramatically to a period of apparent quiescence. This dramatic quiescence associated with dynamic failure is thought to be caused by saturation of the monitoring system due to a cascade of events which become indistinguishable from each other at critical crack coalescence. Similar apparent quiescence has been reported to occur close to dynamic failure in laboratory experiments by a number of other workers (Gowd 1980; Kikuchi *et al.* 1981; Sondergeld *et al.* 1984) but *only* when a strain softening phase preceded failure. Finally, following instability the AE rate recovers before decaying in a manner analogous to an earthquake aftershock sequence.

b-value data for this experiment exhibit all of the features predicted by model C (dashed line, Fig. 7). During the early quasi-elastic phase of loading, where there is a low level of AE activity, the b-value remains essentially constant, with a high value close to 1.5 (and by inference $D = 3$). At higher levels of stress, the major trend is of a decreasing b-value, correlated with increasing AE rate and the growth of new microcracks, and which flattens out at around peak stress ($b = 1$, $D = 2$). This is followed by an inflection point leading to a much shorter time-scale b-value anomaly leading to dynamic failure close to the predicted value of $b_c = 0.5$ ($D_c = 1$). Post-failure the b-value recovers as expected. Note that the error in measurement of b-value is proportional to $1/\sqrt{N}$, where N is the total number of AE events in each time period for which a b-value is determined, so that variability in b-values is much greater during the relatively quiet low-stress phase of loading than during the dilatant phase where the AE rate is much increased.

In this experiment we have been able to produce the two-stage b-value anomaly predicted from model C behaviour, but with an inflection point rather than a recovery to a precursory maximum prior to dynamic failure. There are a number of possible explanations for this observation. First, we have already noted that an inflection point would be expected where the dominant strain-softening mechanism was quasi-static slip behind an accelerating crack tip without any significant pore-fluid diffusion. Such precursory slip is the only mechanism which give rise to a reduction in stress combined with an increase in stress intensity. Second, this experiment was conducted at a strain rate some 9 orders of magnitude faster than inferred typical crustal rates. Under such relatively rapid loading, cracks can extend at a faster rate than reactive pore fluids can migrate to their tips. Hence, crack growth in this final phase of loading at high strain rates may not be influenced by the chemical action of environmental fluids (Atkinson & Meredith 1981; 1987). Third, the difference between crustal and laboratory strain rates highlighted above is further exacerbated by the fact that our assumption of constant strain rate loading breaks down during the crucial phase of decreasing stress preceding dynamic failure, due to the rapid release of strain energy stored in the deformation apparatus. This problem will be alleviated in future experiments by means of a newly-installed loading arrangement that will allow a constant strain rate to be maintained throughout by servo-control of the loading system stiffness.

Finally, Figure 10 shows data from a test on water-saturated Darley Dale sandstone conducted under a higher confining pressure of 200 MPa. In this case there is no stress drop, and deformation takes place quasi-statically by cataclastic flow (model D behaviour) rather than by dynamic rupture on a localised fault. Under these conditions, the AE rate again increases with the onset of microcracking, but remains at a consistently high level throughout the phase of cataclastic flow, with no indication of any period of quiescence or decay. As expected for model D behaviour, the *b*-value is negatively correlated with the level of stress, and falls from its background level of 1.5 to reach a constant value of unity (with an inferred instantaneous value of $D = 2$) associated with distributed microcracking during the phase of cataclastic flow. The value never falls to anywhere near the critical value of 0.5, since there is no critical stress concentration.

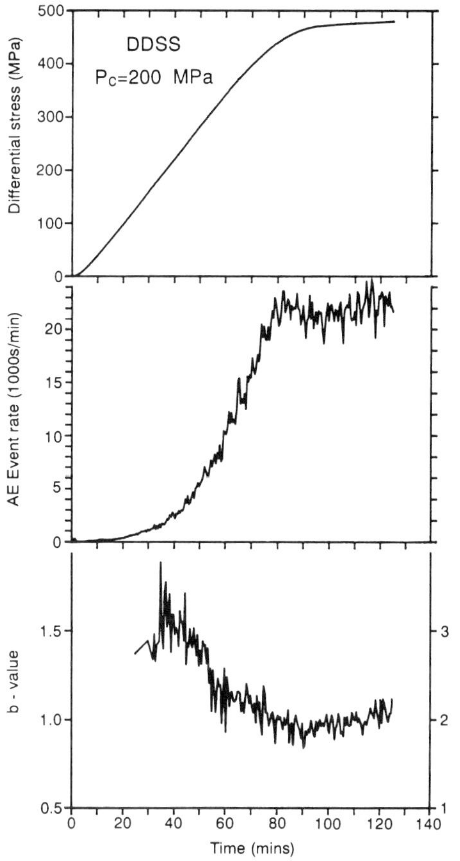

Fig. 10. Contemporaneous measurements of differential stress, acoustic emission event rate, and *b*-value as functions of time for a water saturated sample of Darley Dale sandstone (DDSS) deformed at a nominally constant strain rate of $10^{-5}\,s^{-1}$ under a confining pressure (P_c) of 200 MPa. The stress/time curve shows no stress drop, and deformation occurs quasi-statically by cataclastic flow (model D). The AE rate increases exponentially with the onset of microcracking, but remains at a consistently high level throughout the phase of cataclastic flow. The *b*-value is negatively correlated with the level of stress, but never approaches the critical value of 0.5 ($D = 1$ for $D = 2b$), since there is no critical rupture.

Seismological monitoring of earthquake precursors

Seismicity statistics are notoriously unreliable in terms of the accurate prediction of earthquakes, since their statistical significance is severely limited by the number and quality of the data. Nevertheless, in some cases it is possible to establish a consistent pattern in the event rate and magnitude distribution of smaller earthquakes leading up to a mainshock. We have already noted that $b \approx 1$ for random background seismicity, with $b \approx 0.5$ for foreshock sequences where there is sufficient statistical significance. For several events Smith (1981, 1986) has demonstrated a statistically significant precursory increase in b, followed by a short-term decrease, of respective relative duration 4:1. The only model capable of explaining a precursory *b*-value increase of this kind is the one in which the precursory strain softening is dominated either by stress release off the main fault or a pore pressure increase (see Fig. 7, Model C).

The phenomenon of precursory seismic quiescence in the intermediate term has recently been reviewed by Habermann (1988) and Wyss & Habermann (1988), who defined 'quiescence' as a drop in the seismic event rate by 45–90%. Quiescence has now been established with a statistical significance ranging from 90–99.99% for 17 recorded mainshocks so far (Wyss & Habermann 1988). The false alarm rate, 50%, is not insignificant but is much lower than that of many other empirical precursors. Note that this quiescence is not related to the dramatic drop in event rate in the dynamic failure experiments described above, which is thought to result from the masking effect of crack coalescence at the moment of failure. Instead, quiescence may be more related to the flattening off of event rates in the strain-softening phase of Fig. 9 (Wyss, pers. comm. 1989).

Figure 11 plots the event rate and the b-value prior to and after the $M_s = 6.8$ Western Nagano earthquake, Japan, of 14 September, 1984, after Imoto & Ishiguro (1986). Qualitatively, the b-value follows the form predicted from Model C, with two minima in b or D, and with the mainshock occurring near $b = 0.5$ as predicted. Phases inferred from the model of: (1) elastic loading; (2) strain hardening; (3) strain softening due to, distributed seismicity or crack coalescence, fluid pressure recovery, and or slip-weakening behind a growing crack front; (4) mainshock; (5) aftershocks, are illustrated (Main & Meredith 1989). The statistical significance of the first intermediate-term anomaly is not great, in contrast to that of the short-term decrease to $b_c = 0.58 \pm 0.02$ which is unequivocally established above the noise. However, the inferred peak in stress is followed by a phase of

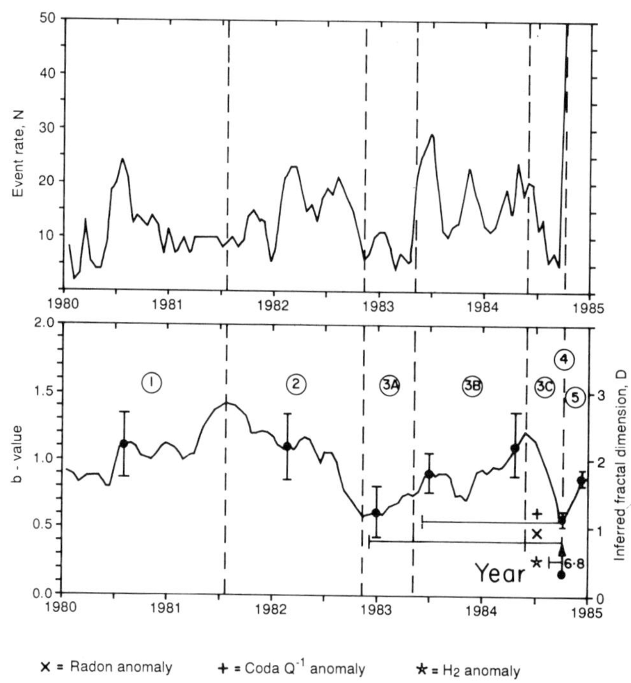

Fig. 11. Temporal variations in the seismic b-value for the Western Nagano earthquake of September 1984 (after Imoto & Ishiguro 1986). All earthquakes with a completeness threshold $M_L > 1.5$, with focal depths shallower than 50 km, and with epicentres within 20 km of the epicentre of the mainshock (35.81°N, 137.55°E) were used to plot the diagram. The upper diagram shows seismic event rates in the previous 72 days plotted as a running average. The b-values were calculated using a Bayesian technique (Imoto & Ishiguro 1986), to which we have added error bars of $\pm b/\sqrt{N}$ at selected points. The major precursory phases inferred from model C of Fig. 7 are: 1, transition from background b-value, 2, strain hardening (decreasing b-value), and 3, strain softening (increasing b-value followed by a short-term decrease near critical rupture). Phase 4 is the mainshock, and phase 5 the aftershock sequence. Phase 3 is split into two quiescent phases (3a and 3c) and a phase of enhanced activity (3b). Horizontal bars indicate the duration of three independent precursors to this earthquake.

marked quiescence during phase 3A. Scholz (1988) has suggested that the stress-memory effect known as the Kaiser effect may be a possible mechanism for intermediate-term quiescence. A definition of the Kaiser effect is that, when a material is subjected to cyclic stressing, no significant number of acoustic emissions is generated on any cycle until the previous maximum stress is exceeded. The phenomenon was first recognized for metals (Kaiser 1953), and has also been observed for Earth materials (Kurita & Fuji 1979; Holcomb 1981). Although this would explain why such a small stress decrease could result in such marked quiescence, care should be taken in extrapolating the Kaiser effect (which is associated with stress *and* strain relaxation) to the earth (stress decrease with monotonically *increasing* remote strain). For example Fig. 9 shows a levelling-off rather than a decrease in event rate during the strain-softening phase.

In contrast, phase 3B is relatively more active, possibly due to the enhancement of the stress corrosion mechanism caused by an influx of pore fluid into the dilatant volume near the nucleation zone. Phase 3C also has a marked short-term quiescence. These phases are all associated with the onset of independent precursors to this event, summarised by Sato (1988), and whose durations are also marked on the diagram. All of these independent precursors occur during the inferred phase of strain softening, with a radon gas anomaly beginning very near the inferred peak stress, and the coda-Q^{-1} anomaly beginning near the phase of inferred fluid influx. Therefore, we conclude that, although the b-value alone is probably not of great enough statistical significance in this case, its correlation with independent precursors adds weight to the interpretation within the context of the model presented here. Also the ratio of b-value increase/decrease prior to the mainshock is 4.7:1, close to Smith's (1986) average of 4:1.

A similar pattern is seen in Fig. 12 for an earthquake of $M_s = 7.8$ in the Tonga zone of the Tonga–Kermadec Trench in 1975, after Carter & Berg (1981). Again, both an intermediate- and short-term anomaly are seen, with a relative duration of increase in b followed by a short-term decrease in the ratio 3:1. Figures 11 and 12 also bear out the inference of $1 \leq D < 3$, with b fluctuating between limiting values of 0.5 and 1.5. Note also the longer time scale associated with the b-value of the larger earthquake, consistent with the scaling of other intermediate-term precursors (Rikitake 1987).

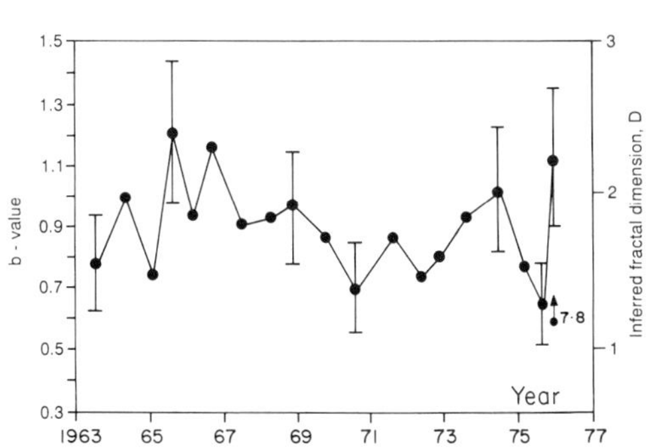

Fig. 12. Temporal variations in the seismic b-value for the December 1975 earthquake in the Tonga zone of the Tonga-Kermadec Trench (after Carter & Berg 1981). All earthquakes reported by the NEIS for the Tonga region with a completeness threshold $M_s > 4.5$ were used to plot this diagram. The intermediate-term and short-term b-value anomalies are consistent with the predictions for model C.

Comparison of laboratory and field results: concluding discussion

The common features of laboratory and field observations are as follows:

(i) $1 \leq D < 3$;
(ii) $D \approx 1$ at dynamic failure;
(iii) separate intermediate-term and short-term variations in D can be seen in both seismic and acoustic emission data;
(iv) the shape of all of the anomalies follow qualitatively one of the variants of the model proposed in Fig. 7;

The differences are as follows:

(i) the laboratory results show no precursory increase in b or D.
(ii) the starting value for D in the laboratory is $b = 1.5$, $D = 3$, compared to $b \approx 1$, $D \approx 2$ in the field;
(iii) the laboratory results show the inferred value of $D < 2$ only in the strain softening phase, and $D > 2$ only in the inferred strain hardening phase. These correlations are not seen in the field examples.

The first discrepancy is obviously important, since the b-value recovery first pointed out by Smith (1981) is a significant and well-documented feature of precursory seismicity. One possible explanation is the obvious differences in strain rate between laboratory and field conditions, possibly allowing a much greater influence of pore fluids during precursory deformation in earthquake sequences (e.g. Sibson 1981). Another technical explanation already mentioned above is the lack of stiffness in the loading system used for the laboratory experiments, so that the shear crack acceleration has been artificially forced due to the release of elastic strain energy stored outside the rock sample. This is reflected in a massive increase in measured strain rate in the strain-softening phase (Meredith *et al.* 1989).

However, the biggest difference in mechanical terms between laboratory rock samples and the Earth's continental crust is that the crust is inherently more heterogeneous, and already contains major faults and sutures. This may explain why the starting value for D is different. In particular, the first phase of dilatancy in intact rocks occurs when $D = 3$ and ends when $D = 2$, the value associated with random seismicity. In this phase, further crack growth is inhibited by local stress relaxation or dilatancy hardening. Random seismicity occurs when the probability of fracture at a particular site is independent of the location of the previous event ($D = 2$), and foreshocks occur when the probability of subsequent events is increased locally by stress concentration due to a previous event ($D < 2$). Thus the value of D is correlated with the idea of positive or negative feedback in the fracture process at various levels of stress concentration. In particular $D > 2$ is associated with dilatancy hardening and $D < 2$ with strain softening by crack acceleration in the intact laboratory samples described by the present work. In the Appendix, this change from negative to positive feedback at $D = 2$ is explained by a change in the dominance of total stress drop from relatively smaller ($D > 2$) to relatively larger ($D < 2$) events. For example, the experimental observation of $D > 2$ for cataclasic flow (Fig. 10) coincides with the inhibition of shear localization or strain softening at high confining pressures. Fluid influx or stress relief off the dominant fault results in an increase in D according to the model presented here, equivalent to a temporary reduction in positive feedback in the fracture process. Values of D near 1 imply a crack system with one dominant fault which takes up most of the deformation (representing a high degree of strain softening), whereas D near 2 corresponds to a more random process of distributed fracture without significant strain softening. A continuum of values between these limits can exist for strain-softened systems. One of our main aims in writing this paper is to encourage the routine measurement of crack distributions, so that as large a database as possible can be made available for comparison with laboratory results. In the meantime, a series of experiments using artificial pre-cut shear faults, with simulated asperities and barriers are planned to investigate the effect of a dominant pre-existing fault on the initial value of D.

Despite the differences in scale and the qualifications raised by the above discussion, there is a great deal of similarity between the laboratory and field observations of the time-dependence of the power-law exponent D. This can be seen as a direct consequence of the system striving to maximise the average fault length, and the presence of positive or negative feedback in the fracture process.

The picture for natural events is made more complicated because of the existence of large pre-existing faults and sutures, but both laboratory and field studies show a distinct short-term drop in b-value from $b \approx 1$ to $b \approx 0.5$ ($D \approx 2$ to $D \approx 1$) in the final phase of acceleration to dynamic rupture. Also, both laboratory and field results qualitatively show an intermediate-term and a short-term anomaly, and although these have different forms, the general model is capable of explaining both equally well.

Despite the obvious differences of scale and strain rate, the major differences between the laboratory and field results appears to be related to the precise mechanism of strain softening, and perhaps in particular with the activity of fluids being more predominant in the field results.

The results presented here give some hope of making inferences about the stress intensity of geological fracture systems, since stress intensity is negatively correlated to D. However, the length distribution in natural systems result from the cumulative effect over a longer time period, and will tend to produce D near 2 for a random process involving distributed damage, or 1 when a single throughgoing fault is produced. Nevertheless, comparative studies could yield important evidence of similarities or differences between different localities, particularly in the relative degree of distributed damage or strain softening.

We thank H. Shaw in particular for sending preprints, for permission to reproduce Figs 1–3, and for many stimulating correspondences on the subject of the fractal geometry of fault systems. The paper also benefitted significantly from suggestions made by two anonymous referees. Financial support was provided by the Natural Environmental Research Council (grant GR3/6812).

Appendix

The total seismic moment from a discrete frequency distribution of magnitudes m_j, $j = 1, n$ is

$$\sum_{j=1}^{n} M_{0_j} = \sum_{j=1}^{n} N_j\, 10^{cm_j+d} \qquad \text{(A1)}$$

N_j is the number of times a magnitude of size m_j is recorded, m_1 is the smallest and m_n is the largest magnitude, with the index j corresponding to a particular bin in the discrete frequency–magnitude distribution histogram. Formulae for c and d are given for a constant stress drop dislocation model by Kanamori & Anderson (1975) and Kanamori (1978). The discrete Gutenberg–Richter frequency–magnitude relation has the form

$$\log N_j = a - bm_j;\ a = \log N_1 \qquad \text{(A2)}$$

whence

$$\sum_{j=1}^{n} M_{0j} = 10^{a+d} \sum_{j=1}^{n} 10^{(c-b)m_j} \qquad \text{(A3)}$$

The contribution to the total moment is therefore dominated by higher magnitudes for $b < c$, or $D < 3$ from equation (2). This result is independent of the actual value of c. Thus the largest events always dominate the seismic moment in any magnitude catalogue.

However, the global stress drop, defined by

$$\Delta\sigma_G = \Delta\sigma l^2/A \qquad \text{(A4)}$$

has a different dependence on fault length l from that of the seismic moment

$$M_0 = C_0 \Delta\sigma l^3. \qquad \text{(A5)}$$

$\Delta\sigma$ here is a local stress drop, A is the area of the mainshock or the size of the laboratory sample, and C_0 is a constant which depends on fault type (Kanamori & Anderson 1975). The global stress drop is defined as the stress relief at some remote boundary due to dynamic slip on a fault or faults smaller than or equal to the area of the boundary. For a catalogue of events of different magnitudes

$$\Delta\sigma_G = \left[(\Delta\sigma)^{1/3}\, (10^{d}/C_0)^{2/3}\, 10^{a} \right] \sum_{j=1}^{n} 10^{[(2c/3)-b]m_j}. \qquad \text{(A6)}$$

In this case for $b = 2c/3$ or $D = 2$, the global stress drop is evenly distributed throughout the magnitude range (m_1, m_n); for $b > 2c/3$ or $D > 2$ the large events dominate the global stress drop and for $b < 2c/3$ or $D < 2$ the smaller events dominate. Again this result is independent of the actual value of c.

References

AKI, K. 1965. Maximum likelihood estimates of b in the formula log $N = a - bm$ and its confidence limits, *Bulletin of the Earthquake Research Institute of Tokyo University*, **43**, 237–239.

—— 1981. A probabilistic synthesis of precursory phenomena, *In*: SIMPSON, D. W. & RICHARDS, P. G. (eds) *Earthquake Prediction: an International Review. American Geophysical Union, Maurice Ewing series*, **4**, 566–574.

ANDERSON, O. L. & GREW, P. C. 1977. Stress corrosion theory of crack propagation with applications to geophysics. *Reviews of Geophysics and Space Physics*, **15**, 77–104.

ATKINSON, B. K. 1984. Subcritical crack growth in geological materials. *Journal of Geophysical Research*, **89**, 4077–4114.

—— & MEREDITH, P. G. 1981. Stress corrosion cracking of quartz: a note on the influence of chemical environment, *Tectonophysics*, **77**, T1–T11.

—— & —— 1987. The theory of subcritical crack growth with applications to minerals and rocks. *In*: ATKINSON, B. K. (ed.) *Fracture Mechanics of Rock*, Academic Press, London, 111–166.

AVILES, C. A., SCHOLZ, C. H. & BOATWRIGHT, J. 1987. Fractal analysis applied to characteristic segments of the San Andreas fault. *Journal of Geophysical Research*, **92**, 331–344.

BROWN, S. R. & Scholz, C. H. 1985. Broad bandwidth study of the topography of natural rock surfaces. *Journal of Geophysical Research*, **90**, 12575–12582.

CAPUTO, M. 1976. Model and observed seismicity represented in a two-dimensional space. *Annals of Geophysics* (Rome), **4**, 277–288.

CARTER, J. A. & BERG, E. 1981. Relative stress variations as determined by b-values from earthquakes in circum-Pacific subduction zones. *Tectonophysics*, **76**, 257–271.

GOWD, T. N. 1980. Factors affecting the acoustic emission response for triaxially compressed rock, *International Journal of Rock Mechanics and Mining Science*, **17**, 219–223.

HABERMANN, R. E. 1988. Precursory seismic quiescence: Past, present and future. *PAGEOPH*, **126**, 279–318.

HOLCOMB, D. J. 1981. Memory, relaxation and microfracturing in dilatant rock, *Journal of Geophysical Research*, **86**, 6235–6248.

HOWARD, K. A., AARON, J. M., BRABB, E. E. & BROCK, M. R. 1978. *Preliminary map of young faults in the U.S. as a possible guide to fault activity*. USGS field studies map MF-916.

IMOTO, M. & ISHIGURO, M. 1986. A Bayesian approach to the detection of changes in the magnitude-frequency relation of earthquakes. *Journal of Physics of the Earth*, **34**, 441–455.

KAISER, J. 1953. Erkenntnisse und folgerungen aus der Messung von gerauschen bei Zugbeanspruchung von Metallischen Werkstoffen, *Archiv für das Eisenhüttenwesen*, **24**, 43–45.

KANAMORI, H. 1978. Quantification of earthquakes. *Nature*, **271**, 411–414.

—— & ANDERSON, D. L. 1975. Theoretical bases of some empirical relations in seismology. *Bulletin of the Seismological Society of America*, **65**, 1073–1095.

KIKUCHI, M., MCNALLY, K. & TITTMAN, B. R. 1981. Machine stiffness appropriate for experimental simulation of earthquake processes, *Geophysical Research Letters*, **8**, 321–323.

KING, G. C. P. 1983. The accommodation of large strains in the upper lithosphere of the earth and other solids by self-similar fault systems: the geometrical origin of b-value. *PAGEOPH*, **121**, 761–815.

KURITA, K. & FUJII, N. 1979. Stress memory of crystalline rocks in acoustic emission, *Geophysical Research Letters*, **9**, 9–12.

LAWN, B. R. & WILSHAW, T. R. 1975. *Fracture of brittle solids*. Cambridge University Press, Cambridge.

MAIN, I. G., 1988. Prediction of failure times in the earth for a time-varying stress. *Geophysical Journal*, **92**, 455–464.

—— 1990. Quasi-static modelling of stress histories during the earthquake cycle: precursory seismic and aseismic stress release. *Geophysical Journal International*, **102**, 195–204.

—— & BURTON, P. W. 1984. Information theory and the earthquake frequency-magnitude distribution. *Bulletin of the Seismological Society of America*, **74**, 1409–1426.

—— & —— 1986. Long-term earthquake recurrence constrained by tectonic seismic moment release rates, *Bulletin of the Seismological Society of America*, **76**, 297–304.

—— & MEREDITH, P. G. 1989. Classification of earthquake precursors from a fracture mechanics model. *Tectonophysics*, **167**, 273–283.

——, —— & JONES, C. 1989. A reinterpretation of the precursory seismic b-value anomaly from fracture mechanics. *Geophysical Journal*, **96**, 131–138.

MANDELBROT, B. B. 1977. *Fractals: form, chance and dimension*. W. H. Freeman, San Francisco.

—— 1982. *The fractal geometry of nature*. W. H. Freeman, San Francisco.

MEREDITH, P. G. & ATKINSON, B. K. 1983. Stress corrosion and acoustic emission during tensile crack propagation in Whin sill dolerite and other basic rocks. *Geophysical Journal of the Royal Astronomical Society*, **75**, 1–21.

——, MAIN, I. G. & JONES, C. 1990. Temporal variations in seismicity during quasi-static and dynamic rock failure. *Tectonophysics*, **175**, 249–268.

PATERSON, M. S. 1978. *Experimental Rock Deformation – The Brittle Field*. Springer Verlag.

RICHTER, C. F. 1958. *Elementary seismology*. W. H. Freeman, San Francisco.

RIKITAKE, T. 1987. Earthquake precursors in Japan: precursor time and detectability. *Tectonophysics*, **136**, 265–282.

SATO, H. 1988. Temporal change in scattering and attenuation associated with earthquake occurrence, *PAGEOPH*, **126**, 465–497.

SCHOLZ, C. H. 1968. The frequency-magnitude relation of microfracturing in rock and its relation to earthquakes, *Bulletin of the Seismological Society of America*, **58**, 399–415.

SCHOLZ, C. H. 1988. Mechanisms of seismic quiescences. *PAGEOPH*, **126**, 701–718.

SCHWARTZ, D. P. & COPPERSMITH, K. J. 1984. Fault behaviour and characteristic earthquakes: examples from the Wasatch and the San Andreas fault zones. *Journal of Geophysical Research*, **89**, 5681–5698.

SHAW, H. R. & GARTNER, A. E. 1986. *On the graphical interpretation of palaeoseismic data*. United States Geological Survey Open File Report, 86-394.

SIBSON, R. H. 1981. Fluid flow accompanying faulting: field evidence and models. *In*: SIMPSON, D. W. & RICHARDS, P. G. (eds) *Earthquake prediction: an international review*. American Geophysical Union, Maurice Ewing Series, **4**, 593–603.

SMITH, W. D. 1981. The *b*-value as an earthquake precursor. *Nature*, **289**, 136–139.

—— 1986. Evidence for precursory changes in the frequency-magnitude *b*-value. *Geophysical Journal of the Royal Astronomical Society*, **86**, 815–838.

SONDERGELD, C. H., GRANRYD, L. A. & ESTEY, L. H. 1984. Acoustic emissions during compression testing of rock, *In*: *Proceedings of the 3rd Conference on Acoustic Emission/Microseismic Activity in Geological Structures and Materials*. 3rd Trans Tech, Clausthal, 131–145.

TCHALENKO, J. S. 1970. Similarities between shear zones of different magnitudes. *Geological Society of America Bulletin*, **81**, 1625–1640.

TURCOTTE, D. L. 1989. Fractals in Geology and Geophysics. *PAGEOPH*, **131**, 171–196.

VON SEGGERN, D. 1980. A random stress model for seismicity statistics and earthquake prediction. *Geophysics Research Letters*, **7**, 637–640.

WYSS, M. & HABERMANN, R. E. 1988. Precursory seismic quiescence. *PAGEOPH*, **126**, 319–332.

Brittle deformation and graphitic cataclasites in the pilot research well KTB-VB (Oberpfalz, FRG)

G. ZULAUF, G. KLEINSCHMIDT & O. ONCKEN

Geologisch-Paläontologisches Institut der Johann Wolfgang Goethe-Universität, Senckenberganlage 32–34, 6000 Frankfurt a.M., FRG

Abstract: Within the research well KTB-VB, late- to post-Variscan brittle deformation is widespread, indicating that the well is situated in a broad fault zone. Graphite-enriched foliated cataclasites within paragneisses are probably connected with late-Variscan NW trending reverse faults. These faults were succeeded by further, graphite-free, faults in post-Variscan time. During the formation of the graphitic cataclasites water-consuming retrograde metamorphism was active producing large amounts of phyllosilicates, especially at the expense of feldspar. The increase in phyllosilicates and graphite was associated with a switch of the dominant deformation mechanisms from brittle fracture to frictional sliding, crystal plasticity and diffusion-controlled processes. The cataclastic fabric suggests that the main part of the movement within the graphitic cataclasites was likely to have been aseismic rather than seismic.

The superdeep research well KTB is situated in the northern part of the Zone of Erbendorf–Vohenstrauss (ZEV). The polymetamorphic ZEV is probably a Variscan nappe covering the suture between the Moldanubian and Saxothuringian zones at the western border of the Bohemian Massif in Central Europe (Vollbrecht *et al.* 1989; see Fig. 1).

The pilot borehole KTB-VB (4000.1 m depth) is located 200 m to the west of the projected superdeep well site. The rocks comprise mainly paragneiss, amphibolite and metagabbro that belong to the ZEV (Müller *et al.* 1989; see Fig. 1). All rocks display features of strong brittle deformation indicating that the well penetrates a broad fault zone. Between discrete zones of ultracataclasite, the upper part of the borehole shows cataclastic zones up to 0.6 m thick. There are notable differences between the brittle shear zones in paragneisses and in metabasites (see also Simpson 1986). Within the metabasites, conjugate discrete shear zones are more common than in the paragneisses. Moreover a conspicuous enrichment of graphite within the brittle shear zones of the paragneisses is observed, whereas within the brittle shear zones of the metabasites, enrichment of graphite is lacking.

Occurrence of graphitic cataclasites within the brittle deformation sequence

In order to determine the attitude and slip direction of the faults, reorientation of the cores was necessary. Apart from the direct method of core orientation (a reference knife grooves the core), measurements of structural elements (e.g. foliation, joints, faults) with borehole televiewer (BHTV) and formation micro scanning tool (FMST) were used for the core orientation. By comparing the measured elements with observable structural elements of the core, reliable orientation of the core is possible (indirect method, see Schmitz *et al.* 1989). The reliability is shown by comparing these data with data of directly oriented cores (Schmitz *et al.* 1989).

In addition to macroscopic core observation and measurement of slickensides (slip plane and direction) 13 samples from the upper core interval (see Fig. 1) were investigated by thin-section observation, SEM-images and by X-ray diffraction analyses. By means of cross-cutting relationships of veins, faults, fault striae and type of mineralization within the different faults, we can distinguish several phases of brittle deformation within the upper core interval:

(1) opening of subvertical tension gashes commonly mineralized with prehnite;
(2) late Variscan evolution of a first generation of SW-dipping reverse faults (with graphite-enrichment);
(3) development of a second (more localized) generation of SSW-dipping reverse faults;
(4) subordinate NW-trending dextral strike-slip faults;
(5) development of discrete normal faults frequently along the previous reverse faults

From Knipe, R. J. & Rutter, E. H. (eds), 1990, *Deformation Mechanisms, Rheology and Tectonics*, Geological Society Special Publication No. 54, pp. 97–103.

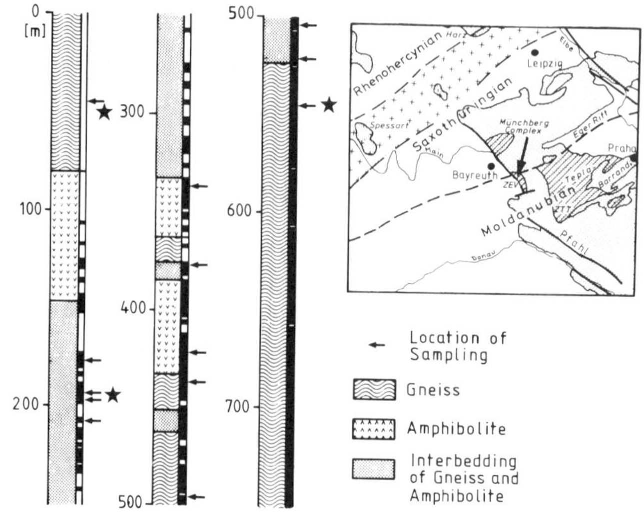

Fig. 1. Locations of investigated samples in the upper part of the well KTB-VB (on the left). Cored intervals are marked black. Stars indicate samples of graphitic cataclasites. A black arrow in the map (in the right-hand top) points to the location of the KTB well site. Simplified lithology of the well section after Müller *et al.* (1989). The north–south edge of the inset map is *c.* 350 km.

(reactivation) or along the foliation, which in most cases dips steeply to SW/SSW;

(6) development of small subhorizontal tension gashes (up to 0.5 cm in width) due to late phase uplift.

Within the nearby Falkenberg Granite, comparable fault populations and veins are observed. All faults therefore seem to be younger than the intrusion of the Falkenberg Granite (311 ± 4 Ma, see Wendt *et al.* 1986). In a few cases, the drilled rocks were intersected by dykes of lamprophyre which are probably of late Variscan age (e.g. Stettner 1981). Within these lamprophyres all fault populations mentioned above are present with the exception of the oldest subvertical prehnite-filled tension gashes and the graphite-enriched reverse faults. Thus, the tension gashes and the graphite-enriched faults seem to be of late Variscan age whereas the younger faults and veins must have developed in post-Variscan time.

Remarkable graphite enrichment is present only within the first generation of reverse faults (2). Among the brittle shear zones listed above, these graphitic reverse faults are the most prominent having accommodated most of the brittle shear strain. Macroscopically the graphitic shear zones are usually millimetre to a few centimetres thick ultracataclasites consisting of minor porphyroclasts (commonly < 10%) that are embedded in a black matrix (Fig. 2). Spacing of the individual graphitic shear zones varies throughout the entire drill-

Fig. 2. Black graphite-enriched cataclasites (lower and central part of picture) in paragneiss. Between the cataclasites synthetic shear planes of the Riedel type can be observed. Sample depth: *c.* 210 m.

hole. Intervals with a high density of graphitic shear zones (spacing ranges from a few centimetres to several decimetres) alternate with intervals with low density (spacing ranges from a few metres to several metres). The thickness of these intervals ranges from several metres to a few tens of metres. The wall rock within zones of high fault density is commonly altered to protocataclasite. In zones of low fault density, the wall rock is not deformed macroscopically or is altered to a crush breccia (*sensu* Sibson 1977). Along the graphitic zones, the core breaks easily and displays black graphitic slip surfaces (slicken-sides). These surfaces are either polished or they show different populations of striae reflecting reactivation of the fault after a change of the stress-field. The striae are wear grooves in the sense of Engelder (1974). Locally it can be observed that from graphitic ultracataclasites, graphite intrudes into foliation-parallel tension gashes, up to 4 mm long, in the paragneiss wall rock.

Mineralogy of the graphitic cataclasites

The intact wall rock (paragneiss) consists of plagioclase, quartz and biotite with subsidiary garnet, hornblende, kyanite and sillimanite (see also Müller *et al.* 1989). Within the brittle shear zones this mineral assemblage is intensely altered. The alteration during cataclastic deformation was associated with water influx contributing to an increase in volume of sericite and chlorite (both > 50 vol%) within the cataclastic zones. Increased volume of phyllosilicates is commonly observed in fine-grained brittle fault rocks (Rutter *et al.* 1986; Sammis *et al.* 1986; Janecke & Evans 1988). Additional new minerals present in the graphitic cataclasites are pyrite and potash feldspar. Laumontite, frequently present in the younger brittle shear zones, has not developed during the formation of the graphitic cataclasites.

The porphyroclasts consist of quartz and minor feldspar, calcite and garnet. Porphyroclast diameters range from 10 μm to a few millimetres. The diameters of the matrix grains are between 1 and 10 μm as can be observed in SEM-images.

Although the most frequent deformation mechanism observed in quartz is fracturing, undulose extinction as well as subgrain formation and grain boundary migration record subordinate crystal plasticity and recovery. Moreover, replacement of quartz by chlorite and calcite reflects diffusion-controlled mechanisms. There are two generations of calcite mineralization. An older one (pre- to synkinematic) displays a high density of twin lamellae and pressure-solution planes. It is found within gashes of the wall rock and as porphyroclasts (Fig. 3). A younger generation of calcite (post-kinematic) is free from twin lamellae and pressure-solution features. It was formed in fault-parallel gashes during late phase extension and opening of the weak cataclastic shear zones. Pressure solution features are also found in quartz, but are less intensely developed than in calcite (see Fig. 3). Plagioclase is commonly fractured. In most cases transcrystalline conjugate fractures were found that are not related to the crystallographic fabric. Furthermore intra-

Fig. 3. Photomicrograph (crossed polarizers) showing porphyroclasts of quartz (Q) and calcite (C) within a matrix of phyllosilicates and graphite. Section is cut perpendicular to the slip plane and parallel to the slip direction. In comparison to quartz, calcite is more intensely corroded by pressure solution (indicated by black arrows). Sample depth: 195 m.

crystalline fractures are present following chiefly the (001) cleavage planes. Crystal plastic deformation is indicated by bent twin lammellae, undulose extinction and mechanical twins. Alteration of plagioclase to sericite within the intact grains and chiefly along fractures leads to a decrease of plagioclase content within the graphitic cataclasites in comparison with the wall rock (see also Janecke & Evans 1988).

Fresh garnets are deformed by fracturing, but chloritized garnets are penetratively deformed and strongly flattened. Biotite is commonly chloritized, kinked and bent.

Fabric of the graphitic cataclasites

Elongate quartz porphyroclasts have a strong preferred orientation leading to a foliation of the graphitic cataclasites (Fig. 4). Foliated cataclasites have been described from numerous localities (Gay & Ortlepp 1979; Stel 1981; House & Gray 1982; Mitra 1984; Chester *et al.* 1985; Rutter *et al.* 1986; Simpson 1986; Wallace & Morris 1986; Chester & Logan 1986, 1987; Blenkinsop & Rutter 1986).

In order to determine the shape and orientation of the porphyroclasts, eight thin sections from four samples (see Fig. 1) were investigated by image analysis. Four thin sections were cut parallel to the fault zone. The other four sections were cut perpendicular to the fault zone and parallel to the slip direction. Measurements of the long and short axes of the porphyroclasts in both sections yield flattened oblate bodies (k = 0.4).

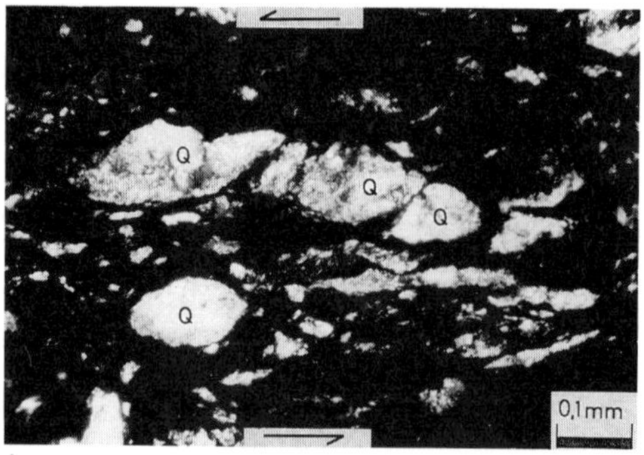

Fig. 4. Photomicrograph (crossed polarizers) of a foliated graphitic cataclasite. Elongated quartz porphyroclasts (Q) show healed fractures. Section is cut perpendicular to the slip plane and parallel to the slip direction. Sample depth: 195 m.

Furthermore the angle between the long axis of the porphyroclast and the slip plane (slip direction) were measured. In all sections cut perpendicular to the fault zone and parallel to the slip direction, a bimodal distribution of the long axes of the quartz porphyroclasts can be observed. The axes deviate symmetrically from the slip plane by an angle of 5° to 20° (Fig. 5). While the ellipticity ranges between 1 and 4, the mean values are between 1.5 and 2.5.

In all sections cut parallel to the slip plane, a similar but less distinct bimodal distribution of the long axes of the porphyroclasts with respect to the shear direction can be observed. Most of the long axes deviate from the slip direction by angles of 15° to 30° (Fig. 6). The ellipticity is notably less compared with sections cut perpendicular to the fault zone. In most cases the values range between approximately 1.2 and 2.0.

In sections cut perpendicular to the slip plane and parallel to the slip direction, secondary planes were frequently observed. There are Y-, R_1- and P-planes (Fig. 7; the terminology of these planes is adapted from Logan *et al.* 1979). Y-planes are synthetic shear planes oriented parallel to the slip direction. R_1-planes are synthetic shear planes (Riedel shears), arranged at inclinations of between 10° and 30° to the direction of relative movement. P-planes are arranged at inclinations of between 150° and 170° to the direction of relative movement (see also schematic sketch in Figs 5 and 6). Y-, R_1- and P-planes are also described from artificial fault gouges (Rutter *et al.* 1986; Logan & Rauenzahn 1987) as well as from natural fault gouges (Rutter *et al.* 1986). Moreover Chester & Logan (1987) observed local bimodal distributions of shear bands within fault gouges. Rutter *et al.*

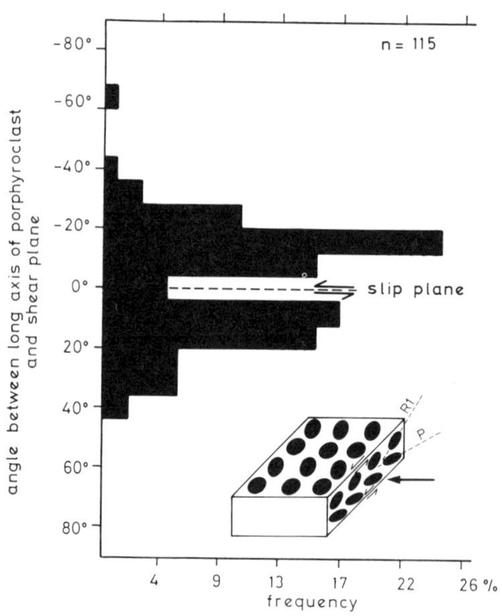

Fig. 5. Orientation distribution of long axes of porphyroclasts in relation to the slip plane. Same sample and section as in Fig. 4. The arrow in the schematic sketch (lower part of the diagram) points at the investigated section.

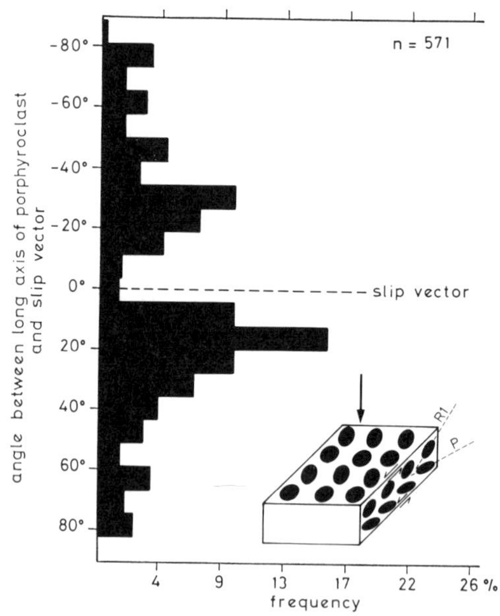

Fig. 6. Orientation distribution of long axes of porphyroclasts in relation to the slip direction. Section is parallel to the slip plane. The arrow in the schematic sketch (lower part of the diagram) points at the investigated section. Same sample as in Fig. 4.

(1986) emphasized that in clay-rich fault gouges a preferred orientation of clay platelets forms in the P-orientation. This can also be observed within the graphitic cataclasites. However, in most cases the P-planes in the graphitic shear zones are marked by pressure-solution planes as well as by quartz layers. The latter often show boudinage or pinch and swell structures (Fig. 7). The P-planes are commonly displaced by R_1-planes.

Possible origin and kinematic significance of cataclastic fabric

To interpret the cataclastic fabric, one must be aware that the dominant deformation mechanisms have changed during the formation of the graphitic cataclasites. In the early stage of the brittle strain path, fracture must have been the dominant mechanism as indicated by abundant cracks in quartz, plagioclase and garnet. Fracture processes were associated with grain-size reduction and dilatancy producing pathways for fluids (e.g. Sibson *et al.* 1975). Thus, alteration of primary minerals to fine-grained phyllosilicates as well as graphite enrichment were possible. The grain-size reduction and mineral change went along with a switch of the dominant deformation mechanisms leading to strain localization and strain softening (fabric and reaction softening; see White *et al.* 1986). The newly formed phyllosilicates were deformed by crystal plasticity and frictional sliding. Pressure solution and reaction fabrics are responsible for the development of the P-planes and the oblateness of the quartz porphyroclasts. The displacement of the P-planes by R_1-planes seems to be the reason for the observed symmetrical deviation of the long axis of porphyroclasts in sections cut perpendicular to the fault zone and parallel to the slip direction. By this R_1-displacement, quartz porphyroclasts, oriented parallel to the P-planes, were rotated into the R_1-direction. The interpretation of the orientation of porphyroclasts in fault-parallel sections is more difficult because shear planes in those sections are not very common.

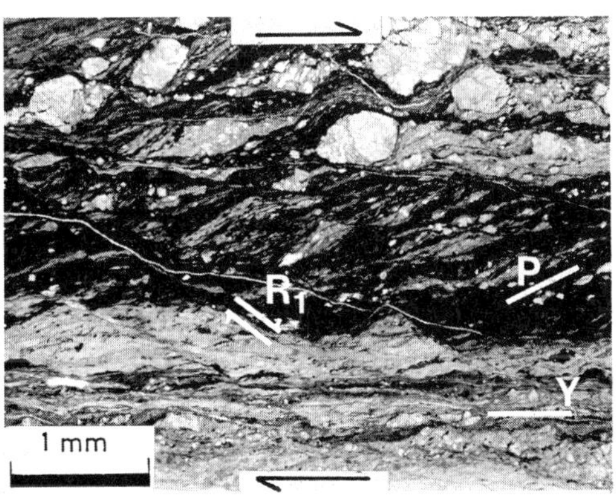

Fig. 7. Photomicrograph of a graphitic cataclasite (parallel nicols). Y-, R_1- and P-planes are indicated. Note boudinage besides pinch and swell structures of quartz layers in the P-orientation. Sample depth: 508 m.

Apart from the distinct symmetric fabric of the graphitic cataclasites, the relatively high portions of phyllosilicates and graphite suggest that the main part of the deformation within these zones was aseismic. The problem of distinguishing cataclasites of seismic and aseismic nature, due to their composition and fabric, was mentioned by Sibson (1986). However, the author emphasized that a preferred shape orientation of grains within cataclasites may rather develop by aseismic than by seismic slip. Engelder *et al.* (1975) as well as Byerlee *et al.* (1978) have deduced experimentally that shear planes in the orientation of R_1-shears originate preferably by aseismic creep within sheared gouge. Thus, the R_1-planes observed in the graphitic cataclasites can probably be interpreted in a similar manner. The kinematic significance of Y-planes is as yet not clear. Rutter *et al.* (1986) describe them as indicators of stick-slip events. On the other hand, Logan & Rauenzahn (1987) relate them to stable sliding of the gouge.

Conclusion

Because of their conspicuous foliation, at first sight the graphitic shear zones resemble mylonites (see Sibson 1977). However, apart from crystal plastic and diffusion-controlled processes during advanced periods of the development of the graphitic shear zones, the microscopical investigations have clearly shown that intense cataclasis were active, especially during the early stage of the brittle strain path.

The relationship between the observed secondary planes (P- and R_1-planes) and the orientation of the porphyroclasts in sections cut perpendicular to the fault plane and parallel to the slip direction is obvious. However, in sections, cut parallel to the slip plane, explanation of the porphyroclast orientation is as yet not possible because of lacking planes. A third section, cut perpendicular to the slip plane and perpendicular to the slip direction, has to be investigated. Results from this section will probably give more information about the spatial distribution of the shear planes and the porphyroclasts. The three-dimensional orientation of shear planes and porphyroclasts seems to be of great kinematic interest, especially for the determination of the kind of movement (stable or unstable) within fine-grained gouges and ultracataclasites.

The lack of laumontite and of recrystallization in quartz indicates a formation temperature of the graphitic cataclasites of between approximately 250° and 300° (see Liou *et al.* 1987; Voll 1976). If we assume an elevated heat flow during late-Variscan time (e.g. Lorenz & Nicholls 1984), the pressures may have ranged between approximately 1.5 and 2.0 kbar.

The research reported here was supported by the Deutsche Forschungsgemeinschaft (DFG) (grant No.: Kl 429/7–2).

References

BLENKINSOP, T. G. & RUTTER, E. H. 1986. Cataclastic deformation of quartzite in the Moine thrust zone. *Journal of Structural Geology*, **8**, 669–681.

BYERLEE, J. D., MJACHKIN, V., SUMMERS, R. & VOEVODA, O. 1978. Structures developed in fault gouge during stable sliding and stick-slip. *Tectonophysics*, **44**, 161–171.

CHESTER, F. M. & LOGAN, J. M. 1987. Composite planar fabrics of gouge from the Punchbowl Fault, California. *Journal of Structural Geology*, **9**, 621–634.

——, FRIEDMAN, M. & LOGAN, J. M. 1985. Foliated cataclasites. *Tectonophysics*, **111**, 139–146.

ENGELDER, J. T. 1974. Microscopic wear grooves on slickensides: indicators of paleoseismicity. *Journal of Geophysical Research*, **79**, 4387–4392.

——, LOGAN, J. M. & HANDIN, J. 1975. The sliding characteristics of sandstone on quartz fault-gouge. *Pure and Applied Geophysics*, **113**, 68–86.

GAY, N. C. & ORTLEPP, W. D. 1979. Anatomy of mining induced fault zone. *Geological Society of America Bulletin*, **90**, 47–58.

HOUSE, W. M. & GRAY, D. R. 1982. Cataclasites along the Saltville thrust, USA and their implications for thrust-sheet emplacement. *Journal of Structural Geology*, **4**, 257–269.

JANECKE, S. U. & EVANS, J. P. 1988. Feldspar-influenced rock rheologies. *Geology*, **16**, 1064–1067.

LIOU, J. G., MARUYAMA, S. & CHO, M. 1987. Very low-grade metamorphism of volcanic and volcaniclastic rocks — mineral assemblages and mineral facies. *In*: FREY, M. (ed.) *Low temperature metamorphism*. Chapman & Hall, New York, 59–113.

LOGAN, J. M. & RAUENZAHN, K. A. 1987. Frictional dependence of gouge mixtures of quartz and montmorillonite on velocity, composition and fabric. *Tectonophysics*, **144**, 87–108.

——, FRIEDMAN, M., HIGGS, N. G., DENGO, C. & SHIMAMOTO, T. 1979. Experimental studies of simulated gouge and their application to studies of natural fault zones. *Proceedings of Conference VIII, Analysis of actual fault zones in bedrock*. US Geological Survey Menlo Park, California, 305–343.

LORENZ, V. & NICHOLLS, J. A. 1984. Plate and intraplate processes of Hercynian Europe during the late Paleozoic. *Tectonophysics*, **107**, 25–56.

MITRA, G. 1984. Brittle to ductile transition due to large strain along the White Rock thrust, Wind River mountains, Wyoming. *Journal of Structural Geology*, **6**, 51–61.

MÜLLER, H., KEYSSNER, S. & RÖHR, C. 1989. Kontinentales Tiefbohrprogramm der Bundesrepublik Deutschland. Geowissenschaften im KTB-Feldlabor — Geologie — Geologisches Profil (0–3900 m) und Entwicklung der Gesteine. *KTB Report 89–3*.

RUTTER, E. H., MADDOCK, R. H., HALL, S. H. & WHITE, S. H. 1986. Comparative microstructures of natural and experimentally produced clay-bearing fault gouges. *Pure and Applied Geophysics*, **124**, 3–30.

SAMMIS, C. G., OSBORNE, R. H., ANDERSON, J. L., MAVONWE, B. & WHITE, P. 1986. Self-similar cataclasis in the formation of fault gouge. *Pure and Applied Geophysics*, **124**, 53–78.

SCHMITZ, D., HIRSCHMANN, G., KOHL, J., RÖHR, C. & WOHLGEMUTH, L. 1989. Kontinentales Tiefbohrprogramm der Bundesrepublik Deutschland. Die Orientierung von Bohrkernen in der KTB-Vorbohrung. *KTB Report 89–3*.

SIBSON, R. H. 1977. Fault rocks and fault mechanisms. *Journal of the Geological Society, London*, **133**, 191–213.

—— 1986. Brecciation processes in fault zones: Inferences from earthquake rupturing. *Pure and Applied Geophysics*, **124**, 159–175.

——, MOORE, J. M. & RANKIN, A. H. 1975. Seismic pumping — a hydrothermal fluid transport mechanism. *Journal of the Geological Society, London*, **131**, 653–659.

SIMPSON, C. 1986. Fabric development in brittle- to ductile shear-zones. *Pure and Applied Geophysics*, **124**, 269–288.

STEL, M. 1981. Crystal growth in cataclasites: diagnostic structure and implications. *Tectonophysics*, **78**, 585–600.

STETTNER, G. 1981. Plutonite und Ganggesteine der variskischen Ära. *In*: BAYERISCHES GEOLOGISCHES LANDESAMT (ed.) *Erläuterungen zur geologischen Karte von Bayern 1:500000*. 28–29.

VOLL, G. 1976. Recrystallization of quartz, biotite and feldspars from Erstfeld to the Leventina Nappe Swiss Alps, and its geological significance. *Schweiz. mineral. petrogr. Mitteilungen*, **56**, 641–647.

VOLLBRECHT, A., WEBER, K. & SCHMOLL, J. 1989. Structural model for the Saxothuringian-Moldanubian suture in the Variscan basement of the Oberpfalz (Northeastern Bavaria, F. R. G.) interpreted from geophysical data. *Tectonophysics*, **157**, 123–133.

WALLACE, R. E. & MORRIS, H. T. 1986. Characteristics of faults and shear zones in deep mines. *Pure and Applied Geophysics*, **124**, 107–125.

WENDT, I., KREUZER, H., MÜLLER, P. & SCHMID, H. 1986. Gesamtgesteins-und Mineraldatierungen des Falkenberger Granits. *Geologisches Jahrbuch*, **E34**, 5–66.

WHITE, S. H., BRETAN, P. G. & RUTTER, E. H. 1986. Fault-zone reactivation: kinematics and mechanisms. *Philosophical Transactions of the Royal Society, London*, **A317**, 81–97.

Brecciation and fracturing within neotectonic normal fault zones in the Aegean region

IAIN S. STEWART & PAUL L. HANCOCK

Department of Geology, University of Bristol, Wills Memorial Building, Queen's Road, Bristol BS8 1RJ, UK

Abstract: Neotectonic normal fault zones cutting bedrock carbonates within the Aegean extensional province contain contrasting breccia types that record different stages of fault zone evolution at shallow (< 500 m) crustal levels. Initial blind faulting at depth produces a near-surface shatter zone. Locally, blocks within the shatter zone become increasingly disorganized as slip planes advancing through the zone, create incohesive breccia belts ahead of propagating tips. Mineralization and frictional wear accompanying fault movement form narrow compact breccia sheets adjacent to slip planes. In some major fault zones, fault movement is concentrated along metre-wide zones of intensely deformed stylobreccia. When slip planes reach the free surface, post-seismic stress release initiates a new fracture network that is superimposed on and restricted to breccia belts and sheets.

Fault breccias are generally regarded, along with microbreccia and fault gouge, as cataclastic products of brittle faulting at high crustal levels (e.g. Sibson 1977). Increasingly, however, it is recognized that fault breccias exhibit a range of textures that reflect the contrasting deformational regimes operating at shallow levels within fault zones (Sibson 1986; Hancock & Barka 1987; Stewart & Hancock 1988). In the Aegean extensional province, investigation of ten neotectonic normal fault zones (three in the Corinth region, four in Crete and three in western Turkey) of varying scales but all in carbonate bedrocks, demonstrates a range of fault zone architectures that reflect differing deformation histories.

Description of breccia types

Shatter zones

While several studies in the Aegean region have recognised the association between neotectonic faults and broad belts of brecciated and fractured bedrock (e.g. Vita-Finzi & King 1985; Hancock & Barka 1987), there has been little investigation of the structure of such zones. Detailed examination of a number of shatter zones accompanying neotectonic fault zones in the Aegean region reveals a characteristic arrangement of regularly spaced (0.2–0.5 m), vertical to near-vertical fractures, striking roughly parallel with and normal to the fault zone (Fig. 1). Lack of consistent abutting relationships between members of the two fault-related fracture sets indicate their contemporaneity.

Although bedding is obscured by fracturing within decametres of a fault, in adjacent areas where bedding is better preserved, fractures are largely dip- and strike-joints, that is, fractures striking parallel with and normal to the dip of the rocks. Where bedding is oblique to a fault zone, fault-parallel and fault-normal fractures occur as short cross-fractures that abut the more dominant dip- and strike-joints. Often, however, the orientation of bedding is fault-controlled, with beds, and the fractures contained within them, being tilted in the direction of downthrow across a normal fault. Within the resulting orthogonal fracture network it is difficult to distinguish between bedding-related and fault-related fractures.

Compact breccia sheets and incohesive breccia belts

Stewart & Hancock (1988) showed that many fault scarps in the Aegean region are underlain by zone-parallel layers of contrasting fault breccia types (Fig. 1). Immediately below slip planes are 3–30 cm thick sheets of well indurated and erosionally resistant fault breccia. These *compact breccia sheets* comprise a mosaic of small (0.1–0.5 cm), sub-rounded to sub-angular carbonate clasts supported in a matrix of comminuted clast fragments and secondary calcite cement.

From Knipe, R. J. & Rutter, E. H. (eds), 1990, *Deformation Mechanisms, Rheology and Tectonics*, Geological Society Special Publication No. 54, pp. 105–112.

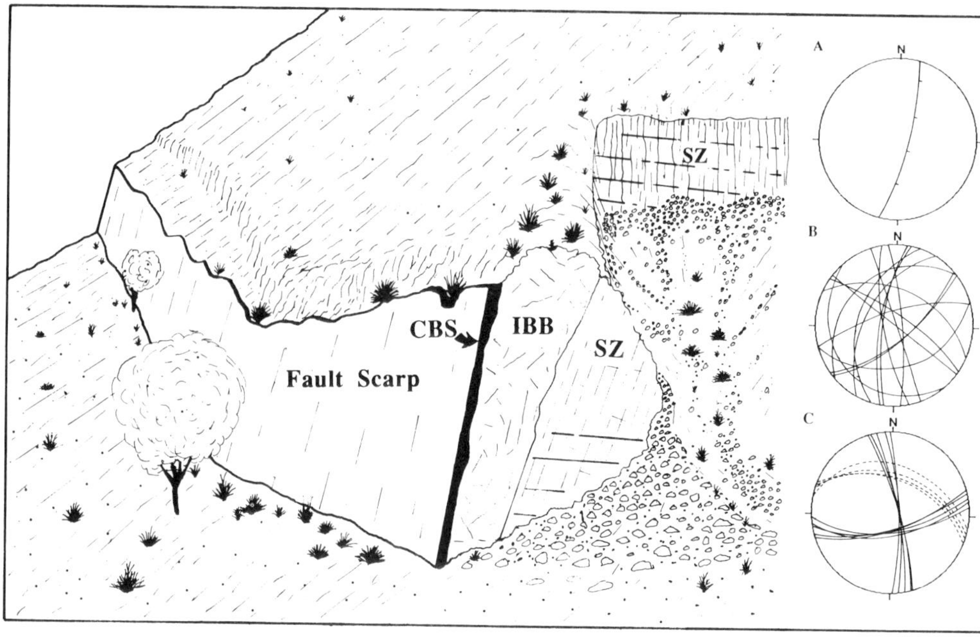

Fig. 1. View, looking southwest, of a section through a 6 m high fault scarp within the Lastros fault zone (Crete) showing the characteristic layered architecture of a compact breccia sheet (CBS), an incohesive breccia belt (IBB) and a broad shatter zone (SZ). Stereoplots are equal-area, lower-hemisphere diagrams that contrast the chaotic arrangement of fractures within incohesive breccias (B) with the orthogonal network of fractures that characterize the shatter zone (C). These regularly arranged shatter zone fractures trend roughly parallel with and normal to the principal slip plane (A). The solid great circle with ticks (downthrown side) depicts the principal slip plane, solid great circles represent fractures, and pecked great circles show bedding.

Underlying compact breccia sheets, and separated from them by a sharp but irregular surface, are 1–2 m wide belts of disaggregated and relatively erodible fault breccia. These *incohesive breccia belts* comprise coarse (0.5–8.0 cm), angular carbonate clasts cut by a network of chaotic and closely spaced (0.5–5.0 cm) fractures. The chaotically arranged fractures within incohesive breccia belts do not penetrate overlying compact breccia sheets, indicating that compact breccias formed after incohesive breccias.

Stylobreccias

Although the longer the history of tectonic activity within a fault zone the more complex its architecture, many major fault zones are only larger-scale equivalents of minor fault scarps. At the base of some major fault zones, however, decametre-high fault scarps are underlain by moderately dipping zones of highly resistant fault rock, here called *stylobreccia*. A section through a 12 m wide stylobreccia zone in the Pisia fault zone (Fig. 2) shows it that it comprises a network of closely-spaced, sub-parallel slip planes, dipping at 38–43° and bounding 0.5–2.5 m thick sheets of stylobreccia. Stylolitic sutures and widespread calcite recrystallization within the uppermost stylobreccia sheets indicate that there has been intense deformation. A 1–2 m thick layered marble complex, lying in the footwall of the slip plane reactivated in the 1981 Corinth earthquake sequence, is interpreted here as a product of fault movement having been concentrated along this relatively narrow zone. Overlying this complex, blocks of layered marble have been incorporated into stylobreccias, presumably as a consequence of slip plane activity migrating towards the active slip plane (reactivated in 1981). Later deformation has overprinted a regular network of near-vertical extension fractures on the uppermost stylobreccia fabric. This regular network contrasts with the chaotic fracture pattern in the incohesive breccia belt underlying the stylobreccia zone.

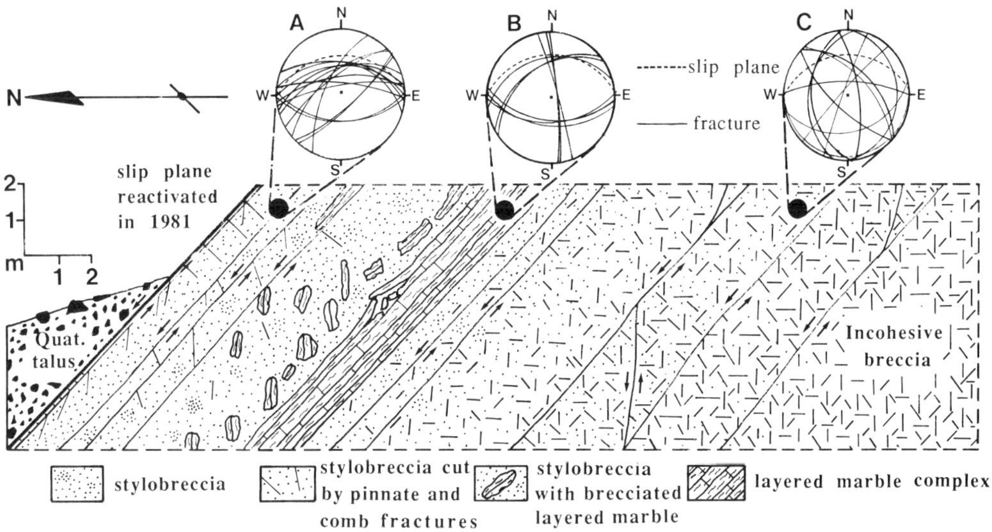

Fig. 2. Section through a stylobreccia zone at Pisia (Corinth). Stereoplots (equal-area, lower-hemisphere) of great circles of slip planes and other fractures, demonstrate how fault rock fabric within the zone grades from chaotically fractured incohesive breccia (C) in the footwall to increasingly recrystallized and deformed stylobreccias underlying the active slip plane. Both a highly deformed 1−2 m thick layered marble complex (B), interpreted as a former zone of fault movement, and the uppermost stylobreccia sheets (A) are cut by a regular arrangement of near-vertical extension fractures, produced by deformation along slip planes at the Quaternary sediment/bedrock contact. Modified from Stewart & Hancock (1988, fig. 11).

Comb fractures and slip-parallel fractures

Commonly, the uppermost stylobreccia or compact breccia carapace on many fault scarps in the Aegean region is cut by a regular network of extension fractures. Hancock & Barka (1987) called closely-spaced slip-normal fractures cutting compact breccia sheets *comb fractures* because, in profile, they subtend slip planes at 75° or more and, therefore, resemble the teeth of a hand-comb. Such small faults, joints and fissures are also common in other neotectonic normal fault zones cutting carbonate bedrocks in the Aegean region (Stewart & Hancock 1988).

In addition to comb fractures, slip planes are also cut by sub-vertical fractures that strike normal to fault zones and give rise to intersection lineations on slip planes that are roughly parallel to the slip vector. Like some comb fractures, some of these *slip-parallel fractures* offset principal slip planes and cut brecciated colluvium in their hangingwalls, indicating that fracture propagation and displacement post-dates the most recent fault movement.

Commonly, comb fractures and slip-parallel fractures occur together although frequently one set is dominant and the other poorly developed or absent (Fig. 3). In particular, where slip planes are corrugated (undulations of a slip plane, the long axes of which are oriented parallel to the direction of slip) comb fractures are preferentially developed on corrugation crests while slip-parallel fractures are largely

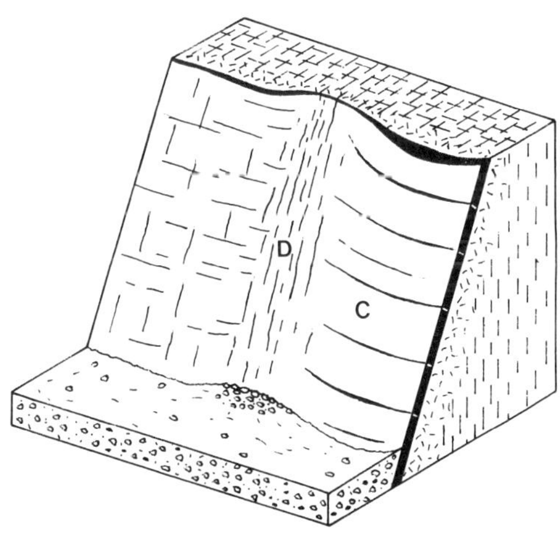

Fig. 3. Geometry of fractures developed as a consequence of post-seismic stress release. Comb fractures (C) preferentially occur on corrugation crests while slip parallel fractures (D) are best developed in corrugation troughs. See text for explanation.

restricted to corrugation troughs. On uncorrugated segments of fault scarps mutually abutting relations indicate that the orthogonal fractures belong to roughly coeval sets.

Model of brecciation and fracturing within 'Aegean-type' fault zones

Aegean-type fault zones, that is, neotectonic normal fault zones cutting carbonate bedrocks in the Aegean region, are characterized by a range of fault rock types which indicate the operation of contrasting deformation processes. Following Vita-Finzi & King (1985) and Petit & Barquins (1988) we interpret broad belts of brecciated and shattered rocks as products of distributed deformation above and ahead of upwards-propagating tip lines of normal faults (Fig. 4a). According to Hancock & Barka (1987) such belts of fault-precursor deformation are likely to be developed in hangingwalls, which preferentially experience dilational deformation. Observations in shatter zones (Petit & Barquins 1988; this study) indicate that initial fault-precursor fracturing reactivates preexisting dip- and strike-joints in the carbonate cover to produce an orthogonal network of well-developed, sub-vertical fractures. Subsequent deformation, however, may warp this fault-precursor fracture pattern and locally overprint a fault-parallel/fault-normal fracture pattern. Although fractures striking parallel to fault zones are common and express extension perpendicular to a fault zone (Angelier & Colleta 1983), orthogonal networks of roughly contemporaneous extension fractures or faults have been described from only a few neotectonic domains (e.g. Hancock *et al.* 1987; Doutos *et al.* 1988) where it is thought they reflect repeated and short-lived 90° switches in the direction of greatest extension.

Later motion on a fault as it propagates towards the free surface leads to additional deformation in the immediate vicinity of the fault tip (Fig. 4b). This motion brecciates the already fractured bedrocks, giving rise to a complex of closely-spaced and randomly oriented fractures and ultimately creating incohesive breccia belts. Once a slip plane reaches the free surface, by propagating through its own incohesive breccia belt and shatter zone, coseismic deformation is restricted to a relatively narrow zone of attrition and mineralisation adjacent to the slip plane, thus forming compact breccia sheets. Sibson (1986) and Petit & Barquins (1988) have also described cataclastic zones developing as slip planes propagate through bedrocks cut by preexisting subsidiary shear and extension fractures of varying size and orientation.

The observation that in other regions, such as central Arabia (Hancock *et al.* 1987), there are no shatter zones or incohesive breccia belts accompanying high crustal level but older Cenozoic normal faults in carbonates from which about 500 m of cover has been denuded, is taken as indicating that fault-precursor fracturing and brecciation occurs only in the epidermis of the brittle crust. Compact breccia sheets accompanying the exhumed Arabian faults indicate that attrition and mineralization continue to accompany fault movement at depths greater than 500 m.

Because the active or youngest slip plane within Aegean normal fault zones is commonly the uppermost one separating footwall brecciated carbonates from hangingwall Quaternary deposits, it is inferred that there is a hangingwall-directed migration of slip plane activity. This tendency, called *intrafault-zone hangingwall-collapse* by Stewart & Hancock (1988), is initiated by an upward-propagating normal fault tip seeking a pathway through its own fault-precursor breccia belt (preferentially developed in hangingwalls). The process results in a fault scarp developing a stepped profile and a fault zone developing a layered architecture (Fig. 4c).

Although many range fronts display a similar *step-type* morphology, those that display moderately-dipping, planar stylobreccia zones at their bases exhibit a *ramp-type* morphology (Stewart & Hancock 1990). According to Stewart & Hancock (1990), the presence or absence of stylobreccias within a fault zone is related to whether or not there is a salient block of uneroded bedrock cover remaining in the hangingwall at the time the free surface is being displaced. Thus it is noteworthy that 20–80 m thick blocks of Mesozoic flysch overlie stylobreccia zones in the western parts of the Pisia (Corinth) and Furnofarango (Crete) fault zones, whereas such blocks are absent in the eastern parts of these fault zones where step-type scarps occur. It is suggested that such *hangingwall salients* inhibit the tendency of slip planes to propagate into the more brecciated hangingwall and, instead, concentrate deformation along the main fault plane, thereby creating metre-wide stylobreccia zones (Fig. 4d). In contrast, where a pre-faulting cover has been eroded, the relative lack of overburden in the uppermost part of the hangingwall leads to the pathway to the surface being along high-angle faults that splay into the hangingwall from the main fault

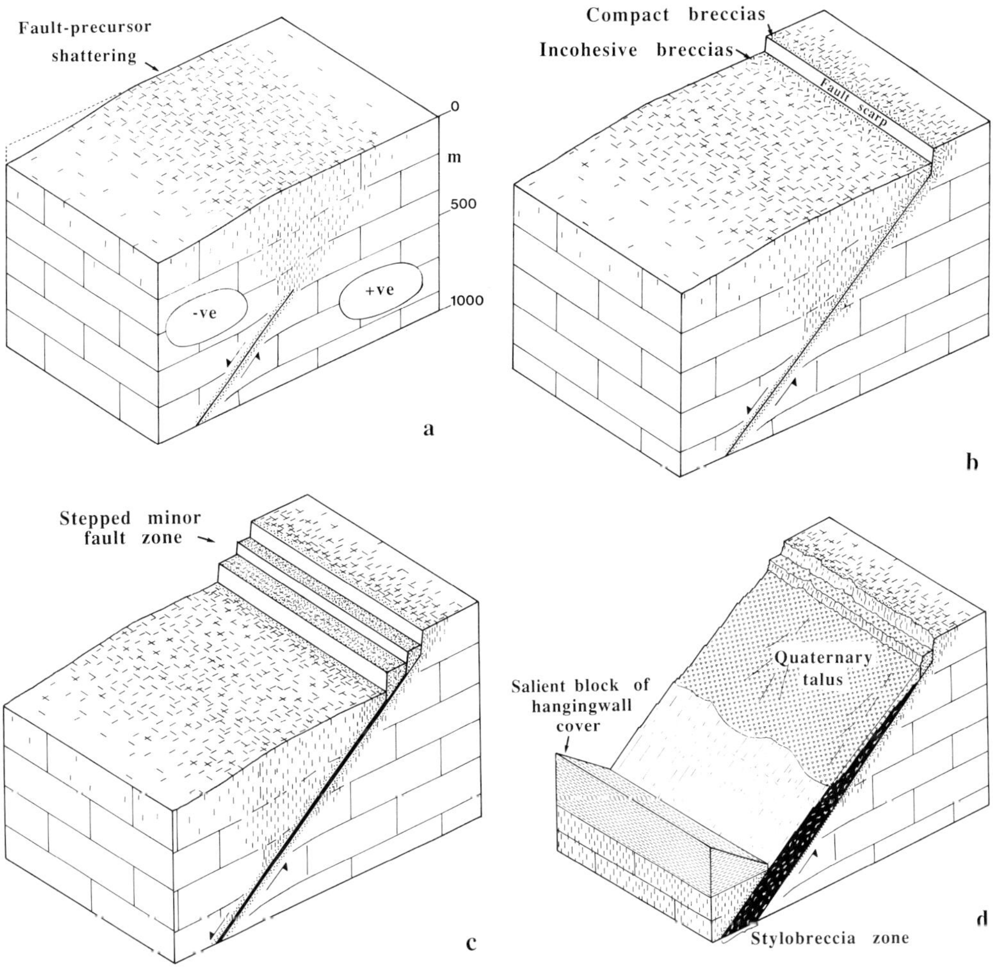

Fig. 4. (**a**) Deformation ahead of an upward-propagating normal fault tip-line. Vertical to near vertical fault-precursor fractures striking parallel with and normal to the trend of a blind fault at depth, are developed in the uppermost 200–500 m of the hangingwall, which experiences dilational (negative) deformation. (**b**) Propagation of a slip plane to the surface through its own incohesive breccia belt, developed from a shatter zone. A narrow zone of attrition and mineralization, defined by a compact breccia sheet, occurs adjacent to the slip plane. (**c**) The progressive migration of slip plane activity into the hangingwall (intrafault-zone hangingwall-collapse) in a minor fault zone results in a stepped scarp underlain by alternating, zone-parallel attrition breccia sheets and fault-precursor breccia belts. (**d**) Evolution of a ramp-type range front as a consequence of a salient block of pre-faulting cover in the hangingwall inhibiting intrafault-zone hangingwall-collapse and guiding the main fault to emerge at the surface along an intensely deformed stylobreccia zone.

plane. The resulting distributed deformation produces less intensely reconstituted fault rocks.

Although shatter belts are regarded as preceding the main development of a fault zone, it is clear that distributed fracturing continues to accompany faulting during later stages of fault zone evolution. For example, the layered marble complex within the Pisia stylobreccia zone is cut by a fracture network comparable to that developed in shatter belts (Fig. 2). In addition, when a slip plane locally defines the free surface some preexisting fractures are reactivated as small faults or fissures that offset the slip plane a few centimetres. While comb fractures are regarded as tension cracks reflecting down-dip stretching of a compact breccia carapace during localised post-slip stress reorientation (Hancock & Barka 1987), slip-parallel fractures are indistinguishable in orientation from fault-normal fractures developed in the footwall block. It is suggested that because all fault zones are subjected to some degree of intrafault-zone hangingwall-collapse, distributed fracturing continues to characterise fault zones throughout their evolution.

Conclusions

Different stages of neotectonic fault zone evolution in carbonate bedrocks in the Aegean region are characterized by contrasting deformational processes and products. Initial deformation ahead of an upward-propagating normal fault tip-line is accommodated by the formation of a broad fault-precursor shatter belt containing an orthogonal fracture network. Later motion propagates the slip plane through the shattered bedrock, locally brecciating the shatter belt ahead of the slip plane and forming a fault-precursor breccia belt. Frictional wear and mineralisation accompanying increments of motion on throughgoing slip planes comminute and cement fault-precursor breccias into indurated attrition breccia sheets immediately adjacent to slip planes. The progressive migration of slip plane activity into the hangingwall (i.e., intra-fault-zone hangingwall-collapse) results in all minor, and some major, fault zones possessing a layered architecture, comprising alternating zone-parallel attrition breccia sheets and fault precursor breccia belts. Within some major fault zones, however, repeated movement along the same slip plane, as a result of confinement by an uneroded hangingwall salient of pre-faulting cover, produces a metre to decametre-wide zone of intensely deformed stylobreccia. Finally, post-seismic stress release overprints an orthogonal fracture network on the co-seismic deformation structures.

Both authors thank NERC and the University of Bristol for financial support.

References

Angelier, J. & Colletta, B. 1983. Tension fractures and extensional tectonics. *Nature*, **301**, 49–51.

Doutos, T., Kontopoulos, N. & Poulimenos, G. 1988. The Corinth-Patras rift as the initial stages of continental fragmentation behind an active island arc (Greece). *Basin Research*, **1**, 177–190.

Hancock, P. L. & Barka, A. A. 1987. Kinematic indicators on active normal faults in western Turkey. *Journal of Structural Geology*, **9**, 573–584.

——, Al-kadhi, A., Barka, A. A. & Bevan, T. G. 1987. Aspects of analysing brittle structures. *Annales Tectonicæ*, **1**, 5–19.

Petit, J. P. & Barquin, M. 1988. Can natural faults propagate under mode II conditions? *Tectonics*, **7**, 1243–1256.

Sibson, R. H. 1977. Fault rocks and fault mechanisms. *Journal of the Geological Society, London*, **133**, 191–213.

—— 1986. Brecciation processes in fault zones: inferences from earthquake rupturing. *Pure and Applied Geophysics*, **124**, 156–175.

Stewart, I. S. & Hancock, P. L. 1988. Normal fault zone evolution and fault scarp degradation in the Aegean region. *Basin Research*, **1**, 137–153.

—— & —— 1990. Neotectonic range-front fault scarps in the Aegean region. *In*: *Neotectonics and Resources: Proceedings of the Conference on Economic Geology and Geotechnics of Active Tectonic Regions, London*, (In Press)

Vita-finzi, C. & *King*, G. C. P. 1985. The seismicity, geomorphology and structural evolution of the Corinth area of Greece. *Philosophical Transactions of the Royal Society of London*, **A314**, 379–407.

Mechanical compaction and the brittle–ductile transition in porous sandstones

TENG-FONG WONG

Department of Earth and Space Sciences, State University of New York, Stony Brook, NY 11794, USA

Abstract: Grain scale brittle fracture is an important mechanism for porosity reduction and tectonic deformation in rocks under diagenetic and low grade metamorphic conditions. Microstructural observations and acoustic emission measurements were used to elucidate the micromechanical processes involved (1) in pressure-induced grain crushing, and (2) in the brittle-ductile transition in porous sandstones. Experimental observations show that the critical effective pressure for the onset of grain crushing decreases as a function of increasing grain size and of increasing porosity. The correlation among grain crushing pressure, porosity and grain size can be interpreted using a Hertzian fracture mechanics model. Experimental observations show that both porosity and effective pressure have profound influence on the inelasticity and failure mode. Based on such observations, a brittle–ductile transition map for porous rocks can be constructed in the effective pressure-porosity space. The theoretical predictions of a fracture mechanics model of a pore-emanated crack are in qualitative agreement with experimental observations on the effect of pressure and porosity on inelastic behaviour and failure mode. However, such a model overestimates the crack stabilization effect due to confining stresses since it neglects grain rotation and pore collapse which are important deformation mechanisms in porous rocks. Implications of the experimental and theoretical results on depth–porosity relation, aseismicity at the shallow region of subduction zones, and faulting in sandstones are discussed.

Brittle fracture on the grain scale has been recognized as an important micromechanical process for porosity reduction and tectonic deformation in rocks under diagenetic and low grade metamorphic conditions. Traditionally the concepts of soil mechanics are used for the mechanical analysis of the deformation processes operative before sedimentary rocks were lithified, whereas post-lithification deformation is considered to be a typical rock mechanics problem (Jones & Preston 1987). The phenomenological study of inelastic deformation and failure modes of porous rocks have been actively pursued in rock mechanics laboratories for more than two decades. Some of the principal similarities and differences between the brittle behaviour of low porosity crystalline rocks and of porous rocks were discussed by Paterson (1978).

Fracture mechanics has been established as a very powerful technique for the understanding of brittle failure processes in materials science (Kanninen & Popelar 1985). Recently the technique has been applied to crustal scale tectonic processes, providing important physical insights on earthquake mechanics (e.g. Rice 1980) and structural geology (e.g. Pollard 1987). It has also been used with some success to model micromechanical processes in low porosity crystalline rocks, such as stress-induced dilatancy (e.g. Horii & Nemat-Nasser 1986), thermal cracking (Fredrich & Wong 1986), and fluid inclusion decrepitation (Wanamaker *et al.* 1990). In contrast, grain scale brittle cracking processes in porous rocks have not been analyzed using fracture mechanics in a systematic manner.

This paper summarizes our recent experimental results on the inelastic deformation and failure modes of porous sandstones over a wide range of pressures and porosities. In addition to the standard phenomenological measurements, we also focus on (i) the use of nondestructive test techniques (such as acoustic emission measurement) and microstructural observations to characterize the micromechanics, and (ii) the use of fracture mechanics to model the micromechanical processes. Two specific aspects of porous rock deformation will be discussed: (i) the hydrostatic compaction by grain crushing, and (ii) the brittle-ductile transition.

From Knipe, R. J. & Rutter, E. H. (eds), 1990, *Deformation Mechanisms, Rheology and Tectonics*, Geological Society Special Publication No. 54, pp. 111–122.

Grain crushing as a compaction mechanism

Sands and sandstones lose their primary porosity by three principal compaction mechanisms: mechanical compaction, intergranular pressure solution, and cementation. During initial stages of burial, mechanical compaction can reduce primary porosity by as much as one-third in quartzose sands and by even more in sands rich in lithic fragments (Houseknecht 1984). If the effective pressure (i.e. overburden pressure minus pore pressure) is relatively low, then mechanical compaction is principally from grain scale frictional slip and rotation, a process which has been studied comprehensively in soil mechanics (Lambe & Whitman 1969; Mitchell 1976). If the effective pressure is sufficiently high, then significant porosity reduction can be induced by grain crushing (e.g. Maxwell 1960). Since the critical pressure for the onset of grain crushing is higher than that relevant to geotechnical applications, grain crushing as a compaction mechanism has not received much attention in soil mechanics research.

Micromechanical processes revealed by microscopy and acoustic emission measurements

In hydrostatic compression experiments, an inflection point on the hydrostat is usually interpreted to be due to the onset of grain crushing (Brace 1978). Figure 1 shows the hydrostats of five quartzose sandstones with porosity ranging from 5% to 30% deformed at a fixed pore pressure of 10 MPa under fully 'drained' conditions. For the Weber sandstone which had the lowest porosity and smallest grain size, the sample stiffened monotonically as a function of pressure. There was appreciable hysteresis, but not much permanent porosity reduction after a loading-unloading cycle. For all the other sandstones with porosities greater than 10%, the sample initially stiffened as a function of pressure, and suddenly became extremely compressible when the hydrostat reached an inflection point, after which the sample gradually stiffened again. There was appreciable permanent porosity reduction. The critical pressure at the inflection point has a negative correlation with porosity and grain size. The Weber sandstone hydrostat did not show an inflection point possibly because the maximum pressure we used was not sufficiently high.

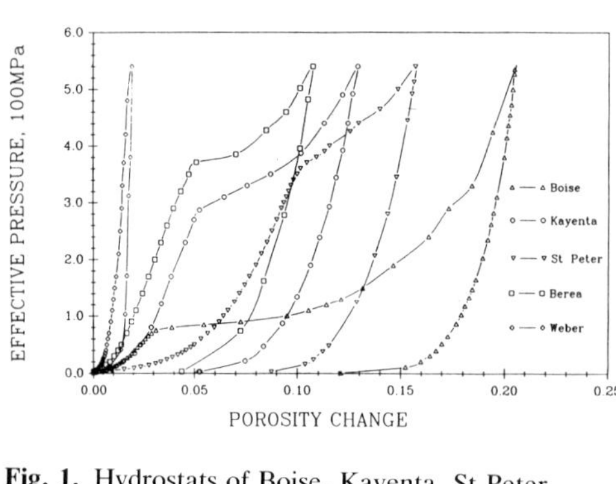

Fig. 1. Hydrostats of Boise, Kayenta, St Peter, Berea, and Weber sandstones in terms of effective pressure versus porosity change. All the hydrostatic tests were conducted at a pore pressure of 10 MPa under fully drained conditions.

Zhang *et al.* (1990) conducted detailed microstructural observations on the compacted samples. They concluded that although other deformation mechanisms (such as distributed cleavage cracking in calcite and intensive kinking in mica) were operative, the major micromechanical process was brittle grain crushing which initiated at a critical pressure corresponding to the inflection point. Microscopy observations of crushed quartz and feldspar grains reveal that the microfractures radiated from grain contacts and 'cone crack' patterns observed near the grain contacts are reminiscent of tensile indentation fractures usually referred to as 'Hertzian fractures' in materials science (Wilshaw 1971). Similar observations were reported previously by Gallagher *et al.* (1974), McEwan (1981), and Gallagher (1987). The grain crushing process introduces a complex network of intragranular cracks and provides additional degrees of freedom for grain rotation allowing some of the crushed grains to move into the pore space.

Although it can provide many details on the micromechanical processes, microscopy techniques have the disadvantage that observations can only be performed on samples after they have been unloaded and possibly had damage introduced by the unloading process. A non-destructive test technique such as acoustic emission (AE) measurement avoids such ambiguities. To complement our microstructural observations, we monitored the AE activity as a function of the effective pressure (Zhang *et al.* 1990). Our AE data for Berea sandstone are shown in Fig. 2a and b. It can be seen from Fig. 2a that a dramatic increase in AE rate occurred at a critical effective pressure (of 380 MPa) corresponding to that at the inflection point of the hydrostat. In total, more than 1 million AE events were counted. In compari-

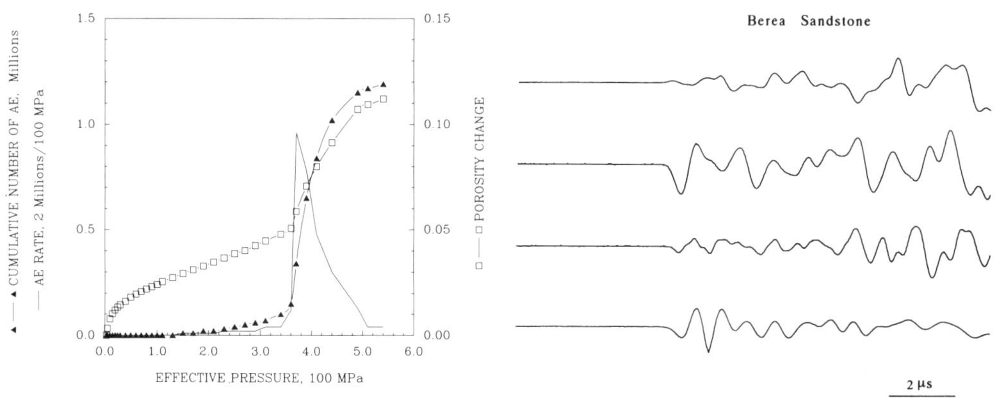

Fig. 2. Acoustic emission (AE) and porosity change measurements of Berea sandstone hydrostatically compressed at a pore pressure of 10 MPa under fully drained condition. (**a**) Cumulative AE count, AE rate, and porosity change as functions of effective pressures. Note the peak in AE rate at the inflection point of the hydrostat. (**b**) Representative waveforms of AE induced by Hertzian fracture and pore collapse.

son, the total number of grains in our specimen was on the order of 1 million, suggesting that brittle cracking was pervasive throughout the sample. Some representative AE waveforms are shown in Fig. 2b. Our preliminary first motion analysis of the waveforms indicates that both extensional and implosion sources were involved, in agreement with the microstructural observations that both tensile 'Hertzian fracture' and pore collapse processes were operative.

Hertzian fracture mechanics model

We use the theory of Hertzian fracture mechanics to analyse the onset of grain crushing. As is true for any micromechanical model, idealized assumptions have to be made so that the theoretical analysis is tractable. Following Brandt (1955) we model the porous rock as a randomly packed assemblage of spherical particles of several distinct sizes. In the vicinity of the grain contact region, the two particles are subjected to a local normal force F (Fig. 3) which is related to the effective pressure, porosity Φ, and grain radius R. The local stress field is governed by the magnitude of F, the grain radius and the elastic moduli (Young's modulus E and Poisson's ratio ν). The maximum hoop stress is attained at the perimeter of the contact area, and the stress intensity factor of a micro-crack located at the contact region can be approximated by that of an edge crack (Frank & Lawn 1967). Hertzian fracture can initiate if the stress intensity factor K_{I} at the tip of a preexisting crack (of length c) reaches a critical value given by the fracture toughness K_{Ic}

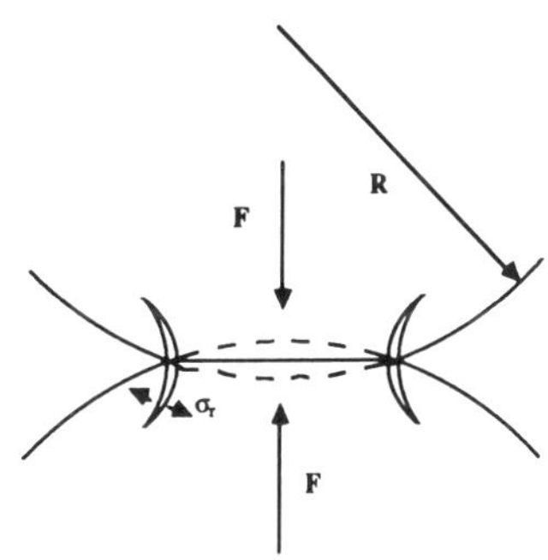

Fig. 3. Schematic diagram of a grain contact between two spherical elastic particles. F is the local normal force acting on the grain contact of two spherical particles of radius R. The maximum tensile hoop stress is denoted by σ_r.

(Wilshaw 1971). Using these assumptions, Zhang *et al.* (1990) derived the following expression for the critical effective pressure for the onset of grain crushing:

$$P_{\mathrm{cr}} = 2.2\,[(1-\nu^2)^2/(1-2\nu)^3]\,[K_{\mathrm{Ic}}^{\,3}/E^2]\,(\alpha\Phi R)^{-3/2} \quad (1)$$

where $\alpha = c/R$ is the ratio of the initial crack length to grain radius.

It is plausible that the preexisting defect dimension c should scale as the grain dimension R, and if the materials have comparable elastic moduli and fracture toughness, then a log–log plot of P_{cr} versus ΦR falls on a straight line of slope $-3/2$ and the ratio α can be estimated from the intercept. A comparison of experimental data with our Hertzian fracture mech-

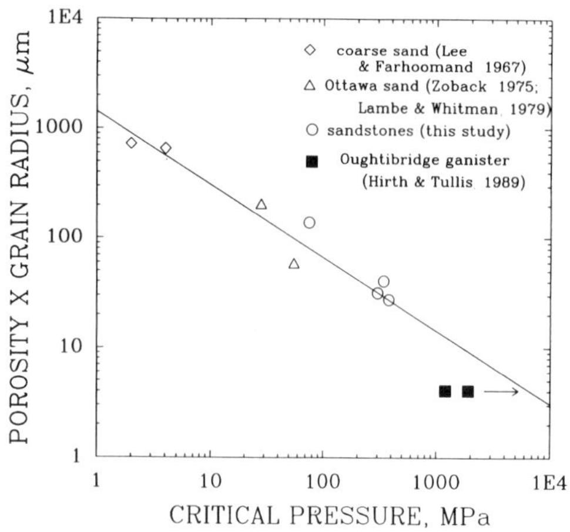

Fig. 4. Experimental data on the critical effective pressure for grain crushing as a function of the product of grain radius and porosity. The straight line corresponds to the theoretical relation given by equation (1).

anics model is shown in Fig. 4. Our sandstone data (Fig. 1) cover an intermediate range of pressure. Sand data obtained by Lee & Farhoomand (1967), Zoback (1975), and Lambe & Whitman (1969) cover the low pressure range, whereas the deformation data of Oughtibridge ganister (Hirth & Tullis 1989) cover the very high pressure range. The experimental data follow reasonably a linear trend over a critical pressure range of 3 orders of magnitude. For reference, the straight line shown in Fig. 4 corresponds to $\alpha = 3.6 \times 10^{-5}$ for quartz grains and $\alpha = 1.7 \times 10^{-4}$ for feldspar grains respectively. Hence the preexisting flaws for Hertzian fracture are inferred to have submicron dimension. Although such short cracks are difficult to resolve even with the use of a scanning electron microscope, they play a key role in the initiation of Hertzian fracture which ultimately leads to distributed grain crushing and irreversible compaction.

Geological implications

The experimental observations summarized above show that there is not a unique pressure at which a porous rock or a granular material undergoes grain crushing. The Hertzian fracture mechanics model suggests that the critical effective pressure for the onset of grain crushing is a function of the porosity and the grain size, as well as the elastic moduli and fracture toughness. For a coarse sand with a relatively high porosity, the critical pressure can be as low as several mega pascals, implying that grain crushing as a compaction mechanism can be quite common in geological settings.

Since grain crushing is associated with an inflection point in the hydrostat, one may expect that the process results in a sudden decrease in porosity as a function of depth. That this is not evident from depth-porosity profiles implies that time-dependent, thermally activated chemical compaction processes are coupled to mechanical compaction processes (Schmoker & Gautier 1989). A mechanical compaction mechanism such as grain crushing decreases the porosity and grain size, which in turn enhances chemical compaction by a mechanism such as pressure solution, the rate of which decreases as a function of grain size (Renton *et al.* 1969; Houseknecht 1983; Tada & Siever 1989). On the other hand, chemical compaction processes modify the grain contact geometry by smoothing out grain angularity and by precipitating mechanically compliant cement into the pore space. Consequently, the local stress concentration is alleviated and an elevated critical pressure is necessary for grain crushing to initiate.

Brittle–ductile transition in porous rock

Other than the overburden pressure, the tectonic stress can also influence the inelastic behaviour. Laboratory studies have shown that in addition to the mean pressure, nonhydrostatic stresses can also cause appreciable compaction (Schock *et al.* 1973). Alternatively, the deviatoric stress can induce dilatancy, which may ultimately lead to failure by shear localization (Edmond & Paterson 1972; Jamison & Teufel 1979). Whether the inelastic behaviour and failure mode is brittle or ductile depends on the porosity and the effective pressure (Jones 1980; Zhang *et al.* 1987; Hirth & Tullis 1989).

The brittle–ductile transition in rock is characterized by qualitative changes in both the inelastic behaviour and failure mode. In the brittle field, pressure-sensitive inelastic deformation accompanied by dilatancy is generally observed and under overall compressive loading, failure usually occurs by shear localization. In the ductile field, dilatancy and shear localization are inhibited. In rocks with relatively low porosities (of say, several percent), the deformation in the ductile field is usually due to a crystal plasticity mechanism (e.g. Evans *et al.* 1990) and requires elevated temperature. In calcite rocks, this type of ductility can be accomplished by an increase in confining pressure since inelastic deformation by homogeneously distributed crystal plasticity can occur at room temperature (e.g. Fredrich *et al.* 1989). In low porosity silicate rocks at effective pressures accessible in the laboratory, elevated temperatures are required to activate the crystal plasticity processes which control quartz and feldspar deformation in the ductile field (e.g. Tullis & Yund 1977). In high porosity silicate rocks, a brittle–ductile transition (not involving crystal-plastic flow) can occur at room temperature (e.g. Mogi 1966). The macroscopically ductile deformation is due to homogeneously distributed grain scale brittle cracking. The inelastic deformation is pressure-sensitive; in addition it is also sensitively dependent on the porosity (Byerlee & Brace 1969; Jones 1980).

Effects of pressure and porosity on inelastic behaviour and failure mode

Previous studies have focused on the effect of pressure. The brittle–ductile transition in Berea sandstone as a function of pressure has been studied in some detail (Handin *et al.* 1963; Jamison & Teufel 1979; Bernabe & Brace 1990). Figure 5 shows our data for Berea sandstone deformed at effective pressures of 50 MPa, 100 MPa and 200 MPa respectively. The strain rate was $2.6 \times 10^{-5}/s^{-1}$ under fully drained conditions. At an effective pressure of 50 MPa, the peak deviatoric stress was attained at a relatively small axial strain (of less than 2%) and brittle failure occurred by shear localization and strain softening. It is as yet unclear whether strain softening preceded the onset of shear localization in these experiments. Both dilatancy and shear localization were inhibited at effective pressures of 100 MPa and 200 MPa. At 100 MPa, the stress–strain curve showed a small amount of softening followed by hardening which persisted to an axial strain of more than 20%. At 200 MPa, no strain softening was observed.

To investigate the effect of porosity on the brittle-ductile transition, we overconsolidated samples of Kayenta sandstone by hydrostatically compacting it beyond the inflection point (Fig. 1). Figure 6 show the axial stress–strain curves of a normally consolidated Kayenta sandstone sample and that of an overconsolidated sample respectively, both deformed at an effective pressure of 150 MPa under fully

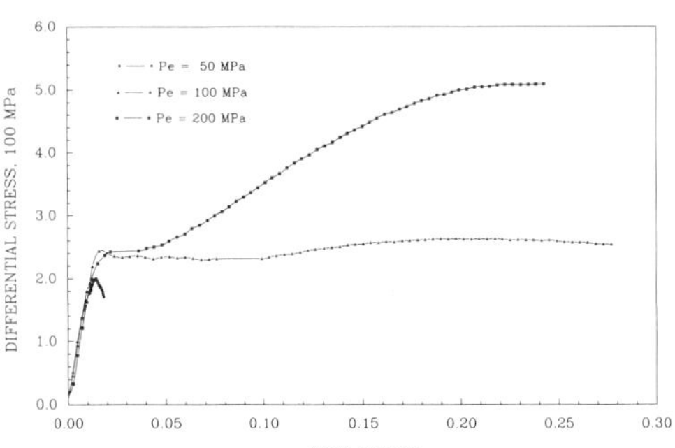

Fig. 5. Axial stress–strain data of Berea sandstone at three effective pressures (50 MPa, 100 MPa, and 200 MPa). The pore pressure was 10 MPa, and the strain rate was 2.6×10^{-5} s^{-1}. The deformation was fully drained.

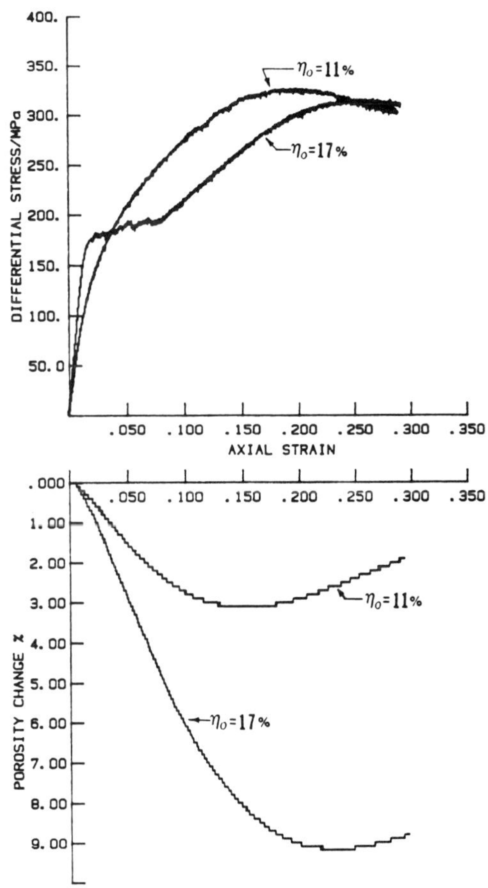

Fig. 6. The differential stress and porosity change as functions of the axial strain of two samples of Kayenta sandstone deformed at confining pressures of 160 MPa and pore pressures of 10 MPa under fully drained condition. The strain rate was 2.6×10^{-5} s^{-1}. The overconsolidated sample had an initial porosity of 11%, and the normally consolidated sample had an initial porosity of 17%.

drained condition at a strain rate of 2.6 × 10^{-5}/s^{-1}. The former had an initial porosity (i.e. the porosity attained after the application of the confining pressure but before the application of the differential stress) of 17% and the latter 11%. The normally consolidated sample first underwent some grain crushing when the mean pressure reached the critical pressure, then it strain hardened up to an axial strain of about 30%. It can also be seen from Fig. 6 that the porosity change was basically all-compaction. The deformed sample had a barrelled shape, and the microstructure was characterized by homogeneously distributed cataclasis. On the other hand, the overconsolidated sample showed dilatancy fairly early in the deformation process and the sample failed by development of a shear band at the strain softening stage. When observed under the microscope, the grains inside the shear band showed intense comminution.

Preliminary acoustic emission measurements by Zhang *et al.* (1989) showed substantial AE activity for all samples whether they were macroscopically brittle or ductile. A surge in AE activity usually marked the onset of shear-enhanced compaction. Surprisingly, no marked changes in AE activity was observed at the onset of shear localization in the brittle field. Details of the AE measurements will be presented in a future publication.

Zhang *et al.* (1987) proposed three criteria to separate the inelastic behavior and the failure mode into two separate fields in a plot of effective pressure versus initial porosity. The three criteria were: (1) the inelastic behaviour—

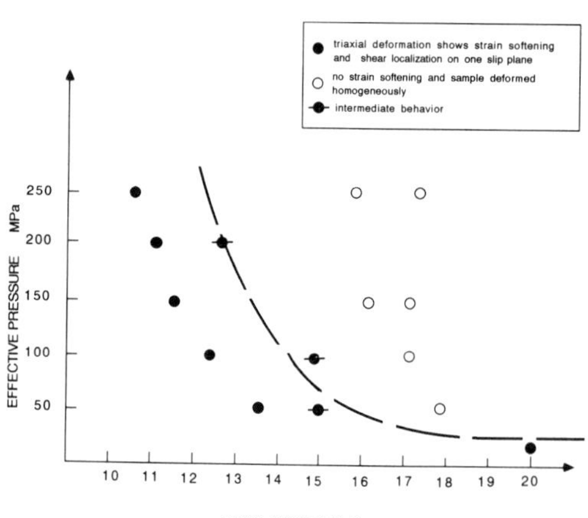

Fig. 7. A brittle–ductile transition map for Kayenta sandstone in the effective pressure–initial porosity space.

whether the sample showed dilatancy or persistent compaction; (2) whether it showed strain softening or persistent hardening; (3) the failure mode—whether it developed shear localization or homogeneously distributed cataclastic flow. Such a map separating the brittle field from the ductile field of Kayenta sandstone is shown in Fig. 7. This brittle–ductile transition map was constructed on the basis of our deformation data (Zhang *et al.* 1987). A similar map has also been constructed for Berea sandstone. It is evident from the maps that both pressure and porosity exert important controls over the brittle-ductile transition in porous sandstones, and that the development of dilatancy and shear localization are both promoted by a decrease in porosity and in confining pressure (or equivalently, an increase in pore pressure).

Micromechanical model based on linear elastic fracture mechanics

Although the effects of pressure and porosity on Kayenta sandstone shown in Fig. 7 are qualitatively similar to that for soil (Lambe & Whitman 1969; Schofield & Wroth 1968), the micromechanical processes may differ significantly. Stress-induced microcracking subparallel to the maximum compression direction is an important mechanism for dilatancy in porous rocks at elevated pressures. The coalescence of the stress-induced cracks causes shear localization in the brittle field (Hoshino & Koide 1970; Jamison & Teufel 1979; Hirth & Tullis 1989; Bernabe & Brace 1990). On the other hand, dilatancy and strain softening in granular materials of very high porosities (say, more than 30%) deformed under relatively low pressures (on the order of 1 MPa) can occur through stress-induced fabric change by grain rotation without any grain crushing. A micromechanical model which considers the change in coordination number and the development of anisotropy in the orientational distribution of the contact normals in an assembly of rigid spheres can capture some of the important attributes of the inelastic behavior and failure mode (e.g. Cundall 1988).

Hence, a micromechanical model of porous rock deformation has to consider the important role of stress-induced microcracking. A micromechanical model based on fracture mechanics which provides useful physical insights on the effects of pressure and porosity on the inelasticity and failure of porous rocks was formulated recently by Sammis & Ashby (1986). This model considers the initiation, propagation and co-

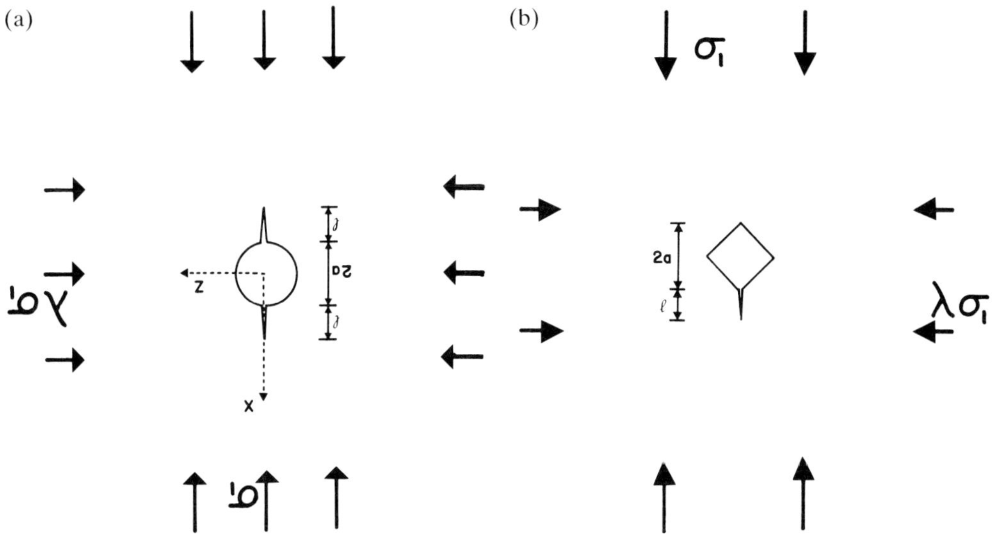

Fig. 8. (**a**) The pore-emanated cracking model analyzed by Sammis & Ashby (1986). A cylindrical pore of radius a is subjected to a remotely applied biaxial stress field with principal components given by $-\sigma_1$ and $-\sigma_2$ $(= -\lambda\sigma_1)$. The stress-induced cracks have length l. (**b**) The model of a crack emanating from a corner of a square hole analyzed by Hasebe & Ueda (1980). The square has dimension $\sqrt{2}a$, and hence the half-length of its diagonals is a. The remotely applied biaxial stress field is given by principal components $-\sigma_1$ and $-\sigma_2$ $(= -\lambda\sigma_1)$ parallel to the diagonals. The stress-induced crack has length l.

alescence of microcracks emanating from cylindrical holes or spherical pores (Fig. 8a). Stable growth of the pore-emanated cracks causes strain hardening. As the cracks extend, they begin to interact and if their mean spacing is sufficiently small, then crack propagation is unstable, possibly leading to crack coalescence and shear localization.

I will first summarize the pertinent results from Sammis & Ashby's (1986) fracture mechanics analysis and then compare their theoretical predictions with our experimental observations (Figs 5, 6, and 7). Consider a cylindrical hole of radius a under a remotely applied biaxial stress field. The maximum and minimum compressive principal stresses are $-\sigma_1$ and $-\sigma_2$ respectively. Tension is taken to be positive here and $\sigma_2/\sigma_1 = \lambda \leq 1$. Since the maximum tensile hoop stress is attained at $(x, \mathbf{z}) = (\pm a, 0)$, stress-induced microcracks nucleate from these positions on the void surface (Fig. 8a). An analytic estimate of the mode-I stress intensity factor at the tip of such a crack of length l is:

$$K_1 = \sqrt{L}\,[1.1(1-2.1\lambda)/(1+L)^{3.3} - \lambda]\sigma_1\,(\pi a)^{1/2} \qquad (2a)$$

where $L = l/a$. For $\lambda \geq 0$, the stress intensity factor increases with crack extension and reaches a maximum at a value of $l < a$, and then decreases monotonically as a function of L (Fig. 9a). Hence, immediately after propagation initiates, the crack extends unstably to a length $l < a$, and the propagation behavior is stable after this 'pop-in' stage. As the pore-emanated cracks extend, their stress fields interact with one another, and enhance the stress intensity factor. Sammis & Ashby (1986) used a beam flexure analysis to derive an approximate analytic estimate of the enhancement of stress intensity factor due to the crack interaction effect:

$$K_1' = (\sqrt{2}/\pi)\,[\Phi(1+L)]\sigma_1(\pi a)^{1/2} \{[1-(8/\pi)\Phi\lambda\,(1+L)^3] [1-(2/\pi)\Phi\lambda(1+L)^3]\}^{1/2} \qquad (2b)$$

where Φ is the initial porosity. The term outside the curly bracket is the contribution from the bending effect due to the axial compression σ_1, whereas the term inside is the contribution from the crack closing effect due to the confining stress σ_2. The first term increases whereas the second term decreases as a function of L. If the critical stress intensity factor or the fracture toughness is denoted by K_c, then the condition for crack extension is:

$$K_1 + K_1' = K_c \qquad (3)$$

with K_1 and K_1' given by equations (2a) and (2b) respectively. It should be noted that in their recent study, Isida & Nemat-Nasser (1987)

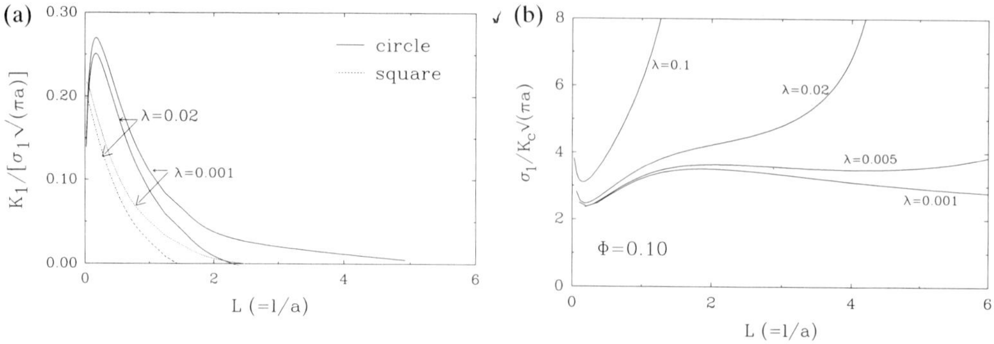

Fig. 9. (**a**) Normalized stress intensity factor as a function of normalized pore-emanated crack length (l/a). Crack interaction is ignored and results for the square hole and circular hole cases are computed based on the work of Hasebe & Ueda (1980) and Sammis & Ashby (1986) respectively. (**b**) Normalized maximum principal compressive stress as a function of normalized pore-emanated crack length (l/a) at a fixed porosity $\Phi = 10\%$ for four different values of the stress ratio λ. Crack interaction effect is estimated by equation 2b.

carried out rigorous computations which showed that equation (2b) overestimates the destabilization effect due to crack interaction. The analytic estimate is used here for mathematical convenience, and the stress intensity factors so calculated should be considered as upper bounds only.

Since Sammis & Ashby (1986) did not show numerical computation results for the range of porosities and pressures of interest here, I perform some relevant calculations based on equations (2) and (3). The crack propagation behaviour is characterized by the normalized stress $\sigma_1/[K_c(\pi a)^{1/2}]$ as a function of the normalized crack extension $L\ (= l/a)$. Figure 9b shows the effect of confining stress ($\lambda = 0.001$ up to $\lambda = 0.1$) on crack propagation behaviour at a fixed initial porosity of 10%. For all values of λ, the normalized stress for crack extension initially decreases with L, corresponding to a small amount of unstable crack extension at the 'pop-in' stage. As noted above, the extension l at the 'pop-in' stage is generally less than a (corresponding to $l < 1$). The stability of crack propagation behaviour after the 'pop-in' stage depends on the ratio λ. For λ relatively small (i.e. confining stress relatively low), the behaviour is stable up to a certain point beyond which the crack driving stress decreases with crack growth. Hence this point is the onset of crack growth instability. If the deformation behaviour of a porous rock can be modelled as the collective behaviour of an ensemble of such pore-emanated cracks, then the onset of instability corresponds to a transition from strain hardening to strain softening. On the other hand, if λ is increased by increasing the confining stress, then the crack growth will be stabilized. As a matter of fact, the crack growth stops beyond a certain length for $\lambda = 0.02$ and $\lambda = 0.1$ (Fig. 9b). Therefore the collective behaviour of an ensemble of pore-emanated cracks will be strain hardening.

Figure 10a and b illustrates the effect of porosity on crack propagation behaviour at $\lambda = 0.001$ and $\lambda = 0.005$ respectively. It can be seen that with an increase in porosity, there is a transition from unstable to stable crack growth behaviour, implying that an increase in porosity at a fixed confining stress can ultimately result in overall hardening and homogeneously distributed cataclastic flow.

In summary, the important parameters in the pore-emanated crack model are λ (the ratio of confining stress to maximum compressive stress), l/a (the ratio of crack extension to pore radius), Φ (the porosity), and K_c (the fracture toughness). The model predicts that the development of strain softening (and possibly shear localization) is promoted by a decrease in porosity and a decrease in confining pressure, in qualitative agreement with experimental observations (Fig. 7). However, the model has an important shortcoming in that it predicts too large a crack stabilization effect due to pressure. According to Fig. 9b, strain softening is inhibited in a rock of 10% porosity if $\lambda > 0.005$ which corresponds to a very small confining pressure (as can be seen from Fig. 6). This discrepancy can not be due to the approximation procedures adopted by Sammis & Ashby (1986) since the rigorous computation of Isida & Nemat-Nasser (1987) showed that Sammis & Ashby (1986) probably overestimated the destabilization effect due to crack interaction, therefore implying that the confining stress is

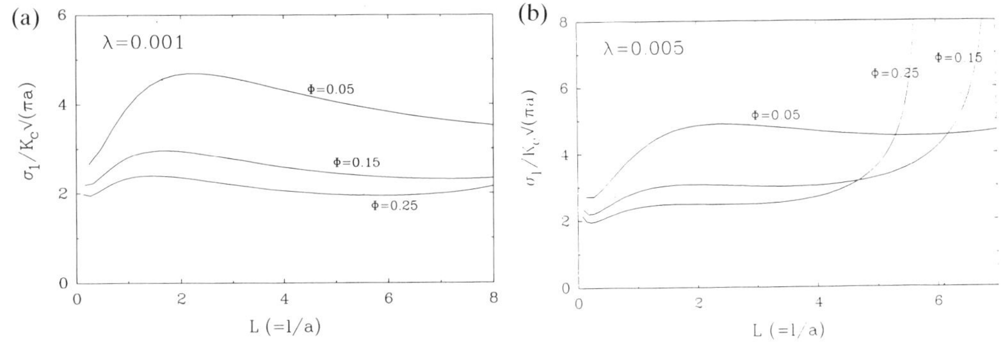

Fig. 10. Normalized maximum principal compressive stress as a function of normalized pore-emanated crack length (l/a) at fixed stress ratios λ of (**a**) 0.001, and (**b**) 0.005 respectively for three different values of porosity. Crack interaction effect is estimated by equation 2b.

even more effective in inhibiting the growth of a pore-emanated crack than predicted by equations (2) and (3).

Since the pore geometry in rocks ranges in shape from nearly equidimensional to elongated with acute corners, one may conjecture that the discrepancy between theory and experiment is because the pore-emanated crack model underestimates the local stress concentration by using a smooth cylindrical geometry. However, it can be shown that this geometric effect is relatively small. The effect of an acute geometry was analysed by Hasebe & Ueda (1980) who calculated the stress intensity factor of a microcrack emanating from a corner of a square hole using conformal mapping and complex potential techniques. I use Hasebe & Ueda's (1980) tabulated values to calculate the stress intensity factor of microcrack of length l subjected to remotely applied principal stresses parallel to the diagonals of the square (Fig. 8b). The square has dimension $\sqrt{2}a$ and hence the half-length of its diagonal is a. The calculated K_1 values are shown in Fig. 9a. Since the sharp corner of a square hole accentuates the stress in its immediate vicinity, the stress intensity factor of a crack emanating from a square hole at the initiation stage is somewhat higher than that of a crack emanating from a circular hole. Nevertheless, K_1 for a crack emanating from a square hole decreases monotonically as a function of crack extension, implying that strain softening is not possible unless crack interaction results in significant increase in the stress intensity factor. Since the interaction effect can still be estimated by the K_1' in equation (2b), the square hole model should predict mechanical behaviour which is qualitatively similar to the circular hole model.

This model focuses on a single micromechanical process, stress-induced microcracking which is expected to be a key deformation mechanism responsible for dilatancy and shear localization in the brittle field. However, for a very high porosity rock deformed under elevated pressures, compaction mechanisms (including pore collapse and grain rotation) are probably operative. Our preliminary measurements (Zhang *et al.* 1989) have shown that the AE activity is intimately related to shear-enhanced compaction. Therefore a micromechanical model for the ductile field must also consider the important roles played by these compaction mechanisms. It should be noted that although equidimensional pores are also found in low porosity rocks, stress-induced microcracking emanating from such pores does not play a dominant role in the development of brittle fracture (e.g. Wong 1982). Furthermore the brittle failure behaviour in these rocks does not show a strong dependence on porosity (Paterson 1978). Hence Sammis & Ashby's (1986) model is not appropriate for low porosity rock.

Geological implications

Because the seismogenic layer is usually confined to the upper 15 km or so, it is usually accepted that crustal faults must be brittle at shallow depths and ductile at greater depths. The brittle deformation is pressure-sensitive, and the transition to ductile behaviour at depth is held to be due to the onset of thermally activated crystal plasticity processes (e.g. Brace & Kohlstedt 1980). Hence the transition depth is sensitive to pressure, temperature, and strain rate. However, our study also highlights the important role of porosity in controlling the brittle–ductile transition in porous rocks. Macroscopically ductile deformation is possible in very high porosity rocks under elevated pressures even if crystal plasticity processes are not operative. At very shallow depths, rocks may be highly porous and probably deform in a macroscopically ductile manner. At somewhat greater depths, a transition to macroscopically brittle behaviour comes about due to porosity reduction by the various compaction mechanisms discussed above. This implies that there is an aseismic layer at the very shallow part of the crust. The plate boundaries at subduction zones are the loci of the largest known earthquakes and of many smaller magnitude events as well. An interesting seismological observation recently pointed out by Byrne *et al.* (1988) is that the shallowest parts of the same plate boundaries are areas of persistent aseismicity. This transition from aseismicity to seismicity as a function of depth can partially be explained by the porosity-sensitivity of the inelastic behavior and failure mode of porous rocks (Fig. 7). Recently Marone & Scholz (1988) summarized similar seismological observations in various tectonic settings.

A common field observation in faulted high porosity sandstones is that fault zones characteristically evolved from many subparallel, closely spaced sets of granulated fault strands, sometimes referred to as 'deformation bands' (Aydin & Johnson 1978) or 'microfaults' (Jamison & Stearns 1982). Underhill & Woodcock (1987) argued that the development of shear localization in the sandstones is relatively stable, involving a stress–strain path that oscillates between strain softening and hardening. It should be noted that this type of stress–strain curve is indeed observed in laboratory experiments of porous sandstones (e.g. Berea sandstone deformed at effective pressure of 100 MPa shown in Fig. 5). The micromechanical process may involve a combination of 'pop-in' cracking following by stable crack propagation as we discussed above (e.g. the curves for $\lambda = 0.02$ and $\lambda = 0.1$ shown in Fig. 9) although other processes (such as pore collapse, grain rotation, and cement deformation) suggested by Underhill & Woodcock (1987) are also possible mechanisms.

Conclusion

Our experimental investigation of the inelasticity and failure mode of porous sandstones shows that a nondestructive testing technique (such as acoustic emission measurement) in conjunction with microscopic observations can be very useful in elucidating micromechanical processes involving grain scale cracking in porous rocks. Furthermore, linear elastic fracture mechanics is a powerful tool for the analysis of the micromechanical processes. We have formulated a Hertzian fracture mechanics model for the onset of pressure-induced grain crushing. The model predicts that the critical effective pressure for grain crushing correlates inversely with grain size and porosity. The theoretical prediction is in reasonable agreement with laboratory measurements over a critical pressure range of 3 orders of magnitude. We conducted a systematic investigation of the effect of porosity and pressure on the development of dilatancy and shear localization in porous sandstones. We showed how a brittle-ductile transition map can be constructed in the effective pressure-porosity space. A fracture mechanics model of a pore-emanated crack is in qualitative agreement with experimental observations. However, such a model overestimates the crack stabilization effect of confining stress, probably because it focuses on a single deformation mechanisms and neglects the significant effects of grain rotation and pore collapse. The significant control of porosity over the brittle-ductile transition implies that there is an aseismic crustal layer at the very shallow depths.

This project was conducted in close collaboration with D. M. Davis. The acoustic emission measurement system was designed and set up by T. Yanagidani, and the deformation experiments were performed by J. Zhang. This paper was prepared while I was a visiting professor at the Massachusetts Institute of Technology. I would like to thank W. F. Brace and B. Evans for their hospitality during my visit. I have benefited from several useful discussions with S. Hickman and J. Fredrich. This research was partially supported by the National Science Foundation through grant EAR-8721045 and the US Geological Survey through grant 14–08–001–G1352.

References

Aydin, A. & Johnson, A. M. 1978. Small faults formed as deformation bands in sandstones. *Pure and Applied Geophysics*, **116**, 913–930.

Bernabe, Y. & Brace, W. F. 1990. *Deformation and fracture of Berea sandstone*. Geophysical Monograph Series, The Heard Volume, American Geophysical Union (in press).

Brace, W. F. & Kohlstedt, D. L. 1980. Limits on lithospheric stress imposed by laboratory experiments. *Journal of Geophysical Research*, **85**, 6248–6252.

—— W. F. 1978. Volume changes during fracture and frictional sliding: a review. *Pure and Applied*

Geophysics, **116**, 603–614.

BRANDT, H. 1955. A study of the speed of sound in porous granular media. *Transaction of the American Society of Mechanical Engineers*, **22**, 479–486.

BYERLEE, J. D. & BRACE, W. F. 1969. High pressure mechanical instability in rocks. *Science*, **164**, 713–715.

BYRNE, D. E., DAVIS, D. M. & SYKES, L. R. 1988. Loci and maximum size of thrust earthquakes and the mechanics of the shallow region of subduction zones. *Tectonics*, **7**, 833–857.

CUNDALL, P. A. 1988. Computer simulations of dense sphere assemblies. *In*: SATAKE M. & JENKINS, J. T. (eds) *Micromechanics of Granular Materials*. Elsevier, Amsterdam, 113–123.

EDMOND, J. M. & PATERSON, M. S. 1972. Volume change during the deformation of rocks at high pressure. *International Journal of Rock Mechanics and Mining Science*, **9**, 161–182.

EVANS, B., FREDRICH, F. T. & T-f. WONG. 1990. *The brittle–ductile transition in rocks: recent experimental and theoretical progress*. Geophysical Monograph Series, The Heard Volume, American Geophysical Union, (in press).

FRANK, F. C. & LAWN, B. R. 1967. On the theory of Hertzian fracture. *Proceedings of the Royal Society of London*, **A229**, 291–306.

FREDRICH, J. T. & WONG, T.-f. 1986. Micromechanics of thermally induced cracking in three crustal rocks. *Journal of Geophysical Research*, **91**, 12743–12764.

——, EVANS, B. & WONG, T.-f. 1989. Micromechanics of the brittle to plastic transition in Carrara marble. *Journal of Geophysical Research*, **94**, 4129–4145.

GALLAGHER, J. J. 1987. Fractography of sand grains broken by uniaxial compression. *In*: MARSHALL, J. R. (ed.) *Clastic Particles: Scanning Electron Microscopy and Shape Analysis in Sedimentary and Volcanic Clasts*, Von Nostrand Reinhold, NY, 189–228.

——, FRIEDMAN, M., HANDIN, J. & SOWERS, G. M. 1974. Experimental studies relating to mircofracture in sandstone. *Tectonophysics*, **21**, 203–247.

HANDIN, J., HAGER, R. V., FRIEDMAN, M. & FEATHER, J. N. 1963. Experimental deformation of sedimentary rock under confining pressure: pore pressure effects. *American Association of Petroleum Geologists Bulletin*, **47**, 717–755.

HASEBE, N. & UEDA, M. 1980. Crack originating from a corner of a square hole. *Engineering and Fracture Mechanics*, **13**, 913–923.

HIRTH, G. & TULLIS, J. 1989. The effects of pressure and porosity on the micromechanics of the brittle-ductile transition. *Journal of Geophysical Research*, **94**, 17825–17838.

HORII, H. & NEMAT-NASSER, S. 1986. Brittle failure in compression: splitting, faulting and brittle-ductile transition. *Philosophical Transactions of Royal Society of London*, **319**, 337–374.

HOSHINO, K. & KOIDE, H. 1970. Process of deformation of the sedimentary rocks. *Proceedings of the Congress of the International Society of Rock Mechanics*, **2(1)**, 353–359.

HOUSEKNECHT, D. W. 1984. Influence of grain size and temperature on intergranular pressure solution, quartz cementation, and porosity in a quartzose sandstone. *Journal of Sedimentary Petrology*, **54**, 348–361.

ISIDA, M. & NEMAT-NASSER, S. 1987. On mechanics of crack growth and its effects on the overall response of brittle porous solids. *Acta metallurgica*, **35**, 2887–2898.

JAMISON, W. R. & STEARNS, D. W. 1982. Tectonic deformation of Wingate Sandstone, Colorado National Monument. *American Association of Petroleum Geologists Bulletin*, **66**, 2584–2608.

—— & TEUFEL, L. W. 1979. Pore volume changes associated with failure and frictional sliding of a porous sandstone. *Proceedings of the US Symposium on Rock Mechanics*, **20**, 163–170.

JONES, L. M. 1980. Cyclic loading of simulated fault gouge to large strains. *Journal of Geophysical Research*, **85**, 1826–1832.

JONES, M. E. & PRESTON, R. M. F. 1987. Introduction. *In*: JONES, M. E. & PRESTON, R. M. F. (eds) *Deformation of Sediments and Sedimentary Rocks*. Geological Society, London, Special Publication, **29**, 1–8.

KANNINEN, M. F. & POPELAR, C. H. 1985. *Advanced Fracture Mechanics*. Oxford University Press, NY.

LAMBE, T. W. & WHITMAN, R. V. 1969. *Soil Mechanics*. John Wiley, NY.

LEE, K. L. & FARHOOMAND, I. 1967. Compressibility and crushing of granular soil in anisotropic triaxial compression. *Canadian Geotechnical Journal*, **4**, 68–99.

MARONE, C. & SCHOLZ, C. H. 1988. The depth of seismic faulting and the upper transition from unstable to stable slip regimes. *Geophysical Research Letters*, **15**, 621–624.

MAXWELL, J. C. 1960. Experiments on compaction and cementation of sand. *Geological Society of America Memoir*, **79**, 105–132.

MCEWEN, T. J. 1981. Brittle deformation in pitted pebble conglomerates. *Journal of Structural Geology*, **5**, 25–37.

MITCHELL, J. K. 1976. *Fundamentals of Soil Behaviour*. John Wiley, NY.

MOGI, K. 1966. Pressure dependence of rock strength and transition from brittle fracture to ductile flow. *Bulletin of the Earthquake Research Institute, Tokyo University*, **44**, 215–232.

PATERSON, M. S. 1978. *Experimental Rock Deformation—The Brittle Field*. Spinger Verlag, New York.

POLLARD, D. D. 1987. Elementary fracture mechanics applied to the structural interpretation of dykes. *In*: HALL, H. C. & FAHRIG, W. F. (eds) *Mafic Dyke Swarms*. Geological Society of Canada Special Paper, **34**, 5–24.

RENTON, J. J. HEALD, M. T. & CECIL, C. B. 1969. Experimental investigation of pressure solution in quartz. *Journal of Sedimentary Petrology*, **39**, 1107–1117.

RICE, J. R. 1980. The mechanics of earthquake rupture. *In*: Dziewonski, A. M. & Boschi, E. (eds) *Physics of the Earth's Interior*. Italian Physical Society, Bologna, 555–649.

Sammis, C. G. & Ashby, M. F. 1986. The failure of brittle porous solids under compressive stress states. *Acta metallurgica*, **34**, 511–526.

Schock, R. N. Heard, H. C. & Stephens, D. R. 1973. Stress–strain behavior of a granodiorite and two graywackes on compression to 20 kilobars. *Journal of Geophysical Research*, **78**, 5922–5941.

Schmoker, J. W. & Gautier, D. L. 1989. Compaction of basin sediments: modelling based on time–temperature history. *Journal of Geophysical Research*, **94**, 7379–7386.

Schofield, A. & Wroth, P. 1968. *Critical State Soil Mechanics*, McGraw-Hill.

Tada, R. & Siever, R. 1989. Pressure solution during diagenesis. *Annual Review of Earth and Planetary Science*, **17**, 89–118.

Tullis, J. & Yund, R. A. 1977. Experimental deformation of dry Westerly granite. *Journal of Geophysical Research*, **82**, 5705–5718.

Underhill, J. R. & Woodcock, N. H. 1987. Faulting mechanisms in high-porosity sandstones; New Red Sandstone, Arran, Scotland. *In*: Jones, M. E. & Preston, R. M. F. (eds) *Deformation of Sediments and Sedimentary Rocks*. Geological Society, London Special Publication, **29**, 91–105.

Wanamaker, B. J., Wong, T.-f. & Evans, B. 1990. Decrepitation and crack healing of fluid inclusions in San Carlos olivine. *Journal of Geophysical Research*, (in press).

Wilshaw, T. R. 1971. The Hertzian fracture test. *Journal of Physics*, **D4**, 1567–1581.

Wong, T.-f. 1982. Micromechanics of faulting in Westerly granite. *International Journal of Rock Mechanics and Mining Science*, **19**, 49–64.

Zhang, J., Wong, T.-f. & Davis, D. M. 1987. Failure mode as a function of porosity and effective pressure in porous sandstones. *Geological Society of America Abstracts with Programs*, **19**, 904.

——, —— & —— 1989. The brittle to ductile transition in porous sandstones. *EOS, Transactions of the American Geophysical Union*, **70**, 1360.

——, —— & —— 1990. Micromechanics of pressure-induced grain crushing in porous rocks. *Journal of Geophysical Research*, **95**, 341–352.

——, ——, Yanagidani, T. & Davis, D. M. 1990. Pressure-induced microcracking and grain crushing in Berea and Boise sandstones: acoustic emission and quantitative microscopy measurements. *Mechanics of Materials*, (in press).

Zoback, M. D. 1975. *High Pressure Deformation and Fluid Flow in Sandstone, Granite and Granular Materials*, PhD thesis, Stanford University.

Shear bands in a plastic layer at yield under combined shortening and shear: a model for the fault array in a duplex

RUSSELL K. DAVIES & RAYMOND C. FLETCHER

Center for Tectonophysics, Departments of Geology and Geophysics, Texas A&M University, College Station, TX 77843, USA

Abstract: To model the regular spacing of the thrust faults in duplexes we subject a single layer embedded in an overlying viscous fluid, and an underlying rigid medium, to combined layer-parallel shortening and shear. The layer is a rigid–plastic solid at yield. A cusp introduced as a small perturbation on the upper layer interface perturbs the flow. The lower layer interface is weak and remains flat throughout the deformation, whereas the upper interface is deformable and can assimilate both weak and strong conditions. If the upper interface is strong it folds into a broad open fold. If the interface is weak the flow localizes into narrow shear bands that cut into the layer from the cusp, and reflect from the layer interfaces at a constant spacing. Material is displaced across the shear bands in a fault-like manner. If the layer is subjected solely to layer-parallel shortening these shear bands are symmetric. Superposing a shear component on the layer interface reorients the shear bands. The shear band dipping opposite to the sense of shear is more strongly developed and intersects the interface at a lower angle than the other. This is in agreement with the observations of duplex configurations.

In a duplex (Boyer & Elliott 1982; Mitra 1986), the thrust faults bounding the horses are regularly spaced as measured along the bedding. They dip opposite to the direction of translation of the hanging wall in the sense appropriate for thrust ramps linking the active portions of the floor thrust and the roof thrust. This regular, asymmetric pattern of faults is assumed *a priori* in kinematic models of duplex development (Boyer & Elliott 1982; Groshong & Usdansky, 1988). In this paper, we seek a mechanical rationalization for it.

Mechanical analysis and interpretation of faulting has been pursued along several lines. As an empirical basis for predicting regions of faulting and fault orientations, Hubbert (1951) adopted the Mohr–Coulomb failure criterion together with the observation that faults in isotropic rocks are oriented at $\pm(\pi/4-\phi/2)$ to the direction of maximum compression, where ϕ is the angle of internal friction. The scheme proposed by Hubbert (1951), and used by Hafner (1951) and Sanford (1959) and others to estimate the region of failure by faulting, consists of evaluating the stress distribution from a purely elastic solution and assigning the region of faulting as that in which the stress state satisfies or would violate the failure criterion. Hubbert (1951) had suggested that this procedure is apt to be reliable if the faults forming are isolated from each other, but this notion has no basis in observation or in physical principles.

Hubbert recognized a key issue in the interpretation of fault arrays. While the Mohr–Coulomb criterion does not distinguish between the two conjugate fault sets that may develop at any point, frequently only a single set is present or predominates. In speaking of faulting under conditions of asymmetric loading in a sandbox Hubbert (1951, pp. 370–371) says the following.

> At any given point, the stresses on each of these conjugate surfaces are the same; so that if they are at the critical value for fracturing on one surface, the same is true for the other. However, slippage along a finite surface involves an integral of conditions along the entire surface. Whatever this integral should be, in homogeneous materials with symmetrical stress distributions, its value over each of a pair of conjugate surfaces of potential slippage will be the same, so that equal slip should occur on each family of surfaces.
>
> In asymmetrical cases, the integral of any stress quantity over one surface or family of surfaces will in general be different form that over the conjugate surface or family. Because of this inequality, slippage should occur on one set before on the other, or on only one of the two sets. In the

From Knipe, R. J. & Rutter, E. H. (eds), 1990, *Deformation Mechanisms, Rheology and Tectonics*, Geological Society Special Publication No. 54, pp. 123–131.

asymmetrical geological system here considered, faulting develops almost exclusively on the concave-upward set of surfaces. The same was true for the reverse faulting in the sandbox experiment.

In experiments with symmetrical systems, slippage on both sets of surfaces is observed.

These remarks are not concrete enough to provide an explanation for why a single fault set forms, since neither the integral nor the physical principle involved are specified. Hubbert clearly excludes, however, an explanation which involves only the local behaviour.

Odé (1960) assumed that the process of faulting in rock is equivalent to that in sand and clay, and it is chiefly from experiments in these materials that he draws the observations that motivate his treatment. He notes that visible faulting in these materials occurs well beyond the elastic limit, and that the faults propagate slowly with imposition of continued deformation. He concludes that faulting is not dependent on elastic behavior. He chooses rigid-plastic behavior with a pressure-dependent Coulomb yield condition and an associated flow law as constitutive equations. The characteristic directions in plane strain, which are the same for both the stress equations and the velocity equations for the choice of an associated flow law, are identified with fault orientations because they are surfaces across which discontinuity in velocity is possible.

A major result from more recent studies of natural fault configurations is that faults or fault zones have characteristic structures that can be interpreted in terms of the mechanical interaction between discrete fault surfaces (Segall & Pollard 1980, 1983), or between segments of fault surfaces undergoing simultaneous slip (Martel & Pollard 1989). Segall & Pollard (1980) estimate stress distributions in a body containing a set of pre-existing fault segments by assuming that the body behaves purely elastically during an episode of slip on the segments in response to a far-field stress. Solutions are obtained which satisfy, for example, an appropriate relation between normal and shear tractions on the fault surfaces. The resulting stress field may be used to estimate the subsequent propagation of the faults or the inelastic deformation in their vicinity. To this extent, the evolution of faults and fault zones can be followed. The large difference in mechanical properties between a set of weak fault surfaces or fault zones and the surrounding rock justifies to some extent the approximation of an elastic response to an episode of slip. The origin of the faults themselves is not treated. This contrasts with Odé's (1960) formulation, in which the material is treated as uniform, and the fault surfaces do not have properties which distinguish them from the rest of the rock.

In the present study, we return to Odé's (1960) suggestion that faulting in rock may be treated in terms of the deformation of an idealized rigid-plastic solid at yield with constant, strain independent properties. In brittle material with a strain-dependent yield strength, the conditions for faulting as a shear instability are met only after initial inelastic yielding (Rudnicki & Rice 1975). Because of this, an identification of faults or fault zones with weakened surfaces or zones cannot be made in our model. Odé (1960) estimated only the two possible fault orientations at each point of the medium, but in the present study we seek the positions of faults relative to a given pre-existing fault, and their orientations. These new faults are identified with narrow shear zones which occur in a flow that is induced by a perturbation representing the pre-existing fault. If the actual material exhibits weakening in shear, these shear zones in the flow would then act to localize the development of faults.

The model

Model configuration

Typically, duplexes exhibit two kinds of asymmetry. (1) The faults separating the successive horses from the layer dip opposite to the direction of translation of the hanging wall. (2) Horses are only added at the 'front' of the duplex, the end in the direction of hanging wall motion, and are accreted to the hanging wall. The floor thrust propagates forward into the zone of decollement, and as it does so, the roof thrust above it is de-activated. The rocks above the roof thrust undergo some deformation as the successively detached horses are translated along the ramp thrust and the upper flats, but the rock below the floor thrust is not deformed.

The second group of features is tied to the circumstance that the medium below the floor thrust is either significantly stronger or subject to a lower stress then the medium above the roof thrust, or both. The planar form of the floor thrust is then maintained and the other features are a consequence of this. To conform to this, the model configuration will consist of a 'rigid' basement separated from the layer in which the duplex forms by a weak sliding surface. A weak interface also separates the layer from an overlying deformable half-space.

It is beyond the scope of the present study to model the complete structure of a typical duplex with its array of fully-imbricated to partially-imbricated horses. Instead we address the following two-part question. Given a pre-existing fault, what is the position, relative to it, of the next fault to form, and what is the attitude of this fault? We may suppose that the specification of a pre-existing fault with a particular orientation will automatically fix the orientation of any subsequent fault. In the model, however, we only specify beforehand the position of the intersection of the pre-existing fault with the upper surface of the layer, not its dip.

In the model, the region containing the pre-existing fault and several times the layer thickness on either side of it is treated as a portion of a layer of nearly uniform thickness. For tractability, this region is duplicated periodically, so that the analysis deals with a periodic flow within a layer of unbounded extent. The locus of the pre-existing fault is established by introducing a shallow cusp or notch. This form is suggested by the configuration associated with incipient imbrication, but it has also been established empirically that a form of this sort is required to produce a structure like that observed. The perturbed form of the shallow notch on the interface is symmetrical. This avoids one source of bias that might favor one fault orientation over the other.

Rheological behaviour

The layer is taken to be a rigid plastic solid at yield since the flow in such a body admits the development of narrow shear zones that simulate in certain respects faults or fault zones. In the present study, we use the constitutive relations for a von Mises plastic solid with a pressure-independent yield strength K. A pressure-dependent Mohr–Coulomb yield condition would be more suitable for rock, but the proper choice of a flow law remains uncertain. The von Mises material is the simplest kind of plastic solid, and its use leads to some simplifications in the details of formulation and analysis of the model. The yield stress in the layer is taken to be uniform.

Slip is permitted on the interfaces bounding the layer, which simulate the floor and roof thrusts in a duplex. The slip rate is taken to be linearly proportional to the resolved shear stress. This is called a viscous slip condition. For example, if the upper interface is plane, the slip velocity of the half-space relative to the layer for a uniform shear stress σ_{xz} is given by $U = \Gamma\sigma_{xz}$. Slip is easier as the coefficient Γ, which has the dimension of velocity over stress, increases. In the model the value of Γ on the floor thrust will be assigned a large value, so that slip is easy there; its value at the upper surface will be varied.

In the natural process, the rock overlying the duplex may only be weakly deformed. In particular, the thickening of the duplexed layer will be far greater than that in the rock above the roof thrust. The present model cannot incorporate such a difference, and the overlying rock must also be supposed to undergo thickening. To minimize the effect that this will have on the deformation with the duplex layer, the overlying half-space is a uniform, isotropic, incompressible, linear-viscous fluid.

Basic state of flow

The layer and bounding media are subject to the basic state of flow illustrated in Fig. 1. Both are subject to layer-parallel shortening at a uniform rate $\bar{\varepsilon}_{xx}$, and to layer-parallel shear at uniform rates $\bar{\varepsilon}_{xz}$ and $\bar{\varepsilon}'_{xz}$, respectively. Uniform translation on the upper and lower weak surfaces takes place according to the viscous slip condition. The velocity of the rigid substrate is set equal to zero.

Perturbing flow

The basic state described is a solution to the field equations and boundary conditions only if the interface between the layer and the medium is perfectly plane, and if the slip constants on the two surfaces are uniform. The simulation of faulting, or more exactly of fault initiation, corresponds to inducing a perturbing flow in which an array of narrow shear bands is present. Because of the assumed uniformity of material properties and the rigidity of the substrate, the

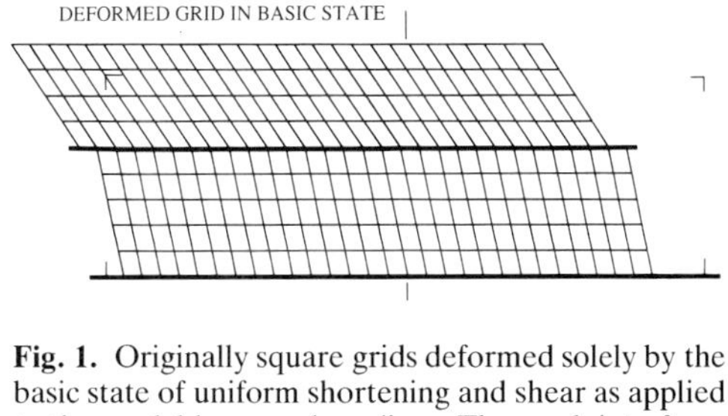

Fig. 1. Originally square grids deformed solely by the basic state of uniform shortening and shear as applied to the model layer and medium. The weak interfaces of the layer are marked by the two heavy lines. The tick marks show the original location of the undeformed grid. Here the layer is stronger than the medium, and is, therefore, sheared by a smaller amount.

only way to induce such a flow is through a perturbation in the shape of the upper interface.

The shape of the perturbed interface that has been selected is a shallow symmetrical cusp or notch. On either side of the cusp the interface is smooth. In evaluating the flow in the vicinity of the perturbing feature on the interface, a truncated Fourier series solution is used, and consequently, the form of the interface is periodically continued beyond the immediate region of interest. The notches are located along the upper surface spaced at the wavelength L. Since each of these gives rise to shear bands, it is necessary to choose L as some integral of the repeat distance determined by geometry of the velocity characteristics in order to avoid a dense array of intersecting shear bands. If L is at least equal to twice the repeat distance, a shear band will be excited forward of the notch at a place where the interface is smooth. This demonstrates that the excitation of shear bands does not require the prior presence of a notch at its point intersection with the weak upper surface.

Mathematical formulation and analysis

A detailed account of the analysis will appear in another paper (Davies & Fletcher, in prep.). Briefly, the perturbing flow is found as an approximate solution obtained by: (1) linearizing the constitutive relations (yield condition and flow law) about the basic state; (2) approximating the boundary conditions to linear terms in the interface slope. The solution for the periodic interface shape is obtained by summing the independent elementary solutions for the Fourier components in the shape. Similar techniques have been used in earlier studies (Fletcher 1974, 1977; Smith 1977; Kilsdonk & Fletcher 1989).

Terms involving the basic flow and the perturbation in interface height about its mean level ζ that enter the boundary conditions at the upper surface drive the perturbing flow. (The height of the interface above the planar floor thrust is $z = H + \zeta$, where H is the layer thickness.) When linearized to first order in the slope of the interface, the boundary conditions for continuity of normal velocity, for the dependence of the slip rate on the shear traction, and for the continuity of shear and normal traction, respectively, take the forms,

$$\tilde{w}'(x,H) - \tilde{w}(x,H) = U\partial\zeta/\partial x$$

$$\tilde{u}'(x,H) - \tilde{u}(x,H) - \Gamma\tilde{\sigma}_{xz}(x,H) = -2\bar{\varepsilon}_{xx}H \{\{\cos 2\delta/(R\theta)\}\,\partial\zeta/\partial x - \sin 2\delta\{(1/R) - (1/\cos 2\delta)\}\,(\zeta/H)\}$$

$$\tilde{\sigma}'_{xz}(x,H) - \tilde{\sigma}_{xz}(x,H) = 2K\{\cos 2\delta - R\}\,\partial\zeta/\partial x$$

$$\tilde{\sigma}'_{zz}(x,H) - \tilde{\sigma}'_{zz}(x,H) = 0$$

where u and w are components of velocity in the directions x and z, parallel and normal to the layer; σ_{xx}, σ_{zz}, and σ_{xz} are components of stress; the tilda denotes quantities in the perturbing flow and a bar denotes quantities in the basic state; a prime denotes quantities in the overlying half-space; and $\bar{\varepsilon}_{xx} < 0$ is the mean rate of layer shortening. The behaviour depends on the three dimensionless parameters in the right-hand sides of the above equations δ, R, and θ, δ is the angle between the positive x-axis and the direction of maximum compression in the layer, and $\bar{\varepsilon}_{xz}/\bar{\varepsilon}_{xx} = \tan 2\delta$. $R = 2\eta' \, \|\bar{\varepsilon}_{xx}\|/K$ is the ratio of the strength of the overlying half-space to that of the layer, where η' is the viscosity of the half-space. $\theta = H/(2\eta'\Gamma)$ is a measure of the relative contribution of deformation in the half-space to the accommodation of a gradient in horizontal velocity at the surface of the layer. If θ is small, this gradient is chiefly accommodated by interfacial slip. A large local gradient in horizontal velocity is associated with the presence of a shear zone.

The effects of the separate terms in the right-hand sides can be assessed by directly examining their contributions to the perturbing flow. It is known for example that the right hand side of the third equation drives interfacial instability such as folding, and that the term in the right hand side of the first equation does not lead to either interface amplification or shear band formation. The excitation of shear bands derives chiefly from the first term in the right-hand side of the second equation. The development of asymmetry derives from the second term. At present, however, a simple physical explanation for the development of the asymmetry has not been worked out.

Results

The results to be presented and discussed here are the velocity fields in the layer for the perturbing flow alone. These are represented either as an array of velocity vectors in a square grid (e.g., Fig. 2) or the deformed grid obtained by connecting the ends of these vectors (e.g., Fig. 3).

We first consider the case in which the shearing stress on the weak surfaces vanishes. This is not the case when duplexes are formed, but it is realized in the shortening by brittle processes in a layer embedded in a medium that behaves in a ductile fashion. Figure 4 shows an example in which a single layer embedded in more clay-rich

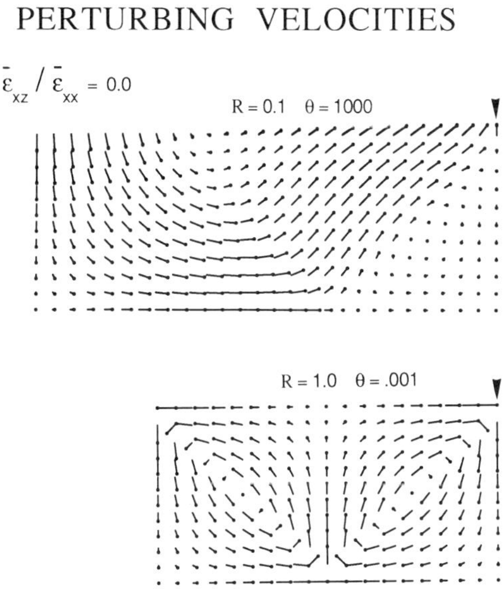

Fig. 2. The perturbing velocities for half the wavelength described in Fig. 3. The magnitude is arbitrary.

rocks is both faulted and folded. This approximately symmetric case provides a reference against which to judge the effect of superposed layer-parallel shear.

The only two factors which may effect the behavior in this case are the relative strength of the overlying medium, as described by the ratio of strengths or effective viscosities, and the magnitude of the slip resistance on the interfaces. If the strengths of the layer and the overlying medium are the same, the contribution from one of these terms vanishes. The remaining term is, in fact, that which is chiefly responsible for the shear band pattern within the layer. Figure 3 shows the deformed grids for two cases: (a) a nearly welded interface (θ = 1000) and a weak overlying medium (R = 0.1); and (b), a weak interface (θ – 0.001) and an overlying medium of equal strength (R = 1). The basal surface is a weak slip surface in both cases. The position of the cusp on the upper interface is indicated by the arrow.

In the case of the welded interface, a choice of equal strengths would result in no perturbing flow, so we have chosen a case where the overlying medium is weak. This leads to a weak interfacial or 'folding' instability and the spacing

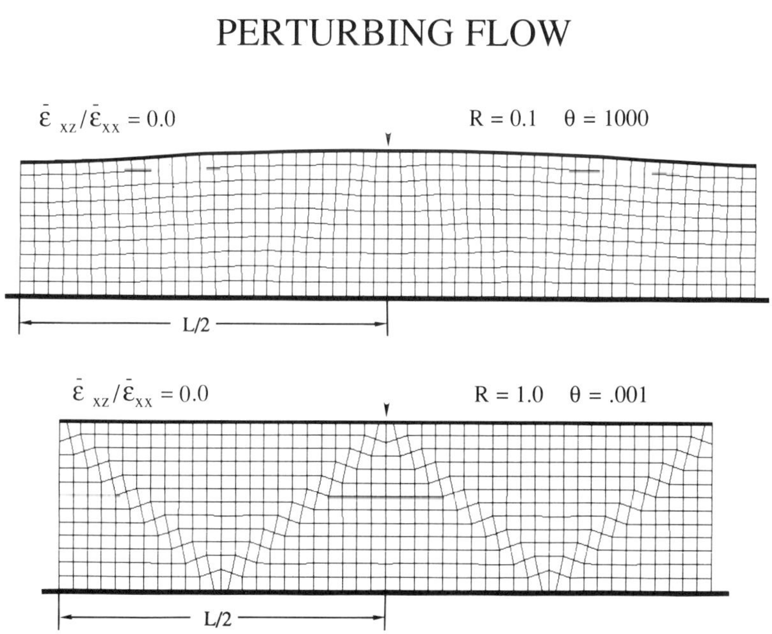

Fig. 3. Originally square grids deformed by connecting the ends of the velocity vectors for the perturbing flow (Fig. 2). Layers are deformed solely in layer-parallel shortening. The small arrowhead marks the position of the notch on the interface. In the top diagram the upper interface is nearly bonded, whereas in the bottom diagram both interfaces are weak.

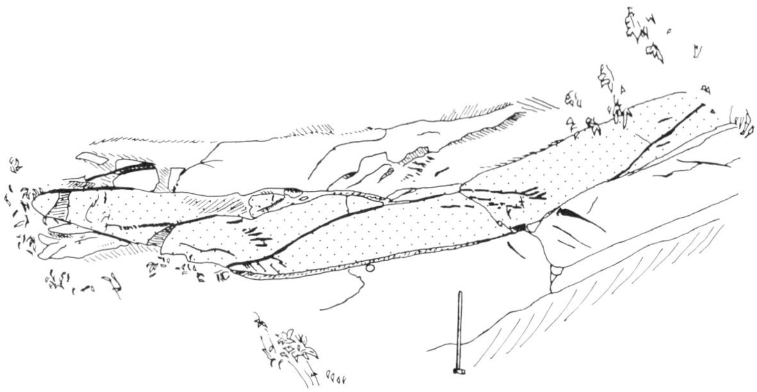

Fig. 4. Single, confined layer (stippled) broken-up by through-going right-dipping faults and left-dipping fault zones. The two sets of shear zones alternate in the layer each reflecting near the termination of a shear zone from the conjugate set. Approximately 1 m long sledge hammer in the foreground for scale.

of the cusps is chosen here to be twice the dominant wavelength for the instability. Instability is indicated by the vertical motions at the upper surface amplifying the initial interface form (Fig. 3). The presence of the cusp also leads to weakly-developed disturbances that extend symmetrically downward from the cusps at 45°. This is the orientations of the characteristic directions. These disturbances reflect at the weak basal surface, but then disappear into regions of diffuse flow.

For the case of a weak interface and strong overlying medium, the spacing of the cusps is chosen to be twice the repeat distance for reflecting shear bands, as determined by the orientation of the characteristic directions, or $L = 4H$. Well-defined shear bands extend from the cusp, reflect from the weak basal surface, and again from the weak upper surface of the layer. Recall that the configuration continues periodically beyond the portion shown. There is no apparent attenuation of the shear bands. As a consequence, shear bands are present not only at the cusp, but at a point on the upper surface where this is smooth. Although an example is not shown here, strongly-developed shear bands are also the dominant feature in models with a weak overlying medium and a weak interface. This implies a locus of values of R and θ which separates cases which show predominantly interfacial instability from those which show predominantly internal flow.

The shear bands of the perturbing flow are more clearly shown in plots of the perturbing velocity fields (Fig. 2). This is particularly true of the more weakly-developed disturbances in the perturbing flow along the characteristic directions for the nearly welded interface case. The velocity field of the case of a weak interface (Fig. 2) shows an additional feature of the flow. The total perturbing flow over the wavelength L must vanish in the sense that the area integral of any component of the velocity vector or of the strain-rate tensor must vanish. Considered separately, the shear bands correspond to a net shortening of the layer. Consequently, the flow within the blocks bounded by the shear bands must show a complementary layer-parallel extension, and this can be seen to be the case. The extension within the blocks is approximately uniform. When the basic-state shortening is superposed, this extension subtracts from it, and so the total flow consists of blocks with diminished shortening and shear bands which balance this to give the prescribed mean rate of shortening.

Consider now the superposition of layer-parallel shear stress. Only models with a weak upper interface ($\theta = .001$) and equal strength of layer and medium are shown (Fig. 5). The spacing between cusps L is taken to be twice the repeat distance for reflecting shear bands, now given by $x_r = 2H/\cos(2\delta)$. We present the results for the three cases, in which the ratio $\bar{\varepsilon}_{xz}/\bar{\varepsilon}_{xx} = 0.2, 0.4$, and 1.0. These correspond to $\delta = 5.5°$, 10.7°, and 22.5° and to $x_r/H = 2.04$, 2.15, and 2.83. In each case, pairs of shear zones emanate from the cusp and reflect off both the weak basal surface and the upper weak interface (Fig. 5). The two sets of shear zones make an angle of 90° with each other, and since they lie along the characteristic directions, they are bisected by the directions of the maximum and minimum compressive stresses. The two shear zones are not equally well-developed. The shallower shear zone is the more strongly

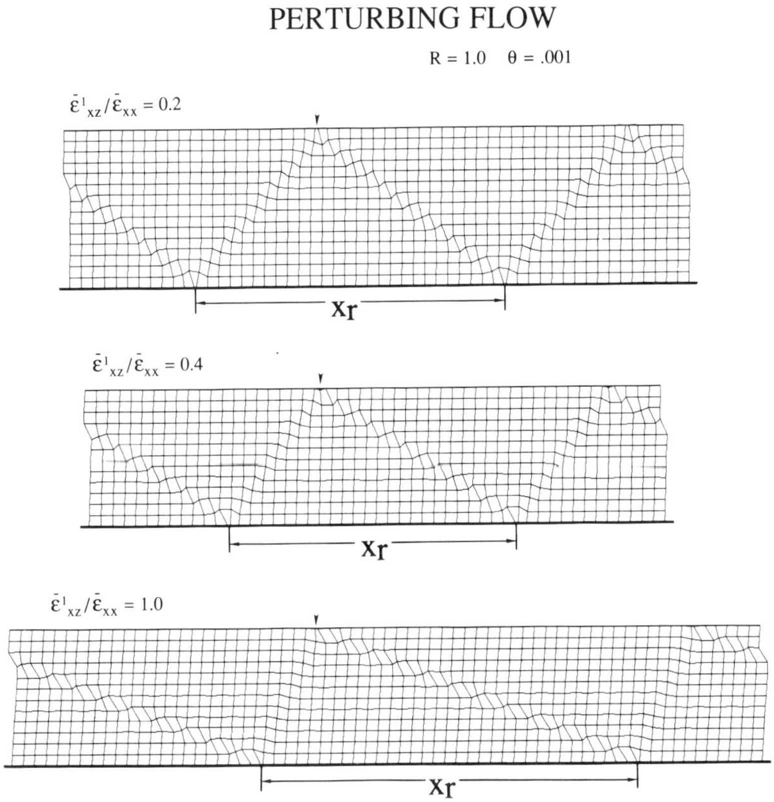

Fig. 5. Initially square, deformed grids in layers with weak interfacial contacts. The layers are deformed in layer-parallel shortening and shear.

developed, and this asymmetry increases with the relative magnitude of the shear stress parallel to layering or with increase in the angle δ. By the time δ has reached 22.5°, the steeper and weaker shear zones are barely discernible.

The total perturbing flow in the layers deformed in layer-parallel shortening and shear, as shown in a plot of the perturbing velocity field (Fig. 6), must also vanish over the distance x_r. In this case, the kinematics of the flow is more complex than that for layer-parallel shortening. The shear bands correspond not only to a net shortening of the layer, but also a counterclockwise (negative) rotation of the flow. Thus, the integral of the perturbing components of velocity, strain rate, and the single vorticity component $\omega = 1/2(\partial u/\partial z - \partial w/\partial x)$ must all vanish over the repeat distance. Consequently, the blocks in between the shear zones show a complementary layer parallel extension and a clockwise (positive) rotation as seen in the velocity field.

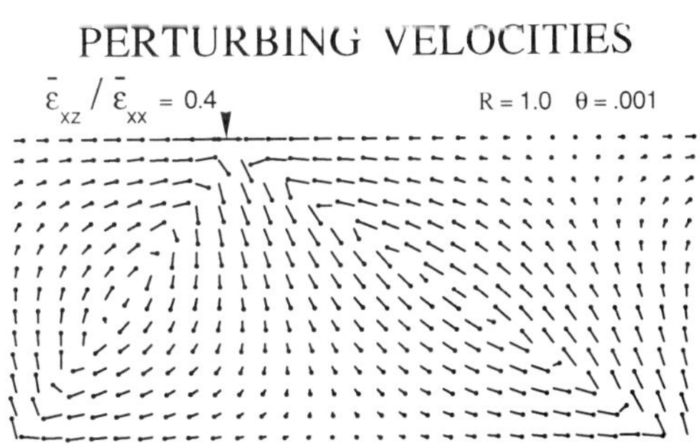

Fig. 6. Perturbing velocity field in a section of the wavelength of the middle diagram in Fig. 5.

Discussion

The actual process that results in a single set of regularly-spaced thrust faults bounding the horses of a duplex is more complex than that in the present model. It may best be described as the initiation and propagation of a fault or fault zone. Qualitatively, the process may be described as follows. A thrust fault cutting the layer is in a state of plastic yielding. The disturbance associated with the intersection of the fault with the weak detachment surface is then propagated downward along the characteristic direction. The process is then repeated at the lower detachment. In the case of a duplex, something prohibits the development of an incipient shear zone along the downward path, but along the upward path, another incipient fault zone is produced which localizes the position of the fault zone demarcating the next horse. The present analysis and results confirm and correct this qualitative picture, but on the other hand, they replace the process of propagation with a simpler 'process' that incompletely represents it.

The perturbation in interface shape simulates in an approximate way the conditions induced by the intersection of a fault with the upper detachment surface. The flow associated with this shape perturbation is shown to correspond to a shear zone leading upwards to the cusp, and, if the layer is subject to shortening alone, to an additional shear zone extending downward from it to the floor thrust. Another shear band is induced at the intersection of the former band and the floor thrust. With superposed layer-parallel shear, the shear bands make shallower and steeper angles to the layer, the shallower set is intensified, and the steeper set is weakened. For a sufficient amount of layer-parallel shear, the steeper set is wholely suppressed. Since the average of the perturbing flow vanishes, the perturbing motions in the regions between the shear zones correspond to layer-parallel extension, and in the case of superposed layer-parallel shear to a rigid-body rotation as well. If the actual material is strain-softening, the additional deformation in the shear bands will lead to locally enhanced degradation of the material and the establishment of a true fault zone. Finally, if the upper surface is not sufficiently weak, only shear bands passing downward from the notch are formed (the lower surface is always weak), and further zones are suppressed by the strong attenuation of the localized flow due to deformation of the overlying viscous medium.

The instantaneous state of motion in the model is evaluated at a single 'time' or more precisely, a single interface configuration. This cannot be viewed as a single step in the actual process. Neither fault initiation nor fault propagation have been simulated. The present results are compelling enough to suggest that the model goes some way towards providing a mechanical explanation for the regular spacing of faults in a duplex, and an answer to the long-standing question of what determines the selection of one fault set from the two permitted by the failure criterion.

Using the present model, it would be possible to follow both the evolution of the interface shape (as long as this remains a low-slope surface), and the internal deformation of the layer. However, since deformation of brittle rock leads to marked strain-hardening or strain-softening, this exercise would not apt to be particularly fruitful.

Conclusions

A model for the regular spacing and dip direction of the thrust faults initially demarcating the horses of a duplex has been based on the incipient flow set up in a layer separated from a rigid substrate and an overlying deformable medium by weak sliding surfaces. The rheological behaviour of the layer is that of a uniform von Mises plastic solid at yield. The direction of the mean compressive stress in the layer is inclined at an angle δ to the layer interface. A secondary flow is induced by perturbing the interface shape to a form that, in the region of interest, has a shallow cusp or notch but is otherwise smooth. The velocity characteristics in the layer lie at approximately $\pm$ 45° to the direction of maximum compression. The principal conclusions are as follows.

(1) The initiation of a perturbing flow requires either that the interface permit slip or, if it is welded, that the overlying medium be weaker than the layer. In either case, shear bands extend from the cusp along the characteristic directions downward to the base of the layer.

(2) The effect of a weak interface is to enhance strongly the development of shear bands and to suppress the broad deformation of the interface (folding). More significantly, shear bands in the layer are induced at a distance from the cusp, where the interface is still smooth. This feature is identified with the initiation of a new thrust fault forward of those already formed.

(3) The effect of shear is to steepen one set of characteristics and shallow the other. Shear bands developed along the shallower set are intensified while those along the steeper set become more diffuse. With significant shear parallel to layering, the more diffuse set becomes hardly discernible. This conforms to the pattern seen in duplexes. It is probable that more subtly developed structures corresponding

to the diffuse conjugate shear bands are generally present in natural structures.

This research was supported by US National Science Foundation Grant EAR-8511842. We thank C. Froidevaux, F. Morrison, and two anonymous reviewers for reviewing the manuscript, and C. Baker for assisting with the typing.

References

BOYER, S. E. & ELLIOTT, D. 1982. Thrust Systems. *American Association of Petroleum Geologists Bulletin*, **66**, 1196–1230.

FLETCHER, R. C. 1974. Wavelength selection in the folding of a single layer with power law rheology. *American Journal of Science*, **274**, 1029–1043.

—— 1977. Folding of a single viscous layer: exact infinitesimal amplitude solution, *Tectonophysics*, **39**, 593–606.

GROSHONG, R. & USDANSKY, S. I. 1988. Kinematic models of plane-roofed duplex styles, *Geological Society of America, Special Paper*, **222**, 197–206.

HAFNER, W. 1951. Stress distributions and faulting. *Geological Society of America Bulletin*, **62**, 373–398.

HUBBERT, M. K. 1951. Mechanical basis for certain familiar geological structures. *Geological Society of America Bulletin*, **62**, 355–372.

KILSDONK, B. & FLETCHER, R. C. 1989. An analytical model of hanging-wall and footwall deformation at ramps on normal and thrust faults. *Tectonophysics*, **163**, 153–168.

MARTEL, S. J. & POLLARD, D. D. 1989. Mechanics of slip and fracture along small faults and simple strike-slip fault zones in granitic rock. *Journal of Geophysical research*, **94**, 9417–9428.

MITRA, S. 1986. Duplex structures and imbricate thrust systems: Geometry, structural position, and hydrocarbon potential. *American Association of Petroleum Geologists Bulletin*, **70**, 1087–1112.

ODÉ, H. 1960. Faulting as a velocity discontinuity in plastic deformation. *Geological Society of America, Memoir*, **79**, 293–321.

RUDNICKI, J. W. & RICE, J. R. 1975. Conditions for the localization of deformation in pressure-sensitive dilatant materials. *Journal of the Mechanics and Physics of Solids*, **23**, 371–394.

SANFORD, A. R. 1959. Analytical and experimental study of simple geological structures. *Geological Society of America Bulletin*, **70**, 19–52.

SEGALL, P. & POLLARD, D. D. 1980. Mechanics of discontinuous faults. *Journal of Geophysical Research*, **85**, 4337–4350.

SEGALL, P. & POLLARD, D. D. 1983. Joint formation in granitic rock of the Sierra Nevada. *Geological Society of America Bulletin*, **94**, 563–575.

SMITH, R. B. 1977. Formation of folds, boudinage, and mullions in non-Newtonian materials. *Geological Society of America Bulletin*, **88**, 312–320.

The failure mechanism for deep-focus earthquakes

H. W. GREEN II & P. C. BURNLEY

Department of Geology, University of California, Davis, CA 95616 USA

Abstract: Experimental deformation of Mg_2GeO_4 olivine at pressures between 1 and 2 GPa in the spinel stability field has led to discovery of a faulting instability that develops at the kinetically-controlled threshold of transformation. Very fine-grained olivine and spinel are found in fault zones. Deformation at lower temperatures is ductile; transformation is inhibited and specimens are very strong. Deformation at higher temperatures also is ductile but transformation is rapid and specimens are much weaker. Detailed examination of the microstructures of specimens deformed in the faulting regime lead to an anticrack theory of faulting that explains the experimental data and provides a fundamentally new mechanism for deep-focus earthquakes. The new mechanism is analogous to the Griffith theory of fracture; nucleation and growth of spinel under stress produces spinel-filled microanticracks normal to the maximum compressive stress that link up to produce faulting. The friction paradox for deep earthquakes is resolved because this faulting process provides a fine-grained, superplastic, 'lubricant' for faults. The temperature distribution within subducting slabs of lithosphere requires that the conditions of instability are reached as a natural consequence of subduction; metastable olivine in the interior of deep slabs warms to a critical temperature where faulting ensues in the presence of a shear stress.

Deep-focus earthquakes have been an enigma since shortly after their discovery nearly 70 years ago (Turner 1922; Wadati 1928) because it was recognized that friction would make rock fracture or sliding on a fault impossible at great depth (Jeffreys 1929; 1936). The Terzaghi effect (reduction of the effective normal stress due to a fluid partial pressure; cf Raleigh & Paterson 1965) might allow this difficulty to be overcome, but it was not until recently that evidence was produced suggesting that the phenomenon may be able to operate at very high pressures (Meade & Jeanloz 1988). Furthermore, there is little reason to expect that significant fluid pressures would be generated at very great depth. It is unlikely, therefore, that the high frequency of earthquakes at depths > 300 km (> 20% of earthquakes wolrdwide; Frohlich 1989) can be attributed to this mechanism.

A considerable variety of alternative mechanisms has been proposed, including plastic instabilities, shear-induced melting and instabilities accompanying recrystallization and/or phase transformation (see Kirby 1987, for discussion of the various hypotheses and their drawbacks). Many recent workers have postulated fundamental involvement of the olivine–spinel phase transformation because it is known to occur in subducting lithosphere under conditions approximating those of deep earthquakes. In particular, Sung & Burns (1976) proposed an implosive instability caused by rapid transformation of significant volumes of material. They based their model in part on studies of the kinetics of the transformation. Vaisnys & Pilbeam (1976) proposed a theortical model that had aspects of both implosive transformation and shearing instability, and Kirby (1987) proposed that a shearing instability could be triggered by incipient transformation. Kirby based his arguments on experimental evidence of anomalous faulting in metastable tremolite and ice, although he and his colleagues found no direct evidence of transformation in either system (Burnley & Kirby 1982; Durham *et al.* 1983). Both Vaisnys & Pilbeam (1976) and Kirby (1987) cited a number of possible reasons for initiation of a shearing instability. In particular, Kirby (1987) collected together all previously proposed suggestions and, in addition, called attention to the stress concentrations to be expected around an elliptical inclusion of the more dense phase formed in a plane of high shear stress within the less dense phase. He suggested that such an inclusion of spinel in olivine could propagate in its own plane and lead to faulting. More recently, Meade & Jeanloz (1989) produced acoustic emissions accompanying martensitic transformations in Si

From Knipe, R. J. & Rutter, E. H. (eds), 1990, *Deformation Mechanisms, Rheology and Tectonics*, Geological Society Special Publication No. 54, pp. 133–141.

and Ge at very high pressure and suggested that their results might have analogy to deep earthquakes.

We have now verified that a faulting instability does develop under certain restricted conditions during the olivine-spinel transformation in Mg_2GeO_4 (Burnley & Green 1990) and we have proposed a new mechanism of instability and deep faulting based on the microstructures present in our experimental specimens (Green & Burnley 1989). Here we summarize both the experimental results and the theory of faulting and we demonstrate how deep earthquakes are an expected consequence of this new mechanism.

Experimental results

Mechanical data

All experiments were conducted on a synthetic Mg_2GeO_4 polycrystal in a modified Griggs apparatus at $T = 900-1400$ K, $P = 1-2$ GPa in a salt confining medium. Deformation experiments were at strain-rates ($\dot{\varepsilon}$) of $10^{-4}-10^{-5}$ s^{-1}, except for a single experiment at 1300 K, 2 GPa, $\dot{\varepsilon} = 10^{-3}$ s^{-1}. Stresses were measured externally with a load cell. Under conditions where the confining salt is molten, this apparatus can measure stresses to an accuracy of a few megapascals and a precision of about 1 MPa (Green & Borch 1989, 1990). When the salt is solid, however, precision of stress measurement remains high but accuracy is reduced to about $\pm$ 100 MPa. Unfortunately, under differential stresses ($\sigma_1-\sigma_3$) greater than about 800 MPa, molten salt induces axial splitting in the Al_2O_3 loading piston which leads to extrusion of the top of the Pt capsule into the cracks, rupture of the capsule and consequent intrusion of salt and axial splitting of the specimens. Most experiments in this study were very strong (>1000 MPa), hence had to be conducted with solid salt. Fortunately, at these extreme strengths, the poorer stress resolution has little effect on interpretation of the results (Burnley & Green 1990).

Conditions under which specimens were deformed are shown in Fig. 1. Specimens deformed below 1100 K were very strong and ductile; the only sign of transformation was production of submicroscopic martensitic lamellae of spinel that did not coarsen (Burnley & Green 1989). With the sole exception of the 10^{-3} s^{-1} experiment, specimens deformed above 1200 K were also ductile, but much weaker; transformation was extensive by incoherent nucleation and growth. In the intervening temperature region, *all* experiments failed by through-going faults which formed at approximately 30^0 to σ_1 (except for specimens that were purposefully stopped early to investigate pre-faulting microstructures). Under these conditions, incipient transformation by both mechanisms was found. Stress–strain curves for representative experiments are given in Fig. 2.

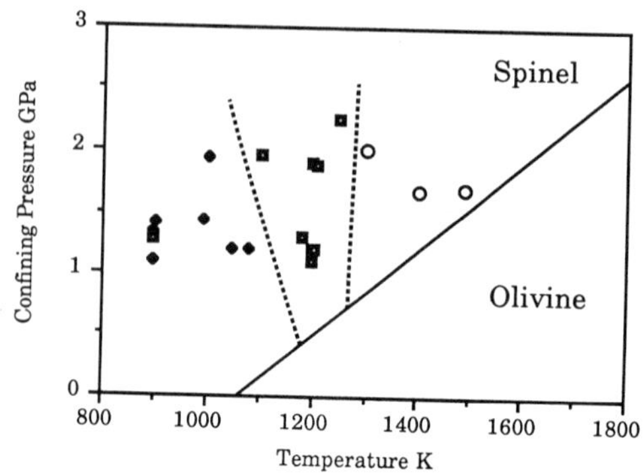

Fig. 1. Conditions of experiments conducted at $10^{-4}-10^{-5}$ s^{-1} on metastable Mg_2GeO_4 olivine. Filled diamonds represent specimens that were very strong and ductile; submicroscopic martensitic nuclei of spinel are produced under these conditions but they do not grow. Open circles represent specimens that were weak and ductile; extensive incoherent nucleation and growth occur at these higher temperatures. Filled squares denote specimens that failed by faulting; in all faulted specimens except the single colder one, very fine-grained spinel occurs in the fault zones and also throughout the specimens in the form of microlenses oriented normal to σ_1. No spinel was found in the cold faulted specimen. We believe that this specimen failed by normal brittle failure (see text for discussion). Phase boundary after Ross & Navrotsky (1987).

A single cold specimen (Fig. 1) developed faulting without any evidence of transformation. That specimen buckled, leading to local stress concentrations in excess of the already very high stress measured during loading. That specimen also showed textures indicative of both plastic deformation and microcracking. We conclude that that specimen failed by normal brittle processes (Burnley & Green 1990). We emphasize that this specimen and the other lower temperature, ductile, specimens were much stronger than the specimens that failed in association with the transformation. Therefore, we know that specimens displaying transformation-associated faulting were loaded to stresses very much lower than would be required to produce ordinary faulting by brittle fracture.

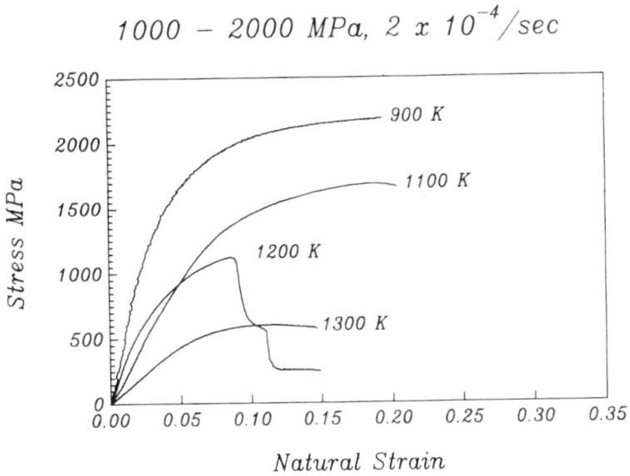

Fig. 2. Stress–strain curves for representative specimens deformed at 10^{-4} s^{-1} in the cold, ductile regime (strong), in the hot, ductile regime (weak), and in the intermediate region where faulting occurs. The 1300 K specimen was conducted at a confining pressure of 2 GPa; the others at 1 GPa.

Petrography

Under the conditions where faulting occurs, transformation is slow compared to the time-scale of the experiment. Although most of the spinel that grows under these conditions is below the resolution of the optical microscope, spinel was easily seen optically along the interface between the specimen and the capsule; apparently the transformation is catalysed on platinum under conditions just slightly colder than it initiates in bulk. This phenomenon was reported previously for Mg_2GeO_4 by Rubie & Champness (1987). In addition, spinel could be identified on olivine grain boundaries by the combination of the reduction of apparent birefringence and the characteristic morphology that develops in connection with the incoherent nucleation mechanism (see below). In several cases, small euhedral spinels nucleated by the martensitic mechanism also were found. Transmission electron microscopy (TEM) has established (Burnley & Green 1990) that the material in fault zones is extremely fine-grained and consists of fragments of olivine (<2 μm) and crystals of spinel ($<0.3\mu$m). Scanning electron microscopy (SEM) of polished specimens etched with HCl (Fig. 3a) confirmed the presence of fine-grained spinel in fault zones (olivine etches preferentially, leaving spinel standing above the general surface and making it easy to identify; spinel also is dissolved slightly, especially at grain boundaries, revealing the very fine grain size previously determined by TEM).

Detailed optical and SEM examination of faulted specimens and others deformed under the same conditions but stopped just before or just past the maximum stress showed that transformation was just beginning throughout the specimens. The majority of this portion of the study has been accomplished through utilization of the SEM at Leeds University, UK, that is equipped for electron channelling. This instrument has allowed imaging of the spinel on polished specimens using the back-scattered-electron mode. The 8% density difference between the two polymorphs results in a brighter image for the spinel phase. An earlier search for this contrast difference on our conventional SEM had failed to produce positive results, but the superior (syton) polishing method and the combination of the back-scatter and channelling modes (Lloyd 1987) at Leeds produced excellent contrast. Under the conditions of faulting (Fig. 3b, c) and also at slightly higher temperatures where strengths were still significant but transformation ran more quickly (Fig. 3d), the spinel developed as small lenticular domains both within olivine crystals and on grain boundaries. In both the specimens that faulted and those stopped just before faulting, the microlenses displayed a very strong preferred orientation normal to the maximum principal stress, σ_1, but in the warmer specimens, the preferred orientation was greatly reduced (Fig. 3; see also histograms in fig. 11 of Burnley & Green 1990). The grain size of the spinel within the microlenses of specimens that faulted was found to be even finer than in fault zones, with many, perhaps most, grains less than 30 nm diameter (Burnley & Green 1990). This texture is in marked contrast to the habit of crystals nucleated by the martensitic mechanism, which grow as cube-modified octahedra both under stress and in post-deformation annealing experiments. The lens morphology is also very different from spinel nucleated and grown under hydrostatic conditions. In the latter case, spinel nucleates by the incoherent mechanism on grain boundaries or on pyroxene inclusions within the olivine, but the crystals grow as octahedra.

Discussion

Anticracks

The striking difference between the morphology of spinel developed under faulting conditions and that developed under hydrostatic conditions suggests a relationship between the spinel microlenses and the instability that leads to faulting. In particular, the crack-like shape and strong preferred orientation of the spinel do-

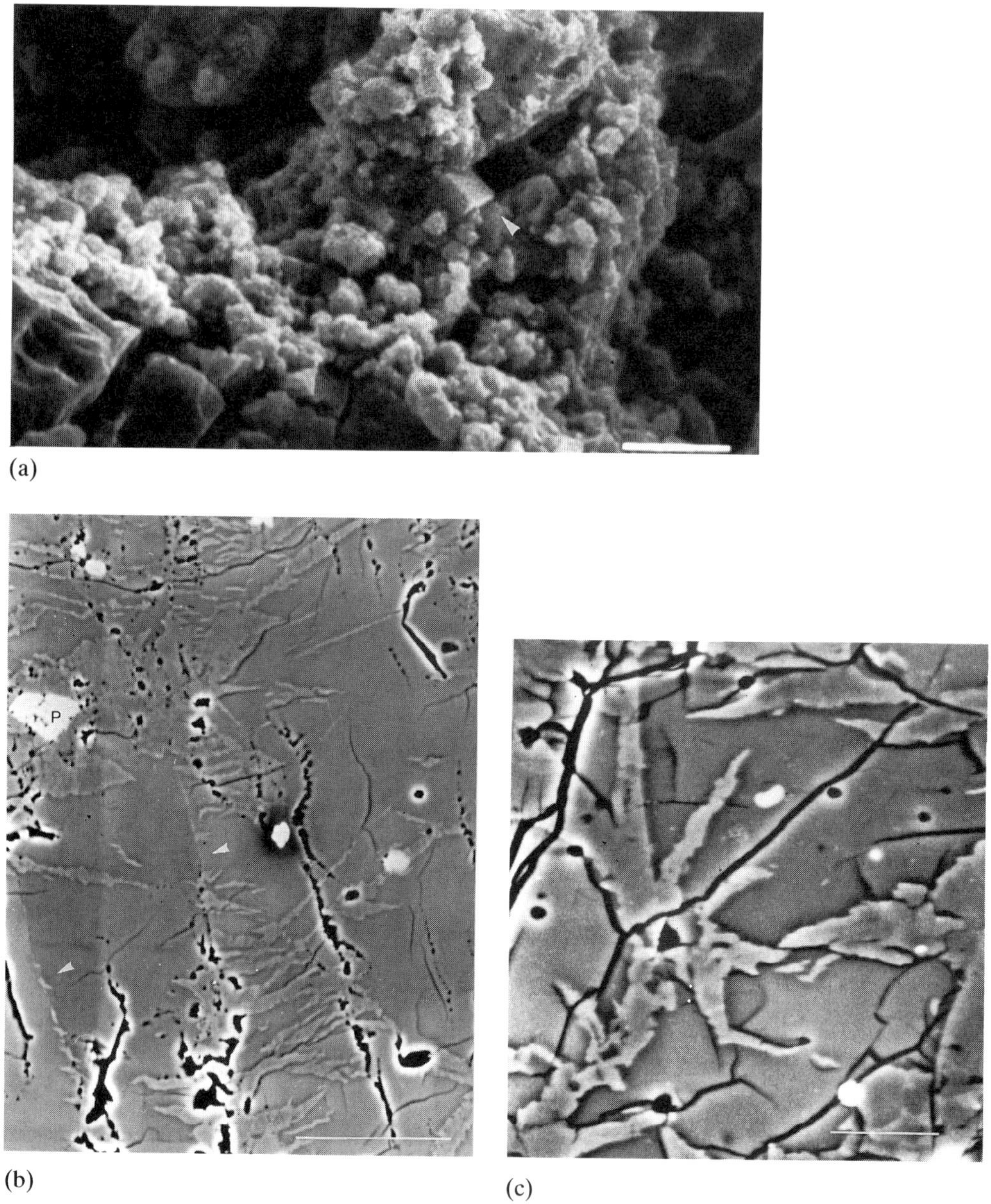

Fig. 3. (**a**) Scanning electron micrograph of a fault zone in a specimen polished and etched with HCl to enhance visibility of spinel. The boundary of the fault zone runs across the lower left of the micrograph. The small, rounded grains are spinel. The elongated crystal fragment (arrowhead) that is more extensively dissolved is olivine. Scale bar is 10 μm; (**b**) Scanning electron micrograph, imaged with combined back-scattered and channelling modes (BSE/CH), of a region distant from faulting. The denser spinel phase occurs as lenses (anticracks) and shows as brighter than the olivine but much less bright than pyroxenes (p). Lenses are especially numerous growing off of grain boundaries (arrowheads). Note the very strong preferred orientation of the spinel lenses normal to σ_1 which is NS; Scale bar is 30 μm; (**c**) BSE/CH SEM micrograph of a specimen deformed at 1300 K, 2 GPa, just above the window of instability, showing development of larger anticracks that lack significant preferred orientation. Scale bar is 10 μm. The black lines and spots in (b) & (c) are, respectively, cracks produced by quenching and reduction of pressure, and plucking during specimen preparation.

mains is remarkably similar to the well-known development of microcracks with a strong preferred orientation that leads to normal brittle failure (cf. Paterson, 1978). Green & Burnley (1989) adopted the concept of anticracks developed by Fletcher & Pollard (1981) in another context and extended the concept to include anticracks filled with a phase more dense than their host. The anticrack model provides a physical basis for interpretation of the orientation and shape of spinel microlenses and, by analogy to brittle fracture, provides a connection between the spinel morphology and the faulting in our experiments.

Briefly, anticracks are the inverse of cracks (hence the name). If a small planar flaw is pulled open by a tensile stress, a (mode I) crack opens. The displacements associated with the crack are outwardly directed and the stress concentrations at the tips of the crack are tensile. In the anticrack case, densification within a crack-shaped region produces inward-directed displacements and compressive stresses at the tips of the region. For similar geometries the displacements and stresses are identical in the two cases except for a change of algebraic sign (Fletcher & Pollard 1981). Microcracks oriented obliquely to σ_1 do not propagate in their own plane (Petit & Barquins 1988). Rather, they propagate such that newly created portions are more parallel to σ_1, leading to the strong preferred orientation of microcracks in brittle materials just before failure. Similarly, oblique microanticracks will propagate by enhanced transformation at their tips in such a way that their orientation becomes progressively more perpendicular to σ_1, leading to the strong preferred orientation of spinel microlenses in our specimens deformed under faulting conditions.

Macroscopic shear failure by fracturing occurs by linking-up of microcracks that are parallel to σ_1 (mode I cracks), but the detailed nature of the instability is still poorly understood (Paterson 1978; Petit & Barquins 1988). We also have little direct evidence of the detailed process by which faults form in our experiments. Specimens stopped during the brief period of strain softening that just preceeds faulting show no evidence of any features in shear orientation. Apparently, the formation of a singular shear feature (the fault) occurs in a very short period of time. Accordingly, we have proposed (Green & Burnley 1989) that shear failure in our specimens occurs analogously to that of brittle fracture, by linking-up of microanticracks that are perpendicular to σ_1 (mode I anticracks). In the former case (cracking), macroscopic shear faulting is controlled by friction and hence is strongly inhibited by pressure; in the latter case (anticracking), the very fine-grained spinel from the mode I features is incorporated into the fault zone as it forms, hence macroscopic shear faulting is controlled by the rheology of this material. Vaughan & Coe (1981) showed that the flow law for fine-grained Mg_2GeO_4 spinel has a stress exponent of *c*. 2 and they concluded that the material was 'superplastic'. Superplasticity (cf. Suery & Mukherjee 1985) is a ductile phenomenon in which the dominant flow mechanism is grain-boundary sliding and the rate-controlling step is either diffusion or dislocation glide. The grain size of spinel in the mode I anticracks of our specimens ranges from less than 10 to a few hundred nanometres making it 1–3 orders of magnitude smaller than that of Vaughan & Coe's (1981) material (1–3 μm) and suggesting that strain rates in our faults could be many orders of magnitude more rapid than in their experiments. Although one intuitively tends to associate extensive grain-boundary sliding with slow strain rates, there is a trade-off between grain-size and strain rate; superplastic creep now has been documented to strain rates in excess of 10 s^{-1} in materials with a stable grain size of 0.5 μm (Mukherjee *et al.* 1990). This also suggests that very much higher strain rates are possible during the faulting in our specimens, given the extremely small grain size of spinel within the anticracks. A further contribution to rapid flow during faulting could be a result of the observed incorporation of olivine fragments into the fault zones. If incompatibilities build up at olivine–spinel interfaces during grain-boundary sliding, local transformation is another mechanism for accomodation of those incompatibilities.

The ductile processes operating during superplastic flow are only weakly dependent on pressure, hence there is no theoretical reason to consider this faulting mechanism to be unreasonable at very high pressures (see note added in proof). Indeed, any relatively coarse-grained material undergoing transformation to a finer-grained aggregate of a denser phase is potentially susceptible to such an instability. We surmise, however, that an exothermic reaction may be necessary so that when the anticrack linking process begins to run, local heating increases the kinetics of the transformation, and leads to catastrophic runaway. Since faulting occurs under conditions where nucleation and growth of the more dense phase are just possible, an endothermic transformation would likely be controlled by flow of heat to the site of reaction and hence catastrophic runaway would be inhibited in these poorly conducting materials. Alternatively, if the runaway can get started without heating, viscous heating could provide the feedback mechanism leading to catastrophic failure.

Deep-focus earthquakes

Virtually all deep earthquakes take place in regions of currently-active subduction and are universally attributed to failure within the slab of oceanic lithosphere as it descends (cf. Frohlich 1989). The models differ in detail, but all of them have in common a cold internal core because the known velocity of subduction is too rapid for conduction of heat from the surrounding mantle to keep pace. As a consequence, all models of active slabs show depression of internal isotherms by hundreds of kilometres. A representative example of such models (Turcotte & Schubert 1982) is presented in Fig. 4. In addition to the temperature distribution within the slab, the equilibrium position of the olivine-spinel phase boundary is shown. The positive slope of the boundary in pressure-temperature space (e.g. Akimoto *et al.* 1976) results in rise of the boundary to shallower depths within the slab. Of course, in the $(Mg,Fe)_2SiO_4$ olivine of the mantle, the equilibrium reaction is $\alpha \rightarrow \beta$ rather than the direct $\alpha \rightarrow \gamma$ reaction of our experiments. The β phase is also significantly more dense than olivine and our anticrack mechanism is driven only by the volume reduction, hence our mechanism should apply equally to either transformation. This is especially true because our evidence suggests that the faulting instability is associated with the incoherent mechanism rather than the martensitic mechanism (Green & Burnley 1989; Burnley & Green 1990).

Our experiments show that transformation at sufficiently high temperature leads to rapid bulk transformation and no faulting instability. We indicate that in Fig. 4 by horizontal steps in isotherms of 1500 K and more, reflecting the exothermic nature of the reaction. For colder temperatures, however, our experiments (and those of Sung 1979, on natural olivine) show that transformation is inhibited and therefore the slab will pass through the equilibrium boundary unaffected. The conditions where the instability develops will depend on the degree of overstepping of the boundary (and hence pressure) and also on kinetic factors (temperature, time, defect density in the olivine) and the level of shear stress. We do not know the detailed nature of these various contributions for our experiments or for natural olivine, nor do we know the detailed temperature distribution

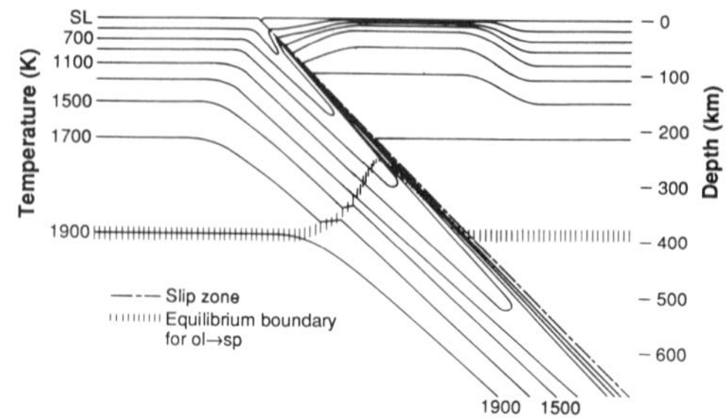

Fig. 4. Thermal model of subducting lithospheric slab (after Turcotte & Schubert 1982) showing the temperature distribution within the slab and the consequent distortion of the equilibrium olivine–spinel phase boundary. Isotherms of 1500 K and higher display horizontal steps at the phase boundary denoting aseismic transformation with negligible overstepping of the boundary. Lower temperature contours pass through the equilibrium boundary unperturbed, implying that the transformation is kineticly inhibited. For the sake of illustration, the conditions of instability can be taken to be the 1300 K isotherm. Any volume of peridotite in the cold interior of the slab must inevitably encounter the conditions of mechanical instability discussed in the text, leading to faulting if shear stresses are present. The result is a scattering of hypocenters over several hundred kilometres depth.

in downgoing slabs, but the evidence from our experiments is that temperature and time should be the most critical factors. Therefore, for the purpose of illustrating the implications of our mechanism for the Earth, we will neglect any pressure effect and assume that the rate of subduction, defect density and temperature distribution are constant. With these simplifications, faulting will be dependent only on temperature and stress; attainment of the critical temperature will initiate faulting if the slab is under sufficient stress.

It is immediately apparent from Fig. 4 that different parts of the slab will reach the critical temperature at different depths and therefore the model predicts faulting to be spread over a considerable depth range unless the critical temperature is very cold (*c*.900 K). Indeed, if the critical temperature is 1300 K, the entire range of earthquakes deeper than *c*.300 km could be a result of anticrack faulting associated with the olivine–spinel transformation (either $\alpha \rightarrow \beta$ or $\alpha \rightarrow \gamma$ depending on depth). The critical temperature for breakdown of natural olivine is not known, of course, but temperatures in the range 1100–1300 K are consistent with the experimental data of Sung (1979). If the critical isotherm is 1100 K or 1200 K, however, some other mechanism or transformation would be necessary for deeper earthquakes. Several other candidate transformations occur in the mantle transition zone, but we hesitate to suggest that they could lead to anticrack faulting because they operate in phases that represent less than half of the volume of the rock, if the rock is peridotite. The most likely possibility would be the various reactions occurring in $(Mg,Fe)SiO_3$, but unless the proportion of orthopyroxene is greater than currently believed for shallow lithosphere, this, too, is unlikely. The breakdown of the γ phase to perovskite + magneiowüstite is also not a strong candidate because the phase boundary has a negative slope, indicating that it should be depressed within the slab and therefore it should occur at depths greater than the deepest earthquakes.

The olivine–spinel transition cannot be associated with earthquakes shallower than *c*.250 km and no other densification transformation is known that could run through an interconnected phase of peridotite, basalt or eclogite. We consider it unlikely, therefore, that anticrack faulting can be responsible for intermediate depth earthquakes.

The remarkable exponential decline of earthquake activity from more than 100 per year of magnitude 5 or greater near the surface to about 1 per year at 300 km, followed by a marked rise in frequency before the total cutoff at *c*.700 km (Sykes 1966; Frohlich 1989) led Sykes (1966) to suggest that the distribution represents two different types of seismic activity. It is universally accepted that shallow earthquakes are the result of brittle failure and stick-slip frictional sliding on existing faults. We expect that any fundamental change of mechanism would alter the depth-dependence of earthquake frequency, hence we propose that all earthquakes shallower than 250–300 km are the result of brittle failure, probably assisted by fluid pressure (Terzaghi effect), due to trapped pore fluid, dehydration or decarbonation reactions or melting. The five-fold rise in frequency of earthquakes at greater depths is initiated at the appropriate depth for correlation with the olivine–spinel transition. Our anticrack mechanism readily explains that correlation as well as the lack of frictional inhibition of faulting and the existence of a narrow band of earthquake foci extending over a considerable depth interval.

The results of this investigation do not prove that deep earthquakes are a consequence of anticrack faulting. It would be an extraordinary coincidence, however, if the discovery of a fundamentally new faulting mechanism in connection with a phase transformation that is known to take place under the conditions of deep earthquakes were to prove to be geologically irrelevant. The simplest explanation of deep earthquakes, then, is that they are due to anticrack faulting accompanying the olivine-spinel transitions ($\alpha \rightarrow \beta$ or $\alpha \rightarrow \gamma$), with a critical temperature corresponding approximately to the 1300 K isotherm of the model in Fig. 4. Of course, Fig. 4 is just one of many models; the actual critical temperature could be somewhat lower or higher than 1300 K. At a depth of 650–680 km, the stable mineral assemblage of the mantle becomes perovskite + magnesiowüstite. This transformation may participate in the deepest earthquakes, but it is endothermic and anticracking may not be unstable. As far as is known, there are no phase transitions affecting the major mantle phases at greater depths, hence earthquakes would not be expected in the lower mantle, whether or not descending slabs penetrate the 650 km seismic discontinuity.

In retrospect, we note that individual aspects of our anticrack theory have appeared previously in various papers on this subject. For example, Vaisnys & Pilbeam (1976) invoked a rapidly-running phase transformation in a metastable phase, as did Sung & Burns (1976) and Sung (1979). McGarr (1977) implicitly foresaw the tendency to form mode I anticracks and Kirby (1987) recognized the potential mechanical instability introduced by thermodynamic metastability and the stress concentrating effects of a densification reaction. Sung & Burns (1976), Vaughan & Coe (1981) and Kirby (1987) also listed structural superplasticity as a possibility for loss of cohesion. Most remarkably, Bridgman (1945) listed the essential elements over 40 years ago when he wrote about a phase transformation: 'It is however, conceivable that *if* the transition is sluggish and *if* it is initiated by the chance formation of a nucleus in a region sufficiently far from thermodynamic stability, and *if* there is a small heating effect so as to bring the material automatically into a region of more rapid progress of the transition, the progress of the transition may be catastrophic' [emphasis added]. However, he went on to say that he considered it unlikely that all of these 'ifs' would be satisfied simultaneously. Over the last 45 years, however, we have learned that olivine undergoes an exothermic phase transformation under the appropriate conditions for deep earthquakes and that cold slabs of lithosphere penetrate to great depths. The results of our experiments now have shown that at a temperature just sufficient for growth of the denser phase, the progress of the transition can be catastrophic. It addition, modern knowledge of the superplastic behavior of fine-grained aggregates provides the mechanism to circumvent the friction paradox and our analogy with the Griffith theory of fracture establishes the self-organizing principle to make such failure an expected rather than improbable event.

The problem of the mechanism responsible for deep earthquakes has a very long history and has had many contributors. We thank all those whose earlier work contributed to the ideas expressed in this paper. In particular, the experiments summarized here are a natural extension of the work begun by R. Coe and P. Vaughan in the mid-seventies to examine the effect of stress on the olivine–spinel transition and the proposal by S. Kirby in 1987 that incipient transformation to a denser assemblage can trigger high pressure faulting. Special thanks go to D. Prior for assistance with the SEM and to J. Abril, whose constant care and development of the apparatus contributed in countless ways. This work was supported by the US National Science Foundation through grants #EAR85-18019, #EAR89-05059 and #EAR89-15938.

References

Akimoto, S., Matsui, Y. & Syono, Y. 1976. High-pressure crystal chemistry of orthosilicates and the formation of the mantle transition zone. *In*: Strens, R. G. S. (ed.) *The Physics and Chemistry of Minerals and Rocks*. Wiley & Sons, London.

Bridgman, P. W. 1945. Polymorphic transitions and geological phenomena. *American Journal of Science*, **243A**, 90–97.

Burnley, P. C. & Green, H. W. 1989. Stress dependence of the mechanism of the olivine–spinel transformation. *Nature*, **338**, 753–756.

—— & —— 1990. Faulting associated with the olivine to spinel transformation in Mg_2GeO_4 and its implications for deep-focus earthquakes. *Journal of Geophysical Research*, (in press).

—— & Kirby, S. H. 1982. Pressure-induced embrittlement of polycrystalline tremolite. *EOS*, **63**, 45.

Durham, W. B., Heard, H. C. & Kirby, S. H. 1983. Experimental deformation of polycrystalline H_2O at high pressure and low temperature: Preliminary results. *Journal of Geophysical Research, Proceedings of the Fourteenth Lunar and Planetary Science Conference*, **88**, b377–b392.

Fletcher, R. C. & Pollard, D. D. 1981. Anticrack model for pressure solution surfaces. *Geology*, **9**, 419–424.

Frohlich, C. 1989. The nature of deep-focus earthquakes. *Annual Reviews of Earth and Planetary Science*, **17**, 227–254.

Green, H. W. & Borch, R. S. 1989. A new molten salt cell for precision stress measurement at high pressure. *European Journal of Mineralogy*, **1**, 213–219.

—— & —— 1990. High pressure and temperature deformation experiments in a liquid confining medium. *American Geophysical Union Geophysical Monographs*, **56**, 195–200.

—— & Burnley, P. C. 1989. A new, self-organizing, mechanism for deep-focus earthquakes. *Nature*, **341**, 733–737.

Jeffreys H. 1929. *The Earth, It's Origin, History and Physical Constitution, 2nd ed.* Cambridge University Press.

—— 1936. The structure of the earth down to the 20° discontinuity. *Proceedings of the Royal Society of Edinburgh*, **56**, 158–163.

Kirby, S. H. 1987. Localized polymorphic phase transformation in high-pressure faults and applications to the physical mechanism of deep focus earthquakes. *Journal of Geophysical Research*, **92**, 13789–13800.

Lloyd, G. E. 1987. Atomic number and crystallographic contrast images with the SEM: a review of backscattered techniques. *Mineralogical Magazine*, **51**, 3–19.

McGarr, A. 1977. Seismic moments of earthquakes beneath island arcs, phase changes, and subduction velocities. *Journal of Geophysical Research*, **82**, 256–264.

Meade, C & Jeanloz, R. 1988. Ultra-high pressure fracturing: First experimental observations of deep focus earthquakes (abs). *EOS*, **69**, 490.

—— & —— 1989. Acoustic emissions and shear instabilities during phase transformation in Si and Ge at ultrahigh pressure. *Nature*, **339**, 616–618.

Mukherjee, A. K., Bieler, T. R. & Chokshi, A. M. 1990. Superplasticity in metals and ceramics, 207–233. *In*: Bilde-Sorensen, J. B. (ed.) *Materials Architecture*. Riso National Laboratory, Roskilde, Denmark (1989).

Paterson, M. S. 1978. *Experimental Rock Deformation – The Brittle Field*. Springer-Verlag, Berlin.

Petit, J. P. & Barquins, M. 1988. Can natural faults propagate under mode II conditions? *Tectonics*, **7**, 1243–1256.

Raleigh, C. B. & Paterson, M. S. 1965. Experimental deformation of serpentinite and its tectonic implications. *Journal of Geophysical Research*, **70**, 3965–3985.

Ross, N. & Navrotsky, A. 1987. The Mg_2GeO_4 olivine-spinel phase transformation. *Physics and Chemistry of Minerals*, **14**, 473–481.

Rubie, D. C. & Champness, P. E. 1987. The evolution of microstructure during the transformation of Mg_2GeO_4 olivine to spinel. *Bulletin de Mineralogie*, **110**, 471–480.

Suery, M. & Mukherjee, A. K. 1985. Superplasticity—correlation between structure and properties. *In*: Wilshire, B. & Evans, R. W. (eds) *Creep Behaviour of Solids*. Pineridge Press, Swansea, UK, 137–200.

Sung, C. M. 1979. Kinetics of the olivine-spinel transition under high pressure and temperature:

Experimental results and geophysical implications. *In*: TIMMERHAUS, K. D. & BARBER, M. S. (eds) *High Pressure Science and Technology*. Plenum Press, New York, 31–41.
—— & BURNS, R. G. 1976. Kinetics of the olivine-spinel transition: Implications to deep-focus earthquake genesis. *Earth and Planetary Science Letters*, **32**, 165–170.
SYKES, L. R. 1966. The seismicity and deep structure of island arcs. *Journal of Geophysical Research*, **71**, 2981–3006.
TURCOTTE, D. L. & SCHUBERT, G. 1982. *Geodynamics*. Wiley and Sons, New York.
TURNER, H. H. 1922. On the arrival of earthquake waves at the antipodes, and on the measurement of the focal depth of an earthquake. *Monograph of the National Royal* Astronomical Society Geophysics Supplement, **1**, 1–13.
VAISNYS, J. R. & PILBEAM, C. C. 1976. Deep-earthquake initiation by phase transformation. *Journal of Geophysical Research*, **81**, 985–988.
VAUGHAN, P. & COE, R. 1981. Creep mechanism in Mg_2GeO_4: Effects of a phase transition. *Journal of Geophysical Research*, **86**, 389–404.
WADATI, K. 1928. Shallow and deep earthquakes. *Geophysical Magazine*, **1**, 162–202.

Note added in proof

We have recently produced anticrack faulting at 14 GPa in association with the $\alpha \rightarrow \beta$ transformation in natural olivine (Green *et al.* 1990; in prep). These results, produced in D. Walker's multianvil apparatus, confirm that this faulting mechanism is not inhibited by pressure and strengthen the likelihood that it is reponsible for deep earthquakes.

Additional reference

GREEN, H. W., YOUNG, T. E., WALKER, D. SCHOLZ, C. & PRIOR, D. 1990. A test of the anticrack theory of deep earthquakes: initial results. *EOS*, **71**, July.

Instability, softening and localization of deformation

B. E. HOBBS, H. -B. MÜHLHAUS & A. ORD

CSIRO Division of Geomechanics, PO Box 54, Mt Waverley, Victoria 3149, Australia

Abstract: Material strain softening is commonly taken as a necessary and sufficient condition for localization in deforming rocks. However, there is a wide range of experimental and theoretical information which shows that localization can occur in sands, brittle rocks and ductile metals under strain-hardening conditions. This paper aims to bring these two contrasted views together. Three separate criteria are necessary in order to understand localization behaviour. The first involves the stability of the deforming system. The second determines whether a deforming system will undergo bifurcation so as to cease deforming in a homogeneous mode and instead deform in an inhomogeneous mode such as barrelling or localization. The stability and bifurcation criteria are independent of each other since barrelling is a stable mode whereas localization is unstable. The third criterion establishes if the unstable bifurcation mode is one of localization or of some other kind. Localization may arise from the presence of vertices on the yield surface (as in the case of pressure insensitive, rate dependent metals and in brittle rocks due to the development of preferred microfractures for slip) or from the constitutive relation being such that the plastic strain-rate vector is not normal to the yield surface (as in the cases of pressure sensitive, dilatant rocks, of materials deforming by crystal-plastic processes involving dislocation cross-slip and/or climb, and of visco-plastic materials in which voids are forming due to diffusive processes). It is important to distinguish between *material* and *system* softening (or hardening) behaviour. The theory for a kinematically unconstrained shortening experiment (that is, rigid, frictionless platens) indicates that localization can occur in strain-hardening materials but the system must strain-soften from then on; that is, localization occurs at peak stress for the system even though the material may continue to harden (or soften). However, the addition of kinematic constraints (such as friction at elastic platens, a constraint to deform in plane strain or at constant volume) means that localization may occur in a system that is monotonically strain hardening. Shear zones in naturally deformed rocks show ample evidence of dilatant behaviour in that evidence for the passage of large volumes of fluid during localization is common as is the development of dilatant vein systems. As such, since shear zones are strongly constrained by the elastic and (limited) plastic response of the relatively undeformed rocks surrounding the shear zones, strain-hardening behaviour of the system is to be expected as the norm, even if the rocks within the shear zones are undergoing material strain-softening.

'.... a geologist may not infer the type of stress-strain curve from the nature of faults or flow.' (Griggs & Handin 1960)

(1) Statement of the problem

Zones of localized shear strain are ubiquitous in deformed rocks and exist on all scales ranging from the scale of individual grains (in the form of deformation bands and kinks), through mesoscopic crenulation cleavages, kink bands and shear zones, to mylonite zones that occur on the scale of the crust (see Ramsay & Graham 1970; Ramsay 1980; Ramsay & Huber 1983 for some examples). There is a prominent theme in both the materials and the geological literature that strain softening, or a loss of load bearing capacity, is a necessary and sufficient condition for localization of deformation (see for example Tullis *et al.* 1982; Argon 1973; Poirier *et al.* 1979; Poirier 1980; White *et al.* 1980; Hobbs & Ord 1988).

In order to explain the loss of load bearing capacity, a number of softening mechanisms have been proposed including crystallographic preferred orientation softening, recrystallization softening, phase-change softening, fluid weakening and thermal softening (see Cobbold 1977; Poirier 1980; White *et al.* 1980). Such softening mechanisms are now firmly entrenched in geological thought and are the basis for many discussions on shear zone formation and evolution (White & Knipe 1978; Yuen *et al.* 1978; Zeuch 1982; Sorensen 1983; Segall & Simpson 1986). Strain-softening, or lack of

From Knipe, R. J. & Rutter, E. H. (eds), 1990, *Deformation Mechanisms, Rheology and Tectonics*, Geological Society Special Publication No. 54, pp. 143–165.

load bearing capacity, and the development of localized zones of deformation are also fundamental concepts in defining brittle as opposed to ductile behaviour of rocks (see Griggs & Handin 1960; Heard 1960; Paterson 1958, 1978; Rutter 1986).

However, and in contrast to this identification of localization with strain-softening, there is a wealth of literature which shows that localization may in fact be associated with strain-hardening. These studies include work on uncemented sands (Arthur *et al.* 1977; Vardoulakis *et al.* 1978; Vardoulakis 1980; Vardoulakis & Graf 1985), Coulomb materials with cohesion and dilatancy (Vermeer 1978; de Borst 1986), brittle rock with dilatancy (Rudnicki & Rice 1975), visco-plastic solids with and without thermal coupling (Anand *et al.* 1987), and plane strain compression of metals (Harren *et al.* 1988).

Indeed, it is not difficult to find examples in the geological literature where localization is associated with strain-hardening (see Griggs & Handin 1960; Byerlee & Brace 1969). Griggs & Handin, for instance, present an example of a dolomite loaded uniaxially at 25°C and 500 MPa confining pressure and another example of a dunite loaded uniaxially at 500°C and 500 MPa confining pressure where localization of deformation occurs under strain hardening conditions. Edmond & Paterson (1972) realized that in dilating materials deforming at non-zero confining pressure, strain-softening as represented by decrease in the applied load during uniaxial shortening is not the criterion to adopt in defining unstable deformation, since a contribution to the total amount of plastic work arises from the dilation working against the confining pressure as well as from the work done by the uniaxial stress. Edmond & Paterson point out that under general stress states no single stress component is a measure of the load bearing capacity of the material. They proposed instead that the increment of work needed for an increment of deformation is a better measure of stability than strain softening as measured by a single stress component. These arguments are continued in a recent paper devoted to dilatancy in the high temperature deformation of marble (Fischer & Paterson 1989).

The purpose of this paper is to bring these diverse views of stability, strain-softening and localization together. There are in fact three important aspects of the problem that in turn involve three quite different criteria. The first criterion is one that defines the *stability* of the deformation history with respect to an infinitesimal perturbation of the system. The second criterion is one that defines the *uniqueness* of the deformation path and whether *bifurcation* in the deformation path is possible. The third criterion is one that defines whether the bifurcation mode is one of *localization*, or of some other kind.

Throughout the discussion it is important to distinguish between the stress-strain response of the *material* undergoing deformation and that of the *system* being deformed. In Section 5 we give an example of a material whose stress-strain behaviour under uniaxial shortening conditions is strain-softening yet when this same material is deformed in a constrained manner (for instance so that the volume of the specimen remains constant) then the system response can be one of strain hardening. It is the system response which is important in structural geology in discussing the dynamics of deforming bodies of rock.

In particular, it arises that localization does not necessarily depend on strain-softening for its occurrence although softening of the system (not necessarily of the material) may result from localization. The important point as far as rock deformation is concerned is that potentially any system which involves dilatancy, or, quite generally, a non-unique relationship between the stress-rate and the strain-rate, may, under some circumstances, become unstable and localize, even in the strain-hardening regime.

In this paper we discuss criteria for stability (Section 2), uniqueness (Section 3), and localization (Section 4); we present some examples of localization in Coulomb materials (Section 5). We conclude with a brief review of conditions for localization in other materials (Section 6).

(2) Stability of the deformation history

(2.1) Summary of this section

In this section we discuss the conditions for a deforming body to continue deforming in the next increment of deformation in a stable or in an unstable manner. A deforming body is said to be stable if a vanishingly small perturbation in the deformation remains vanishingly small (Hill 1959; Koiter 1960). As indicated in the Introduction, many authors (see for instance, White *et al.* 1980) propose that the necessary and sufficient condition for instability is that the material *strain soften*. This criterion is a necessary condition only for materials which possess constitutive relations in which a special structural symmetry exists; such symmetry is present in elastic–plastic, pressure insensitive, incompressible materials at infinitesimal strains. The term *plastic* is used here to indicate any *permanent* deformation and is not meant to imply any particular mechanism of deformation. A pressure sensitivity of deformation arises in geological materials from frictional behaviour when grains or microfractures slide past each other, from dilatant behaviour and, in temperature sensitive materials, from a pressure term in the activation enthalpy (Nicolas & Poirier 1976) or from the opening of voids (Anderson & Rice 1985). For elastic-plastic materials with frictional properties and dilatancy, loss of stability is possible at infinitesimal deformations in the hardening regime of the material. The destabilizing influence of the inclusion of finite strain terms will be briefly discussed in Section 4. For pressure dependent materials (and this includes dilatant, frictional materials as well as for crystal-plastic deformation involving cross-slip and climb of dislocations [Rice 1976]) the criterion proposed by Hill (1958), namely, that at each point in the body $\dot{\sigma} \cdot \dot{\varepsilon} > 0$ is the condition for stable material behaviour, where σ is the stress and ε is the total (elastic plus plastic) infinitesimal strain. A dot indicates differentiation with respect to time. Some of the history of this stability criterion is considered by Truesdell & Noll (1965, p. 252) who attribute a special form of the criterion (ensuring strong ellipticity, rather than positive definiteness, of the governing equations) initially to Hadamard (1903). The Hill criterion simply states that the inner product of the next increment of stress with the next increment of strain should be positive for a stable deformation. This criterion therefore takes into account how the stress and the strain are varying with time and how they interact with each other to produce a product $\dot{\sigma} \cdot \dot{\varepsilon}$ which is negative for instability. For quasistatic deformations, the stability criterion adopted by Edmond & Paterson (1972) is approximately equal to $\dot{\sigma} \cdot \dot{\varepsilon}$.

(2.2) The Hill criterion for stability of deformation

In some instances, such as the necking of a specimen in extension or the buckling of a slender beam, the instability may arise for purely geometrical reasons such as a slight irregularity in the cross sectional area of the specimen or a critical slenderness of the specimen. The main areas of interest in geology however arise from material instability where it is the constitutive behaviour of the material itself rather than a geometrical irregularity that is the intrinsic reason for instability. Under such conditions where the instability arises solely from the material properties, a body of rock under dead loading (i.e. prescribed loads and or displacements which are either constant or purely externally controlled) is stable if over the volume, V, of the body,

$$\int_V \dot{\sigma}_{ij}\, \dot{\varepsilon}_{ij}\, \mathrm{d}\, V > 0 \qquad (2.1)$$

where the summation convention is implied for repeated subscripts. Equation (2.1) will certainly be satisfied if the assumed constitutive relation is such that

$$\dot{\sigma}_{ij}\, \dot{\varepsilon}_{ij} > 0, \text{ where } \dot{\sigma}_{ij} = \dot{\sigma}_{ij}\,(\dot{\varepsilon}_{ij}), \qquad (2.2)$$

for all kinematically admissible but otherwise arbitrary $\dot{\varepsilon}_{ij}$'s. In (2.1) and (2.2) σ_{ij} is the Cauchy stress tensor and ε_{ij} is the total (elastic plus plastic) strain. Criteria (2.1) and (2.2) are due to Hill (1958); Drucker (1951, 1956, 1959) proposed a stability criterion $\dot{\sigma} \cdot \dot{\varepsilon}^{p} > 0$ where only the plastic strain-rate $\dot{\varepsilon}^{p}$ was considered.

The consequences of the stability criterion (2.2) are given below. The state of stress at a particular point in a deforming body may be represented by a vector σ in stress space; the components of the vector are σ_{ij}. The surface which bounds all stress states that correspond to elastic deformations is the *yield surface* (Fig. 1a). For any stress state within the yield surface the deformation is elastic. If the material strain-hardens the stress state moves off the yield surface on to a new yield surface which is external to the previous one but which may or may not be of similar shape to it. In Fig. 1a the stress vector, σ, causing plastic deformation is shown as the line OP and the stress-rate vector, $\dot{\sigma}$, is shown parallel to the change in stress during strain-hardening (the line PQ).

The yield surface specifies the state of stress at yield. The evolution of the yield surface with continued deformation specifies the manner in which the material hardens or softens during deformation. In order to determine the plastic strain-rate that arises from a given stress, another relationship is required; this relationship is known as a *flow rule*. For materials that possess a constitutive law in which the principal axes of stress coincide with the principal axes of plastic strain-rate a surface is defined, the gradient of which specifies the orientation of the plastic strain-rate $\dot{\varepsilon}^{p}$ in six dimensional stress space. By analogy with potential theory, such a surface is called the *plastic potential surface* (Fig. 1b) and by definition, the plastic strain-rate is always normal to this surface.

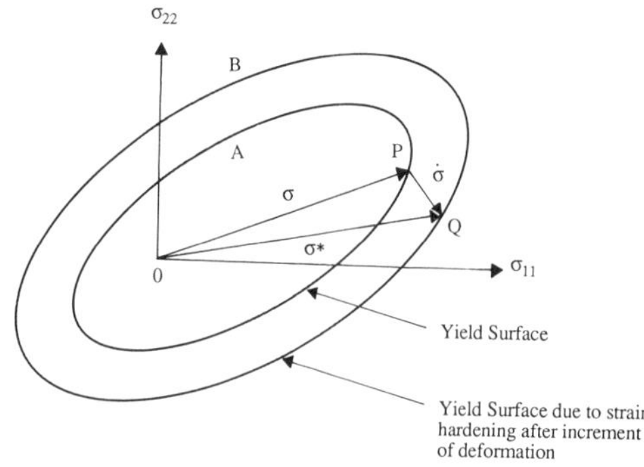

(a)

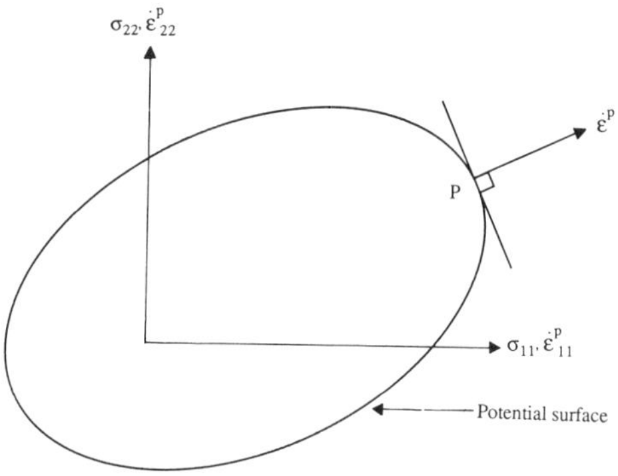

(b)

Fig. 1. (**a**) $\sigma_{11}-\sigma_{22}$ section through yield surfaces. $\overrightarrow{OP}$ represents the stress state σ which corresponds to a stress at yield on the yield surface A. After an increment of deformation the material has strain-hardened with a new yield surface B. σ^* is the new stress state at yield. The change in stress during the increment of deformation corresponds to $\overrightarrow{PQ}$. The stress-rate $\dot{\sigma}$ is parallel to $\overrightarrow{PQ}$. (**b**) $\sigma_{11}-\sigma_{22}$ section through the plastic potential surface. Since the principal axes of stress coincide with the principal axes of strain-rate for the constitutive relations considered here, the strain-rate coordinate axes can be placed in coincidence with the stress coordinate axes. The direction of the plastic strain-rate vector at P is defined as the normal to the plastic potential surface at P.

One can distinguish two kinds of constitutive behaviour. In the first kind, that classically used in metal plasticity, the yield and potential surfaces coincide. In this case, the plastic strain-rate vector is always normal to the yield surface. This is the 'normality rule' of Drucker (1951, 1956, 1959). When such normality exists the flow rule is called an *associated flow rule* and its consequences are illustrated in Fig. 2. Figure 2a shows associated deformation with strain-hardening. At a point P on the yield surface there are three vectors to be considered, namely the stress vector σ, the outwardly pointing stress-rate vector $\dot{\sigma}$, and the plastic strain-rate vector $\dot{\varepsilon}^{\mathrm{p}}$. Since for associated flow, the angles between $\dot{\varepsilon}^{\mathrm{p}}$ and both σ and $\dot{\sigma}$ are always acute (for loading), the plastic energy dissipation during the next increment of deformation, $\sigma \cdot \dot{\varepsilon}^{\mathrm{p}}$, and the plastic part of the stability criterion given by (2.2), $\dot{\sigma} \cdot \dot{\varepsilon}^{\mathrm{p}}$, are always positive. Since the elastic part of $\dot{\sigma} \cdot \dot{\varepsilon}$ is always positive (for loading) the value of $\dot{\sigma} \cdot \dot{\varepsilon}$ is always positive for an associated flow rule with strain hardening.

For strain-softening behaviour (Fig. 2b) the vector $\dot{\sigma}$ may lie at an obtuse angle to $\dot{\varepsilon}^{\mathrm{p}}$ and hence $\dot{\sigma} \cdot \dot{\varepsilon}^{\mathrm{p}}$ can be negative although the plastic energy dissipation $\sigma \cdot \dot{\varepsilon}^{\mathrm{p}}$ is always positive.

Thus for associated flow with a smooth yield surface, the stability criterion $\dot{\sigma} \cdot \dot{\varepsilon}$ can only be negative for a strain-softening situation. This observation is largely the origin of the commonly accepted dogma within geology that instability can only result from strain softening.

In the case of a pressure independent yield surface, the associated flow rule implies that the material is plastically incompressible. In general the associated flow rule applied to Coulomb materials (see Section 5) implies that the angle of dilation is equal to the angle of friction and for granular materials lacking cohesion this is unacceptable (see Vermeer & de Borst 1984) since it implies that zero plastic work is done during plastic deformation. This leads to the second kind of constitutive behaviour where the yield and potential surfaces do not coincide. This is known as a *non-associated flow rule* and the behaviour is characteristic of materials that show frictional and dilational responses to stress (Mandel 1965, 1966). This constitutive behaviour has been applied to soils, uncemented sands, concrete and brittle rock (Rudnicki & Rice 1975; Chen 1982; Vermeer & de Borst 1984).

The consequences of non-associated flow are illustrated in Fig. 3 where at the point P on the yield surface, the stress vector σ and the stress-ıate vector $\dot{\sigma}$ are shown. At P, also, the plastic strain-rate vector, $\dot{\varepsilon}^{\mathrm{p}}$, is shown; it is no longer normal to the yield surface at P but instead is normal to the plastic potential surface. This means that for strain-hardening conditions (Fig. 3a) the angle, θ, between $\dot{\sigma}$ and $\dot{\varepsilon}^{\mathrm{p}}$ can be obtuse and hence the inner product $\dot{\sigma} \cdot \dot{\varepsilon}^{\mathrm{p}}$ can be negative even though the plastic energy dissipation term, $\sigma \cdot \dot{\varepsilon}^{\mathrm{p}}$, is positive. For strain-

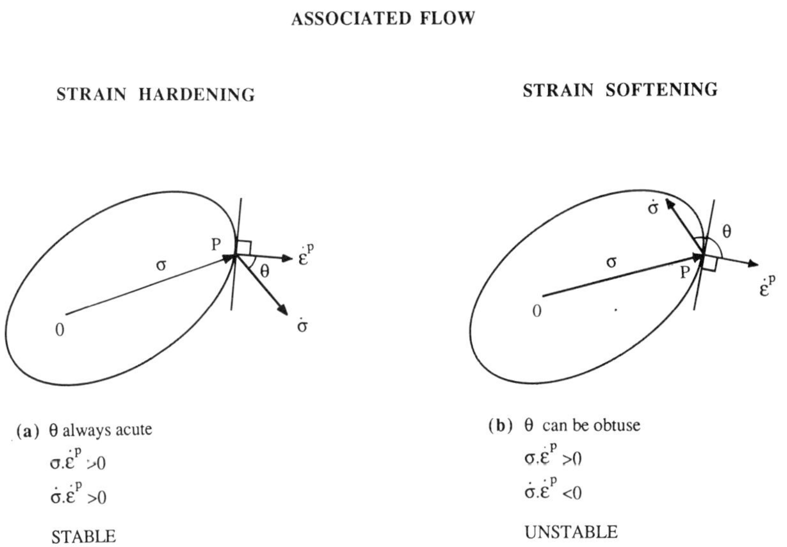

Fig. 2. Yield surfaces for associated flow (the plastic potential surface coincides with the yield surface). σ is the state of stress at P where the stress rate vector is $\dot{\sigma}$ and the plastic strain-rate vector is $\dot{\varepsilon}^p$. In the strain hardening situation (**a**) the angle θ between $\dot{\sigma}$ and $\dot{\varepsilon}^p$ is acute so that $\dot{\sigma} \cdot \dot{\varepsilon}^p$ is positive resulting in stable deformation. In the strain softening situation (**b**) θ can be obtuse so that $\dot{\sigma} \cdot \dot{\varepsilon}^p$ is negative. If $\dot{\sigma} \cdot \dot{\varepsilon}^p > \dot{\sigma} \cdot \dot{\varepsilon}^e$ where $\dot{\varepsilon}^e$ is the elastic strain-rate then $\dot{\sigma} \cdot \dot{\varepsilon}$ in this case is negative leading to unstable deformation.

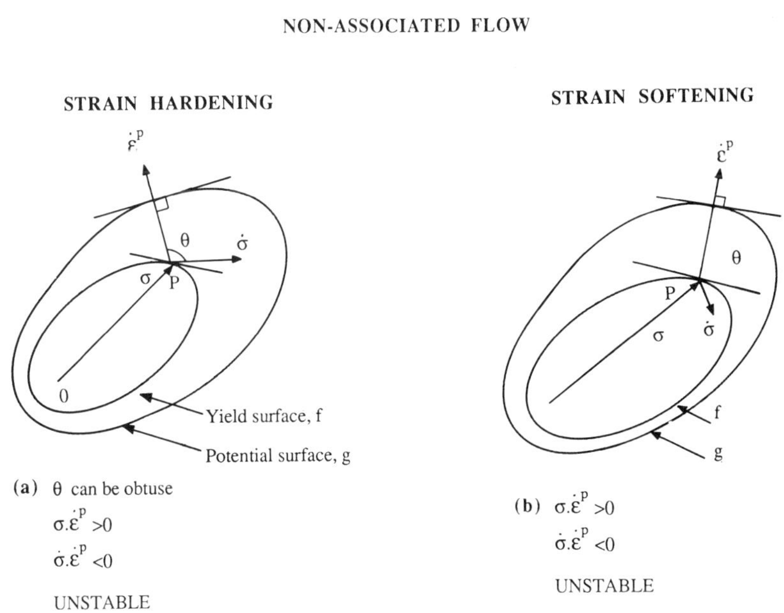

Fig. 3. Yield surfaces for non-associated flow (the plastic potential surface, g, does not coincide with the yield surface, f). At the point P on the yield surface, σ and $\dot{\sigma}$ are the stress state and stress-rate respectively. At P the plastic strain-rate is $\dot{\varepsilon}^p$ and this is normal to the plastic potential surface. In both strain-hardening (**a**) and strain-softening (**b**) situations $\dot{\sigma} \cdot \dot{\varepsilon}^p$ can be negative.

softening conditions (Fig. 3b), $\dot{\sigma} \cdot \dot{\varepsilon}^p$ can be negative although this is not essential.

The important aspect of non-associated flow for a material with a smooth yield surface is that instability is now possible even for a strain-hardening material.

For some materials the yield surface may not be smooth as shown in Fig. 1, 2 and 3 but instead 'corners' or 'vertices' may be present as shown in Fig. 4. This is true for single crystals deforming solely by crystallographic slip governed by the Schmid Law (Bishop & Hill, 1951; Lister *et al.*, 1978; Lister & Hobbs 1980; Hobbs 1985), and ductile polycrystalline materials deforming by crystal plastic processes may develop vertices on an initially smooth yield surface during deformation (Asaro & Needleman 1985). Rudnicki & Rice (1975) have proposed that vertices may develop on the yield surface of a rock deforming in a brittle manner due to the preferred development of potentially active sliding fracture planes within the rock mass.

The consequences of the presence of a vertex on the yield surface for an associated flow rule are illustrated in Fig. 4. The orientation of the plastic strain-rate vector $\dot{\varepsilon}^p$, corresponding to the stress state σ (Fig. 4), is not determined by the orientation of a single gradient but is constrained to lie within the cone of normals at the vertex. This means that the plastic energy dissipation, $\sigma \cdot \dot{\varepsilon}^p$, is always positive. It also carries with it the connotation that localization is possible for hardening conditions; the reasoning here is beyond the scope of this paper (see Hutchinson 1970; Rudnicki & Rice 1975; Rice 1976) but is based on the reduced elastic–plastic stiffness arising from a change in loading path at a vertex.

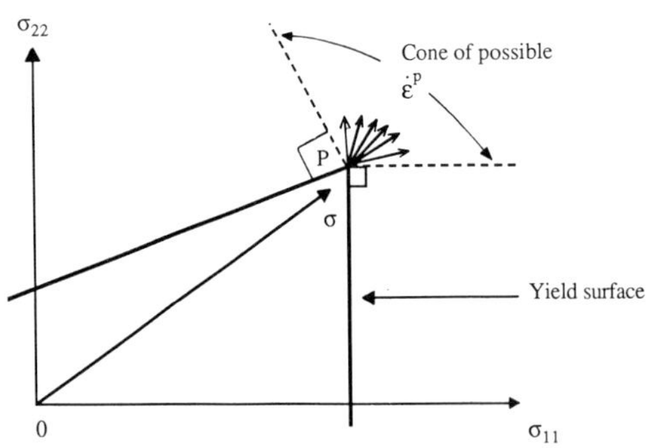

Fig. 4. Yield surface with vertex at P. At the vertex there is a non-unique relation between the stress σ and the plastic strain-rate $\dot{\varepsilon}^p$ which can lie anywhere in the cone defined by the normals to the yield surface at P.

For crystals that deform by mechanisms involving cross-slip or climb (in which case a volume change is involved; Groves & Kelly 1969) the flow rule is non-associated (Rice 1976). Similarly, slip on fracture surfaces during brittle deformation of solids also leads to a non-associated flow rule (Rudnicki & Rice 1975). Similar arguments to those put forward for smooth yield surfaces with non-associated flow also apply to yield surfaces with vertices so that, again, instability is possible during strain-hardening.

The criterion for stability proposed by Edmond & Paterson (1972) is $(d^2W/d\varepsilon^2_1) > 0$ for stability, where W is the total work done during deformation and ε_1 is the shortening strain. In the practical application of this criterion, however, Edmond & Paterson delete the contribution due to the elastic volume change due to the application of the differential stress. Under the conditions of the triaxial compression test, with constant confining pressure, the deformation of the specimen is in fact fully controlled by ε_1 and under these conditions the Edmond & Paterson criterion is a special case of the criterion (2.2). To see this first we note that $d\varepsilon_1 = \dot{\varepsilon}_1 dt$ so that (for constant $\dot{\varepsilon}_1$) $d^2W/d\varepsilon_1{}^2 > 0$ is equivalent to $\ddot{W} = d^2W/dt^2 > 0$. Thus if the stability of the material is only tested with respect to those quasistatic homogeneous deformations possible under triaxial conditions then (2.2) and the Edmond & Paterson criterion are approximately equivalent. The Edmond & Paterson criterion however, as with the Hill criterion, tests only for stability and not for bifurcation. Thus, these criteria may indicate stability for a particular increment of deformation when in reality a bifurcation to an unstable branch is about to take place.

(2.3) Conclusions to be drawn from this section

In classical deformation theory where the yield surface is smooth and coincides with the potential surface (associated flow rule), instability in the deformation is only possible for strain-softening conditions. This is the situation in classical metal plasticity where the constitutive behaviour is insensitive to pressure and is the basis for the entrenched thought in the geological literature that localization only develops under strain-softening conditions. Vertices on the yield surface have a destabilizing effect and enable localization to occur during hardening.

If the yield surface does not coincide with the potential surface (non-associated flow rule) as is the case in constitutive behaviour characterized by friction or by a strong pressure sensitivity for deformation then instability is possible under strain-hardening conditions.

It is important to note that these conclusions apply only to the *material* constitutive behaviour. We now proceed to distinguish between the *material* and the *system* response to deformation.

(3) Uniqueness of the deformation path

Close to the undeformed state, the solution of any reasonably posed deformation problem is well behaved. A one to one correspondence between the loading history and the displacement field exists; that is, the solution is unique. As the load is increased, no abrupt change occurs in the pattern of deformation; that is, the strain remains homogeneous. However, further from the undeformed state more than one solution may exist so that the deformation may continue to be homogeneous or it may become inhomogeneous resulting in localization or shear band formation. Such a situation, where two (or more) solutions to the deformation problem exist, is known as a *bifurcation*. Notice that bifurcation does not necessarily lead to localization (which is an unstable mode of deformation); it may lead to barrelling of the specimen (which is a stable mode of deformation). This is in agreement with practical experience: abrupt changes in the pattern of deformation, for instance due to the sudden formation of a shear band, are events which are expected at or in the vicinity of peak load conditions and not at the very beginning of a loading process.

We wish therefore to understand the conditions of deformation which lead to bifurcation or non-uniqueness of the deformation path. Suppose now that the deformation history is specified by some loading parameter, ξ; in strain controlled deformation histories this may be the shear strain or the shortening strain, whereas in load controlled deformation histories ξ may be the shear stress or the uniaxial compressive stress or some other convenient measure. Consider the case where ξ has reached a value ξ_c which is critical in the sense that, if ξ is further increased, or has to be decreased, more than one solution exists. The deformation problem has ceased to be uniquely solvable and bifurcation of the solution takes place.

So how can we evaluate critical values of the loading parameters and the changes of the deformation pattern quantitatively? The basic premise is of course that a mathematical formulation of each deformation problem is available, that is, we have a mathematical description of the material behaviour (constitutive relation) which together with the equilibrium conditions and the boundary conditions define the boundary value problem corresponding to our deformation problem. Now, in the vicinity of a critical value of the loading parameter we must have at least two displacement fields, each of which satisfies the equilibrium conditions and the boundary conditions. Then the difference between any two of these displacement fields satisfies the homogeneous form of the equilibrium and boundary conditions. This difference field vanishes at $\xi = \xi_c$ and its norm (magnitude) can be made arbitrarily small as we let $\xi \rightarrow \xi_c$. Thus a bifurcation takes place at $\xi = \xi_c$ if the incremental, homogeneous form (zero loading) of our boundary value problem is non-trivially solvable.

Consider the situation shown in Fig. 5 where the axial force F is plotted against the axial displacement u (ξ for this problem) for a homogeneous and perfect specimen. Two branches to the loading curve are recognized, one for continuing homogeneous deformation and one where some form of inhomogeneous deformation has occurred. Typically at various points on the loading curve starting at a critical displacement u_B bifurcation modes may develop in the deformation. These modes may be symmetrical or asymmetrical but commonly are transient in the sense that they do not grow with continued increase in the displacement. Examples are given by de Borst (1986, Fig. 6.9 and 1988 for hardening material; 1989 for softening materials). If we test the stability at u_B then typically $\dot{\sigma} \cdot \dot{\varepsilon} > 0$ and the deformation remains homogeneous. One of two things commonly occurs after u_B: a symmetrical bifurcation mode may begin to grow, resulting in a diffuse bifurcation or barrelling of the specimen; this is a stable deformation mode and $\dot{\sigma} \cdot \dot{\varepsilon}$ remains positive. Alternatively an asymmetrical bifurcation mode may begin to grow resulting in localization or shear band development; this is an unstable deformation mode and $\dot{\sigma} \cdot \dot{\varepsilon}$ becomes zero or negative. In Fig. 5 localization begins at u_p; from this point on, although the homogeneous deformation *material* response would continue to be strain hardening, the inhomogeneous deformation *system* response is one of strain-softening.

After the displacement u_B corresponding to bifurcation more than one incremental displacement field exists satisfying the equilibrium and boundary conditions. If $\dot{\sigma}_1$ and $\dot{\sigma}_2$ are the stress-rate fields corresponding to two such solutions

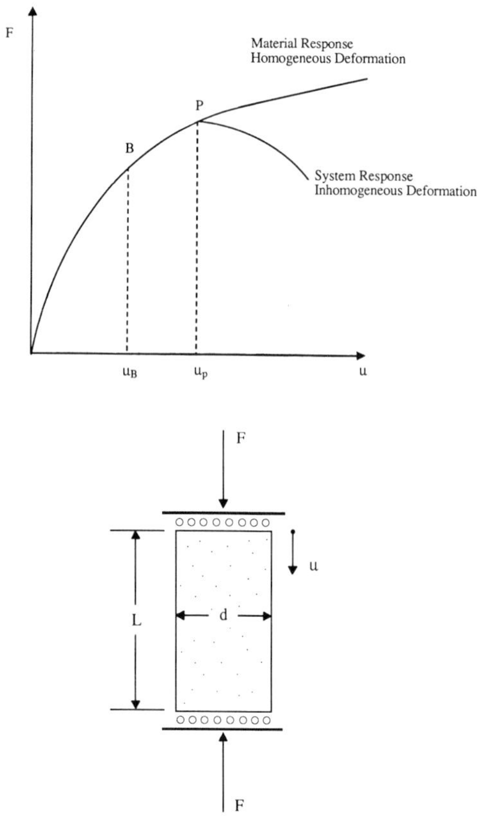

Fig. 5. A perfect homogeneous body is loaded with a force F and undergoes uniaxial displacement u. As the force rises a point B is ultimately reached, corresponding to a critical displacement u_B, where bifurcation occurs (that is, the deformation becomes inhomogeneous). The bifurcation mode may be symmetrical or asymmetrical. The bifurcation mode may amplify or die away. Other bifurcation modes may appear further along the curve until ultimately at P, corresponding to a displacement u_p, a bifurcation mode amplifies strongly and the deformation becomes grossly inhomogeneous. The system response is shown as softening after P and is represented by the lower branch. If such a bifurcation had not amplified so that the deformation remained grossly homogeneous, the material response would be represented by the upper branch.

and $\dot{\varepsilon}_1$, $\dot{\varepsilon}_2$ are the respective strain-rates, then we define $\Delta\dot{\sigma} = \dot{\sigma}_1 - \dot{\sigma}_2$ and $\Delta\dot{\varepsilon} = \dot{\varepsilon}_1 - \dot{\varepsilon}_2$. The necessary condition for a bifurcation to occur is that there exists an incremental displacement field such that

$$\int_v \Delta\dot{\sigma}\, \Delta\dot{\varepsilon} \mathrm{d}V = 0. \qquad (3.1)$$

A simple example will be considered in the next section.

The important point of this section is that the concepts *instability of deformation* and *bifurcation* are not synonymous. A bifurcation mode may or may not be amplified by continued deformation past the bifurcation point. If the bifurcation grows it may be symmetrical or asymmetrical. A symmetrical bifurcation mode results in barrelling of the specimen with no loss in stability (so that $\dot{\sigma} \cdot \dot{\varepsilon}$ remains positive). The system response may continue to be strain-hardening. On the other hand, the bifurcation mode may be asymmetrical resulting in localization of deformation to form a shear band. Here the deformation is unstable (so that $\dot{\sigma} \cdot \dot{\varepsilon} \leq 0$) and, at least for a uniaxial shortening experiment with rigid, frictionless platens (no end constraints) the system response must be strain-softening.

(4) Localization to form a shear band

(4.1) Summary of this section

There is now a very extensive literature on localization in various forms of materials. In this section we first present a simple form of the methods used to investigate localization taking the case of a pressure insensitive, incompressible, elastic-plastic material. We ask, under what conditions can a shear band form at 45° to the loading direction in a uniaxial loading experiment? For such material, using a formulation based on infinitesimal strain, localization is not possible in the hardening regime and must take place at a peak in the stress, followed by softening. However, if a finite strain approach is adopted, then localization is possible in the hardening regime, although the angle between the shear band boundary and the loading direction is no longer constrained to be 45°. If a frictional (that is, pressure sensitive) incompressible material is considered, localization is possible for hardening conditions even for infinitesimal strain; again the angle between the shear band boundary and the loading direction is not restricted to 45°. For these materials, softening of the *system* must always occur after shear bands develop at peak stress, even though the *material* might continue to harden if homogeneously deformed.

(4.2) A simple example: infinitesimal strain

We now give a simple example of the procedure in the solution of a bifurcation problem. For the sake of simplicity we restrict the discussion to a special but nonetheless important type of bifurcation and consider the shear band bifurcation of a rectangular specimen of a pressure insensitive, incompressible, elastic-plastic material. Corresponding results for frictional, dilatant materials will be stated subsequently.

Consider a rectangular block subjected to plane strain biaxial compression (Fig. 6a).

As in the preceding sections σ_{ij} is the stress tensor and u_1, u_2 are the displacements in the x_1, x_2 directions. We assume that compressive stresses have a negative sign. Here the loading parameter ξ is equal to the displacement u of the 'piston', i.e. u_2 $(x_2 = l_2) = u$ and u is assumed to be monotonically increasing. For homogeneous deformation the constitutive relationship can be written as:

$$\tau = f(\gamma), \text{ if } d\gamma = \frac{\partial\gamma}{\partial u} du > 0 \qquad (4.1)_1$$

and

$$\tau = G\gamma \text{ if } d\gamma \leq 0, \qquad (4.1)_2$$

where

$$\tau = 1/2\,(\sigma_{11} - \sigma_{22}),\ \sigma_{11} > \sigma_{22} \qquad (4.2)$$

$$\gamma = \varepsilon_{11} - \varepsilon_{22},\ \varepsilon_{11} > 0 \qquad (4.3)$$

and, since the material is assumed to be incompressible, $\varepsilon_{11} = -\varepsilon_{22}$; ε_{ij} is the strain tensor, which for infinitesimal strain is defined as $\varepsilon_{ij} = \frac{1}{2}\left(\frac{\partial u_i}{\partial x_j} + \frac{\partial u_j}{\partial x_i}\right)$ and G is the shear modulus.

For continued loading $(d\gamma > 0)$ the stress-strain relationship is given by $(4.1)_1$ but in the case of a load reversal $(d\gamma < 0)$ the material is assumed to be Hookean-elastic. The simple relationship (4.1) does not account for Bauschinger effects but for the present purpose where only the response of the material to incipient unloading has to be described it is sufficiently general.

Now for each value of the loading parameter u we can always evaluate the homogeneous strain distribution $\varepsilon_{22} = -\frac{u}{l_2}$ and since the material is incompressible we have $\varepsilon_{11} = -\varepsilon_{22} = \frac{u}{l_2}$. The strain can be inserted into (4.1) to obtain the corresponding deviatoric stress. Thus if σ_{11} is given (confining pressure) there always exists a homogeneous solution of our problem trivially satisfying the equilibrium conditions. The question now is whether there exists a critical value of the ratio G_T/G (where G_T, the tangent shear modulus, equals $\frac{d\tau}{d\gamma}$, see Fig. 7a) which defines a critical value of u for which another solution, and in particular a shear band type solution, exists. To answer this question, we have first to specify an incremental constitutive relationship, which describes the material

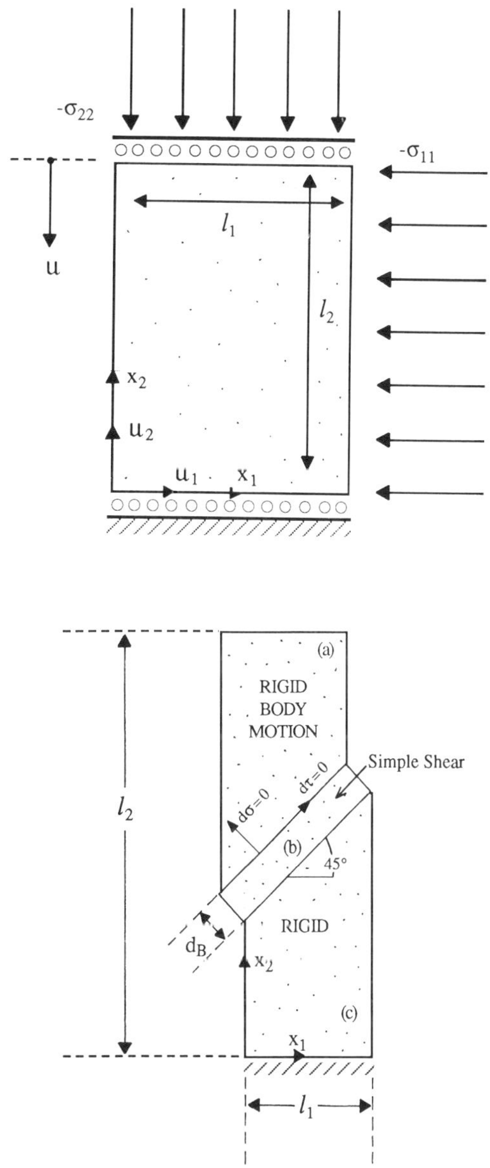

Fig. 6. (**a**) Specimen deformed by axial stress $-\sigma_{22}$ and confining pressure $-\sigma_{11}$. (**b**) Specimen with shear band. At the boundaries of the shear band the stresses $d\sigma$ and $d\tau$ vanish.

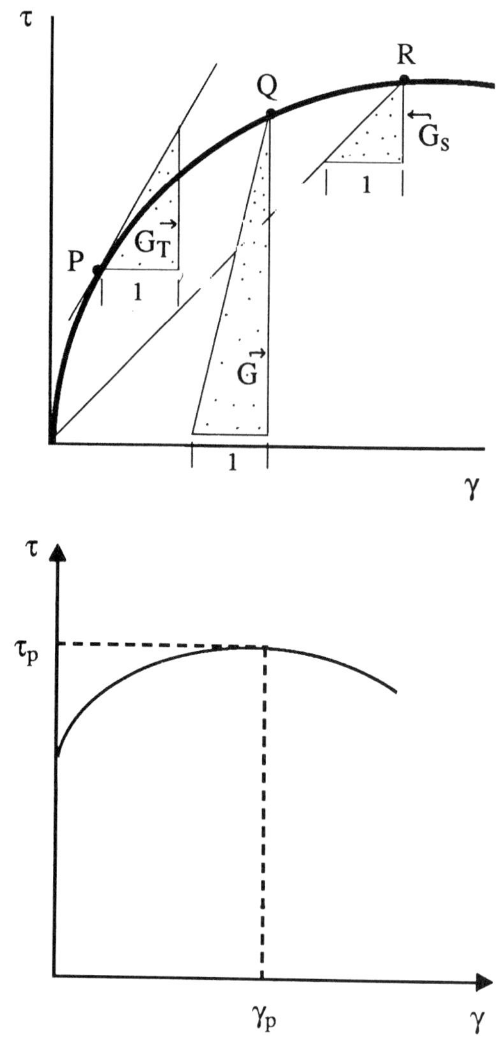

Fig. 7. (**a**) Power law constitutive relation. G_T is the tangent modulus at P, G is the unloading elastic modulus at Q and G_S is the secant modulus at R. (**b**) Strain-softening constitutive law.

behaviour for an infinitesimal deviation from the basic, homogeneously deformed configuration. Use is made of a result of Biot (1965; see also Hill & Hutchinson 1975) which states that this incremental relationship can always (i.e. independent of the specific type of the underlying flow rule) be written in the form

$$d\sigma_{11} = 2G_T d\varepsilon_{11} + dp,\ dp = ½\,(d\sigma_{11} + d\sigma_{22}) \quad (4.4)$$

$$d\sigma_{22} = 2G_T d\varepsilon_{22} + dp, \quad (4.5)$$

$$d\sigma_{12} = 2G^* d\varepsilon_{12} \quad (4.6)$$

The constant G^* takes different values depending on the nature of the yield surface. If we assume that the yield criterion is a smooth surface in the stress space, then we have (e.g. Rudnicki & Rice 1975) $G^* = G$. If the yield surface is not smooth (as in the case of multislip theories, see Batdorf & Budianski 1949) then $G^* < G$. A common practice for yield surfaces with vertices is to approximate the material behaviour by using finite stress-strain relationships (e.g. Budianski 1959; Rudnicki & Rice 1975) and in this case we have $G^* = G_s$, where $G_s = \tau/\gamma$ is the secant shear modulus (see Fig. 7a).

In order to understand the simplest mode of shear band formation, consider the incremental deformation sketched in Fig. 6b. The orientation of the simple shear zone, namely 45° with the x_2 axis, has been chosen for heuristic reasons. The body is subdivided into zones (a), (b) and (c). Zone (a) undergoes a rigid body translation with the incremental displacement vector parallel to a boundary of the shear zone (b). Zone (c) is assumed to be rigid. In both zones (a) and (c) $d\varepsilon_{ij} \equiv 0$, and so we have $d\sigma_{ij} \equiv 0$ there. In zone (b) a simple shearing takes place so that $d\sigma_{ij}$ does not vanish trivially there; however the shear is homogeneous so that local equilibrium is satisfied. To satisfy global equilibrium, the incremental stresses normal ($d\sigma$) and tangential ($d\tau$) to the shear surface have to vanish (Fig. 6b). Strictly speaking, to satisfy the homogeneous stress boundary conditions at the intersections of the shear zone with the boundaries $x_1 = 0$ and $x_1 = l_1$ of the specimen in addition to $d\sigma = d\tau = 0$, we have to satisfy $d\sigma_{11} = d\sigma_{12} = 0$. We assume however that the shear zone is thin compared to l_2 (i.e. $d_B \to 0$, Fig. 6b) so that we can ignore the latter condition. The homogeneous stress boundary conditions then are only violated along line segments of measure $d_B = 0$. Furthermore for the kinematics of Fig. 6b to be possible we must have $l_2 > l_1$. Stress continuity at the interfaces (a)(b) and (b)(c) requires (since the shear zone is at 45° to x_2) that

$$d\sigma = dp - d\sigma_{12} = 0 \quad (4.7)$$

and

$$d\tau = ½(d\sigma_{11} - d\sigma_{22}) = 0 \quad (4.8)$$

where $dp = ½(d\sigma_{11} + d\sigma_{22})$. Equations (4.7) and (4.8) are satisfied if we choose $dp = d\sigma_{12}$ and if according to (4.4, 4.5)

$$G_T = 0 \quad (4.9)$$

The shear band mode of Fig. 6b does not satisfy the homogeneous displacement boundary condition $du_2(x_2 = l_2) = 0$. This can easily be remedied however by combining the shear band mode in an appropriate way with a homogeneous deformation. Thus as soon as the maximum deviatoric stress is reached the present simple model predicts the formation of shear bands which are oriented at $\pm 45°$ to σ_{22}.

Strictly speaking, in this section we should have written $\Delta d\varepsilon$ and $\Delta d\sigma$ to be consistent with the notation used at the end of Section 3. We have omitted the Δs here for convenience. However, we note that in the present simple example of shear zone development $\Delta d\sigma_{11} = \Delta d\sigma_{12} = 0$ and hence $\Delta d\sigma_{22} = 0$ from (4.8). Hence the criterion for bifurcation (3.1) is fulfilled here because the $\Delta d\sigma_{ij}$ are identically zero (even though the $\Delta d\varepsilon_{ij}$ are not).

(4.3) Some simple constitutive models

We now consider two simple constitutive models in order to indicate the application of the above and subsequent calculations.

The simplest possible choice for $f(\gamma)$ in Equation (4.1) is the power law (Fig. 7a)

$$\tau = \tau_0 \, (\gamma/\gamma_0)^n, \tag{4.10}$$

for which the tangent shear modulus $G_T = df/d\gamma$ is given by

$$G_T = n \frac{\tau_0}{\gamma_0} \left(\frac{\gamma}{\gamma_0}\right)^{n-1} = nG_s, \tag{4.11}$$

where G_s is the secant shear modulus (Fig. 7a) and where typically $0 < n < 1$ is the so-called hardening index. The assumption of a power law implies that there is no peak in the stress strain relation $(4.1)_1$; the material is described as monotonically hardening, so that G_T is never equal to zero. This means that, in the case of hardening laws of the power law type, shear bands are never possible at infinitesimal strain.

An example of a constitutive relationship, where a peak exists followed by strain softening (Fig. 7b), has been used by Coleman & Hodgon (1985):

$$\tau = \tau_p - C\,(\gamma - \gamma_p)^2, \tag{4.12}$$

where (τ_p, γ_p) are the peak values of the equivalent stress and strain and $C > 0$ is a material constant.

Here the tangent shear modulus is

$$G_T = -2C(\gamma - \gamma_p) \tag{4.13}$$

Obviously we have $G_T > 0$ for $\gamma < \gamma_p$ and $G_T < 0$ for $\gamma > \gamma_p$. The strain softening indicated by (4.12) may be caused by pore or void growth or if the deformation process is rapid enough to become nearly adiabatic the decrease of the flow stress may be due to thermal softening effects. Clearly, for this material $G_T = 0$ when $\gamma = \gamma_p$ so that a shear band forms at $\pm 45°$ to σ_{22} at peak stress.

(4.4) Analyses involving finite strain and more complicated constitutive models

Hill & Hutchinson (1975) considered pressure insensitive, incompressible, elastic–plastic material in their shear band analysis but included the influence of finite strain; by operations similar to the ones applied here, they arrive at the following critical condition for shear banding:

$$\frac{G_T}{G^*} < \frac{1}{2}\left(1 - \sqrt{1 - \left(\frac{\sigma_{11} - \sigma_{22}}{2G^*}\right)^2}\right) \tag{4.14}$$

which compares with expression (4.9) for an infinitesimal strain formulation. Thus, shear banding is possible as soon as G_T, G^* and the stress deviator $\sigma_{11} - \sigma_{22}$ are such that the inequality (4.14) is satisfied. Equation (4.14) can be simplified if we assume that $(\sigma_{11} - \sigma_{22})/2G^* \ll 1$, so that (4.14) becomes:

$$\frac{G_T}{G^*} < \frac{1}{4}\left(\frac{\sigma_{11} - \sigma_{22}}{2G^*}\right)^2 \tag{4.15}$$

Assuming $G^* = G_s$ (deformation theory) and inserting (4.10, 4.11) into (4.14) we find that no shear banding is possible for n > 0.5 whereas for n ≤ 0.5 shear banding is possible if

$$\gamma > 2\sqrt{n}\,\sqrt{1-n} \tag{4.16}$$

Thus for $n = 0.1$ shear banding is possible if $\gamma > 0.60$. Such magnitudes of strain can of course be reached only if the material is quite ductile. In this case considered by Hill & Hutchinson (1975) the shear bands are oriented at $\pm\,\alpha$ to σ_{22}, where

$$\tan^4 \alpha = \frac{1 - \dfrac{\sigma_{11} - \sigma_{22}}{2G^*}}{1 + \dfrac{\sigma_{11} - \sigma_{22}}{2G^*}}, \tag{4.17}$$

which, in the case of the power law (4.10), can be rewritten as

$$\tan^4\alpha = \frac{1 - \gamma_B}{1 + \gamma_B} \tag{3.18}$$

where γ_B is the value of γ at the onset of shear banding.

Now we again assume infinitesimal deformation but include pressure sensitivity, that is, we consider a frictional material. Assuming incompressibility, Needleman (1979) has derived the critical condition

$$\frac{G_T}{G^*} \leq \frac{1}{2}\left(1 - \sqrt{1 - \delta^2}\right) \qquad (4.19)$$

where

$$\delta = \mu\left(1 - \frac{G^T}{G^*}\right), \; \mu = \sin \varphi^*. \qquad (4.20)$$

Here φ^* is the value of the instantaneous angle of friction at the onset of shear banding which in general is a function of the accumulated plastic strain. The shear band orientation is given by

$$\tan^4 \alpha = \frac{1 - \delta}{1 + \delta} \qquad (4.21)$$

We have seen that under certain circumstances shear banding might precede any limit state of the material, that is, shear banding can occur with $G_T > 0$. Once the band has formed the deformation of the specimen is no longer homogeneous. The question now is whether the average stress measurable below the 'piston' at $x_2 = l_2$ further increases as u is increased. The mathematical condition (see e.g. Hill 1967; Mühlhaus 1982, for more details) for a further stress increase to be possible is that the loading is precisely orthogonal to the eigenmode (that is, the solution of each incremental, homogeneous boundary value problem). A popular example of the latter type of bifurcation is the Euler strut. But when the incremental loading does not have this special orthogonal character then a further increase of the load is not possible. In our present example the 'load' is $F = \int_0^{l_1} \sigma_{22}(u, x_1, x_2 = l_2)\, dx_1$, and the orthogonality condition would require that $\delta F du = 0$, or, since $du > 0$, $\delta F = 0$. Here δF is the load increment due to the eigenmode. The pure shear band mode (Fig. 6b) does not cause any change in stress in zone (a) but as has already been mentioned, the shear band mode has to be supplemented by a homogeneous deformation in order to satisfy the zero displacement condition at $x_2 = l_2$. The homogeneous deformation has to be chosen in such a way that $\delta\sigma_{11} = 0$ but unless $G_T = 0$, $\delta\sigma_{22} \neq 0$ and accordingly $\delta F \neq 0$. Thus, after shear banding a further increase in the load F is not possible.

(4.5) Conclusions to be drawn from this section

If one uses a finite strain formulation of the deformation problem, then localization is possible for strain-hardening materials; however, for a uniaxial shortening experiment with frictionless, rigid platens, the system must show softening after localization occurs. This means that localization forms at peak stress for the system even though the material may continue to harden or soften. In general, the angle that the shear band boundary makes with the maximum principal axis of stress (which in all the constitutive laws used here is parallel to the maximum principal axis of strain-rate) is 45° modified by an angle which depends on the pressure dependence of the yield function. This means that for classical metals the angle is 45°. For frictional materials it is less than 45° (see, for example, expression 4.21).

(5) Localization in Coulomb materials

(5.1) General principles involved in the deformation of Coulomb materials

A Coulomb material is one in which yield occurs if

$$|\tau_s| = c - \sigma_n \tan \phi \qquad (5.1)$$

where τ_s and σ_n are the shear and normal stresses across planes of yield within the material and c and ϕ are material constants known as the cohesion and the angle of friction (Jaeger 1969). Compressive stress components are taken as negative and plane deformation is assumed for simplicity.

Expressed in terms of the Cartesian stress components (5.1) can be rewritten as:

$$\tau^* + \sigma^* \sin \phi - c \cos \phi = 0 \qquad (5.2)$$

where σ^* is the mean stress, giving the centre of the Mohr stress circle in the Mohr plane,

$$\sigma^* = \tfrac{1}{2}(\sigma_{11} + \sigma_{22}) \qquad (5.3)$$

and τ^* is the radius of the Mohr stress circle,

$$\tau^* = \sqrt{\tfrac{1}{2}(\sigma_{11} - \sigma_{22})^2 + \sigma_{12}^2} \qquad (5.4)$$

Thus, the yield function, f, for a Coulomb material may be written

$$f = \tau^* + \sigma^* \sin \phi - c \cos \phi \qquad (5.5)$$

and yield occurs if $f = 0$. The material cannot sustain a stress state for which $f > 0$ and for $f < 0$ the material is elastic.

As a Coulomb material deforms in a plastic manner it may undergo dilation due to sliding of one grain over another, crushing of grains or formation of pore space, opening and closing of fractures, and sliding on fracture surfaces. The angle of dilatancy, ψ, is defined (see Vermeer & de Borst 1984, for a review) by

$$\sin \psi = \dot{\varepsilon}^{\mathrm{p}}_{\mathrm{v}}/\dot{\gamma}^{\mathrm{p}} \tag{5.6}$$

or, the dilation angle is defined by the ratio of the rate of plastic volumetric strain, $\dot{\varepsilon}^{\mathrm{p}}_{\mathrm{v}}$, to the rate of plastic shear strain, $\dot{\gamma}^{\mathrm{p}}$, where

$$\dot{\varepsilon}^{\mathrm{p}}_{\mathrm{v}} = \dot{\varepsilon}^{\mathrm{p}}_{11} + \dot{\varepsilon}^{\mathrm{p}}_{22} \tag{5.7}$$

and $$\dot{\gamma}^{\mathrm{p}} = \sqrt{(\dot{\varepsilon}^{\mathrm{p}}_{11} - \dot{\varepsilon}^{\mathrm{p}}_{22})^2 + (2\dot{\varepsilon}^{\mathrm{p}}_{12})^2} \tag{5.8}$$

We now define the flow rule (that is, the rule that specifies the orientation and magnitude of the strain-rate in terms of the imposed stress state) by

$$\dot{\varepsilon}^{\mathrm{p}}_{ij} = \lambda \frac{\partial g}{\partial \sigma_{\mathrm{ij}}} \tag{5.9}$$

where λ is non-negative for plastic deformation and g is the plastic potential function, often written (see Vermeer & de Borst 1984) as

$$g = \tau^* + \sigma^* \sin \psi - \text{constant} \tag{5.10}$$

Notice that when $\phi = \psi$ the flow rule is associated since then, $f = g$, otherwise it is non-associated. Thus, the commonly accepted use of (5.1) in structural geology as the definition of a Coulomb material together with the assumption of zero dilation implies a non-associated flow rule and hence introduces the possibility of localization under strain hardening conditions (see Section 2).

In what follows, we present four examples of localization in Coulomb materials. The mechanical properties of the materials used in each example are given in Table 1. A number of other examples are presented in Hobbs & Ord (1989) and Ord (this volume). Details of the computational procedure are presented in Cundall (1989) and in Hobbs & Ord (1989). In all of these numerical studies allowance has been made for large deformation effects.

(5.2) Example 1: localization with strain-hardening material and system behaviour

In the first example (Fig. 8) the specimen is loaded uniaxially at constant velocity with zero confining pressure. The specimen may deform subject only to the constraints of the end platens

Table 1. *Constitutive behaviour for localization examples in Coulomb materials*

Example	Constitutive Behaviour	c	ε^c	ϕ	ε^f	ψ	ε^f
1	Friction hardening Dilation hardening Cohesion constant	1 MPa	–	9°	0.05	15°	0.05
2	Cohesion softening Dilation softening Friction constant	10 MPa	0.08	30°	–	20°	0.04
			γ^c		γ^f		γ^f
3	Cohesion softening Dilation softening Friction constant	10 MPa	0.40	30°	–	20°	0.20
4	Cohesion, Dilation, Friction constant	10 MPa	–	30°	–	10°	–

The terms c, ε^c, ϕ, ε^f and ψ define the hardening and softening constitutive behaviour and are given by Vermeer & de Borst (1984, expressions 6.8 and 7.2). The cohesion softening law is $c^* = c \exp[-(\varepsilon^{\mathrm{p}}_1/\varepsilon^c)^2]$ where c^* is the instantaneous value of the cohesion at a uniaxial plastic shortening of $\varepsilon^{\mathrm{p}}_1$. A similar law has been used for dilation softening. The friction (or dilatancy) hardening law is

$$\sin\phi^* \text{ (or } \psi^*) = 2 \sin\phi \text{ (or } \psi). \sqrt{\varepsilon^{\mathrm{p}}_1 \varepsilon^{\mathrm{f}}}/(\varepsilon^{\mathrm{p}}_1 + \varepsilon^{\mathrm{f}}) \text{ for } \varepsilon^{\mathrm{p}}_1 < \varepsilon^{\mathrm{f}}$$

$$\sin\phi^* \text{ (or } \psi^*) = \sin\phi \text{ (or } \psi^*) \text{ for } \varepsilon^{\mathrm{p}}_1 > \varepsilon^{\mathrm{f}}$$

where ϕ^* (or ψ^*) is the instantaneous angle of friction (or dilatancy) at a uniaxial plastic shortening of $\varepsilon^{\mathrm{p}}_1$. In the case of simple shearing experiments γ replaces ε.

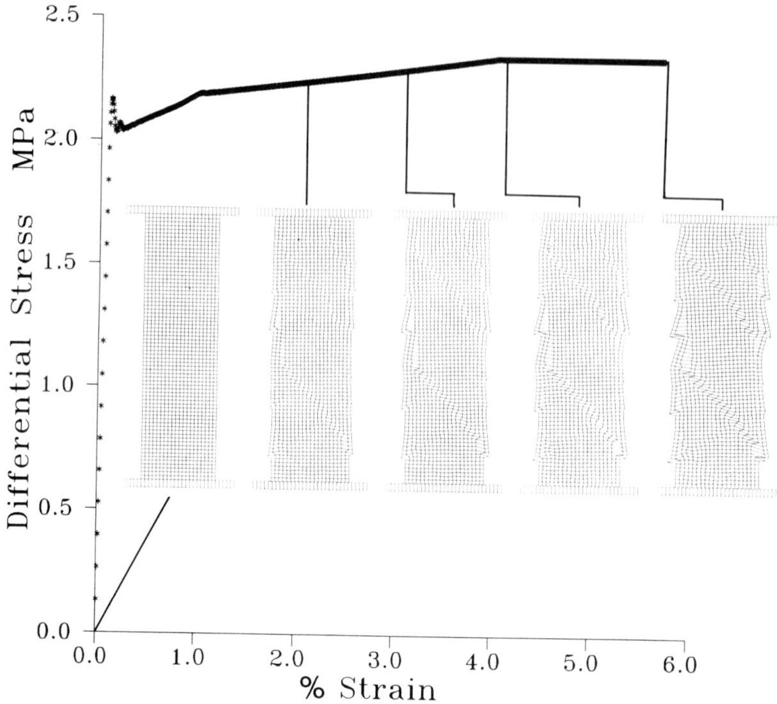

Fig. 8. Shear band development in hardening Coulomb material. Constitutive behaviour given as Example 1 in Table 1.

which are rigid; the friction between the specimen and the platen is approximately 5°. Dilation and friction angles are homogeneous throughout the specimen but the cohesion is inhomogeneous as shown in Fig. 1 of Hobbs & Ord (1989); the cohesion in each element remains constant during deformation whilst the dilation and friction angles harden as indicated in Table 1. The friction hardening results in a strain-hardening stress-strain curve for the material (Fig. 8) since the uniaxial stress σ_2 for homogeneous stress is given (from the geometry of the Mohr circle) by:

$$\sigma_2 = \frac{2c^* \cos \phi^*}{[1 - \sin \phi^*]} + \frac{p[1 + \sin\phi^*]}{[1 - \sin\phi^*]} \quad (5.11)$$

where c^* and ϕ^* are the instantaneous (or 'mobilized') cohesion and angle of friction and p is the confining pressure. The specimen localizes and dilates as shown in Fig. 8 whilst the stress-strain curve for the system continues to harden. This observation, which apparently contradicts the statement at the end of Section 4 (namely that softening should be observed) seems to be the result of the kinematic constraints imposed by low friction at the platens.

(5.3) Example 2: localization with strain-softening material and system behaviour

In the second example the material is again loaded uniaxially at constant velocity subject only to the constraints of the end platens (Fig. 9). In Fig. 9a the confining pressure is zero whereas in Fig. 9b it is 100 MPa. Cohesion, friction angle and dilation angle are homogeneous throughout the specimens but the friction angle remains constant during deformation whilst the cohesion and dilation angle soften during deformation as indicated in Table 1. The cohesion softening leads to a strain-softening material response as given by equation (5.11). At both high and low confining pressures two sets of shear bands develop after yield and in the strain-softening regime although the high pressure specimen ultimately becomes asymmetric (Fig. 9b).

(5.4) Example 3: localization with strain-softening material behaviour but with strain-hardening system behaviour

In the third example the material constitutive

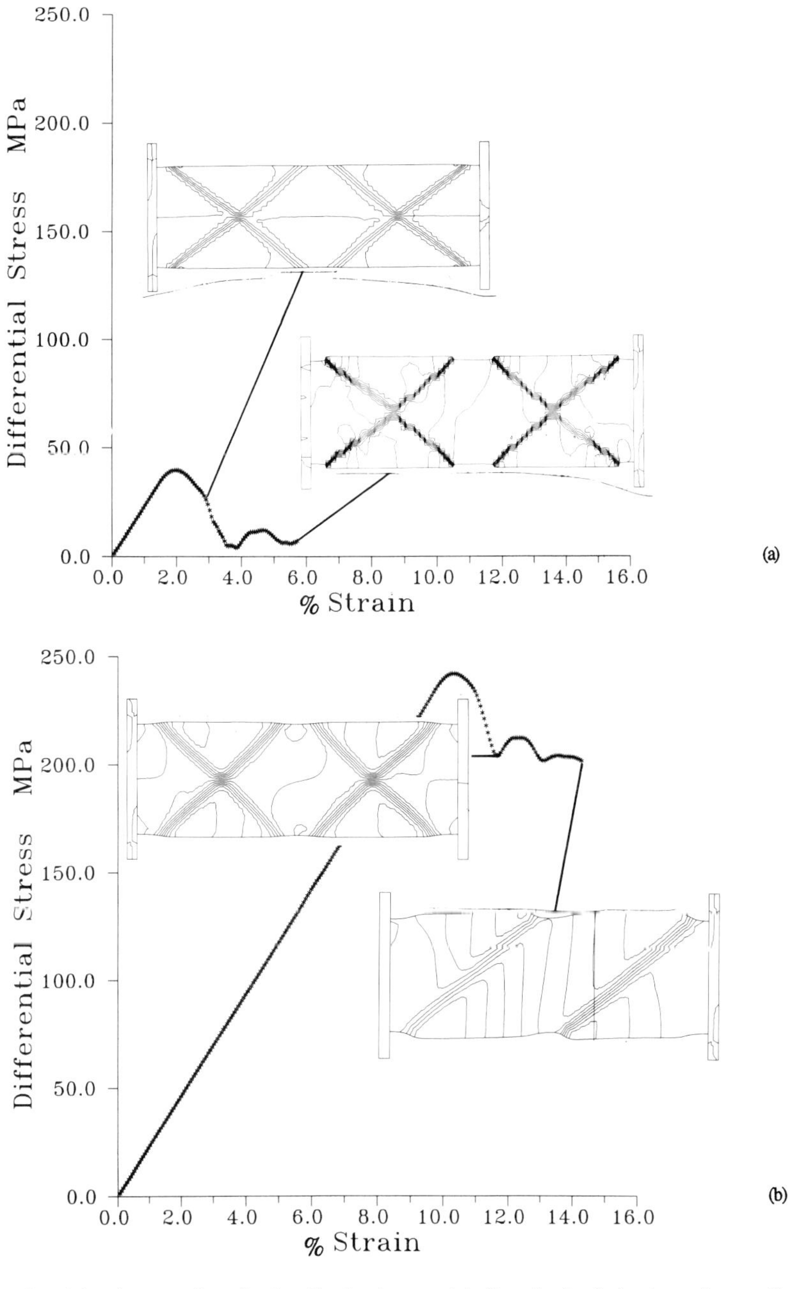

Fig. 9. Shear band development in softening Coulomb material. Constitutive behaviour given as Example 2 in Table 1. (**a**) Confining pressure is zero. (**b**) Confining pressure is 100 MPa.

behaviour is identical (that is, it is strain softening) to that in the second example (see Table 1), but now the specimen is constrained to deform in a homogeneous simple shearing mode and in a strictly isochoric manner as far as the bulk deformation is concerned. The imposed confining pressure is zero. The system response as shown by the various stress strain curves in Fig. 10a, b and c all show hardening but the specimen localizes into a number of shear zones

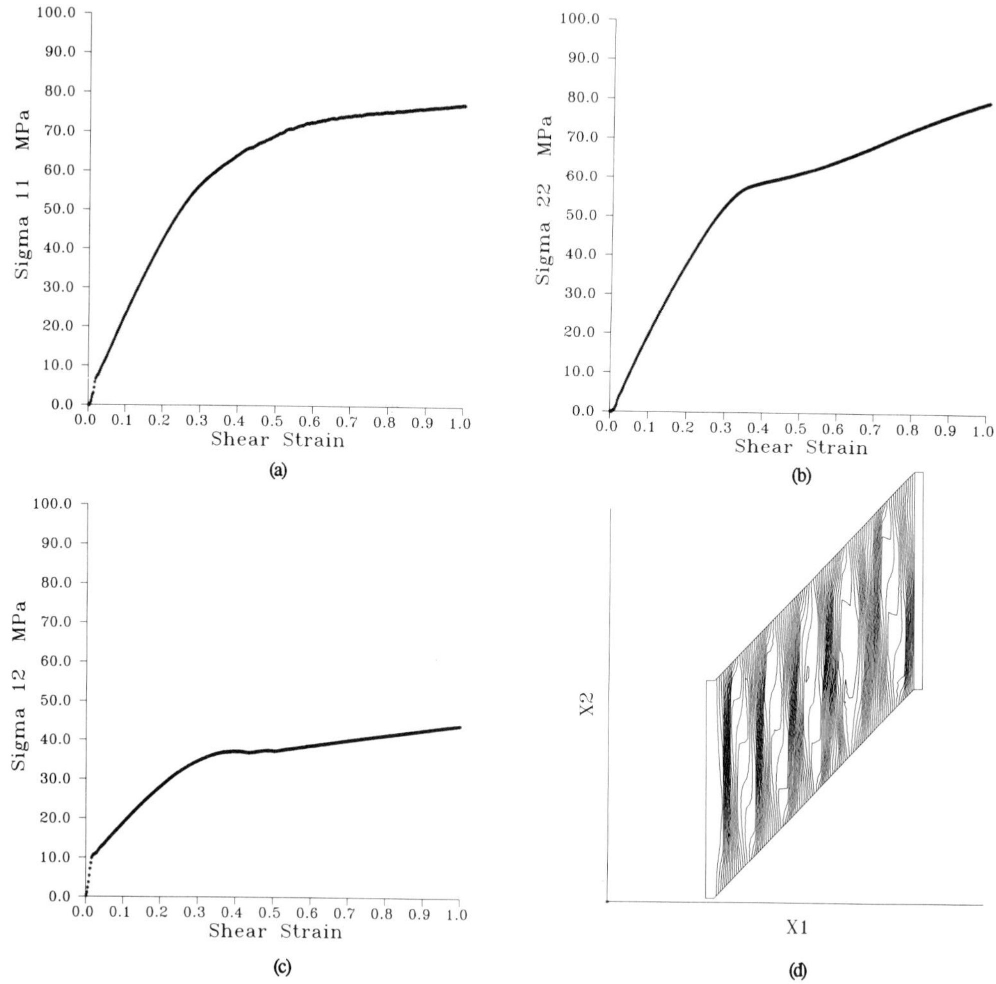

Fig. 10. Strictly isochoric simple shearing deformation history for the bulk specimen. Material constitutive behaviour is strain softening and is given as Example 3 in Table 1. (**a**) σ_{11} measured at the platens; (**b**) σ_{22} measured at the platens; (c) σ_{12} measured at the platens; (**d**) shear band development within the specimen. This is a plot of the x_2 component of the velocity at each point.

spaced in a quasi-periodic manner in a fashion resembling the development of crenulation cleavage (Fig. 10d). This example combined with example two emphasizes the fact that although the material behaviour is one of softening, the system response is one of hardening. Even so, localization is possible in the system-hardening regime.

(5.5) *Example 4: system hardening and instability*

In the fourth example (Fig. 11) the material is non-hardening with cohesion (10 MPa), friction angle (30°), and dilation angle (10°) constant throughout the deformation history and homogeneous throughout the specimen. Two elastically hard inclusions exist at the centre of the specimen (see Table 1 and Ord, this volume). The deformation history for the system is strictly isochoric with zero confining pressure and is hardening throughout for the average shear stress, σ_{12} (Fig. 11b). In order to maintain constant volume overall in this dilating specimen, stresses σ_{11} and σ_{22} must be developed in the platens and these have a form similar to that shown for σ_{12}. Since the deformation history overall is strictly isochoric and simple shearing,

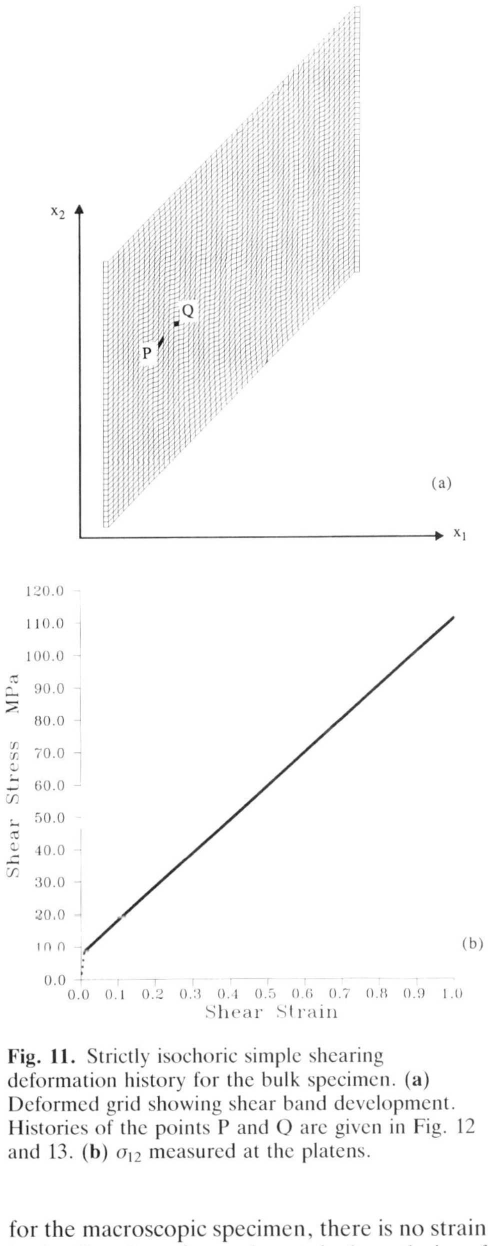

Fig. 11. Strictly isochoric simple shearing deformation history for the bulk specimen. (**a**) Deformed grid showing shear band development. Histories of the points P and Q are given in Fig. 12 and 13. (**b**) σ_{12} measured at the platens.

for the macroscopic specimen, there is no strain in the direction of x_1 and x_2 at the boundaries of the specimen; the product of shear stress-rate on the boundary of the specimen with the macroscopic shear strain-rate is monotonically increasing. Despite this observation, the specimen develops a periodic array of shear zones.

It is instructive to follow the stress and deformation histories of two elements within the body. One of these, P, is highly dilatant and lies within a shear zone; the other, Q, is elastically contractant and lies within a zone that has not experienced as much shearing deformation as P.

The stress and deformation histories for the sheared and dilatant element, P, are given in Fig. 12. These show (Fig. 12h) that the stability measure $\dot{\sigma} \cdot \dot{\varepsilon}$ begins by rising to a positive value but then drops to zero and then becomes negative after a bulk shear strain of 0.34. This contrasts with the stress and deformation history for the contractant element Q (Fig. 13) where the stability measure $\dot{\sigma} \cdot \dot{\varepsilon}$ remains positive for most of the deformation history.

(5.6) Conclusions to be reached from the section

These four examples, for Coulomb materials, illustrate how localization occurs under both strain-softening and strain-hardening material response or under conditions where strain-softening does not accompany the progressive development of the shear zones. In the one example studied in detail, there exist elements within the deforming body where the quantity $\dot{\sigma} \cdot \dot{\varepsilon}$ becomes negative. Localization is associated with such elements. Even though the material constitutive behaviour may be one of softening, the system response can be one of hardening if the deformation is constrained in some manner. Localization still occurs in such hardening systems.

(6) Localization in visco-plastic materials

There is now a large literature on localization in various kinds of visco-plastic materials (see for instance, Pierce *et al.* 1983; Lemonds & Needleman 1986*a*, *b*; Molinari & Clifton, 1987; Anand *et al.* 1987; Needleman 1988; Nemat-Nasser *et al.* 1989). The following brief summary is based largely on the work of Anand *et al.* (1987) who considered a range of materials which could exhibit strain-hardening (or softening), strain-rate hardening, thermal softening and pressure hardening. The constitutive equation for such materials can be written quite generally as

$$\tau = G\,(\dot{\gamma}, \gamma, T, p) \qquad (6.1)$$

where τ is the shear stress, γ and $\dot{\gamma}$ are the shear strain and the shear strain-rate, T is the temperature and p is the pressure. The strain-rate hardening, strain hardening, thermal softening and pressure hardening are defined by

$$\begin{aligned} R &= \partial G/\partial\dot{\gamma} \\ S &= \partial G/\partial\gamma \\ U &= -\partial G/\partial T \\ P &= \partial G/\partial p \end{aligned} \qquad (6.2)$$

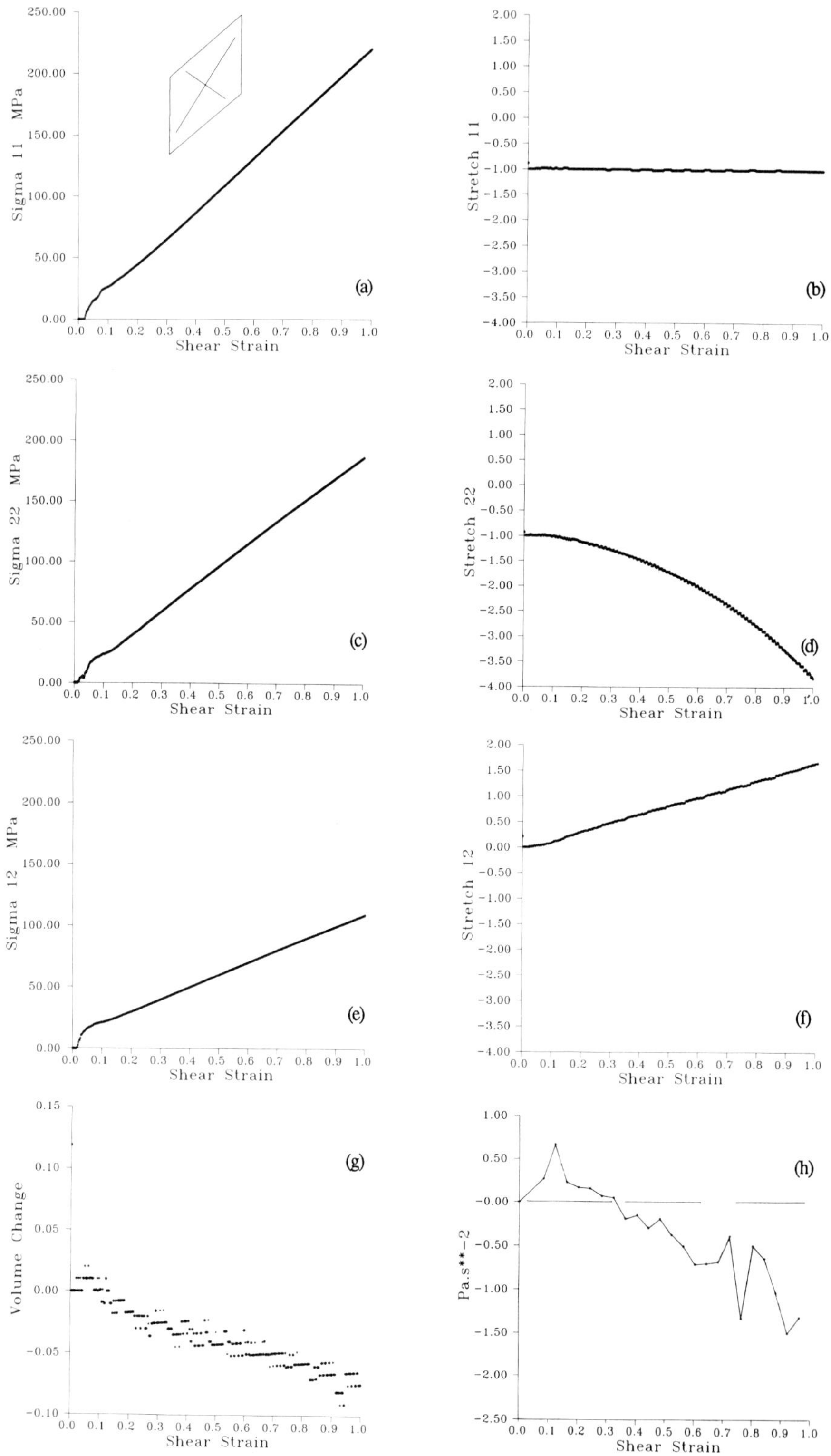

Fig. 12. History of deformation at the point P in Fig. 11. This is a dilatant region within one of the shear bands. **(a)** σ_{11}; **(b)** $c^{-1}{}_{11}$; **(c)** σ_{22}; **(d)** $c^{-1}{}_{22}$; **(e)** σ_{12}; **(f)** $c^{-1}{}_{12}$; **(g)** dilation, (positive for volume decrease); **(h)** $\dot{\sigma} \cdot \dot{\varepsilon}$. c^{-1} is the Finger deformation tensor, shortening is positive.

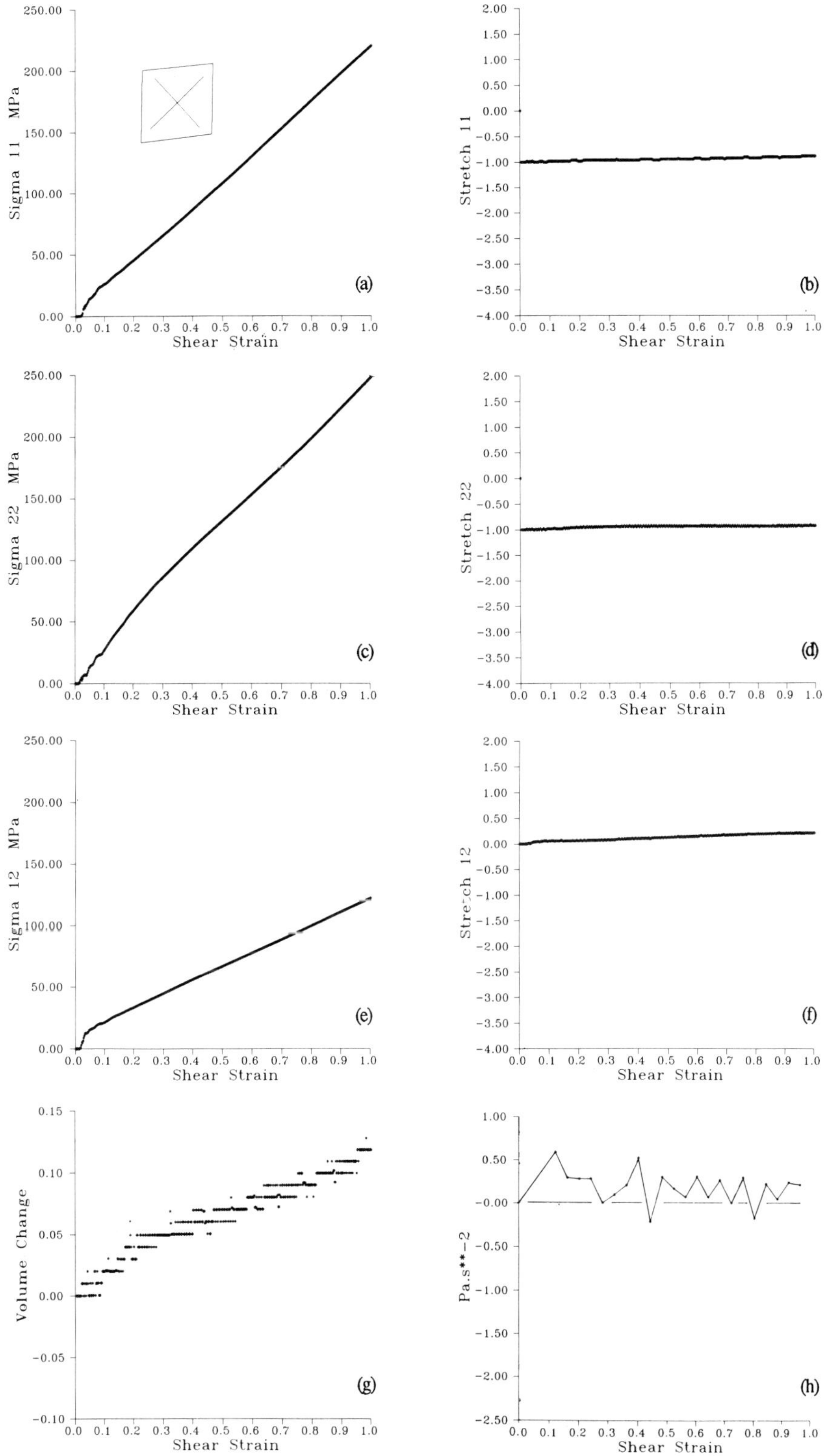

Fig. 13. History of deformation at the point Q in Fig. 11. This is an elastically contractant region between two of the shear bands. **(a)** σ_{11}; **(b)** $c^{-1}{}_{11}$; **(c)** σ_{22}; **(d)** $c^{-1}{}_{22}$; **(e)** σ_{12}; **(f)** $c^{-1}{}_{12}$; **(g)** dilation, (positive for volume decrease); (h) $\dot{\sigma} \cdot \dot{\varepsilon} \cdot c^{-1}$ is the Finger deformation tensor, shortening is positive.

In addition we define the slope of the shear-stress shear-strain curve at constant shearing strain and adiabatic conditions as

$$H = \mathrm{d}G/\mathrm{d}\gamma \qquad (6.3)$$

For deformation histories involving simple shearing under both isothermal and adiabatic conditions, Anand *et al.* (1987) considered perturbations from the simple shearing deformation history and derived conditions under which localized shear zones would grow and remain stable in the deforming body. The results are summarized below.

(6.1) Quasi-static, isothermal deformation histories

(i) For pressure insensitive materials ($P = 0$), localization is possible only for strain-softening ($S \leq 0$) materials.
(ii) For pressure sensitive materials ($P > 0$) there is a critical pressure sensitivity which determines if localization occurs for strain-softening or hardening materials. For a large enough pressure sensitivity shear zones can develop in strain-hardening ($S > 0$) materials.

(6.2) Adiabatic deformation histories

For both pressure insensitive and pressure sensitive materials localization is possible in strain-hardening materials ($S > 0$). For materials with high pressure sensitivity, localization is possible even under conditions where H is positive, that is, the thermal softening, U, does not counterbalance the strain hardening, S, to produce an overall softening response with continued straining.

(6.3) Angle between shear band and principal axis of stress

In all visco-plastic materials with suitable constitutive properties, two sets of shear zones should develop under suitable boundary conditions with the shear zones oriented at α to the axis of maximum principal stress where

$$\alpha = \pm\left[\frac{\pi}{4} - h(P)\right] \qquad (6.4)$$

and $h(P)$ is some function of the pressure hardening. To date there appears to be little treatment in the literature of localization in visco-plastic materials which are dilatant, although some discussion is included in Rice (1976) and in Tvergaard (1982, 1987). Expression (6.4) compares with the similar expression for non-dilatant Coulomb materials:

$$\alpha = \pm\left[\frac{\pi}{4} - \frac{\phi}{2}\right] \qquad (6.5)$$

(see Vardoulakis 1980; Hobbs & Ord 1989).

It appears from this brief summary of the behaviour of visco-plastic materials and the previous discussion in this paper that there is an overall similarity in the localization behaviour of materials, no matter if they be rate dependent or rate independent: localization is possible in the hardening regime so long as the boundary conditions allow this to happen and certain critical constitutive conditions are met. In all cases, the angle between the shear band boundary and the maximum principal axis of stress is 45° minus an angle which depends on the pressure dependence of the yield function. In dilatant materials such as Coulomb materials this angle is further reduced by an angle which depends on the dilatancy (see Anand *et al.* 1987; Vardoulakis 1980, for details).

(7) Discussion and conclusions

The idea that localization in deforming materials is always associated with material softening, as indicated by a loss in load bearing capacity of both the material and the system, arises from two different areas of experience. Firstly, in the classical plasticity of metals shear band formation is commonly observed in strain-softening materials and secondly, in the so-called 'brittle' behaviour of rocks in the laboratory, and in the 'brittle-ductile' transition, shear band formation is again commonly associated with loss of load bearing capacity for both the material and the system. However, the classical metal plasticity arguments are for elastic-plastic materials with very low (or ideally zero) pressure sensitivity for deformation. This means that classically, metals exhibit little if any frictional response within their constitutive behaviour and zero dilation. For such materials, whether they be rate sensitive or rate insensitive, localization must occur in the strain softening regime.

It is wise also to examine the literature on rock deformation carefully for it is not difficult (Griggs & Handin 1960; Byerlee & Brace 1969) to find examples where localization of deformation has been observed in the strain hardening regime. Rocks, unlike classical metals, do show pressure sensitivity within their constitutive behaviour over a wide range of pressure. This is due largely to frictional behaviour as grains and microfractures slide past each other.

In addition, rocks show strong dilational behaviour, even at relatively high temperatures and pressures, even in weak materials such as marble (Fischer & Paterson 1989). We have indicated here that for frictional-dilatant materials, localization can occur in the strain hardening regime. Moreover, systems comprised of dilatant materials, even if their intrinsic material constitutive behaviour is one of softening, can display strain-hardening if the system is constrained in some manner. Such strain-hardening systems still display localization of deformation.

At the present stage of development of the subject there seems to be a substantial gap between theory and experiment. Thus, theory says that for a uniaxial compressive test between rigid, frictionless platens, the system should show weakening after localization no matter what the constitutive behaviour is. Clearly it is not possible to use rigid frictionless platens but there are instances (see Griggs & Handin 1960) where hardening follows localization in what are initially symmetrical shortening experiments: perhaps in the Griggs and Handin instance, the kinematic constraints offered by the jacket material and/or by the friction at the platens is sufficient to change the system response from the theoretical softening to the observed hardening. In the present paper, hardening after localization is shown in Example 1 in Section 5, perhaps instigated by the low friction at the platens. In the case of imposed kinematic constraints that ensure plane strain, hardening has also been observed in metal single crystals after localization (Harren *et al.* 1988). It appears that kinematic constraints are capable of producing system hardening after localization.

Finally, there is now ample evidence that dilatancy is a common feature of naturally occurring shear zones. This includes evidence that large volumes of fluid have passed through shear zones even at conditions up to and including amphibolite facies metamorphism (see for example Kerrich *et al.* 1977; Brodie 1980; Etheridge & Cooper 1981; McCaig 1988). Such observations are consistent with the experimental observation of significant dilatancy even in weak rocks such as marble at temperatures up to 600°C at 300 MPa confining pressure (Fischer & Paterson 1989). Other evidence of dilatancy in shear zones comes from the widespread occurrence of dilatant fracture systems within these zones (see Ramsay & Huber 1983, Figs 2.11, 3.22, for examples).

Since shear zones are kinematically constrained by both elastic and, to a limited extent, plastic deformation in the less deformed rocks surrounding the shear zone, hardening of the system undergoing deformation is to be expected as the norm in nature even if the material constitutive behaviour is one of softening. The onus is on the structural geologist to establish whether softening or hardening is the total response of the system. Softening cannot be assumed in arguments concerning the kinematic and dynamic evolution of the system.

We would like to thank P. Cobbold and P. Warburton for detailed and critical reviews of this paper. ITASCA Consulting Group kindly provided access to the code FLAC.

References

ANAND, K. H. K., KIM, K. H. & SHAWKI, T. G. 1987. Onset of shear localization in viscoplastic solids. *Journal of the Mechanics and Physics of Solids*, **35**, 407–429.

ANDERSON, P. M. & RICE, J. R. 1985. Constrained creep cavitation of grain boundary facets. *Acta Metallurgica*, **33**, 409–422.

ARGON, A. S. 1973. Stability of plastic deformation. *In*: *The Inhomogeneity of Plastic Deformation*, ASM Publication.

ARTHUR, J. R. F., DUNSTAN, T., AL-ANI, A. J. & ASSADI, A. 1977. Plastic deformation and failure in granular media. *Geotechnique*, **27**, 53–74.

ASARO, R. J. & NEEDLEMAN, A. 1985. Texture development and strain hardening in rate dependent polycrystals. *Acta Metallurgica*, **33**, 923–953.

BATDORF, S. B. & BUDIANSKY, B. 1949. *A mathematical theory of plasticity based on the concept of slip*. Technical Note 1871, Nat. Adv. Committee on Aeronautics.

BIOT, M. A. 1965. *Mechanics of incremental deformations*. New York: John Wiley & Sons, Inc.

BISHOP, J. F. W. & HILL, R. 1951. A theory of the plastic distortion of a polycrystalline aggregate under combined stresses. *Philosophical Magazine*, **42**, 414–427.

BRODIE, K. H. 1980. Variations in mineral chemistry across a shear zone in phlogopite peridotite. *Journal of Structural Geology*, **2**, 265–272.

BUDIANSKY, B. 1959. A reassessment of deformation theories of plasticity. *Journal of Applied Mechanics*, **26**, 259–264.

BYERLEE, J. D. & BRACE, W. F. 1969. High-pressure mechanical instability in rocks. *Science*, **164**, 713–715.

CHEN, W. F. 1982. *Plasticity in reinforced concrete*. McGraw-Hill.

COBBOLD, P. R. 1977. Description and origin of banded deformation structures. II Rheology and the growth of banded perturbations. *Canadian Journal of Earth Sciences*, **14**, 2510–2523.

COLEMAN, B. D. & HODGON, M. L. 1985. On shear bands in ductile materials, *Archive for Rational Mechanics and Analysis*, **90**, 219.

CUNDALL, P. A. 1989. Numerical experiments on localization in frictional materials. *Ingenieur-Archiv*, **59**, 148–159.

DE BORST, R. 1986. *Non-linear analysis of frictional materials*. Dissertation, Delft University of Technology.

—— 1988. Bifurcations in finite element models with a non-associated flow law. *International Journal for Numerical and Analytical Methods in Geomechanics*, **12**, 99–116.

—— 1989. Numerical methods for bifurcation analysis in geomechanics. *Ingenieur-Archiv*, **59**, 160–173.

DRUCKER, D. C. 1951. A more fundamental approach to stress-strain relations. *Proceedings of the 1st U.S. National Congress of Applied Mechanics*, 487–491.

—— 1956. On uniqueness in the theory of plasticity. *Quarterly of Applied Mathematics*, **14**, 35–42.

—— 1959. A definition of stable inelastic material. *Journal of Applied Mechanics, Transactions of the American Society of Mechanical Engineers*, **26**, 101–106.

EDMOND, J. M. & PATERSON, M. S. 1972. Volume changes during the deformation of rocks at high pressures. *International Journal of Rock Mechanics and Mining Sciences*, **9**, 161–182.

ETHERIDGE, M. A. & COOPER, J. A. 1981. Rb/Sr isotopic and geochemical evolution of a recrystallized shear (mylonite) zone at Broken Hill. *Contributions to Mineralogy and Petrology*, **78**, 74–84.

FISCHER, G. J. & PATERSON, M. S. 1989. Dilatancy during rock deformation at high temperatures and pressures. *Journal of Geophysical Research*, **94**, 17607–17617.

GRIGGS, D. T. & HANDIN, J. 1960. Observations on fracture and a hypothesis of earthquakes. *In*: GRIGGS, D. & HANDIN J. (eds) *Rock Deformation*. Geological Society of America, Memoir, **79**, 347–364.

GROVES, G. W. & KELLY, A. 1969. Change of shape due to dislocation climb. *Philosophical Magazine*, **19**, 977–986.

HADAMARD, J. 1903. *Lecons sur la Propagation des Ondes et les Equations de l'Hydrodynamique*. Chap. 6. Paris.

HARREN, S. V., DEVE, H. E. & ASARO, R. J. 1988. Shear band formation in plane strain compression. *Acta Metallurgica*, **36**, 2435–2480.

HEARD, H. C. 1960. Transition from brittle fracture to ductile flow in Solenhofen limestone as a function of temperature, confining pressure, and interstitial fluid pressure. *In*: GRIGGS, D. & HANDIN, J. (eds) *Rock Deformation*. Geological Society of America, Memoir, **79**, 193–226.

HILL, R. 1958. A general theory of uniqueness and stability in elastic-plastic models. *Journal of the Mechanics and Physics of Solids*, **6**, 236–249.

—— 1959. Some basic principles in the mechanics of solids without a natural time. *Journal of the Mechanics and Physics of Solids*, **7**, 209–225.

—— 1967. Eigenmodal deformations in elastic/plastic continua. *Journal of the Mechanics and Physics of Solids*, **15**, 371–386.

—— & HUTCHINSON, J. W. 1975. Bifurcation phenomena in the plane tension test. *Journal of the Mechanics and Physics of Solids*, **23**, 239–264.

HOBBS, B. E. 1985. The geological significance of microfabric analysis. *In*: WENK, H-R (ed.). *Preferred Orientation in Deformed Metals and Rocks: An Introduction to Modern Texture Analysis*, Academic Press, Chapter 22.

——, & ORD, A. 1988. Plastic instabilities: Implications for the origin of intermediate and deep focus earthquakes. *Journal of Geophysical Research*, **93**, 10521–10540.

—— & —— 1989. Numerical simulation of shear band formation in a frictional-dilational material. *Ingenieur-Archiv*, **59**, 209–220.

HUTCHINSON, J. W. 1970. Elastic-plastic behaviour of polycrystalline metals and composites. *Proceedings of the Royal Society of London*, **A319**, 247–272.

JAEGER, J. C. 1969. *Elasticity, fracture and flow*. 3rd ed. London: Methuen.

KERRICH, R., FYFE, W. S., GORMAN, B. E. & ALLISON, I. 1977. Local modification of rock chemistry by deformation. *Contributions to Mineralogy and Petrology*, **65**, 183–190.

KOITER, W. T. 1960. General theorems for elastic-plastic solids. *In*: SNEDDON & HILL (eds) *Progress in Solid Mechanics. Vol. 1* North-Holland Publishing Co., Amsterdam, 165–221.

LEMONDS, J. & NEEDLEMAN, A. 1986*a*. Finite element analysis of shear localization in rate and temperature dependent solids. *Mechanics of Materials*, **5**, 339–391.

—— & —— 1986*b*. An analysis of shear band development incorporating heat conduction. *Mechanics of Materials*, **5**, 363–373.

LISTER, G. S. & HOBBS, B. E. 1980. The simulation of fabric development during plastic deformation and its application to quartzite: The influence of deformation history. *Journal of Structural Geology*, **2**, 355–370.

LISTER, G. S., PATERSON, M. S. & HOBBS, B. E. 1978. The simulation of fabric development in plastic deformation and its application to quartzite: The model. *Tectonophysics*, **45**, 107–158.

MANDEL, J. 1965. Généralisation de la théorie de plasticité de W. T. Koiter. *International Journal of Solids and Structures*, **1**, 273–295.

—— 1966. Conditions de stabilité et postulat de Drucker. *Proceedings of the IUTAM Symposium on Rheology and Soil Mechanics, Grenoble*. Springer Verlag, Berlin, 331–341.

MCCAIG, A. M. 1988. Deep fluid circulation in fault zones. *Geology* **16**, 867–870.

MOLINARI, A. & CLIFTON, R. J. 1987. Analytical characterization of shear localization in thermoviscoplastic materials. *Journal of Applied Physics*, **54**, 806–812.

MÜHLHAUS, H-B. 1982. Lösung von Verzweigungsproblemen und Traglastproblemen bei einer Klasse nichtlinearer Randwertprobleme der Kontinuumsmechanik. *Ingenieur-Archiv*, **52**, 379–389.

NEEDLEMAN, A. 1979. Non-normality and bifurcation

in plane strain tension and compression. *Journal of the Mechanics and Physics of Solids*, **27**, 231–254.

—— 1988. Material rate dependence and mesh sensitivity in localization problems. *Computer Methods in Applied Mechanics and Engineering*, **67**, 69–85.

NEMAT-NASSER, S., CHUNG, D-T. & TAYLOR, L. M. 1989. Phenomenological modelling of rate dependent plasticity for high strain rate problems. *Mechanics of Materials*, **7**, 319–344.

NICOLAS, A. & POIRIER, J. P. 1976. *Crystalline Plasticity and Solid State Flow in Metamorphic Rocks*. Wiley, New York.

ORD, A. 1990. Mechanical controls on dilatant shear zones. *This volume*.

PATERSON, M. S. 1958. Experimental deformation and faulting in Wombeyan marble. *Geological Society of America Bulletin*, **69**, 465–476.

—— 1978. *Experimental Rock Deformation – The Brittle Field*. Springer, Berlin.

PIERCE, D., ASARO, R. J. & NEEDLEMAN, A. 1983. Material rate dependence and localized deformation in crystalline solids. *Acta Metallurgica*, **31**, 1951–1976.

POIRIER, J. P. 1980. Shear localization and shear instability in materials in the ductile field. *Journal of Structural Geology*, **2**, 135–142.

——, BOUCHEZ, J. L. & JONAS, J. J. 1979. A dynamic model for aseismic ductile shear zones. *Earth and Planetary Science Letters*, **43**, 441–453.

RAMSAY, J. G. 1980. Shear zone geometry: A review: *Journal of Structural Geology*, **2**, 83–99.

—— & GRAHAM, R. H. 1970. Strain variation in shear belts. *Canadian Journal of Earth Sciences*, **7**, 786–813.

—— & HUBER, M. I. 1983. *The techniques of modern structural geology, Vol. I: Strain Analysis*. Academic Press.

RICE, J. R. 1976. The localization of plastic deformation. *In*: KOITER, W. T. (ed.) *Theoretical and Applied Mechanics*. Proceedings of the 14th IUTAM Conference, Delft, August 30–Sept. 4.

RUDNICKI, J. W. & RICE, J. R. 1975. Conditions for the localization of deformation in pressure-sensitive materials. *Journal of the Mechanics and Physics of Solids*, **23**, 371–394.

RUTTER, E. H. 1986. On the nomenclature of mode of failure transitions in rocks. *Tectonophysics*, **122**, 381–387.

SEGALL, P. & SIMPSON, C. 1986. Nucleation of ductile shear zones on dilatant fractures. *Geology*, **14**, 56–59.

SORENSEN, K. 1983. Growth and dynamics of the Nordre Stromfjord shear zone. *Journal of Geophysical Research*, **88**, 3419–3438.

TRUESDELL, C. & NOLL, W. 1965. *The non-linear field theories of mechanics*. Encyclopedia of Physics, **III/3**.

TULLIS, J., SNOKE, A. W. & TODD, V. R. 1982. Significance and petrogenesis of mylonitic rocks. *Geology*, **10**, 227–230.

TVERGAARD, V. 1982. Ductile fracture by cavity nucleation between larger voids. *Journal of the Mechanics and Physics of Solids*, **30**, 265–286.

—— 1987. *Effect of anisotropic hardening on ductile failure of a material containing two size-scales of particles*. DCAMM Report 354, The Technical University of Denmark.

VARDOULAKIS, I. 1980. Shear band inclination and shear modulus of sand in biaxial tests. *International Journal for Numerical and Analytical Methods in Geomechanics*, **4**, 103–119.

—— & GRAF, B. 1985. Calibration of constitutive models for granular materials using data from biaxial experiments. *Geotechnique*, **35**, 299–317.

——, GOLDSCHEIDER, M. & GUDEHUS, G. 1978. Formation of shear bands in sand bodies as a bifurcation problem. *International Journal for Numerical and Analytical Methods in Geomechanics*, **2**, 99–128.

VERMEER, P. A. 1978. A double hardening model for sand. *Geotechnique*, **28**, 4, 413–433.

—— & DE BORST, R. 1984. Non-associated plasticity for soils, concrete and rock. *Heron*, **29**, 1–62.

WHITE, S. H. & KNIPE, R. J. 1978. Transformation and reaction-enhanced ductility in rocks. *Journal of the Geological Society, London*, **135**, 513–516.

——, BURROWS, S. E., CARRERAS, J., SHAW, N. D. & HUMPHREYS, F. J. 1980. On mylonites in ductile shear zones. *Journal of Structural Geology*, **2**, 175–187.

YUEN, D. A., FLEITOUT, L., SCHUBERT, G. & FROIDEVAUX, C. 1978. Shear deformation zones along major transform faults and subducting slabs. *Geophysical Journal of the Royal Astronomical Society*, **54**, 93–119.

ZEUCH, D. H. 1982. Ductile faulting, dynamic recrystallization, and grain size sensitive flow of olivine. *Tectonophysics*, **83**, 293–308.

Reaction-enhanced formation of eclogite-facies shear zones in granulite-facies anorthosites

EVA M. KLAPER

Institute of Geology, Bern University, Baltzerstr. 1, 3012 Bern, Switzerland

Abstract: Infiltration of a dry granulite-facies anorthosite complex by a hydrous fluid phase under high pressure was crucial to start (a) an eclogitization process which led to the breakdown of plagioclase and the formation of either dense or hydrated minerals and (b) dynamic recrystallization of plagioclase in shear zones. Reaction products of the eclogitization process are related spatially to cracks and shear zones. The reduction in grain size during dynamic recrystallization, the foliation development in shear zones and the occurrence of new 'soft' and/or dense minerals changed the mechanical properties (e.g. ductility) of the rocks considerably.

The initiation of detachment and thrusting in the lower crust may be an important large scale consequence of such localized processes.

Deformation of dry granulite-facies rocks under high-grade conditions may occur in the lower crust during large scale tectonic processes such as subduction or continent–continent collision. However, such dry rocks are able to sustain very large deviatoric stresses and the extent of penetrative deformation will be highly dependent on the presence of a fluid phase. A fluid can act as a trigger for mineral reactions and/or deformational processes both of which will have a strong effect on the mechanical properties of the rocks.

This study describes a representative small scale example of lower crustal deformation localized in an eclogite-facies shear zone cutting a granulite-facies anorthosite. Shear zone formation is ascribed to the infiltration of a dry anorthosite by a hydrous fluid phase. This process had a profound influence on the mineral assemblage and its texture, on the strength of the rocks and, therefore, on the rheological behaviour. An attempt is made to identify the particular mechanisms and processes related to the granulite–eclogite facies transition. Large scale consequences are to be expected.

Geological situation

The Bergen Arcs, western Norway (Fig. 1), represent a series of arcuate Caledonian nappes. One of these nappes, the Precambrian granulite-facies Anorthosite Complex of the Bergen Arcs, is interpreted as a slice of lower continental crust (Sturt & Thon 1978) that had been partially eclogitized (Austrheim & Griffin 1985; Austrheim 1986/87) and emplaced in the upper crust during the Caledonian orogeny (Cohen *et al.* 1988).

The metamorphic conditions (Austrheim & Griffin 1985) during the Grenvillian granulite-facies event (Cohen *et al.* 1988) were determined to be close to T = 800–900°C at P = 10 kbar, with the stable assemblage plagioclase, clinopyroxene and garnet in the anorthositic rocks. The Caledonian eclogitization occurred at T = 750°C and P = 17–18 kbar. The PT-path between the granulite formation and the Caledonian orogeny is poorly constrained. Prior to the Caledonian event the Bergen Arcs granulite terrain was part of the western margin of the Baltic shield. This segment of continental crust probably cooled to a standard shield geotherm during the tectonically inactive 450 Ma interval between the Grenvillian granulite-facies metamorphism and the Caledonian eclogite-facies event. The temperature of the metastable granulites probably did not exceed 450–500°C (shield geotherm at 35 km depth) before the onset of the Caledonian orogeny.

Caledonian structures observed in the Anorthosite Complex are cracks and shear zones of various scales (Austrheim & Griffin 1985; Austrheim 1986/87). Partial eclogitization is confined to these structures. The eclogite-facies shear zones range from a few millimetres to tens of metres in thickness and form an anastomosing network covering an area of 50 km^2 (Austrheim & Mørk 1988). The eclogite-facies rocks replace locally up to 50% of the precursor.

The sample selected for the study of the

From Knipe, R. J. & Rutter, E. H. (eds), 1990, *Deformation Mechanisms, Rheology and Tectonics*, Geological Society Special Publication No. 54, pp. 167–173.

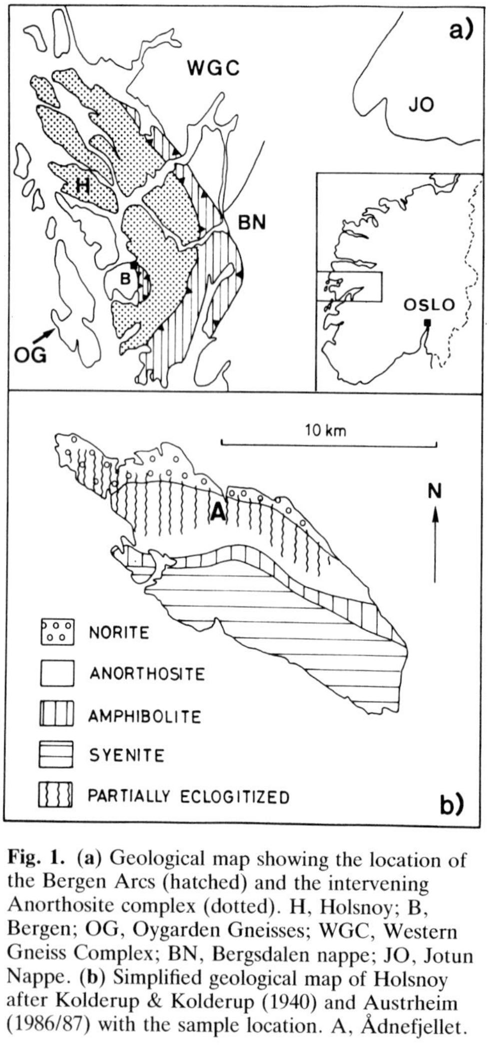

Fig. 1. (**a**) Geological map showing the location of the Bergen Arcs (hatched) and the intervening Anorthosite complex (dotted). H, Holsnoy; B, Bergen; OG, Oygarden Gneisses; WGC, Western Gneiss Complex; BN, Bergsdalen nappe; JO, Jotun Nappe. (**b**) Simplified geological map of Holsnoy after Kolderup & Kolderup (1940) and Austrheim (1986/87) with the sample location. A, Ådnefjellet.

processes related to the granulite-eclogite facies transition is a representative centimetre-scale eclogite-facies shear zone. The field relationships of cracks and shear zones and implications on the regional geology will be described elsewhere.

Macroscopic observations

The granulite-facies sample described in this study has an anorthositic host-rock composition with the stable mineral assemblage plagioclase (An_{55})–garnet–clinopyroxene. The specimen contains the margin of a 5 cm wide representative eclogite-facies shear zone which cross-cuts the anorthosite banding (Fig. 2a and b). The sample was cut perpendicular to the shear zone boundary and within 10° of a very weakly developed mineral lineation. Based on the mineralogy, the sample is divided into three zones described below.

Zone II. Within zone II the plagioclase layers rock is composed of mafic bands consisting of garnet and clinopyroxene in alternation with lilac-coloured plagioclase layers. The grain size ranges from 1–3 mm for garnet and pyroxene and up to 1 cm for plagioclase.

ZoneII. Within zone II the plagioclase layers are white coloured and the mafic bands are green due to a change in the mineral assemblage. The boundary between zones I and II is defined by this colour change and forms a metasomatic reaction front penetrating the host rock.

The shear zone boundary is defined as the line connecting points with the first visible deflection of the wall rock banding. This boundary is located within zone II (Fig. 2b). Within a 4–5 cm wide band the compositional layering has been progressively rotated towards parallelism with the boundary of the shear zone. A new foliation begins to develop and is defined by the shape fabric of clusters of pyroxene + garnet. As the angle between the new foliation and the shear zone boundary decreases, the shape fabric of the (pyroxene-rich) clusters increases and the new foliation becomes dominant. The changing angle between the shear zone boundary and the deflected wall rock banding and developing mylonitic folia is used to determine the shear strain profile (Fig. 2c) (Ramsay & Graham 1970). Zone II grades into zone III.

Zone III. The most strongly sheared part of the sample (shear zone centre) is made up by a green and very fine-grained eclogite-facies rock, with a distinct mylonitic foliation. The original compositional banding is obliterated and the new foliation is parallel to the geometrically defined shear zone boundaries. A strong decrease in grain size can be observed.

Microstructural observations

The three zones described above coincide with three distinct microstructural and mineralogical zones.

Zone I. The unaltered slightly strained anorthosites (An_{55}) (Fig. 3a and b) display a predominantly granoblastic, metamorphic texture with an equigranular to seriate grain-size distribution. Triple junctions with 120° angles and straight grain boundaries are common. Some plagioclase grains, however, show bent (deformation) twins, bulging subgrain boundaries and undulose extinction. The plagioclase grain size varies between 1 and 10 mm.

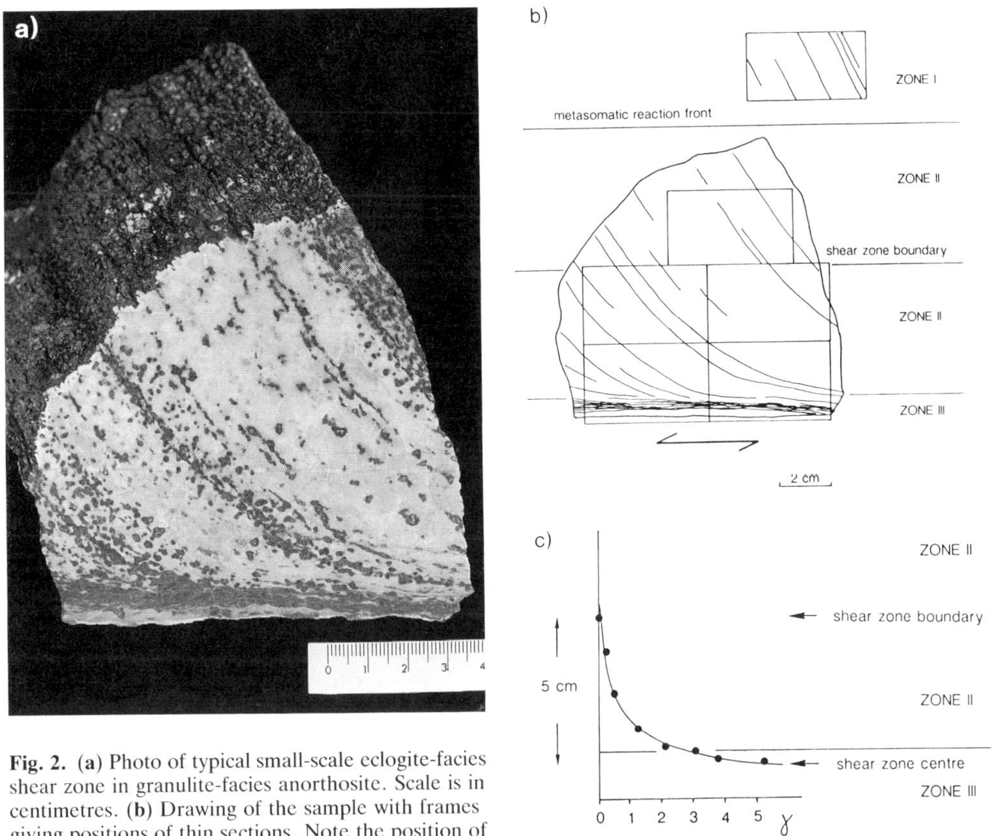

Fig. 2. (**a**) Photo of typical small-scale eclogite-facies shear zone in granulite-facies anorthosite. Scale is in centimetres. (**b**) Drawing of the sample with frames giving positions of thin sections. Note the position of the marker lines: geometrically defined shear zone boundary, metasomatic reaction front. (**c**) Shear strain profile based on the orientation of the deflected compositional banding. Reference line corresponds to the shear zone boundary.

The shear zone boundary as defined geometrically in the previous section is not represented by a distinct microscopic feature.

Zone II. The growth of very fine clinozoisite needles within the plagioclase crystals (Fig. 3c) is the first effect of a beginning textural and/or mineralogical transformation in the anorthosites. The small clinozoisite crystals (up to 0.1 mm) are aligned in two sets which are subperpendicular to each other and are often oriented at approximately 45° to the (albite-) twin lamellae in plagioclase. The long axis of clinozoisite is, therefore, oriented parallel to the (101) and (110) planes or the [010] direction of plagioclase. With increasing number and size of the clinozoisite crystals in plagioclase, the overgrowth loses its crystallographically controlled orientation. Very rare and small nuclei of quartz and kyanite have been observed within plagioclase host grains.

The large clinopyroxene crystals (Fig. 3d) show incipient recrystallization by formation of a mortar texture.

Zone III. The shear zone centre is characterized by a fine grain size of plagioclase (Fig. 3e and f) of 2–6 μm as compared to 1–10 mm in Zone I. Such recrystallized grains occur only at values of $\gamma > 3$. The general microstructure of plagioclase is an equigranular mosaic of grains with straight or lobate grain boundaries and no preferred lattice orientation. About 25% of the recrystallized grains show evidence of strain (undulose extinction or twinning).

Original plagioclase crystals occasionally survived as ribbon shaped grains with a strong preferred crystallographic orientation or as residual clasts. Garnet and omphacite also survived as clasts.

Small cracks which are subparallel to the shear zones (Fig. 3e) are common. Such cracks are always filled with an eclogite-facies paragenesis and may contain omphacite, kyanite, plagioclase (An_{15}) or clinozoisite.

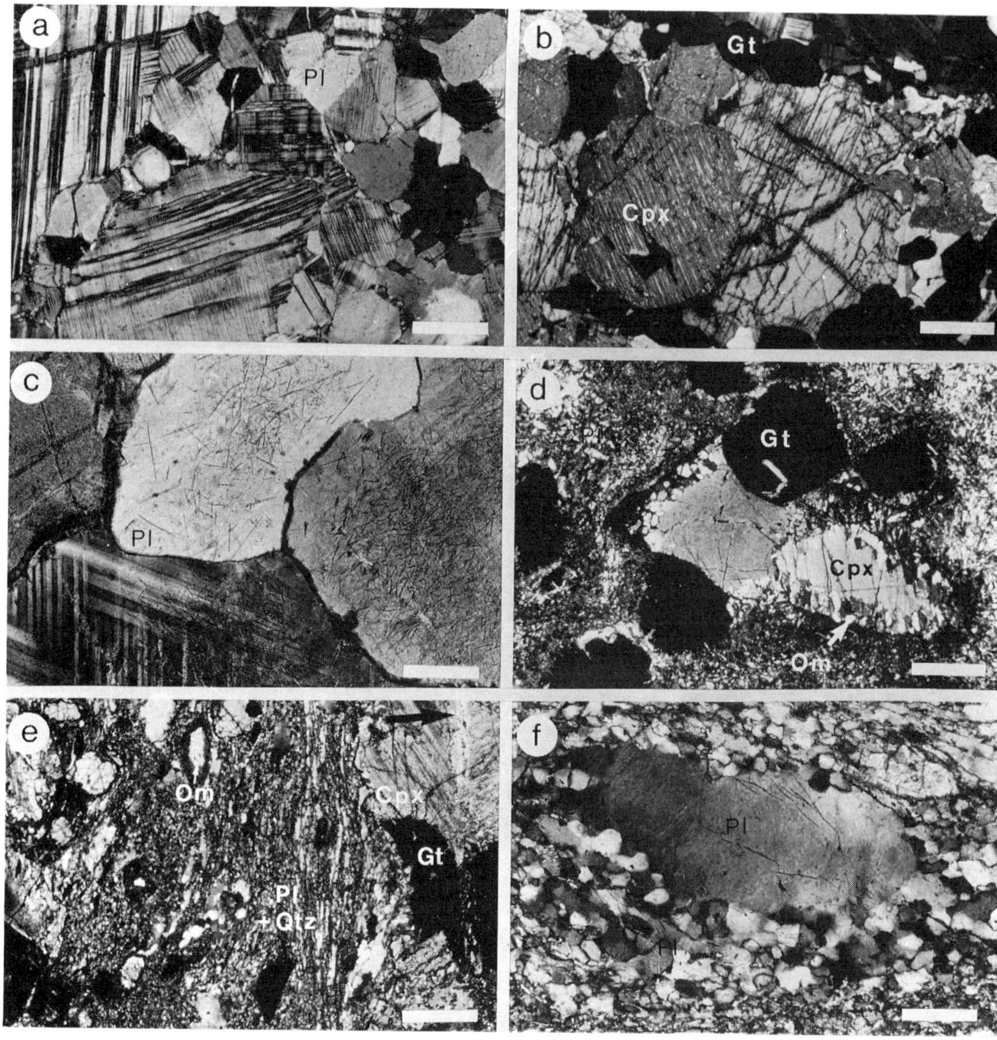

Fig. 3. (**a**) Plagioclase texture in anorthosite unaffected by the Caledonian deformation (scale bar is 0.5 mm). (**b**) Mafic layer in anorthosite. Gt, Garnet; Px, clinopyroxene (scale bar is 0.5 mm). (**c**) oriented clinozoisite needles grow within plagioclase (scale bar is 0.2 mm). (**d**) Recrystallization of clinopyroxene to omphacite in a mortar texture (scale bar is 0.5 mm). (**e**) Microstructure in shear zone with garnet, pyroxene and plagioclase clasts in a fine grained recrystallized matrix. Note the crack (arrow) parallel to the shear zone boundary. The crack cutting the large clinopyroxene is filled with omphacite and albite rich plagioclase. The half garnet within the shear zone may also have formed by cracking a large garnet (scale bar 0.5 mm). (**f**) Plagioclase clast in recrystallized matrix of small grains. Note the bulging of the grain boundaries around subgrains and new grains (scale bar is 0.2 mm).

Plagioclase breakdown reaction and its consequences

Plagioclase is not stable under eclogite-facies conditions (Goldsmith 1982). The granulite-facies assemblage of the anorthosite survived metastably on a regional scale under fluid absent conditions away from cracks and shear zones. Under fluid present conditions and with increasing pressure granulite-facies plagioclase (Plg) is replaced by clinozoisite (Czo) + kyanite (Ky) + clinopyroxene (Cpx) + paragonite (Pg) + quartz (Qtz) according to the schematic reaction:

$$\mathrm{Plg(An_{55}) + fluid \rightarrow Cpx(Jd) + Czo + Ky + Qtz + Pg(in\ pheng.) + Plg(An_{15})}$$

During this process the original plagioclase (An_{55}) is enriched in albite component (An_{15}). The first clinozoisite that can be observed shows the most iron-rich composition with up to 7 wt% ($FeO + Fe_2O_3$). With increasing number and size of the clinozoisite crystals the iron content is reduced to 2–3 wt% total iron. (Microprobe data available from the author on request).

Because hydrated minerals are absent in the original granulite-facies anorthosites it is obvious that a hydrous fluid phase was essential to start the eclogitization process which led to the formation of dense (omphacite) as well as hydrated minerals (clinozoisite, phengite). Garnet does not seem to be a major reactant during the eclogitization process of the anorthosites in opposition to the more mafic granulitic precursors.

The growth of eclogite-facies minerals occurs either at intracrystalline sites (clinozoisite in plagioclase) or at grain boundaries (omphacite around clinopyroxene). In both cases the new minerals grew only close to (within metasomatic reaction front) or within cracks and shear zones. This indicates that the reaction has taken place in response to the infiltration of H_2O along these pathways at high pressure. Several deformation or softening mechanisms may have played an important role on the grain scale during the eclogitization process. As soon as there are traces of fluid in the rock acting as catalyser, they will cause hydrating mineral reactions. Such chemical processes can then lead to reaction softening or reaction enhanced ductility (e.g. White *et al.* 1980; Rubie 1983; Kirby 1985) by the production of soft and fine-grained phases like phengite, quartz and recrystallized plagioclase. Consequently, these processes are responsible for easier accommodation of strain in localized zones. Another mechanism which may be important for the propagation of the metasomatic reaction front into the wall rock of cracks or shear zones is hydrolytic weakening of feldspar (Tullis & Yund 1980) due to diffusion of H_2O. The importance of hydrolytic weakening in the studied sample is not yet clear because it is very difficult to separate the effects of hydrolytic weakening from effects due to mineral reactions enabling the nucleation of clinozoisite needles within plagioclase.

Discussion

The observations presented in this paper suggest a granulite- to eclogite-facies transition model in which fluid availability is crucial for the extent of mineral reactions as well as for the deformation taking place in a specific volume of granulite-facies rocks. Partly eclogitized granulites cover an area of 50 km^2 (Austrheim & Mørk 1988) and it is likely that the volume of altered rocks comprises several cubic kilometres. Thus, the production and subsequent fluid flow is not a local, small scale feature.

The general tectonic framework suggests that the fluid source were crustal rocks which have been subducted together with the granulites at the onset of the Caledonian event. The subduction of a pelitic sequence, for instance, could easily explain the production of the necessary large volume of an aqueous fluid.

Unaltered anorthosites survived metastably on a regional scale which may be due to (a) limited availability of fluid, (b) slow diffusion of NaSi–CaAl even under the presence of a hydrous fluid phase, and (c) slow reaction (nucleation) rates.

The products of the observed plagioclase breakdown reaction are spatially related to shear zones or cracks. This indicates that the reaction has occurred in response to the infiltration or presence of a hydrous fluid phase along these pathways during or after pressure increase. Since cracks with undeformed eclogite-facies fillings can commonly be seen to run subparallel to shear zones (Fig. 3e) they are interpreted as precursor to the ductile deformation or as an initial permeability. The stresses imposed on the anorthosite complex then cause shear strain to localize along such preexisting fractures (Ramsay & Graham 1970; Segall & Simpson 1986) to produce the observed shear zones. Late brittle features are rather uncommon and it seems that the metamorphic effects of the fluid infiltration and reaction softening were more important than the reduction in effective stress.

Original coarse plagioclase grains show little deformation but are gradually reduced in size by progressive dynamic recrystallization of their margins. The increasingly abundant fine grains, only 25% of which show deformation features, are interpreted to undergo cyclic reproduction through grain boundary migration because they take up almost all the strain. As pointed out by Tullis & Yund (1985) the new, strain-free grains are easier to deform, whilst the remaining ribbon grains and clasts show little evidence of recovery (subgrains) and only slight recrystallization along their edges. The existence of relict plagioclase clasts may depend on their original lattice orientation which was unfavorable for the deformation in a given stress field (Olesen 1987).

There are several implications emanating

from the granulite- to eclogite-facies transition. The observed mineral reaction reduces the volume proportion of plagioclase in the rock and favours the growth of the dense phases omphacite and kyanite and of the hydrous phases clinozoisite and phengite. The growth of high density phases as well as field evidence indicate a slight volume decrease during eclogitization. This process may account for an increase in the permeability of the rocks during reaction progress.

Mineralogical alteration also leads to a change in the mechanical properties of the rocks. Deformation in the shear zones is primarily taken up by the recrystallizing matrix of fine-grained plagioclase which governs bulk rheology. Very minor amounts of fine-grained quartz contained in the matrix also show a ductile behaviour with dynamic recrystallization.

The strong minerals like garnet and, to a lesser extent, omphacite form almost undeformed clasts in the deforming matrix. Therefore, deformation partitioning is an important factor in these shear zones and leads to stress concentrations in the matrix. Stress and strain gradients within the shear zone are evident e.g. by the changes in recrystallized grain size. The fine-grained material produced during cyclic dynamic recrystallization, therefore, can accommodate a faster strain rate at a constant stress.

Eclogite-facies rocks containing only minor amounts of hydrated minerals, such as phengite, show a considerably more ductile behaviour even though their density is higher compared to the granulites. Foliation development within the shear zones is important in that it leads to a reduction in rock strength due to the formation of an 'easy' glide plane along oriented layer-silicates. The cyclic production of small strain-free grains during dynamic recrystallization also increases the deformability (Tullis & Yund 1985).

The transformation of a substantial volume proportion of the dry protolith into a network of highly foliated, dense and 'wet' eclogite-facies shear zones of various scales will certainly have an influence on the seismic signature of granulitic lower crust. A transition as discussed in this paper, will certainly have an effect e.g. on the anisotropy and impedance contrasts within a lower crustal granulite-facies complex and may be responsible for the high reflectivity of the lower crust of western Norway (Hurich & Kristoffersen 1988 and references therein).

A very important large scale consequence of the increase in localized ductility during the granulite–eclogite facies transition is the initiation of detachment and thrusting within the lower crust. Only such strain localization in high ductility zones may open the possibility to form shear zones large enough to emplace lower crustal high pressure rocks within the upper crust.

I wish to thank H. Austrheim for his support in the field and during the course of this study. Financial support of field work by the 'Swiss Academy of Science' is acknowledged.

References

AUSTRHEIM, H. 1986/87. Eclogitization of lower crustal granulites by fluid migration through shear zones. *Earth and Planetary Science Letters*, **81**, 221–232.

—— & GRIFFIN, W. L. 1985. Shear deformation and eclogite formation within granulite facies anorthosites of the Bergen Arcs, Western Norway. *Chemical Geology*, **50**, 267–281.

—— & MØRK, M. B. E. 1988. The lower continental crust of the Caledonian mountain chain: evidence from former deep crustal sections in western Norway. *Norges Geologiske Undersokelse, Special Paper*, **3**, 102–113.

COHEN, A. S., O'NIONS, R. K., SIEGENTHALER, R. & GRIFFIN, W. L. 1988. Chronology of the pressure–temperature history recorded by a granulite terrain. *Contributions to Mineralogy and Petrology*, **98**, 303–311.

GOLDSMITH, J. R. 1982, Plagioclase stability at elevated temperatures and water pressures. *American Mineralogist*, **67**, 653–675.

HURICH, C. A. & KRISTOFFERSEN, Y. 1988. Deep structure of the Caledonide orogen in southern Norway: new evidence from marine seismic reflection profiling. *Norges Geologiske Undersøkelse, Special Paper*, **3**, 96–101.

KIRBY, S. H. 1985. Rock mechanics observations pertinent to the rheology of the continental lithosphere and the localization of strain along shear zones. *Tectonophysics*, **119**, 1–27.

KOLDERUP, C. F. & KOLDERUP, N. H. 1940. Geology of the Bergen Arc system. *Bergen Museum Skrifter*, **20**, 1–137.

OLESEN, N. O. 1987: Plagioclase fabric development in a high-grade shear zone, Jotunheimen, Norway. *Tectonophysics*, **142**, 291–308.

RAMSAY, J. G. & GRAHAM, R. H. 1970. Strain variation in shear belts. *Canadian Journal of Earth Sciences*, **7**, 786–813.

RUBIE, D. C. 1983. Reaction-enhanced ductility: the role of solid-solid univariant reactions in deformation of the crust and mantle. *Tectonophysics*, **96**, 331–352.

SEGALL, P. & SIMPSON, C. 1986. Nucleation of ductile shear zones on dilatant fractures. *Geology*, **14**, 56–59.

STURT, B. A. & THON, A. 1978. The Caledonides of southern Norway. Caledonian–Appalachian orogen of the North Atlantic region. *Geological*

Survey of Canada, Special Paper, **78–13**, 39–47.

TULLIS, J. & YUND, R. A. 1980. Hydrolytic weakening of experimentally deformed Westerly granite and Hale albite rock. *Journal of Structural Geology*, **2**, 439–451.

—— 1985. Dynamic recrystallization of feldspar: a mechanism for ductile shear zone formation. *Geology*, **13**, 238–241.

WHITE, S. H., BURROWS, S. E., CARRERAS, J., SHAW, N. D. & HUMPHREYS, F. J. 1980. On mylonites in ductile shear zones. *Journal of Structural Geology*, **2**, 175–187.

The role of second phase in localizing deformation

DAVID L. OLGAARD

Geologisches Institut, ETH, Zürich, CH-8092, Switzerland

Abstract: Small fractions of second phases may play a major role in localizing deformation in rocks. Second phases inhibit the migration of grain boundaries and hence may prevent grain growth. Since flow laws for many deformation mechanisms show a grain size dependence, second phases may indirectly control deformation rate. To illustrate this, a hypothetical sequence of micritic limestones, two impure layers sandwiched between two pure calcite layers, are metamorphosed and then deformed. The initial grain size is assumed to be less than 5 μm. The thermo-mechanical behavior of this sequence is predicted based on three types of experimental data: (1) calcite grain growth kinetics; (2) calcite grain size as a function of second phase; (3) plastic flow laws on natural and synthetic calcite rocks under conditions for grain-size insensitive (GSIC) and grain-size sensitive creep (GSSC). Extrapolations of the results predict that metamorphism at 400°C will increase the grain size in the pure calcite layers to 780 μm in 10000 years. In a layer containing 5% of 0.3 μm particles, however, the grain size will increase to only 6 μm. Under a differential stress of 100 MPa, the coarser-grained layers deform by GSIC at a strain rate of 10^{-14} s^{-1} while the 6 μm layer deforms by GSSC at 10^{-9} s^{-1}. This large difference in deformation rate insures that virtually all of the strain will be localized in the fine-grained layer.

Localized deformation in ductile shear zones and mylonites is a common structure observed in the Earth's crust even in rocks that contain no obvious mineralogical or microstructural heterogeneities. Several processes have been proposed to explain such localization: shear heating, grain size reduction during dynamic recrystallization or cataclasis, reaction-enhanced weakening, increased access of fluids, or geometric softening (e.g. Schmid 1982; Brodie & Rutter 1985). These are all viable mechanisms but strain localization need not be related to syntectonic processes. In this paper, a model is presented which shows that localized deformation can be a consequence of subtle lithological differences. The example described is for calcite rocks because of the abundance of experimental observations of both metamorphic and deformation processes in these rocks. The basic ideas, however, are not limited to calcite rocks but are applicable to any rock type in which grain growth is retarded by second phases and deformation shows a grain size dependence.

Kinetics of grain growth and plastic flow

For the model presented the high temperature, solid state processes of grain growth and plastic flow are considered. The grain growth equations and creep laws applicable to rocks have been reviewed elsewhere (Olgaard & Evans 1988; Schmid *et al.* 1980) and are only briefly summarized below.

The evolution of grain size in nearly monomineralic polycrystals by grain growth is shown in Fig. 1. If the average grain size increases continuously and uniformly, the growth process is called *normal grain growth* and follows the relation:

$$D^p\text{-}D_0{}^p = Kt \qquad (1)$$
$$K - K_0\exp[\ Q_g/(RT)]$$

where D is the grain size after time t, D_0 is the initial grain size, K_0 is an empirical constant dependent on grain boundary mobility, Q_g is the activation energy for normal grain growth, T is temperature, and R is the universal gas constant. The exponent p is theoretically equal to 2 for single-phase materials and 3 when a grain boundary fluid or other mobile second phases are present. Empirical values for p vary from 2 to greater than 5 and may vary with time or temperature (Hu & Rath 1970). Normal grain growth experiments have been conducted on several calcite rocks at temperatures to 1200°C (Tullis & Yund 1982; Olgaard & Evans 1986, 1988; Covey-Crump & Rutter 1989; Olgaard, Evans & Paterson, unpublished data). The results of some of these experiments are summarized in Table 1.

Very large grain sizes may result from normal grain growth at geological conditions unless grain boundary migration is impeded. If a mi-

From Knipe, R. J. & Rutter, E. H. (eds), 1990, *Deformation Mechanisms, Rheology and Tectonics*, Geological Society Special Publication No. 54, pp. 175–181.

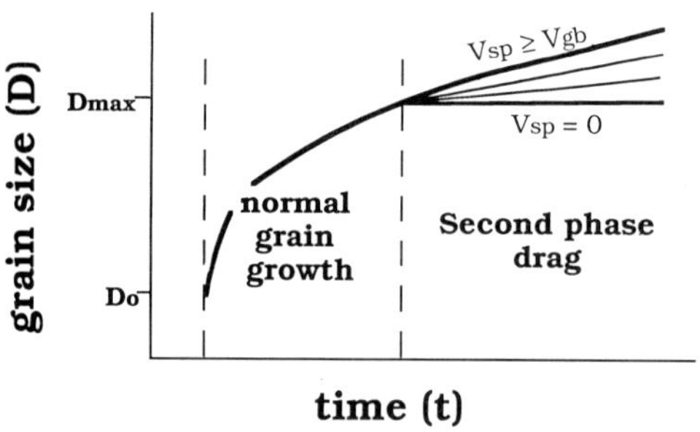

Fig. 1. Grain size evolution by normal grain growth and second phase drag. Metamorphism of an unstrained, dense polycrystal leads first to normal grain growth at a rate defined by Eq. (1). When the grain size reaches the second-phase stabilized grain size, D_{max} further grain growth is controlled by either the maximum velocity of the second phases (V_{sp}) relative to the velocity of the grain boundary (V_{gb}) or the rate at which the second phases coarsen or disappear. If the second phases are immobile and insoluble, D_{max} is a fixed value independent of time or temperature. (modified from Olgaard & Evans 1988)

grating grain boundary intersects second phases such as particles or pores, the second phases exert a drag force proportional to their size, d, and volume fraction, f. In this case, the average grain size increases until a stable size, D_{max}, is reached according to *Zener's (in Smith 1948) equation*:

$$D_{max} = (2/3)(d/f). \quad (2)$$

Olgaard & Evans (1986, 1988) heat-treated two-phase mixtures of calcite plus alumina or mica particles at 800°C for up to 24 hours in the presence of H_2O and compared their results to other studies on metals and ceramics. They found a more general *second-phase drag* relation to be appropriate:

$$D_{max} = C(d/(f^m)). \quad (3)$$

where C and m are empirical constants (Table 2). D_{max} is independent of temperature or time if the second phases are immobile, e.g. alumina and mica particles. If the second phases are mobile, as for pores, equation 3 is still valid but the stable grain size is a function of the rate at which the pores coarsen or migrate out of the material (Fig. 1).

Crystal–plastic flow by recovery-controlled dislocation motion is nominally grain size independent. This *grain-size insensitive creep* (GSIC) follows the power law:

$$\dot{\varepsilon} = A\sigma^n \exp(Q_c/RT) \quad (4)$$

where $\dot{\varepsilon}$ is strain rate, σ is the applied differential stress, Q_c is the activation energy for disloca-

Table 2. *Calcite plus second phase particles*

d(μm)	f	C	m
alumina			
0.29	0.0002–0.01	13.6	0.34
0.40	0.002–0.1	5.0	0.56
2.53	0.01–0.1	2.1	0.39
4.73	0.01–0.1	0.8	0.43
*mica**			
1.3	0.01–0.05	6	0.29

Empirical constants for Eq. 3 (Olgaard & Evans 1986; 1988). d is the particle size, f is the volume fraction. * Biotite and phlogopite.

Table 1. *Normal grain growth in calcite rocks*

	T(°C)	P_c(MPa)	p*	Q_g[§]	Remarks
Synthetic marbles					
Olgaard & Evans (1988)	550–900	300–500	2.6–4.2	150±80	water added
Covey-Crump & Rutter (1989)	650–700	100	2.9	216	water added
Olgaard *et al.* (unpublished)	500–1200	50, 300	3.0	172±60	oven dried
Solnhofen limestone					
Tullis & Yund (1982)	650–1000	1500	3.1–>5	200‡	oven dried
	700–1000	1500	2.5–>5	250‡	water added
Schmid *et al.* (1977)	600–1000	300	2.5–>7	190‡	oven dried
Rutter (1984)	700–800	100	2.2	131	water added
Olgaard & Evans (1988)	700–800	0.1, 300	>5.2		water added
Carrara marble					
Olgaard & Paterson (unpublished)	1200	300	5.6	–	oven dried

Data for Eq. 1. * Not determined at every temperature. [§] Q_g could not always be determined because p varied with temperature. [‡] Calculated assuming $p = 3$.

tion climb and A is an empirical constant. The stress exponent, n, is theoretically equal to 3 to 4.5 (Weertman 1972) and empirically greater than or equal to 3.

Plastic deformation can also depend on the grain size of the deforming material. Grain-size sensitive mechanisms that are believed to be important in rocks include: dislocation pile-ups at grain boundaries (Petch 1953), volume and grain boundary diffusion creep (Herring 1950; Coble 1963, respectively) and grain boundary sliding accommodated by diffusion or dislocation creep (e.g. Ashby & Verrall 1973). High-temperature–low-stress *grain-size sensitive creep* (GSSC) follows a similar power law to that of GSIC but with a grain size dependence and a low stress exponent characteristic of diffusion-controlled creep:

$$\dot{\varepsilon} = B(\sigma^{n'}/D^k)\exp(-Q_c'/RT) \quad (5)$$

where Q_c' is the activation energy for either grain boundary or volume diffusion depending on the accomodation mechanism, D is the grain size, and B is a constant. Theoretically, the stress exponent, n', is equal to 1 and k is equal to 2 or 3. Empirically, n' varies between 1 and 3 implying that both crystal plastic and diffusion creep mechanisms may be operating. The term superplasticity is often given to this regime because of the low stress sensitivity of strain rate and the microstructural similarities of many materials, including rocks, to metals deformed in tension to very high strains without failure.

Deformation experiments have been conducted on both natural and synthetic calcite rocks to explore these GSIC and GSSC fields (e.g. Heard & Raleigh 1972; Rutter 1974; Schmid *et al.* 1977, 1980; Walker *et al.* this volume; Olgaard, Evans & Paterson, unpublished data). The grain growth, second phase drag, GSIC and GSSC relations and the experimentally derived data for calcite will now be used to illustrate that large rheological contrasts can develop from petrographically subtle differences in second phases content.

Initial lithological stratification

In the model discussed below, a four-layer sequence of fine-grained limestones, two less pure layers sandwiched between two pure calcite layers as illustrated in Fig. 2, is considered initially. Fluctuations in sediment source or depositional environment could easily result in such a lithologically stratified sequence. Micritic limestones with a grain less than 5 μm are common in nature. Organic or terrigenous phases such as phyllosilicates, carbon, quartz, pyrite or clay minerals are likely micron and submicron-sized impurities. The lithology and grain size would be similar to Solnhofen limestone.

Similar limestones were made in the laboratory by mixing fine reagent-grade $CaCO_3$ powders with different sizes and concentrations of rigid, non-reacting Al_2O_3 or mica particles (Olgaard & Evans 1986, 1988) and then hot pressing them at 500°–600°C, 200–500 MPa isostatic confining pressure, for a few hours with and without added water. The average size of the polygonal grains in these dense synthetic marbles was 5 to 10 μm.

If this initial sequence is deformed under conditions favoring plastic flow (either GSIC or GSSC), there is little or no rheological contrast because all of the layers have the same initial grain sizes (Fig. 2b). The second phase fractions of 2–5% would cause only small increases in the relative strengths of the layers even if the phases were rigid particles.

Metamorphic stratification

Metamorphism often results in an increase in the grain size and alterations in the petrographic textures of rocks. In the absence of chemical reactions or strain-induced defects such as dislocations, grain boundary migration is driven by the grain boundary energy and boundaries will migrate in a direction to reduce their curvature, i.e. to increase the average grain size. To simulate the effect of metamorphism on grain size, experimental grain growth data is used to predict the grain sizes in the four layer sequence after 100 000 years at 400°C (Fig. 2c)

The kinetics of normal grain growth has been studied in synthetic limestones by heat-treatments at 500°–1200°C for up to 75 days with added water and oven dried (Olgaard & Evans 1988; Olgaard, Evans & Patterson, unpublished data). The results are summarized in Fig. 3. The data are fit linearly to Eq. (1) for the grain size exponent, p, equal to 3, the value expected for fluid or pore-drag controlled grain growth. The slower rate of growth in the wet specimens compared to the dry is probably due to the slightly higher porosities that have been observed in the wet specimens. The activation energy, Q_g, calculated from this data is equal to 172 ± 60 kJ mole^{-1} which is within the range for grain growth as estimated from synthetic and natural calcite rocks (Table 1).

Extrapolations of these data using Eq. (1) with p equal to 3 and Q_g equal to 172 kJ mole^{-1} (therefore, $K_0 = 4 \times 10^{-9}$ m^3 s^{-1}) shows that a grain size of 780 μm would be expected in

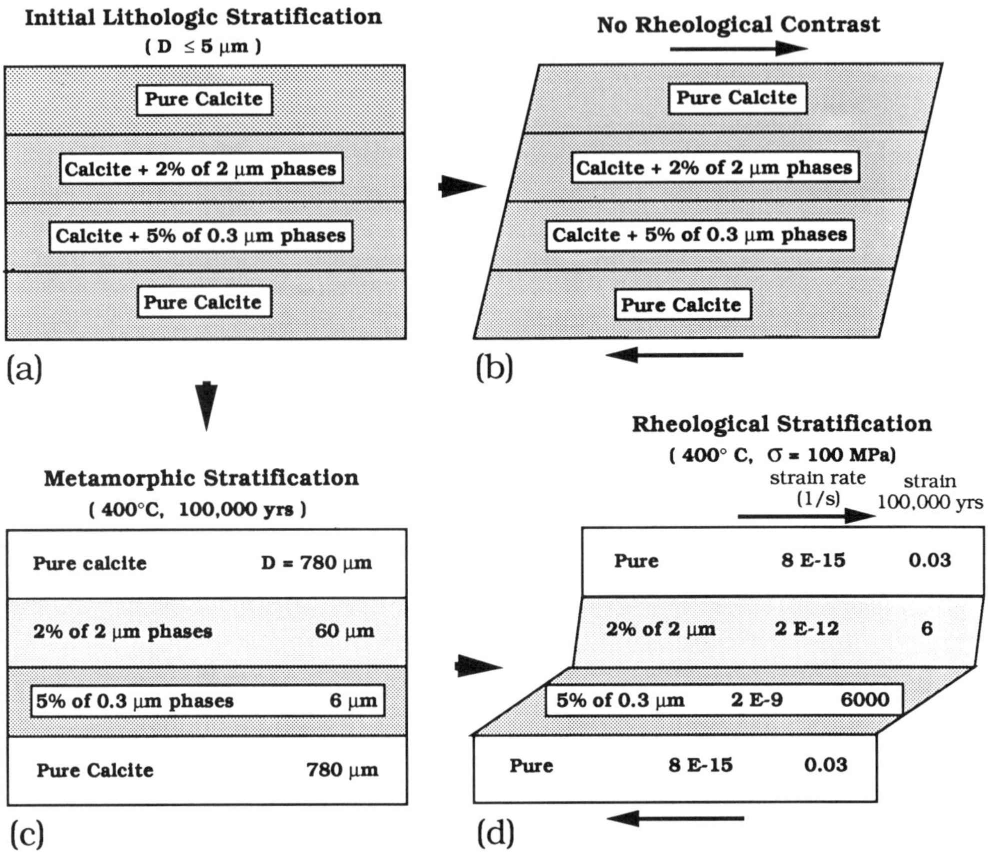

Fig. 2. Synoptic diagram illustrating the evolution of an initially fine-grained sequence through metamorphism and deformation. The thickness of the layers is not specified. If the initial sequence (**a**) is deformed shortly after diagenesis (**b**), there will be no rheological contrast because all of the layers will have similar grain sizes. The second phase fractions of 2–5% will cause only small increases in the strength of the two impure layers. If the sequence is first metamorphosed (**c**), however, the difference in grain sizes will lead to a very pronounced rheological stratification (**d**).

100 000 years at 400°C (3.8 mm after 10 Ma). Such a large increase in grain size in this geologically short period of time implies that small average grain sizes are unstable for even short metamorphic events unless grain growth is inhibited. Larger grain sizes are possible because some small fraction of pores and particles in these nominally pure synthetic marbles probably retard grain growth in the experiments. Of course, smaller grain sizes are also possible, even likely, since natural rocks contain ubiquitous second phases and few, if any, are as clean as the synthetics marbles. For example, the grain growth rate of Carrara marble, a natural rock often used in experiments because of its high purity and low porosity, is slower than in pure calcite synthetic marbles at the same conditions Olgaard & Paterson, unpublished data).

In the two impure layers, grain growth would be inhibited after a very short time by second-phase drag (Fig. 1). The second phases in the middle two layers are assumed to be rigid, non-reacting, immobile particles and therefore the stable grain size, $D_{\max}$, is independent of time or temperature ($V_{sp} = 0$). Mobile second phases ($V_{gb} > V_{sp} > 0$), such as pores, also give a stable grain size defined by Eq. (2), but D_{max} will gradually increase as the pores coarsen (d increases) or disappear (f, decreases). Second phases would be much less effective at retarding migration if the driving energy for grain boundary migration came from chemical reactions or

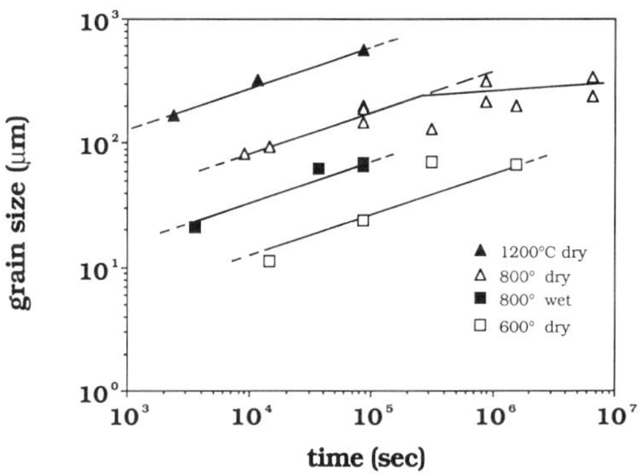

Fig. 3. Grain size v. time for normal grain growth in synthetic marbles with water added (wet) and oven dried (dry). The linear fits through the data are for Eq. (1) with the grain size exponent, p, equal to 3. For 800°C dry and $t > 10^5$ s the growth rate decreases presumably because of porosity. (Olgaard & Evans 1988; Olgaard & Evans unpublished data).

strain induced defects, such as dislocations. Therefore, it is also assumed that the energy driving grain-boundary migration derived only from the grain-boundary area.

The relationship between matrix grain size and second-phase particle size and concentration for calcite is given by Eq. (3) with values from Table 2. Using the average values of C = 5.3 and m = 0.43, respectively in Eq. (3) yields stable grain sizes of 60 μm for the layer containing 2% of 2 μm particles and 6 μm for the layer containing 5% of 0.3 μm particles (Fig. 2c). Obviously, even very small fractions of fine particles may have dramatic effects on the grain size of a metamorphosed rock.

A metamorphic stratification with two orders of magnitude difference in grain size is predicted in 100 000 years at 400°C with only 5% of 0.3 μm particles being necessary to pin the calcite grain size at the initial value. For longer times the stratification becomes even more pronounced as the grain size of the pure layers would continue to increase while the size of the two-phase layers would remain fixed.

Rheological stratification

Now consider the response of the four marble layers to a constant differential stress of 100 MPa at 400°C. Such a stress magnitude, although somewhat arbitrarily chosen, is reasonable for the earth's crust. To illustrate the degree to which deformation may be localized in this grain-size stratified sequence, the stresses applied to each layer are assumed to be equal so that the rheological contrasts are reflected as variations in the strain rates. If strain rates are assumed to be equal, then the stresses across the layers would vary and the competency contrasts could be applied to multi-layer folding. In the earth, neither constant stress nor constant strain rate are strictly applicable, rather stress and strain rate may vary both spatially and temporally.

In Fig. 4, a deformation regime map of stress against grain size with contours of constant strain rate has been constructed by adding Eq. (4) and (5) with appropriate empirical values from Schmid *et al.* (1977, 1980). The GSIC regime is the constant strain rate regime 2 for Carrara marble; the GSSC regime is the constant strain rate superplasticity regime (regime 3) for Solnhofen limestone assuming a grain size exponent, k, of 3. The two coarse-grained layers with grain sizes equal to 780 μm deform by GSIC at a strain rate of $8 \times 10^{-15}\,s^{-1}$; a slow but reasonable estimate for geological rates. The two-phase layer with a grain size of 60 μm is within the GSSC regime and deforms at a strain rate of $2 \times 10^{-12}\,s^{-1}$. The finest-grained layer with a grain size of 6 μm also deforms by GSSC at a strain rate of $2 \times 10^{-9}\,s^{-1}$, or over five orders of magnitude faster than the pure layers.

Second phase particles have a dramatic affect on grain boundary migration, but do not appear

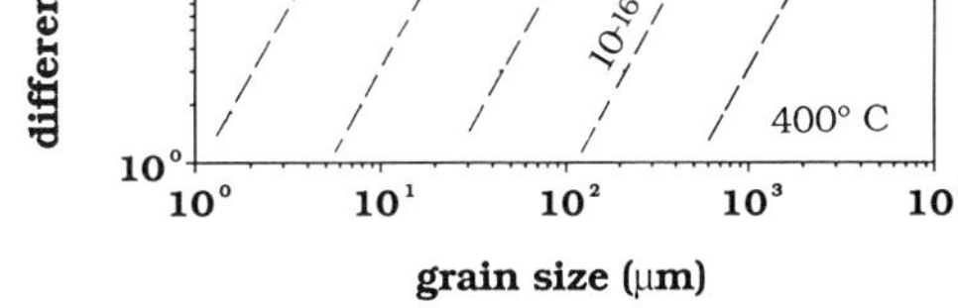

Fig. 4. Deformation regime map for calcite rocks in differential compressive stress v. grain size space (modified from Schmid 1982). The contours are for strain rate (1/s). The map was constructed using the composite flow law: $\dot{\varepsilon} = \dot{\varepsilon}(\mathrm{GSIC}) + \dot{\varepsilon}(\mathrm{GSSC})$. The first term is regime 2 for Carrara marble (Schmid *et al.* 1980) and the second is regime 3 for Solnhofen limestone (Schmid *et al.* 1977), both for constant strain rate. Above 100 MPa the GSIC flow law may have a different form (regime 1 of Schmid *et al.* 1980).

to inhibit grain boundary sliding. Synthetic marbles containing 5 volume percent of 0.4 μm alumina particles (measured D_{max} = 11 μm; calculated D_{max} = 8 μm) were deformed in a triaxial gas-medium apparatus at 600–900°C and strain rates of 10^{-3}–10^{-6} s^{-1} (Olgaard & Paterson, unpublished data); the same conditions define the superplastic regime for Solnhofen limestone (Schmid *et al.* 1977). The synthetic marbles were weaker than Solnhofen limestone at all temperatures even though the synthetic marbles are coarser grained (Fig. 5). Several lines of evidence indicate that the synthetic marbles deformed by GSSC: (1) The stress exponents are between 2 and 3, higher than the 1.7 for Solnhofen limestone but within the range of other superplastic ceramics (e.g. Carry & Mocellin 1986); (2) the final microstructures, e.g. grain boundary morphologies and intragranular defect structures, were similar to the initial microstructures; (3) grain flattening accounted for less than 50% of the total strain. These results suggest qualitatively that GSSC made a significant contribution to the deformation of the synthetic marbles and that the second phases, which did inhibit grain growth (the grain size remained constant in all experiments), did not inhibit grain boundary sliding.

Figure 2d shows the rheological stratification

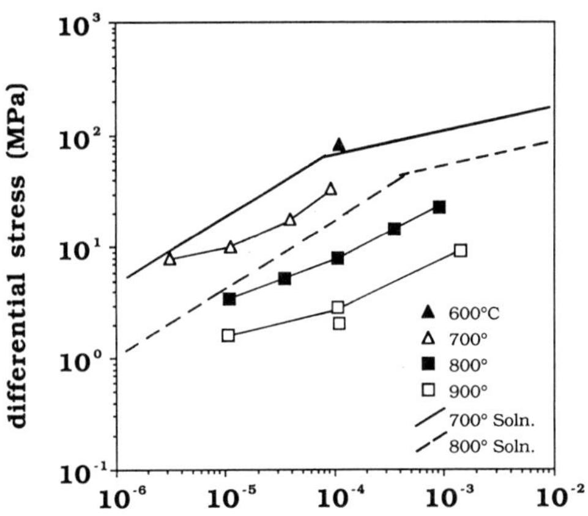

Fig. 5. Steady-state differential compressive stress vs strain rate for two-phase synthetic marbles (D = 11μm) (Olgaard & Patterson, unpublished data). The data are compared to Solnhofen limestone (indicated by the bold lines, Soln.) (D = 6 μm) at similar temperatures (Schmid *et al.* 1977). The low stress exponent (n = 2–3) and microstructural observations strongly suggest that the second phase particles did not inhibit grain boundary sliding and thus did not prevent GSSC.

resulting from a differential stress of 100 MPa imposed at 400°C. After 100 000 years, the pure calcite layers have an almost undetectable strain of 0.03 while the two-phase layer containing 5% of 0.3 μm particles has a strain of 6000! The absolute magnitude of the strain is arguable as these calculations were made by direct extrapolations of data that were determined at much smaller strains. The intention of this exercise was not to predict the magnitude of the strains but to show that an extremely localized zone of deformation is expected in the finest-grained layers after geologically short periods of metamorphism and deformation.

Conclusions

The example presented above demonstrates that localization of deformation into shear zones or mylonites may be due to subtle differences in lithologies and not necessarily the result of syntectonic microstructural or mineralogical changes. Only a few percent of micron-sized particles, sizes and concentrations that may not even be detectable in an optical microscope, are sufficient to keep the grain size of the major phase below 100 μm and, therefore, within the GSSC deformation regime for reasonable geological stresses and strain rates.

The extrapolations are, of course, sensitive to the uncertainties in all measured and calculated parameters in Eqs (1)-(5). Grain growth rates (K) or deformation rates ($\dot{\varepsilon}$) that are an order of magnitude slower, however, do not affect the final stratification because much longer times of metamorphism and deformation are acceptable and even likely. The calculations for the stable grain size do not require extrapolations and the particle sizes and volume fractions used in the experiments and in this illustration are within the range expected in natural rocks. The stable grain size, however, is dependent on the migration velocity of the second phases relative to the grain boundaries. Over geological time, second phases may not be as stable as is assumed but may coarsen, dissolve, precipitate or move. Eq. (3) is, however, still applicable and the relative stratification should remain. If at least part of the stratified sequence remains within the GSSC regime, localization would be expected in the finer-grained layers.

There is abundant evidence that the grain sizes in nearly monomineralic natural rocks are a function of second phase content (e.g. Spry 1969; Evans *et al.* 1980). In at least one case there is evidence that the deformation mechanism changes from GSIC to GSSC between coarse and fine-grained layers (Krabbendam &

Urai 1989). To test the model presented above in the field requires very careful characterization of the second phase size, volume fraction, and dispersion. Special attention needs to be paid to the micron to submicron-sized particles as they are very effective at pinning grain boundaries even in very small volume fractions. It is also necessary to show that at least part of the deformation was by GSSC mechanisms, a task that is often difficult (e.g. Evans *et al.* 1980).

The usefulness of this model is not restricted to only those situations where high temperature crystal plasticity or diffusional flow deformation mechanisms can be identified but is equally applicable to rocks deformed by other grain-size dependent mechanisms such as pressure solution. Furthermore this model is not restricted to calcite rocks. Quantitative predictions of the degree of localization in other rock types, such as quartz or olivine-rich rocks, will depend on grain growth and plastic flow rates which are known to be different than for calcite rocks. However, the qualitative result that second phases may control the grain size and, therefore, deformation rates are applicable. This hypothetical model demonstrates how experimental grain growth and flow law data can be useful in elucidating the possible rheological contrasts that exist in the field.

I wish to thank M. Casey, B. Evans, J. Urai, J. Gilotti, P. Crowley and L. Dell'Angelo for their comments and criticisms of various versions of this manuscript. This work was supported in part by the Swiss National Science Foundation grant No. 2.611–0.87.

References

ASHBY, M. F. & VERRALL, R. A. 1973. Diffusion-accommodated flow and superplasticity. *Acta Metallurgica*, **21**, 149–163.

BRODIE, K. H. & RUTTER, E. H. 1985. On the relationship between deformation and metamorphism, with special reference to the behavior of basic rocks. *In*: THOMPSON, A. B. & RUBIE, D. C. (eds) *Metamorphic Reactions, Kinetics, Textures, and Deformation*. Advances in Physical Geochemistry, **4**, Springer-Verlag, New York, 138–179.

CARRY, C. & MOCELLIN, A. 1985. High ductilities in fine grained ceramics. *In*: BAUDELET, B. & SUÉRY, M. (eds) *Superplasticity. Conference Internationale Grenoble, France*. Centre National de la Recherche Scientifique, Paris, 16.1–16.19.

COBLE, R. L. 1963. A model for boundary-diffusion controlled creep in polycrystalline materials. *Journal of Applied Physics*, **34**, 1679–1682.

COVEY-CRUMP, S. J. & RUTTER, E. H. 1989. Thermally-induced grain growth of calcite marbles on Naxos Island, Greece. *Contributions to Mineralogy and Petrology*, **101**, 69–86.

EVANS, B., ROWAN, M. & BRACE, W. F. 1980. Grain-size sensitive deformation of a stretched conglomerate from Plymouth, Vermont. *Journal of Structural Geology*, **2**, 411–424.

HEARD, H. C. & RALEIGH, C. B. 1972. Steady-state flow in marble at 500–800°C. *Geological Society of America Bulletin*, **83**, 935–956.

HERRING, C. 1950. Diffusional viscosity of a polycrystalline solid. *Journal of Applied Physics*, **21**, 437–445.

HU, H. & RATH, B. B. 1970. On the time exponent in isothermal grain growth. *Metallurgical Transactions*, **1**, 3181–3184.

KRABBENDOM, M. & URAI, J. L. 1989. Deformation mechanism switch due to grainsize stabilization by a dispersed second phase: An example from Naxos (Greece). *Terra abstracts*, **1**, [1], 379.

OLGAARD, D. L. & EVANS, B. 1986. Effect of second-phase particles on grain growth in calcite. *Journal of the American Ceramic Society*, **69**, C–272–C–277.

—— & EVANS, B. 1988. Grain growth in synthetic marbles with added mica and water. *Contributions to Mineralogy and Petrology*, **100**, 246–260.

PETCH, N. J. 1953. The cleavage strength of polycrystals. *Journal of the Iron and Steel Institute*, **174**, 25–28.

RUTTER, E. H. 1974. The influence of temperature, strain rate and interstitial water in the experimental deformation of calcite rocks. *Tectonophysics*, **22**, 311–334.

SCHMID, S. M. 1982 Microfabric studies as indicators of deformation mechanisms and flow laws operative in mountain building. *In*: HSÜ, K. J. (ed.) *Mountain Building Processes, Academic Press*, London, 95–110.

——, BOLAND, J. N. & PATERSON, M. S. 1977. Superplastic flow in finegrained limestone. *Tectonophysics*, **43**, 257–291.

——, PATERSON, M. S. & BOLAND, J. N. 1980. High temperature flow and dynamic recrystallization in Carrara marble. *Tectonophysics*, **65**, 245–280.

SMITH, C. S. 1948. Grains, phases and interfaces: an interpretation of microstructure. *Transactions of the American Institute of Mining and Metallurgical Engineers*, **175**, 15–51.

SPRY, A. 1969. *Metamorphic Textures*. Pergamon Press, Oxford.

TULLIS, J. & YUND, R. A. 1982. Grain growth kinetics of quartz and calcite aggregates. *Journal of Geology*, **90**, 301–318.

WALKER, A. N., RUTTER, E. H. & BRODIE, K. H. Experimental study of grain-size sensitive flow of synthetic, hot-pressed calcite rocks. *This volume*.

WEERTMAN, J. 1975. High temperature creep produced by dislocation motion. *In*: LI, J. C. M. & MUKHERJEE, A. K. (eds) *Rate Processes in Plastic Deformation of Materials*. Proceedings of the J. E. Dorn Memorial Symposium (1972), American Society of Metals, 315–336.

Mechanical controls on dilatant shear zones

A. ORD

CSIRO Division of Geomechanics, PO Box 54, Mt Waverley, Victoria 3149, Australia

Abstract: The finite difference code FLAC is used to examine the distribution of regions of high and low mean normal stress (or pressure) and of maximum dilation around deforming, periodic shear zones. The assumption is made here that the fluid pressure is equal to the mean normal stress. Fluid flow is favoured by large pressure gradients, and is enabled by regions of dilatancy. It is commonly assumed that regions of high dilation are necessarily associated with regions of low pressure. However, it is shown here that this need not be the situation. Cases in which maximum dilation is associated with the maximum pressure may be useful for understanding the presence of periodic melt segregations whereas cases in which maximum dilation is associated with minimum pressure may be useful for understanding metamorphic differentiation during crenulation cleavage development.

Shear band development in a class of frictional-dilatant materials undergoing uniaxial compression during a numerical experiment has been described by Hobbs & Ord (1989). They investigated the use of the computer code FLAC (Fast Lagrangian Analysis of Continua, Cundall & Board 1988) for the study of such localization. They observed that numerical modelling of deformation in a non-hardening Coulomb material, which may be homogeneous or heterogeneous with respect to cohesion, results in localization, compatible with theory (see Vardoulakis 1980) and the results of physical experiments (Arthur *et al.* 1977; Vardoulakis & Graf 1985). The physical conditions required for localization in geological materials are discussed by Hobbs *et al.* (this volume). The behaviour of a similar numerical model subjected to a simple shearing deformation history is investigated here, again changing the friction angle (ϕ) and the dilation angle (ψ) for each numerical experiment. These experiments result in a periodic localization of strain in models for which ϕ and ψ are each between 0° and 30°. The forms of the grids which display these periodic shear bands resemble crenulated rocks so various geological aspects of the deformed specimens were investigated so as to gain greater understanding of their associated geometry, kinematics and dynamics. In structural geology, an additional aim is to assess the consequences of including a dilation angle in the constitutive formulation for a Coulomb material. This results in high dilatancy in regions of high strain, so that there is an obviously greater dilation associated with the shear zones which form in such materials compared to regions outside the shear zones.

This paper therefore concentrates on examining the variations in volume change and the mean normal stress associated with periodic shear bands. Why such patterning occurs is beyond the scope of this paper although a possible approach is outlined in Section 4.

The model

Geometry and boundary conditions

The finite difference grid chosen for these simple shearing deformation history experiments contains 50 × 50 elements. Two columns of elements, one at each side, comprise passive rigid platens. The left hand platen for each diagram described below was given zero velocity while the right hand platen (Fig. 1) was given a velocity of 48×10^{-4} units per time step, where a unit is the length of an initial element. The rows of external nodes between the platens were given velocities varying from 2 to 47×10^{-4} units per time step so that the bulk deformation was constrained to be isochoric (constant volume), resulting initially in an homogeneous simple shear. The 2400 deforming elements are not so constrained; they may dilate or contract and undergo whatever deformation is required so long as the overall boundary conditions are satisfied. A bulk shear strain of 1 is attained at 10 000 computational steps, taking about 17 hours to run on a 386 PC (Toshiba 5200).

Material properties

All elements, except two in the centre of the numerical specimen, have the same shear

From Knipe, R. J. & Rutter, E. H. (eds), 1990, *Deformation Mechanisms, Rheology and Tectonics*, Geological Society Special Publication No. 54, pp. 183–192.

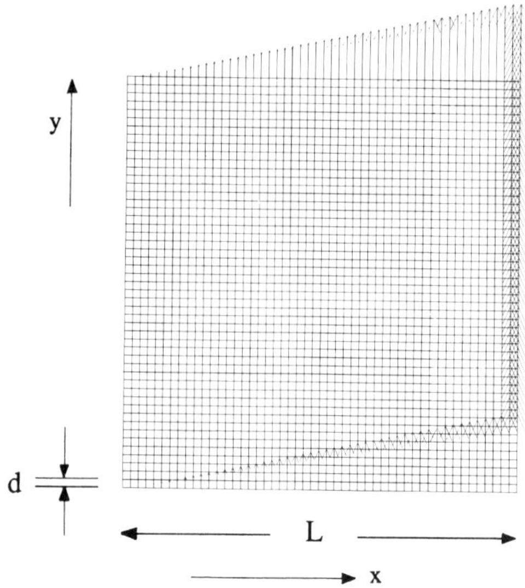

Fig. 1. Finite difference grid containing 50 × 50 elements and showing the imposed boundary conditions. The sinistral nature of the simple shearing deformation history experiments is represented by the velocity vectors in the y direction, parallel to the platen faces.

modulus (1 GPa), Poisson's ratio (0.125), cohesion (10 MPa), friction angle, ϕ, and dilation angle, ψ. The two elements in the middle have higher elastic moduli (shear modulus of 10 GPa), but the same plastic properties. They behave as an elastically hard inclusion within a deforming mass. The shear band formation appears to be triggered by this initial material heterogeneity, as described also by Cundall (1989). The material is equivalent to that described by Hobbs & Ord (1989) in that it is isotropic and elasto-plastic with no hardening. The material follows a non-associated Coulomb constitutive law (see Vermeer & de Borst 1984; Hobbs *et al.*, this volume) with constant values of c, ϕ, and ψ, and it follows an associated flow law whenever $\phi = \psi$.

ϕ is an angle of internal friction analogous to the dry friction between two sliding surfaces and represented on a plot of shear stress against normal stress by the angle of the failure envelope with the axis representing normal stress. ϕ may also be thought of as the angle of repose of a (cohesionless) sand dune.

ψ is a parameter known as the dilation or dilatancy angle which is used for characterising materials which change their volume during plastic deformation. A positive dilation angle leads to dilation while a negative dilation angle results in contraction or a plastic volume decrease. The consequences to a description of the constitutive behaviour of a granular material of recognising that plastic volume changes may occur have recently been reviewed by Vermeer & de Borst (1984). Any process which leads to a change in volume may potentially be incorporated into such a simple elasto-plastic model by way of the dilation angle. Conceptually simple processes which lead to an increase in volume include one layer of grains sliding up over another layer, or a surface with rigid asperities sliding up and over a similar surface. Higher temperature, higher pressure processes may include phase transformations which lead to a volume change. Numerical experiments have therefore been performed for a broad range of friction and dilation angles (0° to 50°) for both uniaxial shortening and simple shearing deformation history experiments, with and without confining pressure, and with boundary conditions constraining the bulk deformation to be isochoric as well as non-isochoric, but we concentrate here on only two dilatant specimens. The first specimen has $\phi = 30°$, $\psi = 10°$, a classical situation in soil mechanics problems, with the dilation angle less than the friction angle. The second specimen is given properties not normally considered in soil mechanics but which may be attained in geological situations with the dilation angle greater than the friction angle, represented here by parameters $\phi = 0°$ and $\psi = 20°$. We suggest here that this might represent a situation close to the liquidus of a deforming system.

Results

Geometrical and mechanical results

The periodic shear-band formation for the two deformed specimens is seen in Fig. 2. The shear band boundaries or boundaries to regions of higher dilation lie at a low angle to the shearing direction; they do not lie parallel to the shearing direction, and the angle is slightly greater for Fig. 2a and 2b than for Fig. 2c and 2d. It therefore appears as though the shear band inclinations relative to the imposed maximum principal axis of stress for a simple shearing deformation history experiment at a shear strain of 1 depend on both the angle of friction and the angle of dilation, as described by Hobbs & Ord (1989) for uniaxial deformation experiments. The shear band boundaries are more sharply defined and also more closely spaced in Fig. 2c and 2d for which $\phi + \psi = 20°$ against

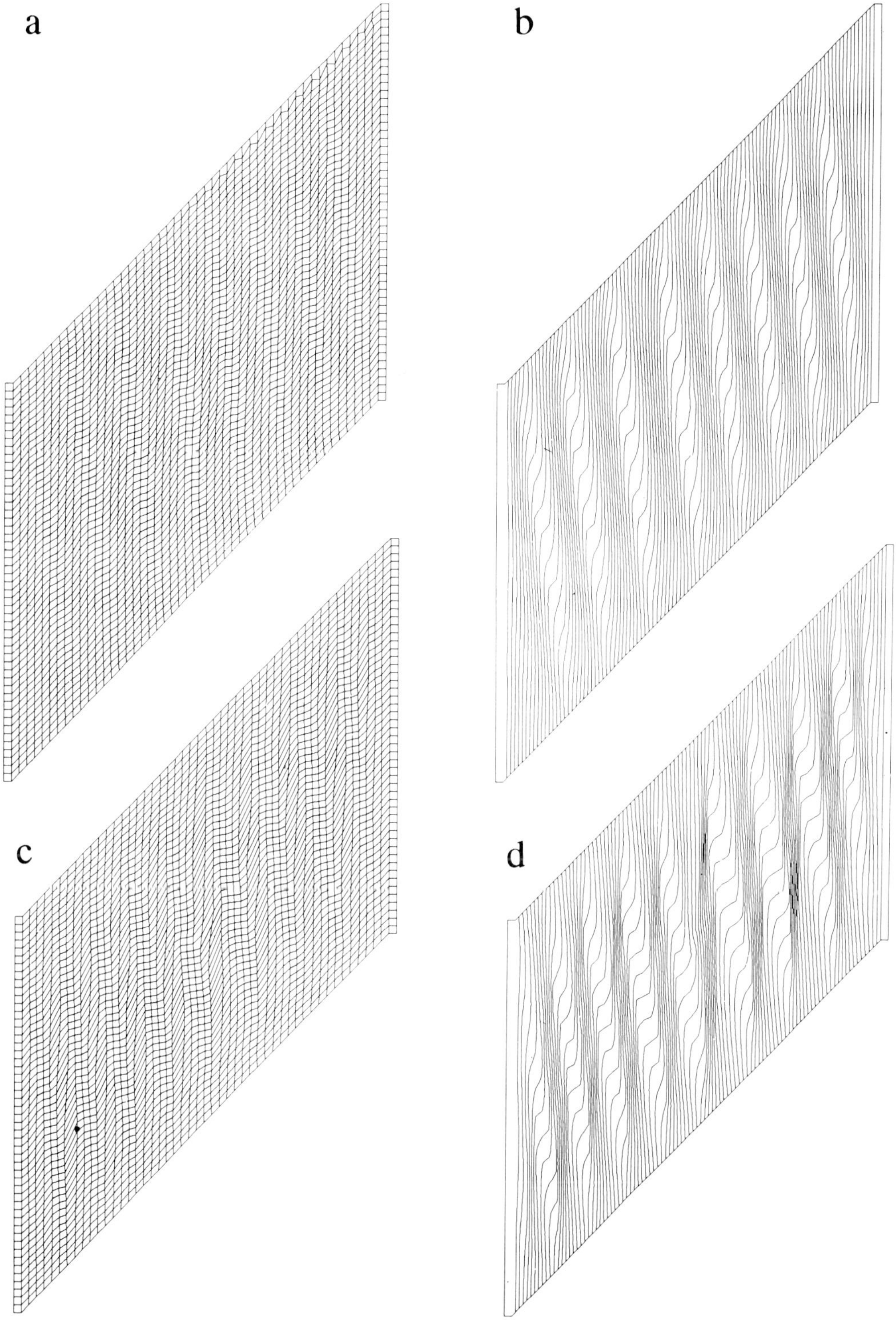

Fig. 2. Deformed specimens at a shear strain of 1. (**a** and **b**) $\phi = 30°$, $\psi = 10°$; (**c** and **d**) $\phi = 0°$, $\psi = 20°$. (a and c) Deformed grid; (b and d) deformation of the specimens as indicated by distribution of contours of the *y*-component of the instantaneous velocity. Contour interval is constant at 5×10^{-5} units.

their greater diffuseness and spacing in Fig. 2a and 2b for which $\phi + \psi = 40°$. This is again in accord with the observations of Hobbs & Ord (1989) for uniaxial shortening deformation experiments. Note also that the elements within the shear bands in Fig. 2c have undergone a greater increase in volume than the equivalent elements in Fig. 2a.

The shear stress (τ)–shear strain (γ) curves for the two experiments are shown in Fig. 3. In both instances, the theoretical gradient of the $\tau-\gamma$ curve for the perfect specimen is reproduced (after about 1000 steps) and is given by

$$\frac{d\tau}{d\gamma} = \frac{K\mu\beta}{1+K\mu\beta/G}$$

where K is the bulk modulus used for a plane strain deformation, G is the shear modulus, and μ and β the sine of the friction and the dilation angles respectively (H-B Muhlhaus, pers. comm., 1988).

a

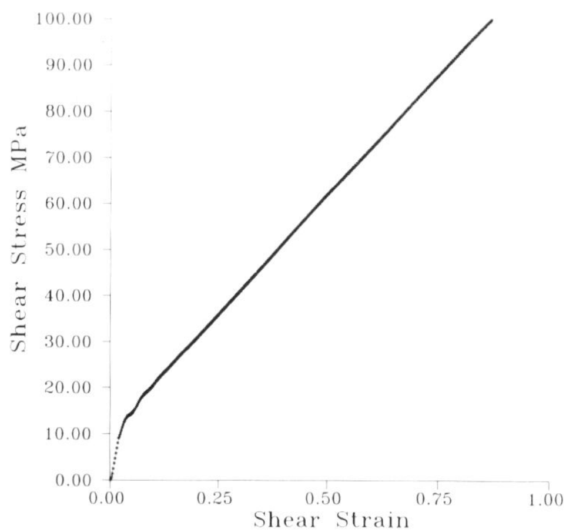

b

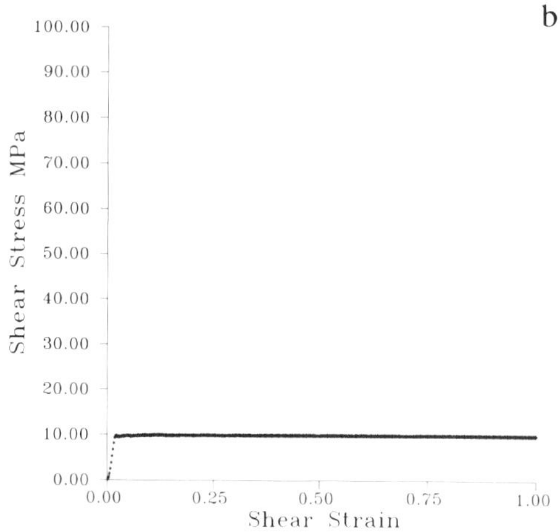

Fig. 3. Shear stress (τ, MPa) versus shear strain (γ). (**a**) $\phi = 30°$, $\psi = 10°$ $d\tau/d\gamma = 104$ MPa. (**b**) $\phi = 0°$, $\psi = 20°$ $d\tau/d\gamma = 0$ MPa. Compression is assumed to be positive.

Volume change and pressure

Fluid flow through a region may be enhanced by an increased pressure gradient across the same region, whereas dilation or an increase in volume in a region represents porosity increase, and therefore enhances the possibility of fluid flow if permeability increases with porosity. The structural representation of the dilation may be on any scale, from micro (fracturing of grains) to macro (dilational fault jogs), since none of the terms of this numerical model contain an absolute length dependance.

Contoured diagrams were therefore constructed for the two specimens of volume change (Figs 4a and 5a) and of mean normal stress (Figs 4b and 5b).

Comparison of the contoured diagrams is facilitated by Figs 4c and 5c where the mean normal stress is plotted against volume change for each of the 2400 deforming elements. The positive volume change in both cases results from elastic contraction of the elements plus a small plastic dilation. This is a result of the bulk deformation being elastically as well as plastically isochoric. In a more natural situation, the bulk of material would be constrained to be plastically isochoric only while the platens absorbed much of the effects of the elastic volume change. A negative volume change for an element means that the plastic dilation is greater than the elastic contraction.

Figure 4 shows an association of a high volume increase of about 20% with low mean normal stress. The maximum pressure difference throughout the specimen and therefore between material in the shear zones and material either side is about 50 MPa, and is about 10 MPa within the high dilation zones at a total pressure of 200 to 210 MPa.

An inverse association is shown in Fig. 5, where the more dilatant regions, with a volume increase of up to 100%, are associated with a high mean normal stress. The maximum pressure difference throughout the specimen is about 7 MPa, and is only about 3 MPa within the high dilation zones at a total pressure of 462 to 465 MPa.

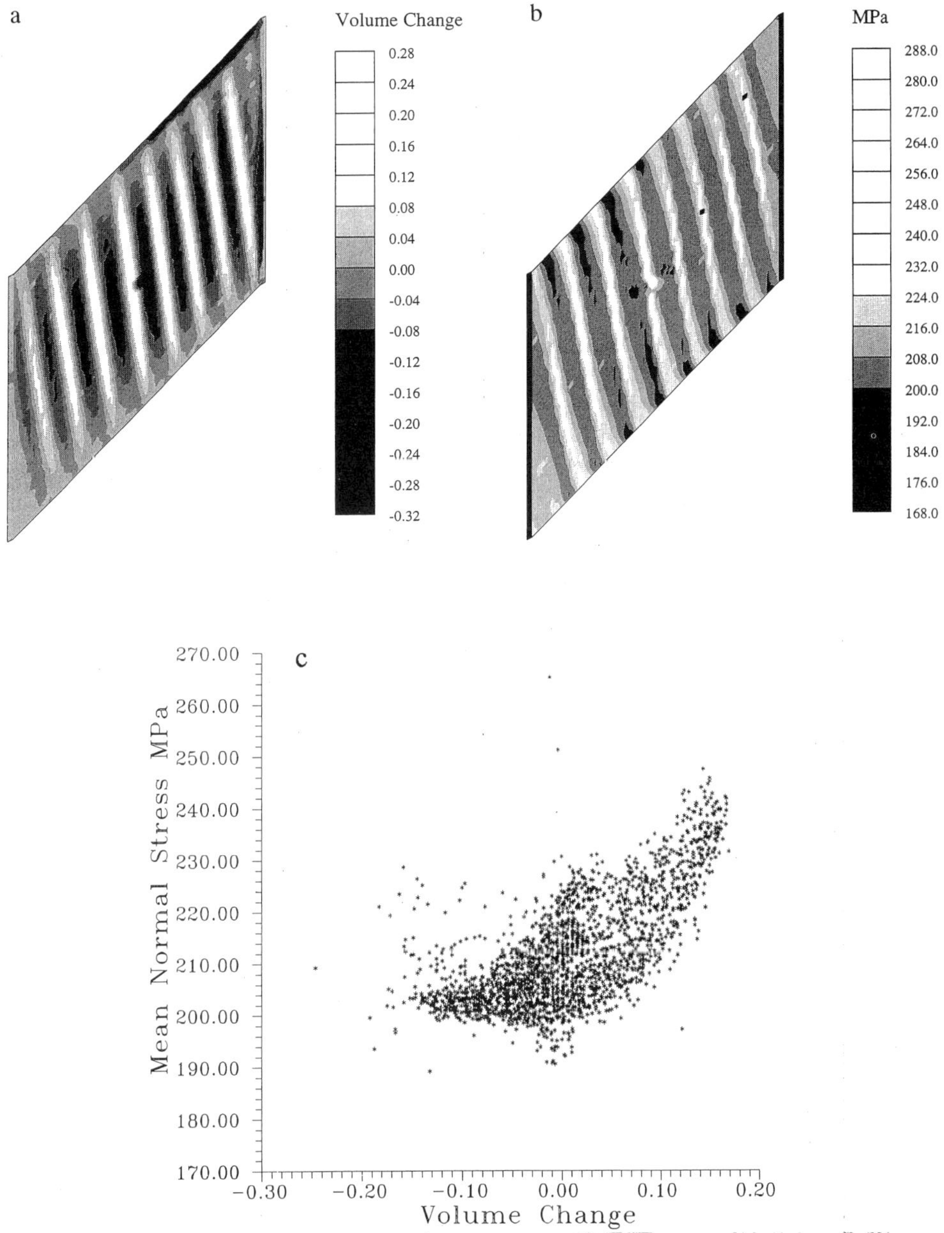

Fig. 4. $\phi = 30°$, $\psi = 10°$. (**a**) Contours of volume change, $\Delta V/V$, for $V = 1$. A decrease in volume is assumed to be positive, compatible with compression being assumed positive. (**b**) Contours of mean normal stress, $(\sigma_{xx} + \sigma_{yy})/2$. (**c**) Mean normal stress versus volume change.

Discussion

Mathematical aspects

The constitutive model used here does not contain a material parameter with the dimensions of length so that the diagrams shown in Fig. 2 could just as well be 1 cm on side as they could be 1 km on side. Clearly it would be highly desirable in many instances in structural geology to have constitutive formulations involving length scales and some progress in this regard has been made by Mühlhaus & Vardoulakis (1987), Mühlhaus (1988) and by Aifantis (1987).

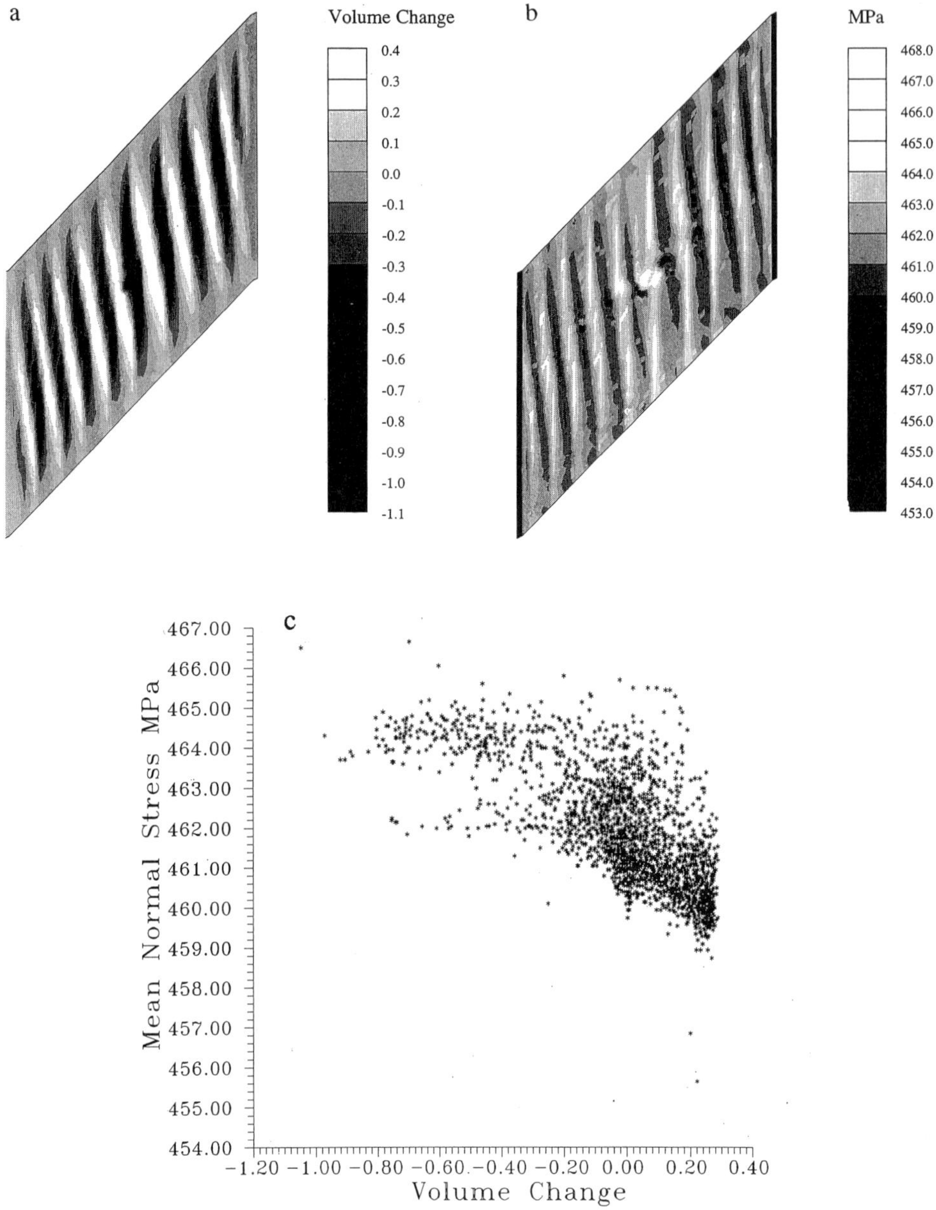

Fig. 5. $\phi = 0°$, $\psi = 20°$. (**a**) Contours of volume change. (**b**) Contours of mean normal stress. (**c**) Mean normal stress versus volume change.

We include here a brief explanation for the patterning of the shear bands observed for these simple shearing deformation experiments motivated by the material presented in these papers.

In an infinite plane under compressive plane strain, shear banding is possible as soon as the governing differential equations change from the elliptic to the hyperbolic type. As soon as the differential equations enter the hyperbolic range, modes of deformation are possible which can be physically interpreted as shear bands. In particular, a discrete shear band type of localized deformation is possible. But in a discretisation procedure such as is followed by finite difference and finite element codes, such a discrete localization can be represented only for vanishingly small element or zone diameters. However, the discrete or single layer type of localization is only one possible solution. Periodic solutions representing non-homogeneous simple shears where the field variables are constant along planes $y = c_1x_1 + c_2x_2$, where c_1, c_2 are constant, are also possible, and are accessible to FLAC. Orthogonal to these planes, the field variables vary according to sin y or cos y. An example is given in Fig. 6. The planes y = constant have the same orientation as the discrete shear band. In a conventional continuum approach, only the ratio c_1/c_2 is determined, so that the wavelengths in the case of the infinite plate problem are arbitrary. In the present case, two length scales are involved, one being the dimensions of the block ($L \times L$) and the other the zone size (d) (see Fig. 1), which induce a wave number selection thus determining the periodicity of the shear bands. The dependance on the mesh size from the physical point of view is of course not very satisfactory. To remedy the undesirable situation, one can employ a so-called higher order continuum approach or a gradient theory of plasticity, where due to the mathematical peculiarities of these equations, internal length scales are induced which could then provide a physically significant scaling of the shear band spacing. In the present case the spacing of the newly formed shear bands clearly depends on ϕ and ψ but is essentially governed by the ratio d/L (Fig. 1); the spacing is therefore strongly determined by the mesh geometry. Further numerical experiments are being conducted to test this assertion.

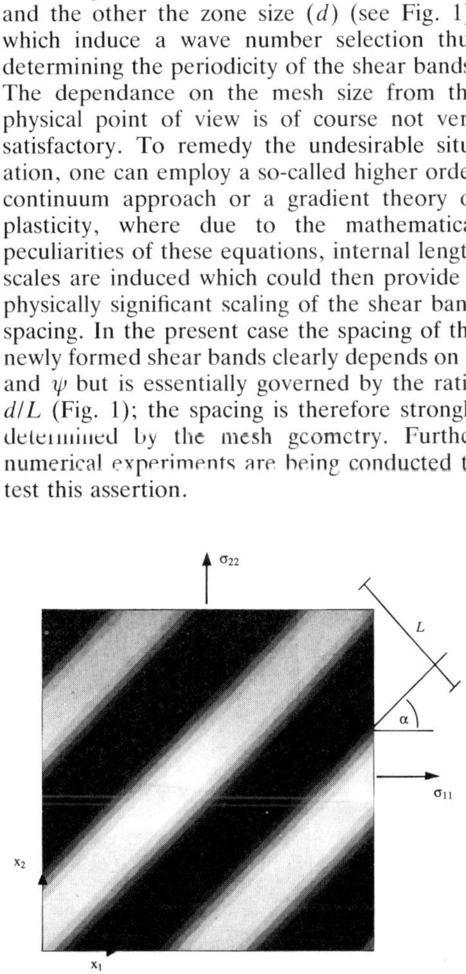

Fig. 6. Non-homogeneous simple shear in an infinite plate. After Mühlhaus & Aifantis (1990).

A new and interesting interpretation of the situation can be made if the finite difference mesh is understood to represent a model of a microscopically inhomogeneous medium. In such a case, d is interpreted as a characteristic fabric length such as a grain diameter whilst L could be a layer thickness. The case of a perfect continuum is obtained in the limit as $d/L \to 0$. In this new interpretation the shear band spacing is determined by the ratio of typical microstructural length to a macrostructural length.

Geological aspects

The association of high dilation with low mean normal stress, observed in these numerical experiments for ϕ greater than ψ, is perhaps the more commonly accepted geological situation, but high dilation may also be associated with high mean normal stress, observed in the numerical experiments for ϕ less than ψ, depending on the constitutive parameters. Both examples are important in that they display a periodic localization of the deformation, and two geological structures are chosen below which may be interpreted in terms of the behaviour of the two numerical specimens described above.

In the first case, consider a classic slate belt, where mica-rich and quartz-rich layers arising from metamorphic differentiation lie parallel to each other, with the mica-rich bands present in the limbs and the quartz-rich bands in the hinges of crenulations (Williams 1968, 1972). The example presented in Fig. 7 is from the low grade (chlorite zone) schists associated with the Cooma granodiorite in south eastern Australia. The metamorphically differentiated (quartz-rich/mica-rich) crenulation layering overprints a schistose foliation (Hopwood 1976; Granath 1980). The suggested pressure for the chlorite zone of between 200 and 250 MPa (Vernon & Hobbs, pers. comm. 1984) is remarkably close to that imposed during the first numerical experiment. The first model described here, for $\phi = 30°$ and $\psi = 10°$, presents a way of explaining this phenomenon. The high dilatancy of the shear zone allows for easy fluid flow and therefore transport of material into and out of the shear zone. The higher pressures in the low dilation zones force any fluids present to flow towards the regions of lower pressure although this flow is hampered by the low dilatancy of the high pressure region. Certainly, flow is towards, not away from, the low pressure zone and is localized along this zone. Such fluid focussing has had many adherents from an observational basis (e.g. Williams 1968, 1972), but the theor-

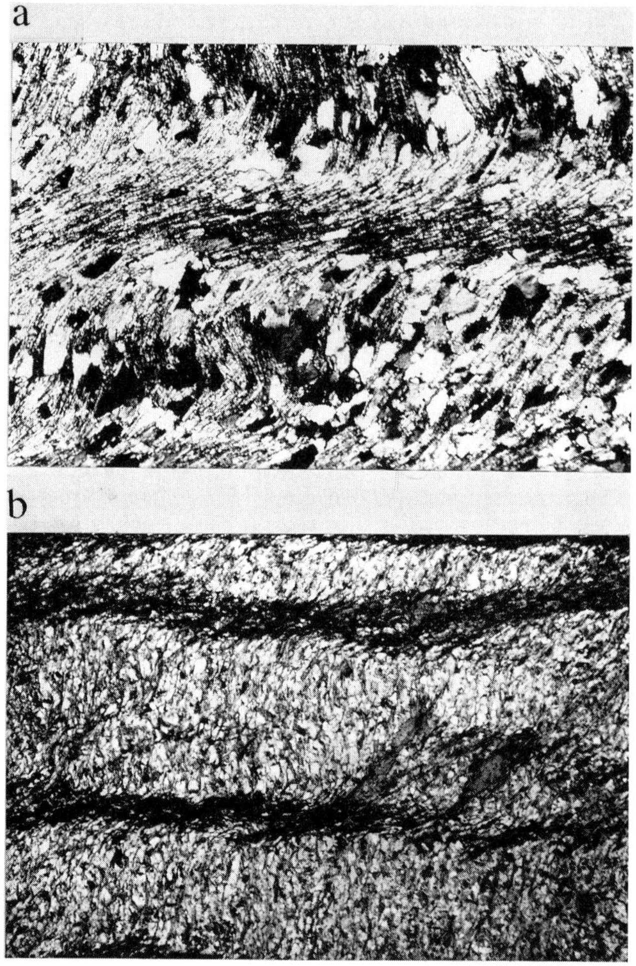

Fig. 7. Crenulations in the low grade schists of the regional-aureole Cooma granodiorite, New South Wales. (**a**) Length of plate is 1.6 mm; (**b**) length of plate is 4.2 mm.

etical and mechanical grounds for such focussing have never been well described.

The second natural example is of gneiss, with periodic, localized segregations of melt. Such pegmatitic segregations are developed in axial planes of F4 crenulations or flexures of the layering of the Archaean Carnot Gneisses (Fanning *et al.* 1979; Parker *et al.* 1988) of the southernmost Eyre Peninsula, South Australia (Fig. 8) and are well displayed in most migmatite terrains. Pressure estimates for the prograde granulite facies metamorphism of the gneisses are 700 to 900 MPa, much higher than the pressure imposed during the second numerical experiment. In a perfectly dry system, where dP/dT for the liquidus is positive (Thompson 1988), for a constant temperature, melt formation is favoured by a decrease in pressure, where the pressure difference is maintained throughout the deformation, and through-going melt segregations could result from a situation as described above. However, in a wet system, where dP/dT for the liquidus is negative (Thompson 1988), an increase in pressure favours melting. This situation is represented by the second numerical model for which $\phi = 0°$ and $\psi = 20°$. The zones of high dilation are also zones of high pressure so that melt may form within them. The pressure gradient is away from this region towards the regions of lower pressure, but since these regions are also regions of low dilation, flow of melt out of the high

a

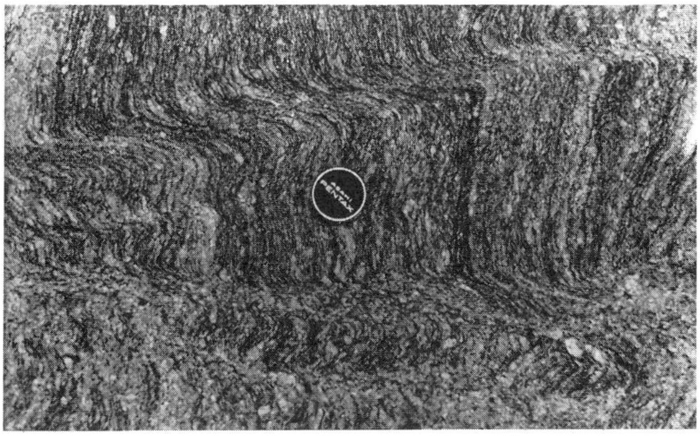

b

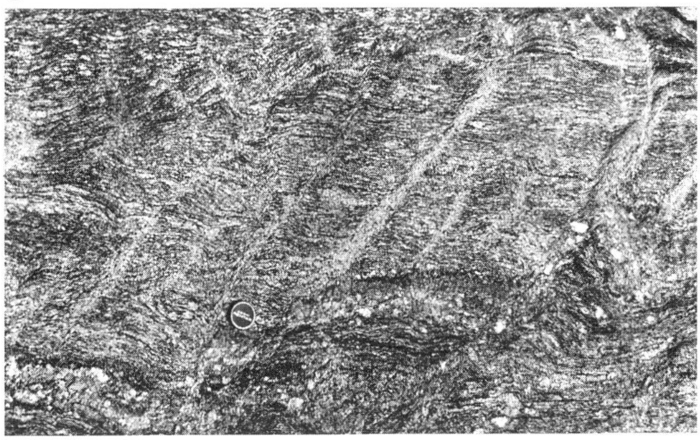

Fig. 8. Crenulations and melt segregations, in the Archaean Carnot Gneisses, South Australia. Diameter of lens cap is 5.4 cm.

dilation zones into the low pressure zones is not favoured until perhaps a perturbation appears in the rock mass which the melt may exploit.

Conclusions

Two examples of numerical models are selected which display periodic localization of the deformation into spaced shear bands. The inclination of the shear bands to the maximum principal stress and the diffuseness of the shear bands increase with both friction and dilation angles.

In one case, for which the dilation angle is less than the friction angle, regions of high dilation are shown to correspond with a lower pressure than surrounding regions, and this is suggested as a feasible mechanism for formation of spaced segregations of mica-rich and quartz-rich layers in a schist, and for transport of quartz far from its origin. In another case, for which the dilation angle is greater than the friction angle, regions of high dilation and high pressure correspond, for which a possible geological phenomenon is that of localized segregations of melt.

I thank B. Hobbs and H. B. Mühlhaus for their illumination of the topic of shear zones; and ITASCA for the numerical code FLAC. I acknowledge a CSIRO/Curtin University of Technology collaborative research grant with A. Duncan for partial financial support.

References

Aifantis, E. C. 1987. The physics of plastic deformation. *International Journal of Plasticity*, **3**, 211–247.

Arthur, J. R. F., Dunstan, T., Al-Ani, A. J. & Assadi, A. 1977. Plastic deformation and failure in granular media. *Géotechnique*, **27**, 53–74.

CUNDALL, P. 1989. Numerical experiments on localization in frictional materials. *Ingenieur Archiv*, **59**, 148–159.

—— BOARD, M. 1988. A microcomputer program for modelling large-strain plasticity problems. *6th International Conference on Numerical Methods in Geomechanics*. Innsbruck, Austria, 11–15 April.

FANNING, C. M., OLIVER, R. L. & COOPER, J. A. 1979. The Carnot gneisses, a metamorphosed Archaean supracrustal sequence in southern Eyre Peninsula. *In*: PARKER, A. J. (Compiler). *Symposium on the Gawler Craton, Extended Abstracts*. Geological Society of Australia, Adelaide, 3–15.

GRANATH, J. W. 1980. Strain, metamorphism, and the development of differentiated crenulation cleavages at Cooma, Australia. *Journal of Geology*, **88**, 589–601.

HOBBS, B. E. & ORD, A. 1989. Numerical simulation of shear band formation in a frictional-dilational material. *Ingenieur Archiv*, **59**, 209–220.

—— MÜHLHAUS, H-B. & ORD, A. 1990. Instability, softening and localization of deformation. *This volume*.

HOPWOOD, T. P. 1976. Stratigraphy and structural summary of the Cooma metamorphic complex. *Journal of the Geological Society of Australia*, **23**, 345–360.

MÜHLHAUS, H-B. 1988. Application of Cosserat theory in numerical solutions of limit load problems. *Ingenieur Archiv*, **59**, 124–137.

—— & AIFANTIS, E. C. 1990. The influence of microstructure-induced gradients on the localization of deformation in viscoplastic materials. *Acta Mechanica* (in press).

—— & VARDOULAKIS, I. 1987. The thickness of shear bands in granular materials. *Géotechnique*, **37**, 271–283.

PARKER, A. J., FANNING, C. M., FLINT, R. B., MARTIN, A. R. & RANKIN, L. R. 1988. *Archaean–Early Proterozoic granitoids, metasediments and mylonites of southern Eyre Peninsula*. Specialist Group in Tectonics and Structural Geology Field Guide Series No 2. Geological Society of Australia.

THOMPSON, A. B. 1988. Dehydration melting of crustal rocks. *Rendiconti della Societa Italiana di Mineralogia e Petrologia*, **43**, 41–60.

VARDOULAKIS, I. 1980. Shear band inclination and shear modulus of sand in biaxial tests. *International Journal for Numerical and Analytical Methods in Geomechanics*, **4**, 103–119.

—— & GRAF, B. 1985. Calibration of constitutive models for granular materials using data from biaxial experiments. *Géotechnique*, **35**, 299–317.

VERMEER, P. A. & DE BORST, R. 1984. Non-associated plasticity for soils, concrete and rock. *Heron*, **29**, 1–62.

WILLIAMS, P. F. 1968. *Tectonic studies of rocks exposed along the south coast of New South Wales*. PhD thesis, University of Sydney.

—— 1972. Development of metamorphic layering and cleavage in low-grade metamorphic rocks at Bermagui, Australia. *American Journal Science*, **272**, 1–47.

Propagation and localization of stylolites in limestones

E. CARRIO-SCHAFFHAUSER,[1] S. RAYNAUD,[2] H. J. LATIÈRE[3] & F. MAZEROLLE[3]

[1] *LGIT, IRIGM, BP53X, 38041 Grenoble Cedex, France*

[2] *Laboratoire de Géologie Structurale et Appliquée, Université de Provence 13331 Marseille Cedex 3 France*

[3] *Laboratoire de Mécanique et d'Acoustique, CNRS, BP71, 13277 Marseille Cedex 9 France*

Abstract: Matrix investigations (X-Ray tomography, porosimetry by mercury injection, SEM analysis) around stylolites revealed major zones that represent different states in the propagation of the pressure solution structure.

Near the stylolite termination, a significant increase of porosity relative to the far-field host rock porosity and variations in the shape of matrix particles are associated with the lateral propagation of the dissolution zone in the plane of the seam. Close to the sides of the seam, this porosity enhancement zone is found again and may be responsible for vertical development of the stylolite style. Above and below the stylolite seam, the rock matrix is less porous than the reference state and this region appears to have been a site of precipitation of diffused solute. These observations imply that the enhanced porosity state around the stylolite tip is a transient one. This zone becomes a site of deposition as the stylolite tip propagates through it.

The mechanism of rock deformation by pressure-solution and deposition involves the dissolution of material at grain boundaries exposed to high normal stress. After diffusion through a fluid phase along the grain contacts, dissolved species are deposited at sites under low normal stress to complete the mass transfer cycle. This deformation appears as a creep mechanism leading to a change in shape (or density) of rocks (Dunnington 1954; Durney 1972; Cosgrove 1976; Gratier 1987; Schwander *et al.* 1981).

Any mechanical stress, whether of gravitational or tectonic origin, can induce this deformation which is found in many geological structures. Stylolites are common expressions of this type of deformation and aspects of their geometry and development are analysed in this paper.

The aim of this work is to provide new data relevant to the problem of the propagation and localization of a stylolitic surface. The use of X-ray tomography (medical scanner, in french: X-ray tomodensitometry) in conjunction with porosimetric measurements and SEM analysis can give a detailed picture of a rock matrix subjected to mass transfer in ways not hitherto possible.

Samples and methods

This study is based on the analysis of tectonic stylolites, sampled (1000 m in average depth) by core-drillings in limestones located in south-eastern France. These upper Cretaceous rocks belong to the Arc syncline (Provence) whose the northern and southern limits are overthrusts. In the syncline, faulting resulted from a mainly compressive deformation (N–S Eocene compression) and a subsequent extensional episode (Oligocene) (Gaviglio 1985). The characteristics of the deformation have been analysed using fault, tension gash and stylolite data. The latter indicate a bulk shortening of 20%. These pressure-solution seams are several centimetres long, vertical and show one or two terminations on the core plugs. They are highlighted by insoluble residue concentrations, consisting essentially of organic matter. The host rock is an homogeneous micritic limestone (98–100% $CaCO_3$), made up of grains a few microns in diameter, with random orientation.

Three major methods of analysis were used to obtain information on the internal structure variations of a rock around a stylolitic surface:

porosimetric measurements by mercury injection;

From Knipe, R. J. & Rutter, E. H. (eds), 1990, *Deformation Mechanisms, Rheology and Tectonics*, Geological Society Special Publication No. 54, pp. 193–199.

SEM analysis of the matrix;
X-ray tomography.

The last mentioned technique, developed by Hounsfield (1973) for medical analysis, has recently been applied to the study of rocks (Dou *et al.* 1985; Raynaud *et al.* 1987, 1989). It is based on the measurement of the attenuation of an X-ray beam through a sample. The radiological density values obtained, expressed in Hounsfield units (H.u.), depend on:

(a) the gravimetric density of the material ($g\ cm^{-3}$), in direct proportionality;
(b) the absorption values per unit mass ($cm^2\ g^{-1}$) induced by the mineralogy of the sample;
(c) the selected width of the cross-section (mm);
(d) apparatus characteristics.

Owing to the great purity of the samples studied, each radiological density variation must therefore be closely related to a change in matrix structure. There is a great importance of this assumption: it allows us to point out accurately each matrix change as a microstructure variation, because no mineralogical change in the mineral species nature can confuse the radiological data.

Radiological images of the sample in the three spatial dimensions can be obtained from the computed data. Changes within the internal structure of the sample can be illustrated as:

(a) density variation profiles along the stylolitic seam, from the undeformed rock to the stylolitized matrix;
(b) radiological maps showing local density variations inside the rock matrix.

On one sample, areas of particular interest were found by this non-destructive method and samples were taken for mercury injection porosimetry and SEM observations.

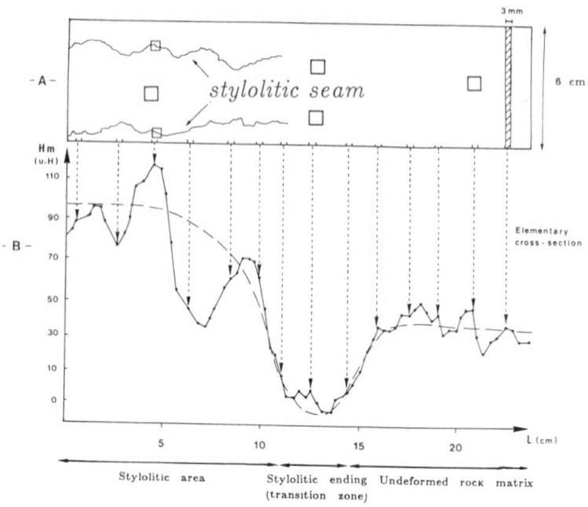

Fig. 1. Average radiological density profile: (**A**) Traces of a single stylolitic seam on the surface of the core. (**B**) Radiological profile obtained from serial cross-sections. Each point on this curve is given by the average density (Hm) of one cross-section, expressed in Hounsfield units u.H. (for comparison, Hm = −700 u.H. for water). Three main zones appear: the undeformed rock matrix (Hm ≈ +35 u.H.); the stylolitic ending or transition zone (Hm ≈ −10 u.H.); the stylolitic area (Hm reaches +115 u.H). The SEM photographs shown in Fig. 3 are localized on the real sample (□). The decrease in Hm, at about 7 cm from the stylolite termination, may be related to the presence of an another small and fine pressure solution seam, too thin to be drawn well on (A).

X-ray tomography results

Figure 1 shows the average radiological density profile, parallel to the stylolite, for the sample studied. Three main zones can be identified on the curve.

(a) Away from the stylolite, in the undeformed rock matrix, the mean radiological densities show only small variations between two successive cross-sections, essentially due to minor local sedimentological changes. There are no signs of mass transfer and this kind of matrix represents the undeformed state of the rock.

(b) In the area around the stylolitic ending (the transition zone), where no evident seam appears, a great decrease in the radiological density values was found which can only be explained by a considerable increase in local porosity.

(c) The stylolitic area is characterized by the highest radiological density values which are indicative of a drastic porosity reduction.

The progressive variation of these data, from the end of the seam to the point of greatest development, might indicate a continuous transformation mechanism in the limestone matrix.

Similar results have been found on four other samples. The variation observed in the radiological density, H_m, in the stylolitic area appears to be due to a meeting with a second stylolite, smaller and finer than the principal pressure solution structure and hardly noticeable on the sample. It seems to appear in the same position on the radiological map (Fig. 2), but it is too close to the core edge to be clearly distinguished.

Passing through natural endings of a stylolite, the contoured radiological density map (Fig. 2) illustrates the distribution of the various kinds of matrix structure around the seam and cor-

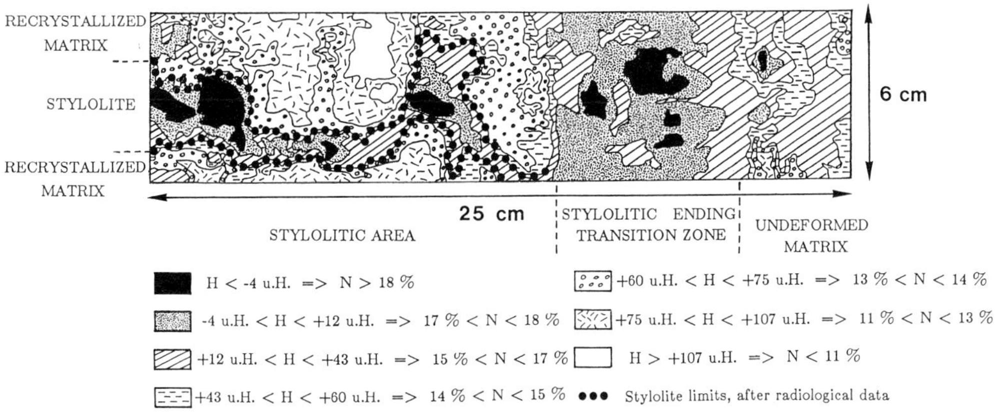

Fig. 2. Radiological map drawn on computerized reconstructed cross-sections obtained on the sample and normal to the plane shown in Fig. 1. The three main zones were found, with the stylolitic area divided into two parts: the stylolite s.s., made up of insoluble residue concentrations and, on either side, part of the rock subjected to dissolution processes; the host recrystallized matrix. The 'radiological' stylolite limits do not coincide with the morphology of the seam on the sample drawn in Fig. 1, but extend beyond it.

roborates the presence of three main areas. At the end of the stylolite (transition zone), a large area of low radiological density can be observed. This area is wider than the stylolitic seam.

The stylolitic area can be divided into two major zones: (a) a large area with high radiological densities is observed near the stylolite ($+ 60$ H.u. $< H_m < + 107$ H.u.) and (b) the stylolite s.s. appears as a sinuous band, with the lowest radiological density. It includes the surface with the insoluble residue concentrations, but also a number of rows of grains on both sides of the stylolitic seam. These density values can be attributed to the nature of the insoluble species (organic matter), but also to the pressure-solution mechanism which is probably not restricted only to the grains fringing the seam, but may modify part of the rock on either side of the stylolite.

The differences of the stylolite morphology between the sample and the radiological map can be explained by the fact that the scanner reveals the zone where the pressure solution mechanism acts, a zone that cannot be observed directly on the sample.

The reduction in the radiological density observed in the stylolitic area on the radiological profile can be explained as follows. As with the major stylolite, the second minor pressure solution seam must also show lower radiological densities at its ending (at the intersection with the first stylolite). Therefore we can observe a relative decrease of the average radiological density in this zone.

Petrophysical analysis

Complementary petrophysical data on these three major zones were obtained from porosimetric measurements and SEM observations. The manner in which the matrix changes through the pressure-solution process can therefore be defined at the grain size scale (SEM photographs, Fig. 3, localization on the sample Fig. 1).

The first zone, far from the stylolite structure, seems to represent the undeformed state of the rock matrix, in accordance with previous petrophysical studies of these limestones (Carrio-Schaffhauser 1987). SEM views show a tight or interlocking structure of micritic particles (Loreau 1972), ovoid to elongated in shape, 0.6 to 1 μm wide, without crystalline faces. The average porosity reaches 15% ($\pm$ 0.5%), with micropores of 0.2 μm average width. Therefore this area can be taken as a reference state for the quantification of petrophysical variations near the stylolite.

The stylolitic end zone, or the so-called 'transition zone', shows considerable matrix transformation, always indicative of calcite removal. The increase in porosity suggested by the lowest radiological densities was confirmed, reaching 18% ($\pm$ 1%) and associated with an increase in the main pore family, always 0.2 μm in size. SEM analysis illustrated this growth of the porous network by a micritic structure made up by smaller spheroidal particles (0.4 to 0.6 μm wide) with variable (weak to tight) contacts.

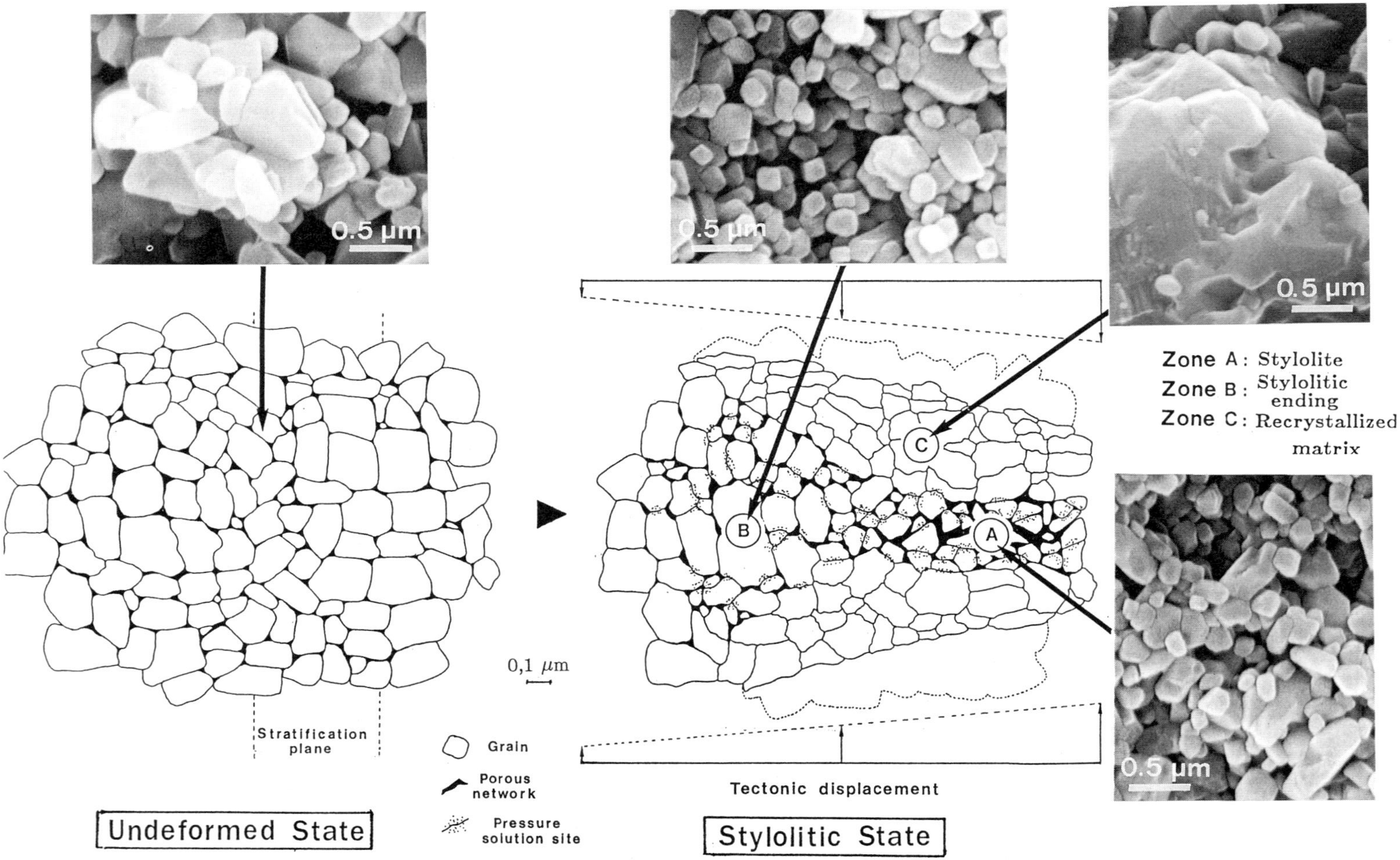

Fig. 3. Interpretation of tomography, porosity and SEM data, showing the microstructural contrast between undeformed and stylolitized rock. SEM views illustrate the successive states in the matrix transformation from a tight to interlocking structure (undeformed state) to a coalescent structure (Zone C in the stylolitic state). Variations in the grain sizes, shapes and contacts, together with pore dimensions and morphology can be distinguished.

Partial dissolution of calcite on the grain surface was considered to be the cause of development of the porous network between particles. Particle size decreases, contact areas become reduced and grain shape more spheroidal.

The stylolitic area itself can be divided into two zones: (a) inside the 'radiological' stylolite limits, the matrix structure has the same features as in the transition zone, with local accentuations; (b) a zone has developed on either side of the seam, where a pervasive densification of the rock has occurred, as shown by tomographic analysis, SEM and porosimetric studies. The porosity decreases (12% ± 0.5%), related to a reduced pore size (0.1 μm). Under SEM, the matrix shows an interlocking to coalescent structure. Micritic grain size growth was noted, either by mass deposition and development of crystalline faces or by accumulation, interlocking and welding of several particles to form new, larger 'grains' (1 to 1.5 μm). Solutes diffusing to the regions adjacent to the stylolite, cementing them and lowering their porosity can explain this rock matrix transformation (Merino 1987). These processes explain the 'densification' of the rock in these sites by mass transfer, probably over very short distances, in closed systems, owing to the impermeability of these limestones (Carrio-Schaffhauser 1987).

The material lost and hence the stylolite normal displacement could not be estimated without geochemical data.

Discussion

Stylolites have been the subject of a great deal of study, especially with regard to parameters activating their occurrence in the rock matrix. Several hypotheses have been suggested to explain the initiation of stylolitic structures in particular zones of the rock and their method of propagation.

(a) Stylolites develop from pre-existing planar (microfractures, argillaceous layers) or point (rigid object, large pore) heterogeneities in the rock matrix (Dunnington 1954; Park & Shot 1968; Sellier & Morlier 1976; Gratier 1987),

(b) Interactions occur between local porosity and local solubility in a rock subjected to a deviatoric stress. Higher-than-average local porosity causes higher-than-average local solubility, because the grain-to-grain contact area is lower. This higher solubility may drive the porosity even higher (Merino *et al.* 1973; Merino 1987).

(c) Fletcher & Pollard (1981) proposed that discrete solution surfaces originate at stress concentrations and propagate through rocks as 'anticracks'.

In view of the matrix structure zoning found around a stylolite (Fig. 3), some new arguments can be brought to bear on the mechanism of propagation of a stylolite (Fig. 4).

At the stylolite tip, porosity enhancement occurs in a 'process zone'. Porous network growth seems to be due to a pressure-solution mechanism, characterized by the formation of smaller and more ovoid grain shapes in this matrix zone. This process induces a diffusive transfer to the both parts of the stylolite where reduced porosity is developing. This latter zone must be the crystallization site of soluble species generated in the 'process zone'. A similar 'process zone' must develop in the small strip fringing the stylolitic seam from which diffusive transfer must also occur in order to feed the stress concentration at the stylolite tip.

In a later state, after an increment of the stylolite growth, the stylolite tip 'process zone' has migrated into pristine material ahead of the stylolite tip. The previous enhanced porosity zone has now become cemented. The diffusive transfer provides mass deposits in the host rock, on either side of the seam (Fig. 4).

The diffusive transfer process inferred to take place during the tip propagation, in the *opposite* direction to the direction of the stylolite growth, now requires more detailed study. Several questions have to be clarified regarding this process.

(a) What is the mass lost out of the dissolution zone?

(b) What is the nature of the fluids that support this flux of diffusive matter?

(c) What is the driving force that leads this flux in the opposite direction to the tip advance, and at what rate does the process proceed?

(d) What is the behaviour of the trapped fluids when the 'process zone' becomes a cemented zone?

Geochemical analyses carried out on these different zones in the rock matrix (and on fluid inclusions?) may provide some answers to these questions.

Conclusions

The results of these petrophysical studies provide new data relevant to the mechanism of the stylolitic propagation. Stylolites seems to grow laterally by a slow, but continuous propagation of the dissolution that induces the presence of an enhanced high porosity area at the ends and on the both sides of the seam, the solute species depositing in the surrounding

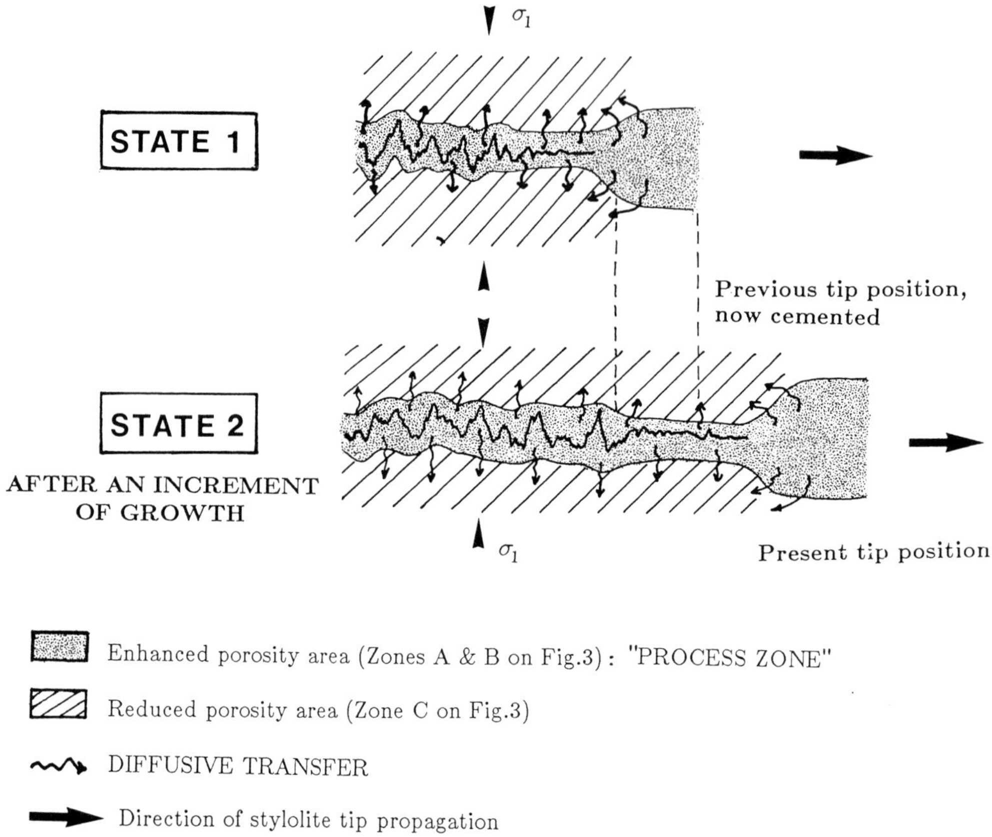

Fig. 4. Implications of the petrophysical results for the diffusive transfer process during the stylolite tip propagation.

material on either side of the stylolite.

A pre-existing heterogeneity appears to be necessary to initiate the process, but this need only be a point source. The planar morphology of a stylolite is due to the dissolution propagation processes and is independent of the shape of the nucleating heterogeneity.

However, further investigations are still necessary relevant to build upon these preliminary observations, and in particular geochemical data must now be acquired to help refine a description of the stylolite propagation process.

References

CÁRRIO-SCHAFFHAUSER, E. 1987. *Evolution des propriétés pétrophysiques d'un calcaire: le rôle de la dissolution-cristallisation dans une déformation cassante*. Thèse de Doctorat de l'Université de Provence, Marseille.

COSGROVE, J. W. 1976. The formation of crenulation cleavage. *Journal of Geological Society, London*, **132**, 155–178.

DOU, M., FLESIA, E., CAGNASSO, A. & LATIERE, H. J. 1985. Etude du charbon de Gardanne par tomodensitométrie. *Analusis*, **13**, 69–75.

DUNNINGTON, H. V. 1954. Stylolite development postdates rock induration. *Journal of Sedimentary Petrology*, **24**, 27–49.

DURNEY, D. W. 1972. Solution transfer, an important geological deformation mechanism. *Nature*, **235**, 315–317.

FLETCHER, R. C. & POLLARD, D. D. 1981. Anticrack model for pressure solution surfaces. *Geology*, **5**, 185–187.

GAVIGLIO, P. 1985. *La déformation cassante dans les calcaires fuvéliens du bassin de l'Arc (Provence, France. Comportement des terrains et exploitation minière*. Thèse de Doctorat d'Etat, Aix-Marseille I.

GRATIER, J. P. 1987. Pressure solution – deposition creep and associated tectonic differenciation in sedimentary rocks. *In*: JONES, M. E. & PRESTON, R. M. F. (eds) *Deformation of Sediments and*

Sedimentary Rocks. Geological Society, London, Special Publication, **29**, 25–38.

HOUNSFIELD, G. N. 1973. Computerized transverse axial scanning (tomography). *British Journal of Radiology*, **46**, 1016–1022.

LOREAU, J. P. 1972. Pétrographie des calcaires fins au microscope électronique à balayage. Introduction à une classification des micrites. *Comptes Rendus Académie Sciences, Paris*, **274D**, 810–813.

MERINO, E. 1987. Textures of low temperature self-organization. *In*: RODRIGUEZ-CLEMENTE, R. & TARDY, T. (eds) *Geochemistry of the Earth's Surface*. Cons. Sup. Investigaciones Cientificos (Spain) and Centre National Recherches Scientifique (France), Madrid, 597–610.

——, ORTOLEVA, P. & STRICKHOLM, P. 1973. Generation of evenly-spaced pressure-solution seams during (late) diagenesis: a kinetic theory. *Contributions to Mineralogy and Petrology*, **82**, 360–370.

PARK, W. C. & SHOT, E. H. 1968. Stylolites: their origin and nature. *Journal of Sedimentary Petrology*, **38**, 175–191.

RAYNAUD, S., MAZEROLLE, F., FABRE, D. & LATIERE, H.J. 1987. Analyse de la structure interne d'un échantillon de roche à l'aide d'une méthode non destructive: la tomodensitométrie (Scanner). Perspectives pour le suivi de la genèse expérimentale des failles. *Comptes Rendus Académie Sciences, Paris*, **305 II**, 707–710.

——, FABRE, D., MAZEROLLE, F., GERAUD, Y. & LATIERE, H. J. 1989. Analysis of the internal structure of rocks and characterization of mechanical deformation by a non-destructive method: the X-ray tomodensitometry. *Tectonophysics*, **159**, 149–159.

SCHWANDER, H. W., BURGIN, A. & STERN, W. B. 1981. Some geochemical data on stylolites and their lost rocks. *Eclogae Geologicae Helveticae*, **74**, 217–224.

SELLIER, E. & MORLIER, P. 1976. Les stylolites: Approche expérimentale du processus de déformation. *Comptes Rendus Académie Sciences, Paris*, **282D**, 953–956.

Deformation of polycrystalline salt in compression and in shear at 250–350°C

R. C. M. W. FRANSSEN[1,2] & C. J. SPIERS[1]

[1] *HPT Laboratory, Department of Geology, Institute of Earth Sciences, University of Utrecht, PO Box 80.021, 3508 TA Utrecht, The Netherlands*

[2] *Present address: Koninklijke/Shell Exploratie en Produktie Laboratorium, 2288 GD Rijswijk ZH, The Netherlands*

Abstract: Dry synthetic polycrystalline salt (NaCl) has been deformed in uniaxial compression and in shear in order to gain insight into the influence of deformation mode on the development of crystallographic preferred orientation, microstructural evolution and mechanical behaviour (flow strength). The experiments were carried out between 250 and 350°C, at strain rates between 10^{-5} and 10^{-7} s^{-1}. Under these conditions halite deforms by climb-controlled dislocation creep. In our samples, the presence of deformed grains containing cellular networks of subgrains, and the development of texture are consistent with such mechanisms. The weaker {110} <1$\bar{1}$0> slip plane appears to align parallel to the shear plane and perpendicular to the uniaxial compression direction. In the sheared samples, a <111> maximum is observed parallel to the shear direction. The textures obtained agree favourably with texture simulations. When the mechanical behaviour seen in uniaxial compression and shear is compared in terms of equivalent stress and strain, as defined in the von Mises theory of isotropic plasticity, the uniaxially deformed samples appear to be stronger. From a detailed consideration of the data, it is inferred that the difference in flow strength is largely due to anisotropy resulting from texture development. The results suggest that the assumptions underlying the von Mises theory of isotropic plasticity are not applicable in the present case, and that the associated flow rule does not offer an accurate method of generalizing creep laws for salt under the conditions investigated.

It is widely accepted that most natural rock deformation phenomena of interest in structural geology and geophysics are non-coaxial (e.g. flow within shear zones, diapiric flow and mantle convection). Despite this, most rock deformation experiments have focused on coaxial configurations, and relatively few non-coaxial deformation experiments have been reported (examples include Bouchez & Duval 1982; Kern & Wenk 1983; Shimamoto & Logan 1986; Schmid *et al.* 1987; Borradaile & Alford 1988; Price & Torok 1989; Williams & Price 1990). This discrepancy is mainly due to experimental difficulties in achieving a well-constrained non-coaxial deformation path while still being able to measure and/or control experimental variables such as the state of stress in the sample. For these reasons, almost all *quantitative* rheological data for rocks have been obtained from axi-symmetric compression experiments.

To predict the behaviour of a body of rock under a given stress state or given boundary conditions, it is necessary to generalize the axi-symmetrically derived flow law, in which stress and strain rate are treated as scalars, to three dimensions, in which stress and strain rate are treated as tensors (see for example Stocker & Ashby 1973). This is usually done using the theory of perfect isotropic plasticity to relate stress and strain rate components (Nye 1953; Paterson 1976). The assumptions underlying this method of generalization (often referred to as the associated flow rule) are that the material is isotropic and stays isotropic during deformation, the material has no other 'memory' for incremental strain history, and that deformation is isovolumetric. However there is no *a priori* justification that flow laws derived from coaxial experiments can be generalized in this way (Ferguson 1979). Indeed, axi-symmetric experiments on most materials with 'memory' are insufficient to fully characterize the material behaviour (see Hobbs 1972). Thus, the generalization of coaxial flow laws is not straightforward, and the validity of the associated flow rule approach (assuming perfect isotropic plasticity) is not known for most geological materials.

An understanding of the effect of deformation mode on flow behaviour and an adequate 3-D

From Knipe, R. J. & Rutter, E. H. (eds), 1990, *Deformation Mechanisms, Rheology and Tectonics*, Geological Society Special Publication No. 45, pp. 201–213.

flow law are of crucial importance in considering (or modelling) deformation localization phenomena. Mechanical weakening, thought to greatly enhance the development of localized shear zones, can be caused by the development of crystallographic preferred orientations within a deforming material (as suggested by texture simulations – see Tomé *et al.* 1984; Wenk *et al.* 1986, 1987) or by microstructural changes such as dynamic recrystallization or grain boundary alignment (White *et al.* 1980). Rotational and irrotational deformation paths have different effects on the development of textures and on microstructural evolution (White *et al.* 1980; Tomé *et al.* 1984; Wenk *et al.* 1986). Hence deformation path (or geometry) can be expected to have a significant influence on the mechanical anisotropy and response of deforming materials.

In this paper we report deformation experiments on polycrystalline NaCl deformed in uniaxial compression and in 'simple shear' at intermediate temperatures (T/T_m between 0.5 and 0.6). The aim of the study was to gain insight into the influence of deformation mode on the development of crystallographic preferred orientation (or texture), on microstructural evolution and on the mechanical behaviour (flow strength) of the material. Polycrystalline NaCl was chosen as a suitable material to investigate for the following reasons. Firstly, polycrystalline salt is ductile and relatively weak at temperatures as low as 200°C. This allows simple shear to be carried out without confining pressure, using relatively simple deformation apparatus. Secondly, NaCl exhibits well understood dislocation creep behaviour at intermediate temperatures and is a useful mineral analogue in a general sense (see Guillopé & Poirier 1979; Banerdt & Sammis 1985; Shimamoto & Logan 1986). Lastly, NaCl is extremely suitable for microstructural and textural studies since dislocation substructures are readily made visible and texture measurements can be carried out relatively easily.

Experimental method

In this study, synthetic polycrystalline NaCl samples were used. These were prepared as follows. First, analytical grade NaCl powder, jacketed in rubber tubes, was hydrostatically cold-pressed into irregular billets. From these billets, right cylindrical samples were machined. For the compression experiments, the samples produced measured 20 mm in length by 10 mm in diameter. For the shear experiments, cylindrical samples with reduced sections were prepared. These samples were 75 mm long and two diameters were used, namely 25 and 15 mm. The reduced sections were cut using a file plus slotted template. Figure 1a illustrates the final geometry. After machining, all samples were annealed for 12–14 hours in an argon atmosphere at 720–725°C (i.e. *c.* 0.9 T_m). The material obtained possessed a recrystallized polygonal 'foam' microstructure with a grainsize 100–400 μm. Grain boundaries contained gas-filled tubular and spherical inclusions of about 10 μm in diameter. The grains contained no optically visible substructure. The porosity of the starting material ranged from 2% to 2.5%. Neutron diffraction analysis showed that no significant texture was developed.

The uniaxial compression experiments were carried out in an Instron 1193 constant displacement rate apparatus equipped with a high temperature furnace and superalloy loading pistons (see de Bresser & Spiers, this volume). The axial force applied to the sample was measured using a 5 kN load cell, with an absolute accuracy better than 0.5%. The maximum strains attained were *c.* 20%.

The shear experiments were carried out in the deformation apparatus shown in Fig. 1b, mounted in an Instron 1362 loading frame. Basically, the shear rig consists of three parallel loading bars. The two outer bars are fixed to end-platens and are immobile. The middle bar is free to move in the vertical direction only, i.e. parallel to the fixed outer bars. The sample (Fig. 1a) is fitted lengthwise through aligned holes in the three loading bars. Advancing the middle bar downwards imposes a shear deformation on the reduced sections of the sample. The shear load and hence the total shear stress supported by the sample were measured using a 10 kN load cell with an absolute accuracy better than 0.5% and a resolution better than 0.01%. Normal stresses acting parallel to the length of the sample could not be measured. Displacement (hence shear strain γ) was measured using the linear variable differential transformer (LVDT) located in the drive unit of the Instron 1362.

Both the uniaxial and shear experiments were carried out at constant displacement rate in the temperature range 250–350°C under unconfined (i.e. atmospheric pressure) conditions. Axial and shear strain rates varied from test to test between 10^{-5} s^{-1} and 10^{-7} s^{-1}. The shear strain rates were chosen to be comparable with the compression strain rates according to the equivalent strain rate concept introduced later.

Stress-strain curves were calculated from strip-chart records of force and displacement v. time (shear tests) or of force v. time alone

(a)

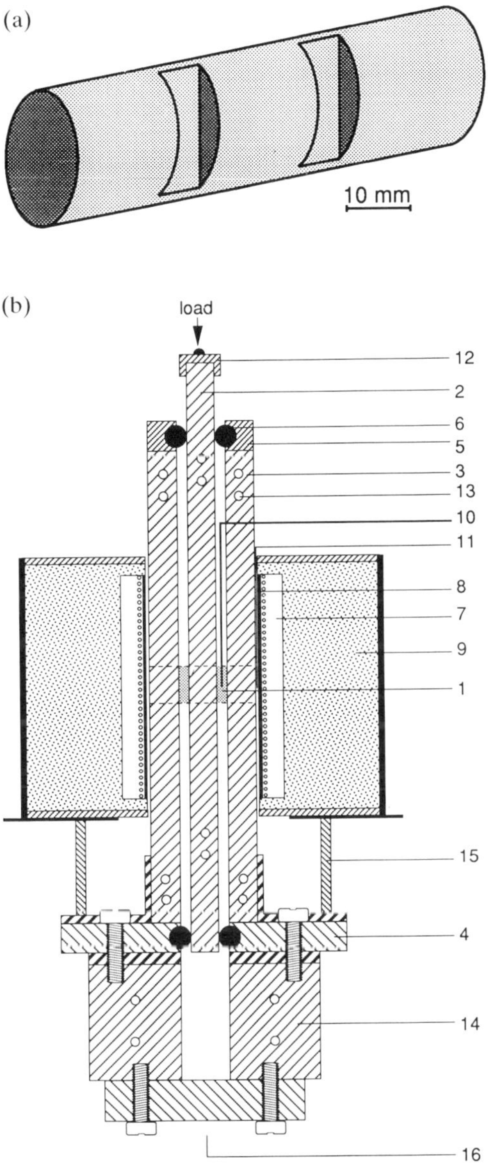

Fig. 1. (**a**) Configuration of cylindrical sample with reduced sections used for shear experiments. This sketch illustrates the final sample geometry. (**b**) Schematic diagram of the shear apparatus. Key: 1, cylindrical sample with reduced sections; 2, mobile middle bar; 3, fixed outer bars; 4, bottom platen supporting outer bars; 5, idem, top platen; 6, roller bearings reducing friction and constraining movement of the middle bar to the vertical; 7, furnace windings; 8, insulating mica sheets; 9, ceramic fibre insulation; 10, sample thermocouple; 11, control thermocouple; 12, spherical seat; 13, water-cooling access; 14, spacers; 15, furnace support; 16, Instron loading ram.

(uniaxial tests). The stresses were corrected for dimensional changes of the sample and, in the case of shear, for the weight of the middle bar. Displacement measurements were corrected for apparatus stiffness and for thermal expansion of the sample. Strain and strain rates were calculated with respect to the starting dimensions of the sample.

The microstructures of both the deformed and undeformed samples were studied by means of reflected light microscopy carried out on polished sections. Grain boundaries and dislocation substructures were made visible by etching the polished sections. Details of section preparation techniques and etching procedures are given in Spiers *et al.* (1986). All micrographs presented here were obtained from polished and etched sections.

Experimental results

Mechanical data

Table 1 shows details of the experiments reported in the present study. The corresponding applied stress v. axial strain curves for the uniaxial experiments are shown in Fig. 2. All of these curves show considerable work hardening during the first 10% shortening, reaching stress levels between *c*.5 and 10 MPa at 10–20% strain. With increasing temperature and decreasing strain rate, the flow stresses decrease. Above 300°C, steady state behaviour is closely approached, particularly at low strain rates.

The applied shear stress v. shear strain curves obtained from the shear tests are shown in Fig. 3 (refer to Table 1 for details). These curves show some work hardening during the first 0.2 shear strain, then attain more or less steady state flow at shear stresses in the range 2–4 MPa. As in the compression tests, the flow stresses in shear decrease substantially towards higher temperatures.

Optical microstructures

The microstructures developed during the uniaxial compression tests are illustrated in Figs 4 and 5. The samples showed a well-defined grain flattening fabric, with clear dislocation substructures. As low strains ($\varepsilon < 10\%$), an irregular pattern of subgrains develops with incomplete boundaries, indicating incipient polygonization involving climb (Friedman *et al.* 1981). With increasing strain, a well-defined cellular network of subgrains forms, and the flattening fabric

Table 1. *Summary of experimental conditions*

Testcode	Temperature (°C)	Strain rate (s^{-1})	Final stress (MPa)	Final strain (%)
Uniaxial compression experiments				
C35	250	2.9×10^{-7}	9.8	6.6
C15*	250	1.8×10^{-6}	14.8	14.5
C6	300	3.6×10^{-6}	9.4	24.0
C53	300	1.8×10^{-6}	7.7	14.2
C42	350	1.8×10^{-6}	7.3	13.9
C46	350	1.8×10^{-6}	6.9	14.6
C45	350	1.8×10^{-7}	5.0	8.9
C52*	350	1.8×10^{-5}	7.4	39.0
Shear experiments				
120s	250	5.8×10^{-7}	4.0	0.66
118s	300	3.6×10^{-6}	2.9	1.22
123	300	3.6×10^{-6}	3.5	0.56
124	365	7.2×10^{-6}	2.2	0.99
121	365	3.6×10^{-6}	2.5	0.34

* Microstructural/textural data only.

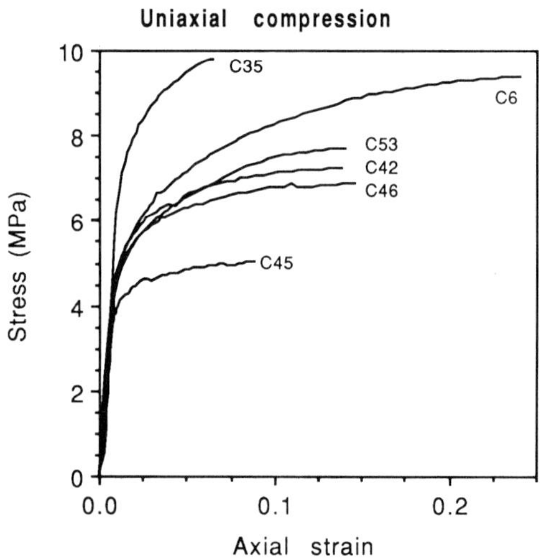

Fig. 2. Plots of applied uniaxial stress ($\sigma_{applied}$) versus axial strain for the present uniaxial compression tests on polycrystalline salt.

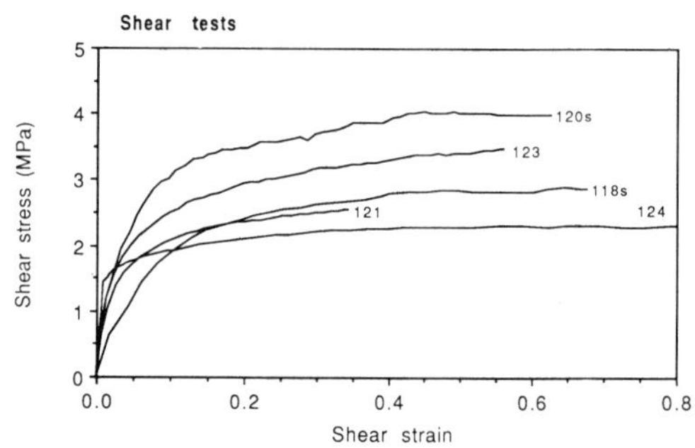

Fig. 3. Applied shear stress ($\tau_{applied}$) versus shear strain plots for the polycrystalline halite samples deformed in shear. The shear stress is determined by the vertical load transmitted through the sample (see text).

becomes more intense. Additionally, in the samples deformed at strain rates above 10^{-6} s^{-1}, characteristic linear arrangements of blocky subgrains are developed (Fig. 4). These linear substructures are oriented at 30–60° to the compression direction. A marked change in grain boundary morphology is linked with the development of subgrains. The straight grain boundaries observed in the undeformed material become bulged where intersected by subgrains (Fig. 5). However, extensive migration of these bulged grain boundaries was not observed.

The microstructures typical of shear deformation are presented in Figs 6 and 7. The transition from undeformed 'wall rock' into the shear zone is illustrated in Fig. 6. The lower part of the micrograph reveals the undeformed part of the sample. The upper half of the micrograph shows the dextral shear zone, containing strongly deformed grains. The transition zone between the 'wall rock' and the shear zone is less than 400 μm (less than one grain diameter). Within typical shear zones, flattened and elongated grains define a clear foliation (Fig. 7). The orientation of the foliation was found to be constant across the width of the shear zone in all samples examined, and corresponded closely to the orientation of the plane of finite flattening of the strain ellipse calculated for simple shear (i.e. a simple shear deformation

Fig. 4. Optical micrograph illustrating the development of linear arrangements of subgrains typical for uniaxial deformation at 250°C. In the two dark grains, just below the centre of the micrograph, two intersecting sets of linear substructures give rise to the formation of blocky subgrains. Thin etch lines present in grain A resemble wavy slip lines. Incipient recovery leads to the development of incomplete subgrains. Close inspection of grain B, which is relatively unaffected by the etching procedure, reveals the presence of a very faint network of subgrains (Sample C15, 250°C, ε = 14.5%, compression direction vertical).

corresponding to the imposed displacement). Good agreement was also obtained between the strain ellipse calculated from grain boundary configurations and that expected for simple shear. These observations indicate that the deformation was close to simple shear. Turning now to the dislocation substructure, the sheared grains within the shear zones show cellular networks of subgrains (Figs 6 and 7). The subgrain walls are preferentially oriented parallel and subnormal to the shear plane. Once again, grain boundaries tend to be irregular due to small bulges developing where subgrains intersect grain boundaries. Grain flattening and polygonization dominate the microstructure. However, evidence for grain boundary migration on the grain scale was observed locally.

In Fig. 8, subgrain size data obtained from both the compression and shear tests are plotted against applied stress (σ_{applied}, τ_{applied}). In most materials, subgrain size is empirically found to be an inverse function of stress (Servi & Grant 1951), and for salt various relationships have been obtained. The subgrain size v. stress relationship for natural polycrystalline salt is given by $d(\mu\text{m}) = 190\ \sigma^{-1}$ (MPa) (Carter *et al.* 1982). For synthetic polycrystalline salt, Burke *et al.* (1981) obtained $d(\mu\text{m}) = 75\ \sigma^{-0.9}$ (MPa). Both relationships are shown in Fig. 8. All previously reported data for salt fall between these two

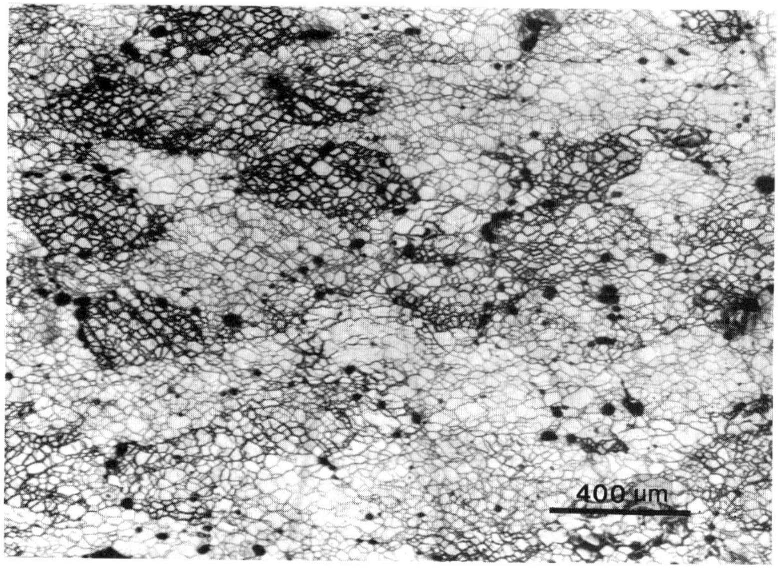

Fig. 5. Optical micrograph of a sample shortened 39% in uniaxial compression revealing the development of a cellular network of subgrains in all grains. The grains are strongly flattened. The subgrains also show a dimensional fabric. Note the scalloped grain boundaries visible even at this low magnification. (Sample C52, 350°C, $\varepsilon = 39\%$, compression direction vertical).

lines (see Carter *et al.* 1982). Thus the present subgrain size v. applied stress data seem reasonably consistent with previous findings although the significance of plotting τ_{applied} is not yet clear.

Textures

Pole figures have been determined for samples deformed in both deformation modes. The pole figures were measured using neutron diffraction goniometry at the GKSS Research Centre, Geesthacht, F.R.G. From the {220} and the {200} pole figures obtained for each sample, the orientation distribution function was calculated. From this, inverse pole figures were determined (Bunge 1969, Dahms 1987). The results presented here have been found to be representative of the two test types, for all temperatures in the range 250–550°C.

Figure 9 shows an inverse pole figure for the compression direction of a uniaxially shortened sample deformed at 250°C to 18% strain (C15). This texture consists of a maximum around the <110> direction spreading towards the <100> and <115> directions. The maximum around <110> is about 1.5 times uniform.

Figure 10 shows the {220} and the {200} pole figures for sample 123 deformed in shear (at 300°C) to a shear strain of 0.6. The {220} pole figure (Fig. 10a) exhibits a maximum perpendicular to the shear plane. In contrast, the {200} pole figure (Fig. 10b) shows a crossed girdle distribution with less intense maxima. Taking into account observations from other samples, no consistent relationship between the sense of shear and the symmetry of the pole figures could be determined. With regard to the inverse pole figures for sample 123, that for the shear direction (Fig. 11a) shows a maximum around the <111> direction and a submaximum around <100>. The inverse pole figure for the shear plane normal (Fig. 11b) shows a maximum around the <110> direction (cf. Fig. 10a).

Discussion

Deformation mechanisms and microstructure

The observed subgrain microstructures plus the presence of crystallographic and shape preferred orientation in compression and in shear are consistent with a climb-controlled dislocation creep mechanism. Furthermore, additional experimental data obtained by Franssen (unpublished) demonstrate power law creep behaviour in both deformation modes under the present conditions, with the power law

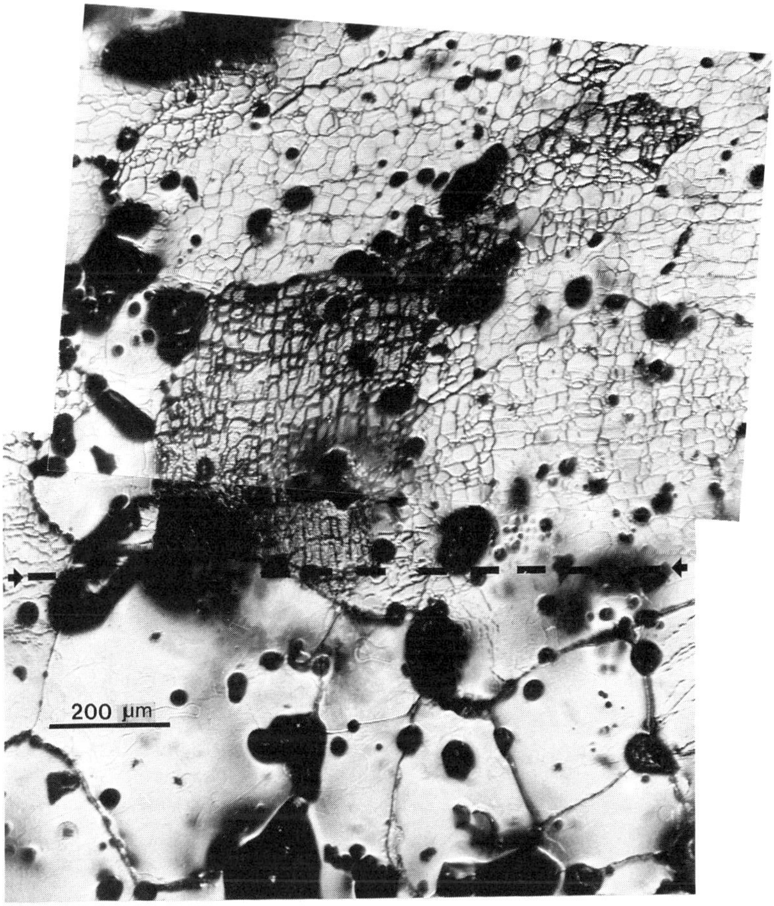

Fig. 6. Photomicrograph revealing the transition from the undeformed wall rock to the shear zone in the reduced section of a sheared sample. The lower half shows the wall rock containing undeformed substructure-free grains. The upper half of the micrograph is well within the shear zone and contains sheared grains with a cellular network of subgrains. The shear zone boundary is less then 400 μm wide (i.e. the average grain size of the starting material). (Sample 118s, 300°C, $\gamma = 1.2$, right lateral shear zone).

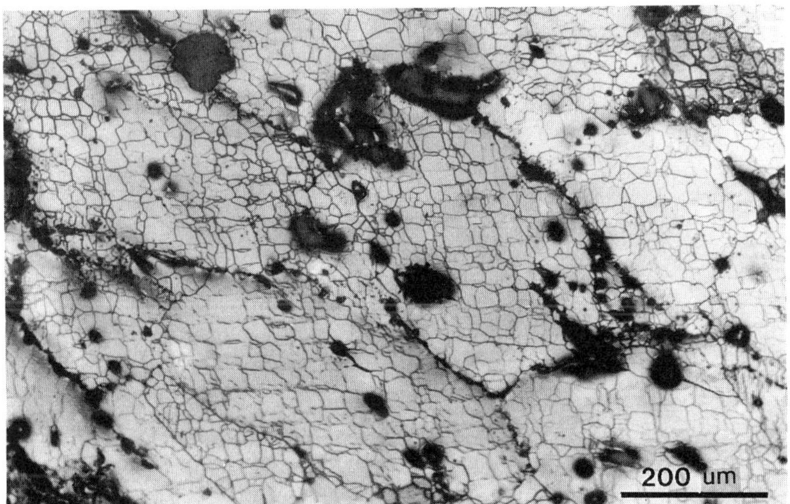

Fig. 7. Sheared sample containing a clear grain shape fabric defining the foliation apparent in this micrograph. The orientation of the foliation, as calculated from the grain boundary markers, corresponds closely to the orientation of the long axis of the strain ellipse. Note the scalloping of grain boundaries developed from small bulges at subgrains. (Sample 123, 300°C, $\gamma = 0.6$, left lateral shear zone).

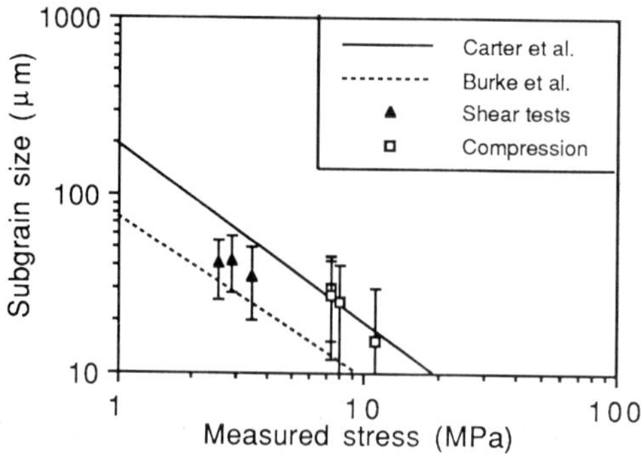

Fig. 8. Subgrain size versus applied (σ_{applied}, τ_{applied}) stress for synthetic rocksalt experimentally deformed in compression (squares) and shear (triangles). The solid lines represent the empirical relationships obtained by Carter *et al.* (1982) and by Burke *et al.* (1981).

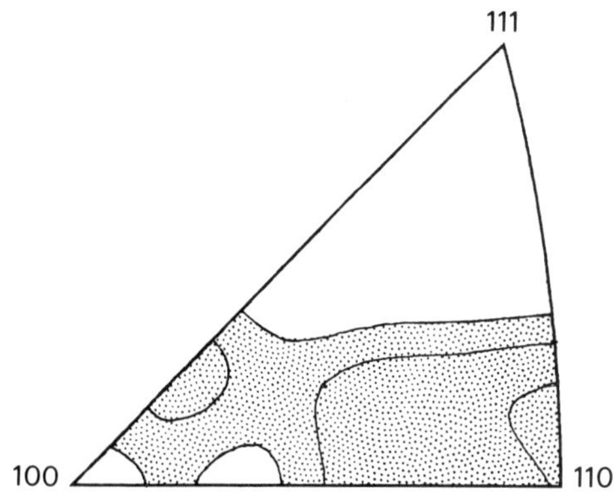

Fig. 9. Inverse pole figure for compression direction obtained from a uniaxially deformed sample (C15) shortened 18% at 250°C. Contour intervals are 0.2 times uniform. The area with an intensity greater than 1 is shaded. Maximum intensity observed is 1.4 times uniform at <110>.

exponent (n) ranging between 4 and 6 and the activation energy for creep (ΔH) ranging from 70 to 139 kJ mole^{-1}. These creep parameters are fully consistent with previous data on diffusion and on dislocation creep in dry salt (Heard 1972; Arieli *et al.* 1982; Heard & Ryerson 1986). It is concluded that deformation occurred by climb-controlled dislocation creep in both deformation geometries. The observation that grain boundary migration is more extensive in shear than in compression is thought to reflect the fact that higher strains are achieved in the shear experiments (Table 1).

Textures

The present texture data for uniaxial compression closely resemble the results obtained by Kern & Braun (1973) for salt deformed in axi-symmetric compression under comparable conditions. We are not aware of any previous experimental data on texture development in salt during simple shear under the present conditions. Recently, however, the Taylor theory and the self-consistent viscoplastic theory were applied to texture development in polycrystalline halite by Wenk *et al.* (1989). Modelling was carried out for uniaxial compression and for simple shear assuming easy $\{110\}<1\bar{1}0>$ slip in accordance with the low/intermediate temperature single crystal yield data of Carter & Heard (1970). The Taylor simulation for uniaxial compression yields inverse pole figures which correspond closely with the present inverse pole figures for compression. The self-consistent simulation gives inverse pole figures with a maximum around <100>. Both the Taylor and the self-consistent texture simulations for shear deformation yield pole figures for {220} which are very similar to our results (Fig. 10), allowing for typical experimental errors. This broad agreement between our tests and the simulations supports the idea that texture development was dominated by the weaker $\{110\}<1\bar{1}0>$ systems in the present experiments. Since our data show a tendency for the {110} planes to align parallel to the shear plane in the shear tests, and normal to the compression direction in the uniaxial tests, at least some difference in mechanical behaviour can be expected in the two modes. Note, however, that our inverse pole figure for shear direction (Fig. 11a) shows a maximum parallel to <111>. Thus in shear, the slip direction of the weakest slip system is not aligned with the macroscopic shear direction.

Mechanical behaviour

An attempt is now made to compare the mechanical behaviour (strength) obtained in compression and shear in the framework of the von Mises theory of isotropic plasticity or the associated flow rule discussed in the introduction. In order to do this, we must first define a number of quantities required for such a comparison (see also Schmid *et al.* 1987).

Definitions. We start by introducing a quantity known as the equivalent stress, σ_{eq} (see McClintock & Argon 1966). This is defined as

$$\sigma_{eq} = \sqrt{\frac{3}{2}\sigma^*_{ij}\sigma^*_{ij}} = \sqrt{3J^*_2} \quad (1)$$

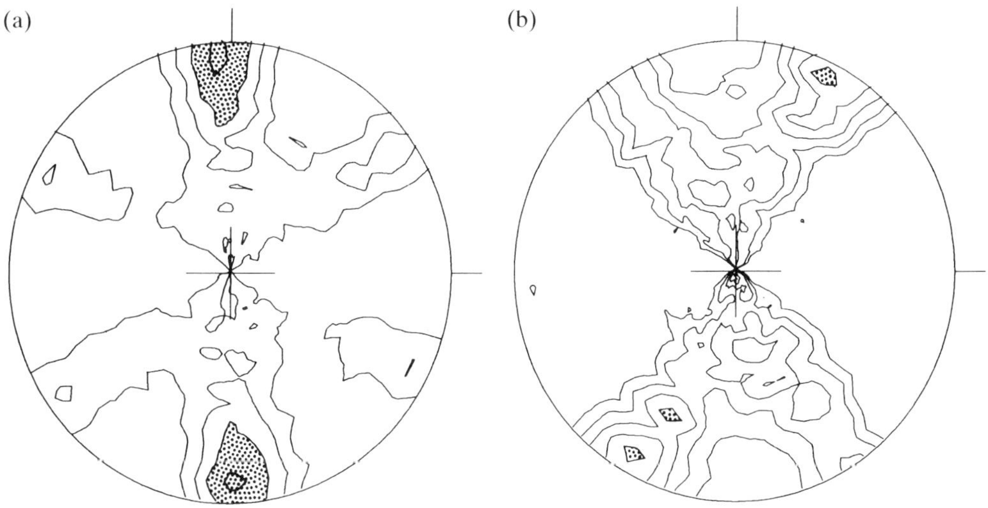

Fig. 10. Pole figures for the (**a**) {220} and the (**b**) (200) crystallographic directions for sample 123, deformed in shear at 300°C to a shear strain of 0.6. Contour interval for (a) is 0.25 and for (b) 0.125 times uniform. Areas with intensities greater than 1.5 times uniform are shaded. The orientation of the shear plane is E–W and the sense of shear is sinistral.

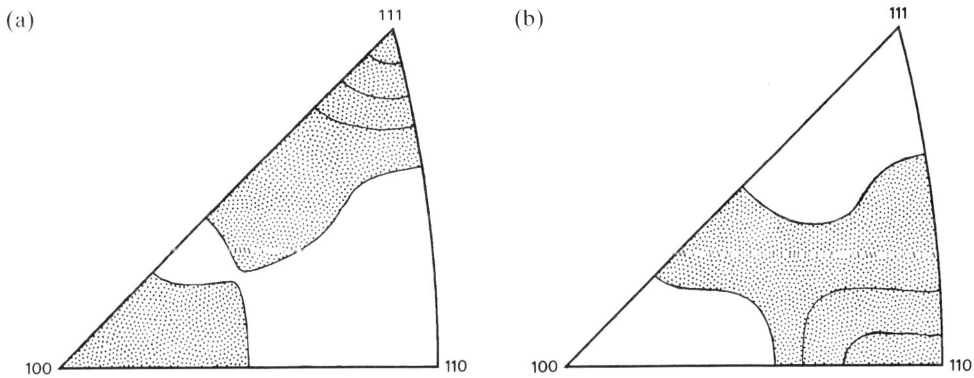

Fig. 11. Inverse pole figures for (**a**) the shear direction and (**b**) the shear plane normal, derived from the pole figures shown in Fig. 10. Contours intervals are 0.2 times uniform. Areas with intensities greater than 1 time uniform are shaded. Sample 123.

(where σ_{ij}^* is the deviatoric stress tensor and J_2^* is its second invariant) and represents a shear stress component equal to the octahedral shear stress of Nadai (1963) multiplied by a factor $\frac{3}{\sqrt{2}}$ (see Schmid *et al.* 1987). Now, the von Mises yield criterion states that flow in an isotropic, perfectly plastic solid occurs when σ_{eq} attains a value equal to the yield (i.e. flow) strength in uniaxial compression (see McClintock & Argon 1966). This holds for all stress states (σ_{ij}^*) causing flow in such materials. For the case of an isotropic, perfectly plastic material undergoing flow, σ_{eq} can thus be regarded as a measure of flow strength, or flow stress, taking the same value independently of deformation geometry. Hence σ_{eq} forms a useful quantity for comparing the flow strength of real materials in different deformation modes. Note that for uniaxial stress ($\sigma_{11} = \sigma_{applied}$) equation (1) yields $\sigma_{eq} = \sigma_{11}$ whereas for a pure shear stress of magnitude τ, $\sigma_{eq} = (\sqrt{3})\tau$.

In order to compare strain rates in different deformation modes, we follow Schmid *et al.* (1987) in using a quantity generally referred to as the equivalent strain rate (see also Stocker & Ashby 1973). For constant volume deformation, this is defined

$$S_{eq} = \sqrt{\frac{2}{3}S_{ij}S_{ij}} = \sqrt{\frac{4}{3}I_2} \qquad (2)$$

where S_{ij} is the strain rate tensor and I_2 is the second invariant of S_{ij}. It can be viewed as an octahedral shear strain rate in which the numerical factor is chosen such that the product $\sigma_{eq}S_{eq}$ equals the mechanical work rate $\sigma_{ij}{}^*S_{ij}$. In uniaxial shortening, S_{eq} is equal to the shortening strain rate (S_{11}). In simple shear at a shearing rate $\dot{\gamma}$, $S_{eq} = \dot{\gamma}/\sqrt{3}$.

To compare finite strains in different deformation geometries, we use the equivalent logarithmic strain. This is equal to the Nadai measure of octahedral strain magnitude (used by Schmid *et al.* 1987) multiplied by the factor $\sqrt{\frac{2}{3}}$. For deformation at constant volume it is defined

$$\varepsilon_{eq} = \sqrt{\frac{2}{3}\varepsilon_{ij}\varepsilon_{ij}} \qquad (3)$$

where ε_{ij} is the logarithmic strain tensor (Hill 1950). This gives $\varepsilon_{eq} = \varepsilon_{11}$ in uniaxial compression and $\varepsilon_{eq} = \sqrt{4/3}\ln\left[\frac{1}{2}(\gamma + \sqrt{\gamma^2+4})\right]$ in simple shear. Note, however, that the physical significance of ε_{eq} is not clear.

Comparison. We begin our comparison of the mechanical behaviour observed in the uniaxial and shear tests by obtaining expressions for the state of deviatoric stress (σ^*_{ij}) in the samples assuming isotropic, perfectly plastic behaviour. We make the additional assumptions (a) that the applied stresses are uniformly transmitted throughout the samples, and (b) that the samples can be considered to be in static equilibrium with no couple stresses. For the uniaxial case, the results obtained for the non-zero components of σ^*_{ij} are $\sigma^*_{11} = (\frac{2}{3})\sigma_{applied}$ and $\sigma^*_{22} = \sigma^*_{33} = (\frac{1}{3})\sigma_{applied}$, whereas in shear we have $\sigma^*_{12} = \sigma^*_{21} = \tau_{applied}$. Using these results and assuming true uniaxial compression and simple shear deformations (at constant volume), equations (1) and (3) can be applied to derive σ_{eq} v. ε_{eq} curves for the samples tested. This has been done for all experiments performed at comparable strain rates, i.e. at similar equivalent strain rates ($S_{eq} \approx 2 \times 10^{-6}\ s^{-1}$, refer to Table 1 and equation 2). The curves obtained are shown in Fig. 12a. From these, it appears that at constant temperature the uniaxially deformed samples support substantially higher equivalent stresses (σ_{eq}) than the sheared samples, for all values of ε_{eq}. Because of ambiguity in the physical meaning of ε_{eq}, however, the most significant observation is that the (near) *steady state* values of σ_{eq} are *c.* 1.5 times larger in compression than in shear (under comparable conditions). This behaviour is clearly not consistent with that expected for an isotropic, perfectly plastic material under the assumptions made above. The implication is: (a) that the samples (though initially free of crystallographic or dimensional preferred

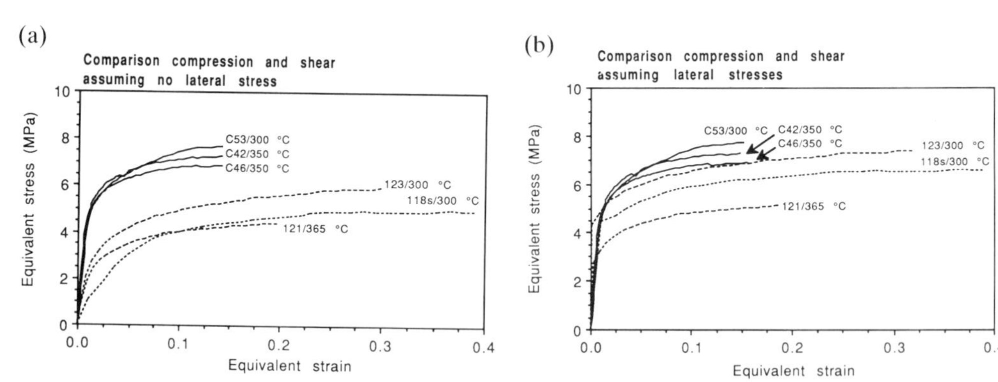

Fig. 12. Comparison of mechanical data from the compression tests (solid lines) and the shear tests (dashed lines) in terms of σ_{eq} and ε_{eq}. In (**a**), σ_{eq} is calculated directly from the applied stress ($\sigma_{applied}$, $\tau_{applied}$). In (**b**) lateral stresses of 3.5 MPa for 300°C and 2.15 MPa for 365°C are superposed for the calculation of the equivalent stress in shear. See text for discussion.

orientation) become mechanically anisotropic during deformation, exhibiting truly lower strength in shear than in compression; or (b) if the samples remain more or less mechanically isotropic, that the calculated values of σ^*_{ij} and σ_{eq} are in serious error because of imperfectly imposed boundary conditions or effects such as volume changes occurring within the samples.

Now, taking into account the lack of texture in the starting material and the symmetry of deformation, the state of deviatoric stress (and hence σ_{eq}) in the compression tests can be considered reasonably well determined $\left(\sigma^*_{11} = \frac{2}{3}\sigma_{\text{applied}}, \sigma^*_{22}=\sigma^*_{33} = \frac{-\sigma^*_{11}}{2}\right)$. In shear deformation, however, the full state of deviatoric stress in the samples is not known, since only vertical stresses are measured. Thus, errors may arise in calculating σ_{eq} in shear. Nonetheless, *for the case that the samples remain roughly isotropic during deformation* (see (b) above), the range in magnitude of σ_{eq} can be constrained. Two extreme constraints can be envisaged.

(1) If the sample in the shear zone dilates during deformation, compressive stresses will be generated in the 'wall rock' of the sample. Density measurements on the deformed samples show dilation during shear was less than 0.2%. If this volume change is translated into 'virtual displacements' occurring perpendicular to the shear plane at a constant rate throughout deformation, then the maximum compressive stress generated normal to the shear zone can be estimated (in a time-averaged sense) from the creep law derived from uniaxial tests. From these calculations, it has been found that the maximum normal stresses induced by dilatancy during the shear experiments take average values between 2.15 and 3.5 MPa, depending on temperature. The $\sigma_{eq}-\varepsilon_{eq}$ curves for shear have been recalculated using these values of normal stress superposed on the measured stresses. The result is shown in Fig. 12b. Still the salt appears weaker in shear than in the compression experiments.

(2) In the present shear, apparatus tensile stresses may develop in the sample due to imperfectly imposed boundary conditions as the middle bar is displaced downwards. However, no evidence of tensile or extensional failure was observed in the sheared samples. It follows that the tensile stresses acting in the shear samples were almost certainly less than the room temperature tensile strength of 1–3 MPa (Gessler 1983). Introducing tensile stresses of 3 MPa; perpendicular or parallel to the shear zone, into the equivalent stress calculation yields closely similar curves to those shown in Fig. 12b.

On the basis of the above, it seems that the 'weak' behaviour observed in shear cannot be accounted for by isotropic behaviour of the samples coupled with errors in determining stress. We therefore infer that the samples became *mechanically anisotropic* during deformation, leading to truly weaker behaviour in shear than in compression when viewed in the framework of the von Mises theory. Thus the assumptions underlying the theory of perfect isotropic plasticity and the associated flow rule (for generalizing creep laws to 3-D) do not seem to be applicable to the present experiments. Since the volume changes are smaller than 0.2% in both shear and compression, and since the same deformation mechanisms (dislocation creep) and microstructural processes are operative, we infer that the observed differences in mechanical behaviour between shear and compression are largely due to the textures developed in the two modes. In future, this explanation should be tested by means of Taylor or self-consistent simulations including the single cystal yield parameters and creep properties appropriate to the present experimental conditions (see Wenk *et al.* 1989). Finally, we note that the 'texture weakening' effect inferred here is quite different from the weakening reported by Burrows *et al.* (1979) and Drury *et al.* (1985). Their weakening was associated with grain-scale shear localization, recrystallization and internal strain softening. The present effect involves no grain-scale localization process.

Conclusions

Dry polycrystalline salt has been deformed in compression and in near simple shear, at temperatures in the range 250–350°C. The microstructure developed in compression and in shear is characterized by strongly deformed grains containing cellular networks of subgrains produced by polygonization. The deformation mechanism is of a climb-controlled dislocation creep type. The textures developed in compression are characterized by a tendency for the {110} planes of the relatively weak {110}<1$\bar{1}$0> slip systems to rotate perpendicular to the compression direction. In shear, the {110} planes tend to align with the flow plane, but the <1$\bar{1}$0> direction does not align with the flow direction. Instead the <111> direction tends to lie parallel to the shear direction. A difference in mechanical behaviour has been observed between shear and compression. In terms of equivalent stress (calculated assuming isotropic behaviour), significantly lower flow stresses were obtained in shear than in compression. It is inferred that this difference in flow strength is largely due to anisotropy, i.e. to the development of the observed textures. Our results suggest that the associated flow rule does not offer an accurate method of generalizing creep laws for salt to 3-D (under the deformation conditions investigated here).

G. Kastelein constructed the deformation rigs at our Institute's workshop. H.-G. Brokmeier is thanked for measuring the textures and introducing R.C.M.W.F. to neutron diffraction techniques. M. Dahms performed the texture calculations. C. J. Peach is thanked for technical assistance and discussions. We are also grateful to an anonymous reviewer for suggestions which substantially improved the manuscript. A grant from the Netherlands Organization for the Advancement of Pure Research (N.W.O.) to visit the GKSS Research Centre at Geesthacht (F.R.G.) is gratefully acknowledged.

References

ARIELI, A., HEARD, H. C. & MUKHERJEE, A. K. 1982. Deformation modeling in sodium chloride at intermediate and elevated temperatures. *In*: ROHDE, R. W. & SWEARENGEN, J. C. (eds) *Mechanical testing for deformation model development*. American Society for Testing of Materials Special Technical Publication, **765**, 342–365.

BANERDT, W. B. & SAMMIS, C. G. 1985. Low stress high temperature creep in single crystal NaCl. *Physics of the Earth and Planetary Interiors*, **41**, 108–124.

BORRADAILE, G. J. & ALFORD, C. 1988. Experimental shear zones and magnetic fabric. Journal of Structural Geology, **10**, 895–904.

BOUCHEZ, J. L. & DUVAL, P. 1982. The fabric of polycrystalline ice deformed in simple shear: experiments in torsion, natural deformation and geometrical interpretation. *Textures and Microstructures*, **5**, 171–190.

BUNGE, H. J. 1969. *Mathematische Methoden der Texturanalyse*. Akademie-Verlag Berlin.

BURKE, P. M., CANNON, W. R. & SHERBY, O. D. 1981. Intermediate and high temperature creep of polycrystalline sodium chloride. *Workshop on structural behaviour of repository materials. Sandia Nat. Laboratories, Albuquerque N.M., April 29.*

BURROWS, S. E., HUMPHREYS, F. J. & WHITE, S. H. 1979. Dynamic recrystallisation. A comparison between magnesium and quartz. *Bulletin Minéralogie*, **102**, 75–70.

CARTER, N. L. & HEARD, H. C. 1970. Temperature and rate dependent deformation of halite. *American Journal of Science*, **269**, 193–249.

——, HANSEN, F. D. & SENSENY, P. E. 1982. Stress magnitudes in natural rocksalt. *Journal of Geophysical Research*, **B87**, 9289–9300.

DAHMS, M. 1987. *Spezielle mathematische Methoden der Texturanalyse und ihre Anwendugen unter Berüchsichtingung der intermetallischen Phasen*. PhD Thesis, T. U. Clausthal.

DE BRESSER, J. H. P. & SPIERS, C. J. 1990. High temperature deformation of calcite single crystals by r+ and f+ slip. *This volume*.

DRURY, M. R., HUMPHREYS, F. J. & WHITE, S. H. 1985. Large strain deformation studies using polycrystalline magnesium as a rock analogue. Part II: dynamic recrystallisation mechanisms at high temperatures. *Physics of the Earth and Planetary Interiors*, **40**, 208–222.

FERGUSON, C. C. 1979. The simple fluid with fading memory as a rheological model for steady-state flow of rocks. *In*: EASTERLING, K. E. (ed.) *Proceedings of the International Conference on the Mechanics of Deformation and Fracture. Luleå, Sweden*. 371–383.

FRIEDMAN, M., DULA, W. F., GANGI, A. F. & GAZONAS, G. A. 1981. Structural petrology of experimentally deformed synthetic rocksalt. *In*: LANGER, M. & HARDY, R. (eds) *First Conference on the Mechanical Behaviour of Rocksalt*, 19–36.

GUILLOPÉ, M. & POIRIER, J. P. 1979. Dynamic recrystallization during creep of single crystalline halite: an experimental study. *Journal of Geophysical Reseach*, **B84**, 5557–5567.

GESSLER, K. 1983. Vergleich der einaxialen Zugfestigkeit mit der Drei-Punkt- Biege Zugfestigkeit und unterschiedlichen Spaltzugfestigkeiten. *Kali und Steinsalz*, **8**, 416–423.

HEARD, H. C. 1972. Steady state flow of polycrystalline halite at a pressure of 2 kilobar. *In*: HEARD, H. C., BORG, I. Y., CARTER, N. L. & RALEIGH, C. B. (eds) *Flow and fracture of rocks*. Geophysical Monograph, **16**, 191–209.

—— & RYERSON, F. J. 1986. Effect of cation impurities on steady state flow of salt. *In*: HOBBS, B. E. & HEARD, H. C. (eds) *Mineral and rock deformation: laboratory studies*. Geophysical Monograph, **36**, 99–115.

HILL, R. 1950. *The mathematical theory of plasticity*. Oxford University Press, Oxford.

HOBBS, B. E. 1972. Deformation of non-newtonian fluids in simple shear. *In*: HEARD, H. C., BORG, I. Y., CARTER, N. L. & RALEIGH, C. B. (eds) *Flow and fracture of rocks*. Geophysical Monograph, **16**, 243–258.

KERN, H. & BRAUN, G. 1973. Deformation und Gefügeregelung von Steinsalz im Temperaturbereich 20–200°C. *Contributions Mineralogy and Petrology*, **40**, 169–181.

—— & WENK, H. R. 1983. Texture development in experimentally induced shear zones. *Contributions to Mineralogy and Petrology*, **83**, 231–236.

MCCLINTOCK, F. A. & ARGON, A. S. 1966. *Mechanical behaviour of materials*. Addison-Wesley, Reading Massachusetts, U.S.A.

NADAI, A. 1963. *Theory of flow and fracture of solids*.

McGraw-Hill, New York.

NYE, J. F. 1953. The mechanics of glacier flow. *Journal of Glaciology*, **2**, 82–93.

PATERSON, W. S. B. 1976. *The physics of glaciers*. Pergamon press, Oxford.

PRICE, G. P. & TOROK, P. A. 1989. A new simple shear deformation apparatus for rocks and soils. *Tectonophysics*, **158**, 291–309.

SCHMID, S. M., PANOZZO, R. & BAUER, S. 1987. Simple shear experiments on calcite rocks: rheology and microfabrics. *Journal of Structural Geology*, **9**, 747–778.

SERVI, I. S. & GRANT, N. J. 1951. Creep and stress behavior of aluminium as a function of purity. *Transactions AIME*, **191**, 909–916.

SHIMAMOTO, T. & LOGAN, J. M. 1986. Velocity-dependent behavior of simulated halite shear zones: an analog for silicates. *In*: DAS, S., BOATWRIGHT, J. & SCHOLZ, C. H. (eds) *Earthquake source mechanics*. Geophysical Monograph, **37**, 49–63.

STOCKER, R. L. & ASHBY, M. F. 1973. On the rheology of the upper mantle. *Reviews in Geophysics and Space Physics*, **11**, 391–426.

SPIERS, C. J., URAI, J. L. , LISTER, G. S., BOLAND, J. N. & ZWART, H. J. 1986. *The influence of fluid–rock interaction on the rheology of salt*. Office for Official Publications of the European Community, Luxembourg.

TOMÉ, C., CANOVA, G. R., KOCKS, U. F., CHRISTODOULOU, N. & JONAS, J. J. 1984. The relation between macroscopic and microscopic strain hardening in F.C.C. polycrystals. *Acta metallurgica*, **32**, 1637–1653.

WENK, H. R., TAKESHITA, T., VAN HOUTTE, P. & WAGNER, F. 1986. Plastic anisotropy and texture development in calcite polycrystals. *Journal of Geophysical Research*, **B91**, 3861-3869.

——, ——, BECHLER, E., ERSKINE, B. G. & MATTHIES, S. 1987. Pure shear and simple shear calcite textures. Comparison of experimental, theoretical and natural data. *Journal of Structural Geology*, **9**, 731–745.

——, CANOVA, G. R., MOLINARI, A. & MECKING, H. 1989. Texture development in halite: comparison of Taylor model and self-consistent theory. *Acta metallurgica*, **37**, 2017–2029.

WILLIAMS, P. F. & Price, G. P. 1990. Origin of kinkbands and shear band cleavage in shear zones: an experimental study. *Journal of Structural Geology*, **12**, 145–164.

WHITE, S. H., BURROWS, S. E., CARRERAS, J., SHAW, N. D. & HUMPHREYS, F. J. 1980. On mylonites in ductile shear zones. *Journal of Structural Geology*, **2**, 175–187.

Experimental determination of constitutive parameters governing creep of rocksalt by pressure solution

C. J. SPIERS, P. M. T. M. SCHUTJENS, R. H. BRZESOWSKY, C. J. PEACH, J. L. LIEZENBERG & H. J. ZWART

HPT Laboratory, Department of Geology, Institute of Earth Sciences, University of Utrecht, P.O. Box 80.021, 3508 TA Utrecht, The Netherlands

Abstract: Theoretical models for compaction creep of porous aggregates, and for conventional creep of dense aggregates, by grain boundary diffusion controlled pressure solution are examined. In both models, the absolute rate of creep is determined by the phenomenological coefficient $Z^* = Z_0 \exp(-\Delta H/RT)$, a thermally activated term representing effective diffusivity along grain boundaries. With the aim of determining Z_0, ΔH and hence Z^* for pressure solution creep in rocksalt, compaction creep experiments have been performed on wet granular salt. Compaction experiments were chosen since theory indicates that pressure solution creep is accelerated in this mode. The tests were performed on brine-saturated NaCl powder (grainsize 100–275 μm) at temperatures of 20–90°C and applied stresses of 0.5–2.2 MPa. The mechanical data obtained show excellent agreement with the theoretical equation for compaction creep. In addition, all samples exhibited well-developed indentation, truncation and overgrowth microstructures. We infer that compaction did indeed occur by diffusion controlled pressure solution, and best fitting of our data to the theoretical equation yields $Z_0 = (2.79 \pm 1.40) \times 10^{-15}$ m^3s^{-1}, $\Delta H = 24.53$ kJ mol^{-1}. Insertion of these values into the theoretical model for conventional creep by pressure solution leads to a preliminary constitutive law for pressure solution in dense salt. Incorporation of this creep law into a deformation map suggests that flow of rocksalt in nature will tend to occur in the transition between the dislocation-dominated and pressure solution fields.

The rheological or creep properties of rocksalt form fundamental input for the modelling of salt tectonic processes and the development of related hydrocarbon traps. They also represent the basic input required for modelling the long-term performance of salt-based repository systems (radioactive or chemical waste repositories), the evolution of solution-mined cavities and storage caverns, and the creep closure of conventional salt mines.

In the last 20 years, a great deal of experimental deformation work has been done on salt, and it has become widely accepted that flow is dominated by cross-slip- and/or climb-controlled dislocation creep mechanisms under most long-term engineering and geologically relevant conditions (Heard 1972; Albrecht & Hunsche 1980; Carter & Hansen 1983; Wawersik & Zeuch 1986; Skrotzki & Haasen 1988). Recently, however, experiments reported by Spiers *et al.* (1986, 1988, 1989) and by Urai *et al.* (1986) have shown that when trace brine is present (always the case in natural salt), the creep behaviour of salt can be strongly influenced by processes such as fluid-enhanced dynamic recrystallization (Urai 1983) and pressure solution creep (i.e. fluid-enhanced grain boundary diffusional creep). The limited data presented to date suggest that pressure solution creep may become important or even dominant over dislocation mechanisms at natural strain rates ($< 10^{-10}$ s^{-1}), particularly in finer-grained salts (Spiers *et al.* 1986; Urai *et al.* 1986). Naturally deformed rocksalts do show microstructural evidence for the operation of solution–precipitation processes (Urai *et al.* 1986, 1987). However, most seem to be dominated by microstructures characteristic of intracrystalline dislocation mechanisms (i.e. lattice preferred orientations and dislocation substructures; see Carter & Hansen 1983). The extent to which pressure solution is important in determining the creep behaviour of salt in nature is therefore unclear. To resolve this question, a fundamental understanding and a constitutive description of pressure solution creep in salt are needed.

In this article, we combine a consideration of *theoretical models* for pressure solution creep with *compaction creep experiments* on wet

From Knipe, R. J. & Rutter, E. H. (eds), 1990, *Deformation Mechanisms, Rheology and Tectonics*, Geological Society Special Publication No. 54, pp. 215–227.

granular salt in an attempt to determine the fundamental phenomenological coefficients or constitutive parameters governing creep of NaCl by pressure solution. The results are used to develop a preliminary constitutive law for pressure solution creep of dense polycrystalline salt. This is applied to plot a deformation map from which the importance of pressure solution during deformation of rocksalt in nature is assessed. Throughout the paper, the term 'pressure solution creep' is used to refer specifically to fluid-enhanced grain boundary diffusional creep of the type often considered analogous to Coble creep (Rutter 1976; Raj 1982). Fluid-enhanced creep mechanisms involving solid state deformation at grain-to-grain contacts plus solution transfer from contact margins to free pore walls, with grain boundary diffusion playing no role (see Pharr & Ashby 1983; Rutter 1983), are referred to as dissolution-coupled mechanisms.

Theoretical background

Theoretical models for pressure solution creep have been developed by numerous authors (e.g. Stocker & Ashby 1973; Rutter 1976; Raj 1982; Spiers & Schutjens 1990). In these models, grain boundaries are considered to contain fluid in some interconnected form which cannot be squeezed out, i.e. in a strongly adsorbed thin film (Rutter 1976, 1983; Robin 1978), *or* in a dynamically stable, island–channel network containing free fluid at uniform pressure (e.g. Raj 1982; Spiers & Schutjens 1990). Creep is viewed to occur by dissolution of material at grain boundary interfaces under high mean normal stress (σ_n), diffusion through the grain boundary solvent phase, and precipitation at interfaces under low mean normal stress. This pattern of mass transfer is illustrated in Fig. 1.

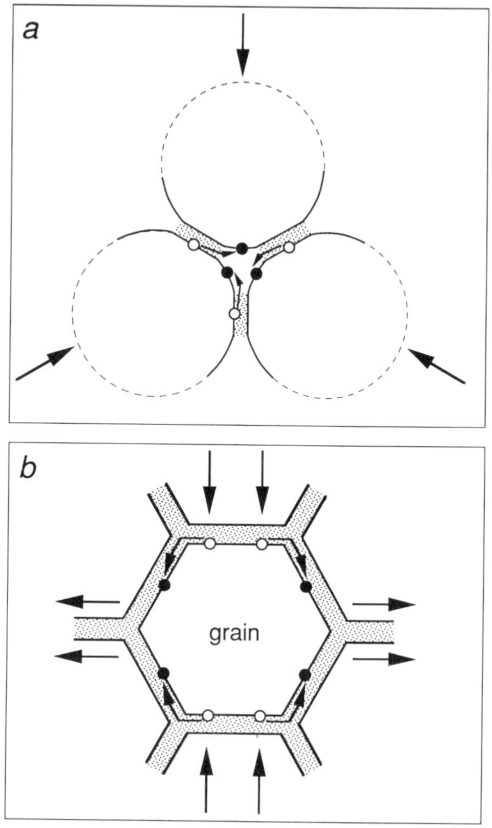

Fig. 1. Schematic diagrams illustrating pattern of mass transfer involved in creep by pressure solution. (**a**) Compaction creep in a porous or granular polycrystalline aggregate saturated with fluid. (**b**) Conventional creep in dense polycrystalline material. In both cases, grain boundaries contain fluid in an adsorbed film or island–channel form.

In most theoretical treatments, the thermodynamic driving force for mass transfer is considered to be provided by gradients in the 'PV' term $\sigma_n\Omega$, where Ω is the molecular volume of the solid phase (Paterson 1973; Rutter 1976; Robin 1978; Green 1980, 1984). While this represents a quantitatively reasonable approximation in most cases of interest (see Lehner 1990), it is important to recognize that the true (i.e. physically correct) driving force for mass transfer is one of stress-induced gradients in the normal component of the chemical potential μ_{se} of the solid at the solid/'fluid' phase boundary (see Lehner & Bataille 1984/85; Heidug & Lehner 1985; Lehner 1990; Spiers & Schutjens 1990). For given boundary conditions in μ_{se} (or in $\mu \approx \sigma_n\Omega$) at grain surfaces, the mass flux field around individual grains is determined by the kinetics of dissolution, diffusion and precipitation, and the *rate* of creep is thus controlled by whichever of these three serial processes forms the slowest overall step (Raj 1982). Constitutive equations for creep rate are obtained either from a consideration of the underlying grain-scale boundary value problem (e.g. Rutter 1976; Raj 1982; cf. Coble 1963), or from thermodynamic dissipation considerations involving solution of a macro-scale Gibbs equation for the deforming aggregate (Lehner 1990; Spiers & Schutjens 1990).

In the NaCl–brine system, the kinetics of dissolution and precipitation are extremely rapid (see Langer & Offerman 1982). For this reason, we now restrict attention to models for grain boundary diffusion controlled pressure solution. When diffusion is taken as rate-controlling, constitutive models developed using the well-known $\sigma_n\Omega$-formulation for driving force produce essentially identical results to those developed using more rigorous thermodynamic dissipation methods. Following recent derivations (Rutter 1976; Raj 1982; Spiers *et al.* 1989; Spiers & Schutjens 1990; Lehner 1990), the results obtained can be expressed as follows. Firstly, for compaction creep of porous aggregates (Fig. 1a) the compaction creep rate can be written

$$\dot{\beta} = AV_m \cdot \frac{Z^*}{T} \cdot \frac{\sigma_e}{d^3 e_v^a} \tag{1}$$

where the various terms appearing are defined in Table 1. This result applies for volumetric strains (e_v) up to *c.* 15% (Spiers *et al.* 1989), and is valid for both true hydrostatic and one-dimensional (1-D) compaction provided grain boundary sliding (GBS) is relatively easy and that no shape fabric develops. In the case of dense polycrystals (Fig. 1b), the conventional creep rate (axi-symmetric case) is given

$$\dot{\varepsilon} = 5V_m \cdot \frac{Z^*}{T} \cdot \frac{\sigma}{d^3} \tag{2}$$

(refer Table 1) which is directly equivalent to Coble's result for solid state grain boundary diffusional creep with non-dissipative accommodation by GBS (Coble 1963, Raj 1982).

As indicated in Table 1, Z^* represents a phenomenological coefficient defining grain boundary diffusivity in terms of the product

$$Z^* = D \cdot C \cdot S \tag{3}$$

where D and C are the diffusivity and average concentration (solubility) of the dissolved solid in the grain boundary fluid, and S is the average thickness of the grain boundary fluid 'film'. This definition of Z^* applies for both the adsorbed film and island–channel grain boundary models (see Rutter 1983; Spiers *et al.* 1989). However, from the theory of solutions, D and C can be expected to exhibit an Arrhenius dependence on temperature. Thus Z^* can be rewritten

$$Z^* = D_0\, C_0 \exp(-\Delta H/RT) \cdot S \tag{4}$$

where D_0 and C_0 are reference values of D and C in the limit $1/T \rightarrow 0$, ΔH is an activation energy for grain boundary diffusion, and R is the gas constant. Substituting this expression into equations (1) and (2) clarifies the temperature dependence of creep rate predicted by these equations. In general, of course, S may

Table 1. *Definition of terms appearing in equations (1) and (2)*

Symbol	Definition	Units
$\dot{\beta}$	Volumetric strain rate (compaction rate) defined $\dot{\beta} = -\dot{V}/V$ where V represents instantaneous volume.	s^{-1}
A	Grain shape and packing factor divided by gas constant. For spherical grains of uniform diameter, $A \approx 22 \pm 11$ (6-fold coordination $\rightarrow A \approx 33$, 12 fold coordination $\rightarrow A \approx 11$).	mol.K.J.$^{-1}$
V_m	Molar volume of solid phase ($2.693 \times 10^{-5} m^3$).	m^3
Z^*	Phenomenological coefficient representing effective grain boundary diffusivity, defined $Z^* = D.C.S$ (see below and in text).	$m^3 s^{-1}$
σ_e	Applied effective stress (hydrostatic or 1-D; see text).	Pa
T	Absolute temperature.	K
d	Grainsize (diameter).	m
e_v	Volumetric strain defined $e_v = -\Delta V/V_0$ where ΔV is total volume change and V_0 is initial volume.	*
a	Numerical exponent dependent (primarily) on grain shape. $a \approx 2$ for spheres, $a \approx 4$ for cubes.	–
$\dot{\varepsilon}$	Axial strain rate (axi-symmetric deformation).	s^{-1}
σ	Applied differential stress (axi-symmetric deformation).	Pa
D	Diffusivity of dissolved solid in grain boundary fluid.	$m^2 s^{-1}$
C	Average solubility of solid in grain boundary fluid.	mol fraction
S	Average thickness of grain boundary fluid.	m

Symbols listed in order of appearance. Expressions for Z^* and values for A and a taken from Spiers *et al.* (1989) and Spiers & Schutjens (1990). Effective stress refers to applied stress minus pore pressure.

* Dimensionless in formulae, % where indicated.

itself be temperature dependent, as well as being potentially dependent on applied stress or strain (Rutter 1983).

It is now clear that for given values of temperature (T), grainsize (d), volumetric strain (e_v) and applied stress (σ_e, σ), the absolute rate of creep predicted by equations (1) and (2) is determined by Z^*, or more specifically by the constitutive parameters ΔH and $Z_0 = D_0 C_0 S$. Assuming that these parameters are more or less independent of deformation geometry, it is also apparent that the creep rates predicted by the compaction model (eqn. 1) are much faster than those predicted by the conventional creep model (eqn. 2), for all values of e_v for which (1) is valid (i.e. for $e_v < 0.15$). This reflects the influence of the decreased area of grain-to-grain contacts in porous aggregates compared with dense polycrystals (i.e. the influence of increased intergranular stresses and shortened grain boundary diffusion paths). The implication is that pressure solution creep can be very substantially accelerated in the laboratory using compaction rather than conventional creep experiments. When powdered samples are used, compaction experiments possess the additional advantage that grainsize can be accurately controlled. Thus, provided other deformation mechanisms are not activated at grain contacts, compaction experiments offer a powerful method of investigating pressure solution phenomena, and of determining the fundamental parameters ΔH and Z_0 (hence Z^*) for any material exhibiting behaviour consistent with equation (1). The present study is based upon this methodology.

Experiments

We now proceed to report compaction creep experiments performed on brine-saturated salt powder. The aim of these experiments was to test the applicability of the above compaction creep model (i.e. eqn. 1) to wet granular salt and, given good agreement, to evaluate Z^* in terms of ΔH and Z_0. The approach adopted involved 1-D compaction experiments in which the applied stress (i.e. the effective axial stress, σ_e), the sample grainsize (d), and temperature (T) were independently varied from test to test.

Starting material, experimental conditions, apparatus, and data acquisition

Starting material was prepared by sieving analytical grade NaCl powder into controlled grainsize fractions of *c.* 100 μm (98 ± 8 μm), *c.* 200 μm (196 ± 16 μm) and *c.* 275 μm (± 25 μm). Individual samples of these fractions (nominal mass *c.* 115 g) were tested at temperatures in the range 20–90°C, and at applied effective stresses (σ_e) of 0.5–2.2 MPa, using the conventional 1-D compaction apparatus shown in Fig. 2. Load was applied to this apparatus using an INSTRON 1362 servo-controlled testing machine operated with a 10 kN load cell in 'load control' mode. This system allowed applied stresses to be measured and controlled to within 500 Pa. Displacement of the loading ram was measured using the LVDT located in the drive unit of the testing machine. Since the entire system was extremely stiff, this yielded a direct measurement of sample compaction, with a resolution of *c.* 1 μm. In tests performed at elevated temperature, the temperature of the compaction vessel and sample was controlled to within *c.* 1°C by means of the electrically heated oil bath shown in Fig. 2. Sample temperature was measured using a type K thermocouple embedded in the vessel wall. Raw data signals (load, displacement, temperature) were relayed to an analogue chart recorder. The chart records were digitized after completion of the tests, and the data processed pointwise to yield discrete volumetric strain (e_v) and compaction rate ($\dot{\beta}$) data versus time. Strain rates ($\dot{\beta}$) were calculated using the 3-point central difference method.

Testing procedure

In setting up each test, the compaction vessel was first brought to the required test temperature. The loose salt sample was then funnelled into the open vessel, and a light load (< 10 N) was applied by inserting and gently advancing the loading piston (Fig. 2). The sample was then loaded dry at an axial stress of 2.1 MPa for *c.* 15 minutes, venting the pores to atmosphere. In each test, this led to an essentially time-independent (instantaneous) volumetric compaction of 2–3%, producing a well controlled "starting aggregate" with a porosity of 42 ± 1%. The applied axial stress was subsequently reduced to *c.* 0.01 MPa (load ≈ 20 N) and the dry-compacted sample was rapidly flooded with saturated NaCl solution (i.e. pre-saturated at test temperature and at 1 atm. pressure) via the brine inlet shown in Fig. 2. The sample was then reloaded by increasing the applied stress to the desired test value, and the compaction creep behaviour was monitored. Volumetric strains of up to 35% were achieved over periods up to 10 days. In all experiments, the pore brine was maintained at 1 atmosphere pressure by means

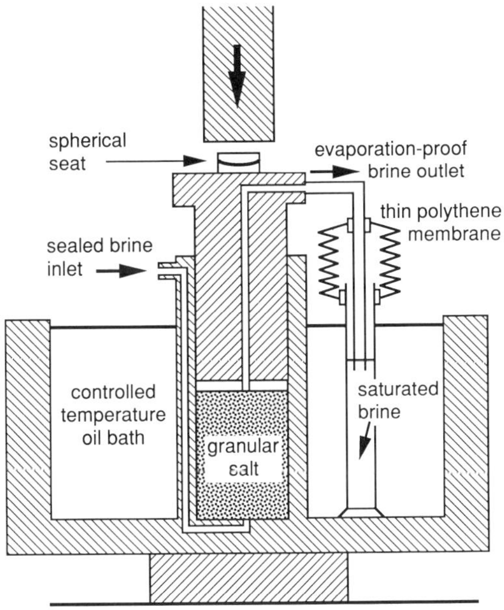

Fig. 2. Semi-schematic diagram illustrating the 1-D compaction apparatus (odometer) used in the present experiments. Piston diameter = 50 mm.

of the evaporation-proof link 'to air' shown in Fig. 2. Tests were terminated by flushing the pore brine out of the sample (in the loaded condition and at test temperature) using compressed air. This was done to minimize corruption of the microstructure by post-test evaporation and precipitation effects. Finally, the sample was unloaded, pressed gently from the compaction vessel, flushed once again (using compressed air and trichlorethane), and resin impregnated to allow sectioning and subsequent microstructural analysis.

Mechanical data

As described above, the dry loading stage of the experiments produced a time-independent (i.e. instantaneous) reduction in the volume of the samples, with little or no on-going compaction creep. In the wet condition, however, all samples exhibited extremely rapid creep. Typical compaction creep curves illustrating the effects of increasing stress, decreasing grainsize, and increasing temperature on the behaviour of wet material are shown in Fig. 3a, b and c respectively. The corresponding numerical data were used to construct plots of compaction rate ($\dot{\beta}$) versus applied stress (σ_e), grainsize (d), and volumetric strain (e_v), as well as an Arrhenius-type diagram in which the quantity $\dot{\beta}T$ is plotted

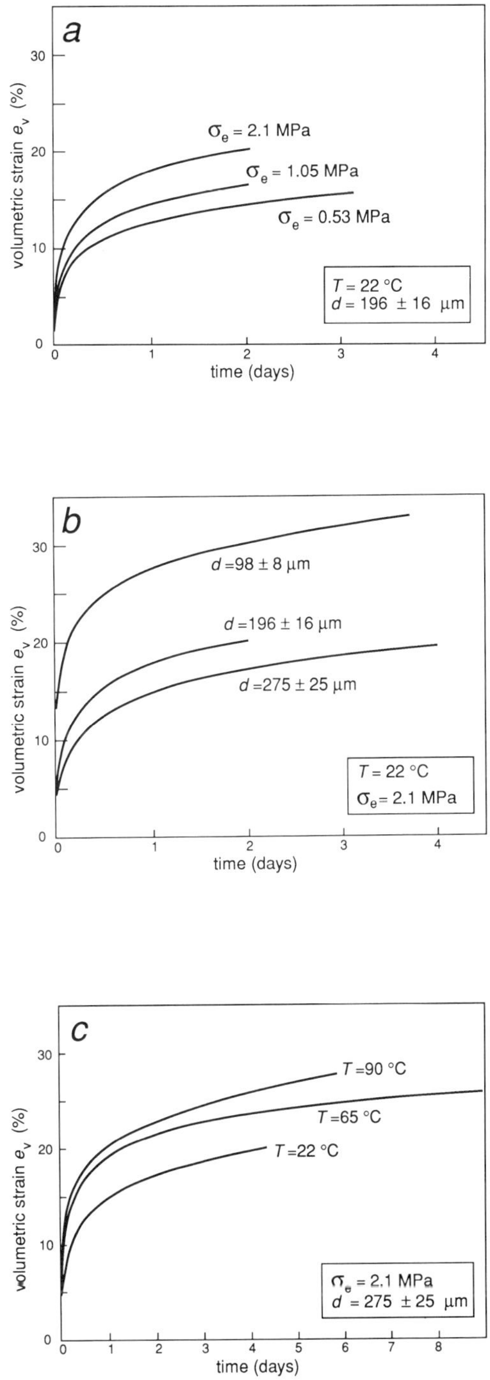

Fig. 3. Typical compaction creep curves obtained for brine-saturated samples. (**a**) Influence of applied effective stress (σ_e) at constant temperature (T) and grainsize (d). (**b**) Influence of grainsize. (**c**) Influence of temperature.

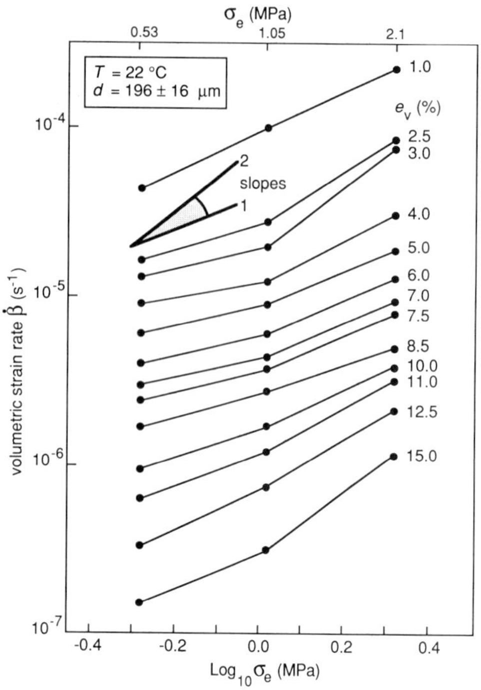

Fig. 4. Log-log plot of volumetric compaction rate ($\dot{\beta}$) v. applied effective stress (σ_e) constructed from the compaction creep data of Fig. 3a for the values of volumetric strain (e_v) shown. Note that the data show a slope of around 1, demonstrating an approximately linear dependence of $\dot{\beta}$ on σ_e.

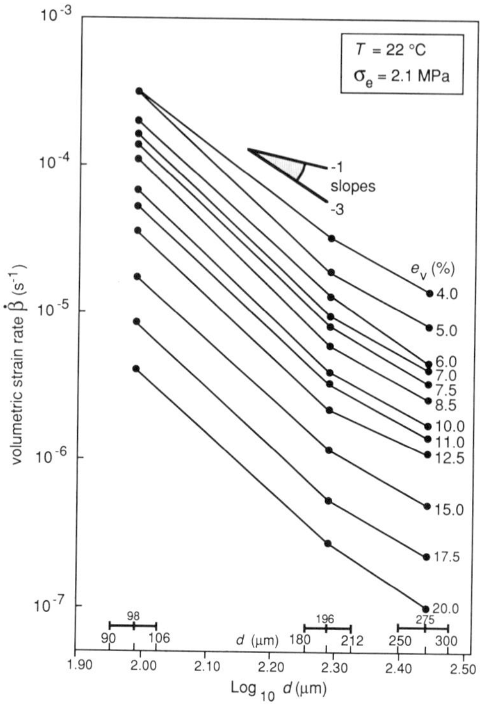

Fig. 5. Log-log plot of compaction rate ($\dot{\beta}$) v. grainsize (d) constructed from the compaction creep data of Fig. 3b for the volumetric strains (e_v) shown. The data show a slope close to -3, demonstrating that $\dot{\beta}$ is approximately proportional to d^{-3}.

against reciprocal temperature. These plots are presented in Figs 4, 5, 6 and 7. Figure 4 shows that for the values of volumetric strain (e_v), grainsize (d) and temperature (T) indicated, $\dot{\beta}$ is more or less linearly related to σ_e (slope of *c.* 1 in log–log- space). In Fig. 5, $\dot{\beta}$ is seen to be approximately proportional to d^{-3}. In Fig. 6, the quantity log ($\dot{\beta}T$) increase 'linearly with $-1/T$ indicating that the dependence of $\dot{\beta}$ upon temperature can be expressed via a relation of the form $\dot{\beta} \propto \exp(-M/T)/T$ where M is approximately constant. Lastly, Fig. 7 shows that when σ_e, d and T are fixed, $\dot{\beta}$ can be viewed as roughly proportional to $1/e_v^n$, with n taking values ranging from around 2 at volumetric strains (e_v) below 10% to 4 or 5 at 20–25%.

With regard to the quality of our data, conventional methods of analysis have shown that the standard relative errors in e_v and $\dot{\beta}$ were less than 0.7% and 4.5% respectively in all experiments. For this reason, no attempt was made to plot error bars in Figs 3–7. Individual creep curves were found to be reproducible to within 5% (see Spiers *et al.* 1989).

Microstructural observations

Microstructural analysis was performed on the sieved salt starting powder, on dry-compacted material (i.e. material produced from the dry loading stage of the tests) and on the wet-compacted samples, using optical microscopy (transmission and reflection methods) and scanning electron microscopy (SEM). The optical work on compacted material was carried out using chemically polished and etched sections prepared in the manner described by Spiers *et al.* (1986).

Sieved salt powder. The various grainsize fractions of this material were found to consist of near-cubic grains of remarkably uniform size and shape. The general nature of the sieved fractions is illustrated in Fig. 8a.

Dry-compacted material. This exhibited a highly porous aggregate structure consisting of a more or less randomly packed array of cubes (Fig. 8b). No clear evidence was found for plastic deformation or for the development of indentation/truncation structures at grain contact points.

Wet-compacted samples. In comparison with the dry-compacted material, these samples showed a marked decrease in porosity, plus a tendency to develop a polygonal texture at high strains. All samples examined showed abundant grain-to-grain indentations, contact truncations,

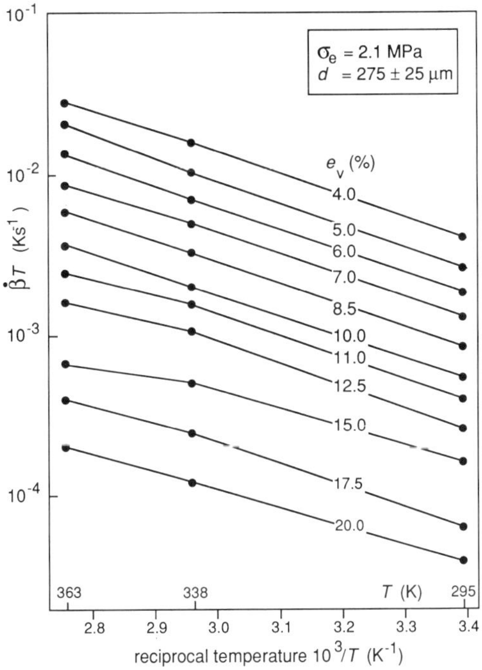

Fig. 6. Arrhenius-type plot of the quantity ($\dot{\beta}T$) v. reciprocal temperature (1000/T) constructed from the compaction creep data of Fig. 3c.

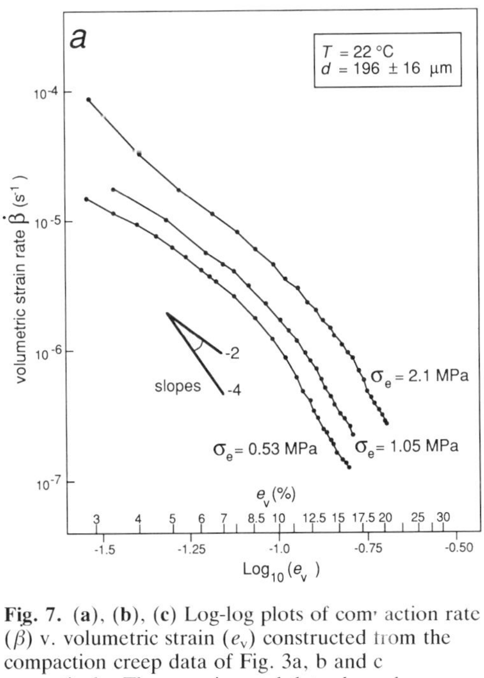

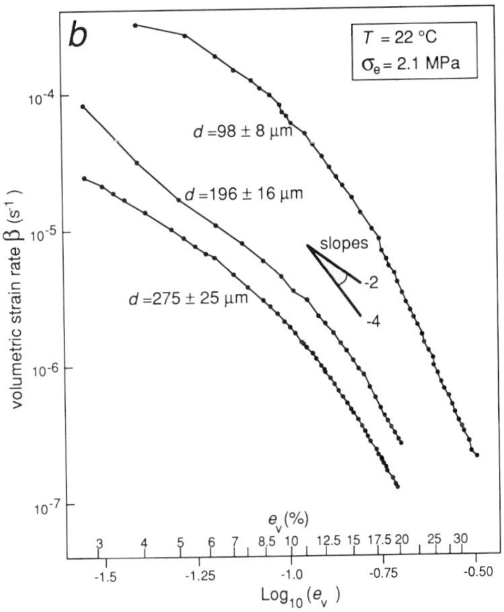

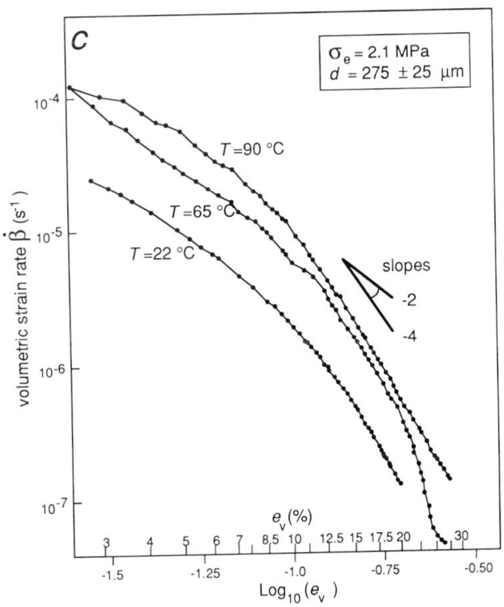

Fig. 7. (**a**), (**b**), (**c**) Log-log plots of compaction rate ($\dot{\beta}$) v. volumetric strain (e_v) constructed from the compaction creep data of Fig. 3a, b and c respectively. The experimental data show slopes between −2 and −5, demonstrating that $\dot{\beta}$ can be considered proportional to $1/e_v^n$ where n = 2 to 5.

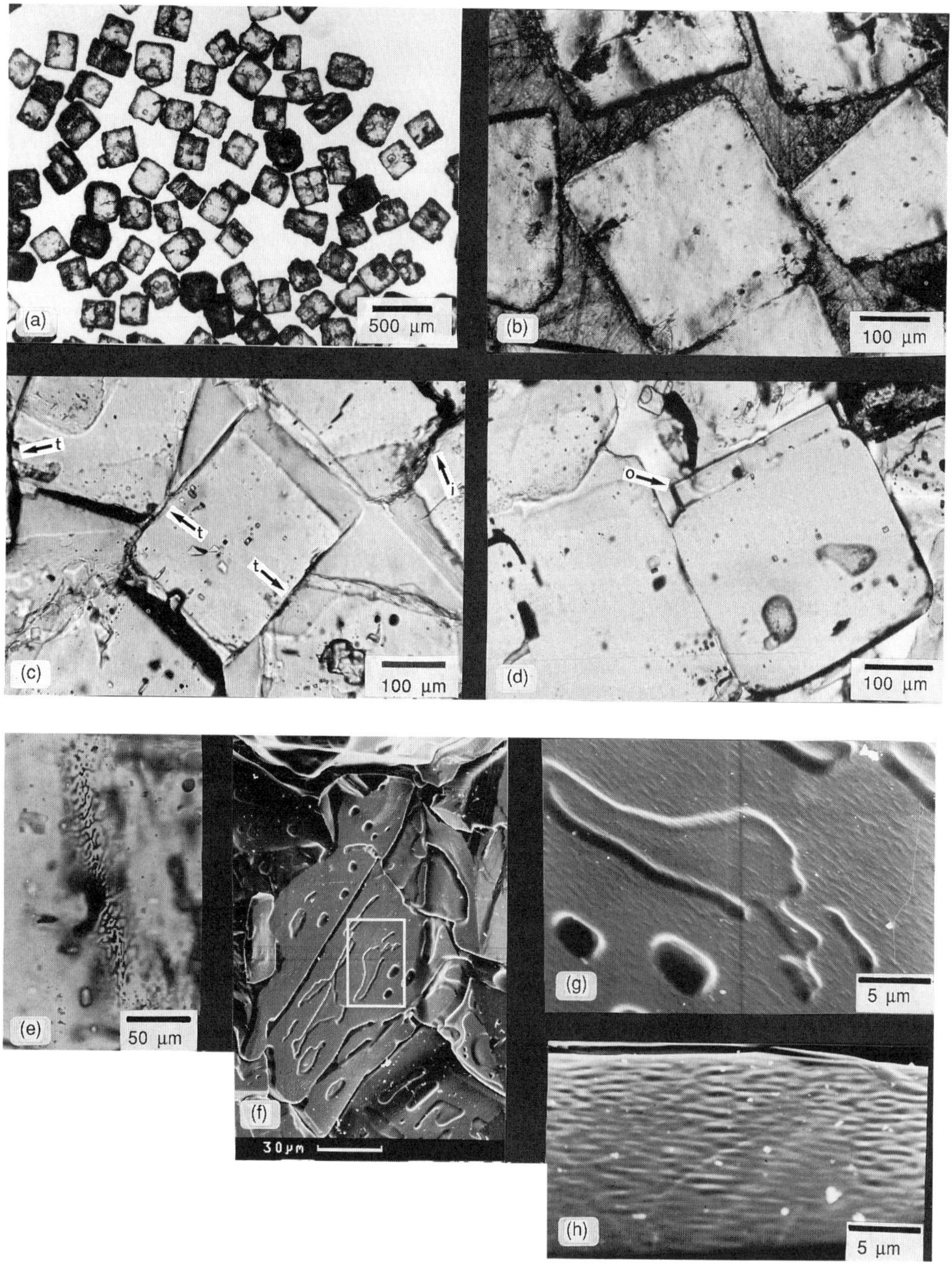

Fig. 8. Microstructures. (**a**) Optical micrograph illustrating the nature of the sieved salt powders used in the present experiments (275 μm fraction shown here). (**b**) Optical micrograph showing the microstructure of dry compacted samples (σ_e = 2.1 MPa). (**c**) & (**d**) Typical indentation, truncation and overgrowth microstructures (labelled i, t, o respectively) seen in wet-compacted samples. (**e**) Grain contact structure in wet-compacted material: transmission optical micrograph of grain contact in oblique view, showing channel-like arrays of (gas-filled) inclusions. (**f**), (**g**) & (**h**) Structure of grain contacts in wet-compacted material: SEM micrographs of mechanically separated grain boundaries showing crystallographically controlled, channel-like inclusions (f, g), plus intervening finer scale structure (g, h). Note that (g) shows detail from (f).

and euhedral overgrowths developed in the pores (see Fig. 8c, d). Little microstructural evidence was found for plastic deformation or microcracking, though material deformed at 90°C did show local, fluid-enhanced grain boundary migration (Urai *et al.* 1986) suggesting that some plastic deformation may have occurred. No clear evidence was found for the development of "necked" grain contacts or near-equilibrium pore configurations of the type expected if dissolution-coupled creep mechanisms (Tada & Siever 1986; Pharr & Ashby 1983) had played an important role. However, examination of thick sections did reveal the presence of channel-like arrays of (gas-filled) fluid inclusions within almost all grain contacts (Fig. 8e). These channels were occasionally interconnected (see Fig. 8e, centre). Most, however, showed a strong tendency to be crystallographically controlled, and consisted of apparently isolated inclusions of tubular or negative crystal form. SEM work on disaggregated samples confirmed the presence of these structures on parted grain contacts (Fig. 8f, g), but also revealed a range of finer, island–channel and terrace-like structures in the intervening regions (Fig. 8g, h). Typically around 80–90% of the area of individual grain contacts possessed the fine-scale structure shown in Fig. 8h. From examination of numerous micrographs, the average relief of this structure was estimated to be around 200 nm. The average thickness of grain contact zones as a whole (i.e. including larger inclusions) was estimated to be *c.* 500 nm.

Discussion

Experimental results v. theory

The present experiments have demonstrated that compaction creep of granular salt is dramatically enhanced by the addition of saturated brine. The data reported show that for values of e_v in the range $3 \leq e_v \leq 20\%$, the compaction creep behaviour of our wet samples can be described by an empirical constitutive relation of the form

$$\dot{\beta} = B \cdot \frac{\exp(-M/T)}{T} \cdot \frac{\sigma_e^k}{d^m\, e_v^n} \quad (5)$$

where B and M are approximately constant, $k \approx 1$, $m \approx 3$, and n takes values ranging from *c.* 2, at volumetric strains (e_v) of 3–10%, to *c.* 4 or 5 when e_v = 15–20%. This agrees closely with the theoretical relation for compaction creep by diffusion controlled pressure solution given by equations (1) and (4), rewritten here in the form

$$\dot{\beta} = A V_m \cdot \frac{Z_0 \exp(-\Delta H/RT)}{T} \cdot \frac{\sigma_e}{d^3\, e_v^a} \quad (6)$$

where $Z_0 \doteq D_0 C_0 S$ and $2 \leq a \leq 4$.

In addition, the widespread grain-to-grain indentation and truncation microstructures, coupled with the observed overgrowths, provide strong evidence that deformation of the wet samples occurred largely by dissolution of material at contacts, diffusion through the grain boundary region, and precipitation within the pores. The observed grain boundary structure suggests that grain contacts contained brine in some kind of non-equilibrium island–channel network during compaction. It is important to recognize, however, that the extent to which the contact structure may have become modified on or after unloading is unknown. In recent in-situ work on salt–salt contacts loaded under brine, we have observed the formation of a contact structure similar to the finer structure reported here (e.g. Fig. 8h). However, the possible existence of some other structure or continuous film during the present tests cannot be ruled-out. The in-situ observation of Hickman & Evans (1988) that salt–salt contacts stressed under brine assume a smooth structure free of liquid, is considered unlikely to be relevant for our material, since it appears to relate to an equilibrium state certainly not achieved in our tests.

Further to the above, the lack of microstructural evidence for plastic deformation, microcracking, and dissolution-coupled creep or contact margin dissolution effects indicates that these were or little or no importance in our wet tests. Pure particulate flow is thought to have been unimportant since all samples were dry-compacted into an essentially 'locked' state at 2.1 MPa prior to wet testing (all wet tests were performed at stresses less than or equal to this value). The implication that solution transfer was indeed the dominant mechanism of deformation is strongly supported by rough estimates of sample strain made from measurements of the amount of material removed at truncated/indented grain contacts. These estimates fall within ± 20% of the strain imposed during deformation.

On the basis of these considerations, we infer that compaction of our wet samples occurred by diffusion-controlled pressure solution creep as described by equations (1) plus (4), or (6). However, the nature of the grain contact structure during deformation is not yet clear. Nonetheless, since the experimentally observed constitutive behaviour (see eqn. 5) more or less exactly matches the theoretical model as given by (6), it would appear that Z^*, or ΔH and $Z_0 = D_0C_0S$, are largely independent of σ_e, T and e_v under the (limited) conditions investigated.

Determination of Z_0, ΔH and Z^*

Having demonstrated close agreement between experiment and theory, it is now possible to evaluate the constitutive parameters $Z_0 = D_0C_0S$ and ΔH, and hence Z^*. This has been done by fitting our experimental data ($\dot{\beta}$ v. σ_e, T, d, e_v) to equation (6) using the non-linear regression method of Marquardt (1963), treating Z_0 and ΔH as fitting parameters. Fitting was carried out for the region $e_v \leq 10\%$ (i.e. for the region in which $\dot{\beta}$ was found to be roughly proportional to $1/e_v^2$; see Fig. 7), using $a = 2$ in equation (6) and using $A = 22 \pm 11$ — see Table 1. The results obtained are presented in full in Table 2. Neglecting standard errors, however, they can be summarized in the form

$$Z_0 = D_0C_0S = (2.79 \pm 1.40).\ 10^{-15}\ \mathrm{m^3s^{-1}} \quad (7)$$

$$\Delta H = 24.53\ \mathrm{kJ\ mol^{-1}} \quad (8)$$

where the upper and lower bounds in (7) reflect the extremum uncertainties in A. Substitution of these expressions for Z_0 and ΔH into (4) then yields

$$Z^* = (2.79 \pm 1.40)\ .\ 10^{-15}\ .\ \exp(-24530/RT)\ \mathrm{m^3s^{-1}} \quad (9)$$

The numerical values of Z^* given by this relation are 2–3 orders of magnitude lower than those calculated from equation (3) assuming values of D and C which correspond to bulk NaCl solution (Kestin *et al.* 1981; Spiers *et al.* 1986, p. 17) and assuming an island-channel grain boundary structure with S in the range 200–500 nm (as suggested by microstructural observations). At the same time, the values of Z^* given by (9) are a *minimum* of 2–3 orders of magnitude higher than predicted from (3) for a truly adsorbed grain boundary film of thickness $\leq$ 3 nm and correspondingly reduced diffusivity (10^{-5} × bulk solution value, or lower); see Rutter (1976, 1983), Tada *et al.* (1987). Further, the value of ΔH obtained above is closely similar to the figure of 19.5 kJ mol^{-1} calculated for diffusion of NaCl in *bulk* solution (using the data of Kestin *et al.* 1981; following Spiers *et al.* 1986, p. 16–17). It seems unlikely that the activation energy for diffusion in a strongly adsorbed film would be so closely comparable with the bulk solution value. Combining all of this information with our microstructural observations, we infer that during deformation, grain boundaries probably contained more or less true liquid in some kind of (mostly) fine scale island-channel form. In the framework of this interpretation, the rather low values of Z^* given by (9) are thought to imply either that the product DC is decreased in the grain boundary liquid, due to relatively minor thin film effects, or that the value of S *during* creep is substantially lower than estimated from the grain boundary structure observed after deformation (i.e < 200 nm). These two effects could of course be coupled. Although highly unlikely in our view, there is insufficient data on the properties of brine films (refer Tada *et al.* 1987) to exclude the possibility of a continuous 'thick' film (3 < S < 100 nm) of weakly 'bound' fluid possessing near-bulk diffusion properties but able to support sufficient shear stress not to be squeezed out of grain boundaries.

Creep law for dense salt (rock)

Development

We now make use of the expression for Z^* obtained above to develop a preliminary constitutive law for pressure solution creep of dense salt. This is done by substituting for Z^* from equation (9) into (2). This step involves the implicit assumption that Z_0 and ΔH (hence Z^*)

Table 2. *Results obtained for Z_0 and ΔH by fitting the present experimental data to equation (6) using A = 22 ± 11 (i.e. A = 11, 22, 33). Fitting was carried out for the region $e_v \leq 10\%$ with a = 2. ΔH was determined using the data of Fig. 6 only. Z_0 was determined using all data, taking ΔH = 24.529 kJ.mol^{-1}*

A mol.K.J^{-1}	Z_0 (± standard error) m^{3}s^{-1}	ΔH (± standard error) kJ.mol^{-1}
11	$(4.1893 \pm 0.2059) \times 10^{-15}$	24.529 ± 0.977
22	$(2.0947 \pm 0.1030) \times 10^{-15}$	24.529 ± 0.977
33	$(1.3964 \pm 0.0687) \times 10^{-15}$	24.529 ± 0.977

take the same values in conventional creep of dense salt as in compaction creep of brine-saturated granular salt. Provided that sufficient brine is present in dense material to fill grain boundaries to the average thickness S active in our compaction tests, this assumption seems reasonable, since our experimental data indicate that $Z_0(= D_0C_0S)$ and ΔH are *not* strongly dependent upon σ_e and e_v (hence porosity ϕ). Excess brine can be expected to be squeezed into triple junctions and other sinks. The result obtained for the conventional creep rate is

$$\dot{\varepsilon} = (13.95 \pm 7.00)\ V_m \cdot 10^{-15} \cdot \frac{\exp(-24530/RT)}{T} \cdot \frac{\sigma}{d^3} \qquad (10)$$

In view of the above-mentioned assumption, it is desirable to compare the above expression for $\dot{\varepsilon}$ with the (limited) data previously presented by Spiers *et al.* (1986) and by Urai *et al.* (1986) for pressure solution creep in dense polycrystalline salt (*c.* 1.0% brine). This is done in Fig. 9. From this it is clear that within the specified upper and lower bounds on $\dot{\varepsilon}$, the present constitutive equation is closely consistent with the previous data for dense material. This strongly supports the use of equation (10) as a preliminary constitutive law for pressure solution creep in dense salt or salt rock.

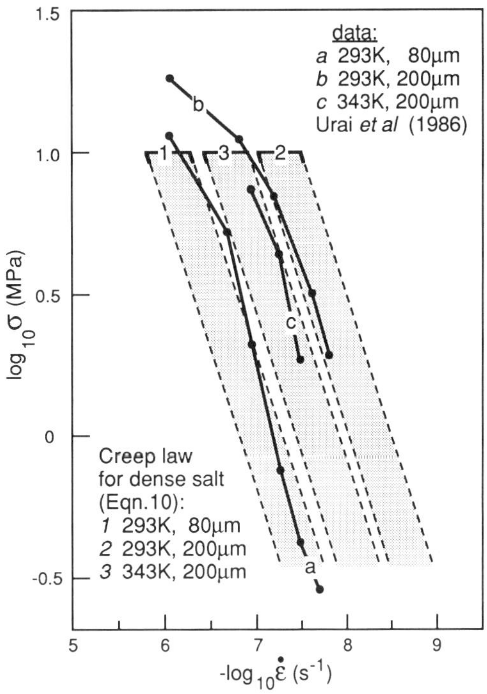

Fig. 9. Log(σ) v. log($\dot{\varepsilon}$) diagram comparing the pressure solution creep law for dense salt developed here with previous experimental data (Spiers *et al.* 1986; Urai *et al.* 1986) for pressure solution creep in dense salt. The previous data sets (a, b and c) represent $\sigma/\dot{\varepsilon}$-stepping data obtained at the temperatures (T) indicated, using samples of grainsize $d \approx 80\ \mu m$ (a) and $d \approx 200\ \mu m$ (b, c). For corresponding values of T and d, the present creep law (eqn. 10) predicts the $\sigma-\dot{\varepsilon}$ behaviour represented by the stippled bands labelled 1, 2 and 3. N. B. σ is differential stress, $\dot{\varepsilon}$ is axial strain rate.

Application

Finally, we combine equation (10) with constitutive laws for other mechanisms to construct a deformation map for natural (brine-bearing) rocksalt of grainsize 10 mm. Three mechanisms are considered, namely dislocation glide (following Skrotzki & Haasen 1988), cross-slip controlled creep (using the cross-slip laws given by Wawersik & Zeuch 1986, for the Asse and West Hackberry salts), and pressure solution. The cross-slip based laws of Wawersik & Zeuch (1986) were chosen in preference to the widely used power law creep model of Carter & Hansen (1983), since most recent evidence points to cross-slip being the true rate controlling mechanism in the low temperature dislocation creep field (see Skrotzki & Haasen 1988; Wawersik & Zeuch 1986). Note, however, that at temperatures < 200°C, the cross-slip law for W. Hackberry salt predicts $\sigma - \dot{\varepsilon}$ behaviour which is closely similar to Carter & Hansen's power law. The deformation map obtained is presented in Fig. 10, along with various constraints on the stresses and strain rates associated with natural salt flow. From this diagram, and taking into account the variability of natural salt (in grainsize and dislocation creep parameters), it is apparent that deformation under natural conditions will tend to occur in the transition zone between dislocation controlled creep and pressure solution. This is consistent with the observation made in the introduction that natural salts seem to show microstructural evidence both for solution transfer and intracrystalline creep phenomena (see Urai *et al.* 1986, 1987). In particular, the Asse rocksalt (FRG) shows overgrowth features indicative of fluid-enhanced recrystallization and possibly pressure solution, as well as dislocation substructures (Urai *et al.* 1987).

Microstructural confirmation is needed from a wider variety of naturally deformed salts. However, viewed overall it seems likely that natural (halokinetic) flow of salt will often occur

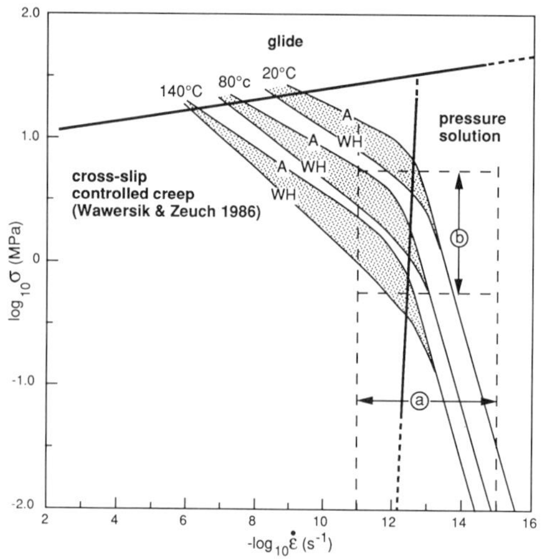

Fig. 10. Deformation map for rocksalt (grainsize d = 10 mm) incorporating the pressure solution creep law (eqn. 10) for dense salt developed in the present study. The map was constructed using the lower bound form of this law (refer eqn. 10). Dislocation creep behaviour is represented using the cross-slip controlled creep models given by Wawersik & Zeuch (1986) for the Asse (A) and West Hackberry (WH) salts. The map indicates that at the strain rates ⓐ, stresses ⓑ, and typical temperatures (50–150°C) thought to be associated with the flow of salt in nature (see Carter & Hansen 1983), deformation will occur in the transition zone between cross-slip controlled creep and pressure solution. *N.B.* Range ⓐ denotes maximum and minimum strain rates estimated from salt mine closure rates and mean salt dome growth rates (Carter & Hansen 1983). Range ⓑ denotes flow stresses inferred from subgrain sizes in naturally deformed salts (Carter & Hansen 1983).

in the cross-slip/pressure-solution transition, probably with accompanying fluid-enhanced dynamic recrystallization (Urai *et al.* 1986, 1987). This latter process may reduce the creep strength of rocksalt in the dislocation dominated regime (cf. Urai 1983). However, on the basis of data reported by Spiers *et al.* (1989), it is not expected significantly to modify Fig. 10. With regard to waste disposal and salt mining, Fig. 10 indicates that pressure solution will be important only for extremely long-term considerations. In this long-term engineering context, it is important to note that solution transfer effects seem to be inhibited in salt which has undergone dilatancy around gallery walls (see Spiers *et al.* 1986, 1989).

Summary and conclusions

(1) Theoretically-based constitutive models for compaction creep of porous polycrystals, and for conventional creep of dense polycrystals, by grain boundary diffusion controlled pressure solution have been examined. These take the form $\dot{\beta} \approx (22 \pm 11)\, V_m Z^* \sigma_e / Td^3 e_v^a$ (when $e_v \leq 0.15$ — eqn. 1) and $\dot{\varepsilon} \approx 5V_m Z^* \sigma / Td^3$ (eqn. 2) respectively. Attention has been drawn to the fact that in these models, the absolute rate of creep is determined by the *effective* grain boundary diffusivity $Z^* = Z_0 \exp(-\Delta H/RT)$.

(2) Compaction creep experiments have been performed on brine-saturated NaCl powder. The mechanical data obtained agree closely with equation 1, and all samples showed well-developed pressure solution microstructures. It is concluded that creep must indeed have occurred by diffusion controlled pressure solution as described by equation (1). Both microstructural and mechanical data suggest that during deformation, grain boundaries possessed a non-equilibrium, island-channel structure containing brine in liquid or near-liquid form (mean thickness ≤ *c.* 500 nm).

(3) Having obtained good agreement between experiment and theory, the compaction creep data were fitted to equation (1) using the full Arrhenius form for Z^*. This yielded $Z_0 = (2.79 \pm 1.40) . 10^{-15}\ m^3s^{-1}$, $\Delta H = 24.53$ kJ/mol, and $Z^* = (2.79 \pm 1.40) . 10^{-15} \exp(-24530/RT)\ m^3s^{-1}$.

(4) Substitution of this expression for Z^* into equations (1) and (2) provides quantitative constitutive laws for both compaction creep of granular salt and conventional creep in dense salt. The law for dense salt is preliminary but agrees closely with previous data.

(5) Incorporation of the creep law for dense salt into a deformation map suggests that deformation of rocksalt in nature will often occur in the transition zone between 'dislocation creep' (plus fluid-enhanced recrystallization) and pressure solution.

This research was supported by the Netherlands Ministry of Economic Affairs (OPLA Project REO-1) and by the Commission of the European Communities (Contract FI1W-0051-NL) in the framework of their respective research programmes on management and storage of radioactive waste. B. K. Smith is thanked for discussions. These were made possible by NATO grant no. AG.86/0148. D. L. Olgaard is thanked for carefully reviewing the manuscript. The authors are also grateful to G. J. Kastelein and E. de Graaff for technical support, and to M. A. T. Mathot-Martens for processing the text.

References

ALBRECHT H. & HUNSCHE, U. 1980. Gebirgsmechanische Aspekte bei der Endlagerung radioaktiver Abfälle in Salzdiapiren unter besonderer Berücksichtigung des Fliessverhaltens von Steinsalz. *Fortschritte der Mineralogie*, **58**, 212–247.

CARTER, N. L. & HANSEN, F. D. 1983. Creep of rocksalt. *Tectonophysics*, **92**, 275–333.

COBLE, R. L. 1963. A model for boundary diffusion controlled creep in polycrystalline materials. *Journal of Applied Physics*, **34**, 1679–1682.

GREEN, H. W. 1980. On the thermodynamics of non hydrostatically stressed solids. *Philosophical Magazine*, **A41**, 637–647.

—— 1984. "Pressure-solution" creep: some causes and mechanisms. *Journal of Geophysical Research*, **89**, 4313–4318.

HEARD, H. C. 1972. Steady-state flow in polycrystalline halite at pressure of 2 kilobars. *In*: HEARD, H. C., BORG, I. Y., CARTER, N. L. & RALEIGH, C. B. (eds) *Flow and Fracture of Rocks*. American Geophysical Union, Geophysical Monograph, **16**, 191–210.

HEIDUG, W. & LEHNER, F. K. 1985. Thermodynamics of coherent phase transformations in nonhydrostatically stressed solids. *Pure and Applied Geophysics*, **123**, 91–98.

HICKMAN, S. H. & EVANS, B. 1988. Influence of normal stress and contact radius on pressure solution along halite silica contacts. *EOS Transactions, American Geophysical Union*, **69**, No. 44, 1426.

KESTIN, J., KHALEFA, H. E. & CORREIA, R. J. 1981. Tables of the dynamic and kinematic viscosity of aqueous NaCl solutions in the temperature range 20–150°C and the pressure range 0.1–35 MPa. *Journal of Physical and Chemical Reference Data*, **10**, 71–86.

LANGER, H. & OFFERMAN, H. 1982. On the solubility of sodium chloride in water. *Journal of Crystal Growth*, **60**, 389–392.

LEHNER, F. K. 1990. Thermodynamics of rock deformation by pressure solution. *In*: BARBER, D. J. & MEREDITH, P. G. (eds) *Deformation Processes in Minerals, Ceramics and Rocks*. Unwin Hyman London, 296–333.

—— & BATAILLE, J. 1984/5. Non equilibrium thermodynamics of pressure solution. *Pure and Applied Geophysics*, **122**, 53–85.

MARQUARDT, D. W. 1963. An algorithm for least squares estimation of non-linear parameters. *Journal of the Society of Industrial and Applied Mathematics*, **11**, 431–441.

PATERSON, M. S. 1973. Non-hydrostatic thermodynamics and its geologic applications. *Reviews of Geophysics and Space Physics*, **11**, 355–389.

PHARR, G. M. & ASHBY, M. F. 1983. On creep enhanced by a liquid phase. *Acta Metallurgica*, **31**, 129–138.

RAJ, R. 1982. Creep in polycrystalline aggregates by matter transport through a liquid phase. *Journal of Geophysical Research*, **87**, 4731–4739.

ROBIN, P.-Y. F. 1978. Pressure-solution at grain-to-grain contacts. *Geochimica et Cosmochimica Acta*, **42**, 1383–1389.

RUTTER, E. H. 1976. The kinetics of rock deformation by pressure-solution. *Philosophical Transactions of the Royal Society of London*, **A283**, 203–219.

—— 1983. Pressure solution in nature, theory and experiment. *Journal of the Geological Society, London*, **140**, 725–740.

SKROTZKI, W. & HAASEN, P. 1988. The role of cross slip in the steady state creep of salt. *In*: HARDY, H. R. & LANGER, M. (eds) *The Mechanical Behaviour of Salt: Proceedings of the Second Conference*. Trans Tech Publications, Clausthal-Zellerfeld, 69–81.

SPIERS, C. J. & SCHUTJENS, P. M. T. M. 1990. Densification of crystalline aggregates by fluid phase diffusional creep. *In*: BARBER, D. J. & MEREDITH, P. G. (eds) *Deformation Processes in Minerals, Ceramics and Rocks*. Unwin Hyman, 334–353.

——, URAI, J. L. & LISTER, G. S. 1988. The effect of brine (inherent or added) on rheology and deformation mechanisms in salt rock. *In*: HARDY, H. R. & LANGER, M. (eds) *The Mechanical Behaviour of Salt: Proceedings of the Second Conference*. Trans Tech Publications, Clausthal-Zellerfeld, 89–102.

——, ——, ——, BOLAND, J. N. & ZWART, H. J. 1986. The influence and fluid-rock interaction on the rheology of salt rock. *Nuclear Science and Technology*, **EUR 10399 EN**. Office for Official Publications of the European Communities, Luxembourg.

——, Peach, C. J., BRZESOWSKY, R. H., SCHUTJENS, P. M. T. M., LIEZENBERG, J. L. & ZWART, H. J. 1989. Long-term rheological and transport properties of dry and wet salt rocks. *Nuclear Science and Technology*, **EUR 11848 EN**. Office for Official Publications of the European Communities, Luxembourg.

STOCKER, R. L. & ASHBY, M. F. 1973. On the rheology of the upper mantle. *Reviews of Geophysics and Space Physics*, **11**, 391–426.

TADA, R. & SIEVER, R. 1986. Experimental knife-edge pressure-solution of halite. *Geochimica et Cosmochimica Acta*, **50**, 29–36.

——, MALIVA, R. & SIEVER, R. 1987. A new mechanism for pressure-solution in porous quartzose sandstone. *Geochimica et Cosmochimica Acta*, **51**, 2295–2301.

URAI, J. L. 1983. Water-assisted dynamic recrystallization and weakening in polycrystalline bischofite. *Tectonophysics*, **96**, 125–157.

——, SPIERS, C. J., ZWART, H. J. & LISTER, G. S. 1986. Water weakening effects in rock salt during long term creep. *Nature*, **324**, 554–557.

——, ——, PEACH, C. J., FRANSSEN, R. C. M. W. & LIEZENBERG, J. L. 1987. Deformation mechanism operating in naturally deformed halite rocks as deduced from microstructural observations. *Geologie en Mijnbouw*, **66**, 165–176.

WAWERSIK, W. R. & ZEUCH, D. H. 1986. Modelling and mechanistic interpretation of creep of rock salt below 200°C *Tectonophysics*, **121**, 125–152.

Phenomenological superplasticity in rocks

JANE A. GILOTTI[1,2] & JOSEPH M. HULL[1]

[1] *Geology Institute, University of Uppsala, Box 555, S-751 22 Uppsala, Sweden*

[2] *Present address: Geological Survey, New York State Museum, The State Education Department, Albany, NY 12230, USA*

Abstract: Superplastic deformation of rocks can be defined as: homogeneous deformation without loss of continuity to very large strains. This phenomenological definition differs from current geological usage that equates superplasticity with either grain boundary sliding, a deformation mechanism, or with a specific set of material parameters. Two examples of natural, phenomenological superplasticity in rocks are presented. A diabase dyke deformed continuously in progressive simple shear to shear strains of at least γ = 100, despite the low grade of metamorphism and relatively coarse grain size of the resultant tectonite. Very coarse-grained pegmatite dykes show continuous deformation to high strains under predominantly pure flattening; intracrystalline plasticity was an important micromechanism of flow. A much wider range of material and mechanical parameters and physical conditions give rise to phenomenological superplasticity in rocks than predicted by experiments on metals. By inference and example, more deformation mechanisms will be associated with superplasticity in nature. The profound differences between the metallurgical experiments and the geological environment prohibit simple rheological interpretations of natural superplastic behaviour.

The phenomenon of superplasticity was clearly demonstrated by C. E. Pearson (1934), a metallurgist, and is defined in the metallurgical literature as the ability of a material to deform to very high strains, without failure, in tensile experiments (Langdon 1982b). The characteristics of superplasticity in metals and metallic alloys have been summarized in recent review papers (Edington 1982; Ghosh 1982; Ghosh & Hamilton 1982; Gifkins 1982; Langdon 1982*b*). The typical requirements for superplasticity in these materials are: a low flow stress; a high strain rate sensitivity *m* to the flow stress, where $m = \partial \log \sigma / \partial \log \dot{\varepsilon}$; high homologous temperatures ($T_H \geq 0.5$); and a fine grain-size, typically less than 10 μm, where grain growth is suppressed. The microstructure usually consists of a stable network of equiaxed grains, and weak crystallographic preferred orientations are common. Grain boundary sliding contributes between *c*. 50–70% of the total strain in superplastic metals and alloys (Vastava & Langdon 1979; see also Shariat *et al.* 1982) and as such is the dominant deformation mechanism in superplasticity. Much of the current research on superplastic metals and metallic alloys is devoted to determining the physical conditions, material parameters, and micromechanisms of flow that produce strongly enhanced ductility in the laboratory.

Many rocks, under a variety of geological conditions, exhibit ductile behavior. Rutter (1986) has emphasized that the concept of ductility refers to homogeneous, continuous deformation, and not a specific deformation micromechanism, yet many geologists incorrectly equate ductile strain with intracrystalline plasticity. As Rutter points out, *ductility* and *crystal plasticity* are fundamentally different terms which describe macroscopic, mechanical behaviour and a micromechanism of flow, respectively. Indeed, a variety of grain scale deformation mechanisms, including cataclasis (e.g. Paterson 1958), can give rise to ductile flow.

The term *superplasticity* has met a similar fate in the geological literature, where it is used in a variety of inconsistent ways. Some authors treat superplasticity as a synonym for grain boundary sliding dominated flow (e.g. Behrmann 1985). The use of *superplasticity* and *grain boundary sliding* as synonyms should be avoided, for many of the same reasons discussed by Rutter (1986). First, these terms represent fundamentally different entities, i.e. a mechanical phenomenon versus a grain scale deformation mechanism. Second, a variety of deformation mechanisms are associated with the phenomenon of superplasticity in metals and alloys (Edington 1982; Gifkins 1982). Indeed, grain boundary sliding must be accommodated by other micromechanisms, such as crystal plas-

From Knipe, R. J. & Rutter, E. H. (eds), 1990, *Deformation Mechanisms, Rheology and Tectonics*, Geological Society Special Publication No. 54, pp. 229–240.

ticity or diffusional mass transfer, to maintain compatibility between sliding grains (Arieli & Mukherjee 1982). Finally, while grain boundary sliding is certainly the dominant mechanism in the superplastic metals studied so far, the physical conditions, material parameters, and testing configurations of the engineering laboratory are extremely narrow compared to the wide range of conditions found in nature.

Other geologists treat rock superplasticity in a less restricted sense as a set of material parameters and physical conditions that are characteristic of superplastic metals (Boullier & Guegen 1975; Schmid *et al.* 1977; Schmid 1982). For example, Schmid (1982) describes superplasticity in rocks as characterized by a stable, equiaxed microstructure; a low stress exponent in the flow law; and grain boundary sliding. However, there is no *a priori* reason to assume that the engineering laboratory has successfully duplicated the variety of natural deformations. We believe that geologists should test this supposition; that is, identify the phenomenon of superplasticity in rocks, then investigate its cause.

There has been very little discussion of superplasticity as a macroscopic phenomenon in the geological literature (but see Boullier & Guegen 1975; Schmid *et al.* 1977; Schmid 1982; Poirier 1985). In this paper, we discuss the phenomenological definition of superplasticity in terms of natural deformation, and extend the definition somewhat to reflect the more general conditions encountered in nature. We then present two cases of natural phenomenological superplasticity and attempt to describe the physical conditions, material parameters, and deformation micromechanisms that led to superplastic flow. Finally, we consider some of the parameters that may influence superplastic deformation in rocks.

Superplasticity in rocks

Superplastic deformation of rocks in nature can be defined as continuous, homogenous deformation to very large strains. This *phenomenological* definition of superplasticity differs from much of the current geological usage that equates superplasticity with grain boundary sliding, or with a set of characteristics derived from experimental deformation of metals. The proposed rock definition also differs from the metallurgical one in that strain paths other than pure extension are included. As others have noted (Schmid *et al.*, 1977; Schmid 1982; Poirier 1985), constriction is not a common strain state in nature nor is it the common mode of most rock deformation experiments. Superductility is synonomous with superplasticity, as defined above, and would be a preferable substitute were it not for the confusion surrounding the term ductile (Rutter 1986).

The concept of continous deformation is fundamental to the phenomenological definition of superplasticity. Structures that produce a loss of continuity, such as through-going faults, shear bands, or boudinage, must be suppressed or inhibited to achieve superplastic deformation. Recognition of superplasticity requires the identification of strain markers whose original shape or form is known. As with other terms involving homogeneity or pervasiveness, phenomenological superplasticity is dependent on the scale of observation. Schmid (1982) has stated that 'any highly strained rock, such as a mylonite' could be considered superplastic in the phenomenological sense; however, this generalization does not incorporate the concept of continuity. For example, S-C mylonites (Lister & Snoke 1984) may contain discrete shear bands which could lead to discontinuous deformation of a particular marker.

The amount of strain attained before loss of continuity ('very large') is purposely vague in our definition of natural superplasticity. Although no precise value is specified for superplastic metals, the minimum tensile strains usually cited by materials scientists are approximately 500%, corresponding to a natural logarithmic strain of $\varepsilon \approx 1.8$. This strain is rather modest in comparison to those commonly measured in metamorphic terrains, and illustrates again the profound differences between the laboratory and nature.

Some Earth scientists may object to a phenomenological definition of superplasticity on the grounds that it dilutes the meaning of the term by including a broader spectrum of geological situations. We feel that the suggested approach will simply focus attention on first recognizing continuous, very high strain deformations in nature, and then the parameters and mechanisms that produced superplasticity in these rocks. Studies of natural, phenomenological superplasticity should be complemented by future laboratory investigations of rocks to much higher strains than currently realized.

Examples

Diabase dykes and arkoses deformed in simple shear

At the base of the Särv thrust sheet (Caledonides, Sweden), diabase dykes intruded into arkoses are rotated and thinned in a nar-

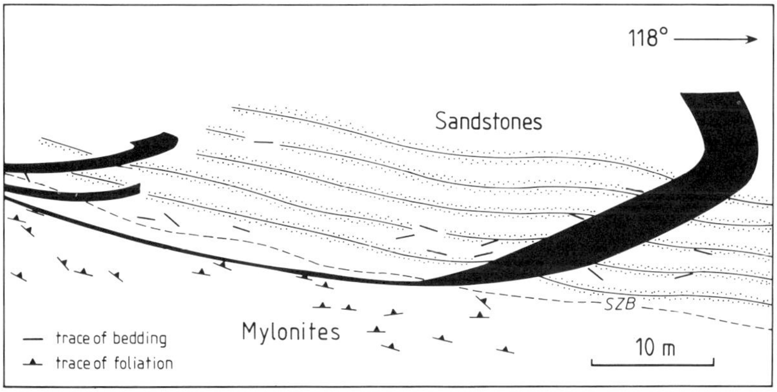

Fig. 1. Strain plane projection (XZ) of a diabase dyke deformed in simple shear from the base of the Särv thrust sheet, Sweden. The upper shear zone boundary (SZB) marks an abrupt change from macroscopically unfoliated rocks above to strongly foliated mylonites beneath. The dike thins continuously from 3.75 m thick above the SZB to 7 cm thick at the last exposure on the NW end of the outcrop, recording a shear strain of $\gamma = 103$ (see Gilotti 1989).

row zone of dominantly simple shear (Gilotti & Kumpulainen 1986; Gilotti 1989). The shear zone consists of interbanded quartzofeldspathic and basaltic mylonites derived from the arkoses and diabase dykes of the thrust sheet. Figure 1 illustrates one such dyke that can be followed for 70 m from the undeformed state in the hangingwall into the basal Särv mylonite zone. This dyke rotates through a minimum angle of 32° into parallelism with the upper shear zone boundary, and thins continuously from an initial thickness of 3.75 m to 0.07 m where it is last seen exposed. The thinnest part of the dyke records a shear strain of $\gamma = 103$, which corresponds to an elongation of 5412% or a natural strain of $\varepsilon = 4$. The mylonites contain numerous centimetre thick metadiabase bands which indicate shear strains in excess of $\gamma = 1000$ ($\varepsilon = 6.6$), if an initial geometry similar to the continuous dike in Fig. 1 is assumed. For comparison, an elongation of 4850% ($\varepsilon = 3.9$) for the Pb–62% Sn alloy is one of the largest experimental values reported (Langdon 1982*b*). The extremely high strains without failure recorded by the Särv dykes demonstrate that deformation of the dykes was phenomenologically superplastic.

The amount of strain in the adjacent quartzofeldspathic mylonites must be comparable to the strain recorded by the dykes, because the interbanded arkosic and basaltic mylonites and ultramylonites do not exhibit great differences in competence. The alternating layers of quartzofeldspathic and mafic mylonites are planar and continuous over long distances, while folds of contact strain (mullions) and boudinage are minor features. One could use the change in thickness and angle of the arkosic screens (i.e. the volume of sandstone between two dikes) as they curl into the shear zone as a strain measure, but poor exposure prevents this. Nevertheless, the quartzofeldspathic mylonites display the same degree of ductility as the deformed dykes and, therefore, are considered to be superplastic as well.

Deformation along the thrust occurred at approximately 450–500°C beneath 15–20 km of overburden (Gilotti & Kumpulainen 1986; Gilotti 1989). The most obvious microstructural feature of the deformed dykes is the syntectonic development of a stable greenschist facies assemblage from the igneous and deuteric mineralogy present outside the shear zone. Gilotti (1989) presents a quantitative set of hydration reactions which describe the development of the metadiabase mylonites. The ultimate product of the syntectonic metamorphism is a biotite–epidote–chlorite–albite schist (Fig. 2a), which forms at $\gamma \approx 100$. K-metasomatism accounts for the abundance of biotite at shear strains of $\gamma > 40$.

The microstructure of the metadiabase tectonite consists of optically unstrained grains with relatively uniform shapes and sizes. Grain size is somewhat coarse, with an abundant 50–200 μm fraction visible in Fig. 2a. Subhedral phyllosilicates with approximately constant aspect ratios display a strong crystallographic preferred orientation which helps define the foliation. Epidote is subrounded with irregular grain boundaries and axial ratios less than 2:1; titanite is euhedral to subhedral with a typical rhombic habit; and both minerals have long-axes parallel to the foliation. Neither the meta-

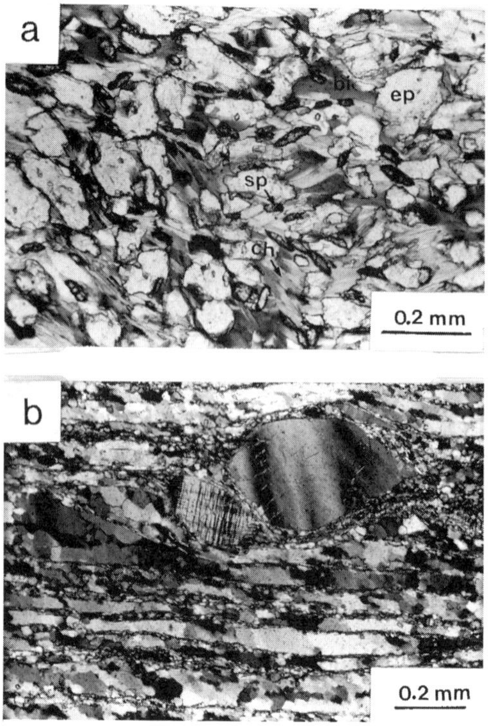

Fig. 2. (**a**) Photomicrograph of metadiabase from the most deformed part of the dyke. The stable greenschist facies assemblage is: ep, epidote; bi, biotite; ch, chlorite; sp, titanite. The grains are optically unstrained with uniform shapes and sizes. Deformation mechanisms were probably dominated by diffusion-accomodated grain boundary sliding (Gilotti 1989). (**b**) Photomicrograph of a quartzofeldspathic augen mylonite derived from the arkoses. Quartz ribbons have developed from clasts via dislocation creep, while the fine-grained matrix between the ribbons and augen has probably deformed via grain size sensitive mechanisms.

morphic assemblage nor the microstructure changes at the higher strains represented by the very thin (< 5 cm) greenschist bands, suggesting that above some critical strain ($\gamma \approx 100$) the deformed dykes attain a steady state microstructure (Means 1981), i.e. a microstructure which is very dynamic, but has a statistically constant grain configuration.

The deformation mechanisms are ambiguous in the mafic tectonites. Aside from a few kinked biotite grains, there is very little evidence for intracrystalline plasticity. Diffusional mass transfer processes are inferred from incongruent pressure shadows around relict phenocrysts, the lack of evidence for cataclasis coupled with the hydration reactions, and metasomatic reactions. No indication of grain boundary sliding could be found (e.g. Drury & Humphreys 1988), but grain boundary sliding rarely leaves a unique microstructure. The creation of a stable microstructure may indicate a large contribution by grain size sensitive deformation mechanisms or dynamic recrystallization. Despite the relatively coarse grain size and moderate temperature of deformation, the metadiabase has deformed superplastically, reflecting reaction-enhanced ductility.

The arkosic sandstones are transformed into feldspar–quartz–muscovite augen mylonites and ultramylonites (Gilotti 1990). Although both lithologies deformed under the same physical conditions, the arkosic mylonites show quite different microstructures (Fig. 2b). Grain size was reduced from an average size of 100 μm in the sandstones to *c*. 20 μm in the ultramylonites, predominantly by dynamic recrystallization. Stable fractures in the feldspar augen account for a small component of grain size reduction. Detrital quartz grains are deformed into ribbons with axial ratios up to 100:1 (X:Z), indicating very large plastic strains. Other microstructures indicative of intracrystalline plasticity are undulatory extinction, deformation bands, twinning, and subgrains.

The arkosic mylonites also show evidence of diffusional mass transport processes. Pressure solution is particularly common in the protomylonites where grains indent one another. Metamorphic reactions produced fine-grained muscovite, quartz and epidote from the incomplete breakdown of the feldspars. These new minerals are added to the fine-grained matrix produced by dynamic recrystallization of the original phases. Quartz ribbons and feldspar augen are eventually replaced by a homogeneous mixture of very fine grains in the ultramylonites. Deformation mechanisms are difficult to evaluate in the fine-grained matrix, but we believe grain size sensitive mechanisms contribute to a large component of the matrix strain. Grain size sensitive flow may even dominate the quartzofeldspathic mylonite rheology, once the fine-grained matrix percentage exceeds *c*. 50% (Gilotti 1990).

The observations along the Särv mylonite zone support our contention that superplasticity in rocks cannot be directly equated with a unique deformation mechanism or a single set of material parameters. Both the metadiabase and arkosic mylonites exhibit macroscopic superplasticity, despite the different phases, microstructures, and deformation mechanisms present in the two rocks. Progressive simple shear of the dykes also demonstrates that superplasticity is not restricted to pure tensile deformations.

Pegmatite dykes deformed in bulk flattening

A swarm of pegmatite dikes intruding Grenville paragneisses (New Jersey Highlands, USA) have been deformed by predominantly bulk flattening (Hull *et al.* 1986). Cross cutting relationships and variability of deformation intensity suggests several generations of pegmatites. We have concentrated our studies on the earliest, most deformed dykes. These dykes predate the regional amphibolite to granulite-facies metamorphism, which produced sillimanite and spinel as the index minerals in the host paragneiss. The highly deformed dykes (Fig. 3a.) may have followed complex strain paths for both boudinaged folds and folded boudinage are seen in the various exposures.

The pegmatites deform continuously by folding when shortened, but exhibit discontinuous deformation by boudinage in extension (Fig. 3a). Measurements of shortening strain perpendicular to the foliation using changes in dyke lengths range from $\varepsilon = -2.0$ to -3.0. Although modest, these strains exceed $\varepsilon = 1.8$, the metallurgical threshold for superplasticity. These strains are also minima, as the original thicknesses of the dykes are unknown, and hence the amount of penetrative strain cannot be determined. The high strains without loss of continuity recorded by the folded pegmatite dykes again illustrate the phenomenon of superplasticity in natural rock deformation. The contrasting behaviour in shortening and extension of the same protolith under the same physical conditions also demonstrates the strain path can determine whether or not a rock behaves ductilely. In this case, necking in extension is much more efficient than the development of shear zones in shortening.

The coarse grained pegmatites (Fig. 3b) are composed mainly of alkali feldspar, quartz, plagioclase, and biotite which show little alteration, except where they are cut by late, low-temperature deformation zones (Hull *et al.* 1986). The original grain size distribution is unknown, but some relict alkali feldspars range up to 2 cm in diameter. The original pegmatites were probably very coarse grained. The homologous temperature during deformation was quite high ($T_H \geq 0.8$), given the granulite facies metamorphism and granitic composition of the

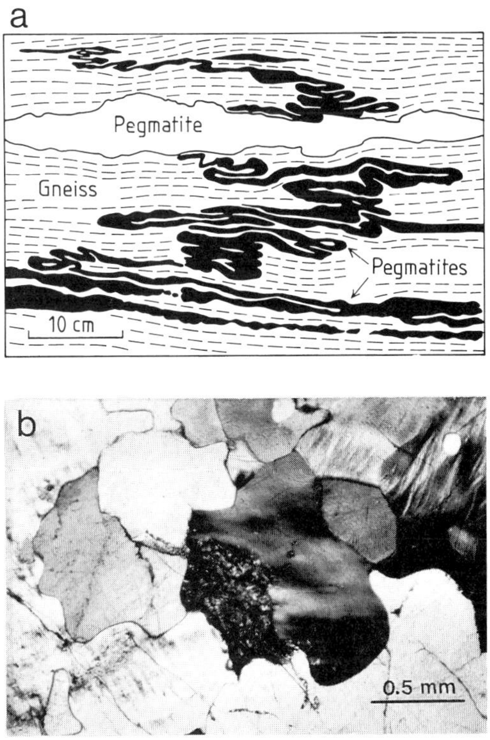

Fig. 3. (**a**) Pegmatite dykes in sillimanite zone paragneisses from Andover, New Jersey, USA show predominantly bulk flattening strain. Strain measurements of shortening recorded by the folded dikes range from $\varepsilon = -2.0$ to $\varepsilon = -3.0$. Note the discontinuously deformed dykes parallel to the foliation (dashed lines) at the bottom of the figure. The unshaded pegmatite represents a later generation. (Sketch drawn from outcrop photograph). (**b**) Coarse-grained pegmatites are partially recrystallized to an equilibrium mosaic modified by grain boundary migration, as exemplified by lobate and interlocking grain boundaries. Both intercrystalline (lower right) and interphase (lower left) grain boundaries show evidence of migration.

pegmatites. The pegmatites are partially recrystallized to an equilibrium mosaic of equant grains with 120° triple junctions at grain boundary intersections. Recrystallized grain size appears fairly uniform, ranging from 100–1000 μm. Although most dyke margins are sharp, some of the pegmatitic dykes show a decrease in grain size towards their margins due to increasing amounts of recrystallization. Many of the quartz and feldspar grains exhibit undulatory extinction, subgrains, deformation lamellae, and other features characteristic of intracrystalline plasticity, but these microstructures were undoubtedly produced later during Palaeozoic, lower temperature events. These optical features are not pervasive and are best developed adjacent to through-going microfaults. Most are products of cold working and not high temperature creep. Away from the microfaults, both quartz and feldspars are optically strain-free.

Many of the grain boundaries are lobate, cuspate, and highly interpenetrating, even along interphase boundaries (Fig. 3b), indicating the ease of grain boundary migration during high-temperature Grenville deformation. Grain boundary mobility is probably enhanced by both the high temperatures during deformation and intergranular fluids in the paragneisses. Partial melts along grain boundaries may have also played a role. Grain boundary configurations similar to the stages of grain switching events illustrated by Ashby & Verrall (1973) are common, indicating grain boundary sliding and/or diffusional creep.

The microstructures in the intensely deformed pegmatites are consistent with both high temperature intracrystalline plasticity and grain size sensitive flow. Unfortunately, we have not been able to quantify the the partitioning of strain among the different micromechanisms, but we believe that crystal plasticity has been important in the deformation of the dykes. Despite the very coarse grain size of both the protolith and the resulting tectonite, the dykes deformed in shortening to very high strains without developing discontinuities. Grain size reduction by recrystallization during deformation may have contributed to the ductility of the pegmatites.

Parameters

A restricted range of grain sizes, phase proportions, and other material parameters are necessary to produce superplastic flow in laboratory experiments on metals and metallic alloys. In natural rock deformation, a much wider range of material parameters may give rise to phenomenological superplasticity, because of the broader range of physical conditions and deformation configurations. For example, deformation path has a strong influence on strain localization, yet there are very few experiments on superplastic alloys deformed with non-coaxial strain histories, such as direct or rotary simple shear. In this section, we discuss the affect of a few common physical variables on rock ductility, drawing on experimental and field work. We have selected just a few of the many variables, simply to illustrate the point that extreme ductility in rocks is expected regardless of the dominant deformation mechanism. An extensive review of the many parameters affecting ductility of superplastic metals and alloys is presented by Gifkins (1982). Comparable discussions for rocks can be found in White *et al.* (1980) and Poirier (1980). We further restrict our discussion to rather homogeneous, small scale protoliths, which are relevant to the examples described above.

The influence of different physical conditions and material parameters on the homogeneity and ductility of rock deformation is summarized in Fig. 4. This diagram is divided into three end member, somewhat idealized micromechanical regimes: fracture or cataclasis (brittle deformation), intracrystalline plasticity, and grain-size sensitive mechanisms (which includes both diffusional mass transfer processes and grain boundary sliding). The influence of the different parameters on uniformity of deformation are shown for each micromechanical regime. Some of these effects are discussed below.

Figure 4 also illustrates some common modes of localized deformation or macroscopic failure for rocks. Heterogeneous deformation can result from initial flaws in the deforming object, grouped in two classes: *mechanical defects*, where the initial strain or rheological behaviour varies along the object, and *geometrical defects*, such as a deviation in initial cross sectional area in a tensile specimen (compare with Kocks *et al.* 1979). In experimental deformation, propagation of surface cracks or other surface imperfections is also important in localizing failure (Griffiths & Hammond 1972). The resulting strain heterogeneities include both microscopic and macroscopic elements (e.g. Lin *et al.* 1981), such as Lüders bands, necks, shear bands, or shear zones (Fig. 4). It is important to note that the presence or development of a macroscopic flaw is not sufficient to produce failure. For example, Mohammed & Langdon (1981) have studied neck formation and growth in Zn–22%Al. Necks form in the superplastic regime but are distributed over large sections of the test specimen. The necks are not sharp and develop slowly.

Figure 4 does not predict whether a rock behaves superplastically or not. Superplasticity depends on the rate of propagation and growth of defects relative to the rate of bulk strain, and thus depends on the magnitude of the deformation as measured by duration or finite strain (e.g. Donath *et al.* 1971, Poirier 1980). The continuity and homogeneity of deformation is also dependent on the scale of observation, as mentioned earlier (see also Poirier 1980). For

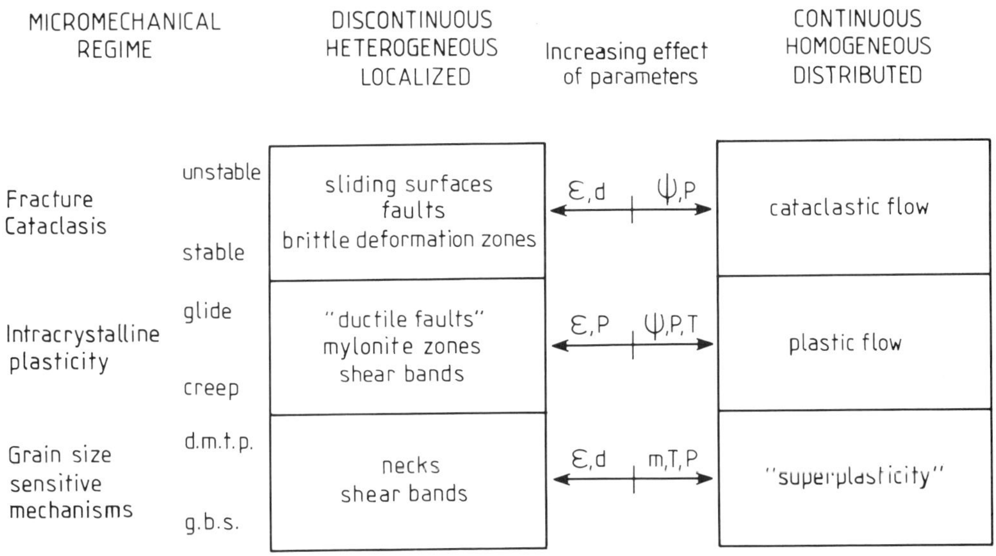

Fig. 4. Diagram illustrating the influence of various material and mechanical parameters on flow localization in three micromechanical regimes: cataclastic, plastic, and grain size sensitive flow. P, pressure; T, temperature; m, strain rate sensitivity; d, grain size; ε, strain; ψ, work hardening coefficient. Increasing magnitudes of parameters to the left (e.g. strain) will favour localized or heterogeneous deformation; increasing parameters to the right (e.g. strain rate sensitivity) will favour ductility. 'Superplasticity' refers to current usage by geologists. Based on a figure in Rutter (1986).

example, a population of faults may produce large, locally homogeneous bulk strains, but each individual fault will show highly localized and discontinuous deformation. To reiterate, superplasticity is not simply a set of physical conditions and material parameters, but rather a phenomenon that must be defined empirically. We discuss how a few parameters *promote* or *inhibit* ductile deformation.

Pressure

For rocks in the brittle regime (Fig. 4), increasing effective pressure decreases the amount of dilatancy and inhibits the propagation of microfractures (Paterson 1958; Griggs & Handin 1960; Sammis *et al.* 1986). Sets of through-going, small-scale faults, brittle 'shear bands', and fault zones give way to more distributed fracturing with increasing pressure, ultimately leading to cataclastic flow and relatively homogeneous deformation at the sample scale (Paterson 1958, Heard 1960, 1976, Donath *et al.* 1971, Tullis & Yund 1987). Some of this trend may be due to an increasing contribution from intracrystalline plasticity and not just modified brittle behaviour (Tobin & Donath 1971), although Tullis & Yund (1977), in low temperature experiments on dry Westerly granite, saw little variation of microstructures with increasing pressure. Ductile flow in very fine-grained cataclasites has also been ascribed to a change in deformation mechanism from fracture to predominantly diffusional mass transfer processes (Mitra 1984). Whether true cataclastic flow at low temperatures can produce relatively continuous, large magnitude deformation of a marker, such as a clast, is unknown. Certainly most cataclastic deformation is discontinuous. 'Smeared out' rock fragments in some fault zone cataclasites might correspond to such large, semi-continuous strain.

The influence of pressure on ductility in the regime of intracrystalline plasticity is variable, but in general, increased confining pressure produces more homogeneous deformation (Heard 1960, 1976; Kronenberg & Tullis 1984; Rutter 1986). 'Ductile faults' and shear bands are suppressed and plastic flow is enhanced at higher pressures, but the actual mechanisms are not well known. Increased pressure can also lower the strength of rocks and minerals deforming predominantly by intracrystalline plasticity. Enhanced recovery by cross slip in perhaps halite, calcite, and amphibole is favoured by high effective pressure (Brodie & Rutter 1985). Hydrolitic weakening in quartz and other silicates (e.g. Tullis & Yund 1980; Kronenberg & Tullis 1984) has also been observed at high pressures, though Kronenberg *et al.* (1986) suggest that high pore pressures (low effective pressures) are responsible. Reducing mineral strength at high pressure would produce more ductility on a small scale, but perhaps less continuity on a large scale, if discontinuous 'ductile' deformation zones formed (i.e. 'ductile faults').

The formation of cavities or voids along grain boundaries is a simple geometric consequence of grain boundary sliding, and is commonly associated with superplastic metals and alloys (e.g. Langdon 1982*a*). However, cavitation failure of nominally superplastic materials can be produced by growth and coalescence of these grain boundary voids. A few experiments on superplastic materials at elevated confining pressures, summarized in Pilling & Ridley (1988), demonstrate that higher effective pressures decrease the rate and amount of cavitation, and enhance ductility. Increased pressure may lower the diffusivity of certain species through its effect on the activation volume (e.g. Karato 1981), and consequently decrease the strain rate. The resulting influence on work hardening is uncertain, but an increased work hardening coefficient ψ seems likely. Pressure may have the same effect on grain boundary sliding if sliding is accompanied by diffusional mass transfer along the grain boundaries.

In summary, there is an overall increase in ductility with increasing effective pressure, regardless of the flow mechanisms. Pressures were high in both of the natural examples of macroscopic superplasticity discussed earlier, and probably made an important contribution to the enhanced ductility. We predict that phenomenological superplasticity will be observed over a wide range of crustal depths, even at relatively low temperatures.

Temperature

No effect of temperature on the ductility of cataclastic deformation is expected. It has long been known that increased temperature will dramatically lower both the strength and the work hardening coefficient in the regime of intracrystalline plasticity, and in general lead to more distributed deformation (e.g. Heard 1960; Griggs *et al.* 1960; Griggs & Handin 1960; Donath & Wood 1976; Tullis & Yund 1980). For example, at relatively high confining pressure, cylinders of dry Westerly granite in compression show more homogeneous deformation and no through going shear bands at higher temperatures (Tullis & Yund 1977). Shear bands or 'ductile faults' were present in these experiments at lower confining pressures. The direct inference is that enhanced plasticity is more distributed at higher temperatures, with fewer undeformed domains.

Temperature may have a dual effect on intracrystalline plasticity, depending upon the mobility of dislocations. Increasing temperature in the dislocation glide regime will influence dislocation interaction by promoting climb, causing fewer pileups which eventually lead to fracture (e.g. Petch 1953; Stroh 1954). Although increasing temperature in the dislocation glide field may allow more distributed deformation, large strains may still be unattainable. In the dislocation creep regime, increasing temperature will enhance recovery and recrystallization, but the effect on ductility will depend upon the specific recrystallization mechanism (Guillopé & Poirier 1979, Drury *et al.* 1985, Urai *et al.* 1986). Recrystallization in the lower temperature portion of the dislocation creep field may have no effect on strength, or actually increase the flow stress (Urai *et al.* 1986, Drury *et al.* 1989), but in general, recrystallization will produce strain softening and thus expedite localized deformation. Again, the ductility will depend strongly on the scale of observation.

In superplastic metals and alloys where grain boundary sliding is known to be a dominant mechanism, one sees an overall increase in the range of strain rates over which superplasticity is observed at elevated temperatures (see, for example, Edington 1982; Gifkins 1982). In contrast, there is an optimum temperature in some alloys, such as aluminium-bronze (Dunlop & Taplin 1972), that is associated with maximum elongation. This unusual response is partially explained by the temperature dependence of the phase proportions of aluminium-bronze, a scenario that is likely among certain rocks.

Temperature is an important variable controlling the ductility of rocks. Mylonites dominated by crystal plasticity can appear very ductile on the thin section scale, but may produce heterogeneous, localized deformation in the outcrop scale. High homologous temperature ($T_H \geq 0.5$) is an oft-cited criterion for superplastic deformation of metals in the grain size sensitive regime. Moderately high homologous temperatures were required for superplasticity in both our natural examples. However, very high temperatures may actually have a deleterious affect on superplastic behaviour by modifying the microstructure. Grain boundary migration and grain growth are probably strongly influenced by diffusion in rocks, which will be greatly enhanced at high temperatures.

Strain history

The influence of various deformation histories on superplasticity in the PbSn eutectoid have been investigated experimentally by Ahmed & Langdon (1983). When the specimens were initially deformed at high strain rates characteristic of the intracrystalline plasticity field, and then deformed in the 'superplastic' regime, the samples showed reduced fracture elongation with increasing amounts of pre-strain. Ahmed & Langdon attribute this behaviour to the formation of strain heterogeneities (incipient necks) in the plasticity field, which are then exploited during the slower deformation.

Some superplastic alloys show *increased* elongation to failure when the tensile tests are interrupted and 'holding times' are introduced into the strain history (Ahmed & Langdon 1983). The mechanism is not well understood but may be related to thermal gradients along the length of the specimen. Ahmed & Langdon (1983) demonstrated that increasing the holding time, following a pre-strain in the plasticity regime, will *decrease* the fracture elongation, mainly because of grain growth.

Episodic deformation and variable strain rates are probably the rule rather than the exception in nature. Inconsistent field relationships may result if the local, episodic strain histories are not recorded by any structural element. Some objects will show discontinuous behaviour, others not, under essentially identical physical conditions and material parameters. Even simple, monotonically varying strain rates can have a profound influence on ductility. For example, early or pre-existing brittle fractures produced during high strain rate deformation will probably localize and promote the growth of ductile shear zones during subsequent low strain rate deformation (see Segall & Simpson 1986).

Stability criteria

A complementary approach to understanding the factors that govern ductility are theoretical stability criteria. There are a large number of theoretical formulations for specific deformation paths and mechanical behaviour that are independent of the deformation mechanism. We briefly consider two, just to illustrate the importance of material parameters and deformation path on the development of instabilities. Whether an instability actually leads to failure depends upon the actual deformation history, in particular the rate of propagation of the flaw, and the duration of deformation.

Tension

The Considère criterion for instability in tension (see e.g. Poirier 1980, 1985) is based upon the load bearing capacity of the material decreasing with increasing strain magnitude. For constant strain rate, the instability will appear when:

$$\mathrm{d}\ln F/\mathrm{d}\,\varepsilon \leq 0$$

where F is the force applied to the cross-sectional area of the specimen.
For constant volume, the criterion can be reformulated to:

$$\left.\frac{\mathrm{d}\ln\sigma}{\mathrm{d}\,\varepsilon}\right|_{\mathrm{d}\dot{\varepsilon}=0} = \psi = 1;$$

that is, instabilities in tension will propagate in materials whose work hardening coefficient ψ is less than unity (Fig. 5). For $0 < \psi < 1$, the material will develop instabilities even though it is work hardening. This behaviour is expected for tensile deformation, as any reduction in cross sectional area will produce an increase in strain rate that must be counterbalanced by some amount of work hardening to prevent necking. Similarly, Griggs & Handin (1960) observed work hardening in dolomite (under compression) concomitant with localized deformation along shear zones. Therefore, strain softening is neither a necessary nor a sufficient criteria for developing discontinuities in tension or compression. Furthermore, there is no simple correlation between steady state flow (where $\psi = 0$) and superplastic behavior, according to the Considère criterion.

More recent theories of instability in tension have been developed by Hart (1967), Kocks *et al.* (1976), and Lin *et al.* (1981), among many others. We will not attempt to summarize this rather contentious subject, but do note that almost all of the theoretical models show a very strong dependence of strain to failure upon two mechanical parameters; the strain rate sensitivity $m = \left(\left.\frac{\mathrm{d}\ln\sigma}{\mathrm{d}\ln\dot{\varepsilon}}\right|_{\varepsilon}\right)$ and the work hardening coefficient.

Pure and simple shear

Poirier (1980, 1985) has modified the Considère criterion for both pure and simple shear deformation, where there is no change in cross sectional area (in addition to constant volume). For these deformation paths, $\psi = 0$, which corresponds to true steady state behaviour (Fig. 5). Therefore, instabilities in pure or simple shear will propagate for $\psi < 0$, the strain

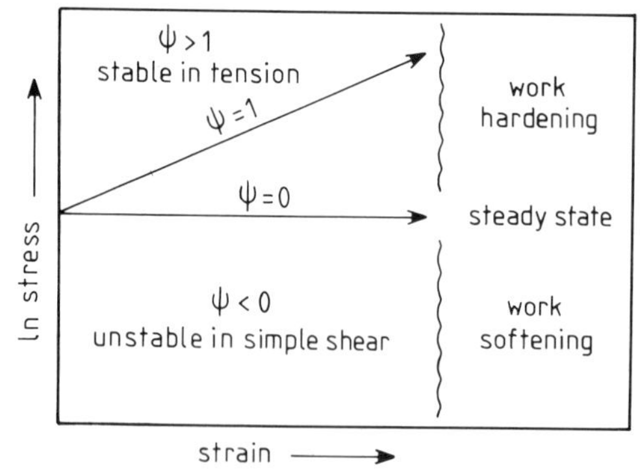

Fig. 5. Stress–strain curve illustrating the effect of varying the work hardening coefficient (ψ) on flow stability. For $\psi > 1$, materials are stable in tension; for $\psi < 0$ materials are unstable in simple shear.

softening regime. Poirier (1980) and Hobbs *et al.* (1986) have coupled this stability criterion with flow laws for plastic deformation to investigate the role of different material parameters on stability.

These two simple formulations illustrate the influence of deformation path or strain style on the continuity of deformation. For the stipulated conditions, a material with a $\psi = 0.5$ (for example) will deform continuously in simple shear but discontinuously in tension (Fig. 5). Both rotational and non-coaxial strains are common in natural deformation, and the resultant strain paths are not simple. We have shown in our second geologic example how strain style affects ductility. Pegmatite dykes in the extension field deformed discontinuously by boudinage, whereas dykes in the shortening field showed more continuous behaviour.

Rheological implications

The above theoretical examples also illustrate the potential for identifying material parameters, such as the work hardening coefficient, from simple field observations; whether continuity of deformation was path dependent or independent, or whether discontinuities propagated or not. The stability criterion cited above has very specific conditions that may not be met by many natural examples, or may be difficult to test. The success of this approach is to find well constrained natural examples, and to develop a broader range of and more generalized stability criteria.

A positive correlation between total elongation to failure (fracture elongation) and the strain rate sensitivity m has been observed for metals and metallic alloys tested in tension. A fairly wide range of values of m and n (the stress exponent) can lead to phenomenological superplasticity in metals and alloys. For $m \geq 0.5$, quasi-stable plastic flow and very diffuse necking can produce very large strains. Yet we believe that it is premature to extrapolate these results to natural deformations, such as those described in this paper, to determine mechanical parameters for rocks. We predict that superplasticity in the geological environment will also be associated with a wide range of mechanical parameters. While a Newtonian viscous rheology, with $m = n = 1$, may favour superplastic behaviour under certain conditions, it is not a requirement for high strains without heterogeneities.

The authors are grateful to D. Olgaard and M. Drury for their constructive comments as reviewers, and to W. Means and C. Talbot for additional comments on the manuscript. J. A. G. would like to thank the Swedish Natural Science Research Council (NFR) for post-doctoral support, as well as travel to Leeds. J. M. H. is supported by a grant from STU (Swedish Council for Technical Development).

References

Ahmed, M. M. I. & Langdon, T. G. 1983. The influence of prestrain ductility in the superplastic Pb-Sn eutectic alloy. *Journal of Materials Science*, **18**, 3535–3542.

Arieli, A. & Mukherjee, A. K. 1982. The rate-controlling deformation mechanisms in superplasticity—a critical assessment. *Metallurgical Transactions*, **13A**, 717–732.

Ashby, M. F. & Verrall, R. A. 1973. Diffusion-accommodated flow and superplasticity. *Acta metallurgica*, **21**, 149–163.

Behrmann, J. H. 1985. Crystal plasticity and superplasticity in quartzite: a natural example. *Tectonophysics*, **115**, 101–129.

Boullier, A. M. & Gueguen, Y. 1975. SP-Mylonites: origin of some mylonites by superplastic flow. *Contributions to Mineralogy and Petrology*, **50**, 93–104.

Brodie, K. H. & Rutter, E. H. 1985. On the relationship between deformation and metamorphism, with special reference to the behavior of basic rocks: *In:* Thompson, A. B. & Rubie, D. C. (eds) *Metamorphic Reactions—Kinetics, Textures, and Deformation*. Springer, New York, Berlin, Heidelberg, 138–179.

Donath F. A. & Wood, D. S. 1976. Experimental evaluation of the deformation path concept. *Philosophical Transactions of the Royal Society London*, **283**, 187–201.

——, Faill, R. T. & Tobin, D. G. 1971. Deformation mode fields in experimentally deformed rock.

Geological Society of America Bulletin, **82**, 1441–1462.

DRURY, M. R. & HUMPHREYS, F. J. 1988. Microstructural and shear criteria associated with grain boundary sliding during ductile deformation. *Journal of Structural Geology*, **10**, 83–89.

——, —— & WHITE, S. H. 1985. Large strain deformation studies using polycrystalline magnesium as a rock analogue. Part II: dynamic recrystallisation mechanisms at high temperatures. *Physics of the Earth and Planetary Interiors*, **40**, 208–222.

——, —— & —— 1989. Effect of dynamic recrystallization on the importance of grain-boundary sliding during creep. *Journal of Materials Science*, **24**, 154–162.

DUNLOP, G. L. & TAPLIN, D. M. R. 1972. The tensile properties of a superplastic aluminum bronze. *Journal of Materials Science*, **7**, 84–92.

EDINGTON, J. W. 1982. Microstructural aspects of superplasticity. *Metallurgical Transactions*, **13A**, 703–715.

GHOSH, A. K. 1982. Characterization of superplasticity in metals. *In*: PATON, N. E. & HAMILTON, C. H. (eds) *Superplastic Forming of Structural Alloys*. The Metallurgical Society of AIME, Warrenville, Pennsylvania, USA, 85–103.

—— & HAMILTON, C. H. 1982. Influences of material parameters and microstructure on superplastic forming. *Metallurgical Transactions*, **13A**, 733–743.

GIFKINS, R. C. 1982. Mechanisms of superplasticity. *In*: PATON, N. E. & HAMILTON, C. H. (eds) *Superplastic Forming of Structural Alloys*. The Metallurgical Society of AIME, Warrenville, Pennsylvania, USA., 3–26.

GILOTTI, J.A. 1989. Reaction progress during mylonitization of basaltic dikes along the Särv thrust, Swedish Caledonides. *Contributions to Mineralogy and Petrology*, **101**, 30–45.

—— 1990. The rheologically critical matrix in quartzofeldspathic mylonites along the särv thrust, Swedish Caledonides. *In*: MITRA, S. & FISHER, G. W. (eds), *Structural Geology of Fold and Thrust Belts*, Johns Hopkins Press, Baltimore, (in press).

—— & KUMPULAINEN, R. 1986. Strain softening induced ductile flow in the Särv thrust sheet, Scandinavian Caledonides. *Journal of Structural Geology*, **8**, 441–455.

GRIFFITHS, P. & HAMMOND, C. 1972. Superplasticity in large grained materials. *Acta metallurgica*, **20**, 935–945.

GRIGGS, D. T. & HANDIN, J. 1960. Observations on fracture and a hypothesis of earthquakes. *Geological Society of America Memoir*, **79**, 347–364.

——, TURNER, F. J. & HEARD, H. C. 1960. Deformation of rocks at 500° to 800°C. *Geological Society of America Memoir*, **79**, 39–104.

GUILLOPÉ, M. & POIRIER, J. P. 1979. Dynamic recrystallization during creep of single crystalline halite: an experimental study. *Journal of Geophysical Research*, **84**, 5557–5567,

HART, E. W. 1967. Theory of the tensile test. *Acta metallurgica*, **15**, 351–355.

HEARD, H. C. 1960. Transition from brittle to ductile flow in Solenhofen limestone as a function of temperature, confining pressure, and interstitial fluid pressure. *Geological Society of America Memoir*, **79**, 193–226.

—— 1976. Comparison of the flow properties of rocks at crustal conditions. *Philosophical Transactions of the Royal Society London* **A283**, 173–186.

HOBBS, B. E., ORD, A. & TEYSSIER, C. 1986. Earthquakes in the ductile regime? *Pure and Applied Geophysics*, **124**, 309–336.

HULL, J. M., KOTO, R. & BIZUB, R. 1986. Deformation zones in the Highlands of New Jersey. *Proceedings of the Geological Association of New Jersey*, **3**, 19–66.

KARATO, S. 1981. Pressure dependence of diffusion in ionic solids. *Physics of the Earth and Planetary Interiors*, **25**, 38–51.

KOCKS, U. F., JONAS, J. J. & MECKING, H. 1979. The development of strain-rate gradients. *Acta metallurgica*, **27**, 419–432.

KRONENBURG, A. K. & TULLIS, J. 1984. Flow strengths of quartz aggregates: grain size and pressure effects due to hydrolytic weakening. *Journal of Geophysical Research*, **89**, 4281–4297.

——, KIRBY, S. H., AINES, R. D. & ROSSMAN, G. R. 1986. Solubility and diffusional uptake of hydrogen in quartz at high water pressures: implications for hydrolytic weakening. *Journal of Geophysical Research*, **91**, 12723–12744.

LANGDON, T. G. 1982*a*. Fracture processes in superplastic flow. *Metal Science*, **16**, 175–183.

—— 1982*b*. The mechanical properties of superplastic materials. *Metallurgical Transactions*, **13A**, 689–701.

LIN, I.-H., HIRTH, J. P. & HART, E. W. 1981. Plastic instability in uniaxial tension tests. *Acta metallurgica*, **29**, 819–827.

LISTER, G. S. & SNOKE, A. W. 1984. S-C mylonites. *Journal of Structural Geology*, **6**, 617–638.

MEANS, W. D. 1981. The concept of steady state foliation. *Tectonophysics*, **78**, 179–199.

MITRA, G. 1984. Brittle to ductile transition due to large strains along the White Rock thrust, Wind River mountains, Wyoming. *Journal of Structural Geology*, **6**, 51–61.

MOHAMMED, F. A & LANGDON, T. G. 1981. Flow localization and neck formation in a superplastic metal. *Acta metallurgica*, **29**, 911–920.

PATERSON, M. S. 1958. Experimental deformation and faulting of Wombeyan marble. *Geological Society of America Bulletin*, **69**, 465–476.

PEARSON, C. E. 1934. The viscous properties of extruded eutectic alloys of lead-tin and bismuth-tin. *Journal of the Institute of Metals*, **54**, 111–124.

PETCH, N. J. 1953. The cleavage strength of polycrystals. *Journal of the Iron and Steel Institute*, **174**, 25–28.

PILLING, J. & RIDLEY, N. 1988. Cavitation in superplastic alloys and the effect of hydrostatic pressure. *Res Mechanica*, **23**, 31–63.

POIRIER, J. P. 1980. Shear localization and shear instability in materials in the ductile field. *Journal of Structural Geology*, **2**, 135–142.

—— 1985. *Creep of Crystals*. Cambridge University Press, London.

RUTTER, E. H. 1986. On the nomenclature of the mode of failure transitions in rocks. *Tectonophysics*, **122**, 381–387.

SAMMIS, C. G., OSBORNE, R. H., ANDERSON, J. L., BANERDT, M. & WHITE, P. 1986. Self-similar cataclasis in the formation of fault gouge. *Pure and Applied Geophysics*, **124**, 53–78.

SCHMID, S. M. 1982. Microfabric studies as indicators of deformation mechanisms and flow laws operative in mountain building. *In*: HSU, K. J. (ed.) *Mountain Building Processes*. Academic Press, New York, 95–110.

——, BOLAND, J. N. & PATERSON, M. S. 1977. Superplastic flow in finegrained limestone. *Tectonophysics*, **43**, 257–291.

SEGALL, P. & SIMPSON, C. 1986. Nucleation of ductile shear zones on dilatant fractures. *Geology*, **14**, 56–59.

SHARIAT, P., VASTAVA, R. B. & LANGDON, T. G. 1982. An evaluation of the roles of intercrystalline and interphase boundary sliding in two-phase superplastic alloys. *Acta metallurgica*, **30**, 285–96.

STROH, A. N. 1954. The formation of cracks as a result of plastic flow. *Proceedings of the Royal Society*, **223**, 404–413.

TOBIN, D. G. & DONATH, F. A. 1971. Microscopic criteria for defining deformational modes in rock. *Geological Society of America Bulletin*, **82**, 1463–1476.

TULLIS, J. & YUND, R. A. 1977. Experimental deformation of dry Westerly granite. *Journal of Geophysical Research*, **82**, 5705–5717.

—— & —— 1980. Hydrolytic weakening of experimentally deformed Westerly granite and Hale albite rock. *Journal of Structural Geology*, **2**, 439–451.

—— & —— 1987. Transition from cataclastic flow to dislocation creep of feldspar: mechanisms and microstructures. *Geology*, **15**, 606–609.

URAI, J. L., MEANS, W. D. & LISTER, G. S. 1986. Dynamic recrystallization of minerals. *American Geophysical Union Monograph*, **36**, 161–199.

VASTAVA, R. B. & LANGDON, T. G. 1979. An investigation of intercrystalline and interphase boundary sliding in the superplastic Pb-62% Sn eutectic. *Acta metallurgica*, **27**, 251–257.

WHITE, S. H., BURROWS, S. E., CARRERAS, J. SHAW, N. D. & HUMPHREYS, F. J. 1980. On mylonites in ductile shear zones. *Journal of Structural Geology*, **2**, 175–187.

Ductile deformation mechanisms in micritic limestones naturally deformed at low temperatures (150–350°C)

MARTIN BURKHARD

Institut de Géologie, rue E. Argand 11. CH-2000 Neuchâtel, Switzerland

Abstract: Deformation microstructures have been examined in a series of micritic limestone samples from the Helvetic nappes of western Switzerland, deformed at temperatures ranging from 150–350°C. Finite strain has been determined with $R_f/\phi-\theta$ and centre–centre techniques applied to pellets and varies between $R_s = 2$ to 10 in samples in fold nappe interiors; much higher strains occur in samples from thrust planes. Optical micrographs from three ultrathin sections per sample and 200 to 500 grains per section were quantified by image analysis. Grain size is fairly constant from diagenesis through to anchizone (3–6 μm) and increases only within the epizone (6–10μm). Grain shape preferred orientations are weak and invariably much smaller than strain ratios R_s. Long axis orientations reflect the finite strain orientation. Roundness measurements are consistent with grains having relatively simple boundaries, frequently meeting in 120° triple junctions independently of temperature, finite strain or grain size. No crystallographic preferred orientation could be detected in the low temperature (up to 300°C) micrites despite large finite strains. Strong *c*-axis preferred orientation is found in one extremely finegrained (< 1 μm) low temperature ($T < 180$°C) faultrock. Some degree of preferred orientation is widespread in the coarser grained epizonal ($T > 300$°C) limestones which all show twinning. Grain Boundary Sliding (GBS) is inferred to be the dominant ductile deformation mechanism in moderately deformed micrites ($R_s < 10$) between 200–300°C. Coarsening of grains with increasing temperature by dynamic grain boundary migration recrystallization leads to a net increase in grain size above 300°C. This favours twinning instead of GBS, which in turn leads to pronounced *c*-axis fabrics.

Deformation behaviour of limestones has been studied by many workers, both in naturally deformed rocks and in experiments (for references see: Turner *et al.* 1954; Rutter 1976; Schmid *et al.* 1981, 1987; Wenk 1985; Evans *et al.* 1986; Olgaard & Evans 1988). The aim of this paper is to quantify optical deformation microstructures and to try to correlate these with finite strain and temperatures of deformation in order to constrain the possible deformation mechanisms.

Naturally deformed rocks are deformed by strain rates several orders of magnitude slower than the slowest rates achieved in laboratory. This and the effect of other parameters make comparisons of nature and experiment difficult (Paterson 1987) and therefore observations on naturally deformed rocks are an important contribution to the understanding of rock deformation. This is especially true for the natural deformation of very fine-grained limestones at relatively low temperatures, which is the subject of this study.

Deformation mechanism maps for calcite predict either pressure solution (Rutter 1976) or some kind of grain size sensitive creep mechanism (e.g. superplasticity, Schmid 1982, fig. 14) for the low temperatures (160–350°C), small grain sizes (2–8 μm) and slow strain rates (10^{-13} to 10^{-15} s^{-1}) of the material studied (Schmid *et al.* 1977, 1980). In fact, brittle fracturing, vein formation and accompanying pressure solution are important in the study area and contribute to the large scale deformation, folding and thrusting in the lower temperature areas. Some plastic deformation, however, occurs even in these limestones and this paper focuses on this ductile low temperature deformation. By following single stratigraphic horizons from non-metamorphic into greenschist facies regions, finite strain and the coarsening of grain size with increasing temperature was monitored. We will examine the interrelationships between dynamic grain growth and the dominant deformation mechanisms.

Even at low temperatures, (estimated at 160°C from vitrinite reflectance by Burkhard & Kalkreuth 1989) micritic limestones do deform plastically, as shown by deformed markers (pellets). If this deformation was achieved by intracrystalline deformation alone, grains should display a grain shape preferred orientation corresponding to the axial ratio of the measured finite strain (Ramsay & Huber 1983; fig. 7.11).

From Knipe, R. J. & Rutter, E. H. (eds), 1990, *Deformation Mechanisms, Rheology and Tectonics*, Geological Society Special Publication No. 54, pp. 241–257.

If not, deformation involved grain boundary sliding or alternatively, recrystallization has obliterated the deformation microstructure. The second case is certainly responsible for the general lack of strong grain shape preferred orientations observed in the extremely deformed tectonites of the Helvetic root zone.

Various authors have examined the crystallographic textures of these epizonal (> 300°C) tectonites and invariably found strong *c*-axis maxima (Schmid *et al.* 1981; Dietrich & Durney 1986) consistent with a significant role being played by twinning in the deformation. These tectonites are not really micrites anymore but have increased their grain size to about 10 μm. This increase in grain size occurs by grain boundary migration and has a strong influence on the deformation behaviour. In the lower temperature regions studied, I expected to find areas where this influence of recrystallization is less important and where the dominant deformation mechanisms can still be inferred from deformation microstructures.

Regional geology and sampling

The Helvetic ranges of western Switzerland expose a variety of Mesozoic limestones and marls within three superposed nappes in continuous profiles (Fig. 1). The metamorphic grade ranges from diagenetic (160°C) in the north and at the top to epizonal (350°C) in the south and at the bottom (Frey 1986; Burkhard 1988). Temperatures were obtained from vitrinite reflectances using Lopatin's modelling technique in the north (Burkhard & Kalkreuth 1989), calcite/dolomite and stable isotope thermometry in the south and at the bottom (Burkhard & Kerrich 1988, table 2). Two particular lithologies have been chosen for this study: massive micritic Malm limestones and lower Cretaceous Öhrli limestones (Fig. 1). The latter consist of micritic pellets in a sparry matrix and are particularly suitable for finite strain determinations. Finite strain in the Helvetic nappes varies mainly as a function of metamorphic grade: upper and northern units are weakly deformed whereas lower and southern units (from the 'root zone') are invariably strongly deformed (Ramsay & Huber 1983, fig. 11.9, 11.10; Burkhard 1986). In order to have a wider spectrum of strain magnitudes, both moderately deformed 'representative limestones' from nappe interiors and extremely deformed 'tectonites' from thrustplanes and faults were sampled. Some of these samples have been described in Burkhard (1986) where finite strain and strain partitioning are discussed in the larger tectonic context. Superposition of different deformation phases is important in the root zone (Masson *et al.* 1980; Dietrich & Durney 1986; Burkhard 1988) but generally it can be assumed that the ductile deformation took place close to peak metamorphic conditions. Strain rates cannot be accurately determined (does a tectonic phase take place within half a million or during fifteen million years?) but can be estimated at 10^{-13} to 10^{-15} s^{-1} (Pfiffner & Ramsay 1982).

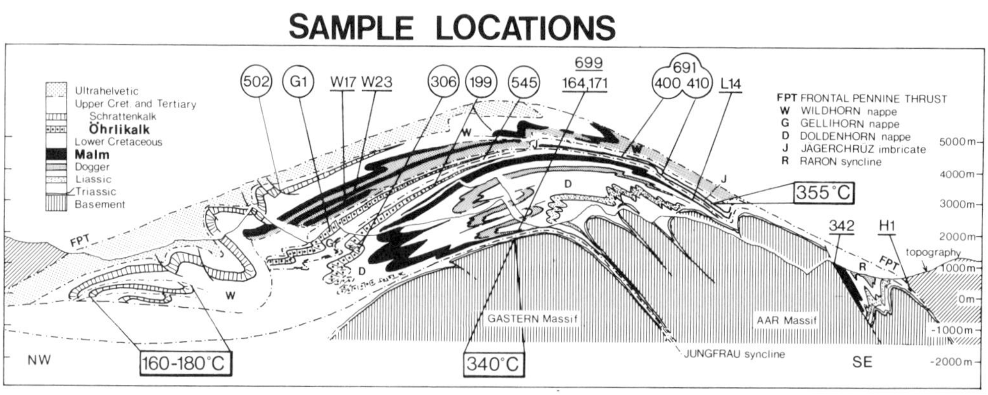

Fig. 1. Sample locations projected onto a synthetic cross section of the Helvetic nappes at the western end of the Aar Massif. The information above current topographic levels was constructed by projecting from the west side along plunging fold-axes. Indicated temperatures are inferred from coal rank data using Lopatin's time depth burial curves (Burkhard & Kalkreuth 1989) in the north, and calcite/dolomite and isotopic thermometry for the higher temperatures (Burkhard & Kerrich 1988, table 2). Malm samples are underscored, Öhrli limestone samples are circled.

These are several orders of magnitudes slower than the slowest experiments and observations of naturally deformed limestones at low temperatures make it possible to study the products of deformation under conditions not accessible to laboratory experiments.

Analytical methods

The analytical procedure followed in this study was designed to determine quantitatively finite strain and various microstructural parameters (Fig. 2).

From the oriented samples three mutually perpendicular thin sections were cut and labelled *.1, *.2, *.3, where * is the sample number. *XZ*, *YZ*, *XY* denote the sections with the largest, intermediate and smallest axial ratio encountered but these do not necessarily coincide with principal planes of the finite strain ellipsoid.

The samples were examined by transmitted light microscopy on ultra-thin sections (Fig. 3). Sections less than 1 μm thick were necessary for the observation of the micritic grains and for semi-quantitative crystallographic texture analyses using the photometric method (Price & Williams 1989). Ultra-thin sections were obtained by grinding the rock slices on successively finer grained waterproof SiC papers (Struers No. 500 to 4000) prior to mounting on glass slides. Standard techniques were applied to produce thin sections approximately 20 μm thick. These were then ground by hand on turning tables by successively finer papers (1000 to 4000) down to about 3 μm thickness and finished with 4, 1 and $\frac{1}{4}$ μm diamond spray on DP-cloth.

Pellet outlines of at least 150 (up to 500) markers were drawn on transparent paper from direct enlargements of the thin sections. The average diameter of pellet contours on the original drawing varied between 1 and 3 cm.

Observations on the grain scale were made using a petrographic microscope with the largest (100 ×) objective, oil immersion. Each analysis is compiled from at least three black and white microphotographs (Fig. 3C, D) taken for three different analyser/polarizer positions (analyser and polarizers were moved in 10–15° steps, keeping the thin section in its original position). This procedure allowed the identification of a maximum number of grain- (and ?subgrain-) boundaries. Complete grain boundary networks of up to 500 grains were traced from the enlarged micrographs. The average diameter of grains on the original drawing was 1–5 cm. The pellet outline and grain boundary network drawings were subsequently analyzed with a QUANTIMET 920 (Cambridge instruments) image analyser with a field of view of 800 × 625 pixels, for the following parameters: area (*A*), longest (*L*) and shortest (*S*) projection-diameter, orientation (ϕ) of the longest diameter, centre coordinates (*X*, *Y*) and Roundness (*Rd*). These parameters were stored on magnetic tape for further data processing on a VAX computer. The following analyses were performed routinely.

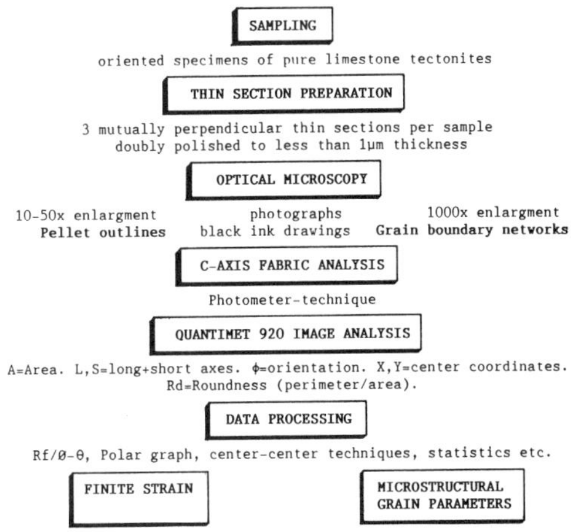

Fig. 2. Schematic diagram of the strategy followed in this study.

R_f/ϕ–*theta*

Such analyses according to Lisle (1977), using *L*, *S* and ϕ allowed a direct determination of both *finite strain* (R_s) from the pellet outlines and mean *grain shape* preferred orientation (*R*) from the grain-boundary networks. The latter result may be questionable because grains are not elliptical particles but complex polygons which are poorly described by *L*, *S* and ϕ alone (Panozzo 1983; Simigian & Starkey 1986). For complex grain shapes (with concave parts), *L*, *S* as measured by QUANTIMET 920 are not necessarily perpendicular to each other. $R_s(R_f/\phi-\theta)$ values nevertheless compared favourably with the R_s values obtained with the more appropriate centre–centre techniques.

Centre–centre

Both classical Fry (1979) and normalized (Erslev 1988) plots were obtained from the centre coordinates *X*, *Y* and $(L + S)/2$ to normalize grain-size for the Erslev technique (Fig. 4). Comparison of the $R_f/\phi-\theta$ (markers) and centre–centre (bulkrock) results further allowed testing for a possible competence contrast between the micritic markers (pellets) and the sparry matrix (Fig. 5A).

Centre-centre methods, applied to the grain boundary networks, are a simple, unbiased way to characterize a mean *grain shape* preferred orientation. Fry-plots are to scale, where the empty space around the centre gives a direct measure of the size of the smallest grains and their preferred orientation. Axial ratios are usually ill-defined even with very large numbers

Fig. 3. Microphotographs of representative samples. All but the SEM pictures have the same scale; scale bars are 10 μm. SEM pictures are made on polished and subsequently HCl (10%, 30 s) etched surfaces. Ultra-thin section micrographs (C–K) are taken with a 100 × objective, oil immersion and crossed polars. (**A**) SEM picture of undeformed, non-metamorphic Malm micrite from the Jura (sample 250). (**B**) SEM picture of weakly deformed, anchizonal Malm micrite (W17) from the northernmost Wildhorn nappe, T estimated at 250°C. (**C, D**) G1.2, micritic pellet interior moderately deformed at *c*. 180°C. C and D are two pictures from exactly the same area but taken with a 30° difference in analyser/polarizer position indicated by + sign within the same grain in each micrograph. (**E**) Micritic pellet interior of sample 400.2, strongly deformed at *c*. 300°C. (**F**) Calc-mylonite (164.2, Doldenhorn thrust), dynamically recrystallized, at *c*. 340°C. (**G**) Extremely deformed and recrystallized Öhrli limestone (sample 545.2) from the Gellihornthrust (T = 300°C); example for 'foam texture'. (**H**) Grain boundary migration recrystallization in Malm limestone (sample 699.1) from Morcles thrust at Saillon (300°C); example for 'sutured grain boundaries'. (**I**) Strongly twinned and recrystallized grains in Malm sample H1 from Raron (350°C). (**J**) Twinning and recrystallization in very fine-grained diagenetic calc-mylonite (sample 502, T < 180°C) led to a strong *c*-axis preferred orientation (vertical in this photograph).

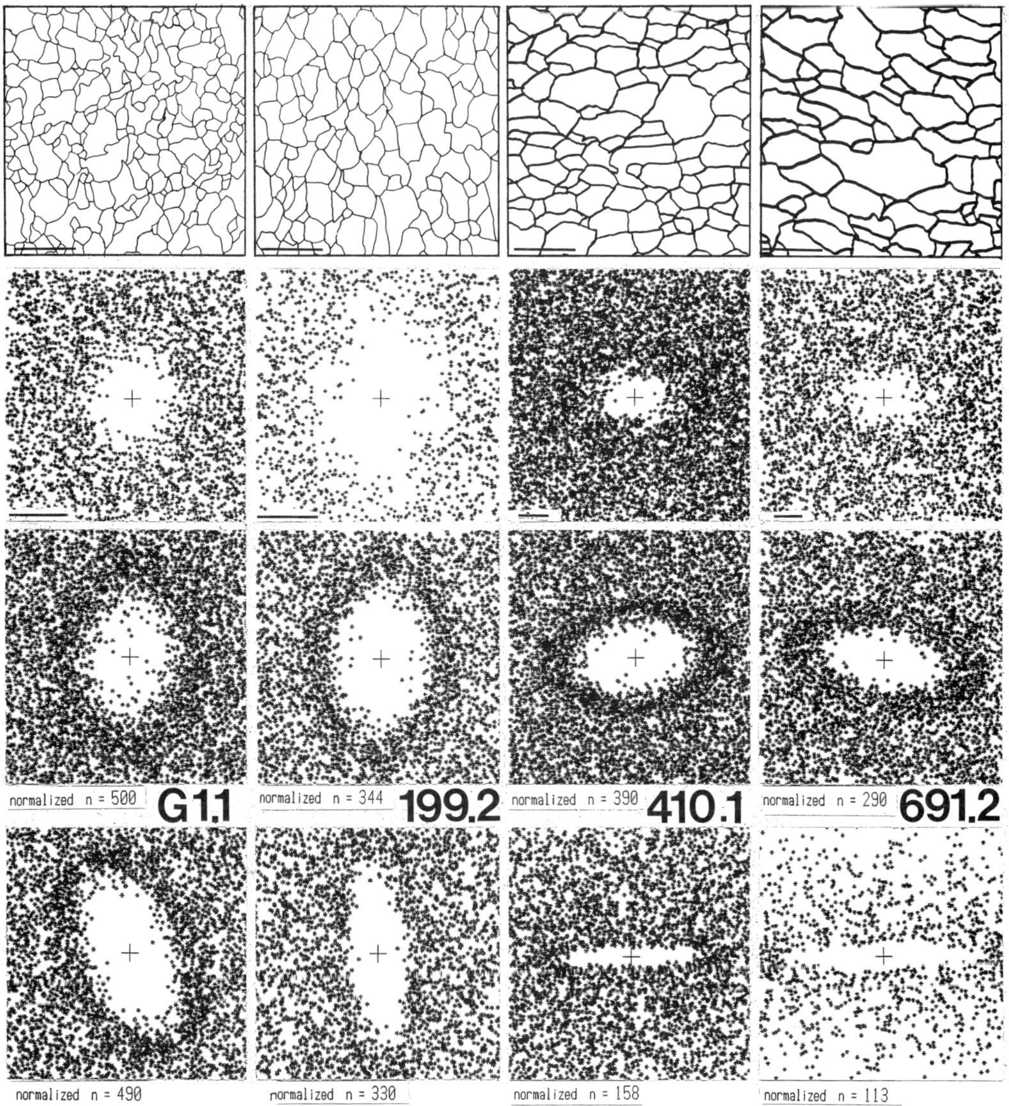

Fig. 4. Different steps of microstructure and strain analysis are illustrated for samples G1.1, 199.2, 410.1 and 691.2 with increasing temperature and finite strain. Scale bars are 5 and 2 μm in the first and second row respectively. Grain boundary networks (first row) traced from enlarged photomicrographs for the use of the image analyser are compared with classical (Fry 1979) centre–centre plots (second row), normalized centre–centre plots (Erslev 1988) (third row) and finite strain determination with the latter technique applied to the pellets (fourth row).

of centres. Normalized (Erslev 1988) plots give much clearer results with pronounced 'high density rings' where ellipses can be determined with a fair level of confidence. Such plots have also been drawn for different classes of grain sizes (and compared to the corresponding $R_f/\phi-\theta$ analyses, Fig. 5c).

Grain shape- and strain-ellipsoids

3-D pellet shape (= finite strain) and grain-shape ellipsoids were calculated from the three faces using PASE5 (Siddans 1971), TRISEC (Milton 1980) and/or TRXL (Genzwill & Stauffer 1981) programs; resulting 3-D para-

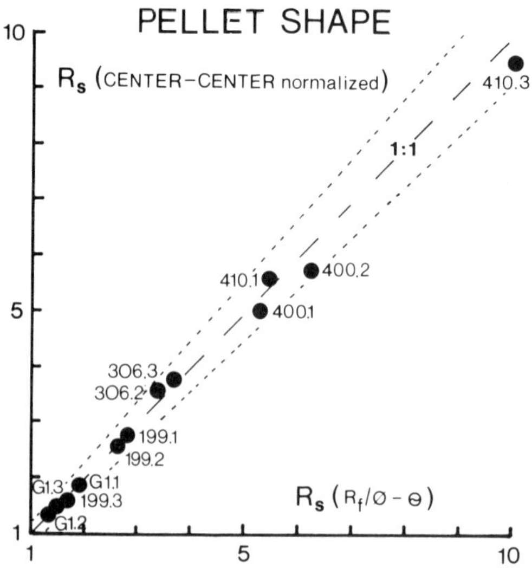

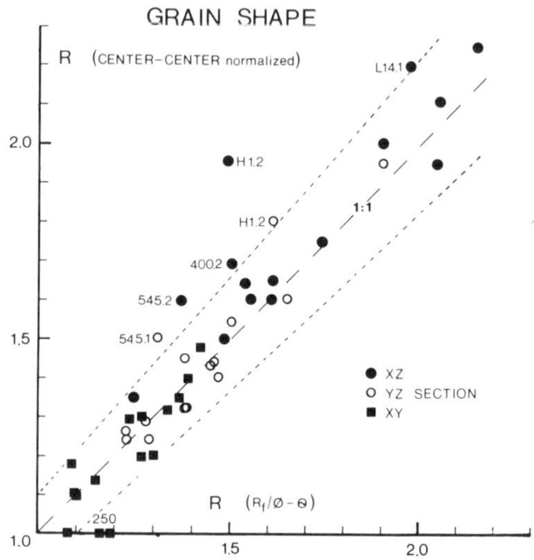

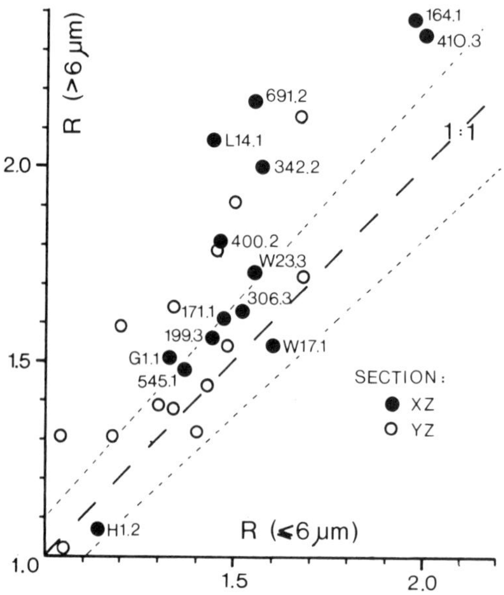

Fig. 5. (**a**) Comparison of results obtained with centre–centre normalized (Erslev 1988) and $R_f/\phi-\theta$ (Lisle 1977) techniques applied to pellet shapes. The generally close compliance of the results suggests that there is no competence contrast between markers and matrix. (**b**) Comparison of centre–centre normalized and $R_f/\phi-\theta$ techniques applied to grain shape. Only samples plotting outside the dotted ± 10% 'error area' are labelled. A large discrepancy exists for sample H1.2 with highly irregular grain shape (H1.2). (**c**) Comparison of grain-shape R of large (> 6 μm) and small (< 6 μm) grains ($R_f/\phi-\theta$ analyses). XZ-sections are labelled. Tectonites (L14, 164, 342, 400, 410, 691) plot clearly above the dotted +10% error line and thus show a tendency for larger grains to be more flattened than the smaller ones. The latter are interpreted as recrystallized grains (or subgrains).

meters are given in Table 1 as X, Y, Z axes, where $X \times Y \times Z = 1$. However, the discussion will mainly deal with axial ratios directly obtained from the XZ-sections.

Statistics

Mean, standard deviation σ and skewness was calculated for each sample together with the following grain-parameters: (a) surface area A; (b) grain size D as defined by $(L + S)/2$; (c) roundness Rd; (d) axial ratio (L/S); (e) log $((L + S)/2$; (f) $\sqrt{A}$; (g) log $(\sqrt{A})$. Histograms were drawn for grain size (D) distributions by number and by surface area and for comparative roundness distributions (see below). Statistical tests have been applied to grain size and roundness distributions in order to decide whether or not the means of any two samples are significantly different from each other (Student's test) and whether or not their distributions are similar (chi^2 test). In either case, $p = 0.05$ has been

Table 1. *Microstructural and finite strain parameters*

Sample no.	n^*	Microstructure: grain shape $R^\dagger$	grain shape $R^\ddagger$ /	grain shape $\phi^\S$	$X^\parallel$ Y Z	grain size (μm) D	grain size σ	grain size skewn.	Finite strain $R_s^\ddagger$ /	Finite strain $\phi^\S$	$X^\parallel$ Y Z	Locality[¶] Tectonic unit, lithology Temperature, tectonic context
G 1.1	500	1.35	1.25	104	1.17	3.0	1.5	1.6	1.92	117	1.64	Gellihorn 616.08/145.9
G 1.2	441	1.18	1.09	81	0.98	3.0	1.4	1.4	1.28	77	.88	Gellihorn nappe, Öhrlifm.
G 1.3	496	1.24	1.23	129	0.87	2.8	1.2	1.3	1.45	108	.70	180°C, hinge of large fold
H 1.1	73	1.24	1.29	171		18.	19.6	2.3	>3.	0		Heidnischbiel 628/129
H 1.1	126	1.8	1.61	175		58.	44.7	1.6	>3.	0		Parautochthonous, Malm
H 1.2	45	1.96	1.49	163		69.	47.8	−.4	>10.	0		350°C, mylonite
L 14.1	123	2.65	1.97	171	1.44	7.9	4.2	1.3	>10.	164		Leuk 614/130.5
L 14.2	150	1.9	1.95	166	1.11	8.5	6.0	1.4	>10.	142		Doldenhorn nappe, Malm
L 14.3	79	1.5	1.42	118	0.63	7.7	5.1	1.4	1−2.	112		330°C, mylonite
W 17.1	376	1.65	1.61	81	1.33	3.5	1.6	1.0	weak			Lämmernh. 610.35/138.40
W 17.2	330	1.32	1.39	95	1.09	3.9	2.2	2.3				Wildhorn nappe, Malm
W 17.3	394	1.37	1.37	112	0.75	3.1	1.6	2.3				250°C, fold limb
W 23.1	430	1.14	1.15	171	1.25	3.0	1.6	1.6	weak			Lämmernh. 610.55/138.80
W 23.2	220		1.34	121	1.04	2.9	1.4	1.5				Wildhorn nappe, Malm
W 23.3	285	1.6	1.61	81	0.76	3.8	2.3	2.4				250°C, fold limb
164.1	236	2.11	2.05	161	1.40	7.6	4.1	1.3	>10.	0		Gasterntal 620.9/143.9
164.2	275	1.44	1.46	5	1.06	5.6	2.9	1.4	>3.	0		Doldenhorn nappe, Malm
164.3	362	1.1	1.1	23	0.68	9.3	5.3	0.7	>3.	0		340°C, thrust mylonite
171.1	113	1.68	1.61	113	1.23	7.9	4.2	1.8	>10.	90		Gasterntal 620.9/143.9
171.2	118	1.0	1.08	170	1.10	7.6	4.2	1.1	>3.	0		Doldenhorn nappe, Malm
171.3	116	1.4	1.47	176	0.73	7.3	5.0	2.4	>3.	0		340°C, thrust mylonite
199.1	332			62		3.5	1.6	1.0	2.78	81	1.68	Rote Chume 612.9/141.1
199.2	344	1.32	1.39	100		3.9	1.5	0.9	2.62	91	1.2	Gellihorn nappe, Öhrlifm.
199.3	249	1.3	1.27	126		4.6	2.2	1.5	1.54	121	.5	250°C, large fold limb
250.1	218	1.0	1.19	118		3.4	1.4	0.9	1.05			Neuchâtel 562/209, Jura, Cret.
250.2	159	1.0	1.16	130		3.0	1.6	1.6	1.06			<60°C, large fold limb

Table 1. *Cont.*

Sample no.	n*	Microstructure							Finite strain			Locality¶
		grain shape			X‖	grain size (μm)					X‖	Tectonic unit, lithology
		R†	R‡ /	ϕ§	Y Z	D	σ	skewn.	R_s‡ /	ϕ§	Y Z	Temperature, tectonic context
306.1	405	1.26	1.23	122	1.29	3.6	1.7	1.3	1.37	116	1.18	Gasteren 616.95/145.35
306.2	401	1.4	1.39	162	1.00	3.1	1.4	0.6	3.24	176	1.30	Doldenhorn nappe, Öhrlifm.
306.3	257	1.5	1.48	171	0.78	3.6	1.7	1.0	3.22	176	.42	280°C, small fold limb
342.1	243	1.7	1.49	160		5.8	2.7	0.9	73.0	0	5.89	Hohtenn 626.05/130.35
342.2	105	2.0	1.90	162		8.2	5.0	1.8	96.0	0	3.34	Parautochthonous, Malm
342.3									1.76	0	.05	350°C, fault mylonite
400.1	272	1.6	1.55	175	1.34	5.3	2.4	0.3	5.26		2.19	Aminona 607.1/130.85
400.2	364	1.7	1.5	9	1.01	4.5	2.1	1.0	6.2	178	1.55	Jägerchrüz imbricate, Öhrlifm.
400.3	223	1.35	1.37	110	0.74	5.8	3.6	1.5	1.35	12	.29	300°C, large fold limb
410.1	390	1.6	1.65	170	1.54	4.3	1.8	1.0	5.41	176	2.42	Aminona 607.9/132.0
410.2	242	1.2	1.27	140	1.02	5.9	2.5	0.9	1.38	159	1.64	Jägerchrüz imbricate, Öhrlifm.
410.3	285	2.25	2.15	14	0.63	4.8	2.5	1.2	7.03	4	.25	300°C, large fold limb
502		<1.2				< 1.0			>10.			Retzligl. 604.6/138.8, Crét. 160°C, fault mylonite
545.1	396	1.5	1.31	4	1.14	4.3	2.0	0.7	>10.	0		Lämmerenboden 611.2/137.5
545.2	258	1.6	1.37	14	1.07	5.3	2.4	0.5	>3.2	0		Gellihorn nappe, Öhrlifm.
545.3	179	1.1	1.10	179	0.82	6.6	3.0	0.9	>3.2	0		300°C, thrust mylonite
691.1	195	1.2	1.30	99	1.37	6.4	3.6	1.4	1−2.	126		Ardon 584.9/117.1
691.2	290	1.75	1.74	14	1.02	5.4	2.9	0.5	>10.	0		Diablerets nappe, Öhrlifm.
691.3	299	1.54	1.5	14	0.72	4.9	2.8	1.5	>10.	0		300°C, thrust mylonite
699.1	216	1.32	1.34	31		12.9	9.8	1.9	>10.			Saillon 579.6/113.3, Morcles thrust, Malm, 300°C

* n, number of grains analysed for the microstructural parameters.
† Centre−centre normalized (Erslev 1988).
‡ $R_f/\phi-\theta$ (Lisle 1977), computer program by C.J. Beach (1977, unpubl.).
§ ϕvalues, in degrees, are with respect to an arbitrary reference line, identical for strain and grain shape.
‖ X, Y, Z = long, intermediate and short axis of 3-D ellipsoid as determined with PASE5, TRISEC and/or TRXL (Siddans 1970; Milton 1980; Genzwill & Stauffer 1981); X, Y, Z are arranged vertically!
¶ Localities and reference grid according to Swiss topographic maps 1:25 000; temperatures interpolated from: Burkhard (1988); B. & Kerrich (1988); B. & Kalkreuth (1989).

chosen as the critical value to accept or reject the testing hypothesis: 'indistinguishable', whenever used in this paper, has this statistical meaning.

Grain size

Grain sizes referred to in this paper are defined by diameter $D = (L + S)/2$ for individual grains. Mean grain size $\bar{D}$ is calculated from a population of n grains. This mean grain size is directly related to the mean grain surface area $\bar{A}$ by $\bar{D} = 1.43 \times \sqrt{\bar{A}}$ (correlation coefficient $r = 1.0$ for all 41 analysed faces). In order to compare different samples, histograms of relative frequency v. grain size have been drawn for each sample together with the calculated mean grain size $\bar{D}$, mean of logarithmic grain size $D' = 10^{\log(D)}$. When examining such histograms it has to be kept in mind that a large grain occupies a larger area than a small grain. Histograms with 'number of grain v. grain size' as well as mean grain size therefore tend to over-accentuate the importance of small grains. In Fig. 6, grain size distributions by surface area have therefore been superimposed (dashed

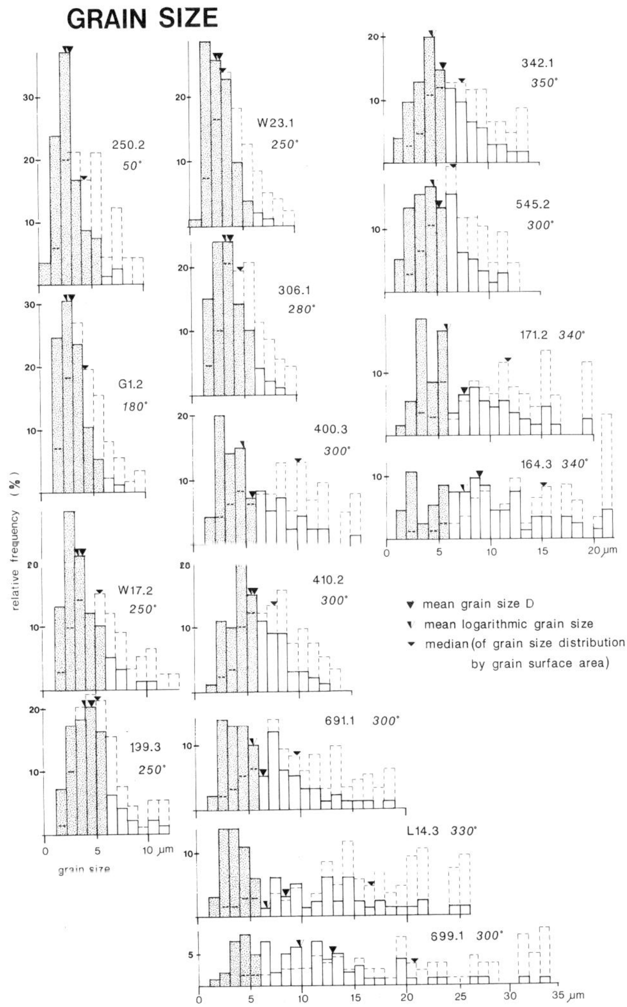

Fig. 6. Grain size distributions by number (plain lines) and by surface area (dashed lines). Sample numbers and estimated temperatures are given according to Table 1. The first two columns are 'representative micrites', the third column are extremely deformed tectonites from thrust planes. Micrite fraction (<6 μm) grains are stippled. This fraction represents either original micrite grains (left column) or subgrains (right column).

histograms). This 'areal' grain size distribution and its median grain size <*D*> more closely reflects the subjective impression of grain size because it emphasizes the presence of large grains, even if these are not very frequent in number.

No attempt at determination of 'real' (3-D) grain size distributions has been made. A priori, two-dimensional observations are not sufficient for the determination of the real grain size distribution. Truncation and sampling effects have to be considered and a real grain size determination requires the knowledge of the exact grain shape in three dimensions (Núñez & Domingo 1987). Conclusions based on differences between grain size distributions are largely independent of such corrections. However, when comparing absolute grain size values with those from other authors, corrections may be necessary to compensate for the different measuring methods applied.

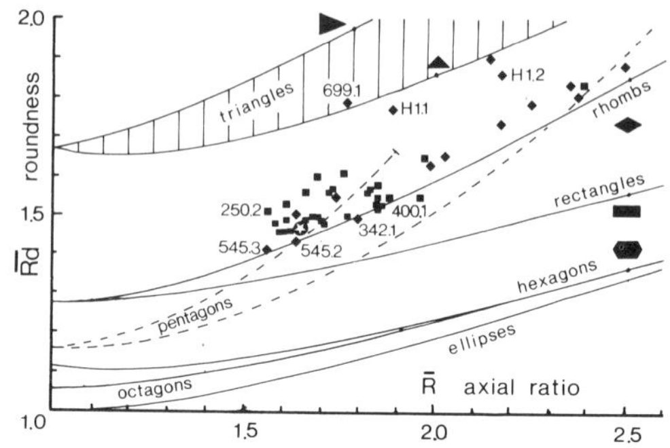

Fig. 7. Mean roundness ($\overline{Rd}$) versus mean grain shape ($\bar{R}$) for all samples analysed. Note that $\bar{R}$ (mean) is not identical to R resulting from $R_f/\phi-\theta$ analyses! Epizonal mylonite samples are represented by diamonds; squares are 'micrites'. Only some outstanding samples referred to in the text are labelled. Curves of $\overline{Rd}$ v. $\bar{R}$ are drawn for a number of regular geometric figures. *denotes a 'typical microstructure for superplastic flow' as determined from Schmid (1982, fig. 18).

Roundness

Roundness is defined as $Rd = P^2/4\pi A$, where P is the perimeter and A the surface area of the grain. This parameter, designed as a measure of the smoothness of grain boundaries, makes it possible to quantify the quality (smooth v. sutured) of grain boundaries. Circles have a roundness of 1, rugged outlines result in higher values, with an unlimited scale. Values higher than 1, however, are also obtained for elongate grains with perfectly straight or smooth boundaries, and there exists a close relationship between Rd and mean axial ratio $\bar{R}$ (Fig. 7). For a better understanding of this effect, Rd has been calculated for different geometric figures as a function of their axial ratio. The resulting curves are plotted on Fig. 7. In order to compensate for this influence of axial ratio, comparative roundness ratios = ($Rd_{\text{measured}}/Rd_{\text{hex}}$) where Rd_{hex} is the roundness of a perfect (flat-lying) hexagon with 120° angles and with the same axial ratio as the grain considered) were separately calculated for each grain in a population and then plotted as histograms for each sample (Fig. 8). Such 'comparative roundness ratio histograms' quantitatively characterize the ruggedness of grain boundaries virtually regardless of axial ratio. Values lower than 1 are in theory possible for figures smoother than a hexagon (ellipse, octagon). Values close to 1 belong to grains that are almost perfect hexagons and thus characterize 'foam' textures. Values up to 1.5 may be either due to more angular figures (rectangles < 1.2, triangles < 1.5) or, alternatively, to rugged grain boundaries. Values higher than 1.5 are positively due to rugged grain boundaries, but the latter is the more probable case for any values higher than 1.3 because triangular grains are rare.

Photometric c-axis fabric analyses

Certain crystallographic textures are easily recognizable under a normal petrographic microscope with the use of a gypsum plate. This is a well known practice in quartz tectonites but works with any uniaxial mineral with first order grey colours under crossed polarizer/analyser. Calcite less than 1 μm thick is comparable to quartz in this respect and allows the use of the photometric method of Price (1973) for semi-quantitative assessment of *c*-axis fabrics. This method was improved for 3-D analyses (Price 1980) and adapted to calcite by Price & Williams (1989) but neither method was available to the author. 2-D analyses (Price 1973, Wallbrecher 1988), however, enable the evaluation of the presence or absence of *c*-axis fabrics from the same small areas of pellet interiors of which the microfabric has been examined. The photometer settings used were the same as for quartz. Since calcite has negative optical characteristics, *c*-axis maxima correspond to minima in transmitted light intensities.

Photometric analyses are an interesting alternative to the X-ray texture goniometer or U-stage which are either not adequate or impossible to use on the samples studied. X-ray

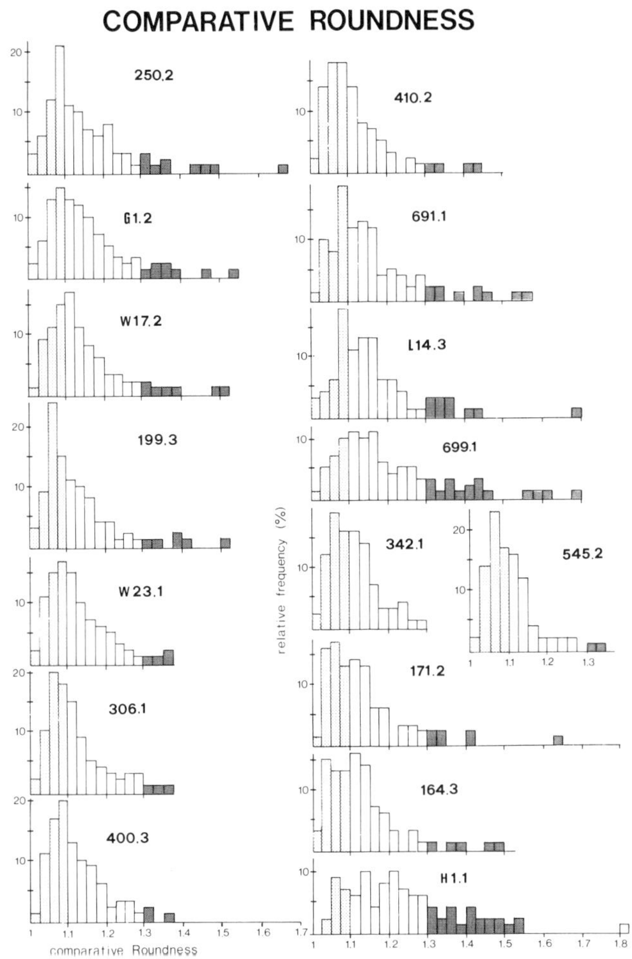

Fig. 8. Histograms of relative frequency (in %) versus comparative roundness ratio. This is a measure of the ruggedness of grain boundaries. Values of 1 to 1.1 (light grey) are almost perfect (120°C angles) hexagons, values higher than 1.3 (dark grey) are due to sutured grain boundaries.

analyses integrate fabrics over a centimetric area of a thick section. Most of our samples have bimodal grain size distributions (micritic pellets in a sparry matrix) and the texture of interest is the micritic part of the sample only. Due to the small grain size, U-stage measurements were impossible.

Results obtained

Grain shape preferred orientation versus finite strain

In Fig. 9 grain shape preferred orientations are plotted versus finite strain (Öhrli limestones). Strain ellipses are drawn for *XZ*-sections, their size corresponding to the respective mean grain size. Highly deformed samples from thrust and fault planes, with finite strain values certainly higher than 10, are represented on the right side of the figure. Visibly elongate grains (*R* up to 2.1) are found in highly deformed samples (410, L14, 164) but *R* values are usually quite small and never correspond to the finite strain R_s with the exception of two *XY* (schistosity parallel) sections (400.3, 410.2).

Grain size

Grain size distributions for the intermediate (YZ) section of each sample are shown in Fig.

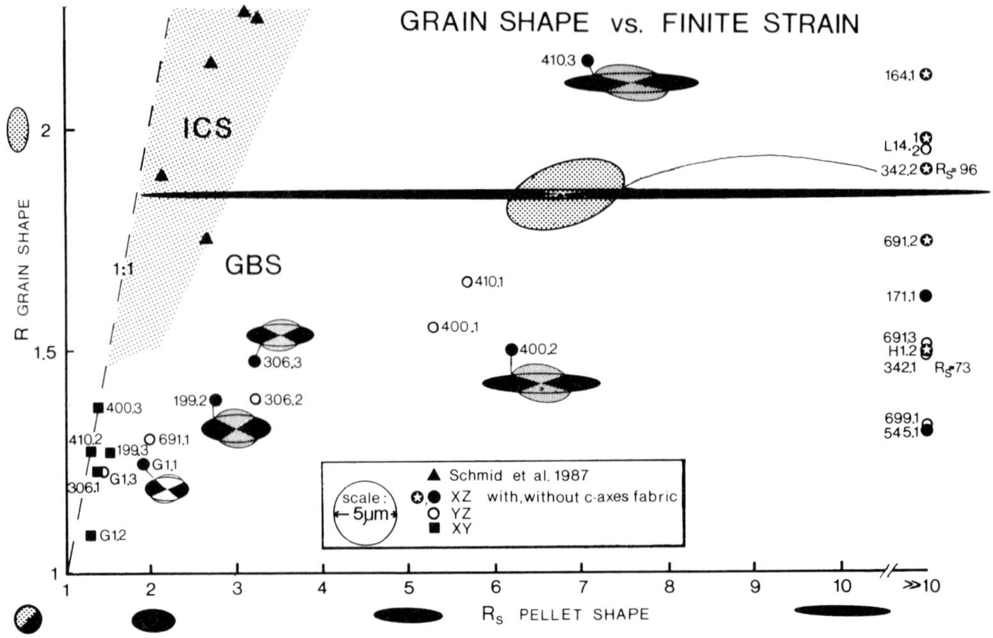

Fig. 9. Grain shape preferred orientations (R) versus finite strain (R_s) from $R_s/\phi-\theta$ analyses of grain boundary networks and pellet outlines respectively. The R-scale is stretched by a factor of 5 with respect to the R_s-scale. Ellipses are drawn for the XZ-sections, where known, with the macroscopically visible schistosity horizontal. The surface area of ellipses corresponds to the measured grain size; note scaled unit sphere. ICS and GBS denote fields of intra-crystalline slip (plus twinning) and grain boundary sliding regimes respectively. Triangles are values from Schmid *et al.* (1987; samples ST1, CT1, CT3, CT5, CT9) deformed in the ICS (twinning) regime.

6. The histograms are arranged with increasing temperature from top to bottom. The first two columns are 'representative micrites' whereas the third column is 'mylonite' from thrust and fault planes. Grain sizes range from 1–35 μm and the lower temperature samples are in the micrite range (1–5 μm) as defined by Folk (1965). Sample 250.2 is an non-metamorphic micritic limestone from the Jura mountains, certainly never heated to more than 60°C.

Up to middle anchizone (250°C), grain size distributions are indistinguishable from this sedimentary standard (250.2). At higher temperatures, larger grains are more frequent but mean grain size only increases slightly. This is in part due to the fact that few large grains contribute much to the area but do not count correspondingly by number. However, the same trends persist in grain size distributions by surface area (dashed histograms). Histograms of the epizonal limestones are characterized by the absence of clearly defined modes, a larger variety of grain sizes and an increasing difference between mean and median grain size (Fig. 6).

In Fig. 6, 'micritic' (1–6 μm) and 'microspar' (6 μm and more) grains are distinguished using an arbitrary, slightly higher, limit than that proposed by Folk (1965: 4–5 μm). This choice was determined by the reference 'starting material' sample 250.2. On the higher temperature side of Fig. 6, the same grain size classes of small grains represent either subgrains or recrystallized grains (or are due to truncation of larger grains). At intermediate temperatures (200–300°C) the question arises whether the observed grain size distributions reflect original grain size, recrystallized grain size or a mixture of both.

Optical criteria for the distinction of subgrains and 'real' grains, notably small angular misorientation (1–8°) and distinct grain boundaries in plain light (Means & Ree 1988, p. 766) are difficult to apply to small micritic calcite grains. Angular misorientation is measured in the plane of the section alone and is always smaller than the real (unmeasurable) 3-D angular difference. Due to the high birefringence of calcite, the appearance of a grain boundary in plain light is strongly dependent on its orientation with respect to the crystallographic orientation of the grains. Some of the grains drawn in the 'grain boundary networks' present some characteristics of and probably are subgrains but since there is no objective way of separating subgrains from grains quantitatively, both were treated as 'grains'. The smaller (1–6 μm) grains in the epizonal samples certainly represent mainly subgrains and recrystallized grains rather than remnant original grains (Fig. 5c).

Comparative roundness ratio

Roundness is a factor which allows the quantitative characterization of textures otherwise subjectively described as 'display straight simple grain boundaries meeting often in 120° triple junctions (foam texture)' or 'irregular, sutured grain boundaries'. Applied to the samples of this study, the two most striking examples for 'foam texture' (545; Fig. 3G) and for 'sutured grain boundaries' (699, H1; Fig. 3H) are clearly distinguished in a roundness v. axial ratio diagram (Fig. 7) and have also quite distinct comparative roundness distributions (Fig. 8). The two extreme samples are characterized by the presence and absence of a clearly defined mode respectively. Most of the analysed samples however, regardless of temperature, finite strain, grain size, grain shape or tectonic context, plot between the two extremes and are also statistically indistinguishable from each other. For instance, the sedimentary reference sample 250.2 is indistinguishable from the epizonal tectonites 410.2, 691.1, 171.2. All these samples have asymmetric comparative roundness ratio distributions with a mode of less than 1.1 which indicates grains with fairly straight boundaries (almost perfect hexagons) but some sutured grains as well. Typical 'foam texture' samples on the other hand are not restricted to the higher temperatures either. Thus, on the basis of grain structure, the epizonal mylonites 342.1, 545.2 are indistinguishable from the weakly deformed, low temperature micrites W23.1 or 306.1.

Photometric c-axis fabrics

Figure 10 illustrates the presence or absence of optically measurable *c*-axis fabrics for *XZ* sections in the different samples analysed on a photometer. No trace of fabric could be detected in the low temperature (up to 300°C) 'micrite' samples despite finite strain R_s values as high as 7 (410.3). In fact, these considerably deformed samples are indistinguishable from the undeformed sedimentary standard 250.2 with respect to *c*-axis fabric. One important exception is the

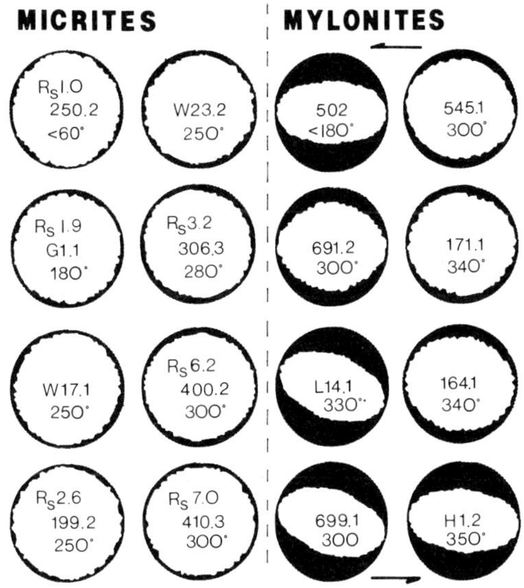

Fig. 10. Rose diagrams of integrally measured azimuths of calcite *c*-axis using the photometric method of Price (1973). Photometer settings are those for quartz as described by Wallbrecher (1988, fig. 15), applied to calcite thin sections of less than 1 μm thickness. *C*-axis maxima correspond to less than maximum light intensities—filled in black. The first two columns correspond to deformed micrite samples. The last two columns are extremely deformed fault and thrust samples. Finite strain ratios R_s (where known), sample numbers and temperatures are given within the rose diagrams. All sections are *XZ* section with horizontal schistosity and sinistral shear sense where known.

low temperature extremely deformed fault sample 502 (Fig. 10). Surprisingly, this sample displays the strongest fabric of all analysed rocks.

Moderate to strong *c*-axis fabrics are widespread in epizonal (T > 300°C) 'mylonite' samples (Schmid *et al.* 1981; Dietrich & Durney 1986) and this was also found in samples 691, L14, 699, H1 with the less powerful photometric method. However, not all epizonal samples show strong *c*-axis fabrics: samples 171, 164 and 545, despite extreme deformation as evidenced from the geological context, have very weak fabrics comparable to those of the low temperature micrites.

Deformation mechanisms inferred from microstructures

In the following discussion, all possible arguments (field relations, tectonic setting, metamorphic grade, microstructures, *c*-axis fabrics) are taken into consideration in an attempt to infer the dominant deformation mechanism in the different samples. The following criteria are considered (with decreasing reliability): (1) *c*-axis preferred orientations due to intracrystalline deformation; (2) grain shape preferred orientations resulting from either intracrystalline deformation or, alternatively, from diffusional mass transfer (pressure solution/crystallization); (3) decrease in grain size (with respect to the protolith) due to the formation of subgrains, related to intracrystalline deformation; (4) increase in grain size due to grain boundary migration either in dynamic or static recrystallization; (5) sutured grain boundaries resulting from dynamic grain boundary migration recrystallization.

At first sight, the great discrepancy between grain shape preferred orientations and finite strain, as illustrated in Fig. 9, would indicate grain boundary sliding (GBS) as the most important deformation mechanism in the analysed limestones. In fact, no sample plots within the stippled intracrystalline slip (ICS) field as expected for deformation dominated by intracrystalline slip plus twinning (compare with laboratory experiments by Schmid *et al.* 1987). A possible grain boundary sliding (GBS) field is also indicated in Fig. 9 below the ICS field (according to Ramsay & Huber 1983; Schmid *et al.* 1987). However, dynamic and/or static recrystallization of the deformed aggregate could produce the same discrepancy and grain shape alone therefore cannot be used to distinguish GBS from ICS deformation. Thus, the key question in our case is whether or not the calcite aggregates are recrystallized.

A first answer to this question can be found from the grain size distributions (Fig. 6). There are clearly two different groups of samples: (a) diagenetic to anchizonal samples, indistinguishable from the sedimentary reference 250; (b) epizonal samples with increased grain sizes. These two groups will be discussed separately below.

Low temperature GBS deformation ($T <$ 300°C)

If grain size distributions of the low temperature samples (notably G1, 199, 306) still reflect original ('detrital') grain size, then these limestones must have been deformed by grain boundary sliding (assisted by some diffusion process, e.g. dissolution/crystallization in the grain boundary region, to accommodate grain boundary misfits, which explains also some minor grain shape fabric). Alternatively, if these samples were deformed by intracrystalline slip (r, f, etc.), the observed grains have to be interpreted as subgrains and not original grains (because of the missing correspondence between grain shape and finite strain). However, since these 'subgrains' are not smaller than the original micrite grains, one would then have to admit that subgrain formation was accompanied by grain boundary migration recrystallization. The two mechanisms (1) grain size reduction by formation of subgrains and (2) grain growth by migration of grain boundaries, would have exactly balanced each other out to result in a final grain size distribution indistinguishable from the original distribution. This seems an unlikely coincidence and we retain GBS deformation as the more reasonable interpretation for the low-T samples. This is also compatible with the lack of *c*-axis fabrics despite large finite strains (Fig. 10). If ICS had been the dominant mechanism and recrystallization had obliterated the corresponding microstructures, crystallographic textures should still reflect the intracrystalline deformation.

Low-temperature ICS deformation

Among the low-T samples, one important exception from the dominantly GBS deformation is found in sample 502: extreme deformation on a fault plane in diagenetic conditions led to a reduction in grain size (Fig. 3K). In fact, this is the only mylonite sample that shows a grain size reduction with respect to the sedimentary protolith. Sample 502 has grains (< 1 μm) too small for quantitative microstructural studies, and are best interpreted as recrystallized (sub-)grains. Both the size of the recrystallized grains and the small size (< 3 μm) of abundantly twinned grains indicate a very high differential stress exceeding 300 MPa using Schmid's (1982, fig. 9) or Rowe & Rutter's (1990, fig. 6) relationships. Twinning also accounts for the pronounced *c*-axis fabric in this specimen (Fig. 10). Thus sample 502 shows that the characteristics of intracrystalline deformation, namely *c*-axis fabric, twinned grains, grain shape preferred orientation? and accompanying grain size reduction due to recrystallization can survive in low temperature environments. This is also indirect evidence for GBS being an important deformation mechanism in the 'representative' low temperature micrites, because none of these features are found in the latter.

Epizonal tectonites ($T > 300°C$): ICS plus recrystallization

The most striking difference between anchi- and epizonal tectonites is the difference in grain size. An increase in final (equilibrium?) grain size in micritic limestones is observed only in the higher anchizone (280°C) and above. Temperature (and/or differential stress?) rather than finite strain seem to be the main factors controlling final grain size. Finite strain had no influence on grain size. It is important to note that the larger grains in these limestones are not original grains (they were all micrites) but the product of dynamic recrystallization. This is further corroborated by the fact that the small (<6 μm) (sub-)grains (samples L14, 164, 342, 400, 410, 691) have clearly lesser axial ratios than the larger grains (Fig. 5c). Furthermore, apart from any deformation-related parameter, the degree of purity of the limestones may also have an influence on the final grain size (Olgaard & Evans 1988). All of the studied samples are quite similar in this respect and have 2–4% optically invisible clay minerals.

Thus, in the case of the epizonal samples, recrystallization gives a plausible explanation for the weak grain shape preferred orientations compared to the finite strain and there is no immediate evidence for the dominant deformation mechanism which may be any combination of ICS, twinning and GBS.

More or less pronounced *c*-axis fabrics in the epizonal samples along with some degree of grain shape preferred orientations give strong evidence for intracrystalline deformation mechanisms. Twinning is known to contribute greatly to the formation of strong *c*-axis point maxima, invariably found in epizonal tectonites from the Helvetic root zone (Schmid *et al.* 1981, 1987; Dietrich & Durney 1986). Twinning alone, however, cannot lead to large strains unless other glide systems are active and/or recrystallization allows a continuous recovery of the highly strained grains. In agreement with Schmid *et al.* (1987) I interpret the strong fabrics (samples 502, 691, L14, 699, H1) as mainly due to twinning, even if most of the twins have been obliterated by subsequent recrystallization. Even though an increase in grain size from 3–10 μm seems small, it has an important effect on the differential stress needed for twinning (Rowe & Rutter, 1990). Thus, at constant differential stress, this small increase in grain size, along with the increasing ease of recovery processes at higher temperatures, allows a shift from dominantly GBS deformation at lower temperatures (with small grain sizes) to ICS and twinning dominated deformation at higher temperatures (with increased grain size).

Dynamic v. static recrystallization

It is tempting to interprete grain forms and microstructures in terms of static v. dynamic recrystallization. However, features like 'frequent triple junctions, straight grain boundaries, foam texture', often cited as evidence for static recrystallization, or 'sutured grain boundaries' as characteristic for grain boundary migration recrystallization are subjective qualities and more than one process may lead to the same end-product. Unequivocal, measurable parameters to characterize 'statically' and 'dynamically' recrystallized textures do not seem to exist (for critical discussions see: Folk 1965; Hobbs *et al.* 1976, p.105ff; Vernon 1981; Urai *et al.* 1986; Groshong 1988; Covey-Crump & Rutter 1989). Here we tried to use a measurable 'comparative Roundness ratio' (Figs 7 & 8) to distinguish different types of recrystallization. However, there is very little difference in grain forms between non-metamorphic micritic limestones and tectonites. 'Foam texture' samples e.g. were found in undeformed low temperature micrites as well as in highly deformed mylonites, the same is true for sutured grain boundaries.

Given the geologic context, 'foam texture' in extremely deformed thrust samples (342, 545, Figs 3G, 7 & 8 and to a lesser extent samples 171, 164) is likely to be due to dynamic recrystallization where an equilibrium grain size has been reached and may reflect superplastic deformation, supported also by the rather weak or absent *c*-axis fabrics (Fig. 10). Alternatively, the microstructures could be due to annealing, whereby former *c*-axis fabrics were erased. The small grain size in these samples, however, puts a limit to the extent of a possible static recrystallization and I therefore favour the former (GBS) interpretation.

Syntectonic grain boundary migration recrystallization obviously caused strongly sutured grain boundaries in samples 699 and H1 (Figs 3H & 8). Here, strong *c*-axis fabrics (Fig. 10) indicate intracrystalline (mainly twinning) deformation and visibly, dynamic recrystallization did not prevent the development of a strong *c*-axis fabric but wiped out most of the direct evidence for twinning.

Conclusions

(1) Micritic limestones do deform plastically at low temperatures (*c.* 200–300°C) without showing much microscopic evidence for this

deformation on the grain scale. The absence of crystallographic preferred orientations and dynamic or static recrystallization features despite finite strain ratios up to 7, indicates grain boundary sliding as the dominant deformation mechanism. For compatibility reasons this has to be assisted by diffusion transport at the grain boundaries. Pressure solution/crystallization at the grain scale could well be operative, and would allow explanation of some minor grain shape preferred orientations, but this mechanism alone cannot accommodate the large finite strains observed.

(2) Grain shape preferred orientations never correspond to the axial ratio of finite strain. No field of particular deformation conditions could be found where intracrystalline deformation alone was responsible for the observed microstructures. At low temperature, grain boundary sliding is an important contributor to finite strain, whereas at higher temperatures dynamic recrystallization continuously obliterates the deformation microstructures.

(3) Within the higher anchizone (above 280°C), grain boundary migration recrystallization leads to an increase of grain size in deformed micrites. This increase of grain size from 3 to 10 μm seems small but it has a very important influence on the deformation regime because twinning is strongly dependent on grain size. Plastic flow by twinning, together with other intracrystalline slips is the dominant deformation mechanism in the epizone resulting in strong crystallographic textures. Microstructures in this regime, however, are dominated by dynamic recrystallization, not deformation processes.

(4) A very strong *c*-axis preferred orientation is found in a low temperature (< 180°C) fault rock with an extremely small grain size (< 1 μm). Recrystallized grain size and smallest twinned grains both indicate differential stress levels exceeding 300 MPa for this mylonite. I believe that this is the first published occurrence of a naturally deformed limestone displaying characteristics of a mylonite (grain size reduction, dynamic recrystallization, crystallographic fabric) at such a low temperature.

(5) Image analysis is not only an effective way of quantifying microstructural parameters such as grain size, grain shape etc., but also enables the assessment of more complex qualities generally described as 'sutured grain boundaries' or 'foam textures'. In the case of the micrites and mylonites analyzed, 'rugged' grain boundaries or 'foam like' textures alone are not reliable indicators of any particular deformation (or recrystallization) regime.

(6) In the case of some mylonite samples from Helvetic thrusts, grain size distributions, optical microstructures and *c*-axis fabrics are not sufficient criteria to distinguish them from moderately deformed micritic limestones. Outcrop scale structures and geologic evidence (important thinning of original bed thickness etc.) are needed in addition to thin section scale microstructures to unravel the deformation history.

Tables with grain parameters, statistics, histograms, stereograms with sample orientation, 3-D finite strain, grain shape results and original grain boundary network drawings have been deposited in the Society library and the British Library Document Supply Centre, Boston Spa, W. Yorkshire, UK as Supplementary Publication No. SUP 18067 (46 pages).

I wish to thank the Institute of Metallurgy at Neuchâtel University for technical assistance, in particular P. A. Girard who introduced me to the metallurgical techniques of thin section polishing, and who carried out the image analyses. I am also grateful to E. Wallbrecher who gave me access to his photometer, S. Simigian, D. L. Olgaard, N. S. Mancktelow and O. A. Pfiffner for stimulating discussions. S. M. Schmid, J. L. Urai and E. H. Rutter provided constructive critiques and reviews; S. Gough and R. Shaw corrected the manuscript. A Swiss National Science Foundation grant (No. 20–5454.87) and support by the Neuchâtel University is gratefully acknowledged. This paper is part of the author's habilitation thesis.

References

BURKHARD, M. 1986. Déformation des calcaires de l'Helvétique de la Suisse occidentale (Phénomènes, mécanismes et interprétations tectoniques). *Revue de Geologie Dynominique et de Géographi Physique*, **27/5**, 281–301.

—— 1988. L'Helvétique de la bordure occidentale du massif de l'Aar (évolution tectonique et métamorphique). *Eclogae Geologae Helvetiae*, **81**, 63–114.

—— & KERRICH, R. 1988. Fluid regimes in the deformation of the Helvetic nappes, Switzerland, as inferred from stable isotope data. *Contributions to Mineralogy and Petrology*, **99**, 416–429.

—— & KALKREUTH, W. 1989. Coalification in the northern Wildhorn nappe and adjacent units, Western Switzerland, Implications for tectonic burial histories. *International Journal of Coal*, **11**, 47–64.

COVEY-CRUMP, S. J. & RUTTER, E. H. 1989. Thermally induced grain growth of calcite marbles on Naxos Island, Greece. *Contributions to Mineralogy and Petrology*, **101**, 69–86.

DIETRICH, D. & DURNEY, D. W. 1986. Change of direction of overthrust shear in the Helvetic nappes of western Switzerland. *Journal of Structural Geology*, **8**, 389–398.

ERSLEV, E. A. 1988. Normalized center to center

strain analysis of packed aggregates. *Journal of Structural Geology*, **10**, 201–209.

EVANS, B., HAY, R. S. & SHIMIZU, N. 1986. Diffusion-induced grain-boundary migration in calcite. *Geology*, **14**, 60–63.

FOLK, R. L. 1965. Some aspects of recrystallization of ancient limestones. *In*: PRAY, L. C. & MURRAY, R. C. (eds) *Dolomitization and Limestone diagenesis*. Society of Economic Paleontologists and Mineralogists Special Publication, **13**, 14–48.

FREY, M. 1986. Very low grade metamorphism of the Alps: an introduction. *Schweizerische Mineralogische Petrographische Mitteilungen*, **66**, 13–27.

FRY, N. 1979. Random point distributions and strain measurement in rocks. *Tectonophysics*, **60**, 89–105.

GENZWILL, D. & STAUFFER, M. 1981. Analysis of triaxial ellipoids, their shapes, plane sections and plane projections. *Mathematical Geology*, **13/2**, 135–152.

GROSHONG, R. 1988. Low temperature deformation mechanisms and their interpretation. *Geological Society of America Bulletin*, **100**, 1329–1360.

HOBBS, B. E., MEANS, W. D. & WILLIAMS, P. F. 1976. *An outline of structural geology*. John Wiley & Sons, New York.

LISLE, R. 1977. Clastic grain shape and orientation in relation to cleavage from the Aberystwyth grits, Wales. *Tectonophysics*, **39**, 381–397.

MASSON, H., HERB, R. & STECK, A. 1980. Helvetic Alps of Western Switzerland, excursion No. 1 *In*: TRÜMPY, R. (ed.) *Geology of Switzerland, part II*. Wepf Basel.

MEANS, W. D. & REE, J. H. 1988. Seven types of subgrain boundaries in octochloropropane. *Journal of Structural Geology*, **10**, 765–770.

MILTON, N. J. 1980. Determination of the strain ellipsoid from measurements on any three sections. *Tectonophysics*, **64**, 19–27.

NÚÑEZ, C. & DOMINGO, S. 1987. Grain shape and its influence on the experimental measurement of grain size. *Metallurgical Transactions A*, **19A**, 933–940.

OLGAARD, D. L. & EVANS, B. 1988. Grain growth in synthetic marbles with added mica and water. *Contributions to Mineralogy and Petrology*, **100**, 246–260.

PATERSON, M. S. 1987. Problems in the extrapolation of laboratory rheological data. *Tectonophysics*, **133**, 33–43.

PFIFFNER, O. A. & RAMSAY, J. G. 1982. Constraints on geological strain rates: arguments from finite strain states of naturally deformed rocks. *Journal of Geophysical Research*, **87/B1**, 311–321.

PANOZZO, R. 1983. Two dimensional analysis of shape fabric using projections of lines in a plane. *Tectonophysics*, **95**, 279–294.

PRICE, G. P. 1973. The photometric method in microstructure analysis. *American Journal of Science*, **273**, 523–537.

—— 1980. The analysis of quartz *c*-axis fabrics by the photometric method. *Journal of Geology*, **88**, 181–195.

—— & WILLIAMS, P. F. 1989. The photometric method of *c*-axis fabric analysis applied to calcite. *Tectonophysics*, **158**, 343–345.

RAMSAY, J. G. & HUBER, M. I. 1983. *The techniques of modern structural geology, vol. 1, strain analysis*. Academic Press, London.

ROWE, K. J. & RUTTER, E. H. 1990. Palaeostress estimation using calcite twinning: experimental calibration and application to nature. *Journal of Structural Geology*, **12**, 1–18.

RUTTER, E. H. 1976. The kinetics of rock deformation by pressure solution. *Philosophical Transactions of the Royal Society of London*, **A283**, 203–219.

SCHMID, S. M. 1982. Laboratory experiments on rheology and deformation mechanisms in calcite rocks and their application to studies in the field. *Mitteilungen Geologisches Institut ETH und Universität Zürich*, **241**, 1–62.

——, BOLAND, J. N. & PATERSON, M. S. 1977. Superplastic flow in finegrained limestone. *Tectonophysics*, **43**, 257–291.

——, PATERSON, M. S. & BOLAND, J. N. 1980. High temperature flow and dynamic recrystallization in Carrara marble. *Tectonophysics*, **65**, 245–280.

——, CASEY, M. & STARKEY, J. 1981. The microfabric of calcite tectonites from the Helvetic nappes (Swiss Alps). *In*: MCCLAY, K. R. & PRICE, N. J. (eds) *Thrust and Nappe Tectonics*. Geological Society, London, Special Publication, **9**, 151–158.

——, PANOZZO, R. & BAUER, S. 1987. Simple shear experiments on calcite rocks: rheology and microfabric. *Journal of Structural Geology*, **9**, 747–778.

SIDDANS, A. W. B. 1971. *The origin of slaty cleavage*. PhD thesis, University of London.

SIMIGIAN, S. & STARKEY, J. 1986. Automated grain shape analysis. *Journal of Structural Geology*, **8/5**, 589–592.

TURNER, F. J., GRIGGS, D. T. & HEARD, H. 1954. Experimental deformation of calcite crystals. *Geological Society of America Bulletin*, **65**, 883–934.

URAI, J. L., MEANS, W. D. & LISTER, G. S. 1986. Dynamic recrystallization of minerals. *In*: HOBBS, B. E. & HEARD, H. C. (eds) *Mineral and Rock deformation: Laboratory studies*. Geophysical Monograph, **36**, 161–199.

VERNON, R. H. 1981. Optical microstructure of partly recrystallized calcite in some naturally deformed marbles. *Tectonophysics*, **78**, 601–612.

WALLBRECHER, E. 1988. A ductile shear zone in the pan-African basement on the northwestern margin of the West African craton. *In*: JACOBSHAGEN, V. H. (ed.) *The Atlas system of Morocco*. Lecture notes in Earth Sciences, Springer, Berlin, 19–42.

WENK, H. R. 1985. Carbonates. *In*: WENK, H. R. (ed.) *Preferred orientation in deformed metals and rocks: an introduction to modern texture analysis*. Academic Press, Orlando, 361–384.

Experimental study of grain-size sensitive flow of synthetic, hot-pressed calcite rocks

A. N. WALKER, E. H. RUTTER[1] & K. H. BRODIE[1]

Geology Department, Imperial College, London SW7 2BP, UK

[1] *Present address: Geology Department, Manchester University, Manchester M13 9PL, UK*

Abstract: Synthetic calcite rocks of controlled grain-size were prepared by crushing large, clear calcite crystals, centrifuged to separate restricted grain-size fractions, cold-pressed and then hot-pressed to produce low-porosity polycrystals of mean grain-size covering the range 2–40 μm. These were deformed dry over the range 400–700°C, mainly at 200 MPa confining pressure, to investigate the onset with decreasing grain-size of grain-size sensitive flow.

The deformation of the fine-grained material can be described by a flow law of the form

$$\dot{e} = A \exp(-H/RT)\, \sigma^n\, d^m$$

in which at low stresses (<25 MPa differential stress) $A = 10^{4.9}$, $H = 190$ kJ mol^{-1}, $n = 1.7$ and $m = -1.9$, when strain-rate, $\dot{e}$, is in s^{-1}, stress, σ, is in MPa and grain-size, d is in μm. At higher stresses (25–250 MPa) $A = 100$, $H = 190$ kJ mol^{-1}, $n = 3.3$ and $m = -1.3$. In the low stress regime, grains remain equidimensional during flow, a pre-existing preferred crystallographic orientation produced during cold-pressing tends to weaken and grain-boundary sliding is important. At higher stresses, a grain flattening fabric develops, the pre-existing preferred crystallographic orientation pattern is preserved, and recovery-accommodated intracrystalline plastic flow is inferred to be the dominant deformation mechanism. Grain-size sensitivity of the flow stress persists even into the intracrystalline plastic flow regime. Grain-size insensitive flow only develops at grain-sizes greater than 40 μm, as demonstrated by reference to mechanical data for the flow of Carrara and Taiwan marbles.

The flow laws which describe most of the experimental data do not extrapolate well through the lowest strain-rate data at the coarser grain-sizes. This indicates that a complete description of the flow, which can be used reliably to extrapolate to geological conditions, requires one or more of the material parameters A, n and m to be a function of one or more of stress, strain-rate and grain-size. This implies variable contributions from different deformation mechanisms as the deformation conditions change.

In fine-grained materials at high temperatures, grain-boundaries form a substantial fraction of the total volume of the material, and thermally-activated processes at grain-boundaries can become dominant controls on the mechanical behaviour of such materials. In such cases the material may exhibit grain-size sensitive flow, in which the mechanical strength decreases rapidly as the grain-size is decreased. At grain-sizes too large for this type of behaviour, the resistance to intracrystalline plastic flow typically decreases slowly as the grain-size is increased (the Hall–Petch effect).

It is commonly observed that localized zones of high shear strain in rocks naturally deformed at medium to high grades of metamorphism are characterized by tectonic grain-size reduction. It is frequently suspected that the shear localization is consequent upon or maintained by weakening associated with grain-size reduction. Although there has been much speculative writing about the possible mechanical properties of fine-grained rocks at high temperatures, very few experimental studies of the mechanical behaviour of such rocks have been made.

When grain-size becomes a significant variable in the constitutive flow law, it is necessary to be able to deform a range of specimens of identical mineralogical and microstructural characteristics but of different grain-sizes. Natural rocks cannot easily be used for such purposes, hence it is necessary to prepare synthetic, hot-pressed rocks having the desired characteristics.

In this paper we report the results of the first phase of a study in which we have deformed dry, hot-pressed synthetic calcite rocks of controlled grain-size ranging upwards from

From Knipe, R. J. & Rutter, E. H. (eds), 1990, *Deformation Mechanisms, Rheology and Tectonics*, Geological Society Special Publication No. 54, pp. 259–284.

1 micron, at temperatures up to 700°C and at confining pressures ranging between 50 and 300 MPa.

Background to the study

The high-temperature creep behaviour of many crystalline solids appears to follow a constitutive flow law of the form

$$\dot{e} = A\ \sigma^{n} \exp(-H/RT)\ d^{m} \qquad 1.$$

in which $\dot{e}$ is the strain-rate, σ is the flow stress, T is the absolute temperature, d is the grain-size, H is the apparent activation enthalpy for flow and A, n and m are material constants. In the regime of intracrystalline plastic flow of relatively coarse-grained materials, n usually lies in the range 3 to 8 and m may range from zero to about +0.5. In fine-grained materials, when grain-boundary processes may become important in the flow, n may range down to perhaps unity (linear-viscous, or Newtonian flow) and m may become as low as −2 or −3. The values of A and H may also be different. With reference to engineering materials, such flow is sometimes termed 'superplastic', because large strains can be accumulated without a strong tendency for strain-rate instabilities to develop.

The microstructures developed in each flow regime are usually strikingly different, reflecting the differences in the dominant deformation mechanisms. In coarse materials, intracrystalline plastic flow results in the formation of a grain-flattening fabric and a crystallographic preferred orientation. Grain-boundary sliding does not contribute significantly to strain. At moderate to large strains the grains in orientations favouring build-up of high dislocation densities tend to recrystallize, either by subgrain rotation or by grain-boundary migration. Except at high temperatures when the material is weak, the recrystallized grain-size is less than the initial grain-size.

In very fine-grained rocks at elevated temperatures and low stresses, grain-surface forces resist the distortion of the grain from an equant shape, and the mobility of the grain-boundaries and grain-boundary dislocations means that grain-boundary sliding may become the dominant deformation mechanism. At a given strain-rate, the finer grained material flows at a lower stress. In this type of flow the microstructure of small, equidimensional grains is insensitive to large changes in strain, but time-dependent grain-growth may occur, causing hardening (e.g. Schmid *et al.* 1977; Karato *et al.* 1986; Rutter & Brodie 1988*b*). If dislocation glide and climb do not provide dominant contributions to the flow, little or no crystallographic preferred orientation develops and any pre-existing fabric may be weakened. To prevent opening of large intergranular voids during grain-boundary sliding, an accommodation process is required, perhaps causing grain-shape to distort during sliding motions. Accommodation may involve stress-induced diffusive mass transfer, intragranular dislocation motion, rearrangement of groups of grains, or some combination of all three (e.g. Kashyap & Mukherjee 1985). It is widely believed (but has yet to be conclusively demontsrated) that dynamic recrystallization during natural rock deformation can reduce grain-size sufficiently to activate this type of flow, thereby producing weakening.

If stresses in fine-grained materials are large enough to activate dislocation motion and large dislocation densities can develop, a cycle of plastic strain, strain-hardening and dynamic recrystallization may develop such that a stable microstructure is dynamically maintained (White 1976; Sellars 1978; Tullis & Yund 1985). Such flow is expected to be characterized by crystallographic fabric formation and an n value which is characteristic of intracrystalline plasticity, but the form of the grain-size sensitivity is unknown.

Both of the above types of flow (with several variations on the basic themes) may be important in the flow of fine-grained rocks in nature. It is important to discover their respective microstructural and mechanical characteristics from experiments.

In previous high temperature studies on rocks, the regime of grain-size sensitive flow appears to have been accessed, but all of the parameters of the flow law have never been determined for a single material. Schmid *et al.* (1977) showed that Solnhofen Limestone (grain-size 5 μm) approached linear-viscous behaviour at low stresses and high temperatures, and demonstrated hardening accompanying grain-growth. Mueller *et al.* (1981) demonstrated that fine-grained anhydrite was weaker than the coarser grained rock.

Schwenn & Goetze (1978) showed that for grain-size sensitivity in the hot-pressing of olivine powders the compaction rate was approximately proportional to the one-third power of the grain-size. Vaughan & Coe (1981) interpreted their experiments on the flow of magnesium germanate olivine under conditions of transformation to fine-grained spinel in terms of superplastic flow of the reaction products. From experiments on hot-pressed olivine rocks of low porosity and controlled grain-size, Chopra

(1986) and Karato *et al.* (1986) demonstrated grain-size sensitive flow and the transition to relatively grain-size insensitive intracrystalline plastic flow with increasing stress and grain-size. They also demonstrated that the presence of intergranular water causes weakening in both flow regimes. Rutter & Brodie (1988*b*) interpreted the weakening of serpentinite during dehydration to olivine + talc + water to grain-size sensitive flow of transiently ultrafine-grained reaction products.

Tullis & Yund (1985) observed strain-softening associated with the dynamic recrystallization and grain-refinement of albite, and showed that dynamic maintenance of the reduced grain-size occurred by cyclic plastic flow and recrystallization. Kirby & Kronenberg (1984) observed the formation of fine-grained shear zones in some of their specimens of clinopyroxenite, which also displayed dramatic strain-softening.

Experimental techniques

All of the deformation experiments reported here were axial compression tests performed on oven-dry (100°C) specimens using two fluid confining medium testing mechines which have been described by Rutter *et al.* (1985) and Rutter & Brodie (1988*b*). Constant strain-rate, constant stress and stress-relaxation testing methods have been used. In all cases specimen pores were vented (subject to their connectivity) to the atmosphere through the hollow, upper loading piston. Some 120 experiments have been performed, including the tests from which the pressure-sintering technique was developed.

In order to assess the effects of intergranular impurity phases on the mechanical behaviour, some of the tests were performed on hot-pressed samples mixed with 1 and 5 vol% of alumina of < 0.25 μm grain-size. A further test was performed on a sample of alumina-doped hot-pressed calcite polycrystal supplied by D. Olgaard.

In addition, some tests were carried out on samples of Solnhofen limestone, to compare its behaviour with that of both the pure and contaminated synthetic polycrystals and with the results on Solnhofen limestone reported by Schmid *et al.* (1977). Solnhofen limestone contains up to 5% of various intergranular phases, mainly illitic clay, oxides and organic matter. The Solnhofen limestone samples were drilled both normal and parallel to bedding from the same block of Solnhofen limestone used by Rutter (1974). Specimens 1 cm in diameter and 2.2 cm long were deformed in annealed copper jackets of 0.25 mm wall thickness.

All experiments were carried out at temperatures between 400° and 700°C and at strain-rates of 7×10^{-4} s^{-1} or slower. At 700°C grain-growth is a problem in long duration experiments in fine-grained calcite polycrystals, hence it was considered pointless to carry out experiments at higher temperatures. Calcite decomposition in pore spaces is insignificant over the temperature range of these experiments. Most runs were carried out at 200 MPa confining pressure, but the effect of pressure on strength in the range 50 to 300 MPa confining pressure was also measured.

All experimental data are presented with stress values corrected for the strength of the copper jacket (obtained from calibration runs) and for changes in cross-sectional area of the specimen with strain. Because in the finest-grained samples at high temperature and low strain-rates the stress supported by the specimen became commensurate with the strength of the copper jacket, special attention was given to a recalibration of the coppper jacket behaviour over the temperature range of interest.

Microstructural studies were made of the starting materials and of the experimentally deformed samples using both flat-stage and universal stage optical microscopy on ultrathin sections, scanning electron microscopy on both broken and polished surfaces, and high-voltage transmission electron microscopy. In order to study the displacements of a planar surface in the interior of the sample, some runs were performed using the split-cylinder technique (Raleigh 1965), in which a composite specimen comprising a cylinder cut longitudinally, with the planar cut faces polished, is separated by a thin sheet of gold.

Measurements of mean grain-size were made from optical or scanning electron micrographs using the method of Abrams (1971), in which intercepts are counted on a series of concentric circles. This method has the advantage that the effects of any shape fabric (for which the section plane is a symmetry plane) of the grains are cancelled out. Particle-size distributions were obtained for the starting materials after hot-pressing from measurements of grain diameters along two orthogonal directions on scanning electron micrographs. Measurements of mean grain shape were also made from some micrographs. The grain diameter was measured at 10° intervals between the shortest and longest dimensions for typically 100 grains in selected samples. In this way the strain accumulated by intragranular distortion could be compared with the distortional strain on the specimen as a whole.

Fabrication of starting materials

The starting material for the synthetic polycrystals was a set of clear calcite rhombs. These were mechanically crushed, then ground in an agate mill for 30 seconds. Milling for longer periods caused clogging through agglomeration of the finest particles. The material was sieved to separate the fraction less than 64 μm. Different grain-size fractions were separated with a centrifuge, following agitation of the sample for 5 minutes in an ultrasonic bath. Nomographs of sedimentation times for soil particles under centrifugal acceleration (Tanner & Jackson 1947) were used as a guide to the theoretical times expected to separate the powder into suitable grain-size fractions.

The loose powder was cold-pressed at about

500 MPa in a split-die into circular cylinders, 1 cm diameter and nominally 2 cm long, so that a porosity of about 15% remained. Prior to deformation, each specimen was hot-pressed under hydrostatic conditions. Optimum hot-pressing conditions were discovered by systematic examination of a range of conditions (70 to 240 MPa hydrostatic pressure, 500–700°C) using a powder batch of *c*. 3.1 μm grain size. Preliminary deformation experiments showed that to obtain reproducible behaviour specimens must have a porosity of no more than 4%, it must be homogeneously distributed, and high coordination-number pores (large pores bounded by the facets of many grains) must be few or absent. These requirements are well known in the production of hot-pressed powder ceramics (e.g. Alford *et al.* 1987).

The shape of the compaction versus time curves was obtained during hot pressing by measuring the displacement of the 'hit-point', the point at which differential load begins to appear on the specimen when the axial loading piston is displaced, with time. In this way the curves shown on Fig. 1 were obtained. They show that at all temperatures densification proceeds rapidly at first, then slows until it has virtually stopped. The density of specimens recovered from these runs was estimated by pyncnometer. Table 1 shows the mean particle sizes of 5 batches (i to v) obtained after hot-pressing the aggregates. Particle-size distributions are shown on Fig. 18. Porosities of less than 3% were obtained in all of the five grainsize batches (Fig. 1c).

Table 1. *Mean grain-sizes obtained (using the method of Abrams 1971) after hot-pressing of the calcite powders (3 hours at 550°C, except batch i, which was usually 72 hours at 700°C, all at 200 MPa confining pressure)*

Batch no.	Grain-size (μm)	Sample size	Standard deviation (μm)
i	38.5	900	1.9
ii	7.5	497	0.8
iii	3.4	1006	0.6
iv	2.5	996	0.2
v	1.9	369	0.1

Grain-size distributions are shown in Fig. 4.

The evolution of microstructure during the hot-pressing was studied by terminating individual runs after different elapsed times (numbered points on Fig. 1). In this way it was discovered that although the most rapid initial densification occurs at 700°C, high coordination-number pores tend to become trapped. Once trapped, they are not easily removed. The slower compaction at 550°C showed less tendency to trap large pores.

Figure 2 shows the characteristics of the hot-pressed samples which were used as starting materials for the deformation experiments. The *c*. 5 μm grain-size

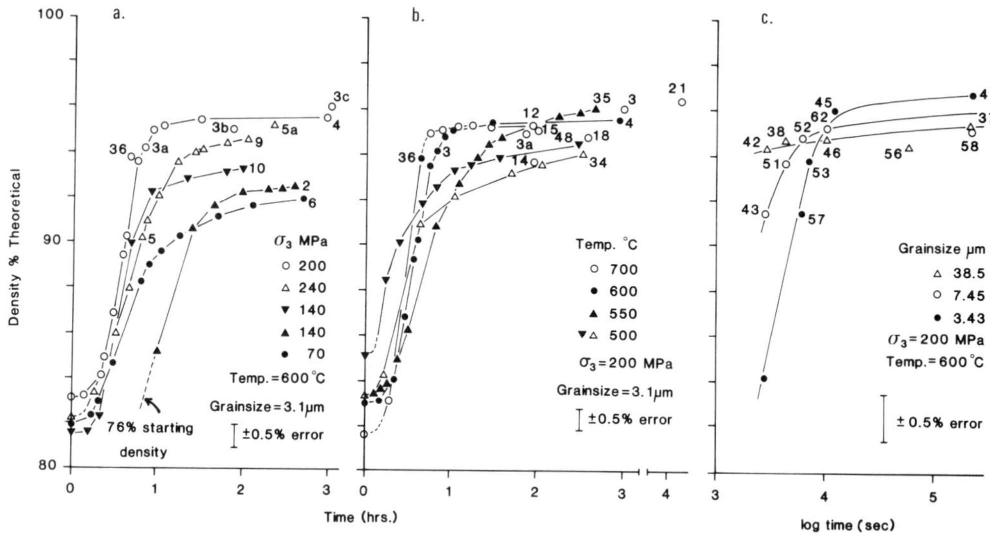

Fig. 1. Densification (percentage of theoretical maximum density) for calcite powder after cold pressing under 500 MPa axial pressure. Symbols represent specimen density calculated from the point at which a differential load starts to appear on the specimen on advancing the loading piston (hit-point). Numbered points correspond to run numbers in which a specimen was removed for microstructural study. (**a**) Runs at different confining pressures but all at 600°C, 3.1 μm grain-size. (**b**) Runs at 200 MPa confining pressure, 3.1 μm grain-size but different temperatures. (**c**) Runs at 200 MPa and 600°C for three of the grain-size batches used for the deformation experiments.

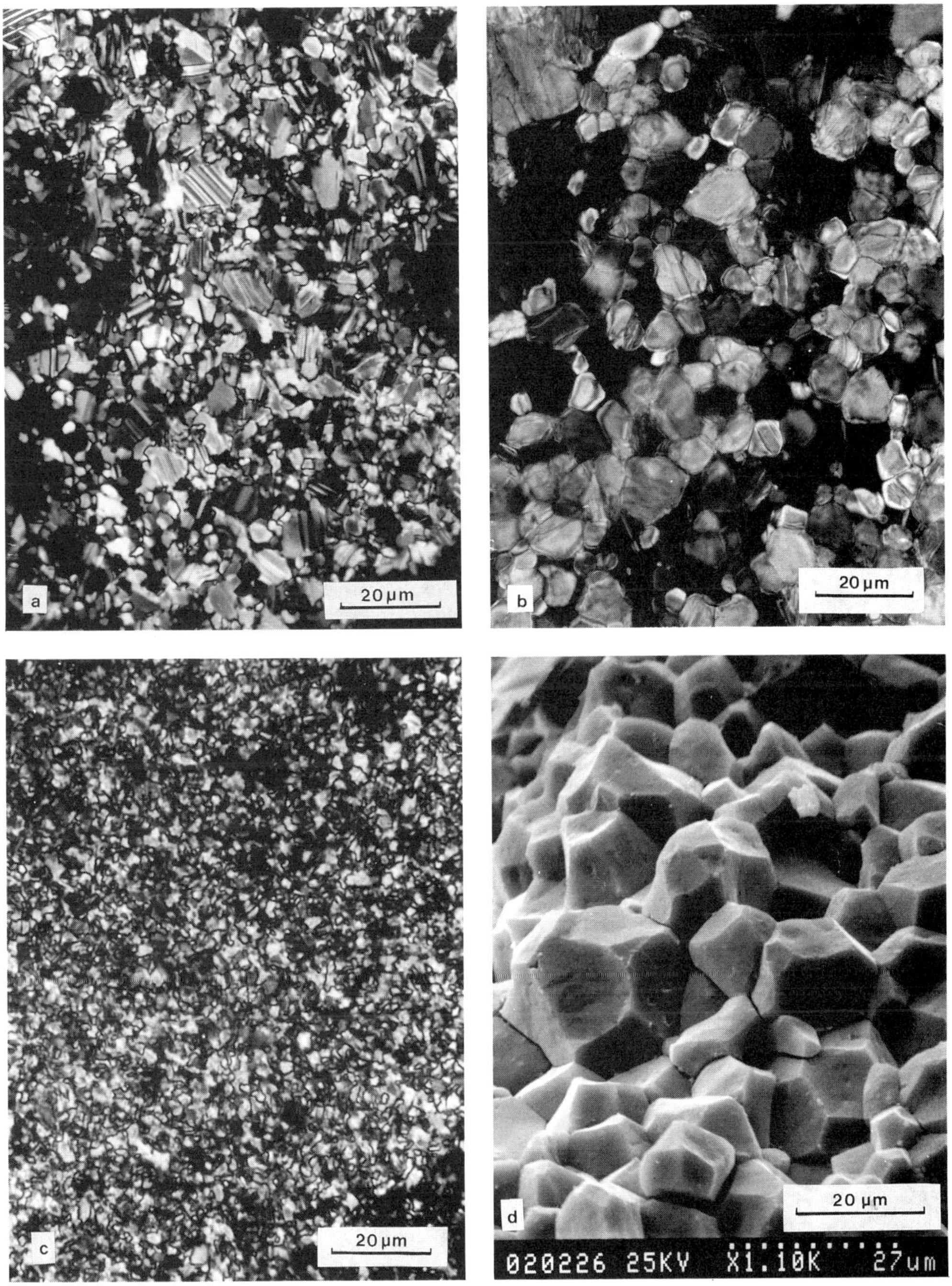

Fig. 2 The hot-pressed starting materials (in all cases after 3 hours at 550°C) and in some cases after further heat treatment.

(**a**) Batch iii (optical micrograph, mean grain-size 3.4 μm) after a further 4 hours at 600°C. Some grain-growth has occurred and the very fine particles which tend to be agglutinated to the larger ones are still present.

(**b**) Batch iii (optical micrograph) after a further 3 days at 600°C. The very fine grains have been eliminated and a unimodal foam texture has developed. There is little further increase in mean grain-size relative to (a).

(**c**) Batch v (optical micrograph, mean grain-size 2.0 μm) after the initial hot pressing treatment. This non-equilibrium texture is very unstable and susceptible to rapid grain-growth at 600 to 700°C.

(**d**) Secondary electron micrograph of a broken surface of an initially 3.1 μm sample after a further heat treatment of 4 hours at 700°C. This shows the low porosity and development of faceted grains which are characteristic of an equilibrium microstructure.

samples (e.g. batch iii) rapidly developed a 'normal grain-growth' arrangement of polyhedra with planar or gently curving faces. Porosity was concentrated mainly in the four-fold grain vertices. The finest-grained samples (batch v) typically developed first an 'interlinked-finger' texture (Fig. 2c) until the regime of normal grain-growth prevailed. At larger grain-sizes (batches i and ii) a bimodal grain-size distribution results from the initial grinding because fine grains tend to stick to the larger ones. Also, the intense cold-working during the cold-pressing leaves the larger grains highly strained and twinned. The hot-press treatment therefore must eliminate the finest grains and anneal the damage done to the larger ones. After the initial treatment at 550°C therefore, a longer sinter at 700°C was applied to the coarser samples. Figure 2a and b shows the microstructural changes observed.

The preparation of polycrystals uniformly mixed with 1 or 5 vol% of ultrafine-grained alumina posed special problems. Various methods were tried with the aim of producing uniform dispersion of the alumina particles. Mechanical mixing in ethanol in a cylindrical bottle rotating slowly about an horizontal axis for several hours produced the most reproducible mechanical behaviour. Figure 14b shows how the alumina was dispersed between the calcite grains. Although a small proportion of the alumina was fairly uniformly dispersed, marked clustering of the alumina is still evident, and further work will be required to solve the problem of producing a more satisfactory starting microstructure in two-phase specimens.

Grain-growth was studied under the pressure–temperature–time conditions of the deformation experiments, with the aim of being able to compensate the mechanical data for the effects of grain-growth. Hydrostatic runs were carried out for various time periods at each temperature to determine the shapes of the grain-growth curves and to compare with the grain-sizes produced during the deformation experiments (Table 2 and Fig. 3). The starting and finishing grain-sizes for some of the deformation runs were also measured (Table 3 and Fig. 3).

Experimental results

Hydrostatic grain-growth tests

Nineteen hydrostatic grain-growth runs were carried out on previously hot-pressed specimens of various initial grain-sizes to assess the grain-growth behaviour. Most were jacketed in copper (as were all of the deformed samples), but five were run in gold jackets, to test the hypothesis that grain-growth might be inhibited through poisoning of grain-boundaries by vapour-deposited copper oxide. Results of these experiments are shown in Table 2 and Fig. 3. There is clearly no significant effect of jacketing material nor of deformation on the growth behaviour. The latter observation is in contrast to

Table 2. *Hydrostatic grain-growth data*

Specimen no.	Temperature (°C)	Batch	Time (s)	Grain-size (μm)	Sample size
(a) Copper-jacketed specimens					
C45	550	iii	10800	3.43±0.61	1006
C62	550	ii	10800	7.45±0.80	497
C95	550	iv	10800	2.52±0.25	996
C97	550	v	10800	1.98±0.13	369
C77	600	iii	25200	4.33±0.23	810
C78	600	ii	24840	11.20±1.7	304
C80	600	iii	25200	4.70±0.40	405
C81	600	iii	85200	6.92±0.28	686
C85	600	iii	14400	3.54±0.27	870
C73	700	iii	22380	8.12±1.28	521
C75	700	ii	24300	11.25±1.77	720
C90	700	iii	14400	4.58±0.26	767
C92	700	i	259200	38.50±1.94	900
C94	700	iii	79200	9.50±0.31	777
(b) Gold-jacketed specimens					
G1	550	iii	10800	3.63±0.46	985
G2	550	iii	262800	9.33±1.01	783
G5	600	iii	18000	3.56±0.17	616
G3	600	iii	23400	7.57±0.27	640
G6	700	iii	626400	14.60±1.28	487

Mean grain-sizes measured using the circular pattern intercept method of Abrams (1971). Experimental times include the time of initial hot pressing, and are measured from the start of heating. Confining pressure was 200 MPa in all cases. Uncertainty values are 1 standard deviation.

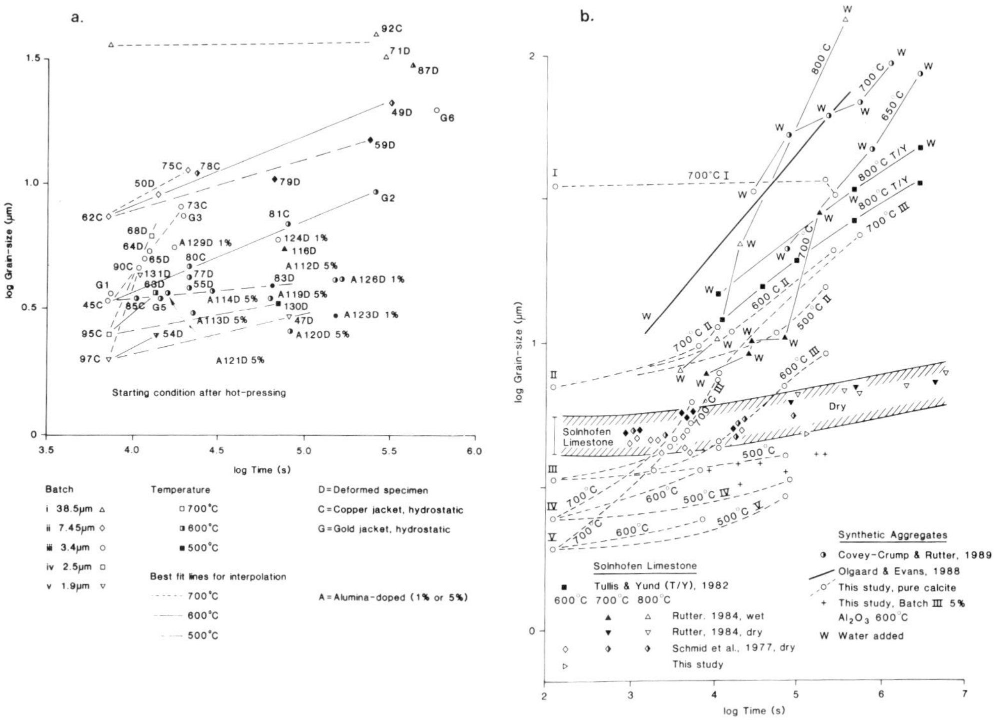

Fig. 3 (**a**) Experimental grain growth data for the synthetic, dry calcite rocks. The powder batches are identified by Roman numerals. The times plotted include the time for the initial hot-pressing at 550°C, so that the lines shown can be used directly to compensate the deformation experiments for the effects of grain-growth. There is no significant difference between growth during deformation and under hydrostatic conditions, neither does the jacketing material (gold or copper) have any discernable effect. Standard deviations on grain-size measurements are listed in Table 2.

(**b**) Compilation of previous grain-growth data for calcite rocks (wet and dry) together with data from the present study. The latter data are replotted so that the grain-growth time starts from the condition obtained at the end of the initial hot-pressing period. Dashed lines indicate trends in data. Note how the addition of alumina (+) effectively stops grain-growth relative to pure calcite at the same temperature (600°C, batch iii).

behaviour typical of metals (e.g. Senkov & Myshlyaev 1986). The trend lines indicated were used to assess the final grain-sizes expected in those deformed specimens which were not studied microstructurally.

There are insufficient data to fit grain-growth laws, but some general conclusion can be drawn by additional reference to other grain-growth data (Fig. 3b). Grain-growth takes place in pure, dry calcite rocks at a rate which is not much slower than when water is present. The behaviour of dry Solnhofen limestone and both dry and wet synthetic aggregates doped with alumina show that small amounts of second phases inhibit grain-growth (Olgaard & Evans 1988). It has previously been noted that grain growth in impure calcite rocks is facilitated by the presence of water because grain-boundary melting is promoted (Rutter 1984).

Clearly, the rate of grain-growth increases with temperature for all initial grain-sizes, but there are anomalies in the behaviour of the synthetic rocks. Batch ii displayed anomalously rapid grain-growth rates (Fig. 3b), which might be attributed to the greater intensity of the initial cold-working of the coarser size fractions. Extrapolation outside the range of the experimental data is unwarranted and unnecessary, for the aim here was simply to compensate the mechanical data by interpolation for the effects of grain-growth over the time-scale of the experiments. Our mechanical data show that grain-size sensitive flow effects can be observed before the hardening effects of grain-growth supervene, but compensation for the effects of grain-growth during the course of the higher temperature tests is clearly worthwhile.

Table 3. *Deformation experiments on dry, hot-pressed calcite cylinders prepared from powder batches i to iv*

Specimen no.		Batch	Temperature (°C)	Total elapsed time (s)	Final grain-size (μm)	Comments
C47	R	ii	600	235800	–	
C49	R	ii	600	324000	21.2	
C50	R	ii	700	18000	9.2	
C54	R	v	600	17400	2.5	
C55	R	iii	600	25200	3.9	
C59	R	ii	500	244800	5.5	
C60	R	iii	600	69000	–	
C63	R	iv	600	17400	3.7	
C64		iv	600	26400	–	Creep test
C65	R	iii	700	15000	5.0	
C66	R	iii	500	241200	–	
C67	R	v	500	86400	2.9	
C68		iv	700	16500	6.3	
C69	R	iii	700	16200	5.6	
C70	R	iv	700	12840	–	
C71	R	i	700	291600	31.6	
C72	R	iii	400	262800	–	
C76	R3	iii	400	73140	2.5	
C79	R2	ii	500	71400	10.6	
C82	R2	iii	520	24000	–	
C83	R	iii	500	76320	4.0	
C87	R	i	600	175680	29.1	HPT 20.5 hrs
C96		iii	700	20800	–	Creep test
C99	R	i	500	1195200	–	HPT 144 hrs
C100	R	i	500	1195200	9.2	
C101	R	iii	700	18000	6.2	
C102		iii	700	20400	–	
C103	R	iii	700	16800	5.7	
C104	R	iii	700	19740	6.9	
C105	R	iii	500	25200	3.7	
C106	R	iii	500	19620	3.7	
C107	R	iii	500	255600	4.3	
C108		i	500	1195620	19.5	
C109		ii	500	27780	–	
C110		ii	700	198000	–	1.8×10^{-6} s^{-1}
C111		ii	500	13500	–	
C115		iii	700	26700	–	
C116		ii	500	81060	5.5	
C118		ii	500	251580	–	
C128		ii	500	21180	–	
C130	R	iv	500	75120	3.4	
C131	R	v	700	14400	4.4	
Confining pressure stepping tests, all stepping through 300, 200, 100 and 50 MPa in that order, at constant strain-rate, establishing steady flow at each pressure (see Fig. 8)						
C84		iii	600	15000	–	
C86		iii	700	15900	–	
C88		iii	500	17100	–	
C89		iii	700	31980	–	
Strain-rate stepping tests (see Fig. 9)						
C91		iii	600	362400	–	4 steps
C93		iii	700	18000	–	3 steps
C98	R	iii	500	1206000	–	5 steps
Two-phase specimens. Volume % of 0.25 μm alumina indicated						
C112		iii	600	158400	4.1	5%
C113		iii	600	19920	2.9	5%
C114		iii	600	29400	4.2	5%
C117		iii	600	18900	–	5%
C119		iii	600	63360	3.5	5%

C120	iii	500	77760	2.4	5%
C121	iii	700	15540	3.5	5%
C122	iii	600	72240	–	1%
C123	iii	500	157620	2.6	1%
C124	iii	700	71640	6.3	1%
C126	iii	600	166200	4.1	1%
C129	iii	700	16320	5.4	1%
DO-1	7.5 μm	700	22380	6.0	5% (Olgaard)
Split cylinder tests					
C125	iii	700	27480	–	
C127	iii	500	19920	–	
C132	iii	500	25080	–	
Solnhofen limestone, cored normal to bedding except SH3 (parallel)					
SH1	5 μm	600	15660	–	bed. normal 3×10^{-5} s^{-1}
SH2	5 μm	600	259200	4.5	bed. normal
SH3	5 μm	600	324000	–	bed. parallel
SH4	5 μm	600	93600	–	bed. normal

All runs at 200 MPa confining pressure unless noted. R denotes a stress relaxation test after initial loading, R*n* denotes *n* relaxation tests on the same specimen. Unless noted, strain-rate lies in range 3 to 7 $\times$ 10^{-4} s^{-1}. Final grain-sizes shown were measured from photomicrographs, others were estimated from grain-growth data. HPT, hot-pressed time for batch i specimens at 700°C when not 72 hours.

Experiments on Solnhofen limestone

Stress–strain curves and stress-relaxation data obtained for Solnhofen limestone at 600°C are summarized on Fig. 4, and compared with the results reported by Schmid *et al.* (1977). It can be seen that there is good agreement with the results of the earlier study, and also that the reproducibility of the results is fairly good.

Experiments on pure, synthetic aggregates

Under most of the experimental conditions in constant strain-rate tests, after about 2% strain the samples exhibited steady-state flow, at least up to 30% shortening (Fig. 5). Reproducibility between stress/strain curves obtained under identical conditions was comparable with that obtained for Solnhofen limestone (Fig. 5). The effects of confining pressure on the flow stress were investigated by means of pressure-stepping tests on 3.4 μm grain-size samples, at a constant strain-rate of 3×10^{-4} s^{-1}, in the sequence 300, 200, 100 and 50 MPa confining pressure and at each of 500, 600 and 700°C. The results are presented in Fig. 6. Pressure stepping in the direction of decreasing pressure minimises the possible effect of progressive porosity reduction with increasing pressure. At 500°C the stength decreases slightly with increasing confining pressure. This is in accordance with the behaviour of Solnhofen Limestone and Carrara Marble at 500°C and 600°C (Heard 1960, Rutter 1972). At 700°C the flow stress increases slightly with increasing confining pressure.

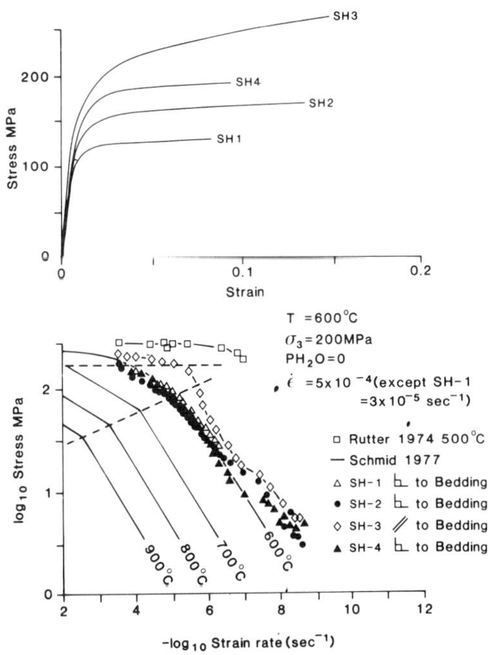

Fig. 4. Results from constant displacement rate and stress relaxations on cylinders of dry Solnhofen limestone. Confining pressure, initial strain-rate and temperature are shown. The constant strain-rate results of Rutter (1974, same block) and Schmid *et al.* (1977) are shown for comparison.

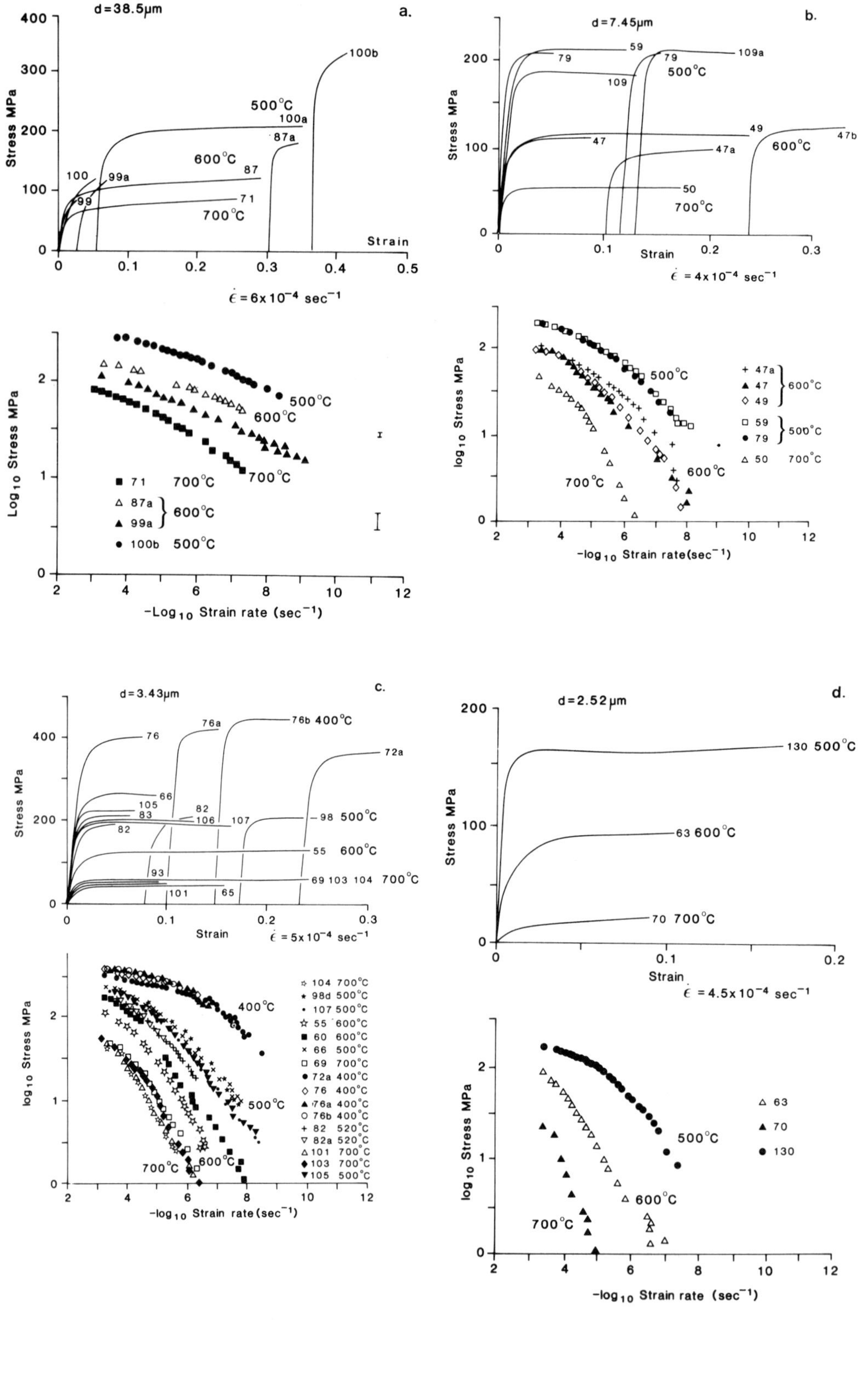
a.
d=38.5μm
Stress MPa
Strain
500°C
600°C
700°C
100b
100a
87a
87
99a
100
99
71
ε̇ = 6x10⁻⁴ sec⁻¹
Log₁₀ Stress MPa
-Log₁₀ Strain rate (sec⁻¹)
71 700°C
87a
99a
600°C
100b 500°C
b.
d=7.45μm
Stress MPa
Strain
59
79
109
109a
500°C
49
47
47a
47b
600°C
50
700°C
ε̇ = 4x10⁻⁴ sec⁻¹
log₁₀ Stress MPa
-log₁₀ Strain rate(sec⁻¹)
47a
47
49
600°C
59
79
500°C
50 700°C
c.
d=3.43μm
Stress MPa
Strain
76
76a
76b 400°C
72a
66
105
83
82
106
107
98 500°C
55 600°C
93
101
65
69 103 104 700°C
ε̇ = 5x10⁻⁴ sec⁻¹
log₁₀ Stress MPa
-log₁₀ Strain rate(sec⁻¹)
400°C
500°C
600°C
700°C
104 700°C
98d 500°C
107 500°C
55 600°C
60 600°C
66 500°C
69 700°C
72a 400°C
76 400°C
76a 400°C
76b 400°C
82 520°C
82a 520°C
101 700°C
103 700°C
105 500°C
d.
d=2.52μm
Stress MPa
Strain
130 500°C
63 600°C
70 700°C
ε̇ = 4.5x10⁻⁴ sec⁻¹
log₁₀ Stress MPa
-log₁₀ Strain rate (sec⁻¹)
500°C
600°C
700°C
63
70
130

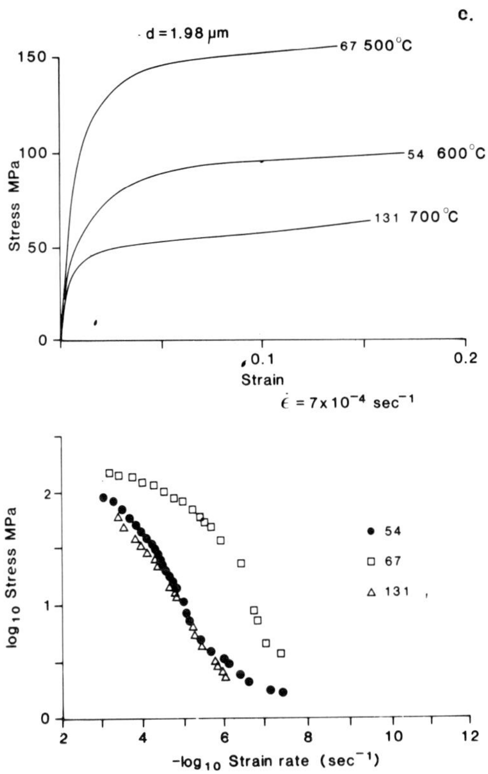

Fig. 5. Summary of results from constant strain-rate tests (strain-rate indicated) and stress relaxations for hot-pressed, dry cylinders of calcite of various grain-sizes (as measured after hot-pressing). (**a**) 38.5 μm grain-size, (batch i); (**b**) 7.45 μm grain-size, (batch ii); (**c**) 3.43 μm grain-size, (batch iii); (**d**) 2.52 μm grain-size, (batch iv); (**e**) 1.98 μm grain-size, (batch v).

Confining pressure = 200 MPa, hot-press treatment = 3 hours at 550°C, followed by at least a further 20 hours at 700°C in the case of batch i samples.

On batch iii samples several repeat runs were done on different specimens, especially at 500°C and 700°C, to establish the reproducibility of the stress relaxation behaviour. The curves shown here indicate the full range of variability obtained between different specimens. Specimen numbers suffixed a, b, c, etc. represent reloadings after stress relaxations. Repeat relaxations on the same specimen (e.g. batch iii) also show a high degree of reproducibility, indicative of the microstructural stability and lack of significant rate-determining substructure changes during stress relaxation.

The weakness during initial loading at 500 and 600°C of some of the batch i (38.5 μm) specimens is thought to be attributable to incomplete pore elimination and persistence of agglomerated finer grains. Densification and coarsening of the agglomerated grains may have advanced by the time they were reloaded, thereby rendering the specimens stronger.

The strain-rate stress grain-size relationship was investigated mainly using stress relaxation experiments (Rutter *et al.* 1978). Figure 7 shows the stress relaxation data obtained over the full range of temperatures and grain-sizes used, in the form of log stress v. −log strain-rate curves. For each test, about 40 to 50 evenly-spaced points from the raw force-time data were selected for data processing. Strain-rates were calculated from the stress difference between pairs of alternate data points, and the corresponding stress was taken as the average of the two points.

At 500, 600 and 700°C constant strain-rate and strain-rate stepping tests were performed over the strain-rate $3 \times 10^{-4}\ s^{-1}$ to $1 \times 10^{-7}\ s^{-1}$, using specimens of grain-size 3.4 μm. The results of these tests were consistent with the stress relaxation data (Fig. 7). As the temperature was increased the rate of change of stress with strain-rate increased, in the same way as was observed for Solnhofen limestone by Schmid *et al.* (1977). The high-temperature, low-stress behaviour approaches linear-viscous.

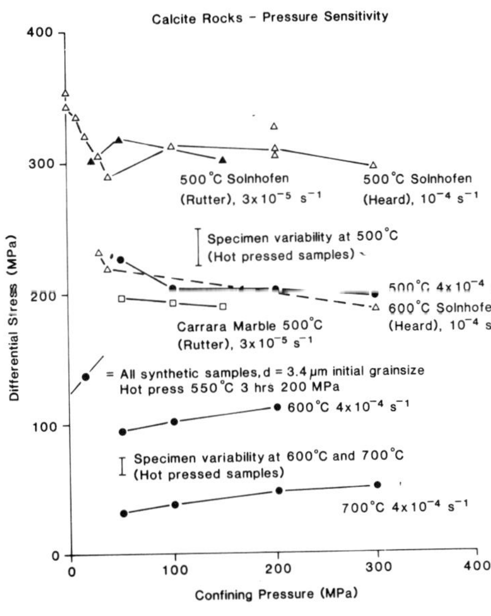

Fig. 6. The influence of confining pressure and temperature on flow stress at constant strain-rate (values indicated). A small negative pressure sensitivity is seen at 500°C, giving way to a small positive pressure sensitivity at 600 and 700°C. Comparative data for Solnhofen Limestone and Carrara marble (from Rutter 1972, 1974 and Heard 1960) are also shown. There is also a tendency to a small negative pressure sensitivity in the latter materials at high stresses.

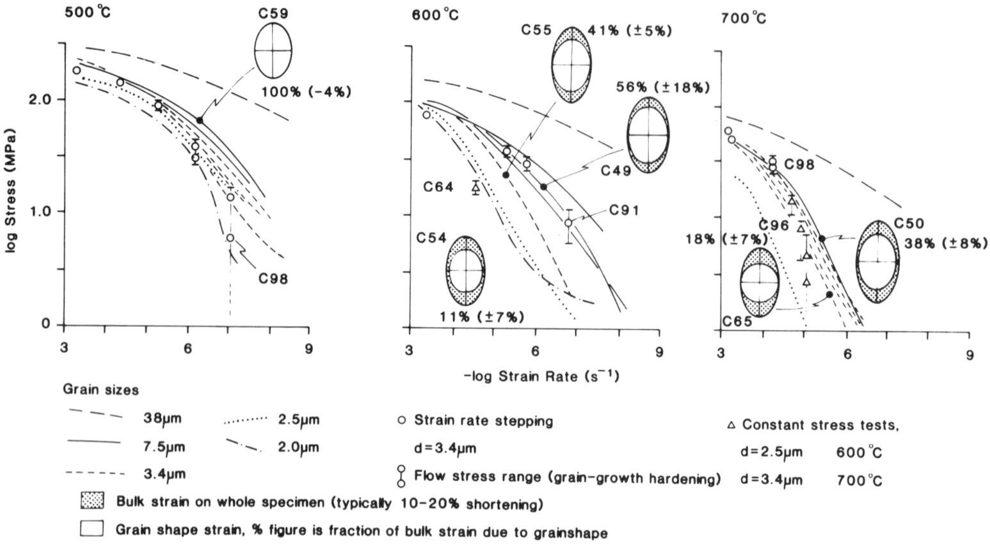

Fig. 7. Summary of stress relaxation behaviour for different grain-size samples at each of three temperatures, showing also the good correlation between the stress relaxation results and creep and constant strain-rate tests. Also shown is the relationship between bulk sample strain and the apparent strain as measured by grain shape. Beside each ellipse is shown the percentage of the bulk strain accounted for by the grain shape. Grain-shape corresponds almost exactly to the bulk strain at high stresses and low temperatures, whereas grains tend to remain equidimensional at high temperatures and low stress levels.

Although in a stresss relaxation test it is possible to estimate rheology without errors arising from variability between specimens, as with the constant strain-rate tests, repeat runs under some sets of conditions were carried out to assess reproducibility and inter-specimen variability. Typical results are shown in Fig. 5c. Of special importance in the present study, the duration of the highest temperature stress relaxation tests was very short (*c.* 1 hour), which minimises the effects of hardening though grain-growth.

In the case of several specimens, repeat stress relaxations were carried out after reloading and further permanent strain (Fig. 5c). This allows the sensitivity of the relaxation behaviour to strain-induced microstructural changes to be investigated. Allowing for the effects of grain-growth, the flow behaviour appears to be insensitive to variations of several percent strain and wide stress variations.

The shape of the log stress vs log strain-rate curves derived from stress relaxation tests is sometimes sigmoidal (Figs 5 and 7), with markedly non-linear flow at high stresses, closer to linear viscous flow at intermediate stresses, apparently giving way to greater non-linearity again at very low stresses (in some 600°C tests). In the regime of very non-linear flow at the highest stresses the flow is almost grain-size insensitive, whilst in the intermediate stress range strength clearly decreases with decreasing grain-size, becoming more grain-size sensitive with decreasing stress. Sigmoidal log stress/log strain-rate curves are well known for superplastic metals (e.g. review by Poirier 1985). However, in our case because the stresses are so low it is not clear whether the non-linearity is real. It will be necessary to employ different experimental apparatus specifically to investigate behaviour in this region in a reliable manner. In the assessment of the form of the constitutive flow law (see below) this region has been ignored.

The assessment of the effect of temperature on the flow behaviour has been carried out in two ways. First, its effect was obtained from the fitting of the flow law to the constant strain-rate and stress relaxation data (see below). Second, activation enthalpy for flow was determined by temperature cycling during a creep (constant stress) test (Fig. 8).

Experiments on aggregates 'doped' with alumina

The 1% alumina samples were nominally slightly weaker than the pure material (but

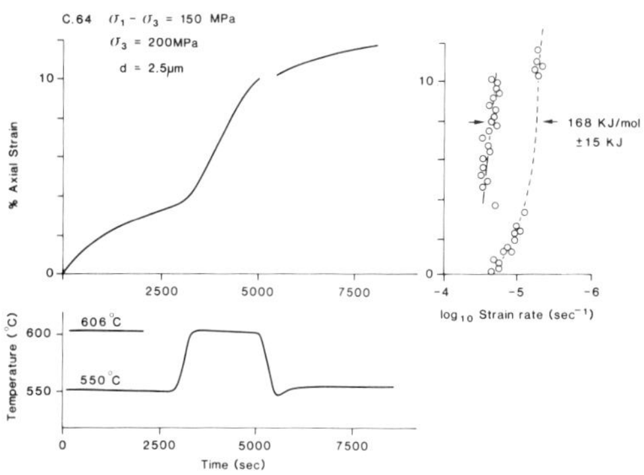

Fig. 8. Results of a temperature-stepping creep test, from which activation enthalpy for flow can be determined. The figure obtained is slightly lower than that from the stress relaxation tests.

identical within the limits of experimental error) whilst the 5% alumina samples were distinctly stronger (Fig. 9). The stress relaxation behaviour of the doped material was comparable in form to that of the pure material, except that the more non-linear high-stress behaviour tended to extend to lower strain-rates. Thus at low strain-rates the more impure material appeared progressively stronger than the pure samples, approaching the strength of Solnhofen limestone, which is stronger than the pure calcite polycrystals under all experimental conditions. There was also a distinctly greater tendency for the low stress parts of the stress relaxation curves to be concave upwards, relative to the behaviour of the pure material.

The stress relaxation data from a specimen supplied by D. Olgaard (grain-size 7.5 μm, 5% volume alumina) was in excellent agreement with his data (Fig. 10), and the pattern of behaviour was consistent with our doped samples.

Microstructural observations

Distinctly different microstructures were observed in synthetic specimens deformed at high flow stress levels (above about 25 MPa) compared to the effects of deformation at lower stresses.

(i) Specimens deformed at high stresses. These

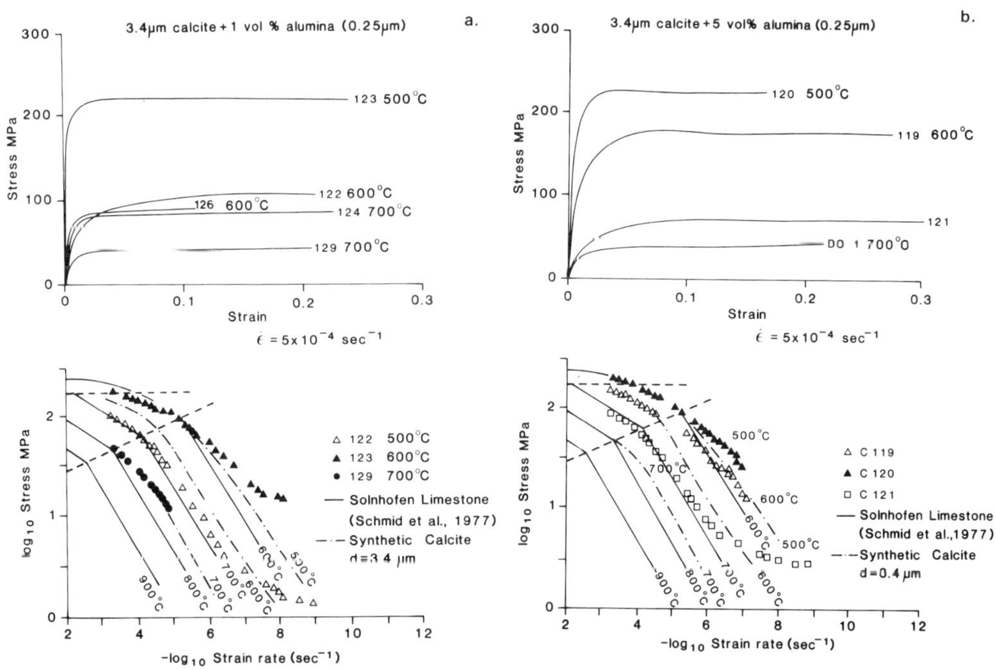

Fig. 9. Constant strain-rate and stress relaxation results for alumina-doped polycrystals (3.4 μm calcite; 0.25 μm alumina), prepared by rolling the mixed powders in ethanol for 4 hours prior to cold pressing. With the stress relaxation curves are shown for comparison the form of curves for Solnhofen limestone and the pure calcite polycrystals. (**a**) 1% volume alumina added. (**b**) 5% volume alumina added. In all tests, confining pressure = 200 MPa, intial strain rate = 5×10^{-4} s^{-1}, hot-pressed at 550°C for 3 hours.

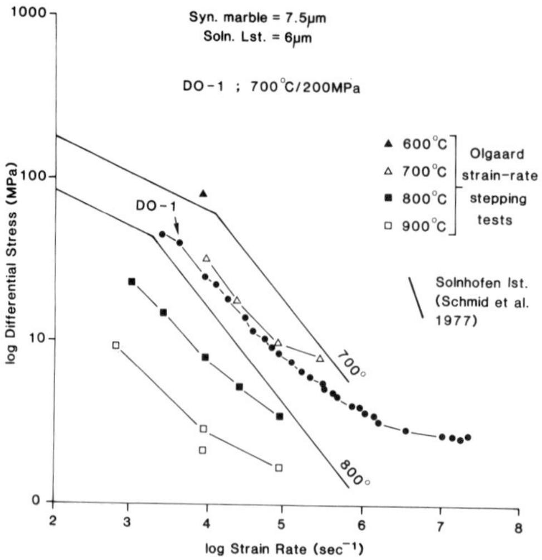

Fig. 10. Comparison between the stress relaxation behaviour of Solnhofen limestone, the behaviour of synthetic hot-pressed aggregates deformed by D. Olgaard using strain-rate stepping, and one of Olgaard's specimens deformed by us (specimen DO-1).

specimens were characterized by a strong grain-flattening fabric (Fig. 11a). The larger grains developed optically discernable twin lamellae. The crystallographic *c*-axis fabric of one such specimen was measured by standard universal stage methods (Fig. 13b). The fine grain-size necessitated the use of an ultrathin section, which meant that twin orientations, and hence the complete grain orientation, could not be measured. The crystallographic fabric observed is the *e*-maximum type (Casey *et al.* 1978; Spiers 1979), which is characteristic of the axisymmetric deformation of calcite rocks when the stress levels are sufficient to activate twinning in many grains. *c*-axis concentrations up to × 10 uniform developed near the compression direction. No indication of an additional, weaker *c*-axis great circle girdle was present, as would be expected from the well-known behaviour of Solnhofen limestone (Casey *et al.* 1978) at high temperatures. It is therefore inferred that the fabric is dominated by that produced during the initial cold-pressing, and that this fabric remains stable during the subsequent high temperature flow.

Measurements of grain-shape also showed that the strain is relatively homogeneous on the grain-scale and corresponds well with the bulk imposed strain (Fig. 7). The development of grain flattening and crystallographic fabric is consistent with flow by intracrystalline deformation involving dislocation motion.

(ii) Specimens deformed at low stresses. In this regime the grain-shape fabric does not develop with the bulk strain. Grains remain approximately equant during the flow (Figs 7 & 11b), even up to strains of *c.* 30%. As previously, the crystallographic fabric of one such specimen (Fig. 13a) is controlled by the cold-pressing deformation, but has become significantly weakened during the high temperature flow. The *c*-axis concentration near the compression direction is more diffuse and small maxima only up to × 6 uniform concentration are preserved. Weakening of a crystallographic fabric is suggestive of flow dominated by high-temperature grain-boundary sliding.

Using the split-cylinder technique of Raleigh (1965), Schmid *et al.* (1977) showed that grain-boundary sliding was important in the 'superplastic' flow of Solnhofen limestone. We have also used the method to infer the importance of grain-boundary sliding (Fig. 12). Deformed material at the centre of a specimen (Fig. 12b,c) is compared with relatively undeformed material (Fig. 15a) adjacent to the loading piston. The deformed material clearly shows evidence of whole grain displacements normal to the polished surface. Oblique views onto the surface also show tilting rotations and curvature of grains as the surface forces between the calcite grains and the gold equilibrate. Such 'diffusional etching' of grain boundaries is also apparent in split-cylinder assemblies deformed in the intracrystalline plasticity (high stress) regime, and it is clear that a certain amount of grain-boundary sliding also occurs under these conditions, although to a lesser degree than in the low-stress flow regime.

Normal-incidence SEM images are not sufficient to evaluate the importance of grain-boundary sliding involving displacements normal to the surface; oblique incidence views provide a better basis for assessment. The normal-incidence micrographs, however, show well the development of grain-boundary voids preferentially in the relatively extended sectors of the grain-boundaries (Fig. 12c). The gold foil intruded these voids, so that their pattern was preserved in relief on the gold surface. The amount of porosity formation is estimated to be on the order of 1%.

(iii) Alumina-doped specimens. Fig. 14b shows the typical microstructure of the 5% volume alumina-doped specimens which displayed slightly higher strength characteristics than the

Fig. 11. Comparative optical microstructures (crossed polars) of specimens deformed in the two grain-size sensitive flow fields. The compression direction is parallel to the short side of each photograph. (**a**) Flow in the high stress regime (specimen C100, 500°C, Batch i, 38 μm grain-size, 30% shortening), characterized by grain-flattening by intracrystalline plasticity. The apparent mean grain-size is reduced owing to the displacement of material out of the plane of section by flattening. A high initial density of twinning produced during cold-pressing has been eliminated by twin and grain-boundary migration, but a strong crystallographic preferred orientation remains. Crossed polars, most grains are close to extinction. (**b**) Flow in the low stress regime (specimen C50, 700°C, Batch ii, 20% shortening, 7.5 μm grain-size, but there has been some grain-growth), characterized by equant grain-shapes and a tendency to weaken the pre-existing crystallographic preferred orientation.

pure polycrystals. A high degree of clustering of the alumina into highly eccentric ellipsoids, flattened normal to the compression direction, is evident. Most of the flattening is probably inherited from the cold-pressing stage. Despite the clustering, there is still an approximately uniform dispersion of a small fraction of the alumina powder, which has apparently significantly suppressed grain-growth relative to the pure calcite specimens (Fig. 3).

(iv) Transmission electron microscope (TEM) observations. Comparative observations of the dislocation microstructure of two samples, deformed in each of the high stress and low stress regimes were made. No significant differences in the dislocation microstructures in the two regimes were seen, as also noted by Schmid *et al.* (1977) in their study of the flow of Solnhofen limestone. A heterogeneous density of dislocations displaying typical hot-creep configurations was seen within individual grains, involving curving dislocations, climb debris and occasional subgrain walls (Fig. 14a). Striking variations in dislocation activity were also seen between grains (cf. Schmid *et al.* 1977). It is noteworthy that there was significant dislocation activity in the weaker sample. It is not clear, however, to what extent the observed dislocation activity has arisen during cooling under pressure (cf. Schmid *et al.* 1980)

Interpretation and discussion of results

Constitutive flow laws for the pure calcite rocks

The flow law (eq. 1) may be written in the form:

$$\log \dot{e} = \log A - H/2.303RT + n \log \sigma + m \log d$$

hence, assuming A, n and m are constant,

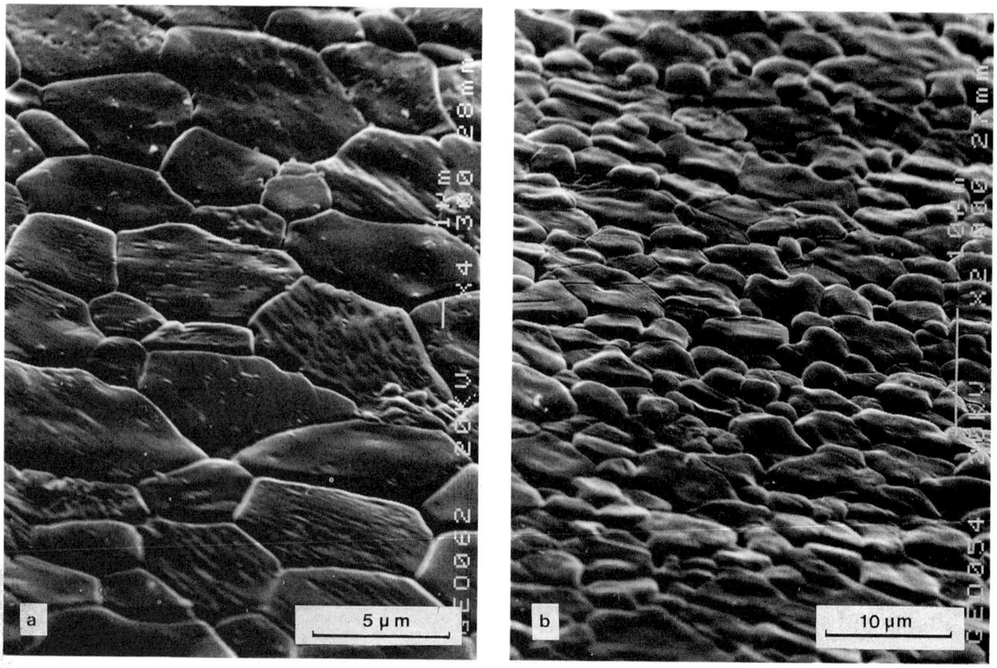

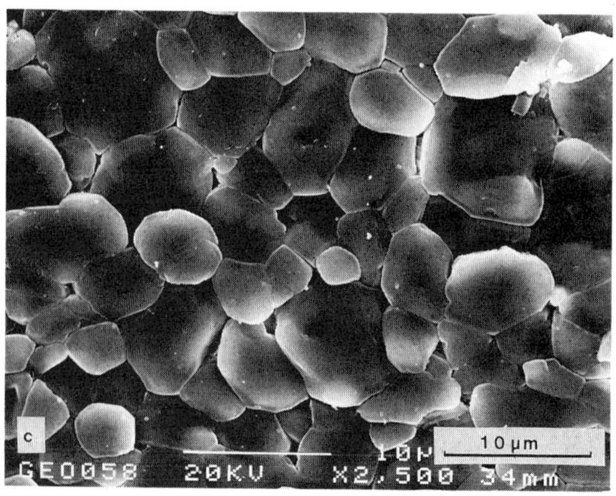

Fig. 12. Split cylinder experiment (C70, shortened 20%, 700°C, grain-size 2.5 μm, but there has been some grain-growth). The secondary electron images (a), (b) and (c) show the originally polished surface of the rock which was in contact with the gold foil, in each case tilted 60° from the specimen surface-normal with the exception of (c), which is a view normal to the surface. In all cases the trace of the compression direction is almost parallel to the short edge of the image, which is also the tilting axis in (a) and (b).

(**a**) The split cylinder surface adjacent to the loading piston, where there has been almost no strain. This provides a reference image for (b) and (c). There is no significant relative motion between grains, but the grain-boundaries have become grooved through equilibration with the gold.

(**b**) The central part of the split cylinder, showing relative motion of individual grains normal to the originally polished surface, rotational (tilting relative to the specimen surface) motion of some grains plus curvature of grain faces through equilibration of surface forces against the gold foil.

(**c**) The central part of the split cylinder showing the equant grain-shapes and the tendency to form voids in the relatively extended intergrain orientations, which are parallel to the compression direction. The shadowing across individual grains arises from the relief created by grain motions normal to the initially polished surface.

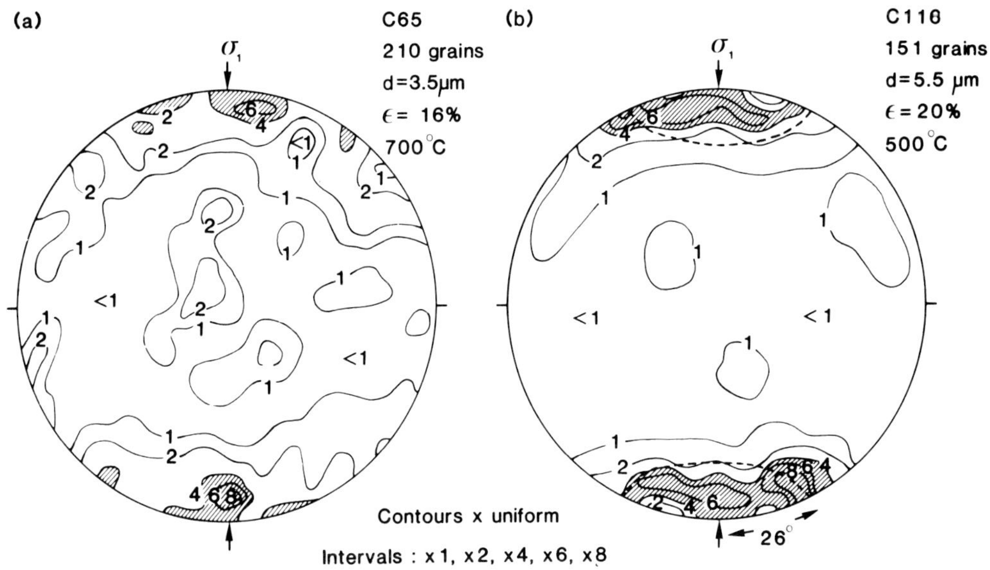

Fig. 13. *C*-axis preferred orientation produced in two samples; (**a**) C-65, at 700°C, grain-size = 3.5 μm, strain = 16%, 210 grains. (**b**) C-116, at 500°C, grain-size = 5.5 μm, strain = 20%, 151 grains. In each case contours are multiples of a uniform distribution, and maximum recorded intensities are indicated. Arrow indicates compression direction. In both cases the influence of the $e(01\bar{1}2)$-maximum fabric induced during initial cold pressing persists. However, at 500°C it remains stable, whereas at 700°C it appears to be weakening, despite similar total high-temperature strains in each case.

experimental data may be fitted by multiple linear regression, and account may be taken of grain-growth during the progression of the highest temperature tests on the finer-grained rocks. The stress relaxation data were used for this analysis and separate fits were carried out on the high stress and low stress parts of the log stress/log strain-rate curves. For each test, the boundary between regimes of 'high stress' and 'low stress' behaviour were selected by visual inspection on the basis of the observed change in the strain-rate sensitivity to stress. For the reasons outlined earlier, very low stress data points were excluded from the analysis, and the low strain-rate data from the 38 μm grain-size specimens were also excluded, as explained below.

In these analyses, each data point is treated as a single measurement, hence the effects of variability between specimens are included in the uncertainty estimates.

The flow laws obtained are:

$$\log \dot{e} = 4.93 - 190\ \text{kJ}/2.303RT + 1.67 \log \sigma - 1.87 \log d$$

at differential stresses up to 25 MPa and

$$\log \dot{e} = 2.00 - 190\ \text{kJ}/2.303RT + 3.33 \log \sigma - 1.34 \log d$$

at differential stresses between 25 and 200 MPa. At higher stresses the flow becomes increasingly non-linear as it passes into the regime of exponential creep (Rutter 1974). For each fit the standard error in log $\dot{e}$ is ± 0.50. Using these parameters, data has been normalized with respect to one or two of the independent variables to show graphically the relationships between the raw data and the best-fit lines (Figs 15 and 16). Fits using stress as the dependent variable are not reported (cf. Schmid *et al.* 1977) because the transitions between the different flow regimes are mainly stress-dependent, and misleading effects would arise from normalization.

It is clear by inspection of the isotherms in log stress/log strain-rate space (Fig. 15) that although these equations describe the bulk of the data reasonably well, the residuals are not always randomly distributed about the best fit lines (e.g. the 400°C data and the 500°C data at low strain-rates). This is probably a reflection of the gradual rather than stepwise changes in the relative contributions of different deformation mechanisms to the total strain as the deformation conditions are varied. It means that extrapolations of the best fits outside the range of experimental conditions will rapidly become invalid. The low strain-rate data from the 38 μm

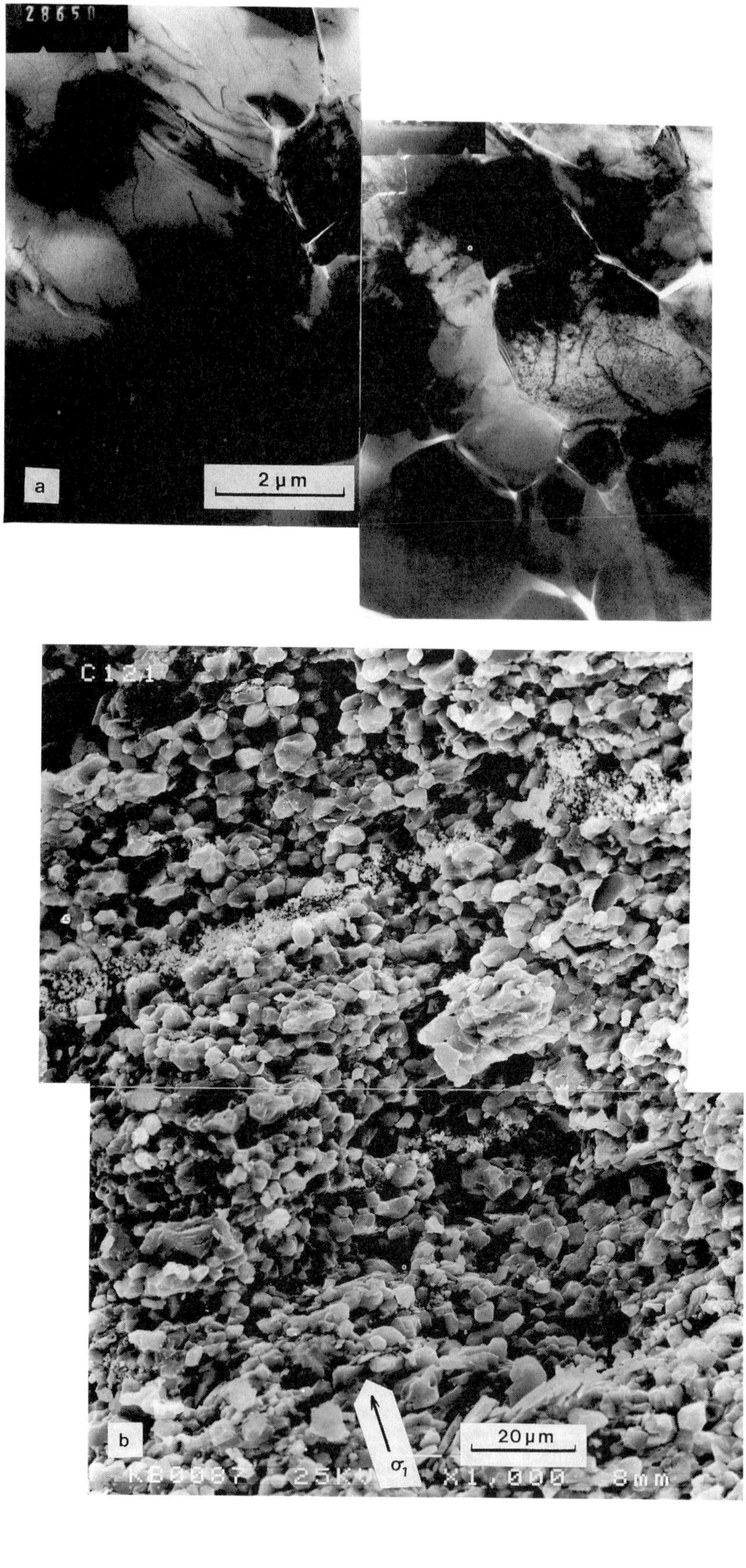

28650
a
2 μm
C121
b
20 μm
σ1
KB0087
X1,000
8mm

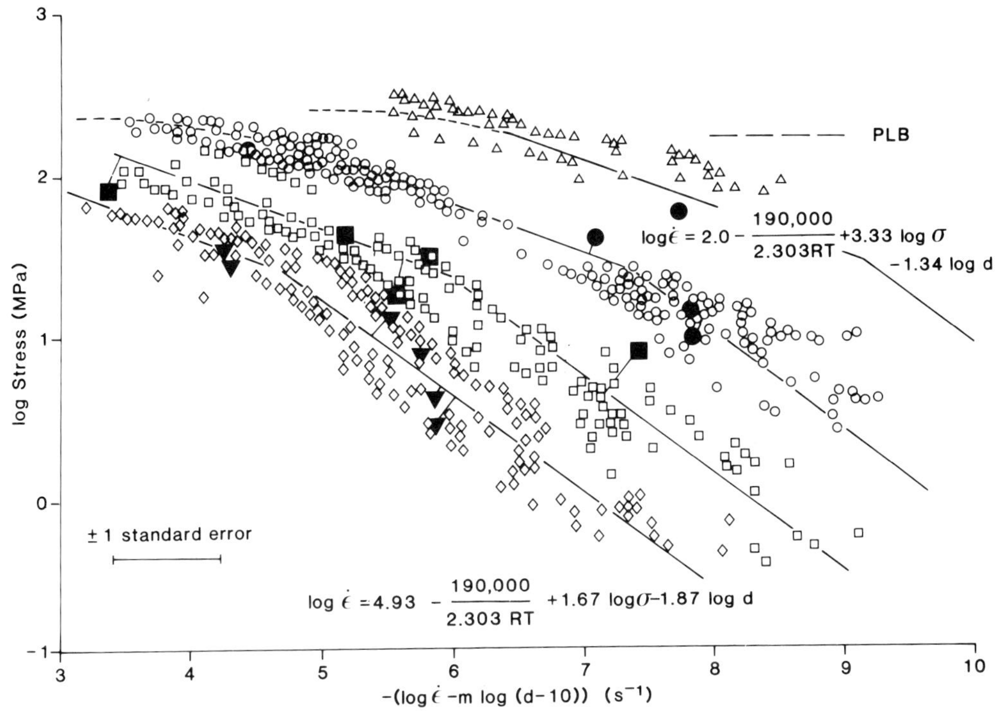

Fig. 15. Results of multiple regression analyses carried out on 34 stress relaxation runs on the pure calcite polycrystals, excluding the low strain-rate data from the 38 μm grain-size samples (because they cannot be satisfactorily described using a common set of fitted parameters) and data at low stress levels commensurate with the strength of the copper jacket. The data have been normalized to a grain-size of 10 μm, using the grain-size sensitivity determined from the fits. Constant strain-rate and constant stress (creep) data are also shown (large, filled symbols), normalized using the same procedure. PLB, power law breakdown stress. Triangles, 400°C; circles, 500°C; squares, 600°C; diamonds, 700°C.

grain-size specimens was clearly stronger and the stress/strain-rate relation more non-linear (Figs 5 and 7) than predicted by the above flow law fits, but there are at present too few data to investigate the change in the flow law parameters. Owing to their different fabrication histories, there may also be significant microstructural differences between (particularly) the 38 μm samples and the finer ones, which might lead to different flow law characteristics, i.e. mean grain-size alone might not be an adequate descriptor of rate-determining geometrical characteristics between the different grain-size specimens. Although alternative methods to that used for fitting flow laws to experimental data exist (e.g. Sotin & Madon 1988), no advantage would accrue in this case owing to the inadequacy of the flow law formulation itself. A better form of the constitutive flow law than the 'standard' forms above will be required, in which the 'constant' parameters of the flow laws are themselves variables.

The grain-size sensitivity of the flow is illustrated in Fig. 17, which shows that there is

Fig. 14. (**a**) High voltage TEM image of a typical field in specimen C55 (25% shortening, 600°C, batch iii, low-stress (30 MPa) flow regime), showing the low and variable dislocation density within grains. It is not clear, however, to what extent the dislocation substructure has been modified during cooling under effective confining pressure. (**b**) Calcite polycrystal (broken surface) doped with 5% volume alumina (specimen C121, 700°C, 31% shortening, secondary electron image). The compression direction is indicated. The image shows the tendency for the alumina (grain-size about 0.25 μm) to form extremely elongate clusters normal to the compression direction, and which are inferred to have acquired their shape largely during the initial cold pressing. Although most of the alumina has clustered, a small amount is dispersed throughout the sample.

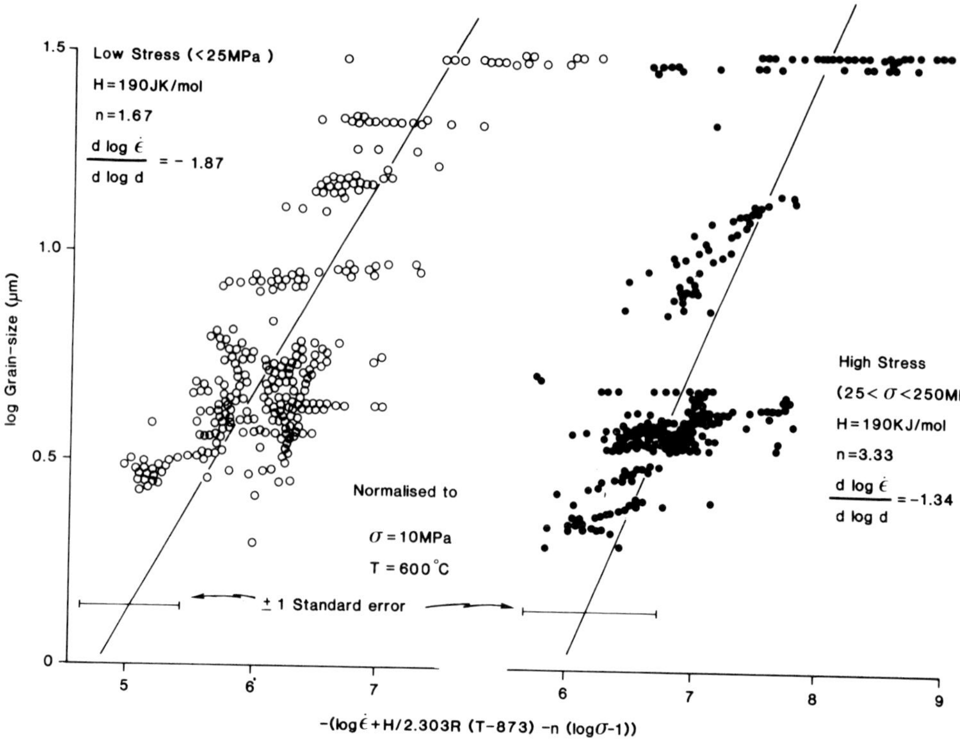

Fig. 16. Results of the multiple regression analyses normalized to a flow stress of 10 MPa and a temperature of 600°C, in order to show the grain-size sensitivity of the flow.

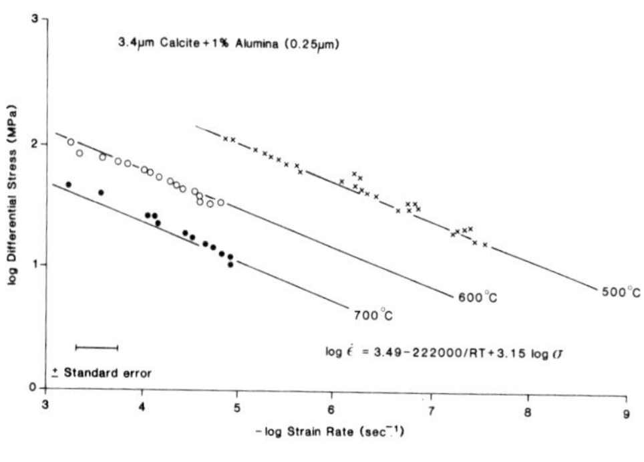

Fig. 17. Multiple regression fit for three stress relaxation runs for 1% volume alumina-doped specimens of 3.4 μm grain-size.

significant grain-size sensitivity even in the high-stress regime of power-law (n = 3.3) creep. In Fig. 18 the transition between grain-size sensitive and grain-size 'insensitive' flow is made clear, using previously obtained results on the coarser-grained Carrara and Taiwan marbles.

Grain-size sensitivity in which strength increases as grain-size increases is characteristic of all of the synthetic calcite rocks, whereas the two natural marbles exhibit the reverse trend. The latter effect may arise from the Hall-Petch effect, but in this case there may be an apparent exacerbation because most of the volume of the natural marbles comprises grains larger than the mean grain-size, which is here estimated on the basis of the *number* of grains falling into each class division. Deformation twinning, which is easier in coarser grained rocks (Rowe & Rutter 1990), probably also enhances the weakening with increasing grain-size.

The accompanying histograms of the grain-size distributions for all sample types shown in Fig. 2 puts the spread of experimental data into better perspective. Mean grain-sizes are shown on the plotted points, but the natural spread of grain-sizes produced during grain-growth, particularly in the finer grained samples, means that different parts of grain population may be behaving differently at any given time (cf. Freeman & Ferguson 1986; Ghosh & Raj 1986).

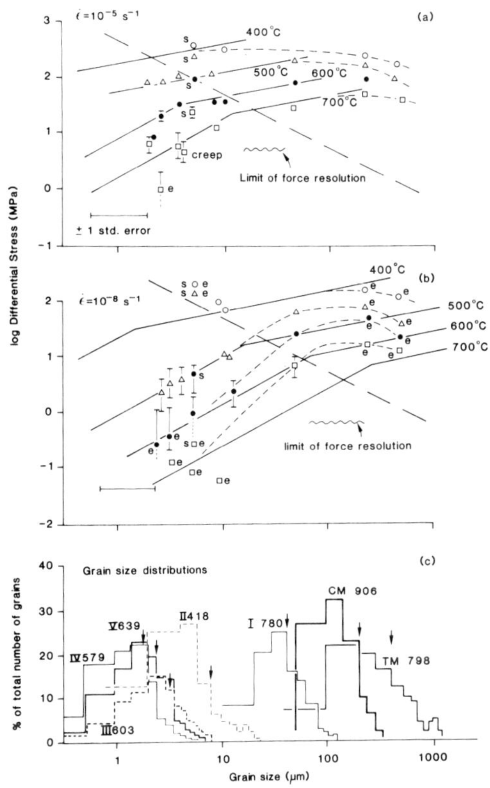

Fig. 18. Log stress v. log grain-size deformation maps showing multiple regression isotherms for constant strain-rates of **(a)** 10^{-5} s^{-1} and **(b)** 10^{-8} s^{-1}. Discrete data points are also shown where they exist on or close to these section planes. Points labelled e represent a (small) extrapolation. S, Solnhofen limestone; open circles 400°C; triangles 500°C; filled circles 600°C; squares 700°C. Dashed straight line, recrystallized grain-size v. stress relationship (Schmid *et al.* 1980). Error bars shown at low stresses are based on the accuracy of force measurements. At the higher strain-rate the fitted lines describe the data quite well, but at the lower strain-rate the fit underestimates the strength of the 38 μm and coarser grained samples. Dashed curves show trends in data. At large grain-sizes, data for Carrara marble (after Rutter 1974 and Schmid *et al.* 1980) and Taiwan marble (after Rowe & Rutter 1990) are also shown. **(c)** shows histograms of grain-size distribution for each sample type (grain-size batches I to V and Carrara (CM) and Taiwan (TM) marbles; numbers indicate grains counted) using the same logarithmic grain-size scale as in (a) and (b), so that the range of grain-sizes is each sample can be fully appreciated. Most of the *volume* of each sample is composed of grains larger than the mean grain-size (indicated by arrow). Mean grain-size is defined on the basis of counted *numbers* of grains.

Figure 18 shows the flow law fits at two strain-rates, 10^{-5} s^{-1} and 10^{-8} s^{-1}. The latter involves a small amount of extrapolation outside the range of experimental observations, from which it is clear that there is divergence between the fitted curves and data points in the high stress regime. Grain-size sensitivity at high stresses appears to become stronger with decreasing strain-rate. For this reason, use of the flow laws to extrapolate to geological strain-rates must be made with due reservation. The presently available data do not cover a sufficiently wide range of conditions for an adequate description of their behaviour to be extracted.

Flow law for the alumina-doped samples and Solnhofen limestone

Only in the case of the 1% volume alumina synthetic polycrystals was a sufficient range of experiments performed to attempt a flow law fit (Fig. 17). The flow law obtained was

$$\log \dot{e} = 3.49 - 222 \text{ kJ}/2.303RT + 3.15 \log \sigma$$

for a single grain size of 3.4 μm, with a standard error in log $\dot{e}$ of ± 0.18. It is likely that a larger number of repeat runs would cause a larger standard error value. The data lie in the 'high-stress' regime. With the possible exception of the activation enthalpy value, the above fit is not significantly different from that for the pure calcite polycrystals in the high-stress regime.

Although only a small number of data are available, it is clear that increasing the concentration of intergranular second phase particles causes hardening and partial inhibition of grain-growth (see also Olgaard, this volume, and Olgaard & Evans 1986). If the latter were not the case it would be difficult to understand the preservation of natural calcite rocks of grain-size less than 10 μm which have been deformed at temperatures on the order of 300°C (e.g Schmid 1975; Burkhard, this volume, see Fig. 19).

We believe that in our alumina-doped samples the observed hardening was less than might have been achieved if the alumina had been more uniformly dispersed, Clustering of second phases causes stress concentrations and hence apparent weakening which may counteract the hardening effects of the dispersion (Kendall *et al.* 1989). The greater strength of Solnhofen limestone is probably a measure of the hardening which can be achieved from a uniform dispersion of about 5% volume of second phases. Except for the higher strength, the flow law parameters for Solnhofen limestone

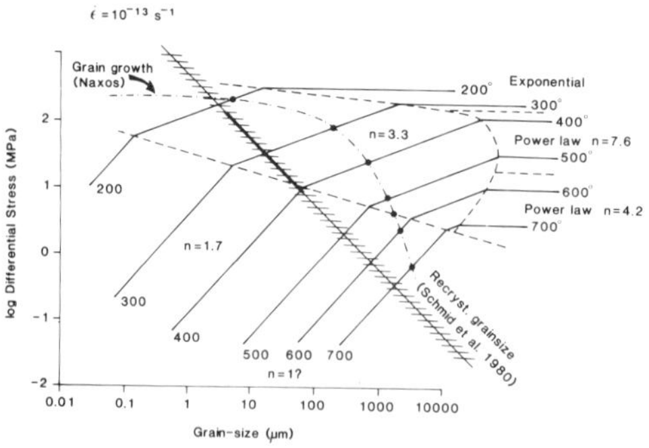

Fig. 19. Deformation map in log stress v. log grain-size showing isotherms for a constant strain-rate of 10^{-13} s^{-1}, constructed by extrapolation of the multiple regression fits, and using Carrara marble data to represent the relatively grain-size insensitive field. Also shown is the empirical relationship between recrystallized grain-size and flow stress of Schmid *et al.* (1980) (emboldened section is range of experimental observations), and the curve showing maximum grain-size *v.* temperature observed for pure calcite marbles on Naxos island, Greece (Covey-Crump & Rutter 1989). Combinations of grain-size and temperature to the left of this curve are expected to be eliminated by grain-growth over the time-scale implied by this strain-rate.

Much of the grain-size sensitive flow field is therefore rendered inaccessible by grain-growth, unless impurity phases inhibit growth. Nevertheless, a sufficient amount is accessible such that a coarse-grained pure marble suffering tectonic grain-size reduction might be expected to be weakened. It appears possible that at high temperatures, the flow of even very coarse marbles may be grain-size sensitive. Given the inadequacies of the multiple regression fits previously mentioned, it is likely that these isotherms *underestimate* the strength of calcite rocks at this strain-rate.

(Schmid *et al.* 1977) are closely comparable to those of the synthetic materials. In the stress range below about 100 MPa the activation enthalpy for Solnhofen limestone is about 210 kJ mol^{-1} and the stress exponent, n, decreases from above 3 towards unity as stress is decreased. At stresses above 100 MPa the differences appear greater.

Deformation mechanisms

The currently available experimental data clearly demonstrate the transition in calcite rocks from relatively grain-size insensitive to grain-size sensitive flow, and the form of the flow laws which describe the data. The transition is best demonstrated by Fig. 18, in which log flow stress is plotted against log grain-size for constant temperature and strain-rate. Over the grain-size range of the synthetic specimens, however, there are two ranges of grain-size sensitivity, characterized by different flow laws and microstructures, and hence corresponding to significantly different deformation mechanisms.

Microstructures in the high-stress ($250 > \sigma > 25$ MPa), grain-size sensitive regime are characterized by grain-flattening by an amount comparable to the imposed bulk strain, a small amount of grain-boundary sliding and the development (or maintenance) of a strong crystallographic preferred orientation. The mechanical behaviour is characterized by steady-state flow to large strains, a distinctly non-linear stress exponent of strain-rate ($n = 3.3$) and a grain-size exponent of -1.3. There is a slightly negative pressure sensitivity of the flow stress, which appears to be characteristic of calcite rocks during high temperature plastic flow (Heard 1960; Rutter 1972). Except for the grain-size sensitivity, these are all expected characteristics of recovery-accommodated flow by intracrystalline plasticity. It is not clear, however, how the grain-size sensitivity arises. It is opposite in sign to that expected of intracrystalline plasticity. We suggest that recovery may be accomplished by dislocation climb into grain boundaries, so that the shorter the climb distance (proportional to grain diameter) the faster the recovery rate.

In the low stress regime ($\sigma < 25$ MPa), grain shapes remain equant over large strain ranges, there is apparent weakening of the crystallographic preferred orientation induced during cold pressing and a slight tendency for dilatation of the least-stressed grain-to-grain interfaces. Split-cylinder tests showed directly the accommodation of at least some strain by grain-boundary sliding. Grain-growth begins to compete effectively with any tendency towards dynamic maintenance of grain-size at the higher temperatures. The mechanical behaviour is characterized by a near-linear viscous behaviour ($n = 1.6$) and a slight positive pressure sensitivity of the flow stress. These are all expected characteristics of flow by grain-boundary sliding with accommodation by a mixture of intragranular dislocation activity solid-state volume diffusion (commonly identified with phenomenological superplasticity).

Apparent activation enthalpies for flow in the two flow regimes for pure calcite rocks are identical, and point to the same rate-controlling process (e.g. diffusion) in both cases. Activation enthalpy for flow in the alumina-doped samples appears to be about 15% higher, but there are insufficient data to be able to say whether the difference is significant.

Some seventeen theoretical models for superplasticity are generally couched in terms of grain-boundary sliding (BGS) with one or more concurrent accommodation processes (review by Kashyap & Mukherjee 1985) involving volume or grain-boundary diffusion, dislocation motion or grain rearrangements. Almost all models predict a flow law of the form

$$\dot{e} = A\,(D_o Gb/kT)\exp(-H/kT)\,(b/d)^2\,(\sigma/G)^n \qquad 2.$$

in which A is a dimensionless number, D_o is a diffusion coefficient, G is the shear modulus, b a Burgers vector, and k is Boltzmann's constant. Most models predict the value of the geometric factor A, which commonly lies in the range 50 to 200. n is expected to be 1 for purely diffusional accommodation and 2 for accommodation by processes involving grain-boundary and intragranular dislocation motion. We can examine how well this law predicts our observed data. Taking $b = 0.7$ nm (Wenk *et al.* 1983), $G = 25$ GPa, $D_o = 5 \times 10^{-6}\ m^2\ s^{-1}$ (Rutter 1976) and our experimentally observed activation enthalpy figure (190 kJ mol^{-1}), at $\sigma = 10$ MPa the diffusional accommodation model overestimates the creep rate by about × 100, and the dislocation accommodation model underestimates the creep rate by up to × 10. Using the observed grain-size exponent of −1.9 and the observed stress exponent of 1.67 in eq. 2, exact agreement with the observed creep rates can be obtained by adjusting the value of A within the above range.

We interpret the above to mean that in the low stress regime flow is by grain-boundary sliding accommodated mainly by dislocation processes but with some diffusional-mass-transfer component. This interpretation is consistent with the microstructural data. As mentioned earlier, the mechanical data show that there is in fact a continuous variation of n with decreasing stress or increasing temperature, rather than the constant value used to describe the observed data by the multiple regression procedure, in agreement with the observations of Schmid *et al.* (1977) on Solnhofen limestone. We interpret this to mean that there is a continuous variation in the contributions of the various processes which accommodate grain-boundary sliding.

Volume changes

Using the split-cylinder technique, evidence was seen for void formation through the separation of the less-stressed grain boundaries. One possible difficulty with the split cylinder technique is that because the gold is so weak and malleable, as demonstrated by its injection into grain-boundary voids, it may behave like a pore fluid under a pressure almost equal to the confining pressure. If this is so we may locally be seeing degrees of dilatation which are more characteristic of deformation under lower confining pressures. An attempt was made to assess the importance of grain-boundary void formation in the bulk of the material by means of an examination of a polished sawcut made on the same specimen but normal to that in contact with the gold. Although some evidence for opening of the least stressed interfaces was seen, it seemed qualitatively to be less important (but not unimportant) than on the gold-covered surface.

Further evidence of dilatation during high temperature flow is provided by the slightly positive pressure sensitivity of the flow stress shown on Fig. 6 (20 MPa increase in flow stress over 300 MPa confining pressure change by pressure stepping on a single sample). The effect is unlikely to be due to more effective elimination of original void space at higher presure because pressure stepping was always carried out in the direction of decreasing pressure.

Void formation during grain-boundary sliding (cavitation) is a common feature of superplastic flow of metals and alloys deformed in tension (e.g. Ridley & Pilling 1985). Behrmann & Mainprice (1987) reported grain-boundary void formation in a quartz–feldspar mylonite and appealed to it as microstructural evidence for grain boundary sliding in a naturally deformed rock. However, void formation requires work to be done against the effective mean stress on the material, so that part of the apparent strength of the material (as measured by differential stress) is attributable to void formation. In axisymmetric compression this is expected to be approximately equal to $\sigma_3 de_v/de_a$ where e_v is volumetric strain and e_a is axial strain. If at 300 MPa confining pressure, 1% volume strain is accumulated over 30% shortening and the flow process is otherwise pressure insensitive, the apparent hardening should be about 10 MPa. This is commensurate with pressure-sensitivity observed in pressure stepping tests at 700°C at high strain-rate (given the perhaps unlikely assumption that rate of dilatation is not strongly pressure-sensitive). It is believed extremely unlikely that the observed pressure sensitivity is attributable to cataclastic processes, because the confining pressure is 5 to 20 times higher than the flow stress, and the flow regime in question is at lower flow stresses than those characteristic of intracrystalline plastic flow, which in calcite appears to be characterized by a slightly *negative* pressure sensitivity.

We have no data on pressure sensitivity of the flow at low strain-rates and high temperatures, but if dilatation becomes progressively less important with decreasing strain-rate then the apparent form of the flow law will be effective-pressure sensitive. This possibility and pressure effects in general remain to be investigated more fully.

Wawersik & Hannum (1979) showed that dilatancy occurs during the hot creep of halite rock and Fischer & Paterson (1989) have demonstrated small amounts of grain-boundary dilatancy during the high temperature flow of marble. These facts, coupled with the possibility of significant dilatation during superplastic flow of fine-grained calcite rocks, have important geological implications despite the amounts of dilatation being small. Carter *et al.* (this volume) point out that if volume increase occurs during high temperature deformation of halite rocks, it is virtually certain to occur during the deformation of silicate rocks. This may explain how plastic shear zones may become preferred conduits for fluid flow, often leading to shear zones becoming loci for metamorphic reactions, even though the grain-size be *reduced* (Rutter & Brodie 1985). If porosity remains constant, permeability is expected to be reduced as a result of grain-size reduction if throat size is reduced in proportion because flux varies as the reciprocal fourth power of throat size. Thus major plastic shear zones transecting the middle and lower crust and possibly lithospheric mantle may provide important pathways for fluid flow if shearing is accompanied by dilatancy. This will only apply whilst they are active, because the void space is dynamically maintained, and it may not be necessary to appeal to high fluid pressures to the extent that is often done to explain fluid flow (e.g. McCaig 1989).

Extrapolation to geological conditions

As mentioned earlier, the multiple regression-fitted flow laws do not provide a good basis for extrapolation outside the range of accessed experimental conditions. However, to give an approximate impression of the pattern of behaviour expected at a low strain-rate, and particularly to speculate on the possible effects of grain-growth, Fig. 19 shows a deformation mechanism map constructed for a strain-rate of 10^{-13} s^{-1}, using the flow law fits given above together with those obtained by Schmid *et al.* (1980) for Carrara marble. We have used the relationship between grain-size and temperature observed for calcite marbles on Naxos island, Greece (Covey-Crump & Rutter 1989) to delimit the inaccessible grain-size/temperature space on the deformation mechanism map. It is not known whether this grain-size is a true, temperature-determined maximum or kinetically frozen-in during cooling, but it is considered typical of grain-size of calcite rocks in nature. The strain-rate chosen is sufficiently slow that grain-size will probably always be maintained at the maximum attainable size. The recrystallized grain-size v. stress relationship obtained by Schmid *et al.* (1980) is also shown. In the plastic flow regime at large grain-sizes, no attempt is made to show the likely effects of weakening with increasing grain-size which arises from the Hall–Petch effect and the increasing importance of twinning. Also, in the $n = 3.3$ regime, the isotherms are more likely to be deflected upwards in the manner shown on Fig. 18, compared to the shapes shown which arise from the strict extrapolation of the regression fit.

Provided that dynamic recrystallization reduces the grain-size of most of the volume of a calcite rock according to the relationship given by Schmid *et al.* (1980) or to a level determined by grain-growth (whichever is the larger), it is expected that tectonic grain-size reduction will lead to significant strain-weakening (e.g. Rutter & Brodie 1988*a*). It seems likely that grain-growth will be an important factor in limiting the degree of grain refinement, but it is not yet clear what role will be played by dynamic grain-boundary migration in determining syntectonic grain-size. The grain-growth curve shown in Fig. 19 applies to one particular geological situation (where the calcite rocks are very pure), and it is not known to what extent it might be shifted to the left through the inhibitory effect of second phases on grain-growth (but see Olgaard, this volume).

The extrapolation also suggests that at high temperatures grain-size sensitive flow may extend to coarse grain-sizes (above 1 mm). This implies that coarse marbles deformed at amphibolite grade may display equigranular textures which are indistinguishable from the products of static grain-growth (e.g. Urai *et al.* 1986).

Conclusions and further work

In the progress of this study we have learned how to make satisfactory synthetic pure calcite polycrystals of controlled grain-size and low porosity, and we have obtained a first approximation to the grain-size sensitive flow behaviour and its transition to (relatively) grain-size insensitive flow. The flow law parameters vary significantly with temperature, grain-size, stress and strain-rate, in ways that we have not been able to investigate fully with the presently available data, making extrapolation outside the accessed conditions unreliable except in a very approximate way. It will be necessary to refine the flow law data by means of new experiments on the best polycrystals that we can make, and to lower stress levels and strain-rates using improved testing equipment. The question of the possible existence of a threshold stress for flow must be addressed, and closer attention given to elucidation of the deformation mechanisms.

The results suggest that it should be possible to weaken coarse-grained calcite rocks through tectonic grain refinement, but this must now be tested through large-strain experiments on coarse marbles.

Improved techniques must be sought for the preparation of doped samples of uniform microstructure, and the effects of doping on the flow law systematically investigated. Tests on doped samples have the advantage that grain-growth is inhibited, so that it may be possible to operate to higher temperatures without serious problems of grain-growth. Second phases are likely to play an important role in stabilizing grain-size during the natural deformation of fine-grained calcite rocks.

The experience of studying grain-size sensitive flow of calcite rocks should prove useful for the eventual experimental synthesis and deformation of fine-grained silicate rocks.

This work was funded through NERC grant GR3/6174, which also provided A. N. W. with a NERC research studentship. Most of the work was carried out at Imperial College, London, but was completed after the transfer of the rock deformation laboratory to the University of Manchester. We thank D. Olgaard (ETH Zurich) for much useful discussion, for the provision of one of his hot-pressed calcite polycrystals for comparative study, together with his experimental data, and for a detailed and constructive review. A second anonymous reviewer provided several constructive suggestions. Experimental officer R. Holloway was as indispensible as always to keeping equipment running and S. Covey-Crump's refinements to data processing procedures and critical reading of the text proved invaluable.

References

ABRAMS, H. 1971. Grain-size measurement by the intercept method. *Metallography*, **4**, 59–78.

ALFORD, N. MCN., BIRCHALL, J. D. & KENDALL, K. 1987. High-strength ceramics through colloidal control to remove defects. *Nature*, **330**, 51–53.

BEHRMANN, J. H. & MAINPRICE, D. H. 1987. Deformation mechanisms in a high temperature quartz-feldspar mylonite: evidence for superplastic flow in the lower continental crust. *Tectonophysics*, **140**, 297–305.

BURKHARD, M. 1990. Ductile deformation mechanisms in micritic limestones naturally deformed at low temperatures (150–350°C). *This volume.*

CARTER, N. L., KRONENBERG, A. K., ROSS, J. V. & WILTSCHKO, D. 1990. Control of fluids on deformation of rocks. *This volume.*

CASEY, M., RUTTER, E. H., SCHMID, S. M., SIDDANS, A. W. B. & WHALLEY, J. S. 1978. Texture development in experimentally deformed calcite rocks., *In*: GOTTSTEIN, G. & LÜCKE, K. (eds) *Proceedings of the Conference on Textures of Materials*, **5**, Springer, Heildelberg, Berlin, New York, 231–240.

CHOPRA, P. N. 1986. The plasticity of some fine-grained aggregates of olivine at high pressure and temperature. *Geophysical Monograph*, **36**, 25–34.

COVEY-CRUMP, S. J. & RUTTER, E. H. 1989. Thermally induced grain-growth of calcite rocks on Naxos Island, Greece. *Contributions to Mineralogy and Petrology*, **101**, 69–86.

FISCHER, G. J. & PATERSON, M. S. 1989. Dilatancy during the deformation of rocks at high temperatures and pressures. *Journal of Geophysical Research*, **94**, 17607–17617.

FREEMAN, B. & FERGUSON, C. 1986. Deformation mechanism maps and micromechanisms of deformation of rocks with distributed grain-sizes. *Journal of Geophysical Research*, **91**, 3849–3860.

GHOSH, A. K. & RAJ, R. 1986. A model for the evolution of grain-size distribution during superplastic deformation. *Acta metallurgica*, **34**, 447–456.

HEARD, H. C. 1960. Transition from brittle fracture to ductile flow in Solnhofen limestone as a function of temperature, confining pressure and interstitial fluid pressure. *Geological Society of America Memoir*, **79**, 193–226.

KARATO, S., PATERSON, M. S. & FITZGERALD, J. D. 1986. Rheology of synthetic olivine aggregates: influence of grain size and water. *Journal of Geophysical Research*, **91**, 8151–8176.

KASHYAP, B. P. & MUKHERJEE, A. K. 1985. On the models for superplastic deformation. *In*: BAUDELET, B. & SUERY, M. (eds) *Superplasticity*. CNRS publications, Paris, 4.1–4.31.

KENDALL, K., ALFORD, A. MCN., CLEGG, W. J. & BIRCHALL, J. D. 1989. Flocculation clustering and weakness of ceramics. *Nature*, **339**, 130–132.

KIRBY, S. H. & KRONENBERG, A. K. 1984. Deformation of clinoyroxenite: evidence for a transition in flow mechanisms and semibrittle behaviour. *Journal of Geophysical Research*, **89**, 3177–3192.

MCCAIG, A. M. 1989. Fluid flow through fault zones. *Nature*, **340**, 600.

MUELLER, W. H., SCHMID, S. M. & BRIEGEL, U. 1981. Deformation experiments on anhydrite rocks of different grain-sizes: rheology and microfabric. *Tectonophysics*, **78**, 527–543.

OLGAARD, D. L. 1990. The role of a second phase in localization of deformation. *This volume*.

—— & EVANS, B. 1986. Effect of second-phase particles on grain-growth in calcite. *Journal of the American Ceramic Society*, **69**, c272–c277.

—— & ——. 1988. Grain-growth in synthetic marbles with added mica and water. *Contributions to Mineralogy and Petrology*, **100**, 246–260.

POIRIER, J-P. 1985. *Creep of Crystals*. Cambridge University Press.

RALEIGH, C. B. 1965. Glide mechanisms in experimentally deformed minerals. *Science*, **150**, 739–741.

RIDLEY, N. & PILLING, J. 1985. Cavitation in superplastic alloys: experimental. *In*: BAUDELET, B. & SUERY, M. (eds) *Superplasticity*. CNRS publications, Paris, 8.1–8.17.

ROWE, K. J. & RUTTER, E. H. 1990. Palaeostress estimation using calcite twinning: experimental calibration and application to nature. *Journal of Structural Geology*, **12**, 1–18.

RUTTER, E. H. 1972. The influence of interstitial water on the rheological behaviour of calcite rocks. *Tectonophysics*, **14**, 13–33.

—— 1974. The influence of temperature, strain rate and interstitial water in the experimental deformation of calcite rocks. *Tectonophysics*, **22**, 311–334.

—— 1976. The kinetics of rock deformation by pressure solution. *Philosophical Transactions Royal Society of London*, **A283**, 203–219.

—— 1984. The kinetics of grain coarsening in calcite rocks. *Progress in Experimental Petrology*, **6**, NERC series D 25, 245–249.

—— & BRODIE, K. H. 1985. The permeation of water into hydrating shear zones. *In*: THOMPSON, A. B. & RUBIE, D. C. (eds) *Metamorphic reactions, kinetics, textures and deformation*. Advances in Physical Geochemistry, **4**, 242–250.

—— & —— 1988*a*. The role of tectonic grain-size reduction in the rheological stratification of the lithosphere. *Geologische Rundschau*, **77**, 295–308.

—— & —— 1988*b*. Experimental syntectonic dehydration of serpentinite under controlled pore water pressure. *Journal of Geophysical Research*, **93**, 4907–4932.

——, ATKINSON, B. K. & MAINPRICE, D. H. 1978. On the use of the stress relaxation testing method in studies of the mechanical behaviour of geological materials. *Geophysical Journal of the Royal Astronomical Society*, **55**, 155–170.

——, PEACH, C. J., WHITE, S. H. & JOHNSON, D. 1985. Experimental syntectonic hydration of basalt. *Journal of Structural Geology*, **7**, 251–266.

SCHMID, S. M. 1975. The Glarus overthrust: field evidence and mechanical model. *Eclogae Geologicae Helvetiae*, **68**, 251–284.

——, BOLAND, J. N. & PATERSON, M. S. 1977. Superplastic flow in fine-grained limestone. *Tectonophysics*, **43**, 257–291.

——, PATERSON, M. S. & BOLAND, J. N. 1980. High temperature flow and dynamic recrystallization in Carrara marble. *Tectonophysics*, **65**, 245–280.

SCHWENN, M. B. & GOETZE, C. 1978. Creep of olivine during hot pressing. *Tectonophysics*, **48**, 41–60.

SELLARS, C. M. 1978. Recrystallization of metals during hot deformation. *Philosophical Transactions of the Royal Society of London*, **A228**, 147–158.

SENKOV, O. N. & MYSHLYAEV, M. M. 1986. Grain-growth in a superplastic Zn-22% Al alloy. *Acta metallurgica*, **34**, 97–106.

SOTIN, C. & MADON, M. 1988. Generalized nonlinear inversion of kinetics data: application to the calcite – aragonite transformation. *Physics of the Earth and Planetary Interiors*, **52**, 159–171.

SPIERS, C. J. 1979. Fabric development in calcite polycrystals experimentally deformed at 400°C. *Bulletin de Minéralogie*, **102**, 282–289.

TANNER, C. B. & JACKSON, M. L. 1947. Nomographs of sedimentation times for soil particles under gravity or centrifugal acceleration. *Proceedings Soil Science of America*, **12**, 60–65.

TULLIS, J. & YUND, R. A. 1982. Grain-growth kinetics of quartz and calcite aggregates. *Journal of Geology*, **90**, 301–318.

—— & —— 1985. Dynamic recrystallization of feldspar: a mechanism for ductile shear zone formation. *Geology*, **13**, 238–241.

URAI, J. L., MEANS, W. D. & LISTER, G. S. 1986. Dynamic recrystallization of minerals. *Geophysical Monograph*, **36**, 161–199.

VAUGHAN, P. J. & COE, R. S. 1981. Creep mechanism in Mg_2GeO_4: effects of a phase transition. *Journal of Geophysical Research*, **86**, 389–404.

WAWERSIK, W. R. & HANNUM, D. W. 1979. *Interim summary of Sandia creep experiments on rock salt from the WIPP study area, southeastern New Mexico*. DOE contract DE-AC04-76DP0079.

WENK, H-R., BARBER, D. J. & REEDER, R. J. 1983. Microstructures in carbonates. *In*: REEDER, R. J. (ed.) *Carbonates: Mineralogy and Chemistry*. Reviews in Mineralogy, **11**, 301–367.

WHITE, S. H. 1976. The effects of strain on the microstructures, fabrics and deformation mechanisms in quartzites. *Philosophical Transactions of the Royal Society of London*, **A283**, 69–86.

High-temperature deformation of calcite single crystals by r^+ and f^+ slip

J. H. P. DE BRESSER & C. J. SPIERS

HPT Laboratory, Department of Geology, Institute of Earth Sciences, University of Utrecht, PO Box 80021, 3508 TA Utrecht, the Netherlands

Abstract: Optical quality calcite single crystals have been uniaxially compressed at constant strain rates and temperatures (T) in the range 3×10^{-4} to 3×10^{-7} s^{-1} and 550–800°C respectively. The tests were performed with the compression direction parallel to $[40\bar{4}1]$, i.e. parallel to the intersection of two cleavage rhombs. At strains above 1% strain, steady state flow was observed at $T \geq 600$°C. The flow stresses at these temperatures were found to be rather insensitive to strain rate, and can be empirically described by a power law creep equation with a stress exponent ranging from *c.* 13 at 600°C to *c.* 9.5 at 700–800°C. The active glide systems were identified by slip line analysis. The crystals were found to deform by *e*-twinning at $T < 600$°C, and by slip on a single $r\langle\bar{2}021\rangle$ system plus a single f system in the so-called positive sense at $T \geq 600$°C. The effective slip direction on the active f-plane was of $\langle 10\bar{1}1\rangle$ type rather than the $\langle 02\bar{2}1\rangle$ type reported previously. The observed creep behaviour in the slip dominated regime has been compared with microphysical models. The deformation behaviour seems best explained by cross slip or glide-controlled creep models, or a combination of these.

It is well established from observations on natural calcite tectonites that intracrystalline plastic mechanisms are important during the deformation of calcite rocks in nature. For this reason, a great deal of work has been done on the rheological behaviour of calcite rocks, and on the development of microstructures and textures (crystallographic preferred orientations) under conditions where 'dislocation creep' mechanisms dominate (e.g. Heard & Raleigh 1972; Schmid *et al.* 1980; Kern & Wenk 1983). On the basis of numerous experimental studies (e.g. Turner *et al.* 1954; Griggs *et al.* 1960; Braillon *et al.* 1972), it is widely accepted that the main glide systems in calcite are twinning on e $\{\bar{1}018\}\langle 40\bar{4}1\rangle$ (3 systems), slip on $r\{10\bar{1}4\}$ $\langle\bar{2}021\rangle$ (3 systems) and slip on f $\{\bar{1}012\}\langle 02\bar{2}1\rangle$ (6 systems). (Note that throughout this paper we use Miller-Bravais indices referred to the hexagonal unit cell with a = 4.99 Å and c = 17.06 Å. Twinning on e occurs in the so-called positive sense, as defined by Turner *et al.* (1954), see Fig. 1, while slip on r and f is generally believed to occur in the negative and positive senses (Wenk 1985). This set of glide systems (e, r, f) has been used in all studies directed at modelling texture development in calcite, using the Taylor method or related techniques (Lister 1978; Wenk *et al.* 1987; Takeshita *et al.* 1987). Such modelling requires detailed specification of the single crystal glide systems and their relative strengths. However, while slip on r and f in the negative sense is widely reported, only very limited data are available for r and f slip in the positive sense (Spiers & Wenk 1980). Furthermore, there are insufficient data on the temperature and strain rate dependence of the absolute strengths of the e, r and f systems to allow their *relative strengths* to be estimated with confidence, particularly under geological conditions. Hence, if meaningful texture modelling is to be carried out for calcite rocks, further fundamental data are needed on the plastic deformation of single crystals and on slip on r and f in the positive sense in particular. Further fundamental data on intracrystalline creep mechanisms in calcite single crystals also offer a basis for developing a better understanding of the rheological behaviour of calcite polycrystals. In this context, single crystal experiments are of particular value, because they provide information on intracrystalline processes independently of grain boundary effects.

In this paper, we report a series of deformation experiments performed on calcite single crystals at temperatures (T) in the range 550–800°C (homologous temperature: 0.5 to 0.7 of the incongruent melting temperature of calcite in the system $CaO-CO_2$ at 1 kbar pressure, after Wyllie & Tuttle 1960). The crystals were compressed in the $[40\bar{4}1]$ direction following

From Knipe, R. J. & Rutter, E. H. (eds), 1990, *Deformation Mechanisms, Rheology and Tectonics*, Geological Society Special Publication No. 54, pp. 285–298.

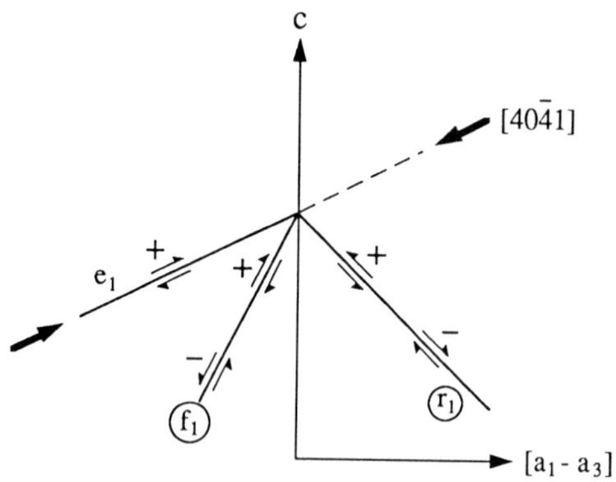

Fig. 1. Definition of sense of gliding on the e_1, r_1 and f_1 planes in calcite, after Turner *et al.* (1954). Loading direction in the present experiments is parallel to [40$\bar{4}$1].

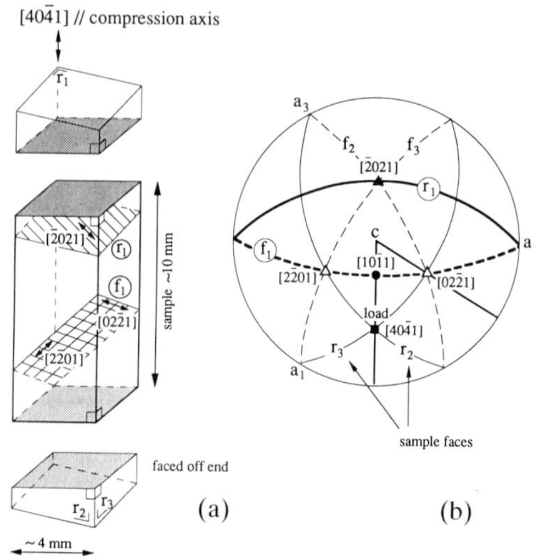

Fig. 2. (**a**) Morphology, dimensions and crystallographic orientation of the samples used in the present study. Shaded areas denote the loaded ends of the cleaved and trimmed sample (cleavage rhombs). (**b**) Stereographic (upper hemisphere) projection of calcite showing relevant planes and directions. Compare with Fig. 2a and Table 1.

Spiers & Wenk (1980), and were found to deform by *e*-twinning ($T < 600°C$) and by slip on *r* and *f* in the positive sense ($T \geq 600°C$). However, the effective slip direction on the active *f* plane was found to be of $\langle 10\bar{1}1\rangle$ type rather than the $\langle 02\bar{2}1\rangle$ type reported previously for *f*-slip (Wenk 1985). In addition to determining the identity of the operative glide systems, we also report a substantial body of data on the influence of temperature and strain rate on the flow strength of our crystals. The creep behaviour observed in the slip-dominated field seems to be best explained in terms of a glide-controlled or cross slip-controlled creep model, or a combination of these.

The samples: preparation and orientation

The present experiments were performed on cleaved calcite 'prisms' compressed in the [4041] direction, i.e. parallel to the intersection of two rhombohedral (*r*) cleavage planes (arbitrarily denoted r_2 and r_3 here). The morphology and dimensions of the samples are illustrated in Fig. 2a, while the orientation of the compression axis with respect to the crystal axes is shown in the stereographic projection of Fig. 2b. All samples were prepared from elongated rhombs of calcite cleaved from a single parent crystal of optical quality 'iceland spar' (total trace element content < 400 ppm, individual trace elements < 100 ppm). The ends of the cleaved rhombs were faced off using a diamond saw, thus producing the final geometry shown in Fig. 2a. The cleaved (vertical) faces of the samples were of high optical quality with only occasional cleavage steps being present. Prior to deformation, all samples were annealed at 500°C for a period of 24 hours to remove dislocation damage. TEM analysis of the annealed samples showed the residual dislocation density to be less than *c.* 10^6 cm^{-2}.

As described above, the samples were compressed parallel to their length, i.e. in [40$\bar{4}$1] direction (Fig. 2a,b). The corresponding Schmid factors for the *e*, *r* and *f* glide systems generally reported in the literature, are listed in Table 1. From this table it is clear that the chosen orientation is unfavourable for *e*-twinning. However, it is relatively favourable for slip on the r_1 (10$\bar{1}$4) [$\bar{2}$021]$^+$ system ($S = 0.31$), on the two symmetrically disposed f_1($\bar{1}$012)[2$\bar{2}$01]$^+$ and f_1($\bar{1}$012)[02$\bar{2}$1]$^+$ systems (S = 0.38), and on f_2 (1$\bar{1}$02)[02$\bar{2}$1]$^-$ and f_3(01$\bar{1}$2) [2$\bar{2}$01]$^-$ ($S = 0.38$; refer Table 1 and Figs. 2a, b). Note that we use the notation (hkil)[uvtw]$^\pm$ here and henceforth, making use of the superscript signs to indicate slip sense.

Experimental method

The samples were deformed in uniaxial compression using an Instron 1193 testing machine equipped with an externally heated, controlled atmosphere cell. The tests were carried out at temperatures in the range 550–800°C and at

Table 1. *Schmid factors* S *for the main twinning and glide systems generally quoted for calcite, for loading in the [40$\bar{4}$1] direction*

e-twinning:	
e_1 ($\bar{1}$018)[40$\bar{4}$1]	$S = 0$
e_2 (1$\bar{1}$08)[$\bar{4}$401]$^+$	$S = 0.12$ positive sense
e_3 (01$\bar{1}$8)[0$\bar{4}$41]$^+$	$S = 0.12$ positive sense
r-slip:	
r_1 (10$\bar{1}$4)[$\bar{2}$021]$^+$	$S = 0.31$ positive sense
r_2 ($\bar{1}$104)[2$\bar{2}$01]	$S = 0$
r_3 (0$\bar{1}$14)[02$\bar{2}$1]	$S = 0$
f-slip:	
f_1 ($\bar{1}$012)[2$\bar{2}$01]$^+$[02$\bar{2}$1]$^+$	$S = 0.38$, 0.38 positive sense both
f_2 (1$\bar{1}$02)[$\bar{2}$021]$^+$[02$\bar{2}$1]$^-$	$S = 0.20$, 0.38 positive and negative sense respectively
f_3 (01$\bar{1}$2)[$\bar{2}$021]$^+$[2$\bar{2}$01]$^-$	$S = 0.20$, 0.38 positive and negative sense respectively

Slip directions after Wenk (1985). All indices refer to the hexagonal cell with a = 4.99 Å and c = 17.06 Å, upper hemisphere coordinates.

approximately constant strain rates in the range 3×10^{-4} to 3×10^{-7} s^{-1}. All tests were carried out using a CO_2 atmosphere maintained at 0.25 MPa overpressure to suppress decomposition of the samples. Maximum axial strains achieved were 5–7%. While most experiments were conducted at fixed displacement rate (i.e. fixed cross-head speed, giving approximately constant strain rate), a few tests were performed in strain rate stepping mode. In these tests flow stresses were found to be independent of stepping history. All tests were terminated by rapidly unloading the sample, with immediate quenching using a blast of cold CO_2 gas (i.e. direct from a CO_2 bottle with the pressure regulator set at 0.25 MPa).

The experimental apparatus allowed temperatures to be kept constant within *c*. 3°C. The temperature drop between the end regions and the centre of the sample was *c*. 4°C. Axial load was measured with an external Instron load cell with an absolute error $\leq$ 0.5% of the measured load. The raw load signal was recorded versus time using a chart recorder. These data were processed to produce true stress-strain curves, calculating displacement from cross-head velocity and elapsed time, and applying appropriate corrections for apparatus stiffness, thermal expansion of the crystal, and change in cross-sectional area of the sample assuming homogeneous deformation and constant volume.

Mechanical data

The complete set of 24 tests reported here is listed in Table 2, and a representative selection of stress-strain curves is shown in Fig. 3. These curves show two broad types of behaviour.

(1) Discontinuous stress–strain behaviour with frequent instantaneous load drops. As in many metals (Reed-Hill *et al.* 1964), the load drop behaviour was found to be associated with deformation twinning (see next section). This was seen in the tests performed at 550°C and in the fastest test at 600°C.
(2) Smooth stress–strain behaviour, with steady state flow being achieved at strains of around 1%. This was seen in all other tests (i.e. at $T \geq$ 600°C).

The steady state flow stresses (or upper bound stresses in the case of twinned samples), arbitrarily measured at 5% strain (see Table 2), are plotted in a standard log–log plot of stress versus strain rate in Fig. 4. From this figure it is clear that the flow stresses are rather insensitive to strain rate. Individually fitted isotherms assuming a power law relationship between strain rate and stress, yield stress exponents in the range 9 to 14 at $T \geq$ 600°C, with a value of 72 at 550°C. The dependence of flow stress on temperature is illustrated in the log stress versus 1/T plot given in Fig. 5. The lines of constant strain rate appearing in this plot show an increase in slope towards lower temperatures, suggesting an increase in apparent activation energy for creep with decreasing temperature.

Glide systems and optical/SEM microstructures

The deformed crystals were studied using a number of techniques, including optical and scanning electron microscopy (SEM). Both were used to carry out 'slip line analysis', of the type frequently used in metallurgy to identify glide systems (Haasen *et al.* 1980). This technique, coupled with morphological observations, revealed that twinning on the e_2 and e_3 systems is important in all tests at 550°C, and in the fastest test performed at 600°C. Occasional twins also developed at $T >$ 600°C, but were clearly associated with anomalous stress concentrations at the corners of the sample. All samples showing load drop behaviour were found to contain twins.

Little or no twinning was observed in samples supporting flow stresses below 75 MPa. Examination of these samples revealed numerous slip

Table 2. *List of experiments reported in this paper*

experiment	temperature [°C]	strain rate $[s^{-1}]$	flow stress at 5% strain [Mpa]
30sc52	550	2.9×10^{-5}	94.2
46sc69	550	2.8×10^{-6}	93.1
40sc42	550	2.9×10^{-7}	88.3*
↘ step	550	2.9×10^{-6}	89.0
51sc68	600	2.9×10^{-4}	85.9
34sc46	600	3.1×10^{-5}	68.3
48sc75	600	2.9×10^{-5}	67.5
42sc37	600	2.8×10^{-6}	54.3
32sc39	600	3.0×10^{-6}	58.7*
↘ step	600	3.0×10^{-7}	46.9*
49sc63	600	2.9×10^{-7}	46.5
↘ step	600	3.2×10^{-8}	42 *
44sc62	650	2.9×10^{-4}	55.3
45sc67	650	2.9×10^{-5}	52.5
41sc43	650	3.1×10^{-6}	37.3
27sc41	650	2.9×10^{-5}	54.3*
↘ step	650	2.9×10^{-6}	44.6
↘ step	650	3.0×10^{-7}	36 *
43sc49	700	2.9×10^{-4}	49.0
21sc45	700	3.1×10^{-5}	42.2
38sc47	700	2.9×10^{-5}	44.9
39sc50	700	3.0×10^{-6}	34.0
26sc36	700	2.7×10^{-6}	32.4*
↘ step	700	2.8×10^{-7}	26.5
37sc51	700	3.0×10^{-7}	22.5
52sc66	800	2.9×10^{-4}	40.8
22sc44	800	1.7×10^{-5}	30.5
24sc34	800	2.7×10^{-6}	25.4
25sc40	800	2.8×10^{-6}	26.8
28sc38	800	3.0×10^{-7}	19.5

* Flow stresses obtained by linear extrapolation of the steady-state portion of the stress–strain curve to 5% strain.
Experiment 27sc41 was stepped both downward and upward.

lines, glide bands and kink bands (Figs 6 and 7), proving intense slip activity. Slip-line analysis, coupled with orientation analysis of glide bands seen in thin section, showed that the operative slip planes were r_1 and f_1 (refer Fig. 2a, b) with slip occurring in the positive sense on both. No evidence was found for slip on f_2 or f_3 (see Fig. 2b, Table 1) in any of the samples tested. In the case of the r_1 and f_1 planes, the observed slip lines could be traced continuously around the entire (cleaved) surface of the deformed samples (Fig. 6d), indicating that the operative slip directions did not lie in the r_2 and r_3 planes making up the sample surface (refer Fig. 2a,b). Taking the observed sense of shear on the r_1 and f_1 slip bands into account, and assuming that slip is confined to rational, low-index directions, this implies that slip occurred in the $[\bar{2}021]^+$ direction on r_1 and in the $[10\bar{1}1]^+$ direction on f_1 (see Fig. 2a,b). Geometric analysis of kink bands (Fig. 6a, b) and c-axis rotations associated with intense f_1 slip (carried out using the method of Turner *et al.* 1954) yielded rotation axes parallel to the a_2 direction (Fig. 2b), thus confirming that slip on f_1 occurred in the $[10\bar{1}1]$ direction.

The above indicates that the main *slip* systems activated in the present experiments were the $r_1(10\bar{1}4)[\bar{2}021]^+$ system and the $f_1(1012)[1011]^+$ system. However, these two systems were not of equal importance under all conditions. Optical examination of the entire suite of samples revealed a transition from $r_1(10\bar{1}4)[\bar{2}021]^+$ slip plus minor f^+ slip (accompanying dominant twinning) at the higher strain rates and lowest temperature (550°C), to dominant $f_1(\bar{1}012)[10\bar{1}1]^+$ slip at the lower strain rates and higher temperatures. This transition is illustrated with a series of optical micrographs in Fig. 7, and is mapped as a function of stress,

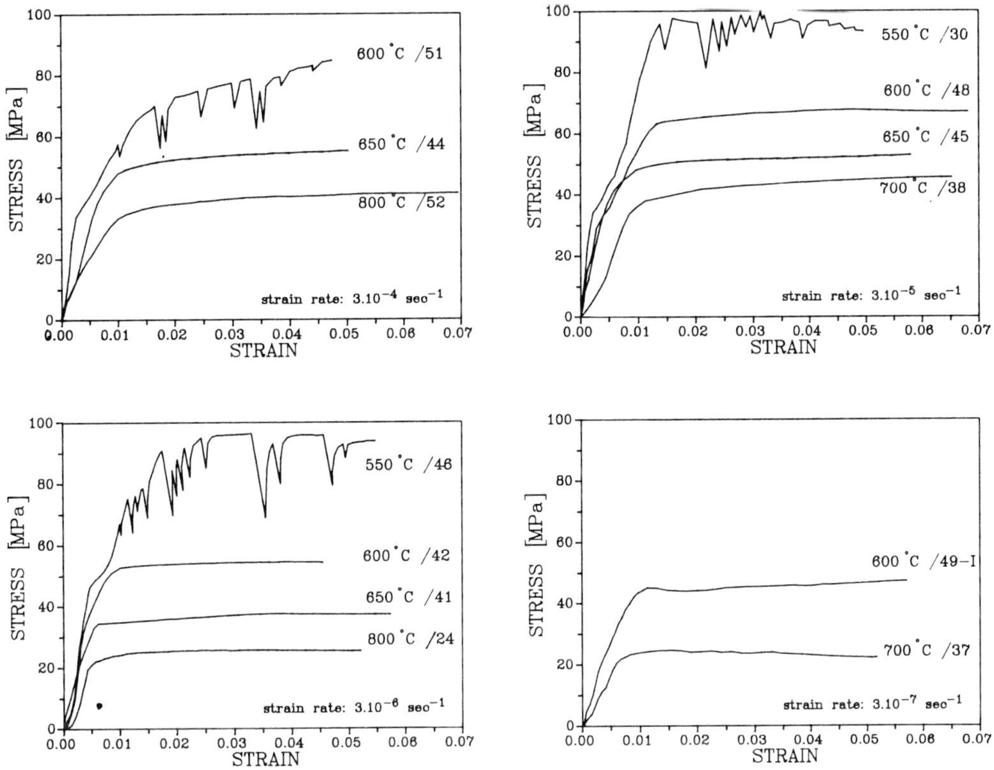

Fig. 3. Typical stress–strain curves obtained for calcite single crystals in the present study. Each curve is labelled with test temperature and test no. (cf. Table 2).

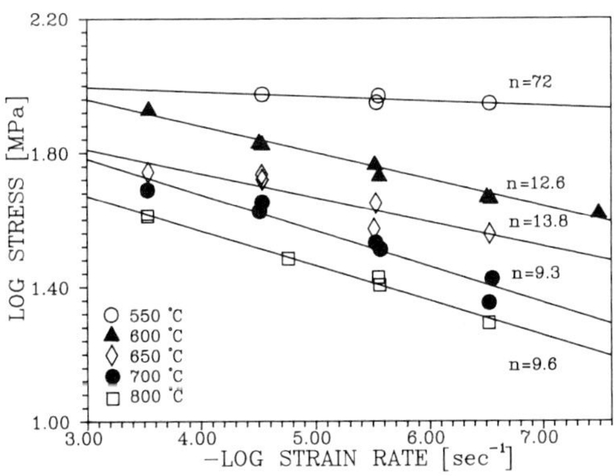

Fig. 4. Log–log plot of strain rate v. differential stress, constructed using the flow stresses (or upper bound stresses in case of twinned samples) supported at 5% strain, see Table 2. Strain rate stepping data are incorporated. Best fit isotherms have been drawn assuming individual power law relationships. The standard error of the stress exponents (n) is c. 1.5.

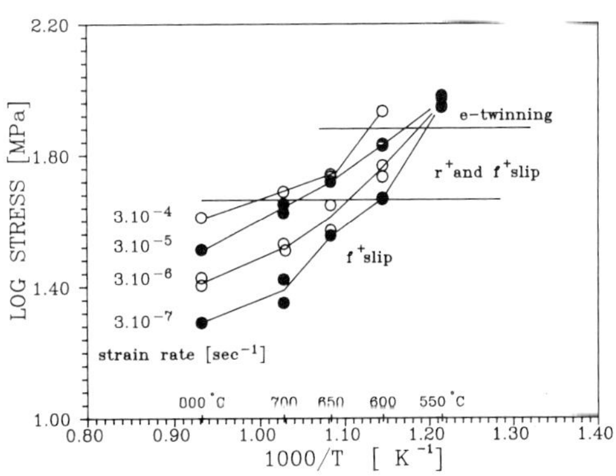

Fig. 5. Log flow stress v. reciprocal temperature for all samples tested. Twinning and slip regimes are based on the data presented in Fig. 8.

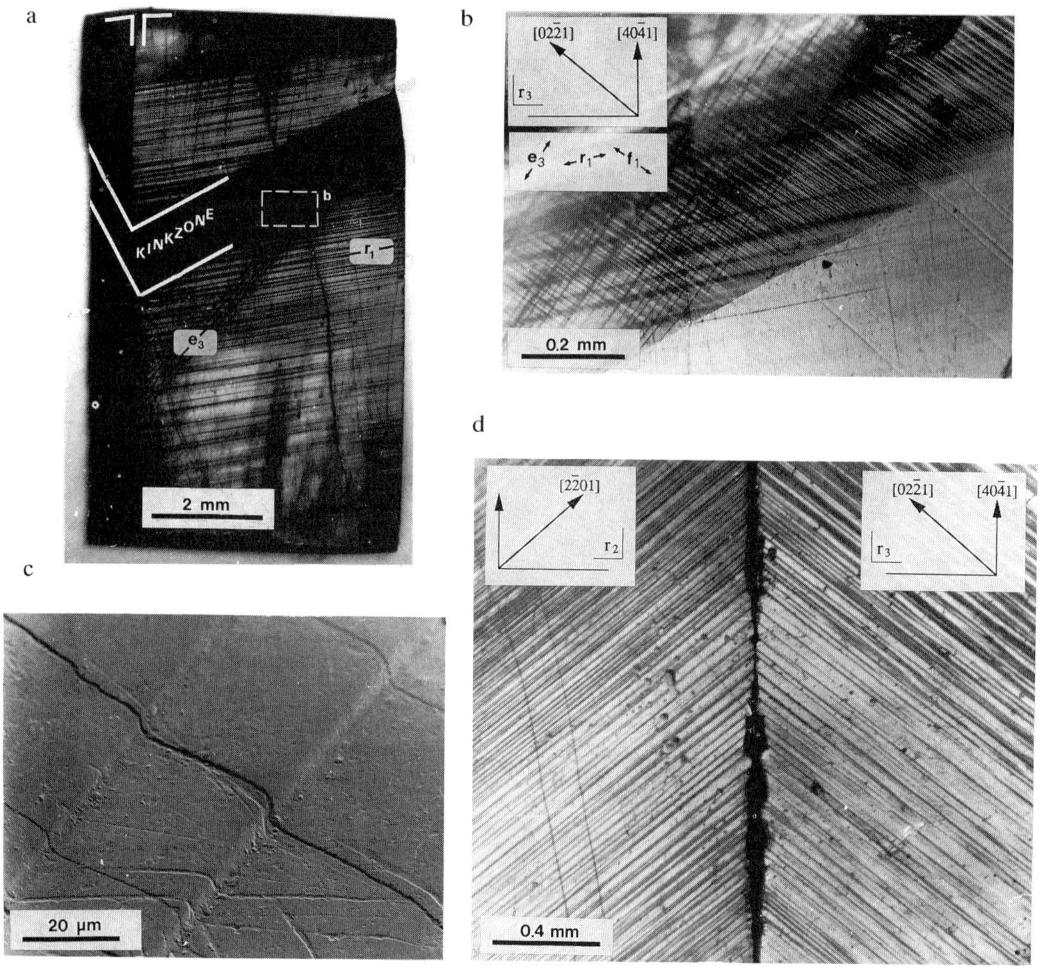

Fig. 6. (**a**) Deformed single crystal of calcite with r_I-slip traces, local e_3-twinning and a kink band dominated by f_I-slip (outlined area refers to b). Load direction vertical.
(**b**) Detail of (a) showing part of the kink zone.
(**c**) SEM (secondary electron) micrograph of f_I-slip lines on a sample face.
(**d**) f_I-slip lines traceable from the r_2 to the r_3 face in a sample deformed at a temperature of 650°C and a strain rate of 3×10^{-5} s^{-1}.

temperature and strain rate in Fig. 8. Note that no r^+ slip was observed at flow stresses below *c*. 47 MPa. The absence of significant twinning at stresses below 75 MPa is also indicated in Fig. 8.

We now report observations on slip band morphology. Firstly, the r_I and f_I slip bands tended to be straight. However, the *f*-slip bands were often seen to terminate in the body of the crystal, transferring their displacement to a neighbouring band in the manner illustrated in Fig. 9a. This type of feature will be referred to henceforth as 'slip band shift'. Locally, *f*-slip lines were found to be arranged in 'en-echelon' packets, making a small angle with the *f*-plane (Fig. 9b), separated by regions of slip band shift.

Estimates of the total displacement accumulated across slip bands indicated that slip activity within these bands was responsible for the bulk of the imposed strain. These estimates were obtained from offsets observed in SEM micrographs. In addition, changes in the external shape of the crystals were found to be consistent with the observed slip systems.

Finally we note that all samples showed evi-

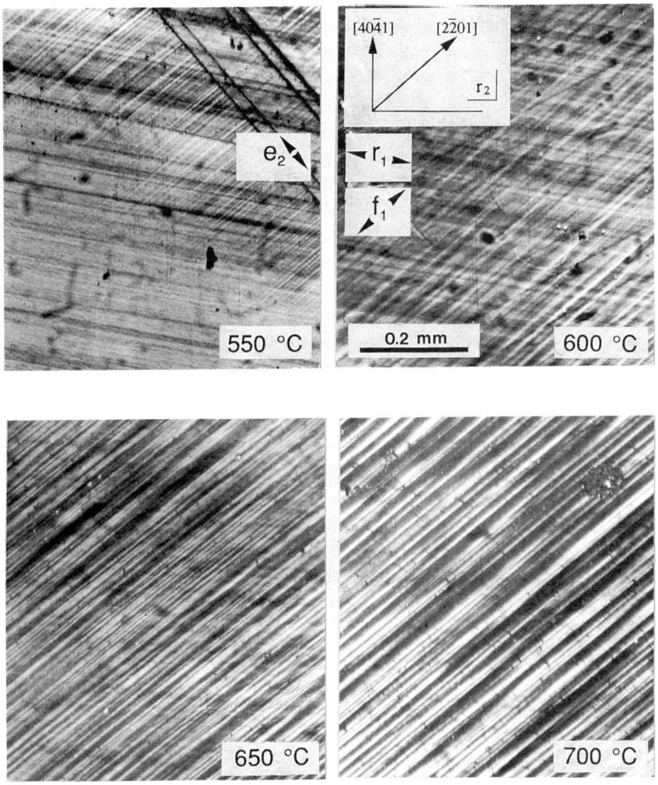

Fig. 7. Optical micrographs showing the transition from r_1^+ slip plus minor f_1^+ slip (accompanying twinning) at low temperature to pure f_1^+ slip at high temperature, seen on a r_2 cleavage plane (load direction vertical). Strain rate 3×10^{-5} s^{-1}, compare with Fig. 8.

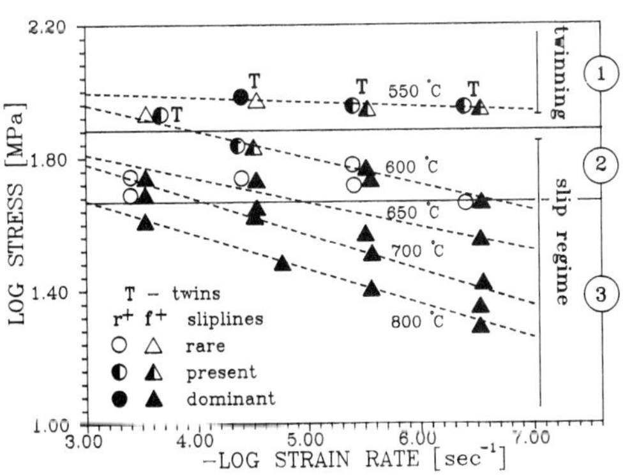

Fig. 8. Log–log plot of strain rate v. differential stress summarizing microstructural observations on twinning and slip activity: (1) regime with dominant *e*-twinning, significant r_1^+ slip and minor f_1^+ slip; (2) regime with f_1^+ slip and minor r_1^+ slip; (3) regime of pure f_1^+ slip.
Note that within the slip regime (2 + 3) *f*-slip dominates overall. Isotherms taken from Fig. 4.

dence of micro cracking. Careful examination of samples during the quenching stage of the tests showed that these cracks were always introduced by the quench treatment.

TEM microstructures

Transmission Electron Microscopy (TEM) was performed using a Jeol 200C microscope operating at 200 kV. The dislocation substructure of the deformed crystals was found to be characterized by both straight and curved, locally helical dislocations, and by irregular pseudo-hexagonal dislocation networks (Fig. 10). Dipoles, jogs, loops and dislocation debris were common. Well defined tilt boundary configurations or subgrains were rare and no evidence was found for dislocation dissociation. Orientation analysis revealed that of the numerous dislocations observed, only relatively few were located in the active *r* or *f* planes. This observation, coupled with the presence of helical dislocations and

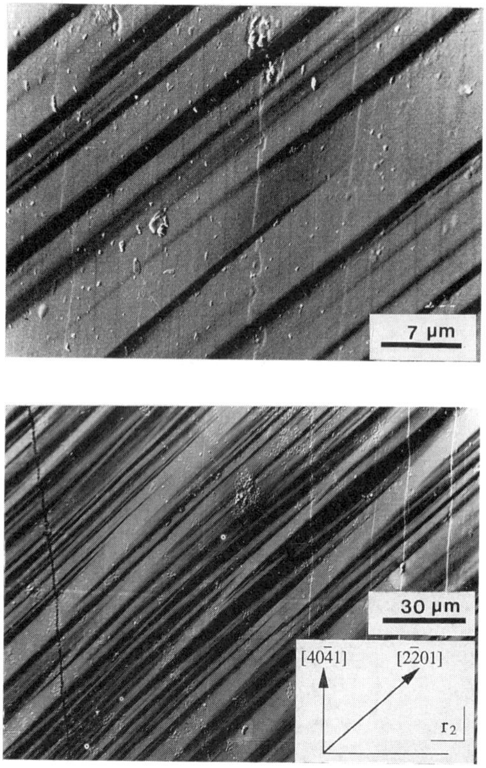

Fig. 9. SEM (secondary electron) micrographs showing:
(**a**) *f*-slip traces with slip band shift microstructure (centre);
(**b**) 'en echelon' packets of *f*-slip lines, making a small angle with the *f*-plane, and separated by regions of slip band shift.
(strain rate 3×10^{-5} s^{-1} and 650°C, sample 45sc67).

(rare) dislocation tilt walls, points to active dislocation climb. The general nature of the dislocation microstructure did not vary substantially with temperature and strain rate. However, the total density of dislocations ρ (cm^{-2}) was found to depend on flow stress σ (MPa) according to the relation

$$\sigma = 4 \times 10^{-4} \cdot \rho^{0.6} \quad (1)$$

reported by De Bresser (1988). No evidence was found for any correlation between the quenching cracks reported above and the observed dislocation substructure and density. The samples deformed at 800°C showed scattered rectangular voids (*c.* 0.1 μm diameter), sometimes developed on dislocations. Most dislocations however were free of these voids.

Preliminary analysis of the pseudohexagonal dislocation networks (Fig. 10b) showed that these lie in planes subparallel to ($0\bar{1}11$) (Fig. 10c), and are made up of three different types of dislocations, with different Burgers vectors. In contrast experiments, these different types of dislocations showed effective invisibility (Edington 1975) for diffracting vectors of g = [$0\bar{1}1\bar{8}$] and [$0\bar{4}48$], g = [$3\bar{2}1\bar{4}$] and [$4\bar{4}0\bar{4}$], and g = [$40\bar{4}4$] respectively. This is consistent with Burgers vectors parallel to [$\bar{2}110$], [$1\bar{1}0\bar{1}$] and [$10\bar{1}1$] (Fig. 10c). In the samples dominated by f_I^+ slip, no contrast conditions were found consistent with Burgers vectors parallel to [$02\bar{2}1$] or [$2\bar{2}01$], the generally accepted directions (e.g. Wenk 1985) for slip on f (in this case f_I).

Discussion

In the remainder of this paper we discuss (a) the evidence presented above for slip on f_I ($\bar{1}012$) in the [$10\bar{1}1$]$^+$ direction (a previously unreported slip system), and (b) the mechanism controlling the *rate* of creep in the regime dominated by r and f slip. We also compare the creep behaviour seen in our single crystals (slip dominated regime) with that of marbles and limestones.

Slip on f *in the [$10\bar{1}1$]$^+$ direction (a new slip direction?)*

The f_I slip line and kink analyses reported in this study indicate an effective slip direction on f_I of [$10\bar{1}1$], i.e. precisely intermediate between the [$2\bar{2}01$] and [$02\bar{2}1$] directions expected for f (i.e. f_I) on the basis of previous literature. This implies either that a truly new f-slip direction has been activated in our tests, or it implies coupled activity in two coplanar <$02\bar{2}1$> directions. In the latter case, any strongly heterogeneous deformation should favour one of the two slip directions relative to the other, at least locally (see Davidge & Pratt 1964). Slip line domains would then develop on opposite faces of the sample, but not on adjacent faces. However, additional deformation experiments performed using samples with aspect ratios (length:width:width) varying between 3:1:1 and 3:5:2:1 (cf. normal ratios 2.5:1:1, Fig. 2), in order to enhance heterogeneous deformation, did not show development of slip line domains on opposite faces. On the contrary, domains of f-lines could be followed around the crystals in all cases.

On this basis we reject the hypothesis of coupled activity involving two <$02\bar{2}1$> directions, and suggest a true slip direction of

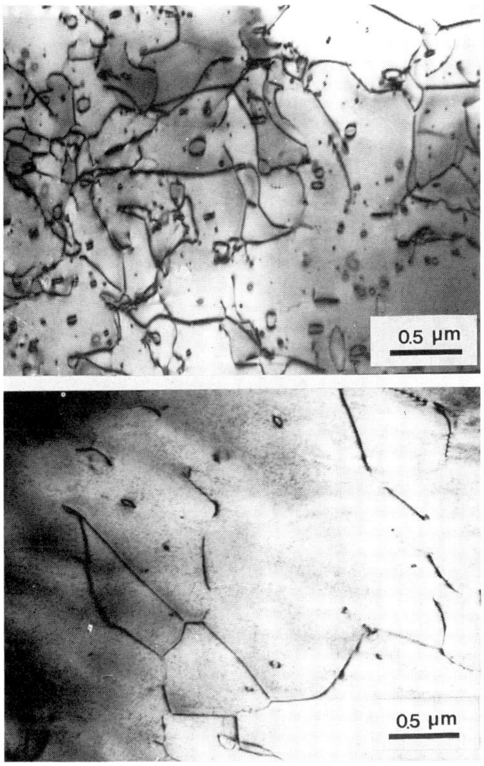

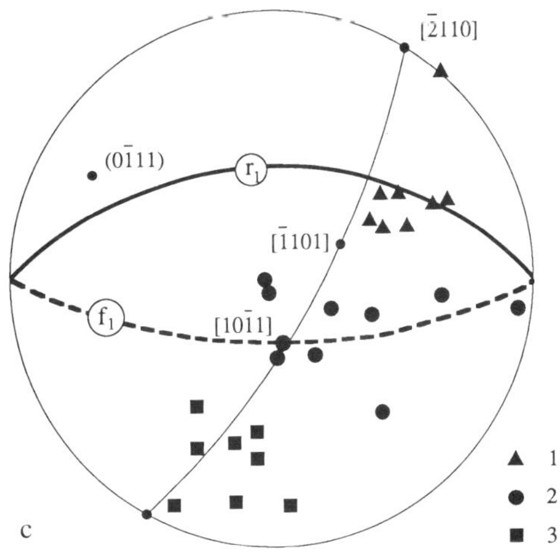

Fig. 10. Bright field (multi-beam conditions) TEM micrographs showing typical dislocation microstructure in the slip regime (**a**) and characteristic irregular hexagonal dislocation network (**b**).
Measured orientations of dislocation lines from networks are given in a stereographic projection of calcite (**c**). 1,2,3 denote the three different types of dislocations joined in a network node. These disclocations correspond to Burgers vectors [1$\bar{1}$0$\bar{1}$], [10$\bar{1}$1] and [$\bar{2}$110], respectively.

$[10\bar{1}1]^+$ for *f*-slip in the present tests. The existence of this new slip direction is strongly supported by dislocation line energy consideration (Motohashi *et al.* 1976; Goetze & Kohlstedt 1977; Paterson 1985), because the length of the Burgers vector parallel to $<10\bar{1}1>$ is only 6.37 Å, compared with a value of 8.09 Å for that parallel to $<02\bar{2}1>$. The proposed slip direction is also consistent with the results of the preliminary Burgers vector analyses by TEM (for network dislocations), described above.

Creep mechanisms in the slip regime

It will be recalled that under conditions dominated by *e*-twinning (550°C tests), the upper bound strength of our samples was found to be extremely insensitive to strain rate (Figs 4 and 8). At the higher temperatures, i.e. in the *slip-dominated regime* ($T \geq 600$°C), our data show an increased but still rather low strain rate sensitivity of the steady state flow stress. In this regime, slip is the main strain accumulating mechanism. Positive slip on r_1 occurs at the higher stresses, but $f_1{}^+$ slip dominates overall (refer Fig. 8). In order to gain insight into the rate controlling mechanism in the slip regime, we now compare the observed creep behaviour with microphysical models for intracrystalline deformation.

We begin by noting that most creep models are derived using Orowan's equation plus the relation $\sigma \propto \rho^{0.5}$ linking dislocation density ρ to the applied stress σ (Kohlstedt & Weathers 1980). Equation (1) reported for our samples (De Bresser 1988) is in reasonable agreement with the latter relation and thus helps justify comparison of our creep data with microphysical models.

For present purposes, we distinguish three broad classes of deformation mechanisms (after Frost & Ashby 1982; Poirier 1985), namely climb-controlled creep, cross slip-controlled creep and barrier-controlled dislocation glide (Table 3). In climb-controlled creep models, thermally activated climb of dislocations is the rate determining step. Although there are many variants on the basic mechanism involved, climb controls recovery and the creep rate is expressed by a power law equation of the general type shown in Table 3, with a temperature independent stress exponent n between 3 and 6 (Weertman 1968, 1972), and a stress independent activation energy Q equivalent to

Table 3. *Creep models and regression results obtained by best fitting of the present single crystal data for the slip dominated regime*

Generalized power law creep equation (Frost & Ashby 1982):

$\dot{\varepsilon} = A \cdot \sigma^{n} \cdot \exp[-Q/RT]$

results of regression analysis:

$LOG(A) = -3.81(\pm 1.22)$, $n = 11.5(\pm 0.9)$ and $Q = 362(\pm 35)$ kJ mole^{-1}, corr. = 0.972

Cross slip equation (Wawersik 1988 references therein):

$\dot{\varepsilon} = B \cdot \exp[(Q_{cs}/RT) \cdot (\ln \sigma_0/\mu_0 - \ln \sigma/\mu)]$

which is equivalent to:

$\dot{\varepsilon} = B \cdot \sigma^{Q_{cs}/RT} \cdot \exp[(Q_{cs}/RT) \cdot (\ln \sigma_0/\mu_0 + \ln \mu)]$

results of regression analysis:

$LOG(B) = 13.51(\pm 2.01)$, $Q_{cs} = 90(\pm 8)$ kJ mole^{-1} and $\sigma_0 \approx 2600$ MPa, corr. = 0.966

Barrier controlled dislocation glide equation

Model 1, obstacle limited (Frost & Ashby 1982)

$\dot{\varepsilon} = C \cdot \exp[(-Q/RT) \cdot (1 - \sigma/\sigma_0)]$

results of regression analysis:

$LOG(C) = 13.05(\pm 2.72)$, $Q = 432(\pm 32)$ kJ mole^{-1} and $\sigma_0 = 202(\pm 15)$ MPa, corr. = 0.937

Model 2, Peierls stress limited (Weertman 1957)

$\dot{\varepsilon} = D \cdot \sigma^{2.5} \cdot \exp[(-Q/RT) \cdot (1 - (\pi/2p) \cdot \sigma)]$

results of regression analysis:

$LOG(D) = 5.98(\pm 1.95$, $Q = 329(\pm 41)$ kJ mole^{-1} and $p = 443(\pm 38)$ MPa, corr. = 0.899

Symbols: $\dot{\varepsilon}$, strain rate; σ, flow stress; Q, apparent activation energy (zero applied stress); A, B, C, D and n, empirical constants; R, gas constant; μ, shear modulus; μ_0, shear modulus at absolute zero; σ_0, flow stress at absolute zero; p, Peierls stress; (ln, natural logarithm; LOG ^{10}logarithm). In the present case, Q_{cs} is the activation barrier to cross slip when the applied stress is approximately equal to $0.3\,\sigma_0$. Temperature dependence of the shear modulus μ is neglected over the temperature range investigated. The quality of fit is measured by the correlation coefficient (corr.).

the activation energy for self-diffusion of the slowest diffusing species. Cross slip can also be viewed as a recovery mechanism (Poirier 1976; Skrotzki & Haasen 1988; Wawersik 1988). However, unlike climb, cross slip is a thermally activated process in which the activation barrier is reduced by the applied stress. Thus cross slip-controlled creep models contain a stress dependent activation energy. For ionic crystals, Skrotzki & Liu (1982) and Skrotzki & Haasen (1988) present a cross slip recovery model in which the free energy of activation is logarithmically dependent on stress. The resulting equation for the strain rate can be written in exponential form or in power law form with a temperature dependent stress exponent (see Table 3). In barrier-controlled dislocation glide, the rate of motion of dislocations is limited by lattice friction resistance (Peierls stress-limited) or by the presence of obstacles such as other dislocations or impurities (obstacle-limited). Recovery processes may operate but are not rate controlling. Rate equations for dislocation glide have an essentially exponential form, but differ in detail depending on the nature of the assumed barriers to glide and the mechanism by which they are overcome (Weertman 1957; Guyot & Dorm 1967; Frost & Ashby 1982). Representative constitutive models for the obstacle and Peierls limited cases are given in Table 3.

The present mechanical data for calcite have been fitted to the equations listed in Table 3, using a non-linear regression method (Marquardt 1963). Note that the temperature dependence of the shear modulus of calcite (which is weak in the temperature range investigated here, see Dandekar 1968) was neglected. The results of the fitting procedures are given in Table 3 and Fig. 11. A comparable quality of fit was obtained for all models, providing no real indication of the rate controlling mechanism.

We now consider the above models in relation to the available microstructural data and constraints on the various fitting parameters. Firstly, the slip band shift features associated with f_I^{+} slip (Fig. 9) are potentially explicable in terms of long range climb of edge dislocations from one slip band to another. In addition, considerable TEM evidence was found for dislocation climb. Furthermore, the general power law fit (Table 3) yields an activation energy of 362($\pm$35) kJ mole^{-1}, which is in good agreement with the values obtained for the activation energy for self-diffusion of carbon and oxygen in calcite, 370 and 420 kJ mole^{-1} respectively (Anderson 1969). However, the high stress exponent of 11.5($\pm$0.9) contrasts strongly with

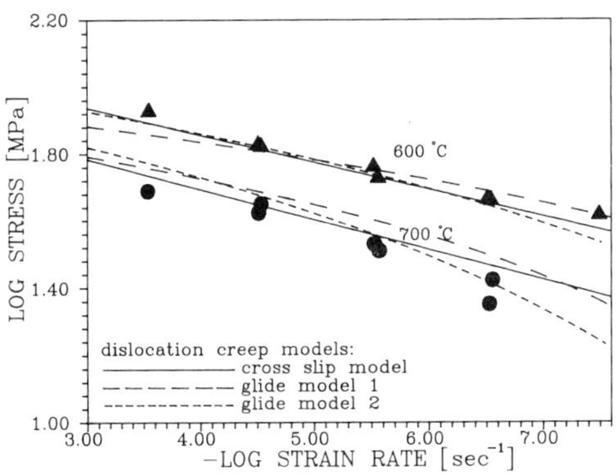

Fig. 11. Log–log plot of strain rate v. differential stress showing data points obtained at 600 and 700°C (see Fig. 4). Best fit lines correspond to the models listed in Table 3.

the values of 3–6 predicted by existing climb-controlled creep models.

The slip band shift features mentioned above in relation to climb, can be equally well explained by cross slip of screw dislocations from one level to another, not on a discrete cross slip plane but homogeneously distributed in the 'shift zone' between the overlapping slip bands (refer Fig. 9). The characteristics of the slip band shift microstructure are not dissimilar to the broad and wavy slip traces reported for involvement of cross slip in creep of salt (Wawersik 1988). In addition, the change from $n \approx 13$ at 600°C to ≈ 9.5 at 800°C seen in the empirical power law fits presented in Fig. 4, shows that n is temperature dependent: one of the principal characteristics of cross slip controlled creep (Table 3). The activation parameter (Q_{cs}) for cross slip obtained by fitting our data to the cross slip model (Table 3) is 90(±8) kJ mole^{-1}. This is 0.25 times that for self-diffusion. Estimates of the activation area (the area swept out by an individual dislocation event) obtained from our data following Skrotzki and Haasen (1988), yield values of $7b^2$ to $30b^2$, where b is the length of the Burgers vector (taken as 6.37 Å). These values for activation energy and area are broadly comparable with data for salt which show an apparent activation for cross slip of 0.1 times that for self-diffusion (Wawersik 1988) and an activation area of $20-400b^2$ (Skrotzki & Haasen 1988). However, no theoretical constraints on these quantities are presently available for calcite. The value of activation area expected for climb-controlled creep is of the order of $1b^2$ (Argon 1966), i.e. about an order of magnitude lower than found here.

Lastly, we consider the evidence that dislocation glide mechanisms might have been rate controlling. The numerous slip bands seen in our samples can certainly be viewed as consistent with the dislocation glide models presented in Table 3. Furthermore, while the dislocation structures observed using TEM are indicative of climb, they do not rule out glide as being rate controlling. On the other hand, glide control is generally expected at lower homologous temperature (0.2–0.4 Tm, Langdon 1985) than in the present tests, and is not considered capable of producing true steady state behaviour (Stocker & Ashby 1973). As far as we are aware, no constraints are available to assess the values of the fitting parameters obtained for the obstacle limited glide model. However, by fitting our data to the Weertman (1957) model for the Peierls mechanism (Table 3), a Peierls stress of 443 (± 38) MPa was obtained, which is c. 0.02 times the shear modulus μ of calcite. This is a high ratio compared with most metals and ionic materials (Guyot & Dorn 1967; Haasen 1985), but is nonetheless of the same order of magnitude as that found for metals such as zinc (Weertman 1957), iron (Arsenault 1975) and iron-alloys (Guyot & Dorn 1967; Christian 1971).

In conclusion, we reject the climb-controlled creep model ($n = 3-6$) because of the very high stress exponent obtained from the power law fit to our data ($n \approx 11.5$). Dislocation climb was clearly active in our tests, but was not a significant strain accumulating mechanism itself (as evidenced by morphological observation), and was apparently too fast to be rate controlling. By contrast, the cross slip and glide models cannot be rejected. They show comparable quality of fit, the values obtained for the various material constants are reasonable (as far as can be assessed), and the observed microstructures are consistent with these models. We infer that the deformation rate was probably cross slip and/or glide controlled, noting that at low stresses the cross slip model is more or less indistinguishable from the obstacle limited glide model (Poirier 1976). However, neither the cross-slip nor the glide model alone can adequately explain the curvature observed in the log stress v. $1/T$ plot (temperature dependent activation energy; Fig. 5). Since the curvature in this plot is apparent in the regime of pure f-slip, it is unlikely that the disappearance of r-slip towards higher temperatures (Figs 7 and 8) can account for it. A possible explanation would be a transition from glide control to cross slip control or vice versa (glide and cross slip are serial processes).

Single crystal behaviour v. creep in polycrystalline calcite

In Fig. 12 we compare the behaviour of our crystals with that of calcite rocks deformed experimentally at similar temperature (700°C). The single crystal data best resemble the results for the coarse-grained Carrara and Yule marbles. Assuming that no other factors were involved (such as composition), this suggests that creep of these marbles at intermediate stresses and temperatures is controlled by the same mechanisms as seen in our single crystal experiments; i.e. glide or cross slip.

Summary and conclusions

(1) Uniaxial compression experiments have been performed on optical quality calcite single crystals at temperatures in the range 550–800°C and at constant strain rates in the range 3×10^{-4} to 3×10^{-7} s^{-1}. The tests were carried out in a controlled atmosphere cell using a CO_2 overpressure of 0.25 MPa to suppress decomposition. The crystals were loaded in the $[40\bar{4}1]$ direction. (i.e. parallel to the intersection of two *r* cleavage rhombs, r_2 and r_3) following earlier experiments of Spiers & Wenk (1980). This orientation was chosen with the intention of activating slip in the (so-called) positive sense on the previously reported $\{10\bar{1}4\}<\bar{2}021>$ and $\{\bar{1}012\}<02\bar{2}1>$ systems.

(2) At temperatures below *c.* 600°C, the samples deformed largely by *e*-twinning. At higher temperatures, the samples exhibited steady state flow behaviour, with deformation occurring by slip on $r_1(10\bar{1}4)$ $[\bar{2}021]$ and $f_1(\bar{1}012)$ $[10\bar{1}1]$ in the positive sense, the f_1 system dominating. Thus, slip on *f* did not occur in the previously reported direction, and the existence of a new set of slip systems, namely $f\{\bar{1}012\}$ $<10\bar{1}1>$ is implied. This possibility is supported by dislocation line energy considerations, and is consistent with the findings of preliminary TEM contrast experiments.

(3) In the slip-dominated, steady-state flow regime ($T \geq 600$°C) the flow stresses supported at 5% strain were found to be relatively insensitive to strain rate, with empirical power law fits to our data yielding a conventional stress exponent *n* ranging from *c.* 13 at 600°C to *c.* 9.5 at 800°C. Although substantial TEM evidence was found for dislocation climb, the observed mechanical behaviour cannot be explained by existing climb-controlled creep models. However, the observed mechanical behaviour and microstructures are consistent with both cross slip- and glide-controlled creep models, with best fitting of our data to these models yielding reasonable values for the various activation and glide resistance parameters. We infer that the deformation of our samples in the slip regime probably occurred by cross slip controlled creep or by glide controlled creep, or a combination of these mechanisms.

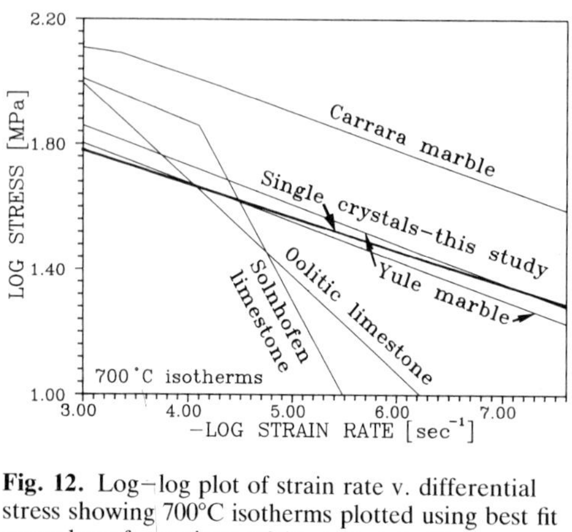

Fig. 12. Log–log plot of strain rate v. differential stress showing 700°C isotherms plotted using best fit creep laws for various calcite materials. Carrara marble isotherm after Schmid *et al.* (1980), Yule marble after Heard and Raleigh (1972), Solnhofen limestone after Schmid *et al.* (1977), Oolitic limestone after Shcmid & Paterson (1977). Single crystal isotherm taken from Fig. 4 of this study.

We thank B. Smith, M. Paterson and J. Boland for critically reading the manuscript. Thanks are also due to C. Peach and G. Kastelein for advice and technical support.

References

ANDERSON, T. F. 1969. Self-diffusion of carbon and oxygen in calcite by isotope exchange with carbon dioxide. *Journal of Geophysical Research*, **74**, 3918–3932.

ARGON, A. S. 1966. Plastic deformation in crystalline materials. *In*: MCCLINTOCK, F. A. & ARGON, A. S. (eds) *Mechanical behavior of materials*. Addison-Wesley Publ. Company, 152–211.

ARSENAULT, R. J. 1975. Low temperature deformation of bcc metals and their solid-solution alloys. *In*: ARSENAULT, R. J. (ed.) *Treatise on materials science and technology (Volume 6); Plastic deformation of materials*. Academic Press, 1–99.

BRAILLON, P., MUGNIER, J. & SERUGHETTI, J. 1972. Deformation plastique de mono-cristaux de calcite, en compression suivant [111]. *Comptes rendus de l'Académie des Sciences, Paris (B)*, **t.275**, 605–608.

CHRISTIAN, J. W. 1971. The strength of martensite. *In*: KELLY, A. & NICHOLSON, R. B. (eds) *Strength-*

ening methods in crystals. Applied Science Publishers ltd. London, 261–329.

DANDEKAR, D. P. 1968. Variation in the elastic constants of calcite with temperature. *Journal of Applied Physics*, **39**, 3694–3699.

DAVIDGE, R. W. & PRATT, P. L. 1964. Plastic deformation and work-hardening in NaCl. *Physica Status Solidi*, **6**, 759–776.

DE BRESSER, J. H. P. 1988. Deformation of calcite crystals by r^+ and f^+ slip: mechanical behaviour and dislocation density vs. stress relation. *EOS* **69**, 1418.

EDINGTON, J. W. 1975. *Interpretation of transmission electron micrographs – Practical electron microscopy in materials science monograph 3*, Macmillan Press Ltd.

FROST, H. J. & ASHBY, M. F. 1982. *Deformation-mechanism maps, the plasticity and creep of metals and ceramics*. Pergamon Press.

GOETZE, C. & KOHLSTEDT, D. L. 1977. The dislocation structure of experimentally deformed marble. *Contributions to Mineralogy and Petrology*, **59**, 293–306.

GRIGGS, D. T., TURNER, F. J. & HEARD, H. C. 1960. Deformation of rocks at 500 to 800°C. *Geological Society of America Memoir*, **79**, 39–105.

GUYOT, P. & DORN, J. E. 1967. A critical review of the Peierls mechanism. *Canadian Journal of Physics*, **45**, 983–1016.

HAASEN, P. 1985. Dislocations and the plasticity of ionic crystals. *In: Dislocations and properties of real materials*. The Institute of Metals, London, 312–332.

——., GEROLD, V. & KOSTORZ, G. (eds) 1980. *Strength of metals and alloys*. Proceedings of the 5th International Conference, Aachen (1979), Pergamon Press.

HEARD, H. C. & RALEIGH, C. B. 1972. Steady-state flow in marble at 500 to 800°C. *Geological Society of America Bulletin*, **83**, 935–956.

KERN, H. & WENK, H. -R. 1983. Calcite texture development in experimentally induced ductile shear zones. *Contributions to Mineralogy and Petrology*, **83**, 231–236.

KOHLSTEDT, D. L. & WEATHERS, M. S. 1980. Deformation-induced microstructures, paleopiezometers, and differential stresses in deeply eroded fault zones. *Journal of Geophysical Research (B)*, **85**, 6269–6285.

LANGDON, T. G. 1985. Regimes of plastic deformation. *In*: WENK, H. -R. (ed.) *Preferred orientation in deformed metals and rocks, an introduction to modern texture analysis*. Academic Press, 219–232

LISTER, G. S. 1978. Texture transitions in plastically deformed calcite rocks. *Proceedings of the 5th International Conference on textures of materials, Aachen*. Springer (Berlin), 199–210.

MARQUARDT, D. W. 1963. An algorithm for least squares estimation of non-linear parameters. *Journal of the Society of Industrial and Applied Mathematics*, **11**, 431–441.

MOTOHASHI, Y., BRAILLON, P. & SERUGHETTI, J. 1976. Elastic energy, stress field of dislocations, and dislocation parameters in calcite crystals. *Physica Status Solidi (a)* **37**, 263–270.

PATERSON, M. S. 1985. Dislocations and geological deformation. *In*: Dislocations and Properties of Real Materials. The Institute of Metals, London, 359–377.

POIRIER, J. P. 1976. On the symmetrical role of cross-slip of screw dislocations and climb of edge dislocations as recovery processes controlling high-temperature creep. *Revue de Physique Appliquée*, **11**, 731–738.

—— 1985. *Creep of crystals*. Cambridge University press.

REED-HILL, R. E., HIRTH, J. P. & ROGERS, H. C. (eds) 1964. *Deformation twinning*. Gordon and Breach, New York.

SCHMID, S. M. & PATERSON, M. S. 1977. Strain analysis in an experimentally deformed oolitic limestone. *In*: SAXENA, K. & BATTACHANJI, S. (eds) *Energetics of geological processes*. Springer N.Y.; 67–93.

—— BOLAND, J. N. & PATERSON, M. S. 1977. Superplastic flow in finegrained limestone. *Tectonophysics*, **43**, 257–291.

—— PATERSON, M. S. & BOLAND, J. N. 1980. High temperature flow and dynamic recrystallization in Carrara marble. *Tectonophysics*, **65**, 245–280.

—— PATERSON, M. S. & BOLAND, J. N. 1980. High temperature flow and dynamic recrystallization in Carrara marble. *Tectonophysics*, **65**, 245–280.

SKROTZKI, W. & HAASEN, P. 1988. The role of cross slip in the steady state creep of salt. *Proceedings second Conference on the Mechanical Behavior of Salt*. Trans Tech Publications Clausthal Zellerfeld, 69–81.

—— & LIU, Z. G. 1982. Analysis of the cross slip process in alkali halides. *Physica Status Solidi (a)*, **73**, k225–k229.

SPIERS, C. J. & WENK, H. R. 1980. Evidence for slip on r and f in the positive sense in deformed calcite single crystals. *EOS* **61**, 1128.

STOCKER, R. L. & ASHBY, M. F. 1973. On the rheology of the upper mantle. *Reviews of Geophysics and Space Physics*, **11**, 391–426.

TAKESHITA, T., TOMÉ, C., WENK, H. -R. & KOCKS, U. F. 1987. Single-Crystal Yield Surface for trigonal lattices: application to texture transitions in calcite polycrystals. *Journal of Geophysical Research (B)*, **92**, 12917–12930.

TURNER, F. J., GRIGGS, D. T. & HEARD, H. C. 1954, Experimental deformation of calcite crystals. *Geological Society of America Bulletin*, **65**, 883–934.

WAWERSIK, W.R. 1988. Alternatives to a power-law creep model for rock salt at temperatures below 160°C. *Proceedings second Conference on the Mechanical Behavior of salt*. Trans Tech Publications, Clausthal-Zellerfeld, 103–128.

WEERTMAN, J. 1957. Steady-state creep of crystals. *Journal of Applied Physics*, **28**, 1185–1189.

—— 1968. Dislocation climb theory of steady-state creep. *Transactions of the American Society of*

Metals, **61**, 681–694.
—— 1972. High temperature creep produced by dislocation motion. *In*: J. E. Dorn Memorial Symposium, Cleveland, Ohio (1972).
WENK, H. -R. 1985. Carbonates. *In*: WENK, H. -R. (ed.) *Preferred orientation in deformed metals and rocks. An introduction to modern texture analysis*. Academic Press: 361–384.
——, TAKESHITA, T., BECHLER, E., ERSKINE, B. G. & MATTHIES, S. 1987. Pure shear and simple shear calcite textures. Comparison of experimental, theoretical and natural data. *Journal of Structural Geology*, **9**, 731–745.
WYLLIE, P. J. & TUTTLE, O. F. 1960. The system $CaO-CO_2-H_2O$ and the origin of carbonatites. *Journal of Petrology*, **1**, 1–17.

Quartzite rheology under geological conditions

M. S. PATERSON & F. C. LUAN

Research School of Earth Sciences, Australian National University, Canberra 2601, Australia

Abstract: The problems associated with extrapolating laboratory measurements on quartzite rheology to geological conditions are discussed, with special reference to the question of equilibration with respect to the effect of water. Some new results on synthetic specimens crystallized from wet amorphous silica are presented in the expectation that water equilibration is more closely attained in these than in natural quartz specimens. These results and earlier ones from the literature are collated and used to arrive at limits on quartzite flow strength under geological conditions.

The steady state flow law for quartzite under geological conditions is of considerable tectonophysical interest because of the widespread occurrence in the Earth's continental crust of deformed quartzites. Thus, knowing the flow law, it may be possible to infer from observations on these quartzites what the mechanical conditions in the crust have been. It has even sometimes been assumed that a flow law for quartzite can be taken as characteristic for continental crustal flow in general (for example, Meissner & Strehlau 1982), although quartzite is not the predominant rock in the continental crust and more complex considerations may be expected to enter (Kirby 1983; Kirby & Kronenberg 1987; Carter & Tsenn 1987). In any case, knowledge of the steady state flow law of quartzite under geological conditions can be expected to be of value eventually in any structural geology studies involving quartzites.

An obvious approach to obtaining a flow law for geological conditions is that of extrapolation from laboratory experiment. However, in the case of quartzite, as pointed out elsewhere (Paterson 1987), extrapolation from laboratory experiments encounters severe difficulties, associated mainly with the role of water in quartz deformation, but also, although possibly to a lesser extent, with the alpha-beta transition. In this paper, we shall attempt to make some progress in overcoming these difficulties by examining more closely the problems surrounding the role of water and by drawing on some new experimental results obtained from synthetic quartzites.

Natural quartzites normally contain substantial amounts of water or water-related species (herein referred to generically as 'water'), as is evident from the ubiquity of fluid inclusions and from the presence of a broad hydroxyl band in the infrared absorption spectrum in the region of 3000–3700 cm^{-1} wavenumber (for interpretation, see Aines & Rossman 1984; Aines *et al.* 1984). Infrared absorption measurements on a number of quartzites indicate water contents ranging from around 1000–4000H/10^6Si or more (Mainprice & Paterson 1984). Following Griggs & Blacic (1965) and Griggs (1967), we assume that the presence of this water gives rise to a mechanical weakening effect and is responsible for natural quartzites being weaker than would be expected for pure dry polycrystalline quartz. This weakening is evident if one compares the high-temperature flow stresses measured in the laboratory on quartzites (Heard & Carter 1968; Parrish *et al.* 1976; Tullis *et al.* 1979; Koch *et al.* 1980; Jaoul *et al.* 1984; Kronenberg & Tullis 1984; Mainprice & Paterson 1984) with those for dry single cyrstals (Griggs & Blacic 1965; Griggs 1967; Heard & Carter 1968; Kekulawala *et al.* 1978; Blacic & Christie 1984; Paterson & Bitmead, quoted in Doukhan & Trepied 1985; Ord & Hobbs 1986).

Recently McLaren *et al.* (1989) have proposed that the water weakening effect can be explained in terms of dislocation generation and climb associated with the presence of water in inclusions, without the need to invoke a specific influence of the water itself on the dislocation mobility, as in the hydrolytic weakening model for glide of Griggs, Blacic and Frank (Griggs & Blacic 1965; Griggs 1967) and in the recovery model of McLaren & Retchford (1969). However, although dislocation generation at pressurized water inclusions appears to be a vital factor in the early stages of deformation in 'wet' synthetic single crystals (cf. Griggs 1974), it is not clear that it plays an essential role in natural quartzite specimens since these can

From Knipe, R. J. & Rutter, E. H. (eds), 1990, *Deformation Mechanisms, Rheology and Tectonics*, Geological Society Special Publication No. 54, pp. 299–307.

already contain substantial dislocation densities initially (McLaren & Hobbs 1972). Moreover, the linear configuration of dislocations observed in 'wet' synthetic crystals at relatively low temperature and high stress and the recovery structures observed at higher temperature and lower stress (Morrison-Smith *et al.* 1976) suggest that water does also play a role, which is strongly temperature dependent, in both overcoming a Peierls stress and facilitating dislocation climb. However, it is to be noted that laboratory tests on natural quartzites can give higher flow stresses than would be expected for polycrystalline specimens from the behaviour of wet synthetic quartz specimens (Mainprice & Paterson 1984), pointing to the existence also of other kinetic factors governing the effectiveness of the water.

In the present paper, we shall therefore accept the postulate of water weakening through an effect on dislocation mobility and recovery, but give close consideration to the question of equilibration with respect to the effectiveness of the water. From this postulate, it can be expected that the flow strength would be a function of the chemical potential of the water, given appropriate equilibration.

Kinetics of equilibration with respect to water effects

A particular difficulty in extrapolating laboratory measurements on quartzite rheology to geological conditions seems to arise from the problem of achieving equilibration with respect to the influence of the water in the experiments, especially in gas-medium apparatus at upper-crustal pressures. This problem is thought to consist essentially of establishing and maintaining a local equilibrium of the chemical potential of water-related components at all points within the specimen under the externally-applied experimental conditions. The local redistribution of water-related components needed for the equilibration can potentially be achieved within the grains by volume diffusion, by pipe diffusion along dislocations, or by penetration along cracks with further local distribution by diffusion. However, the kinetics of the equilibration seem to be very sluggish on the laboratory time scale, as indicated by the following observations.

(1) Annealing studies on wet synthetic quartz crystals, in conjunction with transmission electron microscope observations, show that growth in the size and spacing of water bubbles on the scale of 1 μm or less occurs very slowly at temperatures up to 1200 K (Kekulawala *et al.* 1981; Cordier *et al.* 1988; Gerretsen *et al.* 1989). On the basis that the coarsening of the bubble assemblage is governed by the volume diffusion of water-related species between the bubbles, through the relationship $x/2 \approx \sqrt{2Dt}$ where x is the spacing of the bubbles, D the diffusion coefficient, and t the elapsed time, Cordier *et al.* (1988) have determined the value of D to be approximately

$$D = 10^{-12} \exp(-95000/RT) \text{ m}^2\text{s}^{-1}$$

where R is the gas constant in J mol^{-1} K^{-1} and T the temperature in K. Using this value for D, the spatial range $\sqrt{2Dt}$ for equilibration with respect to the effect of water through volume diffusion can therefore be expected to be of the order of 1 μm during an experiment lasting several hours at 1200–1300 K (assuming that not just the hydrogen component is involved).

(2) Wet polycrystalline specimens made by the hydrothermal isostatic pressing of natural quartz powders with particle sizes of the order of 20–50 μm at 300 MPa are much stronger in plastic deformation at 1200–1300 K than wet single crystals or synthetic polycrystals grown from wet amorphous silica (Luan *et al.* 1986). If we take the lower flow stress in the latter materials to result from water weakening, it appears therefore that the range of action of the water in the natural quartz is less than the order of 10 μm or so. Complementary to this observation, solid medium experiments by Kronenberg & Tullis (1984) at 1500 MPa confining pressure and 1073 K on novaculite, a fine-grained natural quartz aggregate of grain size 1–50 μm, show that the flow stress can be substantially reduced at very high pressure in the presence of water; however, in this case, at least some of the effect may be relatable to microcracking or grain sliding mechanisms. Taken together, these experiments again suggest that the range of action of water in weakening quartz at 1200–1300 K is of the order of 1 μm on the laboratory time scale.

(3) In hydrothermal annealing experiments on dry quartz single crystals at around 1200–1300 K in which special care is taken to avoid microcracking, no water penetration is found to be measurable by infrared absorption at the 100 μm scale (Kronenberg *et al.* 1986: Rovetta *et al.* 1986; Gerretsen *et al.* 1989) or by the H(^{15}N,$\alpha\gamma$)^{12}C nuclear reaction technique at the 100 nm scale (S. Sie, J. Bitmead and M. S. Paterson, unpublished experiments). Thus, while the limited spatial resolution in the former case and limited detection resolution in the latter case leave the diffusive penetration of

water into quartz undetermined, the results are not inconsistent with the other indications of it being limited to the order of 1 μm.

We therefore conclude that the spatial scale over which equilibration with respect to the action of water by volume diffusion occurs is limited to the order of 1 μm in laboratory experiments of durations of a few hours at temperatures up to 1200–1300 K, corresponding to an upper limit of the diffusion coefficient of around 10^{-16} m^{2}s^{-1}.

There is, of course, the possibility to be considered of much faster penetration of water by pipe diffusion along dislocation cores. The effective or bulk diffusion coefficient in this case will be

$$D_{\text{eff}} = D(1 - x) + D_{\text{p}}x$$
$$\approx D\left(1 + \frac{D_{\text{p}}}{D}x\right) \text{ for } D_{\text{p}} \gg D \quad (1)$$

(Le Claire 1976) where D, D_{p} are, respectively, the volume and pipe diffusion coefficients and $x \approx c_{\text{p}}v_{\text{p}}/cv$ is the mole ratio of the amount of the diffusing species in the cores to that in the matrix, c_{p} and c being the respective molar concentrations in cores and matrix and v_{p} and v the respective volumes occupied by cores and matrix. If we take the effective pipe diameter of the dislocation to be of the order of 1 nm, then $v_{\text{p}}/v \approx 10^{-18}\rho$ where ρ is the dislocation density in m^{-2} and, for predominance of pipe diffusion, (1) leads to

$$\rho \gg 10^{18}\frac{D}{D_{\text{p}}}\frac{c}{c_{\text{p}}}$$

As possible examples, if $D_{\text{p}} \approx 10^3 D$ and $c_{\text{p}} \approx c$, an exceedingly high dislocation density of $\gg 10^{15}$ m^{-2} would be required for predominance of pipe diffusion, whereas, if $D_{\text{p}} \approx 10^6 D$ and $c_{\text{p}} \approx 10^2 c$, more moderate dislocation densities of $\gg 10^{10}$ m^{-2} would suffice. Thus, it is conceivable that dislocation pipe diffusion could be significant at achievable, moderately high dislocation densities, especially if water were much more soluble in the dislocation cores than in the matrix. However, its contribution is probably always insignificant at the very low dislocation densities of synthetic crystals (*c*. 10^7m^2; McLaren & Retchford 1969) and, in view of the relatively low dislocation densities observed in crystals deformed at high temperatures (for example, Morrison-Smith *et al.* 1976; Kirby & McCormick 1979), we shall not discuss it further in this paper, although not ruling out that it may eventually turn out to be important in some situations.

Coming now to the question of intragranular equilibration with respect to water by volume diffusion under geological conditions, we need to extrapolate the value of D to lower temperatures, say as low as 600 K (*c*. 300°C). Using the relation of Cordier *et al.* quoted above would lead to a value of D at 600 K of the order of 10^{-20} m^{2}s^{-1} and hence to a diffusion distance $\sqrt{2Dt}$ of the order of 1 mm in 10^{14} s (3 million years). Increasing the activation energy from the value of 95 kJ mol^{-1} of Cordier *et al.* would decrease this distance, while a contribution from pipe diffusion may increase it. Thus we conclude that intragranular equilibration with respect to water effects is probably maintained down to low grade metamorphic conditions (*c*. 300°C) on geological timescales greater than a million years, but only marginally so under the lowest grade conditions, where non-equilibrium conditions could conceivably arise under more rapid deformation or other change. As ambient temperature is approached, the distribution of water within quartz grains will tend to be frozen in on the scale of the grains and no longer respond to changes in external conditions, even on longer geological timescales.

Relevance of rheological extrapolation from experiment

The main question to be settled before extrapolating experimental rheological results to geological timescales is whether, in the experiments, the specimens have achieved a steady state that includes equilibration at all points in the specimen with respect to the influence of water at the experimental pressure and temperature. That is, is the pressure of water in inclusions or pore fluid everywhere the same and in equilibrium with the applied pressure, and is the chemical potential of water-related species within the quartz material itself everywhere the same and in equilibrium with the internal water reservoirs just mentioned?

In the case of experiments on natural quartzites carried out in gas-medium apparatus at moderate confining pressures (*c*. 300 MPa), the observation of flow stresses much higher than those expected on the basis of wet synthetic single crystal behaviour suggests that equilibration with respect to the influence of water is not achieved on the experimental timescale (Heard & Carter 1968; Mainprice & Paterson 1984). In the light of the discussion in the previous section, it may therefore be presumed that the water fugacity frozen in during the late geological history in fluid inclusions or other reservoirs within the grains is too low or the dispersion of the water is on too coarse a scale, to give a low flow stress, and that the effect of the water has not been re-equilibrated to a level corresponding to the applied confining pressure and temperature owing to sluggish kinetics. Presumably also, the redistribution of water through microcracking or pipe diffusion in existing dislocations has been kinetically inadequate as well.

In contrast, in experiments in solid-medium apparatus at higher confining pressures (up to 1500 MPa), the observation that the flow stress in natural quartzites decreases with increasing confining pressure and approaches wet synthetic quartz levels suggests that, unless there is an unknown dilatant defect involved, there is a more effective re-equilibration, presumably driven by the higher pressure (Kronenberg & Tullis 1984) and possibly involving more extensive microcracking and preliminary dislocation generation and pipe diffusion during the pressurizing phase. The flow law reported by Koch *et al.* (1980) from experiments at 1000 MPa also corresponds to lower flow stresses than are observed at the lower pressures in gas apparatus. However, the notable decrease in initial slope of the stress-strain curve with increasing confining pressure (see Kronenberg & Tullis's fig. 3a for the case of added water) requires additional explanation and, taken together with observations of Jaoul *et al.* (1984, table 1) of relatively low values of the stress exponent (1.4 to 2.4), raises the question of whether a component of granular flow is confusing the issue.

It must be concluded that the situation regarding the flow law for natural quartzite under experimental conditions is not very clear. It may be noted that the higher confining pressure experiments are carried out in the alpha-quartz stability field whereas the gas-medium experiments are carried out in the beta-quartz field but it is not clear that this factor is very important in resolving the apparent discrepancies. Thus, Linker & Kirby (1981) found a significant difference in flow stress between alpha and beta fields in wet synthetic crystals of one orientation but not for another, while Cordier *et al.* (1988) found no effect on the diffusion coefficient governing water redistribution in bubbles.

If one accepts that the relatively low flow stress for wet synthetic crystals results from the specimens being close to equilibrium (possibly because of the very fine scale of dispersal of water inclusions), it would appear desirable to measure the flow stress in a polycrystalline aggregate of wet synthetic quartz with a view to using it as a model for natural quartzite that is equilibrated with respect to water effects on the geological time scale. Such an approach will now be described.

New experiments on synthetic quartzite

Polycrystalline aggregates of quartz have been produced by crystallizing wet amorphous silica under 300 MPa confining pressure. Two sources of amorphous silica were used, one designated silica gel (of unknown origin and relatively impure), and the other silicic acid (Mallinckrodt AR Grade 100 mesh, of high purity). Both starting materials were in the form of powders (particle size < 1 μm for the gel and *c.* 150 μm for the silicic acid). After cold-pressing in a piston-cylinder die at around 200 MPa to form pellets, these were heated in the atmosphere at 1100 K in order to remove the greater part of the large water content, reducing it from 12–16% to around 1% or less. The pellets were then isostatically hot-pressed at 1300 K in an iron jacket in the gas-medium deformation apparatus (Paterson 1970) to form cylindrical polycrystalline specimens with water contents of around 1000–10,000 H/10^6Si or somewhat more. Details of the procedures and results will be given in a later publication. The polycrystalline specimens prepared in this way thus consist of quartz grains that have grown rapidly under the experimental 'wet' conditions, in a manner somewhat analogous to that for wet synthetic single crystals except for the growth rates and temperatures being much higher. For comparison, polycrystalline specimens were also prepared under similar conditions from ground quartz sand except that the hotpressing had to be carried out at about 1500 K in order to achieve low porosity.

In order to determine the stress-strain properties, the specimens were deformed immediately after hot pressing, while still in the apparatus at 300 MPa confining pressure, the dimensions being deduced from measurements made after deformation. Typical stress-strain curves are shown in Fig. 1. It will be seen that, in spite of differences in impurity content, including of hydroxyl itself, the two types of polycrystalline specimens synthesized from amorphous silica showed similar flow strengths, which were much lower than that for specimens prepared from natural quartz.

The specimens of natural quartz origin tended to develop a shear failure at larger strains, so that substantial intracrystalline deformation could not be achieved. In contrast, the specimens of amorphous silica origin deformed fairly

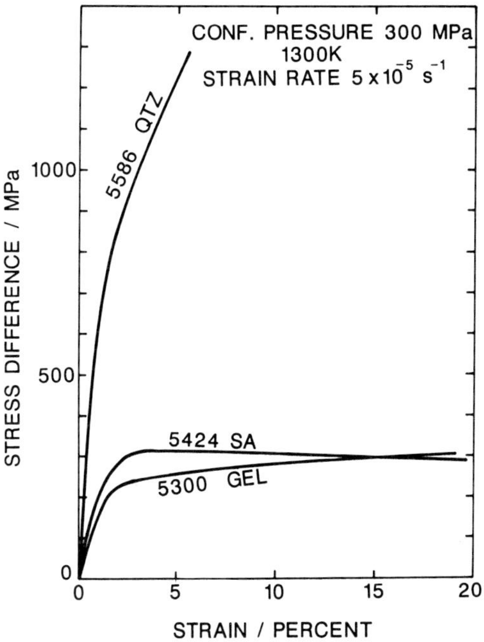

Fig. 1. Stress–strain curves at 1300 K, 5×10^{-5} s^{-1} strain rate, 300 MPa confining pressure for synthetic quartz aggregates previously hot-pressed at 1300 K (1500 K for QTZ). The symbols refer to run numbers, with QTZ signifying quartz-sand origin, SA silicic-acid origin and GEL silica-gel origin. The initial hot-pressing temperatures were 1500, 1300 and 1300 K, the final grain sizes < 38, 14 and 87 μm, the porosities 0.04, 0.03 and 0.01, and the hydroxyl content 1900, 2500 and 12,600 H/10^6Si, respectively, for the QTZ, SA and GEL specimens.

homogeneously and often showed, as well as marked undulatory extinction and flattened grain shapes, microstructural evidence of extensive grain boundary migration (giving highly serrated grain boundaries) and, at larger strains, recrystallization (new small grains). The specimens of gel origin initially contained many spherulitic growth structures but neither these nor the high impurity content obviously affected the deformation response.

We therefore conclude (1) that the deformation mechanism in the specimens of amorphous silica origin is primarily intragranular crystal plasticity, in spite of the grain size in the silicic acid case being near to where a transition to grain-size sensitive flow might be expected, (2) that the mechanism is probably similar to that in natural quartzites under geological conditions, including in respect of the role of water, and (3) that the results from these specimens can therefore be used for extrapolation. The specimens prepared from natural quartz, with addition of water, are presumed to be not equilibrated with respect to the water and so will not be considered further.

Power-law stress exponents (n) for the specimens of amorphous silica origin have been determined in strain-rate stepping experiments in the range 10^{-5} to 10^{-4} s^{-1}, using only upward strain-rate steps. The amount of strain in each step was usually 4–5%, which allows a reasonable approximation to a steady-state stress to be achieved. Specimens of gel origin with grain size in the range 30–80 μm gave n values of 2.3 ± 0.3 at 1200–1300 K, while specimens of silicic acid origin with grain size around 20 μm gave n values of 4.0 ± 0.8. Thus there is a distinct rheological difference between the two types of specimen, suggesting greater grain sliding activity in the more impure material in spite of the larger grain size.

Experimental activation energies Q for the specimens of amorphous silica origin have been determined in temperature stepping experiments in the range 1300–1100 K, using only downward temperature steps. The constant strain rate experiments only give values of Q/n and so the values of Q have been derived from these using the average values of n quoted in the previous paragraph. The available data, for four gel-origin and four silicic-acid-origin specimens, show considerable scatter (148 ± 46 and 152 ± 71 kJ mol^{-1}, respectively) and the two sets overlap considerably. Therefore they have been averaged together, giving a value of Q from the total set of 150 ± 59 kJ mol^{-1}.

A convenient reference stress for comparison and extrapolation is the stress difference at a strain rate of 10^{-5} s^{-1} at 1300 K, which we shall designate herein as σ_0. The values of σ_0 found in the present work are 141 ± 31 MPa for the gel-origin specimens and 222 ± 63 MPa for the silicic-acid-origin specimens, referring to those specimens from which the n and Q values are derived.

Assessment of rheological parameters

We shall now review the range of values for the stress exponent n and experimental activation energy Q as reported by previous workers, taking the results from the collation of Kirby & Kronenberg (1987), as listed in Table 1, but excluding values for quartzite vacuum dried at 1073 K on the grounds that this treatment may seriously compromise the role of water.

Table 1. *Steady-state rheological parameters for quartzites*

Source	Confining Pressure (GPa)	n	Q(kJ mol^{-1})	Comments
Shelton & Tullis (1981)	1.5	2.0	167	No water added
Hansen & Carter (1982)	1.0	1.9	123	
Koch *et al.* (1980)	1.0	2.9	149	
Jaoul *et al.* (1984)	1.5	2.4	167	
Hansen & Carter (1982)	1.0	1.8	167	Water added
Koch *et al.* (1980)	1.0	2.4	160	
Kronenberg & Tullis (1984)	0.9–1.6	2.6	134	
Present experiments	0.3	2.2	149	Gel origin
Present experiments	0.3	3.9	149	Silicic acid origin

Note: The present experiments were carried out in the β-field whereas the others were in the α-field.

If the values of n and Q from the previous workers are averaged within the two groups shown in Table 1, we obtain $n = 2.4 \pm 0.45$ and 2.3 ± 0.3 and $Q = 156 \pm 23$ and 154 ± 14 kJ mol^{-1}, respectively, for the cases of natural quartzite without and with the addition of extra water. We therefore conclude that the two sets of data are statistically indistinguishable from each other and from the results of the present work except for the n value from the silicic-acid-origin synthetic quartzite. This conclusion suggests that, within the present uncertainties, the stress sensitivity and the temperature sensitivity of the strain rate are similar in all cases, with the one exception, although the actual stress level for a given strain rate may vary because of differences in the pre-exponential factor A. That is, the above conclusion suggests that, provided a minimum amount of water is present, whether initially or through addition, the stress exponent and the experimental activation energy will be substantially independent of the total amount or fugacity of the water, but that the amount or fugacity of water (and possibly its dispersion) will have an influence mainly through the pre-exponential factor A.

In the light of the above considerations, we therefore take a global average of all the values listed in Table 1 to arrive at

$$n = 2.5 \pm 0.6$$
$$Q = 152 \pm 15 \text{ kJ mol}^{-1}$$

as the best currently available estimates for application to the rheology of natural quartzites, with the reservation that this value of n may be too low. This selection leaves the laboratory reference stress difference σ_0 or pre-exponential factor A to be chosen according to the actual water situation, probably as characterized by the water fugacity.

Geological extrapolation

Insofar as steady-state, intracrystalline plasticity mechanisms are involved, the most suitable form of flow law for extrapolation is generally agreed to be

$$\dot{\varepsilon} = A\,\sigma^{n} \exp(-Q/RT) \qquad (2)$$

(for example, Frost & Ashby 1982), where $\dot{\varepsilon}$ is the strain rate, σ the stress difference, T the temperature, and A, n, Q constants. The pre-exponential factor A is, in this case, taken to be independent of grain size to a first approximation but, from the discussions above, it is probably a function of water fugacity, of a form at present unknown. It is therefore only feasible to undertake extrapolation at a constant water fugacity similar to that in the laboratory experiments. For the extrapolation, it is convenient to re-write (2) in the relative form

$$\lg \sigma = \lg \sigma_0 + \frac{1}{n}\left[\frac{Q}{2.3R}\left(\frac{1}{T} - \frac{1}{T_0}\right) + \lg \frac{\dot{\varepsilon}}{\dot{\varepsilon}_0}\right] \qquad (3)$$

where σ_0 is the laboratory reference stress difference at temperature T_0 and strain rate $\dot{\varepsilon}_0$ (taken here as 1300 K and 10^{-5} s^{-1}). The value of A can be calculated from (2) using the reference values σ_0, T_0, $\dot{\varepsilon}_0$.

In view of the uncertainties in the rheological parameters, it seems appropriate initially to calculate the extrapolated stresses at a selected geological strain rate only for the limits corresponding to the standard deviations given above. This calculation leads to the shaded band shown in Fig. 2a and b for the geological strain rates of 10^{-12} s^{-1} and 10^{-14} s^{-1}, respectively. Also shown, for comparison, are the lines corresponding to the mean 'wet' quartzite flow law of Koch *et al.* (1980; see also Table 1) and

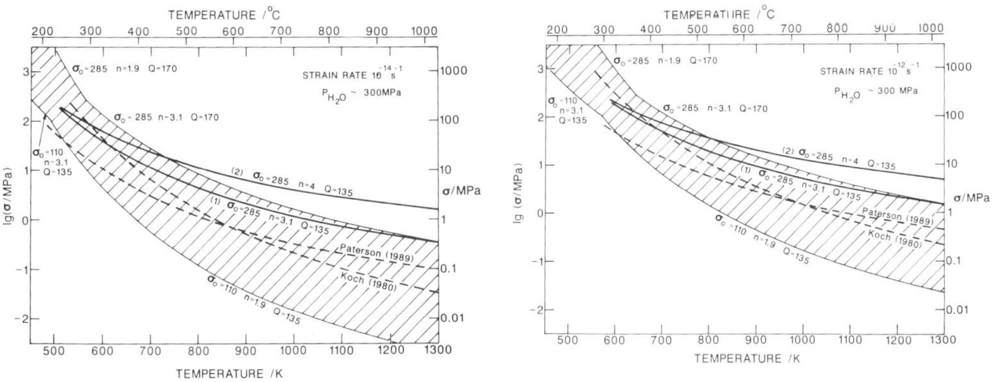

Fig. 2. Calculated range (shaded region) for extrapolated steady state stress difference for dislocation creep of quartzite with $P_{H_2O} \approx 300$ MPa, based on the experimental ranges of $\sigma_0 = 110-285$ MPa (stress at 10^{-5} s^{-1}, 1300 K), $n = 1.9-3.1$ and $Q = 135-170$ kJ mol^{-1}. Also shown are the calculated lines for the flow laws of Koch *et al.* (1980) and Paterson (1989), as well as the two particular cases (1) and (2) discussed in the text. (**a**) For strain rate 10^{-12} s^{-1}. (**b**) For strain rate 10^{-14} s^{-1}. (The marked σ_0 and Q values are in MPa and kJ mol^{-1}, respectively).

the flow law estimated from wet synthetic single crystal data by Paterson (1989).

In the practical application of the extrapolations in Fig. 2, the following qualifications must be borne in mind.

(1) The derived flow stresses only apply for steady state, intracrystalline plastic flow (dislocation creep) and, strictly, for an axisymmetric shortening deformation. A power-law breakdown regime can be expected to intervene at high stresses (Tsenn & Carter, 1987), and a grain-size sensitive regime somewhere at relatively low stresses and high temperatures when the grain size is relatively small (the latter regime has not yet been clearly defined for quartzite but in it the flow stresses would be expected to be lower than those in Fig. 2).

(2) Through the particular choice of σ_0 from the present experiments, the derived flow stresses should apply for water pressures of around 300 MPa. The flow stresses at higher or lower water pressures may be lower or higher, respectively. However, if the speculation above that n and Q are not substantially affected by water fugacity is valid, the effect of varying the water pressure would be mainly to move the curves up or down in Fig. 2, by an amount that cannot yet be predicted.

(3) The laboratory results may be affected to some degree by the presence of partial melt (Mainprice & Paterson 1983; Jaoul *et al.* 1984). The amount and effect of the melt may vary within the experimental range, thus contributing additional temperature dependence, and it may lead to a lower n value because of promoting grain boundary movement. In this connection, the exceptionally high value of $n = 3.9 \pm 0.5$ for the synthetic quartzite prepared from high-purity silicic acid may be noted.

Because the partial melt effect would not be expected to apply at geological temperatures of 900 K and below, it may therefore be expected that, under low to moderate grade geological conditions, relatively low Q and high n values might be appropriate. In this case it is likely that stresses near the upper limits of the bands in Fig. 2 would be the best choice, such as shown by the line (1) for $n = 3.1$ and $Q = 135$ kJ mol^{-1} (for which $A = 6.5 \times 10^{-8}$ MPa^{-n} s^{-1}). The line (2) for $n = 4$ and the same Q could be an even better choice if the n value for silicic-acid-origin specimens is taken more seriously (in this case $A = 4.0 \times 10^{-10}$ MPa^{-n} s^{-1}).

Summary and conclusions

(1) Evidence is reviewed to indicate that, while quartzites seem only to respond on the scale of 1 μm to water diffusion in laboratory times and so may not be equilibrated there, equilibration probably occurs under geological conditions, thus posing a difficulty in extrapolating from the laboratory to the Earth.

(2) New experiments on wet synthetic quartzites, prepared by crystallizing amorphous silica, give results that are taken to be probably representative of equilibration with respect of water.

(3) A review of the various available rhe-

ological results reveals substantial scatter, within which the results for natural quartzites, with or without added water, and for wet synthetic quartzites prepared from amorphous silica cannot be statistically distinguished in respect of stress exponent n and activation energy Q (with the exception of a high $n = 3.9$ for synthetic quartzite prepared from high-purity silicic acid). Within the experimental uncertainties and in view of the wide range of pressures used, it is therefore concluded that n and Q are not strongly dependent on the water fugacity, which may mainly affect the pre-exponential factor in the power-law form.

(4) On this basis, flow stress limits for geological conditions with water pressure around 300 MPa are obtained by extrapolation of the laboratory results, and the limitations on their application discussed.

G. Horwood, P. Brugman and J. Bitmead have given valuable assistance in the experimental work and useful discussions have been held with S. Cox and A. McLaren.

References

AINES, R. D. & ROSSMAN, G. R. 1984. Water in minerals? A peak in the infrared. *Journal of Geophysical Research*, **89**, 4059–4071.

AINES, R. D., KIRBY, S. H. & ROSSMAN, G.R. 1984. Hydrogen speciation in synthetic quartz. *Physics and Chemistry of Minerals*, **11**, 204–212.

BLACIC, J. D. & CHRISTIE, J. M. 1984. Plasticity and hydrolytic weakening of quartz single crystals. *Journal of Geophysical Research*, **89**, 4223–4239.

CARTER, N. L. & TSENN, M. C. 1987. Flow properties of continental lithosphere. *Tectonophysics*, **136**, 27–63.

CORDIER, P., BOULOGNE, B. & DOUKHAN, J. -C. 1988. Water precipitation and diffusion in wet quartz and wet berlinite $A1PO_4$. *Bulletin de Minéralogie*, **111**, 113–137.

DOUKHAN, J. -C. & TREPIED, L. 1984. Plastic deformation of quartz single crystals. *Bulletin de Minéralogie*, **108**, 97–123.

FROST, H. J. & ASHBY, M. F. 1982. *Deformation – Mechanism Maps*. Pergamon Press, Oxford.

GERRETSEN, J., PATERSON, M. S. & MCLAREN, A. C. 1989. The uptake and solubility of water in quartz at elevated pressure and temperature. *Physics and Chemistry of Minerals*, **16**, 334–342.

GRIGGS, D. T. 1967. Hydrolytic weakening of quartz and other silicates. *Geophysical Journal of the Royal Astronomical Society*, **14**, 19–31.

—— 1974. A model of hydrolytic weakening in quartz. *Journal of Geophysical Research*, **79**, 1653–1661.

—— & BLACIC, J. D. 1965. Quartz: anomalous weakness of synthetic crystals. *Science*, **147**, 292–295.

HANSEN, F. D. & CARTER, N. L. 1982. Creep of selected crustal rocks at 1000 MPa (abstract). *EOS Transactions of the American Geophysical Union*, **63**, 437.

HEARD, H. C. & CARTER, N. L. 1968. Experimentally induced "natural" intragranular flow in quartz and quartzite. *American Journal of Science*, **266**, 1–42.

JAOUL, O., TULLIS, J. & KRONENBERG, A. 1984. The effect of varying water contents on the creep behaviour of Heavitree quartzite. *Journal of Geophysical Research*, **89**, 4298–4312.

KEKULAWALA, K. R. S. S., PATERSON, M. S. & BOLAND, J. N. 1978. Hydrolytic weakening in quartz. *Tectonophysics*, **46**, T1–T6.

——, —— & —— 1981. An experimental study of the role of water in quartz deformation. *In*: CARTER, N. L., FRIEDMAN, M., LOGAN, J. M. & STEARNS, D. W. (eds) *Mechanical Behavior of Crustal Rocks*. Geophysical Monograph, **24**, American Geophysical Union, Washington DC, 49–60.

KIRBY, S. H. 1983. Rheology of the lithosphere. *Reviews of Geophysics and Space Physics*, **21**, 1458–1487.

—— & KRONENBERG, A. K. 1987. Rheology of the lithosphere: selected topics. *Reviews of Geophysics*, **25**, 1219–1244, 1680–1681.

—— & MCCORMICK, J. W. 1979. Creep of hydrolytically weakened synthetic quartz crystals oriented to promote $\{2\bar{1}\bar{1}0\}$ <0001> slip: a brief summary of work to date. *Bulletin de Minéralogie*, **102**, 124–137.

KOCH, P. S., CHRISTIE, J. M. & GEORGE, R. P. 1980. Flow law for "wet" quartzite in the α-quartz field (abstract). *EOS Transactions of the American Geophysical Union*, **61**, 376.

KRONENBERG, A. K. & TULLIS, J. 1984. Flow strengths of quartz aggregates: grain size and pressure effects due to hydrolytic weakening. *Journal of Geophysical Research*, **89**, 4281–4297.

——, KIRBY, S. H., AINES, R. D. & ROSSMAN, G. R. 1986. Solubility and diffusional uptake of hydrogen in quartz at high water pressures: implications for hydrolytic weakening. *Journal of Geophysical Research*, **91**, 12723–12744.

LE CLAIRE, A. D. 1976. Diffusion. *In*: HANNAY, N. B. (ed.) *Treatise on Solid State Chemistry, Vol.* **4**: *Reactivity of Solids*, Plenum Press, New York, 1–59.

LINKER, M. F. & KIRBY, S. H. 1981. Anisotropy in the rheology of hydrolytically weakened synthetic quartz crystals. *In*: CARTER, N. L., FRIEDMAN, M., LOGAN, J. M. & STEARNS, D. W. (eds) *Mechanical Behavior of Crustal Rocks*. Geophysical Monograph, **24**, American Geophysical Union, Washington DC, 29–48.

LUAN, F., PATERSON, M. S. & MCLAREN, A. C. 1986. Synthetic quartz aggregates for deformation studies (abstract). *EOS Transactions of the American Geophysical Union*, **67**, 1207.

MAINPRICE, D. H. & PATERSON, M. S. 1984 Experimental studies of the role of water in the plasticity of quartzites. *Journal of Geophysical Research*, **89**, 4257–4269.

MCLAREN, A. C. & HOBBS, B. E. 1972. Transmission

electron microscope investigation of some naturally deformed quartzites. *In*: HEARD, H. C., BORG, I. Y., CARTER, N. L. & RALEIGH, C.B. (eds) *Flow and Fracture of Rocks*. Geophysical Monograph, **16**, American Geophysical Union, Washington, DC, 55–66.

—— & RETCHFORD, J. A. 1969. Transmission electron microscope study of the dislocations in plastically deformed synthetic quartz. *Physica Status Solidi*, **33**, 657–668.

——, FITZ GERALD, J. D. & GERRETSEN, J. 1989. Dislocation nucleation and multiplication in synthetic quartz: relevance to water weakening. *Physics and Chemistry of Minerals*, **16**, 465–482.

MEISSNER, R. & STREHLAU, J. 1982. Limits of stresses in continental crusts and their relationship to depth-frequency distribution of shallow earthquakes. *Tectonics*, **1**, 73–89.

MORRISON-SMITH, D. J., PATERSON, M. S. & HOBBS, B. E. 1976. An electron microscope study of plastic deformation in single crystals of synthetic quartz. *Tectonophysics*, **33**, 43–79.

ORD, A. & HOBBS, B. E. 1986. Experimental control of the water-weakening effect in quartz. *In*: HOBBS, B. E. & HEARD, H. C. (eds) *Mineral and Rock Deformation: Laboratory Studies*. Geophysical Monograph, **36**, American Geophysical Union, Washington DC, 51–72.

PARRISH, D. K., KRIVZ, A. L. & CARTER, N. L. 1976. Finite-element folds of similar geometry. *Tectonophysics*, **32**, 183–207.

PATERSON, M. S. 1970. A high-pressure, high-temperature apparatus for rock deformation. *International Journal of Rock Mechanics and Mining Science*, **7**, 517–526.

—— 1987. Problems in the extrapolation of laboratory rheological data. *Tectonophysics*, **133**, 33–43.

—— 1989 The interaction of water with quartz and its influence in dislocation flow – an overview. *In*: KARATO, S. -I & TORIUMI, M. (eds) *Rheology of Solids and of the Earth*. Oxford University Press, Oxford, 107–142.

ROVETTA, M. R., HOLLOWAY, J. R. & BLACIC, J. D. 1986. Solubility of hydroxyl in natural quartz annealed in water at 900°C and 1.5 GPa. *Geophysical Research Letters*, **13**, 145–148.

SHELTON, G. & TULLIS, J. A. 1981. Experimental flow laws for crustal rocks (abstract). *EOS Transactions of the American Geophysical Union*, **62**, 396.

TSENN, M. C. & CARTER, N. L. 1987. Upper limits of power law creep of rocks. *Tectonophysics*, **136**, 1–26.

TULLIS, J., SHELTON, G. L. & YUND, R. A. 1979. Pressure dependence of rock strength: implications for hydrolytic weakening. *Bulletin de Minéralogie*, **102**, 110–114.

Estimates of the rates of microstructural changes in mylonites

DAVID J. PRIOR[1], ROBERT J. KNIPE[2] & MARK R. HANDY[3]

[1] Department of Earth Sciences, Liverpool University, Liverpool L69 3BX, UK

[2] Department of Earth Sciences, Leeds University, Leeds LS2 9JT, UK

[3] Geologisches Institut, Universität Bern, Switzerland

Abstract: Microstructures developed during dislocation creep such as the dynamically recrystallized grain-size and the subgrain-size are modified in response to increasing differential stresses and shear-strain rates adjacent to rigid minerals in mylonites. In two specimens variations in quartz subgrain-size around rigid minerals has been quantified from orientation contrast SEM images. In a deformed granite from the Pogallo ductile fault zone, quartz two-dimensional subgrain-size is reduced from background values of about 30 μm to 8 μm and then 4 μm in progressively narrower channels between rigid feldspars. In pelitic mylonites from the Alpine fault zone, New Zealand, quartz bands are wrapped around rigid garnets and the subgrain-size is reduced from background values of about 20 μm to 14 μm within 2000 μm of the garnet to 11 μm adjacent to the garnet. In both specimens examined the microstructure responds to both increases and decreases in differential stresses and coupled strain-rates. The analysis of the distribution of subgrain sizes provides information for the assessment of microstructural stability and the estimation of the rates of microstructural changes. The New Zealand specimens have good regional shear strain-rate controls and minimum subgrain boundary migration rates (during reduction in differential stresses and strain-rates) of 1.2×10^{-9} $\mu m\ s^{-1}$ to 1.2×10^{-11} $\mu m\ s^{-1}$ are estimated, for bulk shear strain rates of $10^{-10}\ s^{-1}$ and $10^{12}\ s^{-1}$ respectively. Palaeopiezometry coupled with rheological considerations suggests that these microstructural changes correspond to minimum differential stress-rates of 5×10^{-9} MPa s^{-1} to 5×10^{-11} MPa s^{-1}. Similar or slower regional stress-rates during fault zone evolution, will allow continuous re-equilibration of the microstructure. Thus the time at which any given microstructure is frozen in will be critically dependent upon fault zone deformation history. In the Alpine fault zone rapid uplift and high near surface temperatures combine to generate rapid stress drops as mylonitic deformation gives way to cataclastic faulting during uplift. Mylonite microstructure is frozen in during this rapid stress drop and represents the shallowest level of major crystal-plastic deformation.

Quantification of microstructural stability, specifically the rates at which microstructures are modified due to changes in conditions of stress, strain rate, temperature, depth and fluid composition and activity, is essential if microstructures are to be used to evaluate dynamic processes in the Earth's crust (Knipe 1989). This paper examines one aspect of microstructural stability; the microstructural response of quartz mylonites to changing differential stress and strain-rate conditions.

Quartz mylonites often contain a dynamic equilibrium microstructure (Bell & Etheridge 1973) which can be used to evaluate the kinematics, differential stresses and strain-rates of deformation. Constitutive flow laws for dislocation creep (see Handy 1989; Paterson & Luan, this volume) indicate an interdependence of differential stress and strain rate which cannot be decoupled. For deformation accompanied by dynamic recrystallization, the grain-size and subgrain-size are related to differential stress levels (Twiss 1977, 1986) and coupled strain-rates for a given temperature of deformation. This relationship is documented for quartz in experiments and in natural examples (Mercier *et al.* 1977; Christie *et al.* 1980; Ord & Christie 1984). If conditions of deformation are changing, as might be expected in an uplift path, then it is important to be able to assess whether the observed microstructure developed during the last increments of strain, at the peak stress condition or during some other time period. Recent analysis of mylonite evolution has recognised the importance of cyclic changes in the microstructure (White *et al.* 1980; Means 1981; Knipe 1989). The differential stress and strain-rate fields within a mylonite are likely to be inhomogenous (Lister & Price 1978; Masuda & Ando 1988) and there is a need to identify situations where the influence of changing stress and strain-rate conditions on microstructural

From Knipe, R. J. & Rutter, E. H. (eds), 1990, *Deformation Mechanisms, Rheology and Tectonics*, Geological Society Special Publication No. 54, pp. 309–319.

evolution can be assessed. One such situation is in the proximity of rigid particles. As matrix material flows around and past such particles it experiences an increase followed by a decrease in differential stress and in strain-rate (e.g. Masuda & Ando 1988; Handy 1990). If the geometrical evolution can be modelled and the bulk strain-rates estimated, in these situations, then the rate of microstructural change associated with local changes in differential stress and strain-rates can be constrained. These data can then be applied to large-scale changes in differential stress and strain-rate to infer how mylonitic microstructures will respond to fault zone evolution and changing deformation conditions.

Microstructural environments around rigid phases in mylonites

Mylonites containing phases which behave rigidly are well documented (Boullier 1980; Lister & Price 1978; Mitra 1978; Price 1978; White *et al.* 1980; White 1984). A rigid body in a viscous material will cause a localized perturbation in the differential stress field and strain-rate pattern (Masuda & Ando 1988). In granitic and metasedimentary mylonites; feldspars, garnet, zircon, amphiboles and ore phases usually act as rigid particles and cause the strain in the surrounding matrix (quartz and mica) to accommodate this incompatibility (Lister & Williams 1983; Lister & Price 1978). The microstructure adjacent to these particles may be modified in response to changing differential stress, strain-rate and strain field conditions with or without a change in the dominant deformation mechanism. Such modifications include reduction of the quartz grain and subgrain-sizes in regions adjacent to the rigid bodies (Etheridge & Wilkie 1979; Lister & Price 1978). Knowledge of the flow lines for material around a rigid object combined with constraints on the shear strain rate can be used to quantify the rate at which the material moves around the localized differential stress and strain-rate field of a rigid body. These data coupled with analysis of the quartz microstructures along the flow lines around rigid bodies facilitate quantification of the rates of microstructural change in response to the changing differential stresses and strain-rates.

Changes in quartz microstructure around rigid objects have been investigated in two mylonites: (a) the deformed San Rocco Granite from the Pogallo ductile fault zone, southern European Alps (Handy 1986), and (b) the Alpine fault mylonites, South Island, New Zealand (Sibson *et al.* 1979; Prior 1989). These two examples provide contrasting local differential stress/strain-rate environments and are discussed separately below.

Methods

Specimen blocks were polished to better than 3 μm surface roughness using diamond abrasive on a cloth lap and then put onto a Malvern Instruments Multipol 2 mechanical/chemical polisher (Fynn & Powell 1979; Lloyd 1987) using a polyeurethane lap and SYTON fluid for 4 to 10 hours. Specimens were examined on a Camscan Series 4 scanning electron microscope (SEM) in orientation contrast mode (Lloyd 1987) using a working distance of 7–8 mm, an accelerating voltage of 30 kV and a beam current of 175 nA. Distinct boundaries with lattice rotations $> 0.25°$ can be imaged by this method with sub-micrometre spatial resolution, so that this technique is ideal for imaging subgrains. However it must be emphasized that the change in BSE grey level across a boundary is not simply a function of mis-orientation, so that a 0.25° boundary, a 5° boundary or any other arbitrary boundary mis-orientation are not readily distinguishable without systematic collection of electron channelling patterns (ECPs) from each subgrain (Lloyd 1987). The term subgrain in this paper will always refer to the smallest crystallographic units observed in orientation contrast SEM.

Locating micrographs were taken at 10× to 100× magnification. Micrographs to be used for quantification of subgrain-sizes were all taken at the same magnification in any one given specimen (usually between 500× and 1500×). Areas comprised of pure quartz, as close to an assumed flow line around the rigid object as possible, were chosen for subgrain-size measurement. Where possible ECPs were used to assess typical boundary mis-orientations. Two dimensional subgrain areas (A) were quantified using a digitizing tablet. Two dimensional subgrain-size (d) is calculated as the diameter of a circle of equivalent area (A) to the measured grain. Three dimensional subgrain-size (D) is estimated using the linear stereological correction $D = 4d/\pi$ (Exner 1972). Mean subgrain-sizes and corresponding standard deviation are calculated assuming $\sqrt{d}$ is normally distributed. Because standard deviations are calculated from the $\sqrt{d}$ distribution corresponding errors in d are asymmetrically distributed.

Quartz subgrain-size around feldspars in deformed San Rocco Granite

The Permian San Rocco granite was variably deformed in the Pogallo ductile fault zone under greenschist facies conditions (PDFZ: Handy 1986). Quartz crystallographic fabrics indicate a significant component of simple shear deformation predominated in the PDFZ although shape analyses of quartz ribbons indicate K

values mainly between 0.3 and 0.4. Optical estimates, in transmitted light, of the mean two dimensional dynamically recrystallized quartz grain-sizes of 20 μm to 40 μm suggest differential stresses during deformation of 30 MPa to 50 MPa (using the theoretical recrystallized grain-size palaeopiezometer of Twiss 1977). Rheological considerations bracket shear strain rates between $10^{-8}s^{-1}$ and $10^{-13}s^{-1}$ but shear strain rates are only poorly constrained independently of microstructural criteria.

One specimen of the San Rocco Granite (HD47), cut parallel to lineation, is examined in detail here. The quartz is 100% dynamically recrystallized but its microstructure is different in the areas of constricted flow and pressure shadows created by feldspars 1–5 mm in size. The feldspars have undergone minor fracturing and associated sericitization but have not accommodated significant strain. Flow stresses, estimated from optical measurements of quartz recrystallized grain-size, are amplified by a factor of 2–3 in constrictions relative to pressure shadows (Handy 1990).

An orientation contrast BSE montage of a constriction in which subgrain-sizes have been estimated is shown in Fig. 1. Quartz flow is constrained between a rounded feldspar (*c*. 400 μm radius) and a large flat feldspar. A few millimetres to the right of the imaged area the quartz band thickens to *c*. 1 mm and has a mean subgrain-size in excess of 30 μm. The entire imaged region is a constriction with significantly reduced subgrain-size. The two dimensional subgrain-size is reduced to 8.0 μm [+4.7 −3.6] (mean of cumulated data from areas 1, 4, 5, 6 and 7 in Fig. 1) across most of the imaged area and to 4.5 μm [+1.9 −1.5] (mean of cumulated data from areas 2 and 3 in Fig. 1) in an anvil shaped region above the apex of the rounded feldspar. There is a 40 μm thick arc marginal to the apex of the lower feldspar where subgrain-size is not reduced although there are not enough clearly defined grains in this region for subgrain-size quantification. The microstructural transition into the anvil shaped region is sharp ($\leq$ 10 μm thick) up to 80 μm to 100 μm from the round feldspar, but then widens to an ill-defined transition (over 100 μm to 200 μm) adjacent to the overlying flat faced feldspar (Fig. 1).

Quartz subgrains across the majority of the imaged area are subhedral with axial ratios between 1:1 and 2:1. There are local boundary alignments; broadly top left to bottom right in Fig. 1. Most subgrain boundaries are irregular with serrations of amplitude 1 μm to 5 μm. Subgrain boundary intersections commonly define a 120° dihedral angle. In many cases the 120° angle is only maintained within a few micrometres of the intersection. In the region containing the smaller subgrains there are more euhedral and equant subgrains. Although approximately 25% of subgrains have straight boundaries, the majority of boundaries are serrated, indicating that some subgrain boundary migration has taken place. Subgrain-size histograms (Fig. 1) show that the small fraction which dominates the anvil shaped region also form a significant proportion of the size fraction across the rest of the imaged area. The subgrain-size transition is accommodated by the loss of the broad tail of coarser grains. Crystallographic orientations have not been systematically investigated in specimen HD47 but optical observations indicate that there are strong preferred *c*-axis orientations which are maintained through reduced grain-size regions in flow constrictions. A detailed study of crystallographic preferred orientations (CPO) in other San Rocco Granite specimens shows that constricted regions have strong CPOs but these are distinct from the CPO of pressure shadows (Handy 1990).

Quartz grain-size around garnets in the Alpine fault mylonites

The Alpine fault mylonites (Sibson *et al*. 1979; Prior 1989) developed during Alpine fault motion which probably started no earlier than 22 Ma (Carter & Norris 1976; Kamp 1986). Kinematic indicators (Prior 1989) show that the deformation developed during the transpressive phase of simple shear motion which started about 5 Ma ago (Walcott 1984). Deformation is dominated by simple shear and regional criteria constrain the bulk shear strain rate across the 1 km wide mylonite zone at between 10^{-13} s^{-1} to 10^{-11} s^{-1}. Often an early mylonitic foliation is truncated by a later mylonitic foliation. Kinematic indicators always show that both foliations developed during the same sense of bulk shear. These data suggest that shearing within the Alpine fault mylonites at any one time was restricted to a zone considerably thinner than the total thickness of mylonites. In individual sections which cross the entire width of the mylonites there are typically between 10 and 30 such truncations of one fabric by another. Thus, to account for the localization of mylonitic deformation at any one time the shear strain rate estimates must be increased by at least an order of magnitude to 10^{-12} s^{-1} to 10^{-10} s^{-1}. These estimates are conservative,

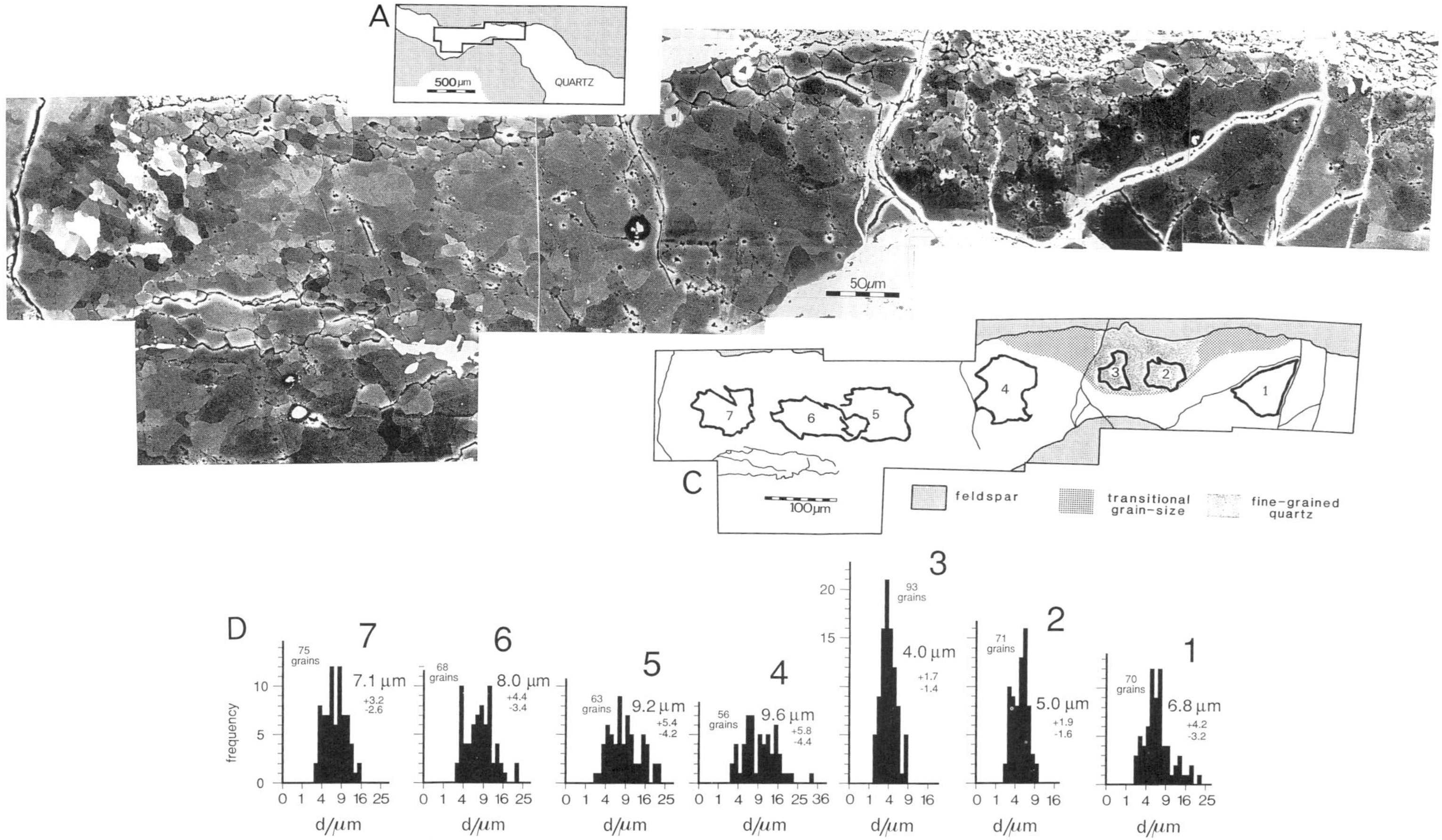

Fig. 1. Microstructure between two feldspars in a specimen of deformed San Rocco Granite (HD47) from the Pogallo ductile fault zone of the southern European Alps. (**A**) Location sketch, traced from a BSE micrograph. Box shows the area examined in detail on the SEM and illustrated in (B) and (C). (**B**) Montage of BSE orientation contrast micrographs showing the constriction of quartz located in (A). Apart from obvious cracks all contrast in the quartz relates to differences in crystallographic orientation. (**C**) Sketch, traced from the montage shown in (B), to locate subgrain-size analysis areas 1 to 7 and to highlight the reduced subgrain-size in the region above the apex of lower rounded feldspar. This anvil shaped region comprises a core of fine-grained quartz and a transitional boundary of variable thickness (see text). Cracks are marked to help location relative to (B). (**D**) Subgrain-size (plotted so that $\sqrt{d}$ is linear) versus frequency histograms for areas 1 to 7. Each plot shows the subgrain-size equivalents of the mean and standard deviation of the $\sqrt{d}$ distribution. Data are not stereologically corrected.

shear strain rates are probably faster than these. Optical estimates in transmitted light of two dimensional dynamically recrystallized quartz grain-sizes of 18 μm away from the rigid particles suggests differential stresses during deformation of 30 MPa to 70 MPa, using the Twiss, (1977) palaeopiezometer (by D. C. Green in 1982; Prior 1989). Mean subgrain-sizes, from orientation contrast SEM images, in regions remote from rigid particles vary from 18 μm to 26 μm.

The mylonite contains phases, including garnet, which developed in the much older Rangitata metamorphism (Landis & Coombs 1967) and are now porphyroclasts. Garnets have not accommodated significant strain although they are affected by fracturing and associated retrogression. The distinctive quartz bands are thinner adjacent to garnets and grain-size reduction in the thinned regions can be observed optically.

Specimen DC3A, from Darnley Creek, a tributary of the Waitangi–Taona river north of Waiho, has been studied in detail. This specimen was cut parallel to the regionally and locally defined movement direction (± 10°) and perpendicular to foliation. The shear sense is clearly

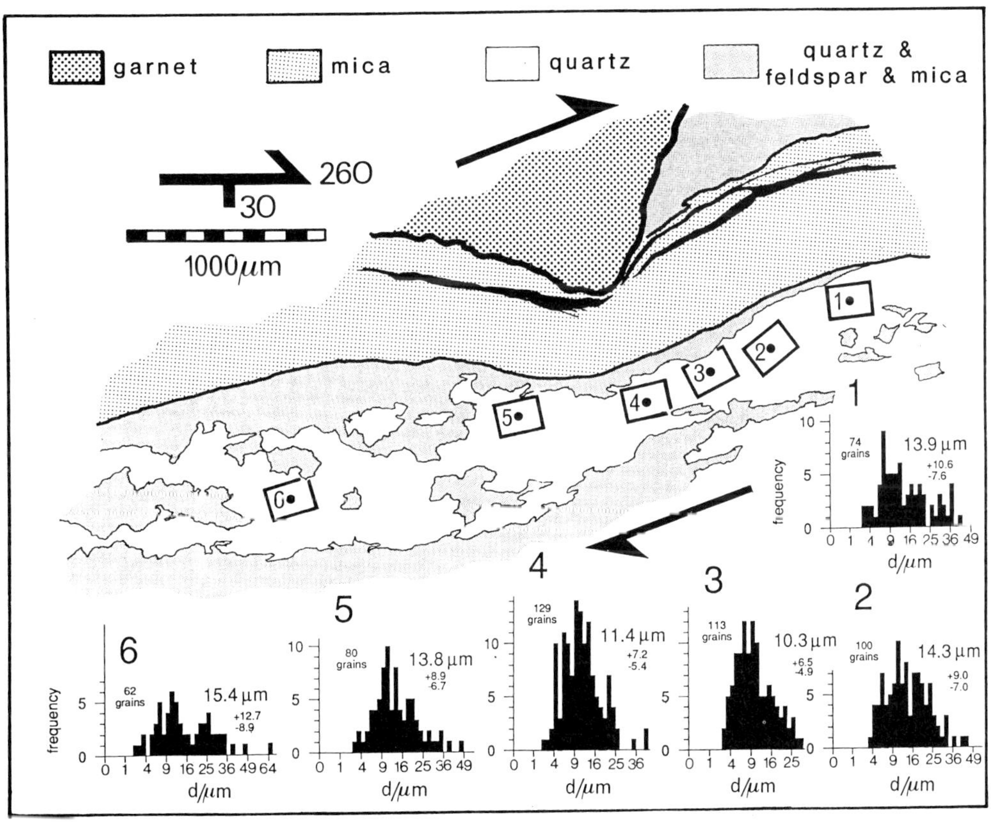

Fig. 2. Microstructure of a quartz band buckled around a garnet in a sample (DC3A) from the Alpine fault mylonites, South Island, New Zealand. Orientation mark relates to the geographical reference frame. Shear sense indicators are based on data from this and neighbouring specimens and are consistent with regional movement senses. The specimen surface is parallel to the movement direction and perpendicular to foliation. A lower magnification sketch of this region is shown in Fig. 4. Some parts of the quartz band contain feldspar and mica impurities. These areas (shaded) are distinguished from pure quartz (no ornament). Numbered boxes show the locations of orientation contrast BSE images used to generate the subgrain-size data. Numbered subgrain size (plotted so that $\sqrt{d}$ is linear) versus frequency histograms 1 to 6 correspond to location boxes 1 to 6. Each plot shows the subgrain-size equivalents of the mean and standard deviation of the $\sqrt{d}$ distribution. Data are not stereologically corrected.

defined by shear bands, asymmetrical microfolds and mica fish in this and neighbouring specimens. Only one of three garnets examined is wrapped by a quartz band sufficiently pure to allow subgrain-size analysis of *c.* 100 grains. A line drawing of the area examined and the subgrain-size data measured in this region is shown in Fig. 2. Flow paths are assumed to be parallel to foliation, here defined by the boundary of quartz/quartz–feldspar–mica bands and mica bands. Five measurements areas (1 to 5) were selected along an interpreted particle flow path which is at its closest 0.6 mm from the garnet margin (2.1 mm from the garnet centre). A sixth measurement (area 6) is not along the same flow path because of the presence of feldspars and is located on a flow path approximately 0.3 mm further from the garnet. Parts of the orientation contrast BSE images used to quantify subgrain-sizes for areas 2 and 3 are shown in Fig. 3. Two dimensional subgrain-size is reduced from measured background values averaging 14.3 μm [+10.1 −7.4] (mean of areas, 1, 2, 5, & 6) to 10.9 μm [+6.9 −5.2] (mean of areas 3 & 4) in the region closest to the garnet. All the areas examined here have subgrain-sizes less than far-field subgrain-size estimates of 18 μm to 26 μm. Subgrain-size histograms (Fig. 2) show that the regions with smaller subgrains contain the same overall range of sizes as the coarser areas but the frequency distribution is altered.

The microstructure of the coarse and fine subgrain regions is similar. Subgrains are euhedral to subhedral with axial ratios from 1:1 to 4:1. About 30% of subgrain boundaries are straight, the others have serrations of amplitude 5 μm to 10 μm. Grain modification occurs by a

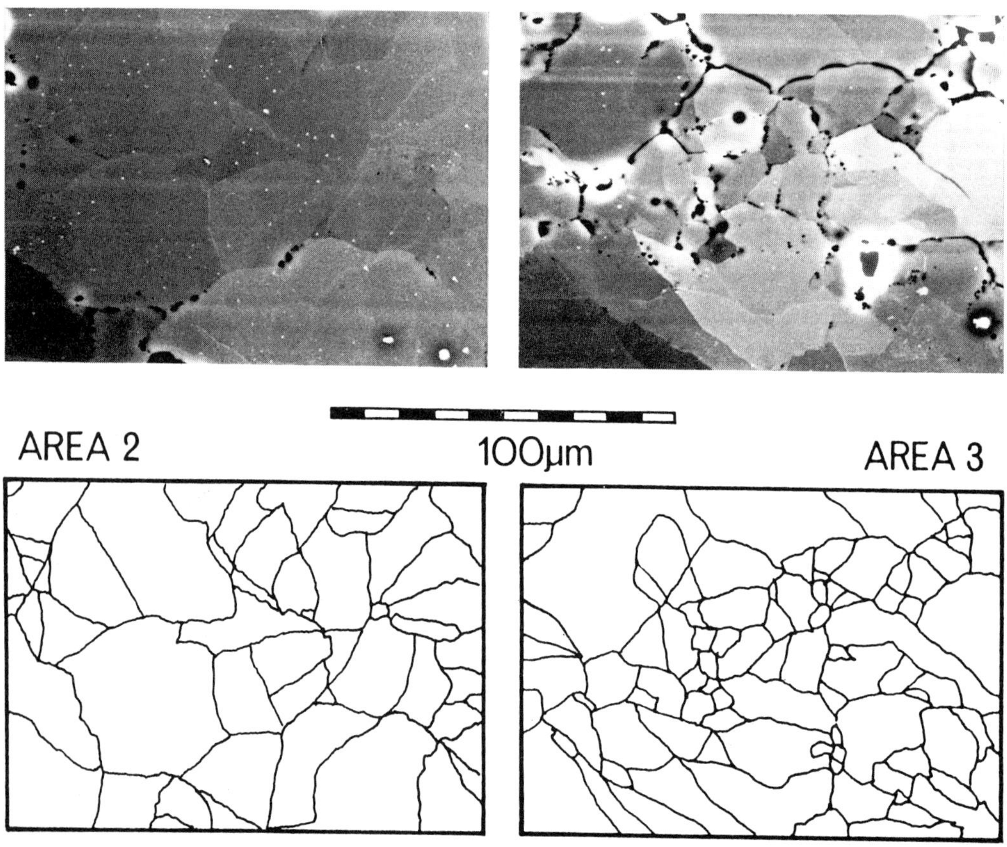

Fig. 3. Part of the BSE orientation contrast micrographs of areas 2 and 3 in Fig. 2. and line drawings of the subgrain boundaries for the same areas.

combination of subgrain rotation recrystallization and grain boundary migration (Prior 1988). ECPs show that a significant number of the subgrain boundaries imaged (from a random sample of about 10) have misorientations of a few degrees, and that very low angle (> 1°) boundaries are rare.

Microstructural stability

In both the mylonites studied the microstructures present arise from the operation of dynamic recrystallization during dislocation creep. There is a well defined decreases in the subgrain-size adjacent to the feldspar and garnet grains studied. This decrease is produced as quartz grains move into the differential stress and strain-rate perturbation adjacent to the rigid particle and from the microstructures preserved probably involves the generation and misorientation of new dislocation walls in addition to subgrain wall migration. It is not possible to assess the relative importance of these two processes during this subgrain-size decrease. The subgrain-size increase which takes place as material flows away from the rigid particle is likely to be controlled by subgrain wall migration.

Edward *et al.* (1982) show that the subgrain-size stress relationship may be perturbed by environmental parameters such as temperature and fluid activity. Temperature must have been constant across the few mm of specimens observed in this study. Fluid activity may vary locally in the vicinity of porphyroclasts of feldspar or garnet (Wintsch & Knipe 1983; Wheeler 1987). However quartz subgrain-size does not vary adjacent to small feldspar and garnets (those the same size as the quartz grains) suggesting that it is the increase in differential stress and strain-rate around relatively large rigid bodies which are the prime control on the local quartz subgrain size.

The data presented in this paper show clearly that the subgrain-size and the recrystallized grain-sizes preserved in mylonites do not necessarily preserve the peak differential stress conditions experienced. Subgrain and grain-sizes can respond to both increases and decreases in differential stress and strain-rate. The microstructures recorded reveal that differential stress and strain rate variations expected during deformation of material containing rigid particles are preserved. However, large scale readjustment of the microstructure after deformation, which would remove such grain-size and subgrain-size variations, has not taken place. The time at which the subgrain sizes have been frozen in is likely to be critically dependent upon the rates of change of deformation conditions (Knipe 1989).

Experimental data (Ross *et al.* 1980) shows that olivine subgrain sizes respond to increases in differential stress and strain rate but not to stress relaxation. Similarly experiments in polycrystalline Mg (White *et al.* 1985) show that peak recrystallized grain-sizes can be preserved. These data contrast with our observation that microstructural change does occur in natural stress relaxation. The conflicting data may reflect different materials and stress relaxation rates. The data presented in this paper are consistent with dynamic models of subgrain development in which subgrain-size is reduced by the formation of new boundaries and may increase by the mobility and mutual annihilation of boundaries (Takeuchi & Argon 1976; Edward *et al.* 1982; Poirer 1985) with substructure modification by the creep of individual dislocations within subgrains and subgrain boundaries. Such models reflect the realistic continuum of processes linking recovery and recrystallization.

There is sufficient geometrical and shear strain rate information to assess the stability of the microstructures and in the case of the Alpine Fault mylonites to estimate the rates at which the microstructural changes take place. Although the shear strain rate is amplified in the constricted region adjacent to the garnet (the magnitude of this increase is estimated later) the mean shear strain rate of the constricted region and the area encompassing the garnet, which does not deform, must be equivalent to the bulk strain-rate. Thus the time taken for flow along a flow line around the garnet can be approximated using the bulk shear strain rate. Figure 4 shows the geometrical model applied to calculate the time it has taken the quartz now at areas 5, 4, 3 and 2 to travel from a position relative to the garnet equivalent to the location of area 1 (see Fig. 2 to locate these areas). In each case the bulk shear strain, relative to a base line bisecting the garnet and parallel to the flow plane, involved in translating from locality one is calculated (Fig. 4) and the time taken estimated using the regional shear strain rate constraints.

The mean subgrain-size of areas 1, 2, 5 and 6 (14.3 μm: Fig. 5) is taken as the background subgrain-size and the mean subgrain-size of areas 3 and 4 (10.9 μm: Fig. 5) is taken for the high stress region adjacent to the garnet. These give a change in two dimensional subgrain-size (d) of 3.4 μm and a change in stereologically corrected three dimensional subgrain-size (D)

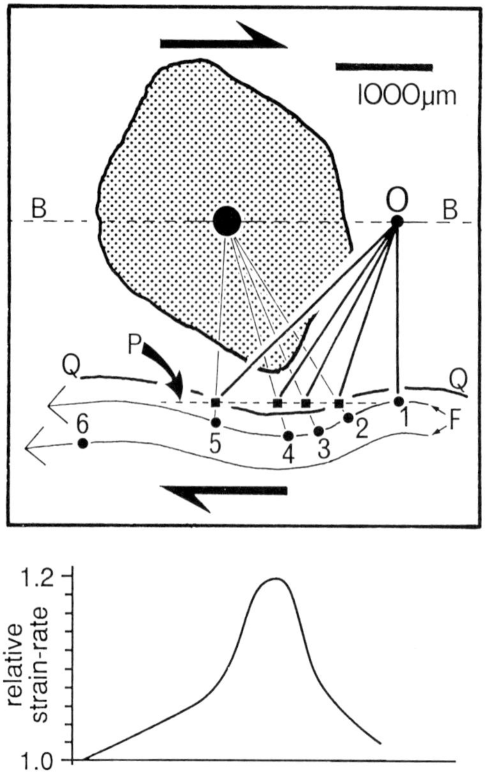

Fig. 4. Geometrical construction used to calculate the bulk shear strain associated with the flow of material from area 1 to areas 2, 3, 4, and 5 in the Alpine fault mylonite specimen. Sketch is traced from a BSE micrograph. Q–Q marks the edge of the quartz band (see Fig. 2). Garnet is stippled. Shear strains are calculated as the change in angle between a baseline (B–B), parallel to foliation and passing through the garnet centre, and a line perpendicular to this and passing through area 1. To make these calculations the deflecting effects of the garnet on particle flow paths (F) must be removed. Deflection due to the garnet is assumed to be radial from the garnet centre. Correction is made by projecting areas 2 to 5 along radial lines onto an 'undisturbed' particle path (P) constructed using the undeflected foliation orientations to the left and right of the image. The bulk shear strain involved in particle translation from area 1 to area 2 is given by tan (O_1 ^OP_2) where P_2 is the projection of 2 onto line P. Similar equations are used for areas 3 to 5. The lower graphs shows the shear-strain rate of points in the constriction relative to the shear strain-rate at locality 6. These data are calculated from the ratio of the flow line separation at each point and locality 6.

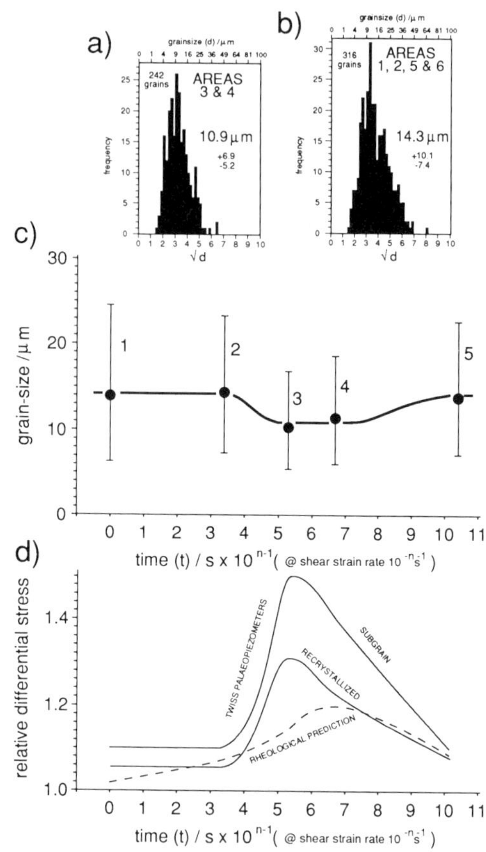

Fig. 5. Cumulation of subgrain-size data and interpretive subgrain-size time relationships for the flow of quartz around garnet in the Alpine fault mylonite specimen (Fig. 2). (**a**) Root subgrain-size ($\sqrt{d}$ in μm) versus frequency histogram for quartz close to the garnet margin (areas 3 & 4). (**b**) Root subgrain-size ($\sqrt{d}$ in μm) versus frequency histogram for background quartz (areas 1, 2, 5 & 6). Histograms show the subgrain-size equivalents of the mean and standard deviation of the $\sqrt{d}$ distribution. Data are not stereologically corrected. (**c**) Graph of subgrain-size (spots show mean and vertical bars standard deviation of the $\sqrt{d}$ distribution) for areas 1 to 5 (see Figs 2 & 4) against the time calculated for flow from area 1 to each of these areas. (**D**) Graphs showing estimates of differential stresses relative to the differential stress at measurement point 6 calculated from two palaeopiezometers and from rheological considerations (see text).

of 4.3 μm. This decrease in the subgrain-size is fully accomplished between areas 2 and 3, indicating a maximum time of 1.9×10^9 s at a shear strain rate of 10^{-10} s^{-1} or 1.9×10^{11} s at a shear strain rate of 10^{-12} s^{-1}. The minimum rate of change in subgrain-size during increasing stress is calculated (for stereologically corrected data) at 2.3×10^{-9} μm s^{-1} (71500 μm Ma^{-1}) at a shear strain rate of 10^{-10} s^{-1} or 2.3×10^{-11}

$\mu m\ s^{-1}$ (715 $\mu m\ Ma^{-1}$) at a shear strain rate of $10^{-12}\ s^{-1}$. This transition occurs during increasing differential stress and strain-rate and reflects the period needed to establish the new microstructure. The equivalent change during decreasing differential stress and strain-rate occurs between areas 4 and 5 and may take twice as long as the up-stress transition although this would be a maximum value since there are no data points between 4 and 5. If, as is likely, this increase in the subgrain-size is controlled by subgrain boundary migration then the minimum net rate of subgrain boundary migration will be equivalent to the rate of change in subgrain-size (between areas 4 and 5) and is calculated (for stereologically corrected data) at $1.2 \times 10^{-9}\ \mu m\ s^{-1}$ (36700 $\mu m\ Ma^{-1}$) at a shear strain rate of 10^{-10} or $1.2 \times 10^{-11}\ \mu m\ s^{-1}$

The rates of microstructural changes calculated here are conservative. The shear strain rates used are minima and the transition in grain-size may occur in only part of the time in the transition time interval. In addition the rates of subgrain migration calculated relate to the net effect of boundary migrations rather than the rate of migration of individual boundaries. Since the microstructure is dynamic, subgrain boundary migration is likely to be occurring to both increase and decrease grain-sizes and the mean rate will be slower that for any individual boundary.

The amplification of shear-strain rate in the constricted region around a rigid body can be estimated from the changing separation of flow lines. The shear-strain rate pattern around the garnet is shown in Fig. 4. By application of a flow law for quartz to the shear-strain rate amplification pattern the differential stress amplification can also be estimated. The differential stress amplification around the garnet is estimated using Paterson de Luan's (this volume) preferred stress exponent of 4 (Fig. 5D). Absolute differential stress magnitudes and amplifications can be estimated by the application of palaeopiezometers to subgrain-size data. The differential stress amplifications calculated using the subgrain and recrystallized grain size palaeopiezometers of Twiss (1977; 1986) are shown in Fig. 5D. The differential stress amplification curve for the recrystallized grain-size palaeopiezometer fits the curve calculated from rheological considerations much more closely than the data from the subgrain palaeopiezometer suggesting, albeit tentatively, that the recrystallized grain-size palaeopiezometer gives more realistic results here. Using the recrystallized grain-size palaeopiezometer (Twiss 1977) the differential stress magnitude calculated for the cumulated, stereologically corrected subgrain-size data of the background region (localities 1, 2, 5 and 6) is 84 MPa [+54 −26] and that for the constriction around the garnet (areas 3 and 4) is 101 MPa [+56 − 29]. These data give differential stress rates during increasing stress of 9×10^{-9} MPa s^{-1} at a bulk shear strain rate of $10^{-10} s^{-1}$ and 9×10^{-11} MPa s^{-1} at $10^{-12}\ s^{-1}$. Similarly differential stress rates during decreasing stress are estimated at 5×10^{-9} MPa s^{-1} at a bulk shear strain rate of $10^{-10} s^{-1}$ and 5×10^{-11} MPa s^{-1} at $10^{-12} s^{-1}$.

In the case of the Alpine fault mylonites the rate of change of the differential stress at the end of the dislocation creep deformation must have been faster than 5×10^{-9} MPa s^{-1} to 5×10^{-11} MPa s^{-1} which is the estimated stress rate range which did allow adjustment of the microstructure during flow around the rigid particle. These results compare well to theoretical predictions of Knipe (1989) which suggest that at 400°C differential stress must drop at a rate faster than 10^{-10} MPa s^{-1} to prevent the subgrain-size maintaining equilibrium with the changing stress conditions.

If the calculated rates of subgrain boundary migration during decreasing stress can be applied to dynamic equilibrium changes at uniform stress then an estimate of quartz microstructural recycling times in the Alpine fault mylonites can be made. For a 20 μm mean subgrain-size (d) the minimum time required to grow a 20 μm subgrain, which will be equivalent to the time needed to recycle the microstructure, will be 540 years and 54 000 years at shear strain rates of $10^{-10}\ s^{-1}$ and $10^{-12}\ s^{-1}$ respectively using the rates calculated for decreasing differential stress and strain rate. These estimated microstructural recycling times are considerably less than the 1–2 Ma required (Wellman 1979) for uplift from the maximum burial depth of *c.* 20 km (Cooper 1980) suggesting that subgrain sizes will maintain equilibrium with fault zone differential stresses and strain rates as they deform during an uplift path from the middle crust to shallow levels. The uplift rate associated with the Alpine Fault Zone is rapid, >10 mm Ma^{-1} (Wellman 1979), and the thermal model of Koons (1987) indicates that the temperatures needed for crystal plastic deformation (300–400°C) may be present within a few kilometres of the surface. Both these features indicate that deformation after the crystal plastic straining and mylonite formation will be short lived and this enhances the chances of preserving the deformation microstructures characteristic of the last stages of dislocation creep. In the final stages of uplift rapid temperature reduction allows major displacement to be accommodated by cataclastic processes and the associated stress drop freezes in the mylonite microstructure.

Similar rapid stress drops and high differential stress and strain-rate gradients in the wall rocks adjacent to active shear zones may have preserved the quartz microstructures in the PDFZ. This crustal scale shear zone was a km wide low angle extensional fault within a hot, uplifting and extending segment of the deep crust (Handy 1986). Amphibolite facies mylonites grade successively into greenschist facies mylonites and cataclasites at the contacts with intermediate depth crustal rocks and the preservation of this zonation of dynamic microstructures is attributed to the very high cooling rates and/or stress drops during the final stages of highly localized deformation.

The rates of microstructural change which we have estimated here provide some constraint with which to assess the period of shear zone evolution a given set of quartz microstructures may represent. In a given shear zone the age of preserved mylonite microstructures will be critically dependent upon the mylonite temperature-differential stress-strain rate history as well as the temperature and strain-rate dependencies of creep of the minerals in the mylonite. As a generalization, slow changes in deformation conditions, particularly differential stress and strain-rate, will favour continuous microstructural re-equilibration, whilst faster changes will allow the microstructure to be frozen in. Microstructures formed under high temperature, deep crustal conditions will require rapid uplift and cooling to avoid equilibration with stresses at shallower levels. This is only possible if the deformation is partitioned into weaker shear zones. Since mechanisms of microstructural recovery are thermally activated, decreasing temperatures during uplift favour higher localized stresses and so tend to increase stress and strain partitioning on all scales.

Work in the Alpine fault mylonites was initiated as part of a NERC studentship to D. J. P. at the University of Leeds. M. R. H. acknowledges support of the Swiss NSF (project 2971–0.88). Staff of the Westland National Park, were most helpful during fieldwork. We thank T. Masuda and E. Rutter for useful discussions in the developmental stages of this work, an anonymous reviewer for helpful comments and B. Smith for a particularly rigourous and perceptive review. SEM work benefitted from the advice of G. Lloyd, J. Harrington and T. Nichells. L. Miralles is thanked for help with collection of digitized data.

References

BELL, T. H. & ETHERIDGE, M. A. 1973. Microstructure of mylonites and their descriptive terminology. *Lithos*, **6**, 337–338.

BOULLIER, A. M. 1980. A preliminary study of the behaviour of brittle minerals in a ductile matrix: example of zircon and feldspars. *Journal of Structural Geology*, **2**, 211–217.

CARTER, R. M. & NORRIS, R. J. 1976. Cainozoic history of Southern New Zealand: an accord between geological observations and plate tectonic predictions. *Earth and Planetary Science Letters*, **31**, 85–94.

CHRISTIE, J. M., ORD, A. & KOCH, P. S. 1980. Relationship between recrystallized grain-size and flow-stress in experimentally deformed quartzite. *Transactions of the American Geophysical Union*, **61**, 377.

COOPER, A. F. 1980. Retrograde alteration of chromian kyanite in metachert and amphibolite whiteschist from the Southern Alps, New Zealand, with implications for uplift on the Alpine Fault. *Contributions to Mineralogy and Petrology*, **75**, 153–164.

EDWARD, G. H., ETHERIDGE, M. A. & HOBBS, B. E. 1982. On the stress dependence of subgrain size. *Textures and microstructures*, **5**, 127–152.

ETHERIDGE, M. A. & WILKIE, J. C. 1979. Grainsize reduction, grain boundary sliding and the flow strength of mylonites. *Tectonophysics*, **58**, 159–178.

EXNER, M. E. 1972. Analysis of grain and particle size distribution in metallic materials. *International Metallurgical Reviews*, **17**, 25–42,

FYNN, G. W. & POWELL, W. J. A. 1979. *The cutting and polishing of electro-optic materials.* Adams Hilger, London.

HANDY, M. R. 1986. *The structure and rheological evolution of the Pogallo Fault Zone. A deep crustal dislocation in the Southern Alps of northwestern Italy.* PhD thesis, Basel.

—— 1989. Deformation regimes and the rheological evolution of fault zones in the lithosphere: the effects of pressure, temperature, grainsize and time. *Tectonophysics*, **163**, 119–152.

—— 1990. The solid state flow of polymineralic rocks. *Journal of Geophysical Research*, **95**, 8647–8661.

KAMP, P. J. 1986. Late Cretaceous – Cenozoic tectonic development of the southwest Pacific region. *Tectonophysics*, **121**, 225–251.

KNIPE, R. J. 1989. Deformation mechanisms, recognition from natural tectonites. *Journal of Structural Geology*, **11**, 127–146.

KOONS, P. O. 1987. Some thermal and mechanical consequences of rapid uplift: the example of the Southern Alps, New Zealand. *Earth and Planetary Science Letters*, **86**, 307–319.

LANDIS, C. A. & COOMBS, D. S. 1967. Metamorphic belts and orogenesis in southern New Zealand. *Tectonophysics*, **4**, 501–518.

LISTER, G. S. & PRICE, G. P. 1978. Fabric development in a quartz feldspar mylonite. *Tectono-*

physics, **49**, 37–38.

—— & WILLIAMS, P. F. 1983. The partitioning of deformation in flowing rock masses. *Tectonophysics*, **92**, 1–33.

LLOYD, G. E. 1987. Atomic number and crystallographic contrast images with the SEM: A review of backscattered techniques. *Mineralogical Magazine*, **51**, 3–19.

MASUDA, T. & ANDO, S. 1988. Viscous flow around a rigid spherical body: A hydrodynamical approach. *Tectonophysics,* **148**, 337–346.

MERCIER, J. C. C. ANDERSON, D. A. & CARTER, N. L. 1977. Stress in the lithosphere. Inferences from steady state flow of rocks. *Pure and Applied Geophysics*, **115**, 199–226.

MEANS, W. D. 1981. The concept of steady state foliation. *Tectonophysics*, **78**, 179–199.

MITRA, S. 1978. Microscopic deformation mechanisms and flow laws in quartzites within the south anticline. *Journal of Geology*, **86**, 129–152.

ORD, A. & CHRISTIE, J. M. 1984. Flow stresses from microstructures in mylonitic quartzites from the Moine Thrust Zone, Assynt area, Scotland. *Journal of Structural Geology*, **6**, 639–654.

PATERSON, M. S. & LUAN, F. C. 1990. Quartz rheology under geological conditions. *This volume*.

POIRIER, J. P. 1985. *Creep of crystals*. Cambridge University Press.

PRICE, G. 1978. Study of the heterogenous fabric and texture within a quartz-feldspar mylonite using the photometric method. *Geological Society of America Bulletin*, **89**, 1359–1372.

PRIOR, D. J. 1988. *Deformation processes in the Alpine Fault Mylonites, South Island, New Zealand*. PhD Thesis, University of Leeds.

—— 1989. Deformation processes in the Alpine Fault Mylonites, South Island, New Zealand: *New Zealand Journal of Geology and Geophysics*, **31**, 526.

ROSS, J. V., AVE LALLEMENT, H. G. & CARTER, N. L. 1980. Stress dependence of recrystallized grain and subgrain size in olivine. *Tectonophysics*, **70**, 39–61.

SIBSON, R. H., WHITE, S. H. & ATKINSON, B. K. 1979. Fault rock distribution and structure within the Alpine Fault Zone: A preliminary report: *In*: WALCOTT, R. I. & CRESSWELL, M. M. (eds) *Origin of the Southern Alps*. Bulletin of the Royal Society of New Zealand, **18**, 55–65.

TAKEUCHI, S. & ARGON, A. S. 1976. Steady state creep of single phase crystalline matter at high temperatures. *Journal of Material Science*, **11**, 1542–1566.

TWISS, R. J. 1977. Theory and applicability of a recrystallized grainsize palaeopiezometer. *Pure and Applied Geophysics*, **115**, 227–244.

—— 1986. Variable sensitivity piezometric equations for dislocation density and subgrain diameter and their relevance to olivine and quartz. *American geophysical Union, Geophysical Monograph*, **36**, 247–263.

WALCOTT, R. I. 1984. Reconstruction of the New Zealand region for the Neogene. *Palaeogeography, Palaeoclimatology, Palaeoecology*, **46**, 217–231.

WELLMAN, H. W. 1979. An uplift map for the South Island of New Zealand and a model for the uplift of the Southern Alps: *In*: WALCOTT, R. I. & CRESSWELL, M. M. (eds) *Origin of the Southern Alps*. Bulletin of the Royal Society of New Zealand, **18**.

WHITE, S. H. 1984. Brittle deformation within ductile fault zones. *Proceedings 27th International Geological Congress*, **7**, 327–350.

——, BURROWS, S. E., CARRERAS, J., SHAW, M. D. & HUMPHREYS, F. J. 1980. On mylonites in ductile shear zones. *Journal of Structural Geology*, **2**, 175–187.

——, DRURY, M. R., ION, S. E. & HUMPHREYS, F. J. 1985. Large strain deformation studies using polycrystalline magnesium as a rock analogue, part I: Grainsize palaeopiezometry in mylonite zones. *Physics of the Earth and Planetary Interiors*, **40**, 201–207.

WHEELER, J. 1987. The significance of grain-scale stresses in the kinetics of metamorphism. *Contributions to Mineralogy and Petrology*, **17**, 397–404.

WINTSCH, R. P. & KNIPE, R. J. 1983. Growth of a zoned plagioclase porphyroblast in a mylonite. *Geology*, **11**, 360–363.

Microstructure in hornblende of a mylonitic amphibolite

WERNER SKROTZKI

Institut für Geologie und Dynamik der Lithosphäre, Universität Göttingen, D-3400 Göttingen, Goldschmidt-Str. 3, FRG

Abstract: The microstructure in a dynamically recrystallized amphibolite of a deep crustal shear zone from the Ivrea Zone, NW Italy, has been investigated by conventional and high resolution transmission electron microscopy. The microstructure in the amphibole phase consists of recrystallized grains, subgrains, free dislocations and various stacking faults bounded by partial dislocations. The Burgers vectors of the dislocations are [001], [100], 1/2 <110> and very likely <101>. Dislocations with [100] Burgers vector and [001] line direction are mainly arranged in low angle tilt boundaries lying parallel to (100) planes. Faceting of these subgrain boundaries is produced by 1/2 <110> edge dislocations dissociated on (010). The free dislocations are mainly non-dissociated [001] and [100] dislocations gliding on (100) and (010) planes, respectively. It is assumed that [001] and [100] dislocations recombined to form <101> dislocations which subsequently dissociate in (010). The stacking faults extended on (010) planes either have pyroxene or sheet silicate character. They act as obstacles for dislocation motion on the main slip system (100)[001]. High resolution imaging of the dislocation cores suggests these areas consist of lower density material providing channelways for fast diffusion which may be responsible for the deformation-induced compositional changes found within amphibolite shear zones.

Deformation-induced microstructures in amphiboles have previously received relatively little attention. Transmission electron microscopic results have been reported by Rooney *et al.* (1975) and Morrison-Smith (1976) on the deformation features in experimentally deformed hornblende. In amphibole single crystals compressed along [001], at temperatures up to 600°C, confining pressures up to 2 GPa and strain rates of about 10^{-5} s^{-1}, they found (101) [10$\bar{1}$] twins and dislocations of the (100)[001] slip system within the twin lamellae. Other slip systems have not been identified unambiguously. On the other hand, in amphibolites naturally deformed at different $P-T$ conditions Brodie (1981), Biermann & van Roermund (1983), Brodie & Rutter (1985) and Cumbest *et al.* (1989*a*, *b*) observed a well-defined subgrain structure parallel to {hk0} planes. Within the subgrains there are free [001] dislocations, mainly screws, gliding on (100), {110} and (010) planes. In addition planar defects parallel to (010) and bounded by partial dislocations are observed. Occasionally in the lower temperature range (< 600°C) also (100) twin lamellae exist (Cumbest *et al.* 1989*b*).

To analyse the dislocation microstructure in naturally deformed hornblende in more detail and to explore further the deformation mechanisms of this type of rock-forming mineral, a study has been made of a mylonitic amphibolite from a shear zone in the Ivrea Zone, NW-Italy (near Premosello, Valle d'Ossola). According to Brodie & Rutter (1985, 1987) the formation of shear zones of this type is due to deep crustal extensional faulting in the Ivrea Zone at relatively high temperatures, > 650°C. ^{40}Ar-^{39}Ar age spectra indicate that this extensional event occurred prior to 280 Ma and predates the tilting of the Ivrea zone (Brodie *et al.* 1989).

The rock is comprised dominantly of brown hornblende. Recrystallization produced a fine-grained porphyroclastic mylonite with porphyroclasts and recrystallized matrix having a grain size in the order of 1 mm and 0.05 mm, respectively. The microstructure of this high temperature mylonite has been studied by conventional and high resolution transmission electron microscopy (CTEM and HRTEM) on ion-milled samples using a Philips EM 400T operating at 120 kV. Preliminary TEM investigations on a similar amphibolite of the Valle d'Ossola (Anzola quarry) have already been reported by Brodie (1981) and Brodie & Rutter (1985).

Microstructural results

The microstructure of the hornblende investigated is characterized by a typical creep microstructure consisting of subgrain boundaries, free dislocations (dislocation density in the order of

From Knipe, R. J. & Rutter, E. H. (eds), 1990, *Deformation Mechanisms, Rheology and Tectonics*, Geological Society Special Publication No. 54, pp. 321–325.

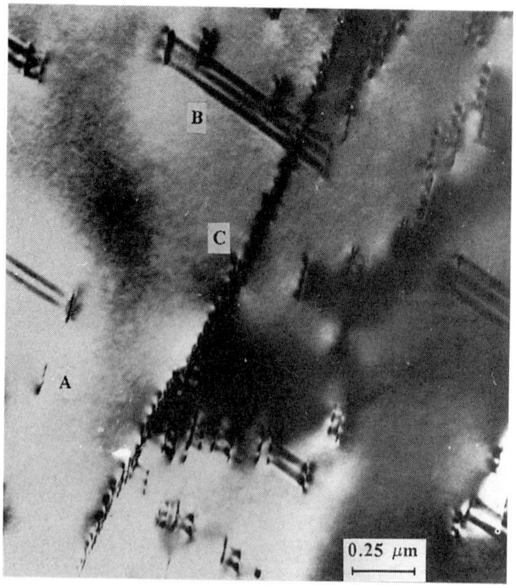

Fig. 1. Microstructure in amphiboles: perfect dislocations (A), stacking faults on (010) (B), subgrain boundary on (100) (C). Bright-field image taken with **g** = 111 reflection.

10^8 cm^{-2}) and various stacking faults bounded by partial dislocations (Fig. 1). No obvious difference has been found between porphyroclasts and matrix, thereby suggesting dynamic recrystallization. The stacking faults have variable width. The Burgers and displacement vectors of these defects have been determined directly from HRTEM images viewed along [001] and by CTEM contrast analyses. Non-dissociated dislocations with [100] Burgers vector (Burgers circuit analysis gives b = [100] sinβ) and [001] line direction are mainly arranged in low-angle tilt boundaries lying parallel to (100) planes (Fig. 2). Faceting of these subgrain boundaries is produced by 1/2 <110> edge dislocations with [001] line direction (Burgers circuit analysis) dissociated on (010). The stacking fault is a single chain multiplicity fault (Czank & Liebau, 1980, Fig. 3). The dissociation reaction and the fault vector are given in Table 1. Only the component of the Burgers and fault vectors normal to [001] can be determined from the HRTEM image taken, the component along [001] is suggested by the fault-type. The free dislocations observed are mainly [001] dislocations with long screw and short edge segments along [001] and [010], respectively (Fig. 4). Hence they belong to the (100)[001] slip system. The Burgers vector of this type of dislocation has been determined by conventional contrast analysis using the $\mathbf{g} \cdot \mathbf{b} = 0$ invisibility criterion. Within the resolution of weak beam TEM [001] dislocations are not dissociated. Other types of free dislocations observed are [100] and most likely <101> dislocations with [001] line direction (Burgers circuit gives only the [100] sin β component), the latter being dissociated on (010) (Figs 5 and 6). The stacking faults produced by dissociation of <101> dislocations are either triple (Fig. 5), two single or quadruple chain multiplicity faults (Fig. 6). Burgers and fault vectors of these defects are given in Table 1. <101> dislocations are assumed to be the reaction product of [001] and [100] dislocations. The slip systems determined from the Burgers vector and line direction of the dislocations observed are (100)[001], (010)[100], (010)<101> and {110} 1/2 <110>, the first having the highest dislocation density and therefore is assumed to be the most active. Because of the predominant screw orientation of [001] dislocations slip on {hk0} planes cannot be excluded. So far [001] glide loops on planes other than (100) have not been observed.

The dislocation cores of dislocations in amphiboles (except those of [001] dislocations) in high resolution are found to have a rather open structure, similar to observations by Carter *et al.* (1978) on a lateral twin boundary in spinel. It is suggested that this results from a lower density in the dislocation core rather than from preferential ion-milling. No changes were observed during electron radiation.

Table 1. *Basic types of chain multiplicity faults in amphiboles, displacement vectors of partial dislocations and reactions of their formation.*

Chain multiplicity fault	Displacement-vector	Dislocation reaction	Figure
1	[0, −1/4, 1/2]	[−1/2, 1/2, 0] → [−1/2, 1/4, 1/2] − [0, −1/4, 1/2]	2c
2 × 1	[1/2, 0, 1/2]	[1, 0, 1] → [1/2, 0, 1/2] + [1/2, 0, 1/2]	6a
3	[1/2, 1/4, 0]	[−1, 0, 1] → [−1/2, 1/4, 1] − [1/2, 1/4, 0]	5
4	[1/2, 0, 1/2]	[1, 0, 1] → [1/2, 0, 1/2] + [1/2, 0, 1/2]	6b

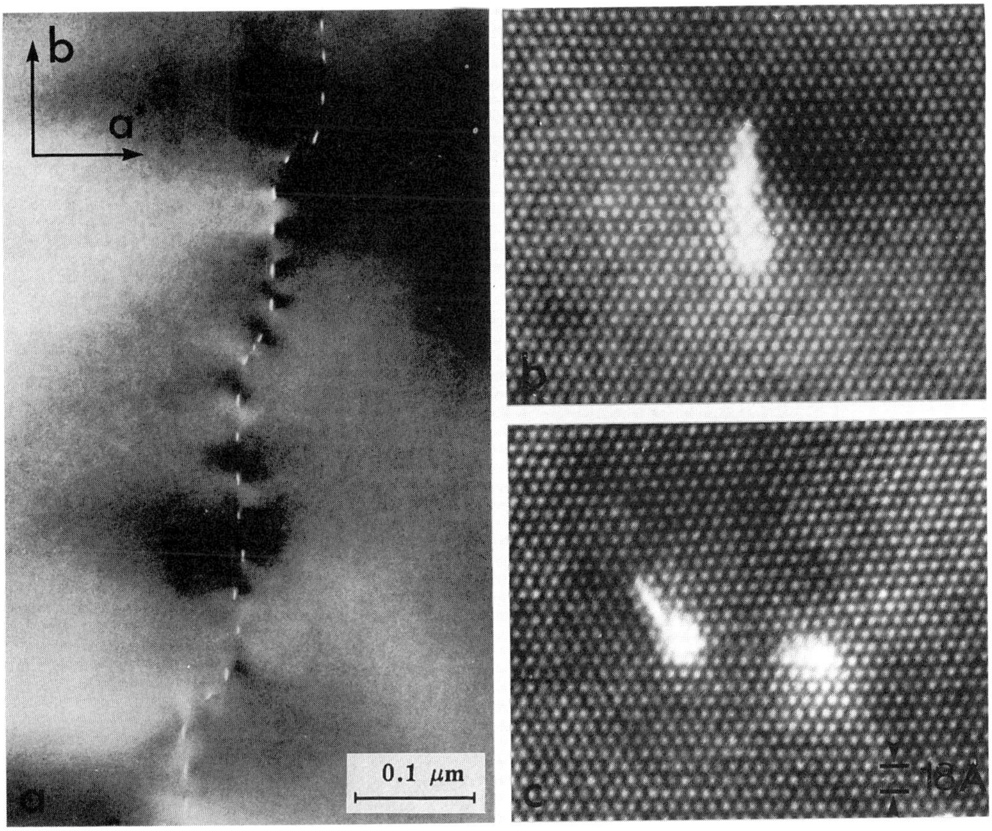

Fig. 2. (**a**) Edge-on view on a faceted (100) low-angle tilt boundary. Multibeam image taken along [001]. (**b**) HRTEM image of a [100] dislocation of the long facets in (a). (**c**) HRTEM image of a [−1/2, 1/2, 0] edge dislocation of the short facets in (a). The dissociation reaction is 1/2 [−1, 1, 0] → [−1/2, 1/4, 0] + [0, 1/4, 0]. The stacking fault between the two partials is a single chain fault. Crystal directions in (b) and (c) are the same as in (a).

Discussion

Slip in high temperature amphibolites mainly takes place on the (100)[001] system with the (010) stacking faults acting as obstacles for dislocation motion (Fig. 4). As a consequence, long [001] screw dislocations are extended along the stacking faults. 'Breaking through' the stacking faults is indicated by bowed-out dislocation segments. During motion [001] screws may react with [100] near edges to [101] dislocations and subsequently dissociate on (010). The reaction products act as obstacles for further slip. Apparently, the stacking faults play an important role in the work-hardening of amphiboles. The nature of the stacking faults is discussed below. Besides (100)[001] in amphiboles at high temperatures other slip systems are active which have not been reported before. All slip planes found have the chain axis as zone axis and therefore tetrahedral (covalent) bonds will not be broken by the movement of dislocations. During high temperature deformation recovery is taking place as is indicated by the subgrain formation. The predominant deformation mechanism is dislocation creep.

Amphiboles may be considered as a transition phase from pyroxenes to sheet silicates. Combining every second single chain in pyroxenes to double chains yields the amphibole structure provided the necessary ion exchange takes place. The combination of double chains finally leads to a sheet structure. The transition from one structure to another is accomplished by partial dislocations with 1/2 <101> Burgers vector moving along (010) planes (Chisholm

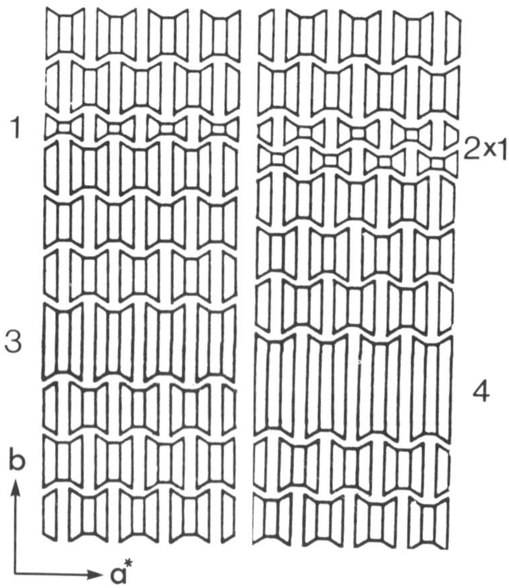

Fig. 3. I-beam representation of the basic chain multiplicity faults in the amphibole structure: single, two single, triple and quadruple chains. (After Thompson 1970).

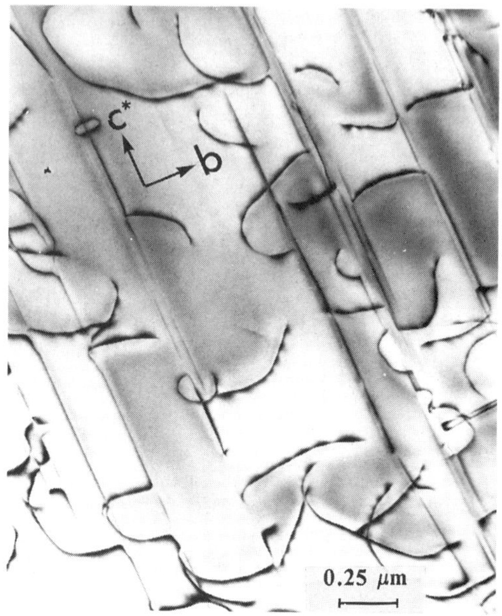

Fig. 4. Dislocation microstructure observed the (100) plane. The interaction of [001] dislocations with stacking faults parallel to (010) is obvious. Bright-field image taken with **g** = 002 reflection.

1973). Because of the transient position of amphiboles, defects are possible in these minerals which have pyroxene and sheet silicate character. The basic types of these defects (chain multiplicity faults) are single, two single, triple and quadruple chains (Fig. 3). Chain multiplicity faults have already been described previously by Veblen & Buseck (1981) and Maresch & Czank (1983). Whilst in the latter instances the faults might be related to growth, here it is demonstrated that another way of forming them is by deformation and subsequent reworking of the dislocation substructure. It should be noted that the faults are not simple stacking faults because there is a change in composition.

The dislocation cores in amphiboles are strongly 'eroded'. If this is due to a region of lower density and not produced by preferential ion-milling, then such an open structure may be a channel for fast diffusion of the ions necessary for phase transitions. It may be also responsible

Fig. 5. HRTEM image of a triple chain multiplicity fault bounded by partial dislocations P and Q with [−1/2, 1/4, 0] and [−1/2, −1/4, 0] Burgers vector components normal to [001], respectively. The total Burgers vector of the defect is assumed to be [101]. Crystal directions and scale are the same as in Fig. 2.

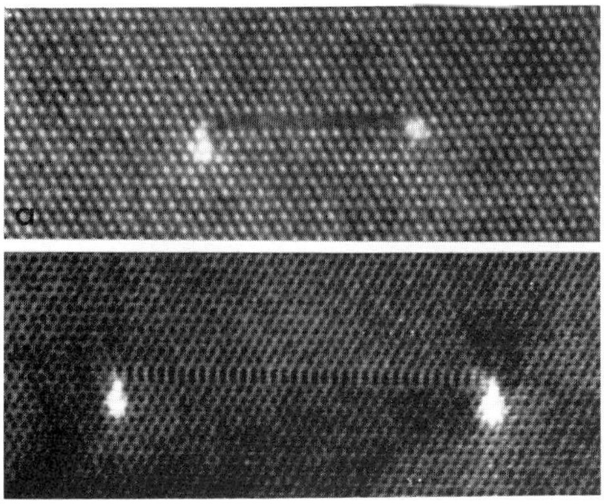

Fig 6. HRTEM images of chain multiplicity faults consisting of two single chains (**a**) and a quadruple chain (**b**). The defects are bounded by partial dislocations with [1/2, 0, 0] Burgers vector component normal to [001]. The total Burgers vector of the defects is assumed to be [101]. Crystal directions and scale are the same as in Fig. 2.

for the deformation-induced compositional changes within shear zones found by Brodie (1981).

References

BIERMANN, C. & VAN ROERMUND, H. L. M. 1983. Defect structures in naturally deformed clinoamphiboles-A TEM, study. *Tectonophysics*, **95**, 267–278.

BRODIE, K. H. 1981. Variation in amphibole and plagioclase composition with deformation. *Tectonophysics*, **78**, 385–402.

—— & RUTTER, E. H. 1985. On the relationship between deformation and metamorphism, with special reference to the behaviour of basic rocks. *In*: THOMPSON, A. B. & RUBIE, D. C. (eds) *Metamorphic Reactions: Kinetics, Textures and Deformation*. Advances in Physical Chemistry, **4**. Springer, Berlin, 138–179.

—— & —— 1987. Deep crustal extensional faulting in the Ivrea zone of northern Italy. *Tectonophysics*, **140**, 193–212.

——, REX, D. & RUTTER, E. H. 1989. On the age of deep crustal extensional faulting in the Ivrea zone, northern Italy. *In*: COWARD, M. P., DIETRICH, D. & PARK, R. G. (eds), *Alpine Tectonics*. Geological Society, London Special Publication, **45**, 203–210.

CARTER, C. B., ELGAT, Z. & SHAW, T. M. (1987). Lateral twin boundaries in spinel. *Philosophical Magazine*, A, **55**, 21–38.

CHISHOLM, J. E. 1973. Planar defects in fibrous amphiboles. *Journal of Materials Science*, **8**, 475–483.

CUMBEST, R. J., DRURY, M. R., VAN ROERMUND, H. L. M. & SIMPSON, C. 1989*a*. Burgers vector determination in clinoamphibole by computer simulation. *American Mineralogist*, **74**, 586–592.

——, ——, —— & —— 1989*b*. Dynamic recrystallization and chemical evolution of clinoamphibole from Senja, Norway. *Contributions to Mineralogy and Petrology*, **101**, 339–349.

CZANK, M. & LIEBAU, F. 1980: Periodicity faults in chain silicates: A new type of planar lattice fault observed with high resolution electron microscopy. *Physics and Chemistry of Minerals*, **6**, 85–93.

MARESCH, W. V. & CZANK, M. 1983. Problems of compositional and structural uncertainty in synthetic hydroxyl-amphiboles; with an annotated atlas of the realbau. *Periodico di Mineralogia – Roma*, **52**, 463–542.

MORRISON-SMITH, D. J. 1976. Transmission electron microscopy of experimentally deformed hornblende. *American Mineralogist*, **61**, 272–280.

ROONEY, T. P., RIECKER, R. E. & GAVASCI, A. T. 1975. Hornblende deformation features. *Geology*, **3**, 364–366.

THOMPSON, J. B., JR. 1970. Geometrical possibilities for amphibole structures: Model biopyriboles (abs) *American Mineralogist*, **55**, 292–293.

VEBLEN, D. R. & BUSECK, P. R. 1981. Hydrous pyriboles and sheet silicates in pyroxenes and uralites: intergrowth microstructures and reaction mechanisms. *American Mineralogist*, **66**, 1107–1134.

Albite deformation within a basal ophiolite shear zone

JOSEPH CLANCY WHITE

Centre for Deformation Studies in the Earth Sciences, Department of Geology, University of New Brunswick, Fredericton, NB Canada E3B 5A3

Abstract: Albite mylonites at the base of the White Hills Peridotite ophiolite fragment at St. Anthony, Newfoundland formed during obduction at lower crustal $P-T$ conditions. Optical microstructures preserve a history of intracrystalline distortion, recovery and cyclic dynamic recrystallization associated with hot-working. The crystallographic fabric is consistent with shear of the polycrystalline aggregate dominated by glide on (010) planes, as is also indicated by TEM observations. Slip systems implied by trace analysis and invisibility experiments using TEM are **b** = [101](010), **b** = [001](010) and **b** = 1/2[112]($20\bar{1}$). Dissociated dislocations have separations ranging from 8.3–15 nm. These separations are notably smaller than in more calcic plagioclases and indicate a relatively higher ductility for albite in the case of cross-slip controlled creep. The observed dislocation densities relate to a high stress/strain rate pulse imposed on the dominant high-temperature creep textures.

A primary goal of structural geology is the characterization of the rheological behaviour of minerals over the widest possible range of natural deformation conditions. A problem inherent to this goal is the difficulty faced in accessing characteristic, deep-level deformation environments. Displaced ophiolitic material from St Anthony, Newfoundland affords such an opportunity to examine deformation associated with lithospheric scale tectonic transport.

The allochthonous St Anthony complex (SAC), located at the northern extremity of the western Newfoundland ophiolite belt is a well-documented example of displaced oceanic lithosphere (Jamieson 1981, 1986 and references therein). The complex comprises the White Hills peridotite (WHP), an ultramafic massif of mantle origin, and a structurally underlying inverted metamorphic sequence (dynamothermal aureole) floored by an emplacement thrust (see Jamieson 1986, p. 16, fig. 3). Within the basal section of the WHP, peridotite mylonites define a high-strain zone that has been associated with the earliest stages of displacement leading to obduction (Calon 1980; Jamieson 1986). Alkaline rocks of igneous origin at the base of the peridotite mylonites (Jamieson & Talkington 1980) include syenitic mylonites that locally are pure albite and are considered part of the basal mylonite structural unit (Jamieson 1981). Syntectonic temperature and pressure estimates for the peridotite mylonites are 850–1050°C and 850–1050 MPa and for the alkaline complex, 775–900°C (Jamieson 1986).

The fortuitous occurrence of polycrystalline, monomineralic albite rocks deformed under the relatively extreme $P-T$ conditions observed in the WHP shear zone provides an opportunity both to characterize the micromechanical behaviour without the complications of additional phases and to extend the field of observed natural behaviour of this feldspar. Previous studies of naturally-deformed albite examined material from substantially lower maximum $P-T$ conditions than are indicated in the WHP shear zone (e.g. Marshall & Wilson 1976; Fitz Gerald *et al.* 1983), while high temperature rheological behaviour and corresponding defect analyses are primarily derived from experimental studies (e.g. Marshall & McLaren 1977; Tullis & Yund 1980).

Study material

The mylonitized albite ranges in composition from An_0-An_3. X-ray powder diffraction data indicate the material now to be highly ordered, low albite. The microstructure of the albite was examined using optical, universal stage and electron microscopy. Transmission electron microscopy (TEM) observations were made using ion-thinned samples in a Philips EM400T operated at 120 keV.

Deformation microstructures and fabric

The optical microstructures are typical of hot-working, recording a history of intracrystalline distortion and progressive grain size reduction

From Knipe, R. J. & Rutter, E. H. (eds), 1990, *Deformation Mechanisms, Rheology and Tectonics*, Geological Society Special Publication No. 54, pp. 327–333.

Fig. 1. Large relict grains in various stages of recrystallization, surrounded by dynamically recrystallized matrix. Scale bar is 1 mm. X-polarized light.

by cyclic dynamic recrystallization (Fig. 1). Large, elongate relict grains up to 3 cm long are surrounded by a matrix of smaller recrystallized grains. Optical subgrains indicative of dislocation recovery are of variable size and are ubiquitous in both relict and recrystallized grains. Recrystallization occurs by strain-induced misorientation across these subgrain boundaries. Large grains are progressively reduced in size to a quasi-equilibrium diameter on the order of 80 μm. The progressive nature of the recrystallization process produces several orders of relict grain size for those grain volumes that have not undergone complete recrystallization.

The foliation within the albite rock is defined by the shape of both elongate relict and recrystallized grains. Although the current high aspect ratio shape of relict grains is largely derived from preferential recrystallization, crystallographic fabric data indicate substantial intracrystalline distortion associated with development of the shape fabric. Optical indicatrix axes (Fig. 2) from both relict and matrix grains show an exceptionally strong preferred orientation. Consideration of the relationships of the albite optical indicatrix to crystallographic elements (Jensen & Starkey 1985; Olesen 1987; Ji *et al.* 1988) places the concentration of (010) planes approximately 10° counterclockwise from the foliation plane defined by the shape fabric. The concentration of (010) planes and the observed fabric asymmetry with respect to the shape fabric is consistent with macroscopic sinistral shear dominated by slip on (010) (Bouchez *et al.* 1983; Ji *et al.* 1988) and rotation of the finite elongation direction toward this plane.

Occasional twin-like lamellar features were observed by polarized light microscopy, and are usually sub-parallel to the shape fabric foliation.

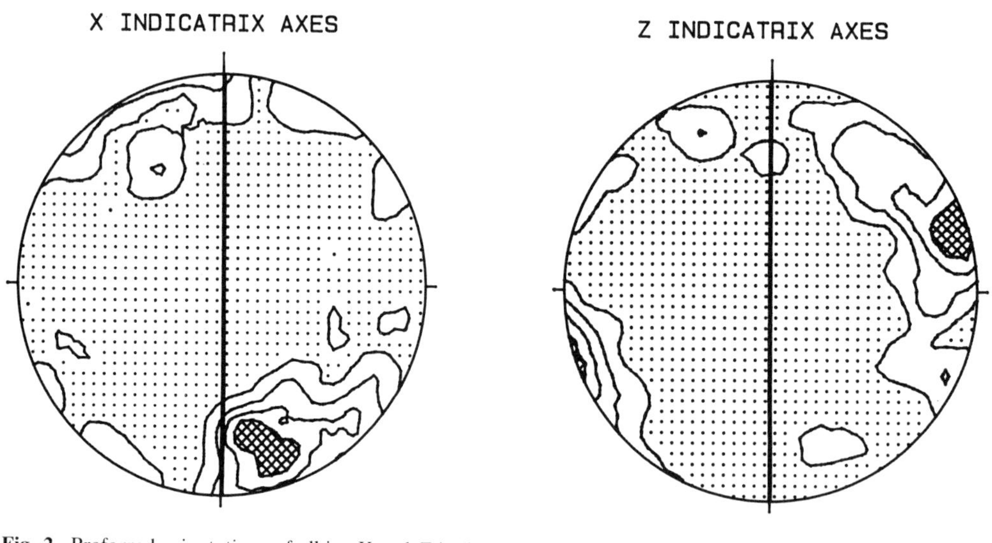

Fig. 2. Preferred orientations of albite X and Z indicatrix axes (lower hemisphere projections) oriented with respect to the grain shape fabric foliation (vertically-oriented N–S solid line) that contains the horizontally-oriented stretching lineation. Details in text. Contours are 1,3,5 and 7 times uniform distribution. Maximum concentration is 12 times uniform. n = 103

Electron microscopy and defect substructures

The essential purity of the grains is indicated in TEM (Fig. 3a) by the modulated texture characteristic of pure albite (McLaren 1974). The limited calcic component of this plagioclase occurs as discrete lamellae which have themselves exsolved into a typical peristerite texture identified by alternating lamellae of albite and oligo-

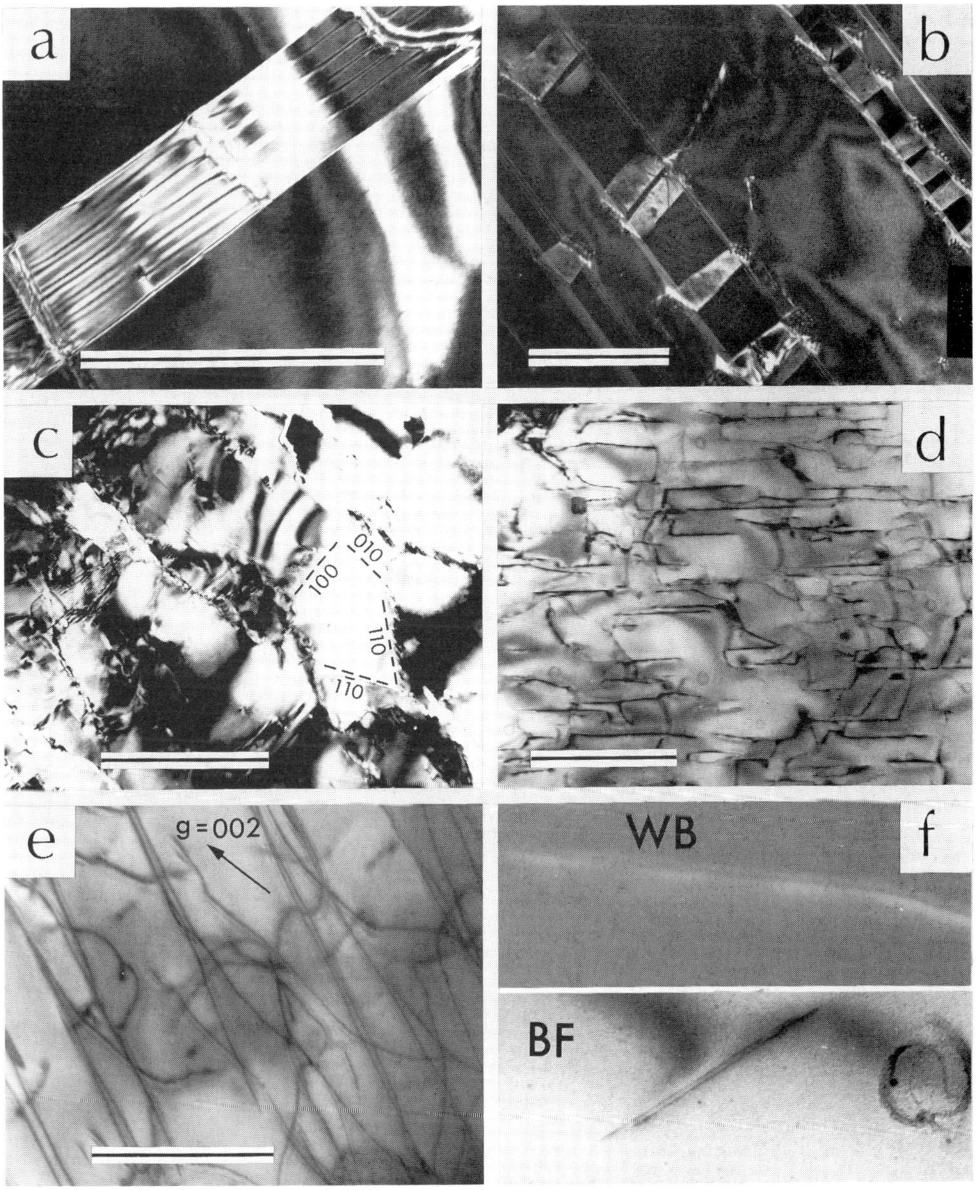

Fig. 3. TEM microstructures (Scale bar is 1 μm). (**a**) Lamellae of exsolved peristerite in albite host. BF $\mathbf{g} = 10\bar{1}$. (**b**) Interfacial dislocations at terminations of planar fringe contrast internal to peristerite lamellae. DF $\mathbf{g} = 20\bar{2}$. (**c**) Subgrains with rational wall orientations in relict grain elongated parallel to the trace of (010). BF $\mathbf{B} \approx [001]$. (**d**) Predominantly $\mathbf{b} = [101]$ dislocations parallel to trace of (010). BF $\mathbf{g} = 002$. (**e**) $\mathbf{b} = [001]$ screw segments of elongated loops imaged near (010) glide plane. BF $\mathbf{B} \approx [1\bar{2}0]$. (**f**) Bright field (BF) and weak beam (WB) images of two different dissociated dislocations with respective separations of 8.3 nm and 15.2 nm.

clase. These lamellae locally extend from dislocation subgrain walls, apparently utilizing such features as initiation points for propagation into the crystals. Although electron diffraction identifies twin reflections, in addition to peristerite reflections, all twins are restricted to the peristerite lamellae boundaries. These are therefore combined phase and twin or twin-like interfaces. In contrast to typical plagioclase twins, pairs of interfacial dislocations occur along the calcic lamellae at the terminations of planar defects (Fig. 3b). Extinction fringe asymmetry and the absence of appropriate twin reflections indicates these are stacking faults.

The different size orders of subgrains observed optically was confirmed in the TEM with subgrains ranging from 1–3 μm in diameter for both host and recrystallized grains. Those in host grains exhibit a strong crystallographic control, with dislocation walls parallel to low order crystal planes, such as (110), ($1\bar{1}0$), (001) and are elongate parallel to the trace of (010) planes (Fig. 3c), whereas those in matrix grains have more equant subgrains with crystallographically irrational walls.

Dislocations are homogeneously distributed throughout most grains (Fig. 2d) with markedly high densities varying from $0.8–3.5 \times 10^{13}$ m^{-2}. Most dislocations observed have traces parallel to (010), which when imaged is clearly the dominant glide plane (Fig. 3e), as was suggested by the crystallographic fabric data. Burgers vectors for dislocations gliding in (010) were identified using the $\mathbf{g}\cdot\mathbf{b} = 0$ invisibility criterion and trace analysis. While recognizing the ambiguities that can arise by applying this criterion to feldspars in the absence of detailed computer simulations (Olsen & Kohlstedt 1984; Montardi & Mainprice 1987), the mutual occurrence of two dislocation types was utilized in combination with effective invisibility of at least one dislocation type for $\mathbf{g} = 11\bar{1}, 020, 1\bar{1}\bar{1}, 110, 1\bar{1}0, 020, 040$ and $\bar{2}00$ to determine $\mathbf{b}$. Additional images were studied using $\mathbf{g} = 002, 20\bar{1}, 101, 00\bar{2}$, 00 10, and $03\bar{1}$. $\mathbf{b} = [101]$ and $\mathbf{b} = [001]$ were identified (indexed using $\mathbf{c} \approx 0.7$ nm) with $\mathbf{b} = [101]$ appearing to be at least as common as $\mathbf{b} = [001]$. These Burgers vectors are typical of plagioclase deformation and have been previously identified (e.g. Olsen & Kohlstedt 1984, An$_{25-48}$). Of the dislocations gliding on planes other than (010), only $\mathbf{b} = 1/2[112]$ ($20\bar{1}$) was identified, as has been reported from albite by Marshall & McLaren (1977). Dislocations are commonly dissociated (Fig. 3f) with the separation between partial dislocations varying between 8.3 nm for $\mathbf{b} = [101]$ and 15.2 nm for $\mathbf{b} = [001]$. These separations respectively approximate 10$\mathbf{b}$ and 21$\mathbf{b}$ for the net slip vectors.

There are few consistent differences between the morphology of the $\mathbf{b} = [001](010)$ and $\mathbf{b} = [101](010)$ dislocations. Locally, both exhibit abundant small and/or irregular loops, whereas more extensive, larger loops are more typical. $\mathbf{b} = [001]$ dislocations appear to be more commonly characterized by narrow loops with much longer screw than edge segments (e.g. aspect ratios of 20:1), parts of which are imaged in Fig. 3e. In contrast to the latter, $\mathbf{b} = [101]$ dislocations exhibited broader loops, more mixed screw/edge segments, long 'crankshaft' morphologies formed by alternating edge and mixed segments and a corresponding abundance of dipoles reflecting the density of edge segments.

Discussion

The microstructural development in the SAC albite mylonite is clearly similar to other monomineralic mylonites such as quartz and calcite (White *et al.* 1980). These textures contrast the more commonly observed behaviour of feldspars as relatively low-ductility silicates with a propensity for semi-brittle deformation (White & White 1983), limited recovery (Tullis & Yund 1980; Fitz Gerald *et al.* 1983) or recrystallization creep (Tullis & Yund 1985) under shallow to mid-crustal conditions. The SAC albite must reflect a deformation regime that supported extensive glide and recovery comparable to that observed in more ductile minerals. The occurrence of these microstructures is a reflection of preservation of deformation from a deeper crustal level than typically observed. Comparison of microstructures from the current study with other examples of extreme single crystal distortion of plagioclase (White & Mawer 1986; Ji *et al.* 1988) shows a mutual association with lithospheric thrusts and/or $P–T$ conditions of 800–1000 MPa and 700–900°C, comparable to the tectonic setting and syntectonic $P–T$ conditions inferred for the SAC albite.

The crystallographic fabric indicates a clear relationship between the hot-working, quasi-steady-state microstructure and the crystal defects observed by TEM. The strong preferred orientation of (010) planes close to the foliation plane reflects significant reorientation of large host grains by intracrystalline glide during creep. Recrystallization of the large grains and subsequent deformation of the matrix grains has produced a strong correspondence in orientation among relict and recrystallized grains. The inferred dominance of (010) glide within the aggregate is matched by TEM observations of a corresponding activity of (010) glide within individual grains.

The absence of work-hardening and/or cold-worked textures supports the introduction of these defects at the high-temperature deformation conditions. The types and multiple activation of slip systems observed are consistent with relaxation of any thermal constraints on glide at high temperature (e.g. Gandais & Willaime 1984 and references therein). The homogeneity of dislocation density can be explained by the strong crystallographic preferred orientation attained at the time dislocations were introduced, whereby the similar orientation of most grains produced an equivalent potential for activation of a given slip system throughout the sample. The latter explanation is analogous to that of Olsen & Kohlstedt (1984) who ascribed highly variable dislocation densities in deformed plagioclase to variations of the Schmid factor for the dominant slip systems as a function of diverse grain orientation.

Cross-slip of dislocations has been proposed as a pressure-sensitive, rate-controlling process for creep for olivine under mantle conditions (Poirier & Vergobbi 1978) and the concept has been extended to plagioclase (Olsen & Kohlstedt 1984; Montardi & Mainprice 1987). The rate control and pressure dependence arises from the need for constriction of dissociated dislocation segments prior to cross-slip. Minerals with low stacking fault energies and corresponding larger separations of partial dislocations would then require greater constriction, making them less amenable to cross-slip. This in turn reduces the recovery rate, leading to less ductile behaviour. Dissociated dislocations in plagioclase show a decrease in separation with decreasing anorthite component: An_{68-70}, 50 nm (Montardi & Mainprice 1987); An_{25-48}, $\geq$ 20 nm (Olsen & Kohlstedt 1984); An_0, $<$ 16 nm (this study). Accordingly, the higher stacking fault energy in albite compared to more calcic plagioclases inferred from the separation of partial dislocations would argue for albite being more ductile, in agreement with generalizations from field and experimental observations (Tullis 1983; Gandais & Willaime 1984).

The deformation of recrystallized grains in these rocks contrasts with highly deformed and recrystallized perthites from a similar $P-T$ environment (White & Mawer) 1988). In the latter case, there is little intracrystalline deformation of matrix grains subsequent to recrystallization, due to the apparent dominance of grain boundary processes that maintain an equant grain shape and that weaken any pre-existing crystallographic fabric by relative grain displacements. However, whereas the albite recrystallizes as a single phase, the perthite recrystallizes as discrete K-feldspar and plagioclase grains (White, 1988). The activation of interface mechanisms may reflect this two-phase nature by analogy with two-phase alloys that exhibit excessive grain boundary sliding (Edington *et al.* 1976), while the absence of such a distinct chemical potential across the grain boundaries of the monomineralic albite rock favours accomodation of deformation in a quasi-uniform manner by dislocation creep.

Despite evidence supporting the high $P-T$ origin of the deformation textures, contradictions remain. Above approximately 700°C, albite exists as a disordered phase (Brown & Parsons 1989), and there is a corresponding expectation that mechanical twinning will be relatively easy (Brown & Macaudiere 1986). The absence of twins in the SAC albite requires either that flow stresses were insufficient to induce twinning at high temperature or that all the microstructures developed below 700°C. The latter is rejected based on the absence of equivalent microstructures in reports of plagioclase deformation at mid-crustal conditions.

For deformation at lower crustal conditions, stresses sufficient to cause twinning are not required to produce the observed mylonite textures. Consideration of power-law creep laws for albite (Ji & Mainprice 1986) shows that for a deformation temperature of 800°C, imposed shear strain rates approaching 10^{-12} s^{-1} can be accommodated by stresses on the order of only 10 MPa, significantly below the estimated minimum critical shear stress for twinning of 100 MPa (Brown & Macaudiere 1986). If, as has been suggested (White & Mawer 1988), intense single-crystal distortion of feldspars is associated with temperatures in excess of 700°C, the SAC albite microstructures could have developed between 700–800°C in partially ordered intermediate albite (Brown & Parsons 1989). This could potentially inhibit twin generation to some degree. Alternatively, the intense preferred crystallographic orientation suggests that any record of twinning in disordered albite may have been obliterated by recrystallization and reorientation for preferred (010) slip. The twins or twin-like features that do occur are dependent on the prior exsolution of the peristerite lamellae, during exhumation and cooling of the albite, that clearly post-date the creep and recovery microstructures.

Estimates of stress based on the observed dislocation densities, as reviewed in Weathers *et al.* (1979), exacerbate the problem in that the determined values in excess of 200 MPa are not appropriate for the hot-working texture and might be expected to induce twinning. The observed dislocation densities, as opposed to the dislocation slip systems, are interpreted to reflect an overprint of the high-temperature microstructure which could have occurred after ordering of the albite.

Calon (1980) has documented a high strain rate/stress pulse in the adjacent peridotite mylonites that cannot be kinematically differentiated from the dominant lower stress event. Such a pulse is conceivably the origin of the high dislocation densities in the albite mylonite, particularly as changes in the dislocation density require strains only on the order of 1% (Kohlstedt *et al.* 1976). The preferred orientation of grains for easy slip on (010) that developed during high-temperature creep would facilitate accommodation of additional dislocations on (010) planes without disruption of the microstructure. An overprint would also help explain the homogeneous distribution of dislocations throughout the grains.

This study has been supported by NSERC through Operating Grant A512 and an Infrastructure Grant to the UNB Electron Microscopy Unit. R. A. Jamieson generously supplied the specimen.

References

BOUCHEZ, J. L., LISTER, G. S. & NICOLAS, A. 1983. Fabric asymmetry and shear sense in movement zones. *Geologische Rundschau*, **72**, 410–419.

BROWN, W. L. & MACAUDIERE, J. 1986. Mechanical twinning of plagioclase in a deformed meta-anorthosite — the production of M-twinning. *Contributions to Mineralogy and Petrology*, **92**, 44–56.

—— & PARSONS, I. 1989. Alkali feldspars: ordering rates, phase transformations and behaviour diagrams for igneous rocks. *Mineralogical Magazine*, **53**, 25–42.

CALON, T. J. 1980. Mylonites at the base of the ophiolitic White Hills Peridotite, northern Newfoundland. *Geological Society of America, Abstracts with Programs*, **12**, 27.

EDINGTON, J. W., MELTON, K. N. & CUTLER, C. P. 1976. Superplasticity. *Progress in Materials Science*, **21**, 61–70.

FITZ GERALD, J. D., ETHERIDGE, M. A. & VERNON, R. H. 1983. Dynamic recrystallization in a naturally deformed albite. *Textures and Microstructures*, **5**, 219–237.

GANDAIS, M. & WILLAIME, C. 1984. Mechanical properties of feldspars. *In*: BROWN, W. L. (ed.) *Feldspars and Feldspathoids*. D. Reidel Publishing Co., 207–246.

JAMIESON, R. A. 1981. Metamorphism during ophiolite emplacement-the petrology of the St. Anthony Complex. *Journal of Petrology*, **22**, 397–443.

—— 1986. P-T paths from high temperature shear zones beneath ophiolites. *Journal of Metamorphic Geology*, **4**, 3–22.

—— & TALKINGTON, R. W. 1980. A jacupirangite-syenite assemblage beneath the White Hills Peridotite, north-western Newfoundland. *American Journal of Science*, **280**, 459–477.

JI, S. & MAINPRICE, D. 1986. Transition from power law to Newtonian creep in experimentally deformed dry albite rock. *Transactions of the American Geophysical Union*, **67**, 1235.

——, —— & BOUDIER, F. 1988. Sense of shear in high-temperature movement zones from the fabric asymmetry of plagioclase feldspars. *Journal of Structural Geology*, **10**, 73–81.

JENSEN, L. N. & STARKEY, J. 1985. Plagioclase microfabrics in a ductile shear zone from the Jotun Nappe, Norway. *Journal of Structual Geology*, **7**, 527–539.

KOHLSTEDT, D. L., GOETZE, C. & DURHAM, W. B. 1976. Experimental deformation of single crystal olivine with application to flow in the mantle. *In*: STRENS, R. G. J. (ed.) *The Physics and Chemistry of Minerals and Rocks*. John Wiley, 35–49.

MARSHALL, D. B. & MCLAREN, A. C. 1977. Deformation mechanisms in experimentally deformed plagioclase feldspars. *Physics and Chemistry of Minerals*, **1**, 351–370.

—— & WILSON, C. J. L. 1976. Recrystallization and peristerite formation in albite. *Contributions to Mineralogy and Petrology*, **57** 55–70.

MCLAREN, A. C. 1974. Transmission electron microscopy of the feldspars. *In*: MACKENZIE, W. S. & ZUSSMAN, J. (eds.) *The Feldspars*. Manchester University Press, 378–423.

MONTARDI, Y. & MAINPRICE, D. 1987. A transmission electron microscopic study of the natural plastic deformation of calcic plagioclases (An 68–70). *Bulletin de Minéralogie*, **110** 1–14.

OLESEN, N. O. 1987. Plagioclase fabric development in a high-grade shear zone, Jotunheimen, Norway. *Tectonophysics*, **142**, 291–308.

OLSEN, T. S. & KOHLSTEDT, D. L. 1984. Analysis of dislocations in some naturally deformed plagioclase feldspars. *Physics and Chemistry of Minerals*, **11**, 153–160.

POIRIER, J-P. & VERGOBBI, B. 1978. Splitting of dislocations in olivine, cross-slip-controlled creep and mantle rheology. *Physics of the Earth and Planetary Interiors*, **16**, 370–378.

TULLIS, J. 1983. Deformation of feldspars. *In*: RIBBE. P. H. (ed.) *Feldspar Mineralogy*. Mineralogical Society of America Reviews in Mineralogy, **2**, 297–323.

—— & YUND, R. A. 1980. Hydrolytic weakening of experimentally deformed Westerly granite and Hale albite rock. *Journal of Structural Geology*, **2**, 439–451.

—— & —— 1985. Dynamic recrystallization of feldspar: a mechanism for ductile shear zone formation. *Geology*, **13**, 238–241.

WEATHERS, M. S., BIRD, J. M., COOPER, R. F. &

KOHLSTEDT, D. L. 1979. Differential stress determined from deformation-induced microstructures of the Moine thrust zone. *Journal of Geophysical Research*, **84**, 7495–7509.

WHITE, J. C. & MAWER, C. K. 1986. Extreme ductility of feldspars from a mylonite, Parry Sound, Canada. *Journal of Structural Geology*, **8**, 133–143.

—— & —— 1988. Dynamic recrystallization and associated exsolution in perthites: evidence of deep crustal thrusting. *Journal of Geophysical Research*, **93**, 325–337.

—— & WHITE, S. H. 1983. Semi-brittle deformation within the Alpine fault zone, New Zealand. *Journal of Structural Geology*, **5**, 579–589.

WHITE, S. H., BURROWS, S. E., CARRERAS, J., SHAW, N. D. & HUMPHREYS, F. J. 1980. On mylonites in ductile shear zones. *In*: CARRERAS, J., COBBOLD, P. R., RAMSAY, J. G. & WHITE, S. H. (eds) *Shear Zones in Rocks, Special Issue of Journal of Structural Geology*, **2**, 175–187.

Crystallographic fabrics: a selective review of their applications to research in structural geology

R.D. LAW

Department of Geological Sciences Virginia Polytechnic Institute & State University, Blacksburg, Virginia 24061, USA

Abstract: In this brief introductory review the potential geological use of crystallographic fabrics is illustrated by considering selected geological problems which, given appropriate conditions, may be investigated in plastically deformed rocks using fabric analysis. Quartz and calcite are taken as the main illustrative examples. Numerical fabric simulations indicate that the imposed strain path (strain symmetry, vorticity etc.) is reflected in the relationship between the fabric pattern, kinematic framework and finite strain axes. Although these fabric patterns are sensitive to the numerical model and combination of crystallographic slip systems chosen, many of the major fabric types have been observed in experimentally and naturally deformed rocks. The fabric pattern itself may contain important information on strain symmetry, orientation of fields of extension and contraction and operative slip systems. Similarly, the angular relationship between the fabric pattern and finite strain features (foliation and lineation) may provide information on shear sense and vorticity of deformation. Spatial transitions with decreasing grain size in naturally deformed rocks, from strongly defined fabrics to a complete lack of crystallographic preferred orientation, have been interpreted as indicating a switch to deformation mechanisms involving grain boundary sliding. Potential problems associated with using the absence of a fabric as an indication of grain boundary sliding (and by inference superplastic flow) are discussed. Experimental studies indicate that geometrical relationships between intracrystalline strain features and the crystal lattice of individual grains may be used to deduce palaeo-stress directions. Results of palaeo-stress analysis techniques based on such relationships are compared.

Since the first pioneering petrofabric study by Schmidt (1925) many thousands of different analyses have clearly shown that the constituent mineral grains of plastically deformed materials commonly display preferred crystallographic orientations (fabrics). Measurement of such fabrics, which was initially confined to the combined optical microscope and universal stage, has now reached a considerable level of sophistication, employing diverse techniques involving X-ray, neutron and electron diffraction (see reviews by Wenk 1979, 1985*a*) as well as electron channelling and electron backscattering (see reviews by Lloyd 1985; Dingley 1984).

Experimental studies (e.g. Green *et al.* 1970; Tullis *et al.* 1973) indicate that there are two dominant mechanisms by which a crystallographic preferred orientation may develop. At low homologous temperatures and/or high strain rates fabrics may develop by rotation of inequant grains (see reviews by Oertel 1985; Mainprice & Nicolas 1989) or by crystallographic slip within individual grains and resultant grain rotation. Secondly, under conditions where recrystallization is dominant, fabrics may be associated with the actual recrystallization process. Wenk *et al.* (1990) have pointed out that whilst there are numerous theories for fabric development associated with deformation by intracrystalline slip (see review by Hobbs 1985), fabric development during recrystallization is poorly understood (but see Jessell 1988). General introductions to the processes by which crystallographic fabrics may develop in deformed rocks are given by Hobbs *et al.* (1976), Nicolas & Poirier (1976) and Mainprice & Nicolas (1989).

One of the most important advances in our understanding of plastic deformation in rocks has been the identification, through single crystal deformation studies, of the exact crystallographic orientation of slip and twin systems active (under different conditions of temperature, strain rate, chemical activity etc.) within the main rock forming minerals. These experimentally detected slip systems form the essential input data for numerical simulations of crystallographic fabric development in mono-

From Knipe, R. J. & Rutter, E. H. (eds), 1990, *Deformation Mechanisms, Rheology and Tectonics*, Geological Society Special Publication No. 54, pp. 335–352.

mineralic rocks (e.g. Lister 1977; Lister *et al.* 1978; Takeshita & Wenk 1988; Wenk *et al.* 1990; Lister 1978; Wenk *et al.* 1987). Listings of identified slip and twin systems have been given for many specific rock forming minerals including quartz (Blacic & Christie 1984; Linker *et al.* 1984; Hobbs 1985), calcite and dolomite (Wenk 1985*b*); less detailed listings for other common rock forming minerals are given by Nicolas & Poirier (1976) and Mainprice & Nicolas (1989).

Of all minerals, perhaps quartz has received the most attention with respect to fabric studies. This is probably due in part to its importance in controlling the rheology of large portions of the Earth's crust, but also partly due to the large variety of fabric types found under different deformational and metamorphic conditions. For example, theoretical studies by Lister and co-workers (e.g. Lister *et al.* 1978; Lister & Paterson 1979; Lister & Hobbs 1980) have indicated that the development of crystallographic fabrics in quartzites during plastic deformation involving intracrystalline slip is controlled by three main factors: (1) the strain path or kinematic framework; (2) the magnitude and symmetry of finite strain; (3) the particular combination of crystallographic glide systems active during deformation. The importance of these factors in controlling fabric development has been confirmed by experimental studies (e.g. Tullis *et al.* 1973, Dell'Angelo & Tullis 1989).

Arguably a point has now been reached in our understanding of the general principles of crystallographic fabric development where, from an observed fabric in a plastically deformed rock, it should be possible to deduce at least some aspects of the deformation history associated with formation of that fabric. This approach, at least philosophically, bears much in common with igneous and metamorphic petrology, where experimental and theoretical studies are used to estimate the geological conditions under which individual rocks have formed. It must be borne in mind, however, that in a complex deformation history early formed fabrics may be overprinted by fabrics which only reflect the later conditions of deformation. Fabric simulation studies by Lister and co-workers indicate that in quartzites 40% shortening is sufficient to result in gross changes to an already existing fabric (Hobbs 1985, p. 477). In addition, it should be noted that all fabric simulations to date have only considered monomineralic crystalline aggregates. The presence of a second mineral phase of different rheology may lead to grain boundary sliding and heterogeneous flow at the grain scale and the formation of either diffuse (Starkey & Cutforth 1978) or domainal fabrics (e.g. Eisbacher 1970; Garcia Celma 1982).

In this brief review the potential geological use of crystallographic fabrics will be illustrated by considering selected geological problems which, given appropriate conditions, may be investigated using fabric analysis of plastically deformed rocks. It is emphasised that whilst attention in this review will be focussed upon quartz and calcite fabrics, many of the general techniques discussed may, with some modification, be applied to other mineral phases.

In the older (pre-mid-1970s) geological literature, fabric data were frequently presented on projection planes orientated perpendicular to both foliation and lineation (e.g. Sylvester & Christie 1968). In such projections, however, the fabric data (at least for quartz and calcite) is generally concentrated near the projection periphery and may easily be misinterpreted. It is therefore recommended that sections for fabric analysis should be cut perpendicular to foliation and parallel to lineation. This approach has the additional advantage that results of most fabric simulations are also presented in a projection plane containing the XZ finite strain axes.

Apparent extensions and strain symmetry

Commonly in deformed rocks, strain analysis indicates that extension has occurred along directions which intuitively one would expect to be associated with either a minimum or zero finite longitudinal strain. For example, analysis of reduction spots in the Cambrian slate belt of North Wales, where measured strain ratios imply a sub-horizontal extension of 35% along the orogenic zone (Wood 1971), obviously present a considerable space problem. Clearly it is important to determine if this is a real or an apparent extension. For the North Wales slates, Ramsay & Wood (1973) were able to demonstrate that tectonic plane strain deformation with progressive volume loss superimposed on previously compacted material would result in a true vertical extension and an apparent horizontal extension.

In other cases, however, volume change cannot explain anomalous apparent extensions. For example, plastically deformed conglomerates in the hinges of folds are commonly observed to have their pebble long axes aligned parallel to the fold hinge. Do such observations indicate that the maximum principal extension direction is orientated parallel to the fold hinge? Ramsay (1967, p. 220) has demonstrated that such pebble alignments could be due to the superimposition of a plane strain deformation (zero extension parallel to fold hinge) upon an initial planar sedimentary fabric. However, such models do not address the specific question of whether clast alignment does, or does not, indicate a real extension.

Theoretical modelling of quartz crystallographic fabric development by Lister and co-workers (e.g. Lister *et al.* 1978; Lister & Hobbs 1980) has, for coaxial deformation, demonstrated a clear set of relationships between strain symmetry, fabric pattern and finite strain axes. These relationships, which are schematically summarised in Fig. 1, are supported by both experimental studies (e.g. Tullis *et al.* 1973; Tullis 1977) and analysis of naturally deformed quartzites (see review by Price 1985). Similar results have been found for calcite (see review by Wenk 1985*b*). The important point here is that the predicted relationship between fabric pattern and the principal extension directions may be used to test whether an individual lineation is indicating a real or an apparent extension. For approximate plane strain deformation, fabric simulations indicate that quartz *c*-axes will either display a cross-girdle (Lister *et al.* 1978; Lister & Hobbs 1980) or single girdle (Etchecopar & Vasseur 1987) pattern intersecting the foliation perpendicular to the lineation (Figs 1 & 2).

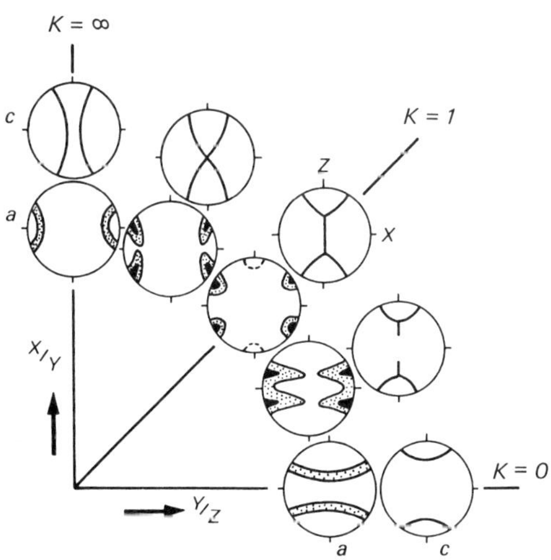

Fig. 1. Theoretical relationships (for coaxial deformation) between strain symmetry (expressed by Flinn Plot) and quartz *c* and *a*-axis fabrics; *c*-axis fabrics represented by fabric skeletons, *a*-axis fabrics represented by contours; in all pole figures foliation (*XY*) is vertical and trends from right to left, lineation (*X*) within foliation is horizontal; adapted from Schmid & Casey (1986).

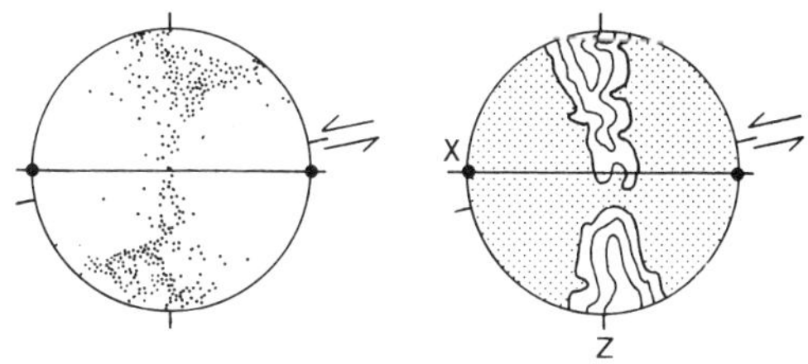

Fig. 2. Simulated quartz *c*-axis fabrics associated with progressive simple shear deformation (shear strain $\gamma = 4.0$) and displayed in the *XZ* section of the finite strain ellipsoid. Orientation of shear plane indicated by opposed arrows. Note whilst full constraints model of Lister & Hobbs (1980, model B) produces a cross-girdle fabric, the model of Etchecopar & Vasseur (1987, fig. 18) with basal and prismatic slip systems only, produces a single girdle fabric (at shear strains greater than approximately 2.5). Note also that whilst single girdle fabric in the Etchecopar & Vasseur model is clearly orientated perpendicular to the shear plane, the relatively diffuse nature of *c*-axis distribution predicted by Lister & Hobbs precludes using fabric to estimate orientation of shear plane.

Such relationships in the Roche Maurice quartzites of western Brittany have been used, for example, by Law (1986) to prove that a horizontal fold hinge parallel lineation defined by plastically deformed detrital grains does in fact indicate the local maximum principal finite strain direction. Similarly, single and cross-girdle *c*-axis fabrics in deformed conglomerates in which pebble outlines define an L–S tectonite (e.g. Strand 1944; Brace 1955; Sylvester & Janecky 1988) frequently indicate that the pebble long axes are orientated parallel to the maximum principal finite extension direction.

Likewise, from Fig. 1, it is clear that the pattern of crystal preferred orientation may potentially, where no strain markers are present, be used broadly to estimate strain symmetry associated with plastic deformation (see reviews by Marjoribanks 1976; Miller & Christie 1981; Law *et al.* 1984; Price 1985; Law 1986; Schmid & Casey 1986; Teyssier *et al.* 1988 for quartz; Wagner *et al.* 1982; Wenk 1985b for calcite). Similarly, for micas (although the exact mechanism by which preferred crystallographic orientation develops remains problematical (e.g. Knipe 1981)), experimental and theoretical studies (e.g. Tullis 1976; Ramsay & Huber 1983, pp. 191–192) indicate a strong correlation between mica fabric patterns and Flinn's strain symmetry parameter (k). These results have been confirmed in analyses of naturally deformed argillaceous rocks (e.g. Tullis & Wood 1975; Le Corre 1979; Wood & Oertel 1980).

Determination of active slip systems

Experimental studies have indicated that the relative activity of different slip systems in a given mineral may be controlled by such variables as temperature, strain rate and water content (see references for slip systems above). Thus, through identifying the crystallographic slip systems in naturally deformed rocks it may ultimately be possible to identify temperature and strain rate regimes (Lister *et al.* 1978, p. 154). For example, the operation of prism <*c*> slip under geological conditions in quartz may be diagnostic of high homologous temperatures, low strain rates and low stress intensity (Lister & Dornsiepen 1982, p. 91) and the increased importance of diffusion in facilitating hydrolytic or water weakening processes (Mainprice & Nicolas 1989, p. 182). When associated with igneous intrusions such as granites (Gapais & Barbarin 1986; Blumenfeld & Bouchez 1988), prism <*c*> slip may indicate plastic deformation at temperatures near the granite solidus (see discussion by Paterson *et al.* 1989, pp. 355–356).

Identification of the active crystallographic slip systems in naturally deformed rocks generally requires TEM analysis of the dislocation structures associated with slip (see Humphreys 1983 for general review of techniques). A listing of recent papers on dislocation analysis of common rock forming minerals is given by White (1985). However, such TEM analyses require extensive operator training and access to sophisticated, highly expensive equipment. In addition, it has been proposed that observed crystal defects in some mylonites may post-date the deformation which produced strong crystallographic fabrics (e.g. Ord & Christie 1984) and that dislocation densities may be radically altered during uplift (e.g. White 1979). Thus it is possible that dislocations imaged in the TEM from at least some naturally deformed rocks displaying a crystallographic fabric may be unrelated to the slip systems responsible for producing that fabric.

Pole figure data

A possible solution to this problem is suggested by numerical fabric simulation models which indicate that the pattern of crystallographic preferred orientation of a given mineral species is directly controlled by the combination of active slip systems. Fabric simulations, such as those reported by Lister *et al.* (1978) which are based on five independent slip systems ('full-constraints' models), indicate that, because of the limited number of available slip systems and their relative dispositions, the crystallographic axes of individual grains in a deforming aggregate tend to rotate towards special orientations relative to the imposed kinematic framework, thus forming a crystallographic fabric. The *c*-axis fabrics produced by these simulations are remarkably similar to fabrics observed in both experimentally deformed quartzites (e.g. Tullis *et al.* 1973; Tullis 1977) and in many naturally deformed quartzites (see review by Price 1985). From this fabric similarity it is tempting to infer that the same slip systems are responsible for producing both the simulated and observed fabrics.

Lister & Paterson (1979, p. 115) suggest that these fabric simulations support the concept put forward by Fairbairn (1949) and Sander (1970) that 'there are several possible discrete and distinct maximum orientations for natural quartz *c*-axis fabrics'. For coaxial deformation these (generally diffuse) fabric maxima would be predicted to bear a simple relationship to the finite strain axes (foliation and lineation).

More clearly defined maxima are generally produced in fabric simulations using less than five independent slip systems (relaxed constraints models) and these simulation models may be more applicable to minerals such as olivine, where only a few slip systems exist (Etchecopar & Vasseur 1987). Such fabric maxima are frequently observed in natural tectonites and are commonly interpreted as indicating the operation of a 'dominant' slip system.

Quartz *c*-axis maxima aligned parallel to the inferred intermediate principal strain direction (Y) have commonly been interpreted as indicating that prism <*a*> is the dominant slip system (e.g. Wilson 1975; Bouchez 1977; Lister & Dornsiepen 1982). Similarly, the much rarer lineation-parallel *c*-axis maximum has been interpreted as indicating prism <*c*> slip; optical and TEM based defect analyses of naturally deformed rocks characterized by such maxima (Blumenfeld *et al.* 1986; Mainprice *et al.* 1986) support this intuitive prism <*c*> slip system interpretation.

Individual 'grain' data

It must be emphasized that in the above

examples the preferred orientation of one crystal direction (i.e. the *c*-axis) was used to infer the orientation of all the other associated crystal directions and hence the alignment of potential crystallographic slip systems with the inferred kinematic framework. Schmid *et al.* (1981) have shown that for individual *c*-axis positions the complete crystallographic orientation may be statistically estimated using the Orientation Distribution Function (ODF) derived, for example, from X-ray data (Casey 1981). Examples of such complete fabric analyses for commonly observed quartz *c*-axis fabrics have been described by Schmid & Casey (1986) and the orientation of potential slip systems relative to specimen finite strain directions used to infer the active slip systems responsible for fabric formation.

Using similar techniques, relative resolved shear stresses (Schmid factors) on potential slip systems have been calculated by Law *et al.* (1990) for grains of different *c*-axis orientations within a geometrically closely constrained quartzose shear zone displaying intense crystallographic fabrics. This Schmid factor data was used to infer operative slip systems responsible for fabric formation assuming: (a) intracrystalline deformation is dominated by a single slip system within each grain, (b) the dominant slip systems have reached stable end orientations; (c) the dominant slip systems become orientated parallel to the inferred simple shear kinematic framework (shear plane and shear direction). Such intuitive models have recently been discussed by Wenk *et al.* (1989).

It should be noted that in general more complex fabrics are observed in minerals where several slip systems operate simultaneously, and therefore fabric interpretation is likely to be most straightforward in minerals where a single slip system is dominant.

Shear sense indicators

Probably the most common problem encountered in structural geology for which crystallographic fabrics are employed is determining shear sense in plastically deformed rocks (see reviews by Bouchez 1978; Lister & Williams 1979; White *et al.* 1980; Behrmann & Platt 1982; Bouchez *et al.* 1983; Passchier 1983; Simpson & Schmid 1983; Simpson 1986). For quartz, shear sense fabric criteria have been based on two slip-induced lattice reorientation models: the relaxed constraints model of Etchecopar and co-workers (e.g. Etchecopar 1977; Etchecopar & Vasseur 1987) and the full constraints model of Lister and co-workers (e.g. Lister & Hobbs 1980). It should be noted that whilst, for simple shear, the Etchecopar model (Etchecopar & Vasseur 1987) predicts a single girdle *c*-axis fabric, the Lister model (Lister & Hobbs 1980) predicts a cross-girdle fabric (Fig. 2). However, although differing in detail, both these fabric simulation models indicate that for simple shear deformation, one *c*-axis girdle develops normal to the bulk shear plane and shear direction (Fig. 2). Thus for a shear zone in which, with increasing shear strain, foliation and lineation is progressively rotated into alignment with the shear zone boundaries, the asymmetry of fabric relative to foliation and lineation can be used to deduce the sense of shear (Figs 2 & 3).

Similar fabric asymmetries have been observed in many naturally deformed quartzites in which the shear sense has been independently indicated by other criteria (see review by Schmid & Casey 1986). Fabric asymmetries have also been observed (Dell'Angelo & Tullis 1989) in experimentally sheared quartzites (Fig. 4). In principle therefore, the asymmetry of, for example, quartz *c*-axis fabrics is a powerful shear sense indicator. It should be kept in mind, however, that in many mylonitic shear zones the margins of the shear zone are not exposed and, even when they are exposed, it may be impossible to decide whether the associated shear plane is parallel to the shear zone margins or parallel to an internal plane of anisotropy.

The essential topological features of a crystallographic fabric may be defined by linking up peaks and crests on the contoured diagram by a

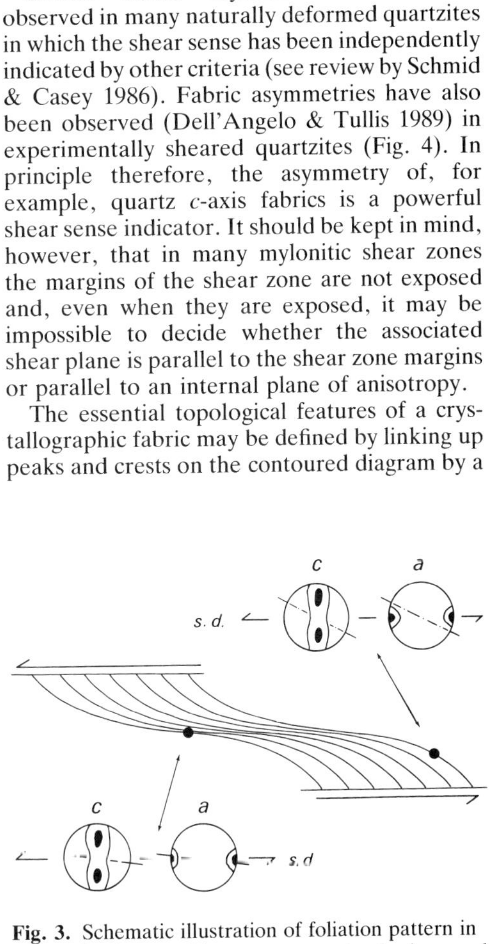

Fig. 3. Schematic illustration of foliation pattern in an idealised zone of heterogeneous simple shear and angular relationships with increasing shear strain between finite strain features (foliation and lineation) and quartz *c* and *a*-axis fabrics; all relationships shown on *XZ* section plane of finite strain ellipsoid; s.d., shear direction.

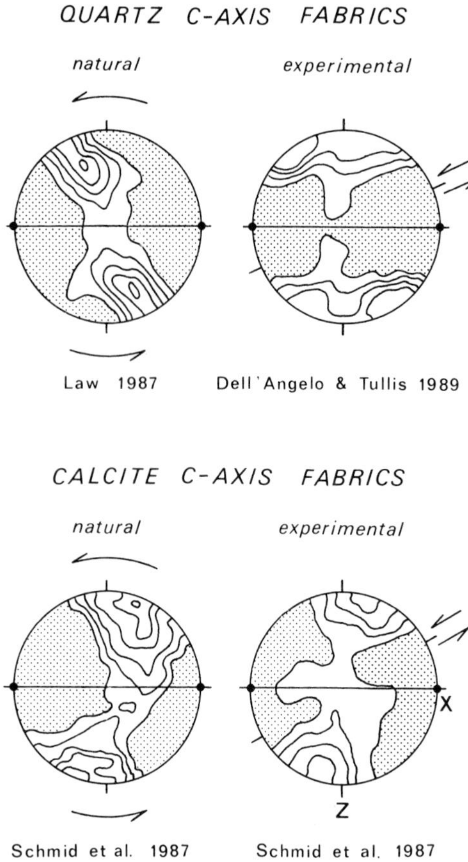

Fig. 4. Selected examples of natural and experimental quartz and calcite *c*-axis fabrics associated with non-coaxial deformation; sinistral shear sense imposed in all examples. Note whilst actual orientation of imposed shear plane is indicated by opposed arrows in experimental examples, general shear sense only is indicated in natural examples. Orientation of lineation (X) and pole to foliation (Z) also indicated. Experimental quartz fabric (Dell'Angelo & Tullis (1989, fig. 5f) produced under plain strain conditions involving 46% shortening and shear strain γ of 2.8. Experimental calcite fabric produced under imposed simple shear conditions in conditions in the twinning regime (Schmid *et al.* 1987, fig. 10).

series of straight line segments. The resulting fabric skelton frequently displays both *external* and *internal* asymmetries (Behrmann & Platt 1982; Platt & Behrmann 1986). External fabric asymmetry is defined by the angle of obliquity between the central segment of the fabric skeleton and the foliation whilst, for cross-girdle *c*-axis fabrics, the unequal inclination of the central fabric segment to the peripheral segments defines the internal asymmetry (Fig. 5). Detailed studies of variation in external and internal fabric asymmetry within thrust related quartz mylonites have been reported by Platt & Behrmann (1986) and Law (1987).

Quartz *c*-axis fabric asymmetry has been used to address many tectonic problems. For example, current crustal extension models for core complex evolution predict that, traced across individual core complexes, a constant shear sense will be associated with mylonite formation (e.g. Lister & Davis 1989). Quartz *c*-axis fabric asymmetry recorded, for example, in the Snake Range of Nevada (Lee *et al.* 1987) support this constant shear sense model. In contrast, older models for core complex evolution involving gravitational spreading (e.g. Compton 1980) predict divergent shear senses traced across individual core complexes. Quartz *c*-axis fabrics consistent with such models are found in the Raft River Mountains of Utah and Idaho (Malavielle 1987).

Bouchez *et al.* (1983) have summarised the conditions required to use the asymmetry method: (1) deformation must result from dislocation creep; (2) the grain-shape fabric (foliation-lineation) must be clearly defined and reflect the imposed finite strain (*XYZ*); (3) deformation must be homogeneous from the thin-section scale up to the scale of observation; (4) the mineral phase being used for shear sense determination must be dominant in volume so as to avoid flow heterogeneities.

In quartz fabric simulations involving simple shear (e.g. Etchecopar & Vasseur 1987) the

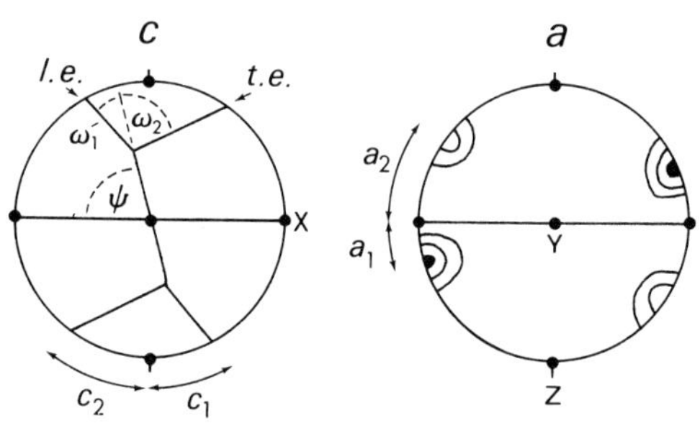

Fig. 5. Parameters used to characterise external and internal fabric asymmetry in quartz *c* and *a*-axis fabrics (adapted from Law 1987). External fabric asymmetry characterised by ψ, c_1, c_2, a_1 and a_2. Internal fabric asymmetry characterised by ω_1 and ω_2. Sinistral shear sense indicated by unequal densities of a-axis point maxima and asymmetry of *c*-axis fabric skeleton and *a*-axis point maxima. Leading and trailing edges of *c*-axis fabric skeleton denoted by l.e and t.e. respectively.

greatest concentration of a-axes develops in the shearing plane parallel to the shearing direction (Fig. 3). Thus the asymmetry between the dominant a-axis maximum and foliation and lineation may, as originally suggested by Bouchez (1978), also be used as a shear sense indicator (Fig. 3). In addition, the angle between lineation and the dominant a-axis maximum may thereoretically be used to estimate bulk shear strain. However, in many quartz mylonite zones the angle between the dominant a-axis point maximum and lineation is too large (i.e. the predicted shear strain is too small) when compared with predictions from field relationships (e.g. Boullier & Quenardel 1981; Law *et al.* 1986; Law 1987; Mancktelow 1987). Possible geological reasons for this, including departure from simple shear, dynamic recrystallisation and fabric overprinting, have been discussed by Schmid & Casey (1986), Mancktelow (1987) and Law (1987). One clearly documented example of such anomalous relationships from the Moine thrust zone at the Stack of Glencoul (NW Scotland) is schematically illustrated in Fig. 6.

Crystallographic fabrics may, under appropriate conditions, also be used to investigate spatial variation in shear sense on a much smaller scale. For example, folded quartz veins in Moine schists located 5 m above the Moine thrust at the Stack of Glencoul yield opposite c-axis fabric asymmetries on adjacent fold limbs (Fig. 7) indicating that mechanically the vein is being actively folded rather than acting as a passive marker within the thrust zone. By analogy with simulation studies (e.g. Lister & Hobbs 1980) these fabrics, which were measured in sections cut perpendicular to foliation and parallel to lineation, also indicate that penetrative deformation is associated with a bulk shearing direction orientated in this section plane rather than perpendicular to the fold hinges (cf. Christie 1963, pp. 382–384). The non-uniform density distributions in these fabrics are probably due to an original crystal preferred orientation in the vein material.

It must be emphasized that, whenever possible, fabric asymmetry should be used in conjunction with microstructural shear sense indicators (see reviews by Bouchez *et al.* 1983; Simpson & Schmid 1983; Simpson 1986). For example, variations in the sense of quartz c-axis fabric asymmetry have in several cases been recorded within individual shear zones (e.g. Passchier 1983), thus casting doubt upon the universal applicability of fabric asymmetry as a shear sense indicator. In such situations microstructural studies are clearly of critical importance in both testing the validity of using fabrics as shear sense indicators and also investigating the possibility of domainal shear sense variation.

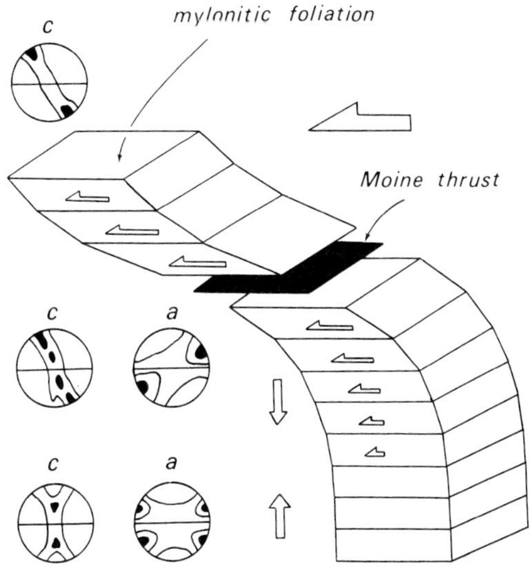

Fig. 6. Schematic illustration (viewed towards the NNE) of quartz c and a-axis fabric variation with distance from the Moine thrust at the Stack of Glencoul, NW Scotland. XZ projection plane used in all fabric diagrams. C-axis fabrics within mylonitic Cambrian quartzites situated beneath the thrust range from asymmetrical kinked single girdles at 0.5 cm beneath the thrust, through asymmetrical cross-girdle fabrics to symmetrical cross-girdle fabrics at 30 cm beneath the thrust. C-axis fabric transition is accompanied by a concomitant transition from asymmetrical single a-axis point maxima fabrics (0.5 cm beneath thrust) through asymmetrical two maxima fabrics to symmetrical two point maxima fabrics (Law *et al.* 1986; Law 1987). Note that although foliation and lineation throughout the mylonite sequence are parallel to the thrust surface, a-axis point maxima are always orientated at c. 25° to the lineation. Deformed quartz veins within phyllosilicate-rich mylonitic Moine metasediments lying above the thrust are all characterised by asymmetrical single girdle c-axis fabrics. See text for interpretation.

It should also be emphasized that, although a particular fabric asymmetry may prove a reliable shear sense indicator for one mineral phase, this does not necessarily mean that a similar fabric asymmetry associated with a different mineral also constitutes a reliable shear sense indicator. This point is illustrated in Fig. 4 for natural and experimental quartz and calcite fabrics associated with sinistral shearing where opposite fabric asymmetries are displayed by the two minerals. The opposite fabric asym-

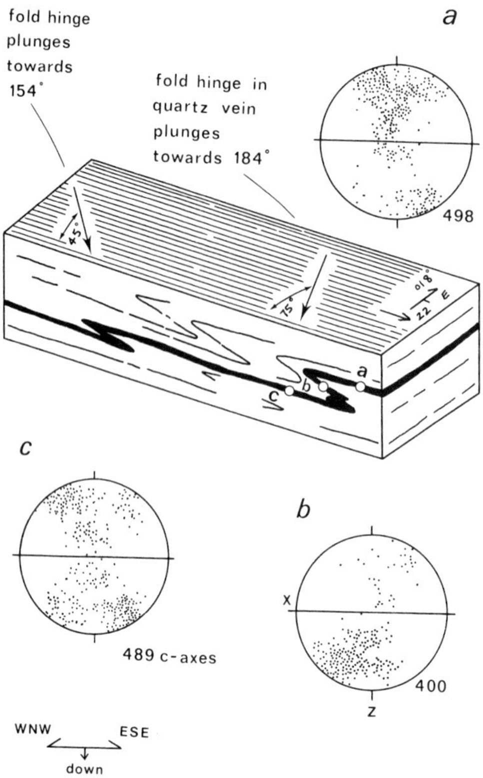

Fig. 7. Schematic sketch of folded quartz veins within Moine mylonites located 5.0 m above the Moine thrust at the Stack of Glencoul, NW Scotland. Surface cut perpendicular to foliation and parallel to lineation is 30 cm in length. *C*-axis fabrics from three small (0.4 × 0.15 cm) domains on adjacent fold limbs displayed on *XZ* projection planes containing lineation (*X*) and pole (*Z*) to foliation; note: (**a**) opposite fabric asymmetries on adjacent fold limbs and; (**b**) non-orthogonal relationship between *XZ* section and fold hinges (orientations indicated by arrows lying within foliation) defined by quartz veins. Specimen and fabrics viewed towards the NNE; movement on Moine thrust associated with WNW directed overthrusting (sinistral shear sense).

metries are explained by the different operative glide systems and the importance of twinning in the development of the calcite fabric, the *c*-axis of the twin moving through a rotation of 52° towards the maximum principal compressive stress axis (Rutter & Rusbridge 1977; Schmid *et al.* 1987, pp. 757–758).

Strain path indicators

Most geological structures owe their formation to heterogeneous flow (see review by Lister & Williams 1983) and hence, in order to understand the detailed processes by which such structures form, it is essential to identify techniques for quantitatively assessing the flow paths (strain paths) associated with their formation. Many authors (e.g. Elliott 1972; Means *et al.* 1980; Pfiffner & Ramsay 1982; Passchier 1988) have pointed out that on theoretical grounds there exists a complete spectrum of strain paths of differing non-coaxiality. For plane strain deformation, such strain paths range between pure shear (coaxial deformation) and simple shear.

Qualitative strain path assessment

Crystallographic fabrics present a potentially useful source of information on strain paths because numerical fabric simulations (e.g. Lister & Hobbs 1980) indicate that the symmetry of deformation (i.e. the strain path) will be reflected in the symmetry of the resultant fabric (see Lister & Williams 1979 for review). Thus internally symmetrical fabrics which are also symmetrical with respect to finite strain axes (foliation and lineation) could be interpreted as indicating coaxial strain paths, whilst asymmetrical fabrics would be interpreted as indicating non-coaxial deformation.

Variation in quartz *c* and *a*-axis fabric symmetry has been used by Law *et al.* (1986) and Law (1987) to infer strain path variations associated with mylonite formation beneath the Moine thrust at the Stack of Glencoul, NW Scotland. These fabric variations, which are summarized in schematic form in Fig. 6, were interpreted by Law (1987) as indicating a constant coaxial deformation component beneath the thrust, and an increasingly important component of a simple shear (non-coaxial) deformation traced towards the thrust. All fabrics above the thrust are indicative of non-coaxial deformation (Law, unpublished data). Strain analysis of mylonites beneath the thrust had previously led Sanderson (1982, p. 215) to suggest that mylonite formation may have involved components of both pure and simple shear deformation; the relative timing of these components could not be determined from the strain data. Inspection of the fabric symmetry variation summarized in Fig. 6 clearly indicates, however, that the component of simple shear deformation must either be contemporaneous with, or have outlasted the component of coaxial deformation. If the coaxial component of deformation had outlasted the simple shear component, then all asymmetrical fabrics would have been overprinted by symmetrical fabrics.

From the above example it is clear that,

asssuming fabric symmetry is related to vorticity of deformation, then crystallographic fabrics can provide important information on the relative timing and spatial distribution of strain paths. However, analyses such as that described above are really only qualitative in that no estimate of the actual degree of vorticity associated with deformation (expressed by the kinematical vorticity number *Wk*; Means *et al.* 1980) is derived. In the quartz fabric simulation studies of Etchecopar (e.g. Etchecopar 1977; Etchecopar & Vasseur 1987) and Lister & coworkers (e.g. Lister & Hobbs 1980) only the cases of coaxial deformation and simple shear deformation were reported. More recently, however, Wenk *et al.* (1987, fig. 12) have modelled calcite *c*-axis fabric evolution for strain paths involving different degrees of noncoaxiality.

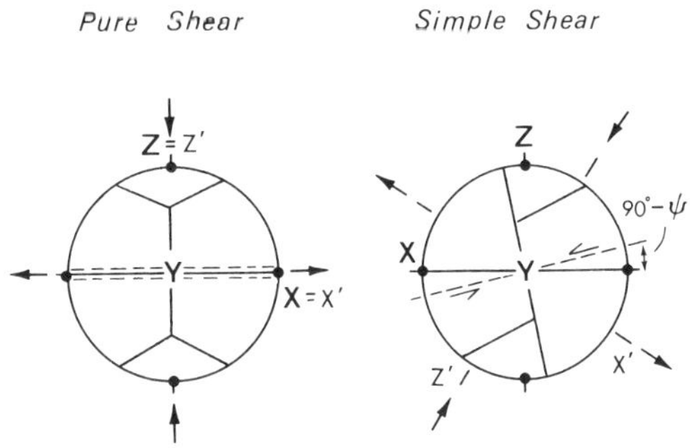

Fig. 8. Kinematic interpretation (adapted from Platt & Behrmann 1986) of quartz *c*-axis fabric skeletons for model quartzite B of Lister & Hobbs (1980) in simulated progressive pure shear (80% shortening) and simple shear ($\gamma = 4$). Principal finite strain axes denoted by X, Y, and Z; maximum and minimum principal axes of instantaneous stretching and shortening denoted by X' and Z' respectively. Broken line represents orientation of particle line of zero angular velocity (extensional apophyses of the flow).

Quantitative strain path assessment

At present very few analytical methods exist for quantitatively estimating vorticity associated with natural deformation in plastically deformed rocks. One method introduced by Platt & Behrmann (1986) makes use of the observation by Lister & Hobbs (1980) that in 'full constraints' quartz *c*-axis fabric simulations the central segment of the cross-girdle fabric establishes itself orthogonal to the shear plane in simple shear and orthogonal to the XY plane of finite deformation in pure shear. Now for simple shear the XY plane of finite strain rotates towards the shear plane as strain increases, so that the measure of external asymmetry (ψ in Fig. 5) approaches 90°. In contrast, for plane strain the central girdle segment of the *c*-axis fabric remains at 90° to the XY plane with increasing strain (Fig. 8). Platt & Behrmann (1986) demonstrated that by plotting 90° − ψ against the finite stretch ($1 + e_x$) along X, curves for simple and pure shear may be constructed (Fig. 9). Platt & Behrmann (1986) have proposed that quartz mylonites deformed in simple shear should be characterized by strain and *c*-axis fabric data which falls on the simple shear curve of this plot. In contrast, mylonites which have followed a strain path intermediate between pure and simple shear should fall between the pure and simple shear curves. Detailed study of a suite of thrust-related quartz mylonites from the Betic Cordilleras by Platt & Berhmann (1986), using this method, revealed that whilst mylonites located adjacent to the thrusts plot close to the simple shear curve, mylonites located further from the thrusts plot closer to the pure shear curve (Fig. 9).

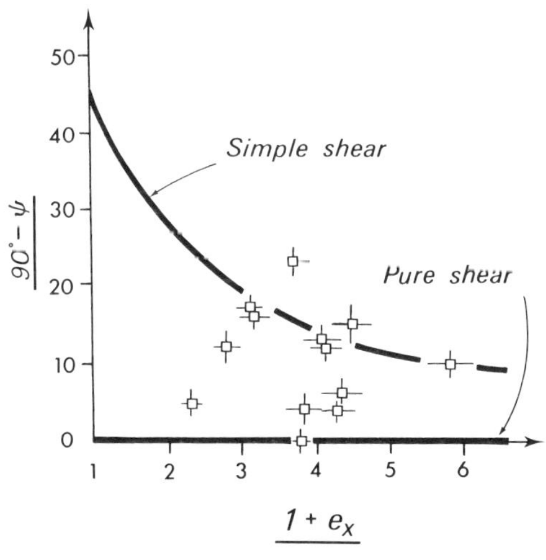

Fig. 9. Relationship between the degree of external asymmetry of quartz *c*-axis fabrics (expressed by 90° − ψ in Fig. 8) and the finite stretch ($1 + e_x$) along X; adapted from Platt & Behrmann (1986). The angle between the XY plane and the line of zero angular velocity (extensional apophyses of the flow) is plotted for progressive simple shear (curve) and progressive pure shear (along horizontal axis). The differing values (with error bars) for thrust related quartz mylonites from the Betic Cordilleras indicate a wide variation in the vorticity of the flow.

It is emphasized that employment of the vorticity estimation method of Platt & Behrmann (1986) is only applicable to plane strain deformation and requires both fabric and strain data of high quality. In addition, suitable strain markers are rarely preserved in dynamically recrystallized tectonites. Therefore the method described by Platt & Behrmann (1986) cannot be applied with confidence to mylonites unless strain markers (such as pebble outlines) are preserved. An alternative method for determining vorticity of deformation associated with plane strain tectonites using quartz *c*-axis fabrics in conjunction with rotated garnet porphyroblasts has recently been described by Vissers (1989).

Several recent studies of natural and experimentally produced crystallographic fabrics have indicated that, with increasing finite strain along a constant strain path, the skeletal outline of a *c*-axis fabric does not (in contradiction to the findings of Lister & Hobbs 1980) remain constant with respect to the kinematic framework. For example, a transition from symmetrical cross-girdle to asymmetrical single girdle *c*-axis fabrics with increasing heterogeneous shear strain of quartz veins has been recorded by Garcia-Celma (1983, p. 78) within the Cap de Creus mylonites of NE Spain. It could be argued (Lister & Williams 1983; pp. 23–24) that in this particular case the fabric transition is due to strain path partitioning followed by fabric (or geometrical) softening. However, a similar transition from double to single fabric point maximum has been observed by Bouchez & Duval (1982) with increasing shear strain in ice subjected to experimental simple shear deformation. A transition from cross-girdle to single girdle quartz *c*-axis fabrics is also indicated for progressive simple shear by the fabric simulation model of Etchecopar & Vasseur (1987, fig. 18). These results must bring into question the universal validity of using fabric asymmetry as an indicator of vorticity of deformation.

Influence of recrystallisation on fabric symmetry

It is a notable feature of many naturally deformed quartzites that the spatial transition from symmetrical cross-girdles to asymmetrical single girdles coincides with a marked increase in the degree of recrystallization. Schmid & Casey (1986) have suggested that such fabrics could all be due to simple shear deformation, the transition from cross-girdle to single girdle fabrics (Fig. 10) marking the bulk finite strain at which grains in unfavourable orientations for continued intracrystalline slip are partially removed by grain boundary migration of more favourably orientated grains, and partially reorientated by selective recrystallization. This model serves to warn us that whenever possible, independent kinematic indicators (e.g. microstructures) should be used in conjunction with crystallographic fabric analysis. A transition from cross-girdle to single girdle *c*-axis fabric

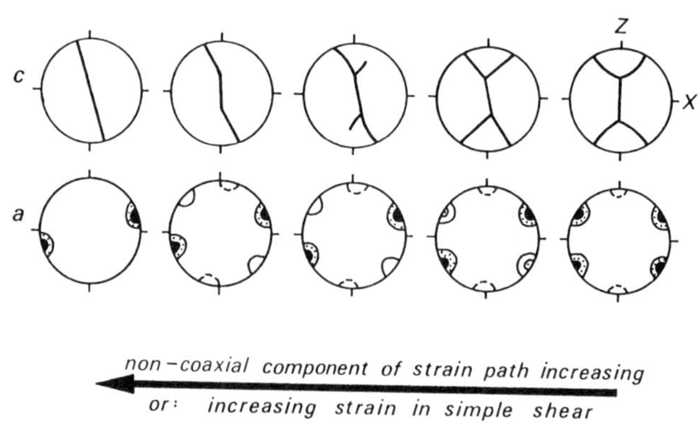

Fig. 10. Two possible interpretations of quartz *c* and *a*-axis fabric transitions produced in plane strain ($k=1$) deformation (sinistral shear sense indicated); *c*-axis fabrics represented by fabric skeletons; *a*-axis fabrics represented by contours. In all stereograms foliation (XY) is vertical and trends from right to left, lineation (X) within foliation is horizontal; adapted from Schmid & Casey (1986).

patterns with increasing strain in simple shear is not predicted by the full constraints fabric simulation model of Lister & co-workers (e.g. Lister & Hobbs 1980) using a given set of active slip systems. Similar fabric pattern transitions are, however, predicted for increasing strain in simple shear by the fabric simulation model of Jessell (1988) involving combined crystallographic slip and dynamic recrystallization (see also Jessell & Lister, this volume), although both cross and partial single-girdle fabrics are asymmetrical with respect to finite strain axes.

Evidence for grain boundary sliding and superplasticity

In metallurgy it has long been known that some polyphase fine-grained alloys can, under certain temperature and strain-rate conditions, be deformed in tension up to strains of more than 1000% without necking or fracture; such materials are then said to behave superplastically. The concept of superplasticity is essentially a phenomological one and does not imply a specific deformation mechanism. However, grain boundary sliding accommodated, for example, by diffusive mass transfer, is generally regarded as the major strain-producing mechanism (Edington *et al.* 1976; Nicolas & Poirier 1976; White 1977; Etheridge & Wilkie 1979; Schmid 1982; Poirier 1985).

Superplastic flow has only recently been observed in experimentally deformed geological materials (e.g. calcite experiments of Schmid 1975, 1976; Schmid *et al.* 1977; Walker *et al.*, this volume). Microstructural characteristics (Boullier & Gueguen 1975, Schmid 1982) of the superplastic regime include: (1) stable microstructure with grains remaining equant even after large bulk strains; (2) very small grain size, typically in the range of 1–10 μm; (3) moderate dislocation densities with no dislocation cells. However, it is generally difficult to find evidence for grain boundary sliding in naturally deformed rocks and a stable microstructure can also be explained by dynamic recrystallization during crystal plastic deformation (White 1977; Schmid 1982).

Dynamic recrystallization is commonly characterized by a crystallographic preferred orientation (e.g. Bell & Etheridge 1976), whilst the absence of a strong crystallographic preferred orientation *may* indicate that grain boundary sliding was the dominant strain producing mechanism (Boullier & Gueguen 1975). By implication, this *lack of crystallographic preferred orientation* in association with the above mentioned microstructural features *may* provide evidence for superplasticity. Possible natural examples of superplasticity in geological materials have been documented for quartz (Behrmann 1985; Behrmann & Mainprice 1987), calcite (Schmid 1975; Behrmann 1983), feldspar (Allison *et al.* 1979), orthopyroxene and hornblende (Boullier & Gueguen 1975). Several recent studies have indicated that a switch from crystal plasticity to superplasticity may occur below a critical grain size (e.g. Schmid *et al.* 1977; Behrmann 1983; Walker *et al.*, this volume). One possible geological example, evidenced by crystallographic fabrics, of grain size controlled millimetre-scale partitioning of crystal plasticity and superplasticity in a quartz-feldspar mylonite (Behrmann & Mainprice 1987) is given in Fig. 11.

Caution must be exercised in using a lack of crystal preferred orientation as evidence for superplasticity. It could, for example, be intuitively argued that annealing (static or post-tectonic) recrystallization has destroyed any pre-existing preferred orientation. Annealing experiments on fine-grained quartz aggregates with pre-existing crystallographic fabrics by Green *et al.* (1970), however, were generally found to result in a strengthening of preferred orientation. Caution must also be exercised in using the presence of a crystal fabric as evidence against superplasticity. For example, Etheridge & Wilkie (1979, p. 175) whilst noting that diffusion accommodated grain boundary sliding

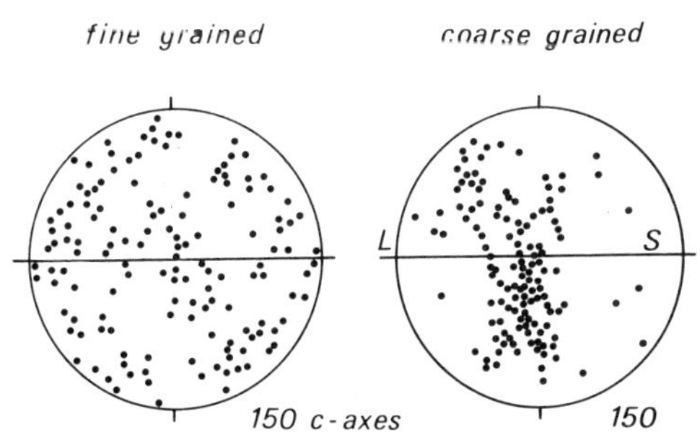

Fig. 11. Domainal variation in degree of *c*-axis preferred orientation within coarse grained (40–100 μm) quartz ribbons and bands of fine-grained (< 10 μm) dynamically recrystallized quartz from a quartz-feldspar mylonite; adapted from Behrmann & Mainprice (1987). Lack of preferred orientation in the fine grained domains was taken (in conjunction with microstructural criteria) to indicate grain boundary sliding and, by implication, superplastic flow. S, foliation; L, lineation.

will result in weakening of any pre-existing fabric, point out that preferred crystallographic orientations may both develop and strengthen in situations where sliding is accommodated by dislocation flow. Possible examples of such fabrics associated with grain boundary sliding have recently been described from quartz mylonites by Mancktelow (1987).

Thus the presence or absence of a crystallographic fabric will not provide totally unequivocal evidence for grain boundary sliding. As pointed out by Behrmann & Mainprice (1987, p. 302), conclusive evidence for grain boundary sliding comes from microstructural evidence for widespread grain boundary failure, and may require TEM analysis. In addition, it should be emphasized that although grain boundary sliding is the dominant strain producing mechanism in superplastic flow, the presence of grain boundary sliding does not provide unequivocal evidence for superplasticity.

Palaeo-stress analysis

Determination of the stress conditions associated with the formation of geological structures is an essential component in our understanding of how such structures evolve. There are a number of methods for determining stress directions and magnitudes in plastically deformed rocks. In general, intracrystalline strain-related petrofabric techniques are most commonly employed in the analysis of palaeo-stress directions (see reviews by Friedman 1964; Carter & Raleigh 1969; Friedman & Sowers 1970; Groshong 1988), whilst microstructural data (e.g. dynamically recrystallized grain size, dislocation density etc.) are used for determining stress magnitudes (see reviews by White 1979; Etheridge & Wilkie 1981; Schmid 1982; Ord & Christie 1984).

Friedman & Sowers (1970) have emphasized that three important principles must be borne in mind when employing crystal fabrics to deduce palaeo-stress directions. First, the principal stress directions deduced from the fabric relate to the state of stress in the rock at the instant that the fabric was formed. Secondly, the principal stress axes can change orientation and/or relative magnitude within a small domain during deformation. And thirdly, for the stress analysis to be valid, the initial fabric of the rock must permit the development of stress-related fabric elements for any orientation of the principal stresses; i.e. the starting material should, ideally, possess no original crystallographic fabric.

Deformation twins

The first crystallographic method for determining stress directions was proposed for calcite by Turner (1953). This method makes use of the fact that twin gliding in calcite on e $\{01\bar{1}2\}$ is a mechanically induced phenomenon, the glide direction being the edge $[e_1{:}r_2]$. The sense of shear (twinning) is positive, the c-axis of the twin moving towards the maximum principal compressive stress axis by a rotation of 52°.

Turner (1953, p. 282) proposed that the applied maximum and minimum principal stresses that would most effectively initiate twin gliding on e_1 in a given grain, lie in the plane containing the c-axis [0001] and the normal to the twin lamella. These stress directions are inclined at 45° to the lamellae pole; the compressive stress axis (σ_1') being oriented within the obtuse angle between the host c-axis [0001] and the lamellae trace at 71° to [0001], whilst the complementary tensional stress (σ_3') is inclined at 19° to [0001]. This analytical method, which was first confirmed in coaxial experimental deformation by Friedman (1963), is summarized in graphical form by Turner & Weiss (1963, p. 243). Subsequent studies have produced refinement in the original Turner (1953) method (e.g. Turner & Weiss 1963, p. 414) and numerical techniques based on this method have been described by Spang (1972), Groshong (1974) and Laurent *et al.* (1981). The Turner (1953) method has been extended, with modifications, to dolomite (Christie 1958), pyroxene (Raleigh & Talbot 1967), olivine (Carter & Raleigh 1969) and plagioclase feldspar (Lawrence 1970).

An alternative three dimensional method for determining the compressive stress direction from e twinning in calcite has been described by Dietrich & Song (1984, p. 31). The validity of the Turner (1953) and Dietrich & Song (1984) methods have been tested by Schmid *et al.* (1987) on calcite rocks subjected to experimental simple shear deformation. Results of this analysis, which indicate that the method of Dietrich & Song (1984) locates the orientation of the maximum principal stress more accurately, are reproduced here as Fig. 12. Rowe & Rutter (1990) have presented a set of experimentally calibrated techniques for the estimation of palaeostress *magnitudes* using calcite twinning.

Deformation lamellae

Three deformation lamellae based methods have been applied to the analysis of palaeo-

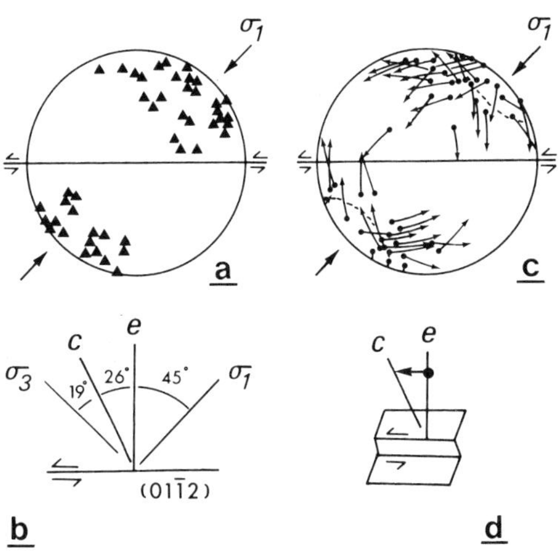

Fig. 12. Experimental comparison between two mechanical twin based methods used to infer the direction of the principal compressive stress in Carrara marble subjected to progressive simple shear; adapted from Schmid *et al.* (1987). Orientation of specimen shear plane and principal compressive stress direction (σ_1) indicated. In (**a**) the orientation of the axis of compressive stress (σ_1') that would be most effective in causing twin gliding on *e* {01$\bar{1}$2} is calculated for individual grains using the method described by Turner (1953). Crystallographic relationships used in this analytical method are summarised in (**b**); the diagram is drawn perpendicular to the glide plane and contains the glide direction (adapted from Friedman 1963). (**c**) & (**d**) method proposed by Dietrich & Song (1984); the arrow joining the pole to the active twin plane (solid circle) with the *c* axis of the host (arrowhead) indicates the direction of translation caused by twinning.

stress directions in quartz (see review by Carter & Raleigh 1969). The first method, regarded by Carter & Raleigh (1969, p. 1245) as the least reliable criterion, assumes that deformation lamellae are statistically inclined at an angle of less than 45° to the maximum principal compression direction. The second method, based on the coaxial experimental work of Carter *et al.* (1964, p. 731) and referred to as to C0-C1 method, assumes that the *c*-axes of the most deformed portion of a quartz grain is rotated closer to the compression direction. Partial great circles drawn between the *c*-axis of the least and most highly strained portions of individual grains should, for coaxial flattening (Carter & Raleigh 1969, p. 1247) converge on the average orientation of the compression direction. The third method, supported by the coaxial experiments of Carter *et al.* (1964) and Heard & Carter (1968) and referred to as the 'arrow method' consists of drawing a partial great circle between the *c*-axis (tail) and lamellae pole (arrow head) of individual grains, the tail being closer to the compression direction. For coaxial flattening, Carter *et al.* (1964) found that poles to lamellae form small-circle girdle patterns centred about the maximum principal compressive stress direction.

As noted by Carter & Raleigh (1969), these methods only apply to quartz containing lamellae inclined at *c.* 10° to 30° to the basal (0001) plane. An experimental comparison for coaxial flattening (Heard & Carter 1968) between the C0-C1 and arrow-head methods is presented in Fig. 13.

In this review, space does not permit a detailed discussion of the voluminous literature on application of these deformation twin and lamellae methods to palaeo-stress analysis of naturally deformed rocks. The reader is referred to Carter & Raleigh (1969), Friedman & Sowers (1970) and references therein, for examples of these applications.

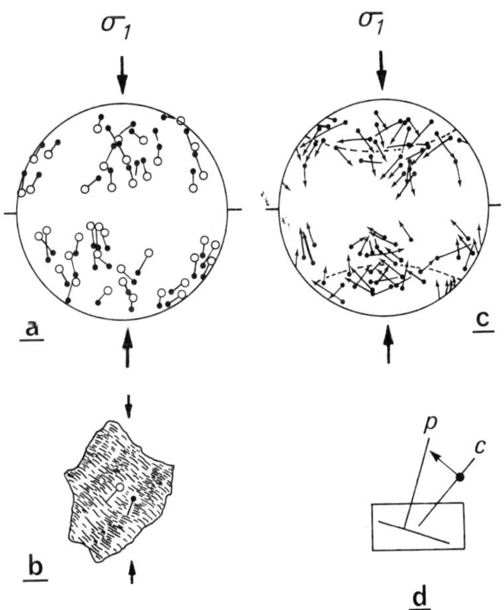

Fig. 13. Experimental comparison of two deformation lamellae based methods used to infer the direction of the principal compressive stress in Simpson quartzite subjected to progressive coaxial flattening; adapted from Heard & Carter (1968). (**a**) & (**b**) C0-C1 method described by Carter *et al.* (1964); *c*-axis in more highly deformed region of grain (solid circle), *c*-axis of less highly deformed region (open circle). (**c**) and (**d**) Arrow method; lamellae pole (arrow head), *c*-axis of grain hosting lamellae (solid circle). See text for details.

Concluding statement

The above review has highlighted seven research topics associated with analysis of strain paths, deformation processes and stress regimes which, given appropriate conditions, may be investigated using crystallographic fabric analysis. It is emphasized that owing to the number of variables potentially controlling crystallographic preferred orientation, fabric analysis in isolation from rock microstructure and structural setting will rarely provide meaningful geological data. However, by integrating fabric studies with all other information available from techniques ranging in scale from field mapping to TEM analysis of crystal dislocations, it is suggested that crystallographic fabrics may provide important data both for constraining deformation histories and identifying new lines of research for individual structural settings.

G. Price and P. Williams are thanked for their detailed reviews of an earlier version of the manuscript.

References

ALLISON, I., BARNETT, R. L., & KERRICH, R. 1979. Superplastic flow and changes in crystal chemistry of feldspars. *Tectonophysics* **53**, T41–T46.

BEHRMANN, J. H. 1983. Microstructure and fabric transitions in calcite tectonites from the Sierra Alhamilla (Spain). *Geologische Rundschau*, **72**, 605–618.

—— 1985. Crystal plasticity and superplasticity in quartzite: a natural example. *Tectonophysics*, **115**, 101–129.

—— & MAINPRICE, D. 1987. Deformation mechanisms in a high temperature quartz-feldspar mylonite: evidence for superplastic flow in the continental crust. *Tectonophysics*, **140**, 297–305.

—— & PLATT, J. P. 1982. Sense of nappe emplacement from quartz *c*-axis fabrics: an example from the Betic Cordilleras (Spain). *Earth and Planetary Science Letters*, **59**, 208–215.

BELL, T. H. & ETHERIDGE, M. A. 1976. The deformation and recrystallisation of quartz in a mylonite zone, central Australia. *Tectonophysics*, **32**, 235–267.

BLACIC, J. D. & CHRISTIE, J. M. 1984. Plasticity and hydrolytic weakening of quartz single crystals. *Journal of Geophysical Research*, **89**, 4223–4239.

BLUMENFELD, P. & BOUCHEZ, J -L. 1988. Shear criteria in granite and migmatite deformed in magmatic and solid states. *Journal of Structural Geology*, **10**, 361–372.

——, MAINPRICE, D. & BOUCHEZ, J -L. 1986. C-slip in quartz from subsolidus deformed granite. *Tectonophysics*, **127**, 97–115.

BOUCHEZ, J -L. 1977. Plastic deformation of quartzites at low temperature in an area of natural strain gradient. *Tectonophysics*, **39**, 25–50.

—— 1978. Preferred orientations of quartz *a*-axes in some tectonites: kinematic inferences. *Tectonophysics*, **49**, T25–T30.

—— & DUVAL, P. 1982. The fabric of polycrystalline ice in simple shear: experiments in torsion, natural deformation and geometrical interpretation. *Textures and Microstructures*, **5**, 1–17.

——, LISTER, G. S. & NICOLAS, A. 1983. Fabric asymmetry and shear sense in movement zones. *Geologische Rundschau*, **72**, 401–419.

BOULLIER, A. M. & GUEGUEN, Y. 1975. SP-mylonites: origin of some mylonites by superplastic flow. *Contributions to Mineralogy and Petrology*, **50**, 93–104.

—— & QUENARDEL, J. -M. 1981. The Caledonides of northern Norway: relation between preferred orientation of quartz lattice, strain and translation of the nappes. *In*: MCCLAY, K. R. & PRICE, N. J. (eds) *Thrust and Nappe Tectonics*. Geological Society, London, Special Publication, **9**, 185–195.

BRACE, W. F. 1955. Quartzite pebble deformation in central Vermont. *American Journal of Science*, **253**, 129–145.

CARTER, N. L. & RALEIGH, B. 1969. Principal stress directions from plastic flow in crystals. *Geological Society of America Bulletin*, **80**, 1231–1264.

——, CHRISTIE, J. M. & GRIGGS, D. T. 1964. Experimental deformation and recrystallisation of quartz. *Journal of Geology*, **72**, 687–733.

CASEY, M. 1981. Numerical analysis of X-ray data: an implementation in Fortran allowing triclinic or axial specimen symmetry and most crystal symmetries. *Tectonophysics*, **78**, 51–64.

CHRISTIE, J. M. 1958. Dynamic interpretation of the fabric of dolomite from the Moine thrust zone, NW Scotland. *American Journal of Science*, **256**, 159–170.

—— 1963. The Moine thrust zone in the Assynt region, northwest Scotland. *University of California Publications in Geological Sciences*, **40**, 345–440.

COMPTON, R. R. 1980. Fabrics and strains in quartzites of a metamorphic core complex, Raft River Mountains, Utah. *In*: CRITTENDEN, M. D. JR, CONEY, P. J. & DAVIS, G. H. (eds) *Cordilleran Metamorphic Core Complexes*. Memoir of the Geological Society of America, **153**, 385–398.

DELL'ANGELO, L. N. & TULLIS, J. 1989. Fabric development in experimentally sheared quartzites. *Tectonophysics*, **169**, 1–22.

DIETRICH, D. & SONG, H. 1984. Calcite fabrics in a natural shear environment, the Helvetic nappes of western Switzerland. *Journal of Structural Geology*, **6**, 19–32.

DINGLEY, D. J. 1984. Diffraction from sub-micron areas using electron backscattering in a scanning electron microscope. *Scanning Electron Microscopy*. SEM Inc., AMF O'Hare (Chicago), 569–575.

EDINGTON, J. W., MELTON, K. N. & CUTLER, C. P. 1976. Superplasticity. *Progress in Materials Science*, **21**, 63–170.

EISBACHER, G. H. 1970. Deformation mechanics

of mylonite rocks and fractured granites in Cobequid Mountains, Nova Scotia, Canada. *Geological Society of America Bulletin*, **81**, 2009–2020.

Elliott, D. 1972. Deformation paths in structural geology. *Geological Society of America Bulletin*, **83**, 2621–2638.

Etchecopar, A. 1977. A plane kinematic model of progressive deformation in a polycrystalline aggregate. *Tectonophysics*, **39**, 121–139.

—— & Vasseur, G. 1987. A 3-D kinematic model of fabric development in polycrystalline aggregates: comparisons with experimental and natural examples. *Journal of Structural Geology*, **9**, 705–717.

Etheridge, M. A. & Wilkie, J. C. 1979, Grainsize reduction, grain boundary sliding and the flow strength of mylonites. *Tectonophysics*, **58**, 159–178.

—— & —— 1981. An assessment of dynamically recrystallised grain size as a palaeopiezometer in quartz bearing mylonite zones. *Tectonophysics*, **78**, 475–508.

Fairbairn, H. W. 1949. *Structural Petrology*. Addison-Wesley, Cambridge, Massachusetts.

Friedman, M. 1963. Petrofabric analysis of experimentally deformed calcite-cemented sandstones. *Journal of Geology*, **71**, 12–37.

—— 1964. Petrofabric techniques for determination of principal stress directions in rocks. *In*: Judd, W. R. (ed.) *State of Stress in the Earth's Crust*, American Elsevier Publishing Co., New York, 451–550.

—— & Sowers, G. M. 1970. Petrofabrics: a critical review. *Canadian Journal of Earth Sciences*, **7**, 447–497.

Gapais, D. & Barbarin, B. 1986. Quartz fabric transition in a cooling syntectonic granite (Hermitage Massif, France). *Tectonophysics*, **125**, 357–370.

Garcia Celma, 1982. Domainal and fabric heterogeneities in the Cap de Creus quartz mylonites. *Journal of Structural Geology*, **4**, 443–455.

—— 1983. *C-axis and shape fabrics in quartz mylonites of Cap de Creus (Spain): their properties and development*. Proefschrift, Rijksuniversiteit, Utrecht.

Green, H. W., Griggs, D. T. & Christie, J. M. 1970. Syntectonic recrystallisation and annealling of quartz aggregates. *In*: Paulitsch, P. (ed.) *Experimental and Natural Rock Deformation*. Springer-Verlag, Berlin, 272–335.

Groshong, R. H. 1974. Experimental test of least-squares strain gage calculation using twinned calcite. *Geological Society America Bulletin*, **85**, 1855–1864.

—— 1988. Low temperature deformation mechanisms and their interpretation. *Geological Society of America Bulletin*, **100**, 1329–1360.

Heard, H. C. & Carter, N. L. 1968. Experimentally induced 'natural' intragranular flow in quartz and quartzite. *American Journal of Science*, **266**, 1–42.

Hobbs, B. E. 1985. The geological significance of microfabric analysis. *In*: Wenk, H.-R. (ed.) *Preferred Orientation in Deformed Metals and Rocks: an Introduction to Modern Texture Analysis*. Academic Press, Orlando, 463–484.

——, Means, W. D. & Williams, P. F. 1976. *An Outline of Structural Geology*. John Wiley & Sons, New York.

Humphreys, C. J. 1983. Imaging dislocations. *Dislocations in Solids*, **5**, 1–56.

Jessell, M. W. 1988. Simulation of fabric development in recrystallising aggregates — II. Example model runs. *Journal of Structural Geology*, **10**, 779–794.

—— & Lister, G. S. 1990. A simulation of the temperature dependence of quartz fabrics. *This volume*.

Kern, H. & Wenk, H.-R. 1983. Calcite texture development in experimentally induced ductile shear zones. *Contributions to Mineralogy and Petrology*, **83**, 231–236.

Knipe, R. J. 1981. The interaction of deformation and metamorphism in slates. *Tectonophysics*, **78**, 249–272.

Laurent, P. 1987. Shear sense determination on striated faults from *e* twin lamellae in calcite. *Journal of Structural Geology*, **9**, 591–596.

——, Bernard, P., Vasseur, G. & Etchecopar, A. 1981. Stress tensor determination from the study of *e* twins in calcite: a linear programming method. *Tectonophysics*, **78**, 651–660.

Law, R. D. 1986. Relationships between strain and quartz crystallographic fabrics in the Roche Maurice quartzites of Plougastel, western Brittany. *Journal of Structural Geology*, **8**, 493–515.

—— 1987. Heterogeneous deformation and quartz crystallographic fabric transitions: natural examples from the Stack of Glencoul, northern Assynt. *Journal of Structural Geology*, **9**, 819–833.

——, Casey, M. & Knipe, R. J. 1986. Kinematic and tectonic significance of microstructures and crystallographic fabrics within quartz mylonites from the Assynt and Eriboll regions of the Moine thrust zone, NW Scotland. *Transactions of the Royal Society of Edinburgh: Earth Sciences*, **77**, 99–123.

——, Knipe, R. J. & Dayan, H. 1984. Strain path partitioning within thrust sheets: microstructural and petrofabric evidence from the Moine thrust zone at Loch Eriboll, North-West Scotland. *Journal of Structural Geology*, **6**, 477–497.

——, Schmid, S. M. & Wheeler, J. 1990. Simple shear deformation and quartz crystallographic fabrics: a possible natural example from the Torridon area of NW Scotland. *Journal of Structural Geology*, **12**, 29–45.

Lawrence, R. D. 1970. Stress analysis based on albite twinning in plagioclase feldspars. *Geological Society of America Bulletin*, **81**, 2507–2512.

Le Corre, C. 1979. L'evolution typologique et texturale des roches argilo-silteuses au cours de la schistogenese. Notion de trajectoire de fabrique. *Bulletin de Minéralogie*, **102**, 273–281.

LEE, J., MILLER, E. M. & SUTTER, J. F. 1987. Ductile strain and metamorphism in an extensional tectonic setting: a case study from the northern Snake Range, Nevada, USA. *In:* COWARD, M. P., DEWEY, J. F. & HANCOCK, P. L. (eds) *Continental Extensional Tectonics*. Geological Society, London, Special Publication, **28**, 267–298.

LINKER, M. F., KIRBY, S. H., ORD, A. & CHRISTIE, J. M. 1984. Effects of compression direction on the plasticity and rheology of hydrolytically weakened synthetic quartz crystals at atmospheric pressure. *Journal of Geophysical Research*, **89**, 4241–455.

LISTER, G. S. 1977. Discussion: Crossed girdle c-axis fabrics in quartzites plastically deformed by plane strain and progressive simple shear. *Tectonophysics*, **39**, 51–54.

—— 1978. Texture transitions in plastically deformed calcite rocks. *In*: Proceedings of the 5th International Conference on Textures of Materials. Springer Verlag, Berlin, 199–200.

—— & DAVIS, G. A. 1989. The origin of metamorphic core complexes and detachment faults during Teriary continental extension in the northern Colorado River region, USA. *Journal of Structural Geology*, **11**, 65–94.

—— & DORNSIEPEN, U. F. 1982. Fabric transitions in the Saxony Granulite Terrain. *Journal of Structural Geology*, **4**, 81–92.

—— & HOBBS, B. E. 1980. The simulation of fabric development during plastic deformation and it's application to quartzite: the influence of deformation history. *Journal of Structural Geology*, **2**, 355–370.

—— & PATERSON, M. S. 1979. The simulation of fabric development during plastic deformation and its application to quartzite: fabric transitions. *Journal of Structural Geology*, **1**, 99–115.

—— & WILLIAMS, P. F. 1979. Fabric development in shear zones: theoretical controls and observed phenomena. *Journal of Structural Geology*, **1**, 283–79.

—— & —— 1983. The partitioning of deformation in flowing rock masses. *Tectonophysics*, **92**, 1–33.

——, PATERSON, M. S. & HOBBS, B. E. 1978. The simulation of fabric development during plastic deformation and its application to quartzite: the model. *Tectonophysics*, **45**, 107–158.

LLOYD, G. E. 1985. Review of instrumentation, techniques and applications of SEM in mineralogy, *In:* WHITE, J. C. (ed.) *Applications of Electron Microscopy in the Earth Sciences*. Mineralogical Association of Canada Short Course, **11**, 151–188.

MAINPRICE, D, BOUCHEZ, J-L, BLUMENFELD, P. & TUBIA, J.M. 1986 Dominant *c* slip in naturally deformed quartz: implications for dramatic plastic softening at high temperature. *Geology*, **14**, 819–822.

—— & NICOLAS, A. 1989. Development of shape and lattice preferred orientations: application to the seismic anisotropy of the lower crust. *Journal of Structural Geology*, **11**, 175–189.

MALAVIELLE, J. 1987. Kinematics of compressional and extensional ductile shearing deformation in a metamorphic core complex of the north-eastern Basin and Range. *Journal of Structural Geology*, **9**, 541–554.

MANCKTELOW, N. S. 1987. Quartz textures from the Simplon Fault Zone, southwest Switzerland and north Italy. *Tectonophysics*, **135**, 133–153.

MARJORIBANKS, R. W. 1976. The relation between microfabric and strain in a progressively deformed quartzite sequence from central Australia. *Tectonophysics*, **32**, 269–293.

MEANS, W. D., HOBBS, B. E., LISTER, G. S. & WILLIAMS, P. F. 1980. Vorticity and non-coaxiality in progressive deformations. *Journal of Structural Geology*, **2**, 371–378.

MILLER, D. M. & CHRISTIE, J. M. 1981. Comparison of quartz microfabric with strain in a recrystallised quartzite. *Journal of Structural Geology*, **3**, 129–142.

NICOLAS, A. & POIRIER, J. P. 1976. *Crystalline Plasticity and Solid State Flow in Metamorphic Rocks*. John Wiley & Sons, London.

OERTEL, G. 1985. Reorientation due to grain shape. *In:* WENK, H.-R. (ed.) *Preferred Orientations in Deformed Metals and Rocks: an Introduction to Modern Texture Analysis*, Academic Press, Orlando, 159–265.

ORD, A. & CHRISTIE, J. M. 1984. Flow stresses from microstructures in mylonitic quartzites of the Moine thrust zone, Assynt area, Scotland. *Journal of Structural Geology*, **6**, 639–654.

PASSCHIER, C. W. 1983. The reliability of asymmetric *c*-axis fabrics of quartz to determine sense of vorticity. *Tectonophysics*, **99**, T9–T18.

—— 1986. Mylonites in the continental crust and their role as seismic reflectors. *Geologie en Mijnbouw*, **65**, 167–176.

—— 1988. Analysis of deformation paths in shear zones. *Geologische Rundschau*, **77**, 309–318.

PATERSON, S. R., VERNON, R. H. & TOBISCH, O. T. 1989. A review of criteria for the identification of magmatic and tectonic foliations in granitoids. *Journal of Structural Geology*, **11**, 349–364.

PFIFFNER, A. & RAMSAY, J. G. 1982. Constraints on geological strain rates, arguments from finite strain states of naturally deformed rocks. *Journal of Geophysical Research*, **87**, 311–321.

PLATT, J. P. & BEHRMANN, J. H. 1986. Structures and fabrics in a crustal scale shear zone, Betic Cordilleras, SE Spain. *Journal of Structural Geology*, **8**, 15–34.

POIRIER, J. P. 1985. *Creep of Crystals*. Cambridge University Press.

PRICE, G. P. 1985. Preferred orientations in quartzites. *In*: WENK, H.-R. (ed.) *Preferred Orientations in Deformed Metals and Rocks: an Introduction to Modern Texture Analysis*. Academic Press, Orlando, 385–406.

RALEIGH, C. B. & TALBOT, J. L. 1967. Mechanical twinning in naturally and experimentally deformed diopside. *American Journal of Science*, **265**, 151–165.

RAMSAY, J. G. 1967. *Folding and Fracturing of Rocks*. McGraw Hill, New York.

—— & Huber, M. I. 1983. *The Techniques of Modern Structural Geology, Volume 1: Strain Analysis.* Academic Press, New York.

—— & Wood, D. S. 1973. The geometric effects of volume change during deformation processes. *Tectonophysics*, **13**, 263–277.

Rowe, K. J. & Rutter, E. H. 1990. Palaeostress estimation using calcite twinning: experimental calibration and application to nature. *Journal of Structural Geology*, **12**, 1–18.

Rutter, E. H. & Rusbridge, H. 1977. The effect of non-coaxial strain paths on cyrstallographic preferred orientation development in the experimental deformation of a marble. *Tectonophysics*, **39**, 73–86.

Sander, B. 1970. *An Introduction to the Study of Fabrics of Geological Bodies*. Pergamon, Oxford, (translated with amendments from the 1950 edition in German).

Sanderson, D. J. 1982. Models of strain variation in nappes and thrust sheets: a review. *Tectonophysics*, **88**, 201–233.

Schmid, S. M. 1975. The Glarus overthrust: Field evidence and mechanical model. *Eclogae Geologicae Helvetiae*, **68**, 247–280.

—— 1976. Rheological evidence for changes in the deformation mechanism of Solenhofen limestone towards low stresses. *Tectonophysics*, **31**, T21–T28.

—— 1982. Microfabric studies as indicators of deformation mechanisms and flow laws operative in mountain building. *In*: Hsu, K. J. (ed.) *Mountain Building Processes*. Academic Press, London, 95–110.

——, Boland, J. N. & Paterson, M. S. 1977. Superplastic flow in fine grained limestone. *Tectonophysics*, **43**, 257–291.

—— & Casey, M. 1986. Complete fabric analysis of some commonly observed quartz c-axis patterns. *In*: Hobbs, B. E. & Heard, H. C. (eds) *Mineral and Rock Deformation: Laboratory Studies - The Paterson Volume*. American Geophysical Union, Geophysical Monograph, **36**, 263–286.

——, —— & Starkey, J. 1981. An illustration of the advantages of a complete texture analysis described by the Orientation Distribution Function (ODF) using quartz pole figure data. *Tectonophysics*, **78**, 101–117.

——, Panozzo, R. & Bauer, S. 1987. Simple shear experiments on calcite rocks: rheology and microfabric. *Journal of Structural Geology*, **9**, 747–778.

Schmidt, W. 1925. Gefugestatistik. *Tschermaks Mineralogische und Petrographische Mitteilungen*, **38**, 392–243.

Simpson, C. 1986. Determination of movement sense in mylonites. *Journal of Geological Education*, **34**, 246–261.

—— & Schmid, S. M. 1983. An evaluation of criteria to deduce the sense of movement in sheared rocks. *Geological Society of America Bulletin*, **94**, 1281–8.

Spang, J. H. 1972. Numerical method for dynamic analysis of calcite twin lamellae. *Geological Society of America Bulletin*, **83**, 467–472.

Starkey, J. Cutforth, C. 1978. A demonstration of the interdependence of the degree of quartz preferred orientation and the quartz content of deformed rocks. *Canadian Journal of Earth Sciences*, **15**, 841–847.

Strand, T. 1944. Structural petrology of the Bygdin conglomerate. *Norsk Geologisk Tidsskrift*, **24**, 14–31.

Sylvester, A. G. & Christie, J. M. 1968. The origin of crossed-girdle orientations of optic axes in deformed quartzites. *Journal of Geology*, **76**, 571–580.

—— & Janecky, D. R. 1988. Structure and petrofabrics of quartzite and elongate pebbles at Sandviksfjell, Bergen, Norway. *Norsk Geologisk Tidsskrift*, **68**, 31–50.

Takeshita, T. & Wenk, H.-R. 1988. Plastic anisotropy and geometrical hardening in quartzites. *Tectonophysics*, **149**, 345–361.

Teyssier, C., Amri, C. & Hobbs, B. E. 1988. South Arunta Block: the internal zones of a Proterozoic overthrust in central Australia. *Precambrian Research*, **40**, 157–173.

Tullis, J. 1977. Preferred orientation of quartz produced by slip during plane strain. *Tectonophysics*, **39**, 87–102.

——, Christie, J. M. & Griggs, D. T. 1973. Microstructure and preferred orientations of experimentally deformed quartzites. *Geological Society of America Bulletin*, **84**, 297–314.

Tullis, T. 1976. Experiments on the origin of slaty cleavage and schistosity. *Geological Society of America Bulletin*, **87**, 745–753.

—— & Wood, D. S. 1975. Correlation of finite strain from both reduction bodies and preferred orientation of mica in slate from Wales. *Geological Society of America Bulletin*, **86**, 632–638.

Turner, F. J. 1953. Nature and dynamic interpretation of deformation lamellae in calcite of three marbles. *American Journal of Science*, **251**, 276–298.

—— & Weiss, L. E. 1963. *Structural Analysis of Metamorphic Tectonites*. McGraw-Hill, New York.

Vissers, R. L. M. 1989. Asymmetric quartz *c*-axis fabrics and flow vorticity: a study using rotated garnets. *Journal of Structural Geology*, **11**, 231–245.

Wagner, F., Wenk, H.-R., Kern, H., Van Houtte, P. & Esling, C. 1982. Development of preferred orientation in plane strain deformed limestone: experiment and theory. *Contributions to Mineralogy and Petrology*, **80**, 132–139.

Walker, A. N., Rutter, E. H. & Brodie, K. H. 1990. Experimental study of grain-size sensitive flow of synthetic, hot-pressed calcite rocks. (*This volume*).

Wenk, H.-R. 1979. Some roots of experimental rock deformation. *Bulletin de Minéralogie* **102**, 195–202.

—— 1985a Measurement of pole figures. *In*: Wenk, H.-R. (ed.) *Preferred Orientations in Deformed Metals and Rocks: an Introduction to Modern*

Texture Analysis, pp. 11–47. Academic Press, Orlando.

—— 1985b Carbonates. *In*: Wenk, H.-R. (ed.) *Preferred Orientations in Deformed Metals and Rocks: an Introduction to Modern Texture Analysis*, pp. 361–84. Academic Press, Orlando.

——, Canova, G., Molinari, A. & Kocks, U. F. 1989. Viscoplastic modeling of texture development in quartzite. *Journal of Geophysical Research*, **94**, 17895–17906.

——, Takeshita, T., Bechler, E., Erskine, B. G. & Matthies, S. 1987. Pure shear and simple shear calcite textures. Comparison of experimental, theoretical and natural data. *Journal of Structural Geology*, **9**, 731–746.

White, S. H. 1977. Geological significance of recovery and recrystallisation processes in quartz. *Tectonophysics*, **39**, 143–170.

—— 1979. Difficulties associated with palaeo-stress analysis. *Bulletin de Minéralogie*, **102**, 210–215.

—— 1985. Defect structures in deformed minerals. *In*: White, J. C. (ed.) *Applications of Electron Microscopy in the Earth Sciences*. Mineralogical Association of Canada, Short Course, **11**, 121–149.

——, Burrows, S. E., Carreras, J., Shaw, N. D. & Humphreys, F. J. 1980. On mylonites in ductile shear zones. *Journal of Structural Geology*, **2**, 175–187.

Wilson. C. J. L. 1975. Preferred orientation in quartz ribbon mylonites. *Geological Society of America Bulletin*, **86**, 968–974.

Wood, D. S. 1971. *Studies of Strain and Slaty Cleavage in the Caledonides of northwest Europe and the eastern United States*. Ph.D thesis, University of Leeds.

—— & Oertel, G. 1980. Deformation in the Cambrian slate belt of Wales. *Journal of Geology*, **88**, 285–308.

A simulation of the temperature dependence of quartz fabrics

M. W. JESSELL & G. S. LISTER

Department of Earth Sciences, Monash University, Clayton, VIC, 3168, Australia

Abstract: This paper presents the predictions of a hybrid model of fabric development in quartzites that combines aspects of previous models based on lattice rotations and dynamic recrystallization. The input parameters for the model have been systematically varied in order to investigate changes in deformation fabrics due to increasing temperature, and predictions of crystallographic preferred orientations and grain shape fabrics are made. The resulting fabrics reflect the combination of both deformation processes over the range of the parameters used. The fabrics evolve continuously with progressive deformation, and produce crossed girdles, single girdles and point maxima for a single deformation geometry. The results suggest that the coupling of dynamic recrystallization with lattice rotations can actually produce a larger variation in crystallographic preferred orientations with temperature than is predicted by using a model that only considers lattice rotations. Using this model point maxima develop even though a multiple slip model has been assumed.

The interpretation of quartz fabrics

It is now well established that crystallographic preferred orientations for a given mineral fit into a relatively small set of patterns, and indeed this has always been one of their attractions to structural geologists, who hope to be able to derive predictive models for fabric development in terms of their conditions of formation. Price (1985) has demonstrated that a part of the variation in quartz crystallographic preferred orientations can be accounted for by considering the geometry and history of deformation using the predictions of the Taylor–Bishop–Hill model (TBH) as formulated for quartz by Lister *et al.* (1978). Theoretical considerations suggest that temperature, strain rate and chemical environment may also play a role in determining crystallographic preferred orientations (see Hobbs 1985 for discussion), via their control of both the relative critical resolved shear stresses of mineral slip system and diffusional constants. Current thinking suggests that a model which only considers a single deformation process will not be applicable to the wide spectrum of metamorphic grades at which rocks deform (Jessell 1986; Knipe & Law 1987; Karato 1987). However, for a broad range of conditions where both dynamic recrystallization and deformation induced lattice rotations are significant deformation processes, it may still be possible to interpret the consequent fabrics in terms of a relatively small number of processes. Etchecopar & Vasseur (1987) have formulated a model based on the combination of lattice rotations with recrystallization, where they reset the grain shape fabric to a foam texture periodically and in an *ad hoc* manner. While this approach fulfills part of the goal of combining the two deformation processes, it cannot be used to investigate the interplay of specific recrystallization processes during deformation. Karato's (1987) analytic model of the coupling of dynamic recrystallization and lattice rotations in olivine assumes that grains with low Schmidt factors will grow preferentially, which as Karato points out implies that the deformation is heterogeneously distributed between grains, since this will lead to grains well oriented for slip deforming at much higher rates than other orientations. We would argue that at the temperatures where grain boundary mobilities are high enough to allow significant grain boundary migration, the aggregate will be deforming more homogeneously, and that grains with low Schmidt factors will in fact be the very grains to be preferentially consumed.

The purpose of this paper is to evaluate a new computer simulation of crystallographic and grain shape preferred orientation development in a quartzite shear-zone, with input parameters for the simulation chosen to reflect a systematic variation in temperature (Lister 1978; Takeshita & Wenk 1988). With the increased use of complete crystallographic fabric analysis (Schmid & Casey 1986; Lloyd & Ferguson 1986; Mancktelow 1987) we hope that these predictions can be tested by detailed fabric analysis of naturally deformed rocks.

From Knipe, R. J. & Rutter, E. H. (eds), 1990, *Deformation Mechanisms, Rheology and Tectonics*, Geological Society Special Publication No. 54, pp. 353–362.

Presentation of the model

The model presented in this paper is essentially a hybrid of models originally described by Lister *et al.* (1978) and Jessell (1988*a*). It is based on a two-dimensional 100 by 100 element discretization of a polycrystal, with each element representing a small area of a crystal, which we shall refer to as a crystallite (Fig. 1). In addition to its spatial location, each crystallite refers to a specific three-dimensional crystallographic orientation. Fabric development in a dextral quartzite shear zone is simulated by repeating four incremental deformation processes. These processes are:

(1) homogeneous strain, simulated by shuffling laterally the positions of entire rows of crystallites;
(2) lattice rotations predicted by the Taylor–Bishop–Hill model;
(3) a Monte Carlo simulation of grain boundary migration driven by the stored energy of deformation;
(4) subgrain formation by a rotation recrystallization mechanism, simulated by nucleating neoblasts with host controlled orientations.

Detailed discussion of the simulation techniques for these deformation processes can be found in Lister *et al.* (1978; point 2) and Jessell (1988*b*; points 1, 3 and 4). The simulation of grain boundary migration has been significantly altered since the work of Jessell (1988*a*), and these changes will be discussed below. The exclusion of other deformation processes, such as kinking, grain boundary sliding and other dynamic recrystallization processes (Urai *et al.* 1986) should be kept in mind when interpreting the results presented here. It should also be emphasised that this is essentially a geometric model of fabric development, not a mechanical model of polycrystalline deformation, and strain heterogeneities, on both the grain scale (Mancktelow 1981) and the scale of the foliation (White *et al.* 1980), are not considered.

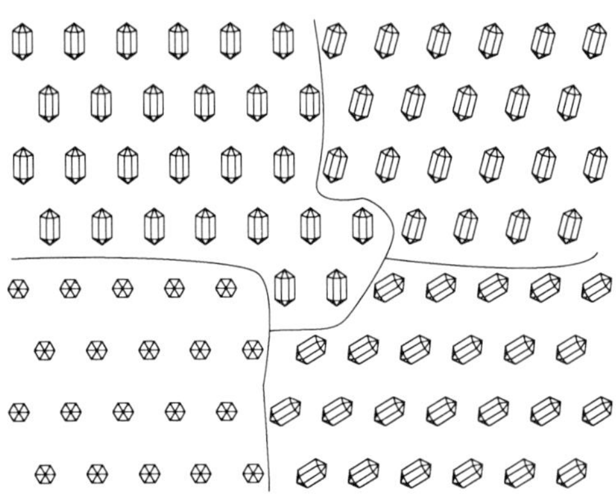

Fig. 1. Schematic basis for simulation. Part of the 100 × 100 array of crystallites arranged on a triangular grid. Each crystallite in this array is associated with a crystallographic orientation, and grain boundaries are inferred features which are defined as lines separating orientation domains in the array.

The driving force for grain boundary migration

In the model formulated by Jessell (1988*a*), the orientation-dependent patterns of stored energy, which are assumed to be the driving force for the grain boundary migration, were defined somewhat arbitrarily. In the present model these patterns are based on the results of the TBH model itself, which uses the critical resolved shear stress (CRSS) values of the available slip systems and the imposed deformation geometry to calculate the work done by each grain. Although we accept it as an inherent drawback, it is fundamental to the model that we assume homogeneous strain, so there is no mechanical coupling between the ease of deformation and the rate of straining. In this model we have assumed that there is a direct correlation between the resistance to deformation and the stored energy of deformation (as dislocations). Since in this model the CRSS values control the patterns of the orientation dependent work term, and this in turn drives the grain boundary migration, we have a coupled variation in the driving force for grain boundary migration and the lattice rotation patterns.

This simulation also considers grain boundary energy as a driving force for grain boundary migration, following the principles used by Anderson *et al.* (1984), so that another driving force for grain boundary migration is provided by the local curvature of grain boundaries. Urai *et al.* (1986) have compared the driving force resulting from dislocations with that from grain boundary energy, and conclude that the driving force from dislocations will be approximately three orders of magnitude higher than that from grain boundary energy. However their experimental results suggest that locally grain boundary energy may still dominate grain shape fabrics, since even in their deformed samples grain boundary angles clustered near 120°. We have chosen to make the driving force for grain boundary energy just one order of magnitude weaker than that of the average work term. Post-tectonic recrystallization, which could clearly have a major impact on the grain shape fabrics developed, has been simulated using this model, and will be the subject of a future paper.

Karato (1987) bases his analytical model of the relative roles of grain boundary migration and lattice rotations on the assumption that olivine crystals swept by grain boundaries will be completely devoid of dislocations immediately following the passage of the grain boundary, and will then progressively work harden until they reach a steady state level. This assumption has been adopted here, so that newly swept crystallites will, like newly formed subgrains, have zero stored energy levels.

Choice of parameter values

As in any simulation, the choice of input parameters is as crucial to the relevance of the model as the formulation of the simulation itself. In the TBH simulations of Lister and coworkers the choice of deformation geometries and CRSS values was intentionally broad, so that in principle the range investigated would encompass all naturally occurring situations. Jessell (1988*a*) took a more limited look at the range of possibilities due to the more simplistic approach of that model. This paper follows Takeshita & Wenk (1988) in choosing a set of parameters that are varied systematically to reflect deformation at different temperatures. This of course requires an understanding and discussion of why these underlying parameters should vary with temperature. The actual values of the parameters chosen for this paper are shown in Table 1.

The variation of CRSS values with temperature

It is well established for metals that the CRSS values for a given slip system vary with temperature (see chapter 2 of Honeycombe 1984 for examples), although there are cases where this does not hold true (Beuers *et al.* 1987). With most geological materials it is inherently more difficult to measure CRSS values for natural conditions and the specific values for quartz can only be extrapolated from laboratory conditions. One important aspect of the temperature dependence is that the CRSSs for each slip system will co-vary. Since there is a limited range of temperatures and strain rates that may be attained within the crust, there are likely to be commonly recurring combinations of CRSS values for a given mineral, as has been recognised by Lister (1978), and Takeshita & Wenk (1988). This means that only a subset of the fabrics which can be produced by these types of simulations will actually occur in nature. This in itself explains why there are a number of commonly recurring patterns of crystallographic preferred orientations, and is analogous to the recurrence of particular metamorphic assemblages. Hobbs (1985) synthesized the existing experimental data to produce a schematic plot of CRSS values v. temperature for three of the slip systems of quartz and we follow the same principle here.

An additional aspect of the assumed driving force for grain boundary migration, the lattice defect energy, is that with increasing temperature the rate of dynamic recovery will also be increasing, so that we have included a dynamic recovery term that reduces the overall stored energy levels for all grains.

Grain boundary mobilities

The choice of mobilities for quartzite is problematic since it is possible that they may vary not only with temperature, but also with the magnitude of the driving force (see Urai *et al.*

Table 1. *The input parameters used to simulate deformation at five temperatures, together with the values used in the comparison with Lister & Hobbs (1980) Model C*

Parameter	L	L-M	M	M-H	H	Model C
CRSS values:						
basal <*a*>	5	4.75	4.5	4.25	4	1
prism <*a*>	6	6	4.5	3.75	3	1
prism <*c*>	11	10	9	9	9	∞
rhomb ± <*c*+2*a*>	9.5	9.25	9	8.75	8.5	3
rhomb ± <*c*+3*a*>	9.5	9.25	9	8.75	8.5	2.5
Grain boundary mobility	0.75	2.5	7.5	17.5	50.0	7.5
Work hardening rate	0.9	0.875	0.85	0.825	0.8	.75
Dynamic recovery rate	1	2	3	5	10	3
Probability of subgrain formation	0.001	0.001	0.001	0.001	0.001	0.001
Shear strain rate	0.1	0.1	0.1	0.1	0.1	0.1

1986 for discussion) and the nature of the grain boundary (Drury & Urai 1988). The results for experimentally annealed quartzites (Tullis & Yund 1982), and for a variety of ceramics (Yan *et al.* 1976), shows that there is a steep positive correlation between temperature and mobility. For the ceramics an increase in temperature of 300°C leads to an increase in grain boundary mobility of over 5 orders of magnitude.

In Table 1 a mobility of 1 means that, on average, each crystallite in the array will be chosen once for testing to see if a grain boundary migration event took place.

Rates of subgrain formation

In this model we only consider one of several models for subgrain formation (Means & Ree 1988), that of the progressive misorientation of subgrains (Guillopé & Poirier 1979). Their model implies that there will be a close relationship between the initial orientation of the subgrain and its host. In our model subgrain formation is a spontaneous nucleation process restricted to grain boundaries. The average size of newly formed subgrains is nine crystallites. The angular relationship between the crystallography of parent grains and their subgrains is not well understood, so this has not been varied with temperature, and subgrains are constrained to form within 10° of the parent orientation. The rate of nucleation is also held constant in this paper. Once a strain-free subgrain has been nucleated it acts just like any other grain and its future behaviour is dependent on its orientation. The rate of subgrain formation in Table 1 refers to the probability that any single Monte Carlo step will be a subgrain nucleation event instead of a grain boundary migration event.

Work hardening rate

In this model the stored energy level in a crystallite (E) reaches its steady state stored energy level (E_{ss}) asymptotically as a function of the age of the crystallite (t) and a hardening term (H), according to the function:

$$E = E_{ss}(1-H^t)$$

and we have assumed that at higher temperatures the rate of work hardening (H) decreases.

Simulation results

By choosing five 'temperatures' which we hope will represent different levels in the Earth's crust, we can derive the appropriate input parameters for this model (Table 1). For the purposes of this paper it is not necessary or wise to equate the five levels we have chosen with specific temperatures or metamorphic grades, especially since the latter also imply pressures. Here we shall simply call them the low, low-medium, medium, medium-high and high temperature models.

Each simulation starts with a regular array of hexagonal grains (each made of 37 crystallites), each grain having a randomly chosen crystallographic orientation. Each deformation increment represents a shear strain of 0.1 and for each increment a 'synthetic thin section' is calculated with a shading intensity chosen to simulate the microstructures as they would appear under a polarising microscope with polarisers at 45° to the sides of the array. The array in fact represents a cylinder of material, flattened out, so that material moved by the dextral simple shear off the left vertical boundary re-enters on the right hand side. The horizontal boundaries are not periodic in this manner.

The effects of dynamic recrystallization on fabric development

In order to show the relationship between this model and the TBH model we will first compare the TBH model using the CRSS parameters described by Lister & Hobbs (1980) for their model C, and this model using the same CRSS values with the addition of parameters relating to dynamic recrystallization. The TBH results have been re-calculated for this study, however the resulting fabrics are identical to those published earlier. The results of the TBH model are displayed as discrete *c*- and *a*-axis orientations plotted on lower hemisphere equal area projections, whereas the Taylor–Bishop–Hill/dynamic recrystallization (TBH/DRX) simulations are displayed as contoured orientation densities, also weighted according to grain area. This technique was used because most measurements of the full crystallographic preferred orientation in quartz are made using x-ray goniometry (Schmid & Casey 1986; Mancktelow 1987), and a grain area weighting mimics this measurement process. In addition it more clearly reflects the impact of the recrystallization on fabric development.

The TBH model predicts the gradual development of three elongate maxima of both *c*-axes and *a*-axes around the perimeter (Fig. 2a). In contrast, the patterns of c-axis preferred orientations predicted by the TBH/DRX simulations evolve from a random distribution through a cross-girdle distribution to a single girdle distribution, and the *a*-axes evolve into a single maximum distribution with two weak secondary maxima. This change from one strong preferred orientation pattern to another with progressive deformation was seen in the simulations of Jessell (1988*b*), however they are even more pronounced here. The overall effect of dynamic recrystallization on the fabric development is systematically to remove those parts of the stable fabric predicted by the TBH model which result in the maximum work done.

The fabrics at the same stages of the deformation are also displayed as a synthetic thin sections (Fig. 2b), with grain colours simulating the effect of using cross polars at 45° to the shear zone boundaries. These synthetic thin sections show the development of a strong grain shape foliation roughly parallel to the orientation of maximum finite elongation. This grain shape foliation at no time lags more than 5° behind the maximum finite elongation direction, and by a shear strain of 3.0 is within a degree of it. Also evident is a marked grain growth and the development of serrated grain boundaries.

Low temperature

The parameters for this simulation were chosen to mimic the low temperature behaviour of quartz (i.e. in the low temperature range of crystalline plasticity), with low grain boundary mobilities, high work hardening rates, low dynamic recovery rates, and CRSS values which favour basal<*a*> slip. By a shear strain of 5.0 the simulation predicts the development of a very pronounced *c*-axis maximum perpendicular to the shear zone boundaries (Fig. 3a). The *a*-axes are distributed along a girdle within the shearing plane. The *c*-axis plot shows that the areal dominance of *c*-axes perpendicular to the flow direction hides a significant proportion of grains in other orientations. The TBH predictions for these CRSS values (equivalent to a grain boundary mobility of 0, and no subgrain formation), look very similar to the predictions of the TBH Model C. The synthetic thin section for a shear strain of 3.0 shows a well developed grain shape foliation, however the same fabric at a shear strain of 5.0 shows very little detail, since most of the grains have a very similar *c*-axis orientation by this shear strain (Fig. 4).

Low–medium temperature

The crystallographic (Fig. 3b) and grain shape preferred orientation predictions for this temperature are essentially indistinguishable from the low temperature model.

Medium temperature

The medium temperature simulation results in the formation a single girdle of *c*-axes containing two orthogonal maxima, although the shear plane normal maximum still dominates the fabric (Fig. 3c). The *a*-axes are still distributed along a girdle within the shearing plane. The synthetic thin section for these conditions shows a clear grain shape foliation, however the grains at a shear strain of 3.0 appear to be somewhat less elongate than the lower temperature runs.

Medium–high temperature

At these conditions the two *c*-axis maxima are almost equally densely populated (Fig. 3d), and the *a*-axis pattern is starting to show strong secondary maxima. The synthetic thin section at a shear strain of 3.0 looks very similar to the medium temperature results, and the prediction at a shear strain of 5.0 is difficult to compare since none of the previous figures have a two-maximum *c*-axis fabric.

High temperature

The simulation predicts the formation of a strong *c*-axis maximum perpendicular to the flow direction and within the shearing plane (Fig. 3e). The *a*-axes distribution is characterised by three equally populated maxima spaced at 120°, lying in a plane parallel to the shear direction and perpendicular to the shear plane. Perhaps surprisingly the TBH predictions for these CRSS values still look very similar to the results produced by the low temperature values. In this simulation the point maximum of *c*-axes develops in an area that is predicted to be empty by the TBH model, however examination of the individual lattice rotations shows that it is an area that is still rotationally stable. The synthetic thin section for these conditions shows the weakest grain shape preferred orientation of all the temperatures, and the largest grain size. The grain shape foliation lags behind the orientation of maximum finite elongation.

Discussion

As we have said the actual values of the CRSSs for slip systems in quartz under geological conditions are not known, however it is the relative values for different slip systems that are important for the TBH model, not absolute values, and even by changing these values conserva-

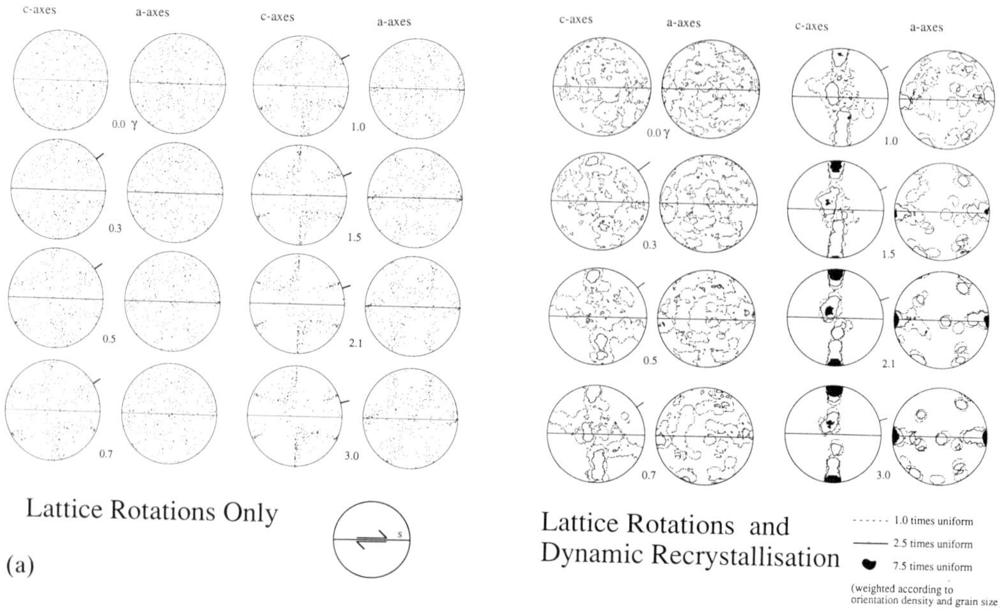

Fig. 2. The effect of coupled lattice rotations and dynamic recrystallization on fabric development. A comparison is made between the fabrics predicted by the Taylor–Bishop–Hill model and those predicted by our model for progressive dextral simple shear. In both cases the CRSS values of Lister & Hobbs (1980) Model C are used. (**a**) Crystallographic preferred orientations: the Taylor–Bishop–Hill fabrics are plotted as *c*- and *a*-axis orientations, while the Taylor–Bishop–Hill/dynamic recrystallization model fabrics are contoured with a weighting scheme combining both crystal axis density and grain area. The selective modification of the lattice rotation fabrics by the recrystallization can be seen, leading to the successive development of a crossed girdle of *c*-axes and then a single girdle with further deformation. For each stage the orientation of the line of maximum finite elongation is shown as a tick mark on the perimeter of the lower hemisphere equal area projections, with the shearing plane (S) horizontal. (**b**) Synthetic thin sections predicted by the Taylor–Bishop–Hill/dynamic recrystallization model, showing the natural looking fabrics predicted by this model. The equivalent Taylor–Bishop–Hill-only model would merely consist of the original hexagonal grains homogeneously deformed to a shear strain of 3.0. The shear plane is horizontal and the shear strain is shown for each stage, together with the orientation of the line of maximum finite elongation.

tively our combined model predicts a large variation in the crystallographic and grain shape preferred orientations. In each case at high strain the predicted crystallographic preferred orientations reflect the combination of the rotationally stable and minimum work orientations predicted by the TBH model, even if the TBH model does not predict that the rotationally stable orientation will be populated, as in the high temperature case. The lack of grains in the centre of the equal area projection in the TBH fabrics is a combination of the initial random distribution of orientations and the tendency in simple shear for *c*-axes to rotate around an axis within the shear plane and perpendicular to the shear direction. At lower strains the TBH fabrics may be only partially modified by the dynamic recrystallization, so that intermediate patterns develop.

These simulations suggest that it is not wise to interpret point maxima as indicators of single slip deformation. It is interesting to note that the fabric that apparently reflects equal CRSS levels for basal<*a*> and prism<*a*> slip is actually the medium–high result, where the CRSS for prism <*a*> slip is actually lower than that for basal<*a*>. This is again because of the lack of grains in the undeformed state with orientations which will rotate into the centre of the equal area projection.

The grain shape fabrics that develop suggest a decrease in the strength of the foliation with increasing temperature, and an increase in grain size although the lower temperature results are hard to interpret at high strains since they develop near single-crystal fabrics.

The progressive evolution of *c*-axis patterns from a crossed to a single girdle in Fig. 2b

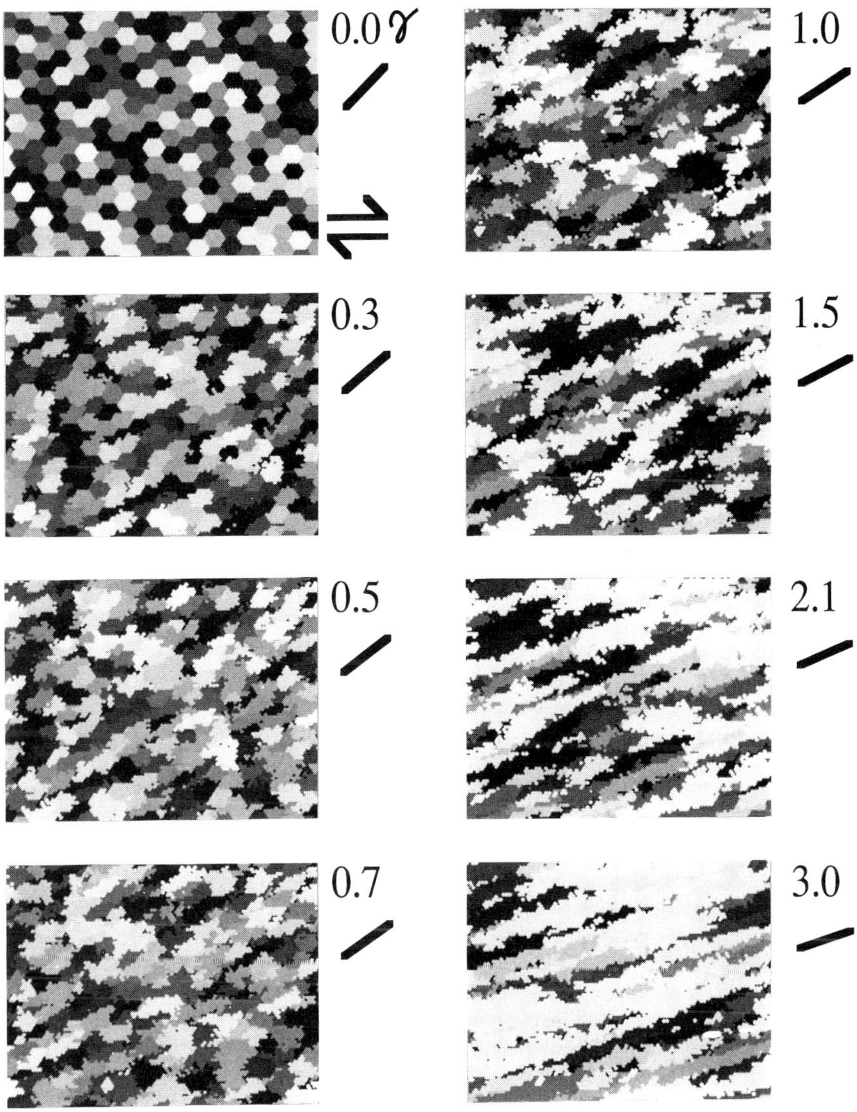

(b)

follows the trend suggested by Schmid & Casey (1986, fig. 14) as being typical of fabric development with progressive simple shear. These simulations provide a possible explanation for this trend, although if we accept this explanation, they also suggest that this is only one of several possible paths, depending on the temperature during deformation.

The analytical study of fabric development in olivine by Karato (1987) concluded that the coupling between lattice rotations and dynamic recrystallization would lead to fabrics that were controlled either by one process or the other. These simulations suggest that the high strain fabrics reflect both aspects of the deformation, the lattice rotations and the dynamic recrystallization. Clearly if there is no dynamic recrystallization it cannot affect the fabrics. However, as soon the grain boundary mobility is high enough to allow significant grain boundary velocities, the coupled deformation processes should always produce coupled fabrics.

These simulations were developed in order to provide a set of predictions for fabric develop-

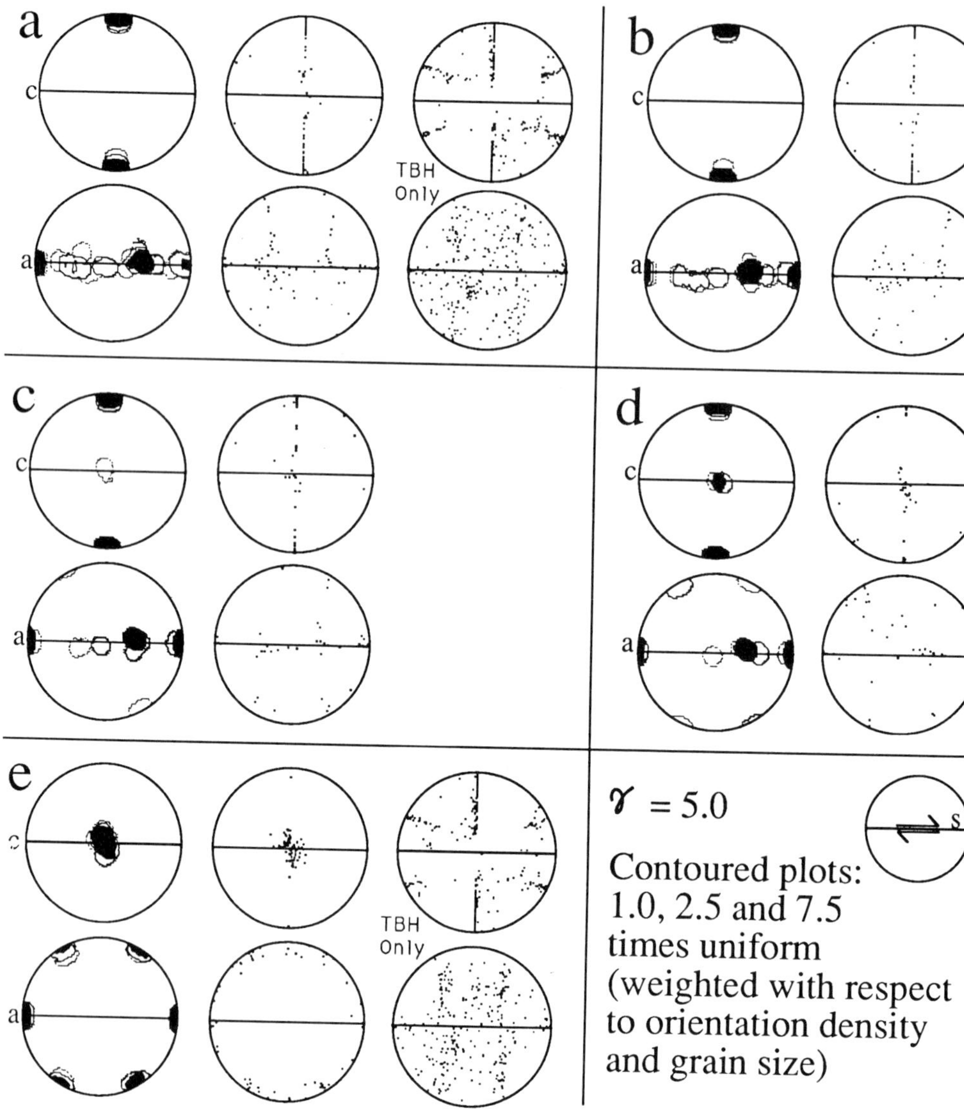

Fig. 3. Crystallographic preferred orientations predicted by this model for a shear strain of 5.0 at conditions set to simulate (*a*) low temperature, (*b*) medium–low temperature, (*c*) medium temperature, (*d*) medium–high temperature and (*e*) high temperature. Each part of the figure consists of a row of *c*-axis projections above a row of *a*-axis projections. The left column is contoured according to axis density and grain size, the right column simply shows the axes. For the low temperature and high temperature cases the equivalent Taylor–Bishop–Hill-only predictions are also presented. All plots are lower hemisphere equal-area plots with the shear plane horizontal. The contour intervals are 1.0, 2.5 and 7.5 times uniform.

ment in a model crust, and it is our hope that they will be tested by the measurement of full crystallographic preferred orientation patterns of quartzites whose metamorphic grade at the time of deformation is well constrained.

Conclusions

(1) A simulation of coupled lattice rotations and dynamic recrystallization predicts the formation of *c*-axis patterns which display point maxima, single girdles and crossed girdles, and

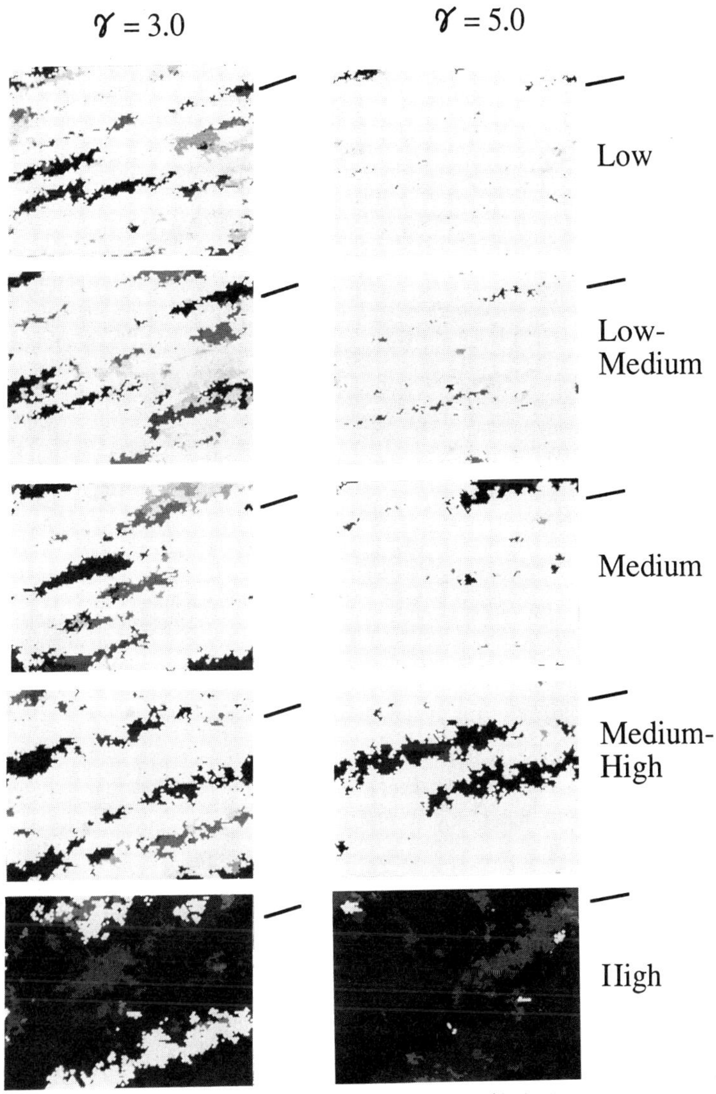

Fig. 4. The variation in grain shape fabrics with temperature predicted by the Taylor–Bishop–Hill/dynamic recrystallization model. Synthetic thin sections showing fabric development at a shear strains of 3.0 and 5.0 with parameters chosen to mimic the five temperature levels. The shear plane is horizontal, and the orientation of maximum finite elongation is shown for each fabric.

that with simple progressive deformation one pattern may evolve into another.

(2) By choosing parameters which reflect a progressive increase in temperature, this model predicts a systematic variation in both grain shape and crystallographic preferred orientations.

(3) The development of point maxima cannot be used as evidence for the activity of a single slip system, since this model is able to produce such patterns assuming multiple slip.

(4) These simulations produce a larger variation in fabrics with temperature than is predicted by using a model which only includes lattice rotations.

(5) The fabrics that develop in natural rocks should be interpreted in terms of a coupled model combining lattice rotations and dynamic recrystallization if their significance is to be properly understood.

References

ANDERSON, M. P., SROLOVITZ, D. J., GREST, G. S. & SAHNI, P. S. 1984. Computer simulation of grain growth. I- Kinetics. *Acta metallurgica*, **32**, 783–791.

BEIJERS, J., JÖNSSON, S. & PETZOW, G. 1987. TEM–In Situ deformation of Beryllium single crystals-a new explanation for the anomolous temperature dependence of the critical resolved shear stress for prismatic slip. *Acta metallurgica*, **35**, 2277–2287.

DRURY, M. R. & URAI, J. L. 1990. Deformation-related recrystallization processes. *Tectonophysics*, **172**, 235–256.

ETCHECOPAR, A. & VASSEUR, G. 1987. A 3-D kinematic model of fabric development in polycrystalline aggregates: comparison with experimental and natural examples. *Journal of Structural Geology*, **9**, 705–718.

GUILLOPÉ, M. & POIRIER, J. P. 1979. Dynamic recrystallization during creep of single-crystalline halite: an experimental study. *Journal of Geophysical Research*, **84**, 5557–5567.

HOBBS, B. E. 1985. The geological significance of microfabric analysis. *In*: WENK, H. -R. (ed.) *Preferred orientation in deformed metals and rocks*. Academic Press, Orlando, 463–484.

HONEYCOMBE, R. W. K. 1984. *The plastic deformation of metals*, (Second Edition). Edward Arnold, London.

JESSELL, M. W. 1986. Grain boundary migration and fabric development in experimentally deformed octachloropropane. *Journal Structural Geology*, **8**, 527–542.

—— 1988*a*. A simulation of fabric development in recrystallising aggregates — I: Description of the model. *Journal of Structural Geology*, **10**, 771–778.

—— 1988b. A simulation of fabric development in recrystallising aggregates — II: Example model runs. *Journal of Structural Geology*, **10**, 779–793.

KARATO, S.-I. 1987. Seismic anisotropy due to lattice preferred orientation of minerals: kinematic or dynamic. *In*: MANGHNANI, M. H. & SYONO, Y. (eds) *High-Pressure Research in Mineral Physics*. Terra Scientific Publishing Company, Tokyo, 455–471.

KNIPE, R. J. & LAW, R. D. 1987. The influence of crystallographic preferred orientation and grain boundary migration on microstructural and textural evolution in an S-C mylonite. *Tectonophysics*, **135**, 155–169.

LISTER, G. S. 1978. Texture transitions in plastically deformed calcite rocks. *In*: GOTTSTEIN, G. & LÜCKE, K. (eds) *Textures of Materials*, **2**. Springer, Berlin, 199–210.

—— & HOBBS, B. E. 1980. The simulation of fabric development during plastic deformation and its application to quartzite: the influence of the deformation history. *Journal of Structural Geology*, **2**, 355–370.

——, PATERSON, M. S. & HOBBS, B. E. 1978. The simulation of fabric development during plastic deformation and its application to quartzite: the model. *Tectonophysics*, **45**, 107–158.

LLOYD, G. E. & FERGUSON, C. C. 1986. A spherical electron-channelling pattern map for use in quartz petrofabric analysis. *Journal of Structural Geology*, **8**, 517–526.

MANCKTELOW, N. S. 1981. Strain variations between quartz grains of different crystallographic orientations in a naturally deformed metasiltstone. *Tectonophysics*, **78**, 73–84.

—— 1987. Quartz textures from the Simplon Fault Zone, southwest Switzerland and north Italy. *Tectonophysics*, **135**, 133–153.

MEANS, W. D. & REE, J. H. 1988. Seven different types of subgrain boundaries in octachloropropane. *Journal Structural Geology*, **10**, 765–770.

PRICE, G. P. 1985. Preferred orientation in quartzites. *In*: WENK, H.-R. (ed.) *Preferred orientation in deformed metals and rocks*. Academic Press, Orlando, 385–406.

SCHMID, S. M. & CASEY, M. 1986. Complete fabric analysis of some commonly observed quartz c-axis patterns. *American Geophysical Union Geophysical Monograph*, **36**, 263–286.

TAKESHITA, T. & WENK, H.-R. 1988. Plastic anisotropy and geometrical hardening in quartzites. *Tectonophysics*, **149**, 345–361.

TULLIS, J. & YUND, R. A. 1982. Grain growth kinetics of quartz and calcite aggregates. *Journal of Geology*, **90**, 301–318.

URAI, J. L., MEANS, W. D. & LISTER, G. S. 1986. Dynamic recrystallisation of minerals. *American Geophysical Union Geophysical Monograph*, **36**, 161–199.

WHITE, S. H., BURROWS, S. E., CARRERAS, J., SHAW, N. D. & HUMPHREYS, F. J. 1980. On mylonites in ductile shear zones. *Journal of Structural Geology*, **2**, 175–188.

YAN, M. F., CANNON, R. M. & BOWEN, H. K. 1976. Grain boundary migration in ceramics. *In*: FULRATHE, R. N. & PASK, J. A. (eds) *Ceramic microstructures '76: with emphasis on energy related problems*. Westview Press, 276–307.

High temperature deformation of octachloropropane: dynamic grain growth and lattice reorientation

JIN-HAN REE

Department of Geological Sciences, State University of New York at Albany, 1400 Washington Avenue, Albany, NY 12222, USA

Abstract: Dynamic grain growth and lattice preferred orientation development in octachloropropane polycrystals simple-sheared at 0.8 Tm using synkinematic microscopy are discussed. Both soft and hard grains are observed to grow. It is suggested that globular grains are not necessarily an indicator of coaxial deformation. Strain heterogeneity is induced by partitioning of deformation into relatively increased components of rigid-body rotation in hard grains and strain in soft grains. With dominant grain rotation, similar lattice preferred orientation to that of a single-slip model is introduced although the unlocking process is different.

Much research on fabric development on crystalline aggregates has been based on either Taylor's model (Taylor 1938; Aernoudt 1978; Lister *et al.* 1978; Jessell 1988*a*, *b*) or Sachs model (Sachs 1928; Leffers 1979) assuming dislocation glide as the only deformation mechanism. In Taylor's model modified to permit grain boundary migration (Jessell 1988*a*, *b*), where the state of strain is assumed homogeneous throughout the aggregates, grains oriented for single slip on weak slip systems ('soft grains') are expected to grow because they accumulate low dislocation energy for a given increment of deformation. On the other hand, in Sachs model for which the stress is assumed to be homogeneously distributed, we can expect that grains unsuitably oriented for single slip on weak systems ('hard grains') will grow because they fail to deform and thereby accumulate low dislocation energy for a given increment of deformation, although the model itself doesn't consider the effect of recrystallization. In a realistic case both stress and strain must be heterogeneous (Karato 1987). Furthermore, other processes such as climb and grain boundary deformation mechanisms are possible in addition to dislocation glide in high temperature deformation. This paper discusses dynamic grain growth and the deformation heterogeneity observed in experimentally simple-sheared octachloropropane (C_3Cl_8, hereafter called OCP) polycrystals at about 0.8 Tm (Tm = absolute melting point). It is shown how rigid-body rotation of grains can contribute to the development of the lattice preferred orientation (LPO). OCP is a uniaxial positive, hexagonal material whose melting point is about 160°C. It has been used in several previous microstructural studies (Beck 1949; McCrone 1949; Means 1983, 1989; Means & Ree 1988; Dong 1985; Jessell 1986; Ree 1988).

Experimental details

The sample to be deformed was prepared with a mixture of OCP and 1000-grit silicon carbide particles as explained in greater detail by Jessell (1986) and Means & Ree (1988). The silicon carbide particles are used to mark material points in the samples. The experiment described here was carried out at 80 ± 5°C (about 0.8 Tm), at a shear strain rate of $4.2 \times 10^{-5}\ s^{-1}$, using a Urai press (see Means 1989 for a picture of this apparatus and reference to its manufacturer). Photomicrographs in plane and cross-polarized light were taken every half-hour so that the strain increment between photographs is approximately constant. In cross-polarized light, four photomicrographs were taken with various rotated positions of the microscope stage every time for a near-complete record of the evolution of subgrain boundaries and other microstructures. *c*-axis orientations of the grains were measured on a universal stage before and after deformation. During the deformation they were measured by using flat stage extinction directions (for the trend of *c*-axis) and birefringence measurement with a Berek compensator (for the plunge of *c*-axis) without interrupting the deformation. Although the measurement of *c*-axis plunge with a Berek compensator is not as accurate as on a universal stage, this gives approximate information on *c*-axis reorientation trajectories. Grain-shape

From Knipe, R. J. & Rutter, E. H. (eds), 1990, *Deformation Mechanisms, Rheology and Tectonics*, Geological Society Special Publication No. 54, pp. 363–368.

foliations were determined by measuring the orientation of short segments of individual grain boundaries from the maps of the sample with a digitizer coupled to a IBM-PC.

At the beginning of the deformation (Fig. 1a), the grain boundaries were straight or slightly wavy with an average grain size of 100 μm. The grain thickness of 50 μm measured normal to the plane of Fig. 1a was constant throughout the experiment. Figure 1b shows the sample immediately after 9.5 hours of deformation during which a total shear strain (γ_t) of about 1.4 accumulated. Although it is weak, there was some local development of grain shape foliation at about 25° to the shear direction, roughly parallel to the direction of finite elongation. Average grain size increased by 50% with the growth of some mega-crystals by grain boundary migration and amalgamation as explained in Means & Ree (1988) and Means (1989). The number and size of subgrain boundaries also increased and most of them are perpendicular to the basal plane of the OCP grains.

The main microstructural changes after 14 hours of static recovery at room temperature (Fig. 1c) include further increase in grain size and the modification of both the intensity and orientation of grain-shape foliation. Grain-shape foliation was somewhat strengthened compared to that immediately after deformation, contrary to what might be expected after static recovery. Grain-shape foliation was also further 'rotated' into the shear direction, past the direction of maximum finite elongation. There was also significant development of type VII subgrain boundaries that occur statically from optically strain-free grains by a process probably similar to classical polygonization (Means & Ree 1988).

Figure 2 shows c-axis fabrics (a) before and (b) after deformation. At the beginning of deformation a LPO had already developed due to the vertical pressing of the specimen during sample preparation. With the deformation, the LPO was significantly modified, to form a broad girdle of c-axes, normal to the shear direction (Fig. 2b). The girdle pattern of c-axes and predominance of prismatic subgrain boundaries suggest that basal slip may have been the dominant slip system.

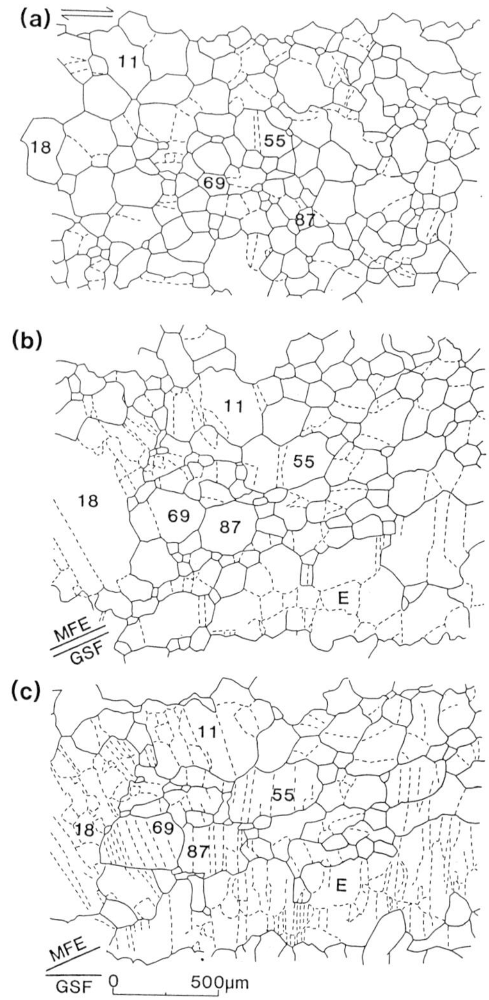

Fig. 1. Maps of OCP sample TO-109, showing grain boundaries (solid lines), and subgrain boundaries (dashed lines). Grains referred to in the text are numbered. Arrows in the left top show the sense and direction of shear in the sample. (**a**) Microstructure immediately before deformation. (**b**) Microstructure 9.5 hours later, immediately after deformation with a bulk shear strain of 1.4. (**c**) Microstructure after 14 hours of static recovery at room temperature. Directions of grain-shape foliation (GSF) and maximum finite elongation (MFE) are indicated in (b) and (c). Uncertainties in these orientations are ± 5° and ± 8° respectively.

Dynamic grain growth and strain heterogeneity

Figure 3a shows the configuration, strain state and c-axis trajectory of four grains that grew 60–500% in area during the deformation, at a bulk shear strain of 1.2. Strain within a grain was measured using displacements of sets of three, widely spaced marker particles within the grain and it is not necessarily homogeneous within a grain. Bulk strain of the specimen was

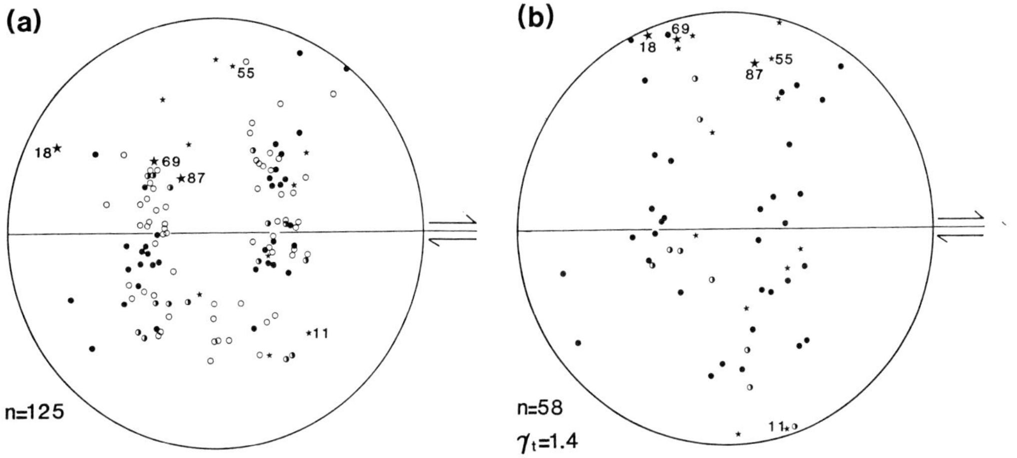

Fig. 2. *c*-axis fabric diagrams of OCP sample TO-109 (**a**) before deformation and (**b**) after deformation. Grains that grew more than five times in area during deformation are represented by big stars, twice to four times by small stars and less than twice by solid circles. Grains that shrunk are indicated by semi-solid circles and those that disappeared by open circles. Lower-hemisphere equal area projection.

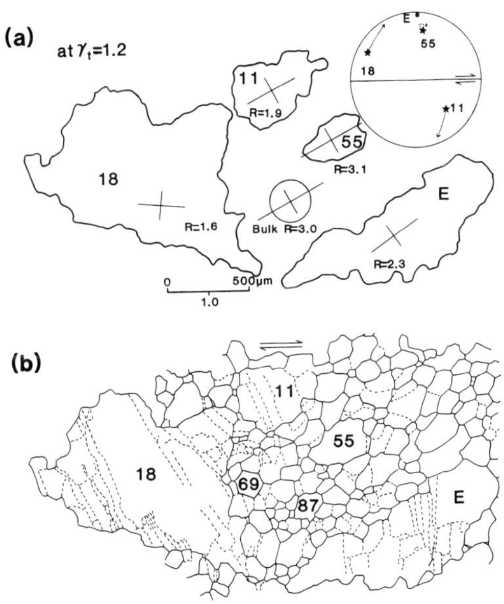

Fig. 3. Maps of TO-109 at bulk shear strain of 1.2. (**a**) Strain state and *c*-axis trajectories of four grains, 11, 18, 55 and E. Maximum and minimum finite stretches are shown in each grain with R value indicating axial ratio of strain ellipse. Scale for stretch is shown in the bottom of scale bar and by unit circle around lines for bulk stretch in the center of the figure. Stereographic inset shows c-axis trajectories of grains 11, 18 and 55 from the beginning to the bulk shear strain of 1.2, and of grain E from the bulk shear strain of 0.2 to 1.2. (**b**) Complete map showing the positions of the grains in (a).

determined by averaging strains indicated by ten sets of three marker particles with average distance of about 1 mm between marker particles in a set.

Two grains (11 and 18) began in hard orientations for basal slip, assuming major and minor principal stress axes to be at about 45° to the shear direction. The ratios of maximum to minimum finite stretch of these grains are 1.6 for grain 18 and 1.9 for grain 11 respectively. Thus they suffered far less strain than the bulk strain with stretch ratio of 3.0. However, their *c*-axes were significantly rotated clockwise. Also they had grown by consuming more strained neighbouring grains probably because strain energies of these hard grains were lower than neighbouring grains. Grain 18 became an especially big mega-crystal with an increase in size of about 500%.

Grain 55 initially had its basal plane about parallel to shear direction, and was thus a 'soft' grain. It experienced much more strain than the 'hard' grains described above, with little rotation of its *c*-axis. The magnitude and direction of stretch are about the same as those of the bulk strain. For another soft grain (E), the initial *c* axis orientation was not measured because it was out of the field of view at the beginning of deformation. It appeared in the field of view at bulk shear strain of 0.2, consuming adjacent grains in the lower right part of the field of view, with its basal plane parallel to shear direction. Its strain was measured only from bulk shear γ_t = 0.7 to γ_t = 1.2 in Fig. 3a.

Nevertheless its magnitude is higher than those of hard grains measured from the beginning. If the stretch ratio of this grain is scaled to represent the total strain from the beginning to $\gamma_t = 1.2$, assuming that it had been strained at the same rate as for $\gamma_t = 0.7-1.2$, its value becomes about 5.5, thus higher than even that of the bulk strain. Its c-axis stayed perpendicular to the bulk shear plane. These two soft grains had grown by migrating their grain boundaries into less favorably oriented neighboring grains, perhaps as a result of fewer dislocation tangles and thus a lower dislocation density than its neighbours as explained by Urai & Humphreys (1981) and Jessell (1986). When hard grains encounter soft grains as in the case of grains 11 and 55 in Fig. 1b, grain boundaries between them were stable.

Lattice reorientation

c-axis trajectories are shown in Fig. 4a for the experiment described here (TO-109) and in Fig. 4b for an experiment not described here but run at the same strain rate and temperature (TO-105). In both cases almost all of the *c*-axes were rotated clockwise toward the normal to the bulk shear plane, that is toward a soft orientation for basal slip. Thus the resulting fabric is similar to that produced in the 'single-slip model' of Etchecopar (1977) and Etchecopar & Vasseur (1987). The major difference is the 'unlocking' of hard grains, which permits them to begin deformation in the easy-slip orientation. In their model unlocking of hard grains occurs by several processes such as forming 'en cornue' crystals, fragmentation, and dynamic recrystallization. In contrast, in the experiment described here, hard grains more or less rigidly and continuously rotated into the easy slip orientation.

To further investigate the rigid- body rotation of hard grains in experiments TO-105 to TO-109, the rotations of their *c*-axes are plotted against bulk shear strain in Fig. 5. Also drawn is a line of $R = \gamma_t/2$ rad (R = rotation, γ_t = bulk shear strain) representing rotation of rigid sphere in a viscous fluid (Rosenfeld 1970). Although there is not a close fit, it can be seen that the general trend of the points fits the line for rigid-sphere rotation in a viscous fluid.

Conclusions

Although these results may not be applicable to the development of microstructures in naturally deformed rocks, they have many potential implications for interpretation of natural microstructures. It has been shown that both hard and soft grains can grow in high temperature deformation of OCP. The two effects of strain and lattice orientation on dynamic grain growth are competing processes with about equal weight. Sample deformation is accommodated by dominant grain rotation in hard grains and by dominant strain in soft grains, resulting in

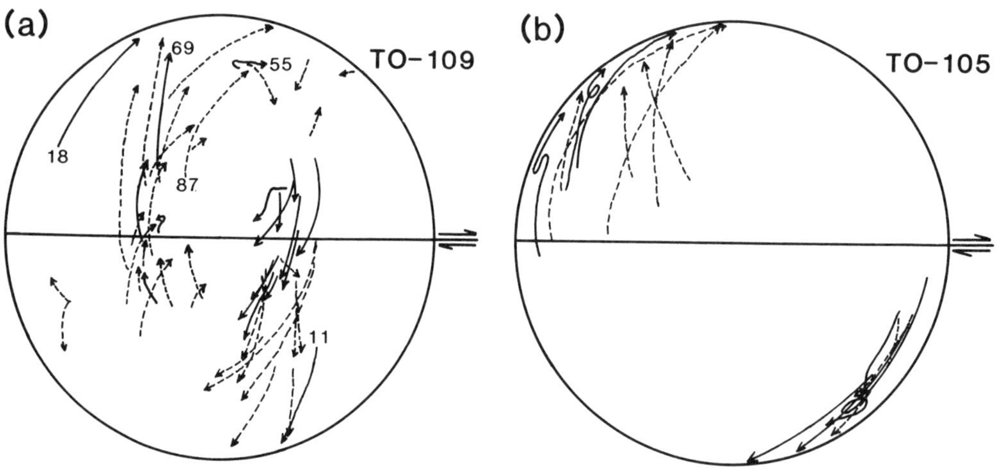

Fig. 4. *c*-axis reorientation trajectories of experiments (**a**) TO-109 and (**b**) TO-105. Solid lines represent trajectories for which c-axes were measured not only before and after deformation but also during deformation twice in TO-109 and six times in TO-105. Those for which *c*-axes were measured only before and after deformation are indicated by dashed lines. Lower-hemisphere equal area projection.

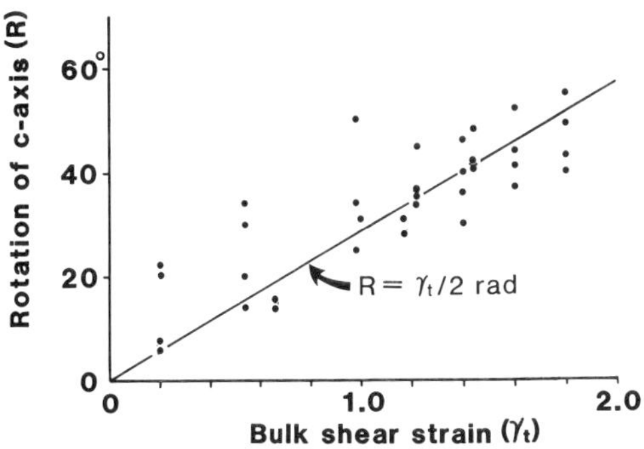

Fig. 5. Graph of rotation of *c*-axis v. bulk shear strain for hard grains. Data from experiments TO-105 and TO-109 are plotted together.

strain heterogeneity. Another point to note is that grains 18 and 11 for example look like porphyro*clasts* in a finer-grained matrix in Fig. 3b despite the fact that they are in fact known to be porphyro*blasts* growing from the beginning.

Globular grains that have their basal planes more or less perpendicular to shortening direction have been interpreted as an indicator of coaxial deformation (Tullis *et al.* 1973; Mancktelow 1981; Law 1986; Law et al. 1984). However, growing globular grains have been frequently observed in the non-coaxial deformation experiments including the one described here. Therefore, care must be taken in interpreting globular grains because their presence is not necessarily an indicator of coaxial deformation.

Grain rotation without much internal straining may be more important than slip-induced rotation in developing LPO in high temperature deformation. With rigid grain rotation a very similar LPO to that by the single-slip model was produced in experiments described in this paper. *c*-axes vigorously rotated toward the normal to the bulk shear plane, i.e. toward an easy-slip orientation. The simulation with a combination of Taylor–Bishop–Hill theory and dynamic recrystallization (Jessell 1988*a*, *b*) also introduces a similar LPO but the fate of hard grains is different. In Jessell's model, hard grains just disappear after being eaten up by soft grains.

For better understanding of LPO development in high temperature deformation, the competition between rigid grain rotation and slip-induced rotation, and between strain and lattice orientation effects on dynamic grain growth should be investigated quantitatively. Although it is not described here, the effect of neighbouring grains may also be important and should be considered.

I thank W. D. Means for valuable suggestions and advice, and Y. -J. Lee for drawings. B. -N. Ree helped with typing. The manuscript was improved by very careful comments from M. Jessell, S. G. Hwang and an anonymous reviewer. This work was supported by US National Science Foundation Grant EAR 8803096 to W. D. Means.

References

AERNOUDT, E. 1978. Calculation of deformation textures according to the Taylor model. *In*: GOTTSTEIN, G. & LUCKE, K. (eds) *Textures of Materials*. Springer, 45–66.

BECK, P. A. 1949. Comments on grain growth in octachloropropane. *Journal of Applied physics*, **20**, 231.

DONG, H. 1985. A possible formational process of mylonite. *Acta Geologica Sinica*, **4**, 286–292.

ETCHECOPAR, A. 1977. A plane kinematic model of progressive deformation in a polycrystalline aggregate. *Tectonophysics*, **39**, 121–139.

—— & VASSEUR, G. 1987. A 3-D kinematic model of fabric development in polycrystalline aggregates: comparisons with experimental and natural examples. *Journal of Structural Geology*, **9**, 705–717.

JESSELL, M. W. 1986. Grain boundary migration and fabric development in experimentally deformed octachloropropane. *Journal of Structural Geology*, **8**, 527–542.

—— 1988*a*. Simulation of fabric development in recrystallizing aggregates — I. Description of the model. *Journal of Structural Geology*, **10**, 771–778.

—— 1988*b*. Simulation of fabric development in recrystallizing aggregates — II. Example model runs. *Journal of Structural Geology*, **10**, 779–793.

KARATO, S. 1987. Seismic anisotropy due to lattice preferred orientation of minerals: kinematic or dynamic? *In*: MANGHNANI, M. H. & SYONO, Y. (eds) *High-pressure Research in Mineral Physics*, 455–471.

LAW, R. D. 1986. Relationship between strain and quartz crystallographic fabrics in the Roche Maurice quartzites of Plougastel, western Brittany. *Journal of Structural Geology*, **8**, 493–515.

——, KNIPE, R. J. & DAYAN, H. 1984. Strain path partitioning within thrust sheets: microstructural and petrofabric evidence from the Moine thrust zone at Loch Eriboll, northwest Scotland. *Journal of Structural Geology*, **6**, 477–497.

LEFFERS, T. 1979. A modified Sachs approach to the plastic deformation of polycrystals as a realistic alternative to the Taylor Model. *In*: HAASEN, P., GEROLD, V. & KOSTORZ, G. (eds) *Strength of Metals and Alloys*. Pergamon Press, 769–774.

LISTER, G. S., PATERSON, M. S. & HOBBS, B. E. 1978. The simulation of fabric development in plastic deformation and its application to quartzite: the model. *Tectonophysics*, **45**, 107–158.

MANCKTELOW, N. S. 1981. Strain variation between quartz grains of different crystallographic orientation in a naturally deformed metasiltstone. *Tectonophysics*, **78**, 73–84.

MCCRONE, W. C. 1949. Boundary migration and grain growth. *Discussions of the Faraday Society*, **5**, 158–166.

MEANS, W. D. 1983. Microstructure and micromotion in recrystallization flow of octachloropropane: a first look. *Geologische Rundschau*, **72**, 511–528.

—— 1989. Synkinematic microscopy of transparent polycrystals. *Journal of Structural Geology*, **11**, 163–174.

—— & Ree, J. -H. 1988. Seven types of subgrain boundaries in octachloropropane. *Journal of Structural Geology*, **10**, 765–770.

REE, J. -H. 1988. Evolution of deformation-induced grain boundary voids in octachloropropane. *Geological Society of America Abstracts with Programs*, **20**, A213.

ROSENFELD, J. L. 1970. Rotated garnets in metamorphic rocks. *Geological Society of America Special Paper*, **129**, 102pp.

SACHS, G. 1928. Zur Ableitung einer Fliessbedingung. *Zeitschrift Verein Deutscher Ingenieure*, **72**, 734–736.

TAYLOR, G. I. 1938. Plastic strain in metals. *Journal of the Institute of Metals*, **62**, 307–324.

TULLIS, J., CHRISTIE, J. M. & GRIGGS, D. T. 1973. Microstructures and preferred orientations of experimentally deformed quartzites. *Geological Society of America Bulletin*, **84**, 297–314.

URAI, J. L. & HUMPHREYS, F. J. 1981. The development of shear zones in polycrystalline camphor. *Tectonophysics*, **78**, 677–685.

The SEM/ECP technique applied on twinned quartz crystals

NIELS Ø. OLESEN[1] & NIELS-HENRIK SCHMIDT[2]

[1] *Geologisk Institut, C. F. Møllers Allé, DK-8000 Aarhus C, Denmark*

[2] *Forskningscenter Risø, Postboks 49, DK-4000 Roskilde, Denmark*

Abstract: The SEM/electron channelling technique, ECP, is useful in fabric studies, since a three-dimensional lattice orientation determination is performed with a precision of *c.* 0.5° from areas a few microns in diameter. Computer simulation of the ECP image (a line diffraction pattern) facilitates analysis, particularly in the case of low-symmetry crystals such as alpha-quartz. The most important types of twinning in quartz are after the Dauphiné and Brazil laws. The latter combines a left- and a right-handed crystal (enantiomorphism), and it is shown that it is *not* possible to discriminate between these two crystals, since the twin plane coincides with mirror planes in the ECP sphere resulting in identical ECPs for the individual Brazil twins. In contrast, Dauphiné (and other) twins are easily identified because their twin planes do not coincide with mirror planes. A specification is given of the point groups which permit a complete lattice orientation determination, and those which involve a loss of information (orientational solubility).

Determination of crystal lattice orientation and microstructure in bulk specimens (thin section size) may be performed either with a Universal Stage mounted on the optical microscope (OM) or with the application of the electron channelling pattern (ECP) technique in the SEM. Basically the ECP is a line diffraction pattern, which is produced by rocking the electron beam around a point on the surface of the specimen and detecting the backscattered electrons from the crystal lattice at the point of incidence (e.g. Joy *et al.* 1982 and references therein). Non-metallic materials, such as most geological specimens, require special polishing techniques (see for example, Lloyd *et al.* 1981). Since the introduction of these techniques, a number of geological studies have been published (Lloyd *et al.* 1988 and references therein). The ECP technique is advantageous compared to the OM technique for two reasons: first it is not restricted to optically transparent and anisotropic minerals; and secondly three-dimensional lattice orientation determination is performed with a precision of *c.* 0.5° from areas a few microns in diameter. The methodology of interpretation of channelling patterns is identical for all minerals and crystal symmetries.

In this paper we discuss the potential of the ECP technique in the identification of twinned crystals, taking quartz as an example. We also address the problem of distinguishing enantiomorphic forms.

Orientation procedure

Determination of lattice orientation is performed by means of one of the following methods.

(1) The ECP image is compared with an ECP map, which is either a computer-drawn pattern, covering all possible orientations (e.g. Young & Lytton 1972) or a mosaic constructed from SEM/ECP micrographs (e.g. Joy *et al.* 1982).

(2) Calculation based on dimensions and angles between the contrast lines in the ECP image (e.g. Kozubowski *et al*, 1987, Weiland & Schwarzer 1986) or dimensions, angles *and* relative diffraction intensities (Schmidt & Olesen 1989).

In the case of low-symmetry crystals such as alpha-quartz (trigonal system) a two-dimensional ECP map is not practical. Lloyd & Ferguson (1986, fig. 5) constructed a micrograph mosaic on the surface of a sphere, whilst Schmidt & Olesen (1989, fig. 6) designed a computer simulation. This is also based on a spherical system, which facilitates comparison. We have studied with the OM, using 'Airy's spiral' technique (e.g. Buchwald 1952), a slab of the quartz single crystal which was used to construct the micrograph mosaic. The crystal is right-handed and homogeneous. The computer simulation is also made on the basis of a right-handed crystal lattice. The two maps are similar in appearance and hence orientation determi-

From Knipe, R. J. & Rutter, E. H. (eds), 1990, *Deformation Mechanisms, Rheology and Tectonics*, Geological Society Special Publication No. 54, pp. 369–373.

nation of quartz with the aid of computer simulation is a reliable procedure. There are certain minor discrepancies between simulation and micrograph mosaic as regards relative intensities (Schmidt & Olesen 1989), particularly at multiple lines, but these are easily incorporated into the computer model, thus facilitating future orientation determination.

It is notable that mirror planes appear on the ECP maps, which are not present in the alpha-quartz lattice (Fig. 1). These mirror planes are products of Friedels Law (Schmidt & Olesen 1989), and we will refer to them as FLMP. The alpha-quartz lattice has no centre of symmetry, but if we cover the whole sphere with a mosaic of ECP images (this is the ECP-sphere) it will display a centre of symmetry. Thus, as a general rule it will not be possible to distinguish between point groups within one Laue group (for exceptions, see below). In the case of alpha-quartz (point group 32) the ECP map displays the symmetry elements of the Laue group $\bar{3}$ m.

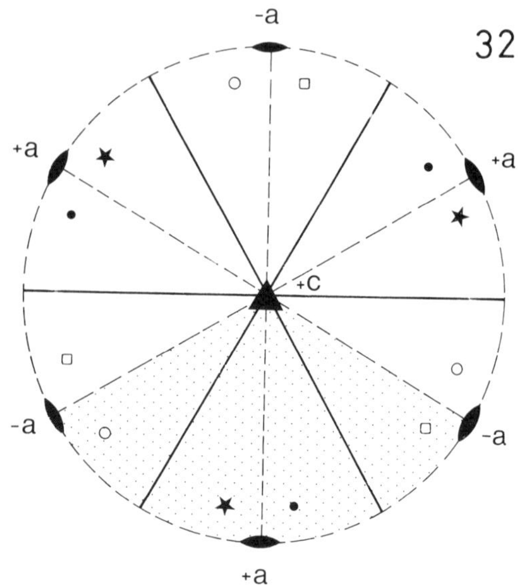

Fig. 1. Schematic projection of the quartz ECP-sphere. Quartz belongs to point group 32, in which the crystal structure only possess rotational axes. In the ECP-sphere additional mirror planes (solid lines) appear, positioned midway between $+a$ and $-a$, containing the c-axis and each of the m-axes. These planes are due to Friedels Law, as explained in the text. The closed circle denotes an image S in the upper hemisphere, and the star an image S in the lower hemisphere. The open square denotes an image Z in the upper hemisphere, and the open circle an image Z in the lower hemisphere. S and Z are mirror images of one another.
The combined symmetry elements acting on an ECP image in a general position on the sphere will generate a set of 6 identical images (S) and a set of 6 mirror images (Z). This is why the size of the elementary pattern of the sphere (EPS) is 1/6 of the whole ECP-sphere (ornamented).

Important twin laws in quartz

Two types of twinning are particularly important in quartz:

(1) Dauphiné law with $\{10\bar{1}0\}$ twin plane, which commonly takes the form of complex interpenetration twins.

(2) Brazil law with $\{11\bar{2}0\}$ twin plane, which combines a left- and right-handed crystal (enantiomorphism) in a complex penetration twin. The composition surface is usually planar and either (0001) or $\{10\bar{1}1\}$ (e.g. Trepied & Doukhan 1978).

The ECP-sphere may be considered as an abstract representation of a crystal. In Fig. 2 we analyse how this ECP-sphere is affected by twinning. Just as two crystals A and B are related through a twinning symmetry operation, so are the ECP-spheres. In the case of Brazil twinning, the twin planes are equivalent to FLMPs and hence the ECP-spheres of crystal A and B are identical. In the case of Dauphiné twinning, the twin planes produce an ECP-sphere of crystal B which is rotated 60° around the c-axis relative to crystal A (equivalent to a 180° rotation).

Thus, it is predicted that a right- and a left-handed crystal related through the Brazil twin law will display identical ECP images and are therefore indistinguishable. We have tested this on a slab of amethyst quartz, cut with an angle of 87° from the c-axis. Amethyst is known to display frequent Brazil twinning, and in this polished slab numerous Brazil twins are optically visible along three of the six prism faces (probably with composition planes on $\{10\bar{1}1\}$). Observation with the 'Airy's spiral' technique demonstrates that composition planes do reach the surface of the slab, and precisely this surface was studied in the SEM. With the SEM in ECP mode, scanning of the electron beam across the surface known to contain Brazil twins displays identical ECP images throughout. We also applied the orientation contrast (OC) mode (e.g. Lloyd 1987) in combination with tilt experiments, in which case subgrains appear as contrast features. A few subgrains were visible along the prism faces, but no contrast features are visible in areas known to contain numerous Brazil twins. This suggests that the two twin-individuals have identical ECP-spheres, in full agreement with theory.

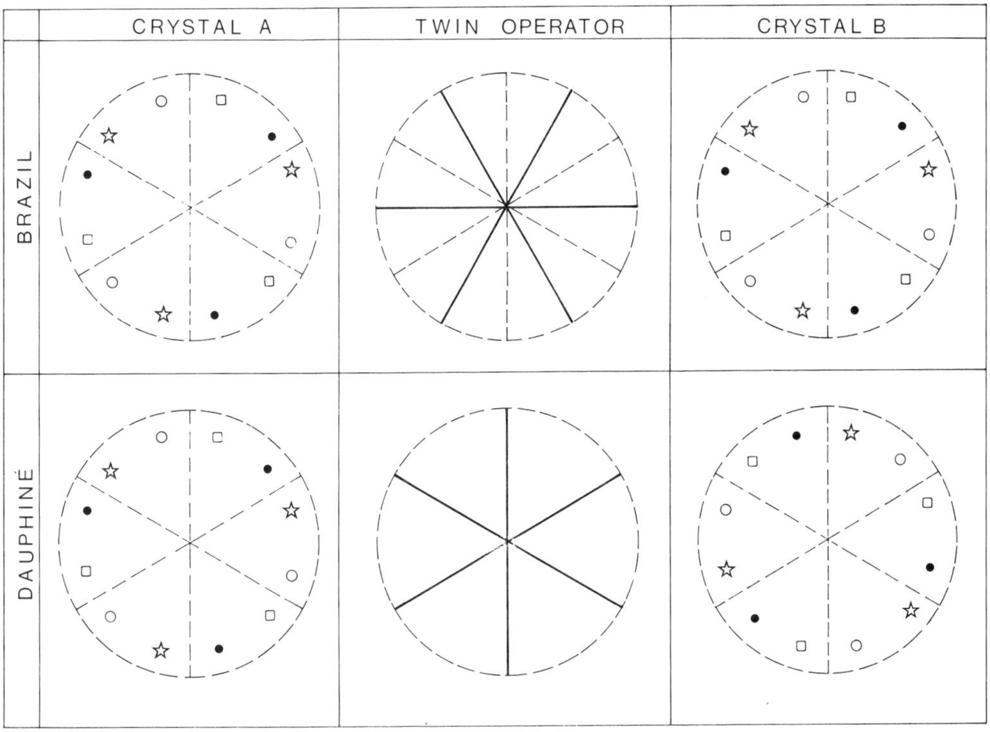

Fig. 2. When twinning takes place between two crystals A and B the two ECP-spheres are related through the particular twin operation. Symbols are defined in Fig. 1, and the 6 S and the 6 Z images represent the ECP-sphere of crystal A. Further explanation in text.

Recently, examples of discrimination of non-centrosymmetric point groups from centrosymmetric ones have been documented (Martinsen & Høier 1988). Fine scale *contrast differences* across the FLMPs demonstrate that Friedel's law in certain cases is violated. As alpha-quartz belongs to the non-centrosymmetric point group 32 it is possible that similar fine-scale differences may exist in the ECP map of alpha-quartz, but it remains to be documented. If this is so, it will be possible to distinguish between right- and left-handed alpha-quartz. However, it must be stressed that this requires images of special orientations, as the image must contain one of the FLMP's, and also possibly specific positions on these planes. It is notable, however, that the *geometry* of the line pattern also in this case is identical in the Brazil twinned individuals.

Choice of crystallographic axes

Based on X-ray texture goniometry, a correction of the indexing of a quartz ECP micrograph mosaic was given by Lloyd *et al.* (1987). They claimed that the positive and negative rhombs had been confused, and hence polarities of the *a*-axes were also confused. The simulation indicates that the positive rhomb (*r*) is positioned to the right in Schmidt & Olesen (1989, fig. 6) where two bands of the highest diffraction intensity are the two r-planes. Therefore it is also to the right in Lloyd & Ferguson (1986, fig. 5), in agreement with the result of the X-ray texture analysis. With this as a starting point two different positions of the *a*-axes are possible, simply by reversing the polarity of these axes. Schmidt & Olesen (1989) follows the crystallographic tradition, given by e.g. Nicolas & Poirier (1976), while Lloyd *et al.* (1987) follows the tradition in X-ray texture analysis (e.g. Schmid *et al.* 1981).

Similarly, a choice must be made on how to position the *a*-axes in the two enantiomorphic forms of quartz in a Brazil-twinned crystal. Evidently the crystal structures are different since the right-handed crystal belongs to space group P3(1)21 and the left-handed to space group P3(2)21. It is convenient, and also crystallographic tradition (e.g. Berry & Mason

1959), to maintain identical indices for any plane and direction in the two crystals. This is done by translating the cell axes. By adopting this convention, a quartz ECP image is unambiguously attached to one crystallographic direction (although we do not know in which structure).

Orientational solubility

Application of a standard diffraction technique such as the SEM/ECP results in a loss of orientational solubility in certain point groups. Following e.g. Wenk *et al.* (1988) the 32 point groups can be divided into three types. Point groups with no symmetry centre and no symmetry planes but only pure rotational axes belong to *type I*. These are the point groups that may display enantiomorphic forms, and they include groups 1, 2, 222, 3, 32, 6, 62, 4, 42, 23, and 43. The point groups which lack a symmetry centre but have at least one symmetry plane or an inversion axis are termed *type II*, and they include groups m, mm, 3m, $\bar{6}$, 6mm, $\bar{6}$m2, $\bar{4}$, 4 mm, $\bar{4}$2m, and $\bar{4}$3m. Note that point group $\bar{3}$ is missing. Those point groups which do possess a symmetry centre are termed *type III* and include groups $\bar{1}$, 2/m, mmm, $\bar{3}$, $\bar{3}$m, 6/m, 6/mmm, 4/m, 4/mmm, m3, and m3m (equivalent to the 11 Laue groups).

If we assume that Friedel's law is not violated, channelling patterns from crystals of type I and II will show reduced orientational solubility, whereas a type III crystal creates channelling patterns that unambiguously give the lattice orientation. We have illustrated above, through the quartz example, that an ECP image from a type I crystal can originate from either of the two possible enantiomorphic structures. In Schmidt & Olesen (1989) a type II structure was illustrated through a crystal with $\bar{4}$3m point group symmetry, and it was shown that an ECP image from this type of structure originates from one of two equally possible orientations.

It is notable, however, that a full structure determination of type I and II crystals can be performed by the application of convergent beam diffraction techniques in the TEM (e.g. Goodman 1975; Goodman & Secomb 1977).

Conclusions

Distinction between the enantiomorphic forms right- and left-handed quartz is not possible in routine fabric studies of quartz performed with the SEM/ECP technique. This is unfortunate, since there are indications that Brazil law twins can be mechanical twins (Trepied & Doukhan 1978) and therefore may be important contributors to the rock deformation. In contrast Dauphiné law twins (and twins after other laws, where the twin plane is not coincident with the Friedels Law Mirror Planes) are easily identified with the SEM/ECP technique.

We thank G. E. Lloyd for constructive criticism of the manuscript and he kindly supplied us with a slab of the quartz crystal, which he used in the construction of the ECP micrograph mosaic map. H. D. Zimmermann and H. Micheelsen gave scientific assistance and technical assistance was received from L. Jans and S. R. Jacobsen. The study was supported by the Danish Natural Science Research Council (11–4715, 11–5333, 11–6543, 81–6881, 81–6895).

References

Berry, L. G. & Mason, B. 1959. *Mineralogy*. W. H. Freeman & Co., San Francisco.

Buchwald, E. 1952. *Einfuhrung in die Kristalloptik*. Walter de Gruyter & Co., Berlin.

Goodman, P. 1975. A Practical Method of Three-Dimensional Space-Group Analysis Using Convergent-Beam Electron Diffraction. *Acta Crystallographica*, **A31**, 804–810.

Goodman, P. & Secomb, T. W. 1977. Identification of Enantiomorphously Related Space Groups by Electron Diffraction. *Acta Crystallographica*, **A33**, 126–133.

Joy, D. C., Newbury, D. E. & Davidson, D. L. 1982. Electron channelling patterns in the scanning electron microscope. *Journal of Applied Physics*, **53**, 81–122.

Kozubowski, J. A., Ji Lii, M. & Gerberich, W. W. 1987. On some practical aspects of orientation determination using electron channelling patterns. *Scanning*, **9**, 237–247.

Lloyd, G. E. 1987. Atomic number and crystallographic contrast images with the SEM: a review of backscattered electron techniques. *Mineralogical Magazine*, **51**, 3–19.

—— & Ferguson, C. C. 1986. A spherical electron channelling pattern map for use in quartz petrofabric analysis. *Journal of Structural Geology*, **8**, 517–526.

——, Hall, M. G., Cockayne, B. & Jones, D. W. 1981. Selected area electron channelling patterns from geological materials: specimen preparation, indexing and representation of patterns and applications. *Canadian Mineralogist*, **19**, 505–518.

——, Law, R. D. & Schmid, S. M. 1987. A spherical electron channelling pattern map for use in quartz petrofabric analysis: correction and verification. *Journal of Structural Geology*, **9/2**, 251–253.

——, Olesen, N. Ø. & Schmidt, N.-H. 1988. Geological use of SEM/ECP technique. *Scanning*, **10**, 163–164.

Martinsen, K. & Høier, R. 1988. Determination of Crystal Symmetry from Electron Channelling

Patterns. *Acta Crystallographica*, **A44**, 693–700.

NICOLAS, A. & POIRIER, J. P. 1976. *Crystalline Plasticity and Solid State Flow in Metamorphic Rocks*. J. Wiley & Sons, London.

SCHMID, S. M., CASEY, M. & STARKEY, J. 1981. An illustration of the advantages of a complete texture analysis described by the orientation distribution function (ODF) using quartz pole figure data. *Tectonophysics*, **78**, 101–117.

SCHMIDT, N.-H. & OLESEN, N. Ø. 1989. Computer-aided determination of crystal-lattice orientation from electron-channelling patterns in the SEM. *Canadian Mineralogist*, **27**, 15–22.

TREPIED, L. & DOUKHAN, J. -C. 1978. Some Precisions on the Brazil twins in quartz. *Physica Status Solidi*, **A50/1**, K37–K41.

WEILAND, H. & SCHWARZER, R. 1986. Online texture determination by Kikuchi or channelling patterns. *In:* BUNGE, H. J. (ed.) *Experimental Techniques of Texture Analysis*. Deutsche Gesellsch. f.Metallkunde, Oberursel, 301–313.

WENK, H. -R., BUNGE, H. J., KALLEND, J. S., LÜCKE, K., MATTHIES, S., POSPIECH, J. & VAN HOUTTE, P. 1988. Orientation distributions: representation and determination. *In*: KALLEND, J. S. & GOTTSTEIN, G. (eds) *Proceedings of the 8th ICOTOM*. The Metallurgical Society, Santa Fe, 17–30.

YOUNG, C. T. & LYTTON, J. L. 1972. Computer generation and identification of Kikuchi projections. *Journal of Applied Physics*, **43**, 1408–1417.

Practical application of entropy optimization in quantitative texture analysis

H. SCHAEBEN[1], H. SIEMES[2], S. HÖFLER[3,4] & G. WILL[3]

[1] *Department of Geology, University of Bonn, FRG; present address: Laboratory of Metallurgy of Polycrystalline Materials, University of Metz, France*

[2] *Department of Mineralogy, University of Aachen, Wüllnerstrasse, 2, 5100 Aachen, FRG*

[3] *Department of Mineralogy, University of Bonn, FRG*

[4] *Nuclear Research Center, Jülich, FRG*

Abstract: Finite series expansion methods are applied to the texture problem of recovering an orientation density function (odf) from its pole density functions (pdf) measured in X-ray or neutron diffraction experiments, the mathematical model of entropy maximization is applied to pole figure data and numerical results are presented as well as an interpretation of their geological significance.

The maximum entropy concept has been explicitly introduced into and constructively adapted to the purposes of texture goniometry in (Schaeben 1988). Its philosophy and justification has been recapitulated in (Schaeben 1990). Thus, the tomographic problem of pole-to-orientation density inversion has completely been restated to maximize

$$S(x) = -\sum_{n=1}^{N} x_n \ln x_n \tag{1}$$

subject to

$$\pi x = y \tag{2}$$

and

$$x_n \geq 0 \qquad n = 1,\ldots,N \tag{3}$$

and

$$\sum_{n=1}^{N} x_n = 1 \tag{4}$$

where π denotes the correspondence matrix relating the discrete orientation density function, or texture vector, x, with the discrete pole density function y, the components y_p of which are identified with the available data, i.e. with the experimental intensities; for more details the reader is referred to Schaeben, (1988).

This model is equivalent to the method of regularization (Tikhonov & Arsenin 1977; Titterington 1985) when the penalty term is chosen to be the entropy functional, (cf. Jaynes 1983; McLaughlin 1983).

Approaching the corresponding primal problem results in a problem of convex programming, and solving both the primal and dual problem simultaneously results in an iterative procedure; the algorithm to be eventually coded reads for the mth component initialization

$$x_m^{(0)} = \begin{cases} 0 & \text{for all } m \text{ for which } \pi_{pm} \neq 0 \text{ and } y_p = 0 \\ \dfrac{1}{N-v} & \text{otherwise} \end{cases} \tag{5}$$

where v denotes the number of *a priori* zero components of $x^{(0)}$,

$(k+1)$th iteration in block step notation

$$x_m^{(k+1)} = \prod_{p \in L_m} \left\{ \left(\frac{y_p}{\sum_{n=1}^{N} \pi_{pn} x_n^{(k)}} \right)^{r_k \pi_{pm}} \right\} x_m^{(k)},$$

$$m = 1,\ldots,N \tag{6}$$

with $L_m = \{p = 1,\ldots,P \,/\, \pi_{pm} \neq 0\}$ denoting the set of numbers referring to the non-zero elements of the mth column vector of π, and relaxation parameters $0 < r_k \leq 1$.

The sequence defined by equation (6) was labeled 'block-iterative' algorithm by Censor & Segman (1987). Recent proofs of convergence of this type of algorithm towards the maximum entropy solution of problem (1) have been given by Censor & Segman (1987), and Lent & Censor (1990).

Some of the more interesting features of this algorithm (6) will be briefly listed. According to the rationale of the maximum entropy concept, it provides the stepwise, i.e. piecewise constant, odf f with the smallest information content consistent with the experimental pdf data y,

From Knipe, R. J. & Rutter, E. H. (eds), 1990, *Deformation Mechanisms, Rheology and Tectonics*, Geological Society Special Publication No. 54, pp. 375–381.

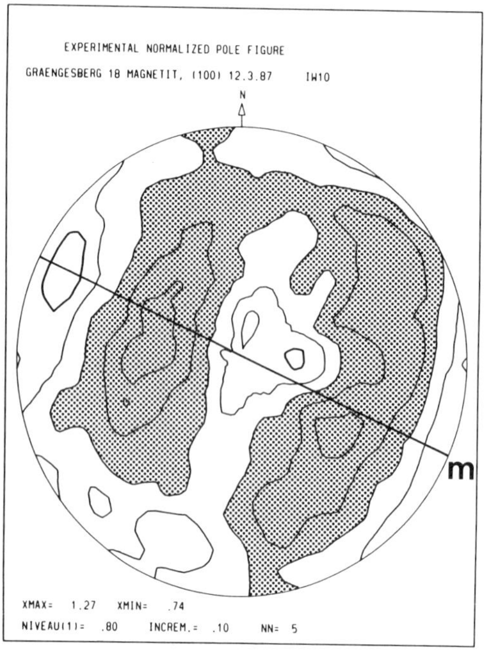

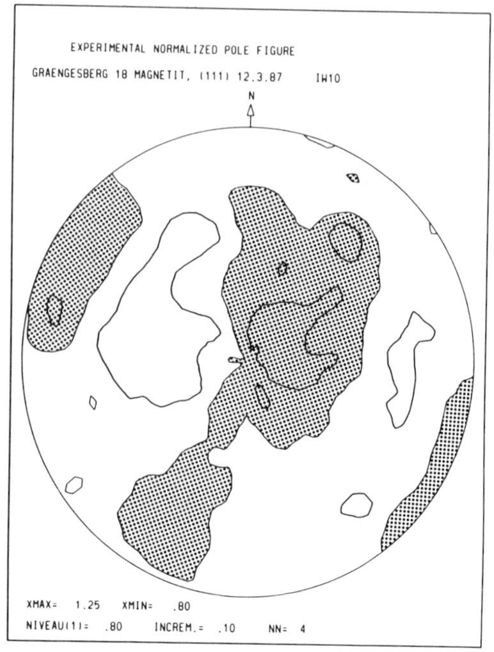

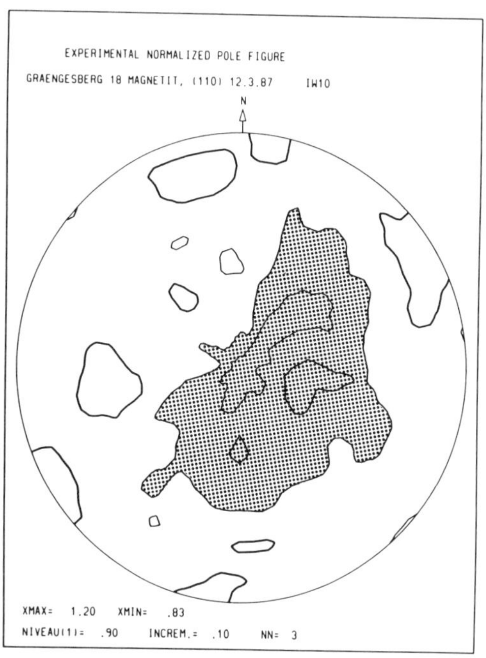

Fig. 1. Contour line diagrams of experimental pdfs of reflections (100), (111), and (110) of magnetite of an ore from Graengesberg, Central Sweden. Magnetite is of cubic crystal symmetry, the sample symmetry is triclinic (monoclinic); the bold line depicts the trace of the mirror plane (m). (**a**) (100) contour intervals: 0.8, 0.9, 1.0, 1.1, 1.2; minimum density: 0.74, maximum density: 1.27; shaded above 1.0. (**b**) (111) contour intervals: 0.9, 1.0, 1.1, 1.2; minimum density: 0.80, maximum density: 1.26; shaded above 1.0. (**c**) (110) contour intervals: 0.9, 1.0, 1.1; minimum density: 0.84, maximum density: 1.20; shaded above 1.0.

thus avoiding artificial textural components for which there is no evidence in the data, also 'ghosts' caused by the specific properties of the diffraction experiment. Synonymously, it may be referred to as the minimal prejudiced, least presumptive, or maximal non-committal solution of equation (2). It is also the most safely balanced solution in the sense of the Backus–Gilbert formalism, i.e. it provides the best resolution given the partitions (G_n) of $G = (SO_3)^+$ and (Z_p) of S^3.

Example

The capability of the maximum entropy method has been shown by means of the cubic/orthorhombic pdf data set MIX2 (Schaeben *et al.* 1990), which has been derived from a mathematical odf model-function (Matthies 1982). In this paper we present an application to a practical geological problem. The complete pdfs of the reflections (100), (111), and (110) of a magnetite ore were measured at the KFA Jülich by means of neutron diffraction (Schäfer *et al.* 1988) on a sample which has been collected at the Graengesberg Mine in Central Sweden by J. Huppertz. The three experimental pdfs (Fig. 1) are projected approximately parallel to the plane of foliation of the ore and they reveal a weak preferred orientation with triclinic specimen symmetry, which may be simplified to monoclinic symmetry as indicated in Fig. 1. The trace of the indicated mirror plane (m) is parallel to the lineation of the ore.

The FORTRAN program package named MENTEX (Maximum ENtropy TEXture) which has been developed to perform the pdf-to-odf inversion by means of finite elements and entropy optimization, is described in more detail by Schaeben *et al.* (1990). Using three complete or two incomplete pdfs the odf has been calculated and the pdfs were recalculated. The complete pdfs of the reflections (100) and (111) were truncated at 75° and used as incomplete pole figures; they have been chosen because Schaeben *et al.* (1990) have shown that pdfs of crystallographic forms with preferably a small number of faces give the most promising results. In Table 1 the performed calculations are summarized. Comparing the values of the mathematical entropy and of the mean relative error in percent (RPO) as well as the maximum and minimum intensities in the pdfs and considering the number of iterations it is obvious that the best results are obtained by analysing three complete pdfs. As an example, part of the calculated odf after 77 iterations is displayed in Fig. 2. In each section two small maximum areas (density > 1.30) are detectable which represent two continuous tubes of preferred orientation. The recalculated pdfs of the larger number of iterations are very similar because the increase of entropy towards its maximum or the decrease of the mean relative error (RPO) becomes progressively slower. The pdfs which were recalculated from the odf of three pdfs after 44 iterations (Fig. 3) and from the odf of two pdfs after 77 iterations (Fig. 4) are presented as examples. The pdfs of both examples are very similar and show convincingly the main features of preferred orientation.

The conclusion of this testing of the entropy method is that in the case of weak preferred orientation and low specimen symmetry already

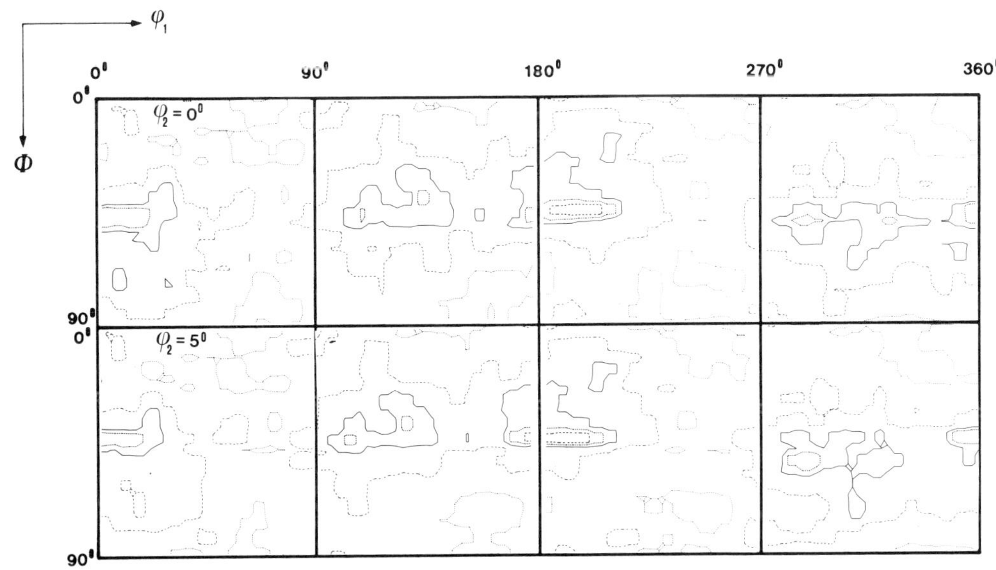

Fig. 2. Calculated odf after 77 iterations using the complete (100), (111), and (110) pdfs of Fig. 1. Contour intervals: 1.1, 1.2, 1.3, 1.4; minimum density: 0.64, maximum density: 1.40;

Table 1. *Calculation of the odf of a magnetite from Graengesberg, Sweden and recalculation of the pdfs by means of MENTEX*

	Orientation distribution function				(100) pole figure				(111) pole figure				(110) pole figure		
	Density				Density		Mean rel. error %		Density		Mean rel. error %		Density		Mean rel. error %
	Min.	Max.	Entropy	Texture Index	Min.	Max.	RPO[1]	RPO[2]	Min.	Max.	RPO[1]	RPO[2]	Min.	Max.	RPO[1]
Normalized experimental pdf															
completed					0.746	1.277			0.810	1.260			0.840	1.205	
incomplete (*)					0.739	1.265			0.805	1.252					
odf calculated from pdfs: (100) (111), (110) complete					*recalculated pdf*				*recalculated pdf*				*recalculated pdf*		
22 iterations	0.69	1.40	70.8	8100.	0.786	1.232	2.89		0.861	1.199	2.57		0.923	1.149	4.18
33 iterations	0.67	1.40	77.7	8100.	0.774	1.239	2.85		0.854	1.202	2.51		0.919	1.151	4.16
66 iterations	0.66	1.40	81.1	8100.	0.769	1.242	2.84		0.850	1.203	2.49		0.918	1.153	4.15
55 iterations	0.65	1.40	83.0	8100.	0.767	1.243	2.83		0.847	1.203	2.49		0.917	1.154	4.15
66 iterations	0.64	1.40	84.5	8100.	0.766	1.245	2.82		0.846	1.203	2.48		0.916	1.155	4.15
77 iterations	0.64	1.40	85.6	8100.	0.766	1.246	2.81		0.844	1.203	2.48		0.915	1.155	4.14
odf calculated from pdfs: (100), (111) incomplete (*)					*recalculated pdf*				*recalculated pdf*				*calculated pdf*		
27 iterations	0.674	1.348	67.9	8000.	0.772	1.234	2.78	2.09	0.867	1.199	2.58	2.28	0.928	1.071	4.84
77 iterations	0.608	1.454	87.3	8100.	0.757	1.258	2.69	1.86	0.840	1.199	2.54	2.14	0.914	1.106	4.88
101 iterations	0.593	1.481	92.0	8100.	0.757	1.261	2.72	1.81	0.835	1.198	2.54	2.12	0.910	1.117	4.94

[1] RPO with reference to the normalized complete pole figure.
[2] RPO with reference to the normalized incomplete pole figure.
* Truncated at 75 degrees pole-distance.

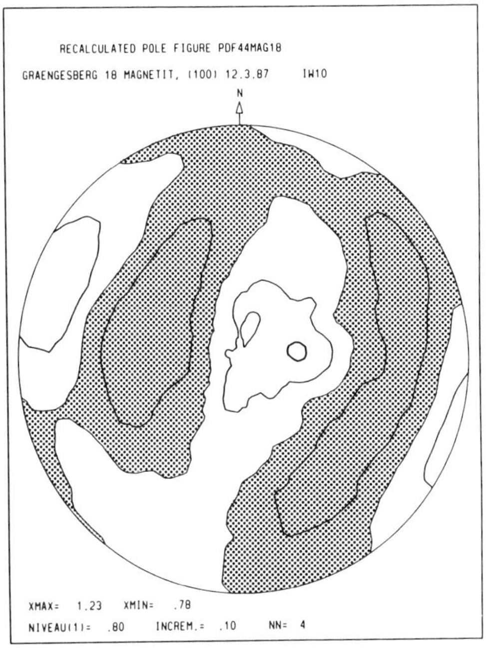

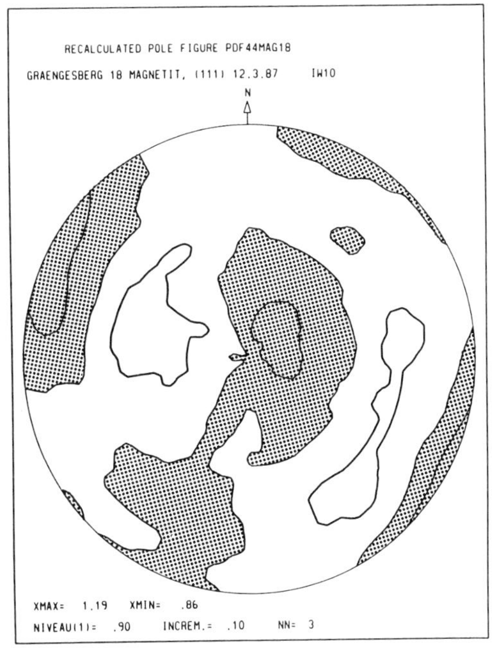

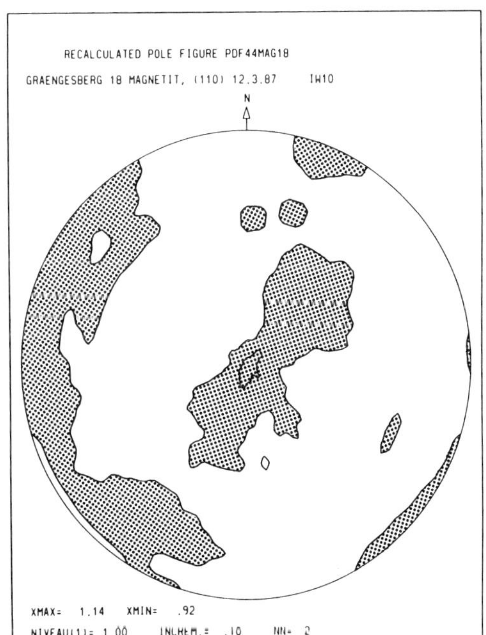

Fig. 3. Recalculated pdfs after 44 iterations corresponding to the three pdfs of Fig. 1. (**a**) (100) contour intervals: 0.8, 0.9, 1.0; 1.1; minimum density: 0.77, maximum density: 1.24; shaded above 1.0. (**b**) (111) contour intervals: 0.9, 1.0, 1.1; minimum density: 0.85, maximum density: 1.20; shaded above 1.0. (**c**) (110) contour intervals: 1.0, 1.1; minimum density: 0.91, maximum density: 1.15; shaded above 1.0.

two (incomplete) pole figures lead to an acceptable and reasonable odf, if the number of iterations is sufficiently large.

The straight forward application of the MENTEX program requiring only a very few and incomplete pole figures makes this method very attractive for those who are measuring preferred orientation on thin, flat samples by means of X-rays and in cases of low crystal-symmetry or overlapping reflections in multi-phase materials like rocks and ores.

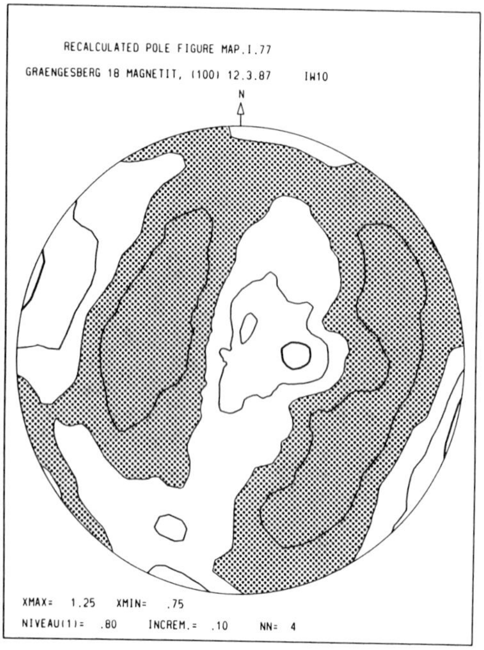

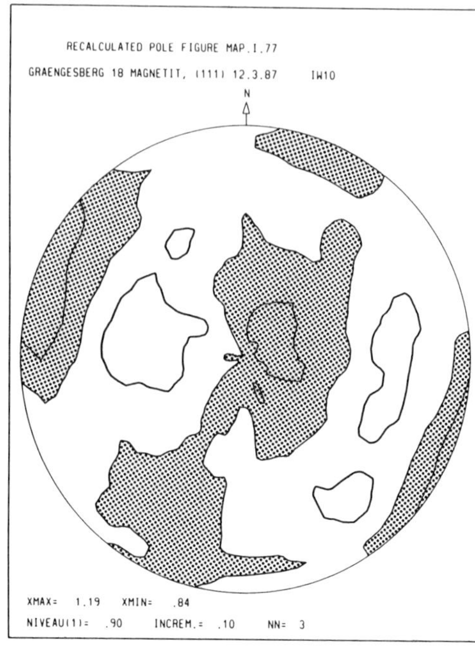

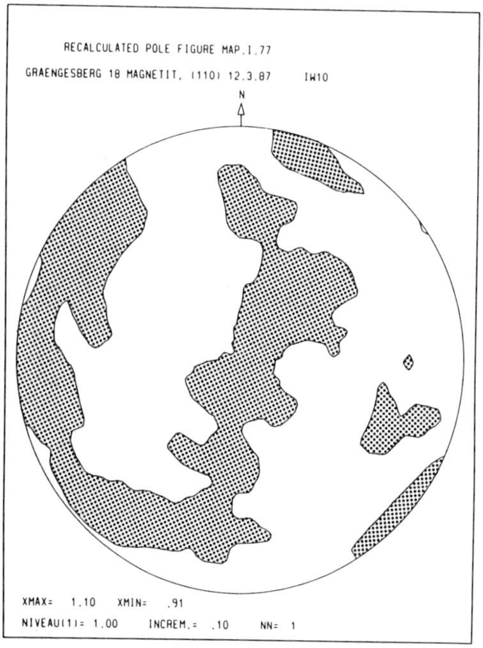

Fig. 4. Recalculated pdfs after 77 iterations corresponding to the two pdfs (100) and (111) of Fig. 1 truncated at 75 degrees of Fig. 1. (**a**) (100) contour intervals: 0.8, 0.9, 1.1, 1.2; minimum density: 0.76, maximum density: 1.26; shaded above 1.0. (**b**) (111) contour intervals: 0.9, 1.0, 1.1; minimum density: 0.84, maximum density: 1.20; shaded above 1.0. (**c**) (110) contour intervals: 1.0 minimum density: 0.91, maximum density: 1.12; shaded above 1.0.

Conclusions

The solution of the tomographic inversion problem of texture goniometry presented here uses finite series expansion, explicitly relates to a well established body of mathematical and information theory, and is given by an algorithm that allows efficient programming, thus liberating the user from confidence in *ad hoc* procedures.

From theoretical considerations, see Schaeben (1988, 1990) as well as from practical applications, see Schaeben *et al.* (1990), including odf reproduction from a single complete or incomplete pdf, and from several complete or (initially not normalized) incomplete pdfs of cubic or trigonal/hexagonal materials, it can be concluded that the technical properties and practical advantages of this new method are:

(1) it immediately applies to complete or incomplete pdfs;
(2) It also applies to simultaneous processing of several pdfs of different crystal forms;
(3) it detects faked pdfs or the erroneous mixture of pdfs of different crystal forms from different samples;
(4) it provides an adjustable relaxation parameter to account for small non-negligible experimental errors in the measured pdfs;
(5) the necessary non-megativity of the odf to be reproduced is a truly constitutive element;
(6) possible zero components in the pdf are exactly and efficiently accounted for, uniform pdfs give a uniform odf;
(7) it provides the most safely balanced solution with respect to the inevitable compromise that must be made between spatial resolution and the accuracy of amplitude measurements;
(8) it is numerically straight-forward and stable;
(9) requires storage for only the vector of iterated approximate solutions which makes parallel processing an attractive possibility;
(10) it uses a property of the odf itself as a stopping rule for the iteration, i.e. the size of the $(k + 1)$th increment of the entropy in accordance with the rate of convergence of the algorithm; stopping rules in all other methods of odf reproduction are based on the residual or simple functions of it thus referring to properties of the recalculated pdfs. Due to the general ambiguity of the texture problem they do not provide sufficient criteria for the physical goodness of the reproduced odf.

Computing work was done with a CDC CYBER 175 at the computer center of Aachen Technical University. This study was partly (H. Siemes) funded by Deutsche Forschungsgemeinschaft (DFG), Bonn.

References

CENSOR, Y. & SEGMAN, J. 1987. On block-iterative entropy maximization. *Journal of Information and Optimization Sciences*, **8**, 275–291.

JAYNES, E. T. 1983. *Papers on probability, statistics, and statistical physics*. D. Reidel Publishing Company, Dordrecht.

LENT, A. & CENSOR, Y. 1990. The primaldual algorithm as a constraint-set-manipulation device. *Mathematical Programming*, (in press).

MATTHIES, S. 1982. *Aktuelle Probleme der quantitativen Texturanalyse*. Akademie der Wissenschaften der DDR, Zentralinstitut für Kernforschung, Rossenddorf bei Dresden, ZfK – 480.

MCLAUGHLIN, D. W. 1983. *Inverse problems, Proceedings of a symposium in applied mathematics*. American Mathematical Society, Providence, R.I.

SCHAEBEN, H. 1988. Entropy optomization in texture goniometry, I: Methodology. *Physica Status Solidi (b)*, **148**, 63–72.

—— 1990. DETERMINATION OF COMPLETE ODF USING THE MAXIMUM ENTROPY METHOD. *In*: BUNGE, H. J. & ESLING, C. (eds) *Quantitative texture analysis. II, Course of advanced instruction* Clausthal-Zellerfeld, Mar 6–10, 1989. Deutsche Gesellschaft für Materialkunde, (in press).

——, SIEMES, H. & AUERBACH, S. 1990. Entropy optimization in texture goniometry. II: Practical applications. *Physica Status Solidi* (b), **158**, 407–425.

SCHÄFER, W., HÖFLER, S. & WILL, G. 1988. Applications of neutron diffraction pole figure measurements on polycrystalline and monocrystalline metallic samples. Examples from the fourcircle neutron diffractometer in Jülich. *Textures and Microstructures*, **8 & 9**, 457–466.

TIKHONOV, A. N. & ARSENIN, V. Y. 1977. *Solutions of ill-posed problems*. J. Wiley & Sons, New York.

TITTERINGTON, D. M. 1985. Common structure of smoothing techniques in statistics. *International Statistical Review*, **53**, 141–170.

Experimental and observational constraints on the mechanical behaviour in the toes of accretionary prisms

DANIEL E. KARIG

Department of Geological Sciences, 2124 Snee Hall, Cornell University, Ithaca, NY 14853, USA

Abstract: The mechanical behaviour of porous sediments in the toes of accretionary prisms can be analyzed by combining the principles of soil mechanics and the results of experimental sediment deformation with the in-situ physical properties observed from deep sea drilling. Porosity is the only physical property that has been extensively collected and even these data have generally been inadequately analysed. After correction for porosity rebound, variations due to lithology, and lack of representative sampling, porosity data must be converted from a spatial or Eulerian description to a Lagrangian description, which follows the behaviour of a given sediment element as it moves through the prism toe. Such an approach shows that sediment elements compact during deformation in prism toes; strongly in the Nankai prism and much less so in the Lesser Antilles. Compactive stress paths are ductile, and there is evidence that much of the deformation, at least in the Nankai toe, is of this type. Nevertheless, there is also a very strong component of brittle deformation, both on diffuse structures and along the major faults. This brittle overprint is explained in the light of experiments as showing that sediments already at a state of ductile failure will shear brittlely if post-failure deformation is associated with a decrease in σ_m'. Such reduction of σ_m' probably occurs in prism toes during fluctuations of pore pressure. Persistent brittle faults are modelled as zones in which porosity and pore pressure are higher than in the surrounding sediments, a condition maintained by fluid channelled up the faults from deeper in the prism. Outward diffusion of fluid from the faults may be responsible for broad zones of brittle failure, characterized by scaly clay.

The toes of accretionary prisms along convergent plate margins are sites where porous sediments are subjected to stress paths that result in large changes in physical properties before, during, and after mechanical failure. In a not too over-simplified way, sediments uniaxially consolidated in the trench wedge can be visualized as deforming as they move through a prism toe having a temporally constant geometry and distribution of physical properties. This distribution is now only sketchily characterized and the nature of stress in the toe is even less well constrained, but as negative as the situation might appear, great progress has been made in the past few years. In part this is a result of better data acquisition, but it stems also from the recognition of the nature of the mechanical response of porous sediments and of the role of pore fluids in affecting mechanical behaviour. Application of principles developed for the mechanics of soil behaviour and from hydrology provide significant constraints and permit semi-quantitative modelling of flow and stress paths. Ironically, this approach, while elucidating the behaviour of prism toes, has also revealed several seeming contradictions. The most striking of these is the existence of discrete fault zones and other manifestations of brittle deformation in compactively, and theoretically ductiley deforming sediments.

In this paper the character of deformation and strain, and the distribution of porosity in prism toes is reviewed and discussed insofar as they pertain to mechanical behaviour of these sediments. Following a brief review of relevant aspects of soil mechanics, much of which is unfamiliar to the geological community, an attempt is made to deduce the stress paths and stress history of sediments in prism toes from their observed physical properties and structural fabric. Finally, a model that appears to reconcile the apparently incompatible observations of persistent brittle faulting in an environment of dominantly ductile deformation is presented.

From Knipe, R. J. & Rutter, E. H. (eds), 1990, *Deformation Mechanisms, Rheology and Tectonics*, Geological Society Special Publication No. 54, pp. 383–398.

Relevant observations in prism toes

Information concerning physical properties has been accumulating over the past decade from a number of prism toes, including those of the Makran, Cascadian, and Middle American arcs, but the largest coherent data sets are from the Lesser Antilles and Nankai prisms (Fig. 1); where deep sea drilling and ancillary geophysical studies have been undertaken. Fortunately, these two study areas represent examples near the end-members of a spectrum of clay rich (Lesser Antilles) to more sand rich (Nankai) prisms, and display quite different responses. The available data consist mostly of physical properties, such as porosity and permeability, and aspects of the structural fabric, which reflect the mode of deformation and deformational mechanisms. Only very limited information concerning the state of stress has been acquired from prism toes, and review of these data seems best incorporated into the discussion of the mechanical behaviour that is deduced from the physical properties.

Structural fabric

The largest data set in both these toes concerns the structural fabric, which includes both discrete faults and diffuse structures, the latter apparently including both ductile and brittle components.

Discrete fault zones, resolved on seismic profiles or by stratigraphic offsets in drill holes, are the most obvious manifestation of brittle deformation within the prism toe. These faults include the sub-horizontal basal decollement and imbricate thrusts that rise upward from this decollement (Fig. 1). Almost all the evidence concerning the deformational structures within these faults has been obtained from drill cores in the Lesser Antilles Arc. Both the decollement and imbricates in the clay-rich lithologies of that setting are marked by zones of scaly clay up to several tens of metres thick (Moore *et al.* 1987). Although Behrmann *et al.* (1988) discuss alternative brittle and ductile interpretations of these fabrics from a mesoscopic perspective, they are clearly discrete fractures that would be termed brittle shears if seen in experimentally deformed samples and would be associated with a dilative response. Dilative calcite veins that accompany and post-date the development of the scaly clay (Shipboard Scientific Party 1988; Behrmann *et al.* 1988) provide further evidence for a brittle response during at least some parts of the strain history in these fault zones.

Only a very fragmentary and poorly preserved

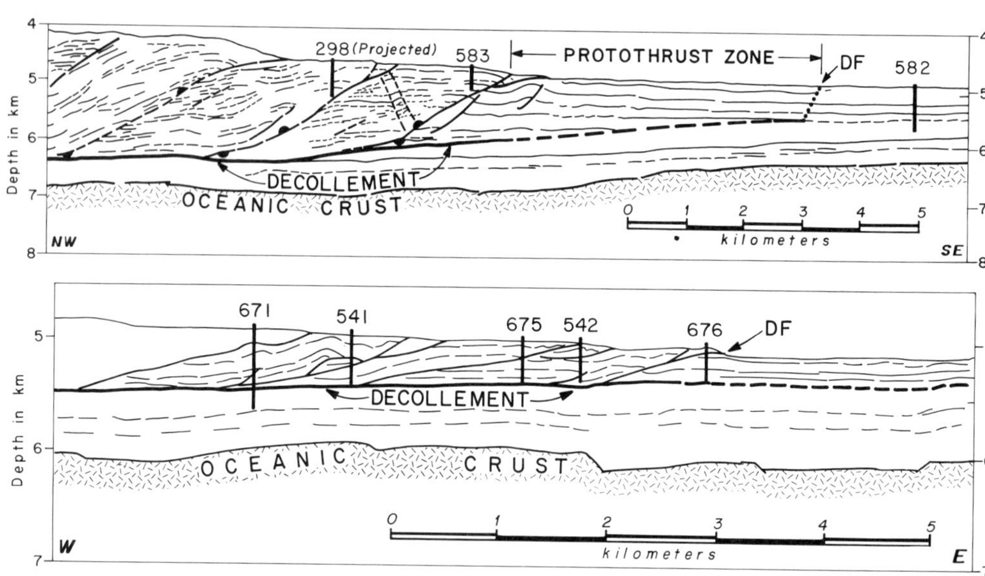

Fig. 1. Interpretive sections of seismic depth profiles across the toes of the Lesser Antilles (upper) and Nankai (lower) prisms (not to same scale). These sections show the basic structural elements and locations of drill sites. The decollement is dashed where nascent or thought to be incipient. The deformation front (DF), defined as the locus of the first observable deformation, may lie a little more to the east in the Lesser Antilles prism. Sections are modified from Shipboard Scientific Party (1988) and Karig (1986*a*, *b*).

record of the frontal imbricate has been obtained in the Nankai toe, because this fault was cut at a shallow depth, where the sediment was poorly consolidated (Kagami, *et al.* 1986). One core (583D-12) suspected to have sampled this fault zone, displayed small lenses of mud in a sand matrix (Shipboard Scientific Party 1986), which, if not a drilling-induced structure, would also suggest a brittle response (cf. Lundberg & Moore 1982). A brittle, Coulomb response is also suggested by the initial 35° angle between these imbricate thrusts and the 5° trenchward orientation of σ_1 deduced from a conjugate set of minor shear fractures (Karig 1986[a]).

Diffuse deformation contributes a significant component of strain in both the Nankai and Lesser Antilles prism toes, but with very different characteristics in the two areas. In the Lesser Antilles prism, significant diffuse deformation is implied by the much greater horizontal shortening calculated with area balancing techniques than with bedding length techniques (Behrmann *et al.* 1988). Structural features observed in drill cores that account for at least some of this deformation include zones of scaly clay and clay-filled 'vein structures', and small faults (Behrmann *et al.* 1988), all of which are dilative features. Ductile features might include folds as well as more uniform flow. Flow is implied seaward of the frontal thrust because movement on the decollement in this setting requires shortening in the upper plate, where discrete failure structures are not observed.

None of the brittle structures noted in the Lesser Antilles cores were observed in cores from the Nankai prism, despite the fact that diffuse deformation accounts for nearly all the horizontal shortening within the protothrust zone and about half of that within the prism toe (Karig 1986*b*). Even in the clay-rich sections of core, where structural features were best preserved, the only structures observed were semi-pervasive deformation bands (Lundberg & Karig 1986) that occur at depths greater than 350 m. These are interpreted as brittle-ductile shear zones with Coulomb geometries.

Displacements on these bands can account for only a small fraction of the observed bulk strain in the prism toe, but apparent passive rotation of these bands to dips as high as 65° implies large ductile strains, at least in the vicinity of the bands (Karig & Lundberg, 1990). The diffuse strain shown by seismic profiles across the Nankai toe can readily be satisfied by such localized and non-uniform ductile strain. In short, the Lesser Antilles toe has a larger component of brittle deformation than does the Nankai despite the much greater clay content in the former and, as will be noted in more detail later, its much higher porosity.

Physical properties: porosity

The measurement of sediment physical properties throughout prism toes is important because some properties, such as porosity and permeability, exert strong controls on the strength and state of stress, and because others, such as seismic velocity, can be empirically correlated with porosity and other mechanical parameters. Variations in other characteristics, such as temperature and pore fluid chemistry, have proved to be important indicators of zones with anomalous porosity and permeability (Shipboard Scientific Party 1988).

Sediment porosity (η) is probably the most critical physical property in prism toes because it is a basic parameter controlling the strength of porous sediments that deform primarily by mechanical rather than mass diffusive processes. At the porosities occurring in that section of the prism toe explored by drilling (> 20%), reduction of porosity is very sensitive to increases in effective mean stress, σ'_m (Fig. 2). This relationship depends as well on the nature of the strain under which compaction is induced, but for all strain paths, $\partial\eta/\partial\sigma$ decreases as σ'_m increases. This stress–porosity relationship is not

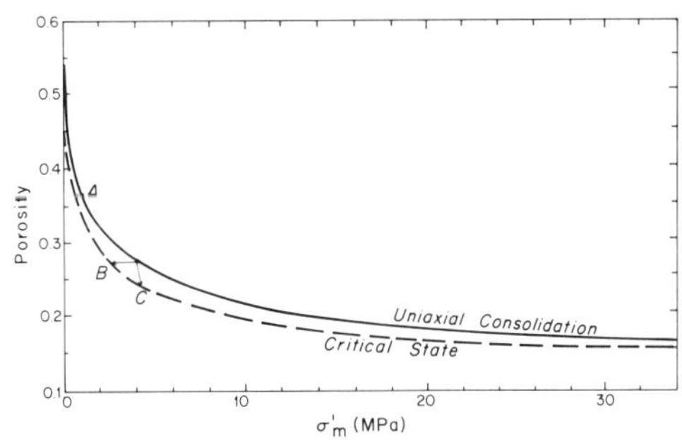

Fig. 2. Porosity as a function of σ'_m for uniaxial consolidation and for critical state failure, based on experimental data using silty clay. The uniaxial consolidation curve is well-constrained but the critical state curve is largely interpolated. Path A shows that, at high porosity, σ'_m can be only slightly decreased before the pre-failure strain becomes dilative. Path B, which approximates the pre-failure stress path of a sediment element in the Lesser Antillies toe originally 400 m deep, shows the much larger reduction of σ'_m possible before the onset of dilatation. Path C represents the situation for a similar element in the Nankai toe, where σ'_m increases slightly and results in a much larger porosity decrease.

reversible, in that the only increase in porosity caused by stress reduction is by elastic expansion at relatively high stress and inelastic expansion at very low stress. Gross increases in porosity are accomplished only by means of dilative brittle failure.

Porosity has been the most frequently measured property in the two prism toes, primarily by laboratory techniques, but also by downhole logging for short sections of drill hole (Kagami *et al.* 1986). However, the laboratory data on which interpretations have been based have not, in general, been adequately reduced or analysed. Laboratory measurements must be corrected for the porosity rebound from in-situ to atmospheric conditions and the different porosity–depth responses of various lithologic components of the sediment section must be taken into account.

In general, sand-rich sediments have much lower porosities than clays at very low stresses, but are also less compressible, so that at stresses corresponding to depths of a kilometre or so, clay porosities often become lower than those of sand (e.g. Plumley 1980). Not only must the lithology of the analysed sample be noted, but the relative in-situ abundance of that lithology must also be estimated before spatially averaged porosities can be determined. This is not a trivial problem in that clay-rich units are very preferentially preserved in rotary drill cores, and even to some extent in hydraulic piston cores.

Porosity rebound, which results from reduction of stress, has both elastic and inelastic components. Rebound varies with lithology, porosity level, and degree of cementation, as well as with the behaviour of pore pressure during stress reduction (Bishop *et al.* 1975), but has not been adequately quantified. A general response, relying on measurements by Hamilton (1976), Hou (1988), and others is that rebound increases with decreasing porosity at high porosity, uncemented conditions, but decreases sharply at lower porosities as cementation increases the values of elastic moduli. Porosity rebound approaching 10% is quite likely for the poorly cemented mudstones with $\eta = 20\%$ in prism toes. Rebound for sand is very much less than for mudstones at these and higher porosities (unpublished experimental data from our lab; see also Lambe & Whitman, 1969, p. 255).

Porosity data from drill holes are augmented and extrapolated throughout the prism toe with velocity data, which can be inverted to porosities using logging and laboratory data for calibration. Expanding spread seismic profiles over the Nankai toe and constant (wide) aperature multichannel profiles across the Lesser Antilles prism (Bangs *et al.* 1990) have greatly improved the velocity structure in these two cases, but inversion of these data to porosities still lacks adequate porosity–velocity correlations.

Even with the problems outlined above, comparison of roughly calculated porosities in the prism toes of both the Nankai and Lesser Antilles arcs reveals both similarities and large differences in their mechanical behaviour during deformation.

At all the three Nankai drill sites, the uncorrected porosities measured on cores decrease monotonically downward to the base of the trench fill turbidites (Bray & Karig 1986). These measurements, primarily on the clay-rich fraction, have been corrected for rebound in this study using estimates from Hamilton (1976) and from our consolidation tests on artificial mudstones (Hou 1988). Estimates from the logging results at Site 583 (Shipboard Scientific Party 1986) suggest a sand/shale ratio of 1:3 in the Nankai prism toe. Porosity and velocity data for sand were available from logging and from consolidation tests in our laboratory. From these, a bulk in-situ porosity–depth curve for conditions seaward of the Nankai deformation front was generated (Fig. 3). These manipulations indicate that the difference between uncorrected and in-situ porosities increases downward to at least 500 m.

This in-situ curve will be merged with the velocity–depth data from this setting to produce a locally applicable velocity–porosity relationship. Application of a preliminary version of this correlation to the velocity distribution of Aoki *et al.* (1986), confirms the modest arcward reduction in porosity between sites 582 and 583 at all depths that was seen earlier by comparison of the core data. However, rather than a simple monotonic arcward porosity decrease there appears to be a pronounced porosity minimum just seaward of the frontal imbricate thrust and thus a remarkably steep lateral reduction in porosity across the protothrust zone. Bulk porosities may be as low as 20% at the base of that zone near the frontal imbricate.

The porosity distribution in the toe of the Lesser Antilles prism in the vicinity of the DSDP and ODP drill holes from core data show only a very small arcward shift (steepening) in gradient (Shipboard Scientific Party 1988). Using constant offset seismic profiles, Bangs *et al.* (1990) developed a porosity distribution in the Lesser Antilles prism that shows a lack of convergence of isoporosity contours in the prism toe and low porosity–depth gradients even

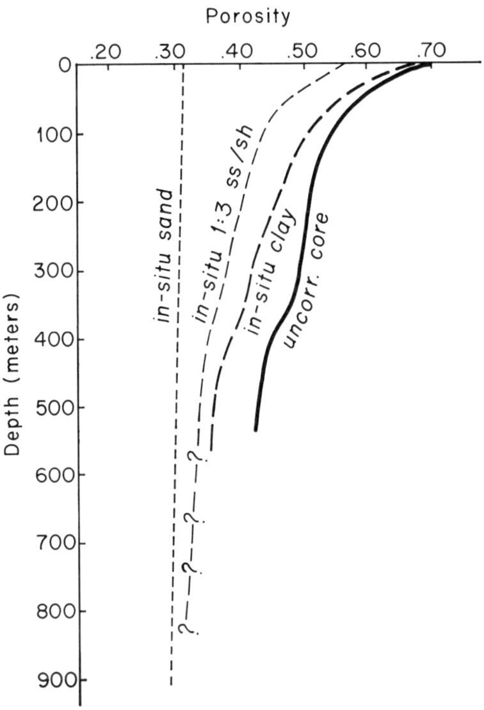

Fig. 3. Various porosity v. depth curves for DSDP Hole 582, illustrating the corrections that must be made to shipboard core measurements before a representative in-situ porosity curve can be developed (see text for details). Uncorrected core data are based on water contents of the clay-rich fraction (Bray & Karig 1986). The in-situ curve is corrected for porosity rebound. The in-situ sand curve is from laboratory measurements, and the in-situ 1:3 curve attempts to match the ratio of sand to shale estimated from well-log data from DSDP Site 583 (Kagami *et al.* 1986).

compared with that of the incoming oceanic sediment column. Clearly porosities have not been reduced as much in the Lesser Antilles as in the Nankai toe, but it is not obvious from the isoporosity contours how much, if any, compaction is occurring in the sediments moving through the toe of the Lesser Antilles prism.

The characterization of porosity by comparison of depth profiles is, in effect, a spatial, or Eulerian description of the porosity field, and is not as relevant to the interpretation of the mechanical behaviour of a progressively deforming sediment element as is the investigation of porosity along the flow trajectories that these sediment elements take through the prism toe. This latter approach would define a Lagrangian distribution of porosity. Before a Lagrangian description of sediment porosity can be made, however, both the spatial Eulerian distribution of porosity and the flow trajectories must be known.

Flow trajectories in a prism toe

Determination of flow trajectories of sediment elements through a prism toe that is assumed to have a fixed shape requires a knowledge of the distribution of strain and displacement throughout that region. Given such problems as the existence of faults, which are moving discontinuities, this displacement field will probably only be approximated with the best of data. However, a conceptual idea can be presented and applied to the Nankai protothrust zone, where deformation occurs without obvious faulting.

In the trench wedge and the protothrust zone, there is little or no displacement parallel to the trench axis, and deformation can be approximated by plane strain. Vertical strains in the Trough and toe of the Nankai prism can be estimated from seismic profiles (e.g. Karig, 1986*b*; Moore *et al.* 1990), which display excellent coherence. With volume strains (Δ) directly obtained from the porosity distribution, the horizontal strains can be calculated: $\varepsilon_h = \Delta - \varepsilon_v$. The reference state for all these strains is an element with having a unit volume at the original near-surface bulk porosity, estimated as 55% (Fig. 3).

Preliminary analysis indicates that the vertical extensional strains acquired as sediment elements move through the protothrust zone are greatest near the base and rear of the zone. Because the largest reduction in porosity also occurs in this region, the horizontal strains in sediment elements near the base of the zone are perhaps as high as 0.3, in contrast to much lower values (0.1) near the rear and top of the protothrust zone.

This Eulerian distribution of strain can then be converted to a Lagrangian distribution, which describes the flow trajectories or velocities of sediments within the prism toe. A very crude approach to this conversion is sequentially to calculate the strain of a column of elements with equal solid (grain framework) volume from the deformation front arcward, using the incompressible basement as a coordinate x-axis. A computerized numerical scheme will be employed for this calculation as the quality of the data improves, but even the initial graphical approach in the Nankai protothrust zone illustrates several points.

Sediment flow trajectories do not parallel

bedding, either in the trench wedge or in the prism toe (Fig. 4). Sediment elements deposited on the trench floor are buried during subsequent sedimentation as they move toward the deformation front. Trajectories are approximately parallel to the facies boundary between wedge fill strata and the underlying basinal strata, being slightly steeper owing to continuing compaction in the sediment section below the trajectory in question. Sediment elements in basinal strata would follow paths that parallel bedding surfaces if no faults existed.

Within the protothrust zone the trajectories fan out in response to the horizontal shortening. Stratification in the protothrust zone behaves as a passive strain marker that rotates toward the direction of maximum extension with progressive deformation (Fig. 4). Intersection of the uppermost strata with the surface of the prism may appear unusual, but is actually observed on the seismic profiles. These terminations mark palaeo-positions of the deformation front, which forms the arcward boundary of sedimentation in the trench. This is consistent with the requirement that, within an accretionary prism, no trench fill reflector can persist laterally for a distance greater than the width of the trench floor at the time of its deposition.

Lagrangian description of porosity

The flow trajectories, as developed in the section above, can be superimposed on the plot of porosity distribution to show the change of porosity in a sediment element along the stress path to which it is being subjected as it moves along the flow trajectory. In the Nankai protothrust zone, and by crude extrapolation across the entire prism toe, all trajectories and stress

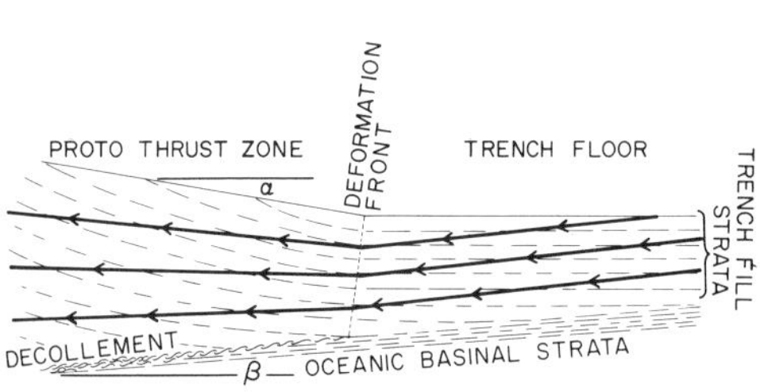

Fig. 4. Diagrammatic section through a prism toe having a simple ductile deformation scheme (see text). The resulting flow trajectories (exemplified by heavy lines) cause the stratification (light dashed lines) to rotate passively toward the axis of maximum elongation. The angles α and β define the surficial and basal slopes of the prism toe. (From Karig & Lundberg 1990).

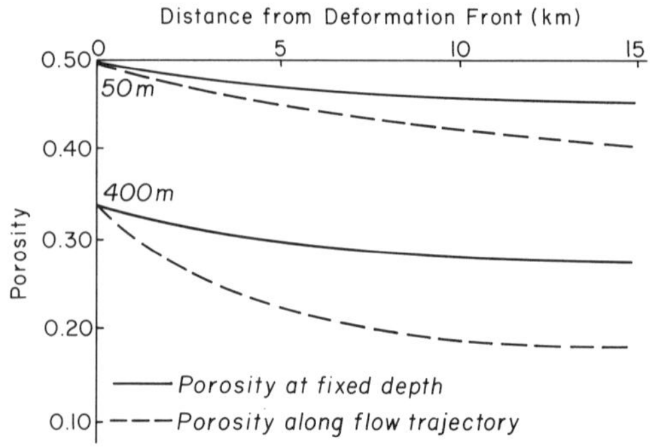

Fig. 5. Comparison of Eulerian (solid lines) and Lagrangian (dashed lines) description of porosity across a prism toe (based on preliminary data from the Nankai prism) for sediment elements at depths of 50 m and 400 m at the deformation front. This display illustrates not only the differences between the two descriptions but also the increase in the difference with the initial depth of sediment element.

paths are compactive. Porosity loss in the toe versus depth can be contrasted with that along trajectories for elements at different depths at the deformation front by plotting η v. distance from the deformation front (Fig. 5). Such a plot shows that, with increasing depth, the difference between Eulerian and Lagrangian porosities increases.

The Lagrangian distribution of porosity loss across the toe of the Lesser Antilles prism can be qualitatively estimated from the porosity distribution of Bangs *et al.* (1990), and with the assumption of a pattern of flow trajectories similar to that in the Nankai protothrust zone. Even though porosity decreases no faster or even slower with depth in this prism that in the ocean basin, most or all flow trajectories describe compactive stress paths. It is certainly the case for the deeper flow trajectories near the decollement, where high excess pore pressures (Moore *et al.* 1982) and large negative velocity anomalies (Bangs *et al.* 1990) have been indicated. It is also clear, however, that there is much less compaction along flow trajectories in the Lesser Antilles than in the Nankai prism.

Porosity and permeability in fault zones

Several lines of evidence indicate that porosity in the large fault zones is quite different from that of the surrounding material in both the Nankai and Lesser Antilles prism toes. In the Nankai prism the first several imbricate thrusts are imaged as seismic reflectors, indicating a strong impedance contrast associated with the faults (Aoki *et al.* 1986). This contrast is more likely to represent conditions in the fault zone than changes across it because there are no obvious differences in lithologies or porosities across the faults. Because the sign of the impedance contrast is determined by the changes in the product of velocity and density, which increase or decrease together, the polarity of the fault plane reflection should discriminate between increased or decreased porosity in the fault zones.

The polarity of the reflections from the imbricate thrusts is not easily determined, and has been interpreted as both positive (Karig 1986*b*) and negative (Aoki *et al.* 1986) from previously available profiles. Additional profiles suggest to Moore *et al.* 1990) that the contrast is negative and that porosities are higher in these faults than in the surrounding prism. A similar but stronger case is made for a negative polarity reflection from the basal decollement (Moore *et al.* 1990), but here the contrast could be in part between the denser trench fill strata and the underlying basinal hemipelagic clays (Bray & Karig 1988).

Faults in the Lesser Antilles toe are not as well imaged as on the Nankai profiles, but implications for anomalous porosities within these zones are provided by drilling data. These faults are marked by zones of scaly clay, which is evidence of brittle, presumably dilative failure at some point along the stress path. The porosity contrast across the fault zones from core data is ambiguous (Shipboard Scientific Party 1988). Within the lenses of clay the porosity may be lower than in unsheared clay, but the many failure surfaces should produce a significant fracture porosity. These zones are conduits for the flow of fluids from deeper within the prism, as evidenced by local maxima in temperature (Fisher & Hounslow 1990) and by chemical species having deeper sources (Shipboard Scientific Party 1988), probably because of enhanced fracture permeability. Fracture permeability does not correlate simply with porosity (e.g. Arch & Maltman 1990), but permeability increase does suggest some porosity increase, as fracturing is a dilative process.

High fault zone permeability in the Nankai prism is less well documented than in the Lesser Antilles but higher fault zone porosities would imply higher permeabilities. However, no indications of anomalous fluid flow were observed in the frontal imbricate where penetrated in DSDP Hole 583D (Shipboard Scientific Party 1986), and no discernable deflection of the temperature-dependent gas hydrate reflector across these faults can be noted. A large positive thermal anomaly over the protothrust zone and inner trench floor along the Nankai system (Kinoshita & Yamano 1986) does imply lateral fluid flow, as does the existence of fluid venting near some of the frontal imbricate faults (Le Pichon *et al.* 1987), but this diffuse anomaly could reflect the relatively high intergranular permeability in these sediments.

Thus, there is evidence for enhanced porosity and permeability in the major faults of both the Nankai and-Lesser Antilles prism toes, but the evidence remains ambiguous. The Nankai faults provide better evidence for higher porosity whereas the Lesser Antilles faults have evidence of higher permeability.

Mechanical behaviour of porous sediments

The study of soils and highly porous sediments by geotechnical engineers has provided the basis for a reasonable model describing the mechanical response of prism toes. One of the most useful concepts is that the mechanical state of such a sediment depends on its differential stress, effective mean stress, and porosity or other measure of specific volume (e.g. Atkinson & Bransby 1978; Jones & Addis 1986). Porous sediments can fail (show a maximum supportable differential stress) either brittlely (with discrete failure surfaces) or ductilely (with diffuse deformation at the grain-size and larger scale). Brittle failure entails a near-constant porosity (slightly dilatant) pre-failure path that basically corresponds to the Coulomb criterion, whereas ductile failure is associated with a compactive pre-failure path. The state of stress associated with ductile failure is roughly equivalent to the state of residual strength after brittle failure at the same σ'_m. This failure state, marked by a specific set of $\Delta\sigma$, σ'_m, and η (at least in the zone of failure) is maintained independent of the amount of strain, and is termed the critical state of failure for those conditions. Critical state conditions form a single valued function relating $\Delta\sigma$, σ'_m, and η that defines a curved line through this 3-D space. For the low-stress conditions typical of geotechnical engineering the projection of this curve on a $\Delta\sigma-\sigma'_m$ plane is a straight line that passes through the origin and has a slope, M (Fig. 6), that is characteristic of the lithology being tested. M is greater for sand than for clay (Atkinson & Bransby 1978), and is about 0.9 at low stress for a silty clay on which we have experimented (Fig. 6).

Although the critical state model provides valuable insights to the mechanical behavior in prism toes, it is oriented toward idealized low stress conditions and has several recognizable

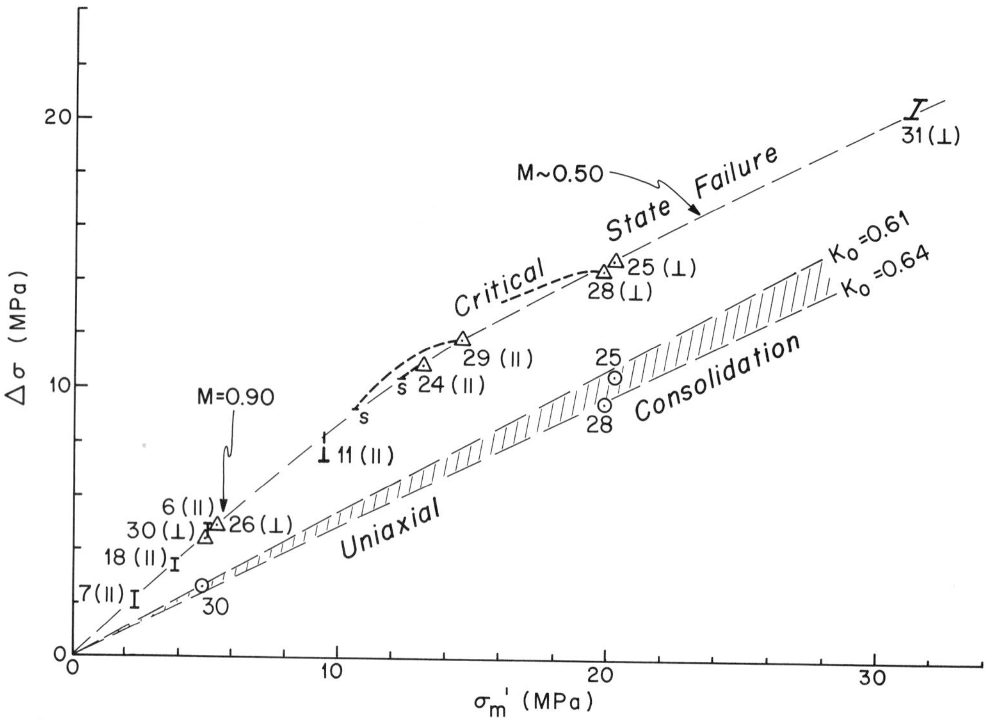

Fig. 6. Plot of differential stress ($\Delta\sigma$) v. effective mean stress (σ'_m) for a silty clay experimentally subjected to uniaxial consolidation and to triaxial stress paths leading to ductile (critical state) failure. All triaxial tests were run at a strain rate of 1.4×10^{-7} s^{-1} and with a constant P = 1.02 MPa ('drained') and all except T-31 were deformed at a constant σ'_m. For uniaxial consolidation $\frac{\Delta\sigma}{\sigma'_m}$ is constant with σ'_m to at least 35 MPa for individual tests, but these values vary slightly among the tests. The slope of the critical state line (M) clearly decreases with increasing σ'_m. To a first approximation samples with σ_1 parallel and perpendicular to bedding have the same critical state relationships. Post failure stress paths for several tests are shown and further explained in Fig. 7.

problems when applied to natural sediments at higher stresses. First, our experimental determination of the critical state line for a silty clay indicates that the value of M decreases with increasing stress above about 15 MPa (Fig. 6).

Second; the fabric anisotropy of natural samples, both from consolidation and from shearing, does impart some strain dependence to the behaviour of porous sediments. Most pertinent to this discussion is the role of strain hardening, defined as the increase in supportable differential stress with progressive shear strain. In experiments in which clay is deformed within a thin pre-cut shear or gouge zone, pronounced strain hardening is reported (Morrow *et al.* 1982), but the determination of effective stresses in this environment is difficult to assess. Such tests have provided the basis for the assumption by a number of workers (e.g. Karig, 1986*a*; Moore & Byrne 1987) that post-failure strain hardening plays a large and very significant role in the progressive deformation of shear zones. To the contrary, triaxial and simple shear tests show that strain hardening is totally a pre-failure phenomenon, associated with porosity decrease and ductile deformation (e.g. Fig. 7 and Skempton 1985). Post-failure behavior in tests in which the confining and/or normal stresses are known to be constant involves a slight strain-softening, due to the development of aligned clay platelets in the shear zone. The amount of strain-induced weakening appears to decrease with increasing σ'_m and decreasing clay content (Skempton 1985). As will be shown later, apparent post-failure strain hardening can be induced by an increase in σ'_m. The importance of large post-failure strain itself in affecting the mechanical behaviour of prism toes is very slight.

The pre-failure strain-hardening noted above

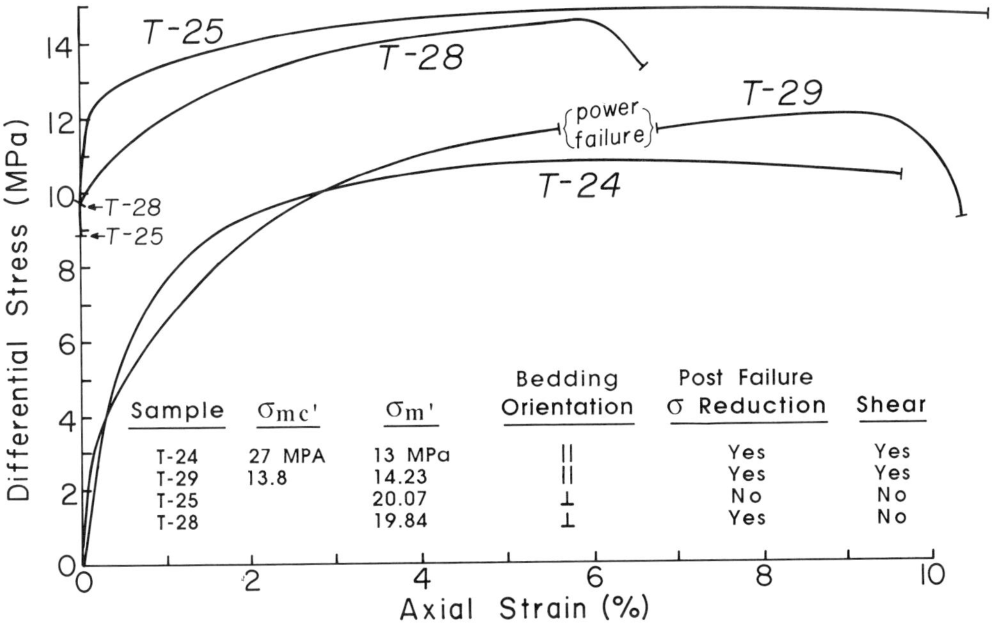

Fig. 7. Differential Stress ($\Delta\sigma$) v, axial strain curves for several specimens of silty clay experimentally consolidated to different σ'_m (see key). Cylindrical triaxial specimens cut perpendicular to bedding were uniaxially reconsolidated to σ'_m (T-25 was not sufficiently reconsolidated), whereas specimens cut parallel to bedding had to be isotropically reconsolidated. All these curves show prefailure strain-hardening, although the amount of strain necessary to reach failure varies as a function of overconsolidation and fabric orientation. Test T-25 shows the lack of post failure strain hardening when σ'_m is constant. The other tests were subjected to post-failure decreases of σ'_m, using different but constant rates of decrease in confining pressure. All initial failures were ductile, but the two tests on samples parallel to bedding resulted in the development of brittle shear zones during post-failure deformation.

is definitely important, and can be understood by analysing curves of η v. σ'_m for uniaxial and for critical state failure (Fig. 2). These curves show that porous sediments are more efficiently compacted when deformation involves shear. Thus, if a sample at some state of unaxial consolidation is subjected to a stress path leading to failure, the pre-failure section of these paths will be compactive for a large range of changes in σ'_m, even for significant decreases. At high porosities, the maximum decrease in σ'_m, that will still permit porosity loss is relatively small (path A on Fig. 2), but at lower porosities, as would be typical for most of the prism toe, the low slope of both curves requires a much larger decrease in σ'_m during pre-failure deformation before non-compactive failure occurs (e.g. path B on Fig. 2). The curves of Fig. 2, generated from experiments on silty clay, indicate that reduction of σ'_m from its consolidation value by as much as one third is necessary to cause initially brittle failure within prism toes. This condition will subsequently be shown to be difficult to achieve during deformation in prism toes.

Post-failure changes in σ'_m (generated by slowly varying the confining pressure at a constant strain rate) produced mechanical responses in several experiments that were initially startling, but in retrospect are readily explainable. An increase in σ'_m at a constant strain rate results in continued ductile failure, but at an increased supportable $\Delta\sigma$ and decreased porosity. In effect the stress path moves up the critical state failure line and strains ductilely.

If, on the other hand, a sediment deforming at critical state is subjected to a reduced σ'_m, as by an increase in pore pressure, its stress path will move down the critical state line, but will develop a dilative brittle shear zone (Fig. 7). This behaviour is explained as the development of a local weaker and more porous zone as an instability, which localizes subsequent strain. The sensitivity of the failure response of a sediment already at critical state to changes in fluid pressure is in sharp contrast with the relative insensitivity to changes in σ'_m along the pre-failure stress path. This difference in stress condition will be shown to have very important implications for the structural fabric of sediments in prism toes.

Implications for mechanical behaviour of sediments in prism toes

Stress paths leading to failure

The most robust observation concerning the mechanical behaviour of sediments in the toe of both the Nankai and Lesser Antilles prisms is that the deformation is compactive except possibly in major fault zones. This observation, together with descriptive and quantitative information on strain, and hints concerning aspects of stress, provide useful constraints for the stress paths of sediments in these prism toes. As for porosity, the analysis of stress paths and stress histories of sediments in prism toes requires the use of a Lagrangian framework, with the determination of flow trajectories. An approximate quantitative estimate of the relative stress changes to failure across the Nankai and Lesser Antilles prism toes is presented below, with recognition that the present data are inadequate to allow a more exact solution.

The stress paths of sediments deformed in prism toes begin with their deposition and inelastic consolidation in the ocean basin or trench wedge. This process approximates uniaxial strain and leads to a ratio of σ'_h / σ'_v that is constant for a given lithology during experiments (e.g. Atkinson & Bransby 1978; also Fig. 6) and may be nearly so for in-situ conditions. This ratio, termed K_0, is 0.6 to 0.7 for clay-rich sediments and decreases with increasing sand content. The magnitudes of the horizontal principal stresses on any sediment element during deposition are determinable when the overburden stress, σ_v, and pore fluid pressure (P) are known. In the following development, all stresses will be stated in terms of σ_v and P, which serve as a basis for comparison of stress at different depths and in different settings.

Pore pressures in the sediments of the Nankai Trough are presumed to be about hydrostatic, except near the deformation front, on the basis of consolidation parameters of these relatively incompressible and more permeable lithologies (Bray & Karig 1988). Seaward of the Lesser Antilles prism toe, excess pore fluid pressures are thought to be generated by lateral flow from sites of higher hydraulic potential within the prism (Westbrook *et al.* 1983). If P increased, the sediments would expand elastically, with $\Delta\sigma_H' = (v/1-v)\Delta P$, where v is the Poisson's ratio, and $\Delta\sigma_v = -\Delta P$. If ΔP were sufficiently large, σ_1' would become horizontal and, in the extreme, could lead to failure by shear or horizontal tensile fracture. All these effects would increase permeability and may predispose this setting for propagation of a décollement.

The uniaxial consolidation caused by sediment loading in the trench wedge or ocean basin is replaced at or near the deformation front by biaxial deformation as the subhorizontal principal stress perpendicular to the trench increases. Because the deformation associated with the increase in stress is compactive in both the Nankai and Lesser Antilles toes, this stress path should lead toward a ductile, critical state failure. This is particularly true for the Nankai, where porosity reduction is pronounced.

It is not possible to quantify the stress paths of sediment elements to failure and beyond without better measurements of in-situ pore pressure and strain, and without data on the mechanical behaviour of the lithologies involved. When these parameters become available, the state of stress throughout the prism toe could be estimated and the location where a failure state is first reached within the toe could be determined. Nevertheless, even the preliminary data from our experiments on sediment behaviour, together with present estimates of pore pressure and strain in the prism toes, provide some plausible and useful results.

The analysis of the stress path to failure can be approached by first estimating the change in σ_m' from the initial uniaxial consolidation state in the ocean basin or trench wedge to that at ductile failure, which is implied by the compactive stress paths. This change in σ_m' can then be related to the change in porosity and to the axial strain with experimental data. These in turn can be compared with porosity changes and strain observed in prism toes to estimate the locus of failure.

Effective mean stresses, σ_m', can be determined in terms of σ_3, or σ_v, using the relationships among the principal stresses at critical state failure and under plane strain conditions. At failure under biaxial strain conditions, σ_2' has been shown experimentally to be about 1.3 σ_3' for sand (Cornforth 1964) and 1.5 σ_3', for clay (Henkel & Wade 1966). The effective maximum principal stress, σ_1', which is about horizontal, can be related to σ_3, the total vertical stress, through the critical state relationship:

$\sigma_1 - \sigma_3 = M\ \sigma'_m$, and the effective stress 'law', $\sigma_1' = (1 - \lambda)\ \sigma_1$, where λ is the modified ratio of pore fluid pressure to total vertical stress (Davis *et al.* 1983). These relationships can be combined and recast into the form:

$$\sigma'_1 = \left(\frac{3 + 2.3\ M}{3 - M}\right)(1 - \lambda)\ \sigma_3. \qquad \text{[eq. 1]}.$$

For the sediments in prism toes, M has a value not far from 1.0, leading to:

$$\sigma'_1 = 2.65\ (1 - \lambda)\ \sigma_v \text{ [eq. 2a] and}$$
$$\sigma_m' = 1.65\ (1 - \lambda)\sigma_v \text{ [eq. 2b]}.$$

To compare σ'_m at failure with that in the trench fill for the same sediment element, Lagrangian flow paths are assumed to diverge uniformly arcward from the deformation front (Fig. 4), which greatly simplifies the analysis and does not significantly degrade these estimates. The geometry of this pattern requires that for any element, which is at a depth z_0 at the deformation front, its depth, z, at any distance, x, from the front will be:

$$z = z_0 \left[1 + \frac{X}{T_0} \tan\ (\alpha + \beta)\right] \text{ [eq. 3]}.$$

where T_0 is the thickness of the accreted section at the deformation front. Tan $(\alpha + \beta)$ is the small angle approximation to the trigonometric function accounting for the apical angle of the protothrust zone (Fig. 4). Because to a first approximation the vertical gradients of density are constant across the prism toe, the total vertical stress, $(\sigma_3)_x$, on a sediment element at distance x, will be related to the vertical stress of that element at the deformation front, $(\sigma_3)_0$ by:

$$(\sigma_v)_x \approx (\sigma_v)_0\ [1 + (x/T_0)\ \tan\ (\alpha + \beta)]. \qquad \text{[eq. 4]}.$$

At this point, the distance, x, behind the deformation front at which the element in question reaches failure must be estimated, but the effect of this estimate on $(\sigma_3)_x$ is small, and the initial estimate can be re-iterated after better control on porosity and strain is acquired.

For the Nankai system, σ_m' in the trench wedge, where pore pressure is assumed hydrostatic, can be calculated:

$$(\sigma_m')_0 = (1 + 2K_0)\ (\sigma'_v)_0/3 \quad \text{[eq. 5.]}$$

With $K_0 \approx 0.6$ and $\lambda \approx 0.55$, $\sigma_m' = 0.4\ (\sigma_v)_0$. Ductile failure in our experiments is reached at axial strains of 10% or less, which would occur in the Nankai protothrust zone at x equals 3 km or so. The thickness of the accreted trench wedge at the deformation front, T_0, is 0.7 km, and the apical angle, $(\alpha + \beta)$, of the protothrust zone is about 4°, leading to σ_v at the failure state, from eq. 4, of $(\sigma_v)_x \approx 1.30\ (\sigma_v)_0$.

With eq. 2b, the value of $(\sigma_m)_x'$ can be determined in terms of $(\sigma_v)_0$, if λ can be evaluated. An admittedly crude estimate of $\lambda = 0.75$ at the rear of the protothrust was obtained from mechanical considerations (Karig 1986*a*), which would lead to $(\sigma'_m)_x = 0.5\ (\sigma_v)_0$ and $(\sigma_m)'_x$ $1.25(\sigma_m)_0'$. Such manipulations suggest that σ'_m is approximately constant, probably slightly increasing, during pre-failure deformation in the Nankai protothrust zone.

Similar manipulations for the toe of the Lesser Antilles prism near the drill sites, where λ within the prism was estimated to be 0.9 or more (Moore *et al.* 1982) and where pore pressures in front of the prism were greater than hydrostatic (Westbrook *et al.* 1983), lead to the conclusion that σ'_m at failure is reduced by between half and one third its value in the initially consolidated state.

The effects of the changes in σ'_m for the two prisms are perhaps more obvious if viewed on plots of porosity v. σ'_m for uniaxial consolidation and for critical state conditions (Fig. 2). The nearly constant σ'_m paths in the Nankai toe for the sediment elements imply prefailure porosity reductions of up to 4% if the naturally and experimentally deformed sediments behaved similarly. When compared with porosity distributions and flow paths in the Nankai prism toe, this suggests that ductile failure initially occurs only a few km behind the deformation front at the deeper levels of the protothrust zone, but much closer to the rear of that zone for the uppermost levels.

The very small Lagrangian reduction of porosity in the toe of the Lesser Antilles prism is shown by the plot of Fig. 2 to be consistent with the very large estimated prefailure reduction in σ'_m. Such a stress path approaches the 'undrained' conditions of soil mechanics, in which pore fluids are prevented from leaving the system. As the reduction in prefailure σ'_m increases, the pre-failure horizontal strain decreases, effectively condensing the width of prism toes across which prefailure strain shortening occurs. It is also important that, despite the very high pore fluid pressure in this prism toe, initial failure is still expected to be ductile.

Stress paths after initial failure: ductile

In both the Nankai and Lesser Antilles prisms, porosity continues to drop along the flow trajectories after a failure state is certain to have been reached. In part this reflects the increase of σ_m' caused by continued tectonic thickening of the prism above each sediment element as it moves along its flow trajectory. This increase in σ_m' is almost undoubtedly not matched by an equivalent increase in pore pressure for several reasons. First, the increase in pore fluid gradient within a saturated sediment element is proportional to the time rate of change in volume strain, from a basic form of the consolidation equation (e.g. Biot 1941). Because the post-failure change in volume (porosity), even with a constant rate $\partial\sigma_m'/\partial t$, is much less than that along the pre-failure stress path (Fig. 1) the pressure gradient will decrease after failure. Second, the rate of horizontal strain is a maximum near the deformation front and decreases rapidly arcward (Karig, 1986*b*) further reducing the rate of pore pressure increase.

Thus, σ_m' should increase after initial failure and, with continuation of shortening, should cause an increase in supportable differential stress (strength), and a decrease in porosity. Clearly the stress state cannot exceed that at critical state because the stress path is compactive. The rate of deformation could be too slow to cause failure, but continuously superposed brittle-ductile failure structures (Karig & Lundberg 1990), indicate that such is not the case. In other words, if the sediments are continously failing and dewatering they must be moving up a stress path closely approximating the critical state line.

Stress path after initial failure: brittle

Although the bulk of both the Nankai and Lesser Antilles prism toes are compactively deforming, and thus expected to strain ductilely, there is obviously an overprint of diffuse but discrete failure surfaces reflecting brittle deformation in both areas. This constitutes one of the apparent mechanical inconsistencies noted initially, but one which appears to have a reasonable solution in light of our experimental results.

In the Nankai toe, the brittle overprint consists of semi-pervasive shear bands that appear to be ephemeral, each forming and deactivating after very small displacements (Karig & Lundberg 1990). As a group, however, these bands appear to have been developing continuously, as shown by crosscutting relationships and by subsequent passive rotation. This behaviour can be explained readily by slight and temporary reductions of σ_m' on sediments already at a state of ductile (critical state) failure. Such fluctuations in σ_m' are more likely to result from variations in pore fluid pressures than from changes of principal stresses, as tectonic conditions and the prism geometry do not appear to change rapidly in this setting. Several causes for fluctuating pore fluid pressure could be surmised, including irregularities in local strain rate related to earthquake slip deeper within the prism and fluctuations in fluid flux along transport zones from depth.

The existence of throughgoing thrust faults with persistent displacements pose a somewhat different problem of brittle deformation in a ductilely deforming prism toe. These faults appear to be not only zones of higher porosity and permeability but, in some cases, fairly wide zones of brittle deformation (Behrmann *et al.* 1988). An earlier explanation of such broad zones of brittle deformation relied on the assumption that preferential dewatering along the faults required higher permeabilities there and a pressure drop from the surrounding material into the fault zone (Karig 1986*a*; Moore *et al.* 1986). Outward migration of failure conditions was attributed to progressive strain hardening in the more central part of the zone (e.g Moore & Bryne 1987).

In a modification of this model applied to the decollement, Moore (1989) continues to interpret this zone as having a lower hydraulic potential than the adjacent sediments, which would engender fluid flow into the decollement. This model assumes a pore fluid pressure gradient, $\frac{\partial P}{\partial z}$, that is greater than hydrostatic above the decollement and is greater than hydrostatic but less than lithostatic below the decollement. Such a pressure profile would not permit downward flow into the decollement, nor is the restriction on the gradient of the fluid pressure below the decollement correct. To engender downward flow of fluids the pressure gradient must be less than hydrostatic, whereas the upward flow beneath the decollement could be associated with pressure gradients even higher than lithostatic if, as is likely, the strength in the decollement is less than that outside the zone. The only constraint is that the pore fluid pressure not exceed the lithostatic pressure.

The data from ODP leg 110 suggests that flow is out of, rather than into the decollement and other major thrust faults in the Lesser Antilles prism toe. Not only fluid temperatures but some chemical species are a maximum in the fault zones and decrease outward from them in both directions (Fisher & Hounslow, 1990; Gieskes *et al.* 1990), a pattern which characterized outward diffusion. Moreover, the source of the excess heat and the chemical species is interpreted as being deeper in the prism (Moore *et al.* 1987), further indicating flow up the fault zones and outward from them, at least in the prism toe.

To explain the dynamics of flow and the zones of scaly clay, Moore (1989) appeals to a deformational pumping scheme generated by dilatation and subsequent brittle failure, which imply a strongly overconsolidated state of stress. The data reviewed and generated during the present study indicate that such an overconsolidated state is not likely to develop in the porous and little cemented sediments of the prism toe. Neither can post-failure strain hardening be called upon to cause the outward migration of a zone of brittle failure. If pore fluid pressures are anomalously high in the fault zones and diffuse outward, and if these fault zones develop in sediments already at or near a ductile failure state, the model of post-failure fluctuations in σ_m' can more reasonably explain the outward diffusion of brittle shear fractures (Fig. 8).

Outward diffusion of high pressure fluids from these faults would increase the width of the zone subjected to reduced σ_m' and could account for the broadening of zones of scaly clay or other brittle failure features with time. If, on the other hand, the supply of high pressure fluid stopped, the fault zones should become almost as strong as the surrounding material and might cease to show localized displacement. This is a possible explanation for the cessation of slip on some of the imbricate thrusts in the prism toe, even before they are rotated out of an orientation producing a high resolved shear stress.

Whereas the application of reduced σ_m' offers a most logical explanation for some of the observations in prism toes, it dodges the questions of how these faults initiate and what conditions would result in fluid flow into the fault zones at greater depth. The first of these might be answered by assuming that faults propagate upward from depth as instabilities driven by high pore fluid pressures. The second is more difficult, as there are almost no data on in-situ conditions at greater depth, and the relationship between porosity and permeability during deformation, which may play a critical role, is only now being explored (e.g. Arch & Maltman 1990).

The probability that the decollement and active imbricate faults are surfaces of low shear strength because of locally higher pore fluid pressures also has several larger scale mechanical implications. These faults effectively divide the accretionary prism into mechanical subunits with trapezohedral cross sections, each of which would behave as a small example of the pressure-dependent wedge model of Dahlen (1984), except that a critical state rather than Coulomb criterion would control the failure state. To transmit the shear forces across these active imbricate faults, discrete changes in the

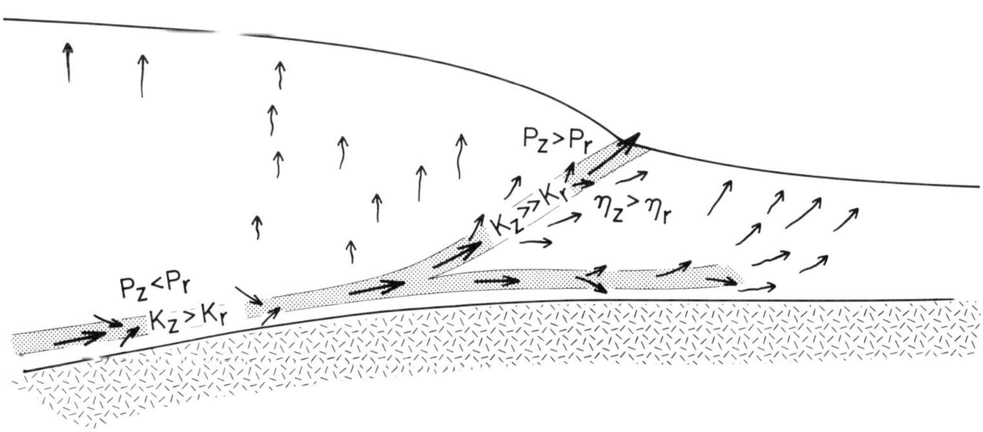

Fig. 8. Model of fluid flow in the toe of a prism, based on the mechanical behaviour of sediments outlined in this study. In this model the properties inside the decollement and imbricate thrusts (with subscript z) are contrasted with those in the bulk of the prism (subscript r). At shallower depths the fault zones have higher porosities (η), permeabilities (K) and pore pressures (P) than surrounding material, leading to outward diffusion of fluid and of brittle failure structures. At greater depth, the pressure gradient is hypothesized to reverse, so that fluids flow into the fault zones.

thickness of the prism at the faults is required, as is observed.

The existence and width of a protothrust zone seems to be related to the behaviour of σ'_m during the pre-failure strain. The Lesser Antilles case, with a very large drop in σ_m', shows very little horizontal shortening strain before failure, and no protothrust zone is observed seaward of the frontal thrust, whereas such a zone is 5 km wide seaward of the frontal thrust in the Nankai case. The relationship of the frontal thrust to the locus of initial ductile failure is not yet apparent, but possibly the closer is the stress path to brittle initial failure, the closer the development of the first thrust is to the failure front.

Conclusions

Although the measurement of sediment physical properties has often appeared to be of questionable significance to structural geologists, it should now be more obvious that, coupled with an understanding of the mechanical behaviour of these materials, they offer powerful constraints for the state and history of stress in sediments. The toes of accretionary prisms provide an illuminating example of this because stress and porosity leading to failure are very sensitively related under these conditions. Moreover, prism toes provide a vivid example of the need to differentiate between the Eulerian distribution of parameters that typifies much of geophysical observation, and a Lagrangian description of behaviour, which is basically the approach of structural geologists who explore the deformation history of a sample. When the transformation between these two frames can be made on the basis of reliable measurements, data-based models of the mechanical behaviour of structural systems can be constructed.

Even the present partial understanding of porous sediments and fragmentary data from prisms provide important insights to the mechanical behaviour of prisms and explain some of the apparent inconsistencies in that behaviour. From experimental studies, dewatering of prism toes implies ductile deformation, yet brittle structures are most commonly recognized from this setting. Ductile deformation leaves only subtle traces in these sediments having few strain markers, but there is evidence to support such a style, particularly in the Nankai prism. Explicit efforts to determine the extent of ductile strain should be a priority objective.

Compactive deformation can lead to significant pre-failure strain, especially in the direction of σ_1. Thus, the deformation front, where the first evidence of tectonic strain is observed, is not generally the locus of initial failure. The deformation front would effectively coincide with failure only if the sediments were deformed without loss of pore fluid, a condition that may be approached in the Lesser Antilles example. In the Nankai prism the physical manifestation of initial ductile failure may be extremely subtle, and appears to lie within the protothrust zone.

Experimental recognition that compaction and ductile deformation can occur despite a significant increase in pore fluid pressure is mirrored by the example of the Lesser Antilles toe. Real sediments, with varying degrees of cementation, may behave somewhat more brittlely than otherwise similar artificially consolidated sediments, but observations in the two prism toes suggests that such differences are small.

The dominance of initial ductile failure, and the difficulty in producing an initial brittle failure have been stressed, but coexistence of brittle failure structures is readily explained as post-failure effects of reduced σ'_m. In this condition the brittle–ductile transition is extremely sensitive to slight changes in pore fluid pressure.

The predictive capabilities of the approach presented here should also be clear. If sediments were deforming at critical state conditions, as they appear to be, and if the functional relationship among η, $\Delta\sigma$, and σ'_m were known, either by laboratory experiments or by in-situ measurements, the state of stress would be largely constrained by knowledge of porosity, at least away from the fault zones.

It is of geological significance to explore the mechanical and structural response of porous sediments, not only in accretionary prisms but where ever porous sediments are subjected to biaxial or triaxial strain. This will require much more information concerning in-situ conditions than has heretofor been available. Not only are the values of such parameters as k, P, η, and their indicators important in critical zones such as faults, but as important, gradients in these parameters throughout the toe are needed. Far from being a mundane exercise, the collection of these measurements is challenging, and potentially one of the most rewarding aspects of structural geology. What is now required are better tools to acquire these measurements and a better understanding of the processes controlling them.

References

Aoki, Y., Kinoshita, H. & Kagami, H. 1986. An

evidence of low velocity layer beneath the accretionary prism of Nankai Trough; inferences from a synthetic sonic log. *Initial Reports of the DSDP*, **87**, U.S. Government Printing Office, Washington, D.C., 727–736.

ARCH, J. & MALTMAN, A. 1990. Anisotropic permeability and tortuosity in deformed wet sediments: *Journal of Geophysical Research*, **95**, 9035–9046.

ATKINSON, J. H. & BRANSBY, P. L. 1978. *The mechanics of soils – an introduction to critical state soil mechanics*. McGraw-Hill.

BANGS, N., WESTBROOK, G. K., LADD, J. W. & BUHL, P. 1990. Seismic velocities from the Barbados Ridge Complex and indications of high pore fluid pressures in an accretionary complex. *Journal of Geophysical Research*, **95**, 8767–8782.

BEHRMAN, J. H., BROWN, K., MOORE, J. C. MASCLE, A., TAYLOR, E. & others 1988. Evolution of structures and fabrics in the Barbados accretionary prism. Insights from Leg 110 of the Ocean Drilling Program. *Journal of Structural Geology*, **10**, 577–591.

BIOT; M. A. 1941. General theory of three-dimensional consolidation. *Journal of Applied Physics*, **12**, 155–164.

BISHOP, A. W., KUMAPLEY, N. K. & EL-RUWAYIH, A. 1975. The influence of pore-water tension on the strength of clay. *Philosophical Transactions of the Royal Society, London*, **278**, 511–554.

BRAY, C. J. & KARIG, D. E. 1986. Physical properties of sediment from the Nankai Trough, DSDP Leg 87A, Sites 582 and 583. *In*: KAGAMI H., KARIG, D. E. ET AL. *Initial Reports of the DSDP*, **87**, 827–892.

—— & —— 1988. Dewatering and extensional deformation of the Shikoku hemipelagic sediments in the Nankai Trough. *Pure and Applied Geophysics*, **128**, 725–747.

CORNFORTH, D. H. 1964. Some experiments on the influence of strain conditions on the strength of sand. *Geotechnique*, **14**, 143–167.

DAHLEN, F. A. 1984. Noncohesive critical Coulomb wedges: An exact solution! *Journal of Geophysical Research*, **89**, 10,125–10,133.

DAVIS, D. M., SUPPE, J. & DAHLEN, F. A. 1983. Mechanics of fold and thrust belts and accretionary wedges. *Journal of Geophysical Research*, **88**, 1153–1172.

FISHER, A. T. & HOUNSLOW, M. W. 1990. Transient fluid flow through the toe of the Barbados accretionary complex: constraints from ODP Leg 110 heat flow studies and simple models. *Journal of Geophysical Research*, **95**, 8845–8856.

GIESKES, J. 1990. Hydrogeochemistry in the Barbados accretionary complex. ODP leg 110. *Tectonophysics*, (in press).

HAMILTON, E. L. 1976. Variations of density and porosity with depth in deep sea sediments. *Journal of Sedimentary Petrology*, **46**, 280–300.

HENKEL, D. J. & WADE, N. H. 1966. Plane strain tests on a saturated remolded clay. *Journal of the Soil Mechanics & Foundations Division, Proceedings of the American Society of Civil Engineers*, **SM6**, 67–80.

HOU, G. 1988. *Physical properties and mechanical state of an artificial silty clay as a simulation of deformation of deep sea sediment at convergent plate boundaries*. MSc Dissertation, Cornell University, Ithaca, New York.

JONES, M. E. & ADDIS, M. A. 1986. The application of stress path and critical state analysis to sediment deformation. *Journal of Structural Geology*, **8**, 575–580.

KAGAMI, H., KARIG, D. E., COULBOURN, W. T. ET AL. 1986. *Initial Reports of the DSDP*, **87**, Washington, D.C., (U.S. Government Printing Office).

KARIG, D. E. 1986*a*. Physical properties and mechanical state of accreted sediments in the Nankai, Trough, Southwest Japan Arc. *In*: MOORE J. C. (ed.) *Structural Fabrics in Deep Sea Drilling Project Cores from Forearc*, Geological Society of America Memoir, **166**, 117–133.

—— 1986*b*. The framework of deformation in the Nankai Trough *In*: KAGAMI, H. KARIG, D. E., COULBOURN, W. ET AL. *Initial Reports of the DSDP*, **87**, Washington, D.C. (U.S. Gov't Printing Office), 927–940.

—— & LUNDBERG, N. 1990. Deformation bands from the toe of the Nankai Accretionary Prism. *Journal of Geophysical Research*, **95**, 9099–9110.

KINOSHITA, H. & YAMANO, M. 1986. The heat flow anomaly in the Nankai Trough area, *In*: KAGAMI, H., KARIG, D. E., COULBOURN, W. T. ET AL. *Initial Reports of the DSDP*, **87**, Washington, D.C., (U.S. Government Printing Office), 737–744.

LAMBE, T. W. & WHITMAN, R. V. 1969. *Soil Mechanics*, J. Wiley & Sons. N.Y.

LE PICHON, X., IIYAMA, T. BOULEGUE, J. CHARVET, J. FAURE, M., KANO, K., LALLEMENT , S., OKADA, H., RANGING, C., TAIRA, A., URABE, T. & UYEDA, S. 1987. Nankai Trough and Zenisu Ridge: A deep-sea submersible survey. *Earth and Planetary Science Letters*, **83**, 285–299.

LUNDBERG, N. & KARIG, D. E. 1986. Structural features from the Nankai Trough lower slope, DSDP sites 582 and 583 *In*: KAGAMI, H., KARIG, D. E. *et al. Initial Reports of the DSDP*, **87**, Washington, D.C., (U.S. Government Printing Office), 797–808.

—— & MOORE, J. C., 1982; Structural features of the Middle America Trench slope off Southern Mexico, Deep Sea Drilling Project Leg 66 *In*: WATKINS, J. S., MOORE, J. C. & OTHERS. *Initial Reports of the DSDP*, **66**, Washington, D.C., U.S. Government Printing Office, 793–805.

MOORE, G. F., SHIPLEY, T. H., STOFFA, P. L., KARIG, D. E., TAIRA, A., KURAMOTO, S. and SUYEHIRO, K. 1990. Structure of the Nankai Trough accretionary zone from multichannel seismic reflection data. *Journal of Geophysical Research*, **95**, 8753–8766.

MOORE, J. C. 1989. Tectonics and hydrogeology of accretionary prisms: role of the décollement zone. *Journal of Structural Geology*, **11**, 95–106.

—— & BRYNE, T. 1987. Thickening of fault zones: a mechanism of melange formation in accreting sediments. *Geology*, **15**, 1040–1043.

——, BIJU-DUVAL, B., BERGEN, J. A., BLACKINGTON, B., CLAY-POOL, G. E., COWAN, D. S., DUENNEBIER, F., GUERRA, R. T., HEMLEBEN, C. H. J., HUSSONG, D., MARLOW, M. S., NATLAND, R. H., PUDSEY, C. J., RENZ, G. W., TARDY, M., WILLIS, M. E., WILSON, D. & WRIGHT, A. A. 1982. Offscraping and underthrusting of sediment at the deformation front of the Barbados Ridge. Deep Sea Drilling Project Leg 78A. *Geological Society of America Bulletin*, **93**, 1065–1077.

——, MASCLE, A., TAYLOR, E., ALVAREZ, F., ANDREIEFF, P., BARNES, R., BECK, C., BEHRMANN, J., BLANC, G., BROWN, K., CLARK, M., DOLAN, J., FISHER, A., GEISKES, J., HOUNSLOW, M., MCCLELLAN, P., MORAN, K., OGAWA, Y., SAKAI, T., SCHOONMAKER, J., VROLIJK, P. J., WILKENS, R. & WILLIAMS, C. 1987. Explusion of fluids from depth along a subduction-zone décollement horizon. *Nature*, **326**, 785–788,

——, ROESKE, S., LUNDBERG, N., SCHOONMAKER, J., COWAN, D. S., GONZALES, E. & LUCAS, S.E. 1986. Scaly fabrics from Deep Sea Drilling Project cores from forearcs *In*: MOORE J. C. (ed.) *Structural fabrics in Deep Seal Drilling Project coves from forearcs* Geological Society of America Memoir, **166**, 55–73.

PLUMLEY W. J. 1980. Abnormally high fluid pressure-survey of some basic principles. *American Association of Petroleum Geologists Bulletin*, **51**, 1240–1254.

SHIPBOARD SCIENTIFIC PARTY 198. SITE 583 *in*: KAGAMI, H., KARIG, D. E., COULBOURN, W. T. *et al. Initial Reports of the DSDP*, **87**, Washington, D.C. (U.S. Government Printing Office), 123–256.

—— 1988. Synthesis of shipboard results: Leg 110 transect of the northern Barbados Ridge *In*: MASCLE, A., MOORE, J. C. *et al. Proceedings of the Ocean Drilling Program, Initial Reports (part A)*, **110**, College Station, TX, (Ocean Drilling Program), 577–591.

SKEMPTON, A. W. 1985. Residual strength of clays in landslides, folded strata and the laboratory. *Geotechnique*, **35**, 3–18.

WESTBROOK, G. K. & SMITH, M. J. 1983. Long décollements and mud volcanos: Evidence from the Barbadoes Ridge Complex for the role of high pore-fluid pressure in the development of an accretionary complex. *Geology*, **11**, 279–283.

The undrained shear behaviour of fine-grained sediments

N. A. YASSIR

Department of Earth Sciences, University of Waterloo, Waterloo, Ontario N2L 3G1, Canada

Formerly at Department of Geological Sciences, University College, London WC1E 6BT, UK

Abstract: A series of high pressure triaxial tests were conducted on three normally consolidated mud volcano silty clays to explore some of the effects on the undrained shear behaviour of argillaceous sediments. The tests followed two different consolidation paths before undrained shear: isotropic, with equal all-around stresses and anisotropic, at a horizontal to vertical effective stress ratio of 0.6. The tests showed that shear stresses produce overpressuring in the fine-grained sediments, and that the behaviour is sensitive to stress path, stress magnitude and grain size. Other important effects on the behaviour are also thought to include natural porosities and cementation.

Thick argillaceous sequences are capable of sustaining high pore fluid pressures and therefore, low effective stresses, due to their thickness and low permeability. They are also often subjected to very high shear stresses, whether imposed by tectonic activity or by the geometry of the depositional basin. It is important to understand the undrained shear behaviour of these sediments for two main reasons:

(a) such sequences are frequently encountered during drilling of oil wells and cause drilling problems;

(b) more generally, clay-rich sequences at high total stresses but low effective stresses can fail catastrophically, forming features such as shale diapirs and mud volcanoes (Barber *et al*. 1986). They can also facilitate structural deformations such as the large decollement zone in the Barbados Ridge accretionary complex (Westbrook & Smith 1983).

This paper will investigate the effect of undrained shear on argillaceous sediments, and the sensitivity of the behaviour to various factors including burial history, stress magnitude, grain size and cementation. Some of these effects are discussed in the light of an experimental study conducted on three mud volcano silty clays: Lagon Bouffe, Trinidad (LB), Taiwan clay (T) and Devil's Woodyard, Trinidad (DW) with clay contents of 10%, 29% and 62% respectively.

Test description

The samples were in a naturally remoulded state with very high water contents. They were therefore consolidated at 2.5 MPa in an oedometer before high pressure triaxial testing. The pre-testing porosities were 28% for the LB samples, 31% for the Taiwan samples and 49 and 53% for the two DW samples (the DW sample tested at 10 MPa had been preconsolidated to 0.6 MPa only).

After isotropic consolidation at 5 MPa in the triaxial cell, the samples were further consolidated either isotropically, or anisotropically, at a horizontal to vertical effective stress ratio of 0.6, approximating Ko conditions for silty clay (Lambe & Whitman 1979). This ratio was not varied, despite the difference in the three materials, for the sake of comparison. Filter strips were placed around the samples to speed up consolidation and equilibrate pore pressures (Bishop & Henkel 1982); these were measured at the top and bottom of the sample. Anisotropic consolidation involved increasing the stresses slowly at a constant ratio. The axial displacement rate varied between 0.0001 and 0.0005 mm per minute, according to the permeability of the material, to avoid pore pressure build-up. Volume change was measured by means of a pressurised volume gauge described in Addis (1987).

When the desired consolidation stress was

From Knipe, R. J. & Rutter, E. H. (eds), 1990, *Deformation Mechanisms, Rheology and Tectonics*, Geological Society Special Publication No. 54, pp. 399–404.

reached, the sample was loaded axially in an undrained state to simulate the effect of shear stresses on a thick, low permeability material. The axial displacement rate used during undrained shear was 0.002 mm per minute. The tests were performed at mean effective stresses of 5 to 60 MPa. This was done to observe any change in behaviour from the low effective stress range, which is expected for overpressured sediments, to the high effective stresses.

The test names will start with the sample initials, LB, T or DW, followed by the letters I or K, to indicate isotropic or anisotropic consolidation respectively, then by a number indicating the value of mean effective stress to which the sample was consolidated.

Results

The results are presented in Figs 1 to 3. The porosities at the beginning of undrained shear are indicated in these figures.

For all three materials, undrained shear results in considerable pore pressure generation, and therefore decreasing mean effective stresses, leading to failure (Figs 1 to 3). The pore pressure response to increasing shear stress, however, tends to decrease with higher stresses (Yassir 1989*b*). Interestingly, sample DWI10 gave a pore pressure response greater than the corresponding shear stress (Fig. 3B&C). The three materials displayed different modes of failure (defined here as the point of maximum shear stress), which also varied within the same test series.

(a) Critical state failure, where the stresses reach a point where they remain constant with increasing strain (e.g. test LBK15 in Fig. 1A&B).

(b) Dilatant failure, where the pore pressures begin to decrease after failure, causing an increase in effective stress and strain hardening (e.g. test LBK60 in Fig. 1A&B). Note here that the overall volume of the sample does not change as the test is undrained; the term is related here to the pore pressure decrease only.

(c) Contractant failure, where the effective stresses decrease after failure (e.g. test TCK25 in Fig. 2A&B). This is often also referred to as brittle failure, although the only sample to have developed a shear surface in association with strain softening is DWI50 (Fig. 3A&B).

The LB series shows a change in failure mode from dilatant, to critical state, and back to dilatant failure at the higher stresses (Fig. 1A). The samples did not display any obvious shear planes, but SEM photography showed increasing particle crushing with increasing stress (Yassir 1989*a*). The Taiwan clays showed both critical state modes of failure as well as contractant failure (Fig. 2A). This failure was not associated with the development of shear planes. Finally, the DW series, which is the richest in clay content, showed both critical state and brittle failures (Fig. 3A&B). Interestingly, both DW samples had well-developed polished shear surfaces, reminiscent of the scaly clay texture, despite the fact that

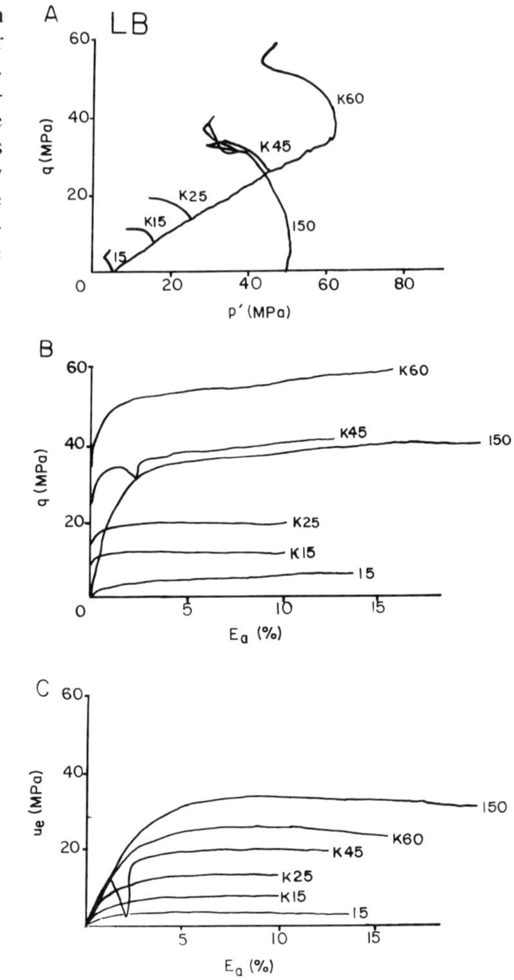

Fig. 1. Results of the Lagon Bouffe, Trinidad (LB) test series. (**A**) Stress path in mean effective (p')–differential (q) stress space. (**B**) Stress (q)–axial strain (E_a) curves. (**C**) Excess pore pressure (u_e)–axial strain (E_a) curves. Porosities at the start of undrained shear: LBI5 (23%), LBK15 (21.5%), LBK25 (20%), LBK45 (16%), LBI50 (17%), LBK60 (15%).

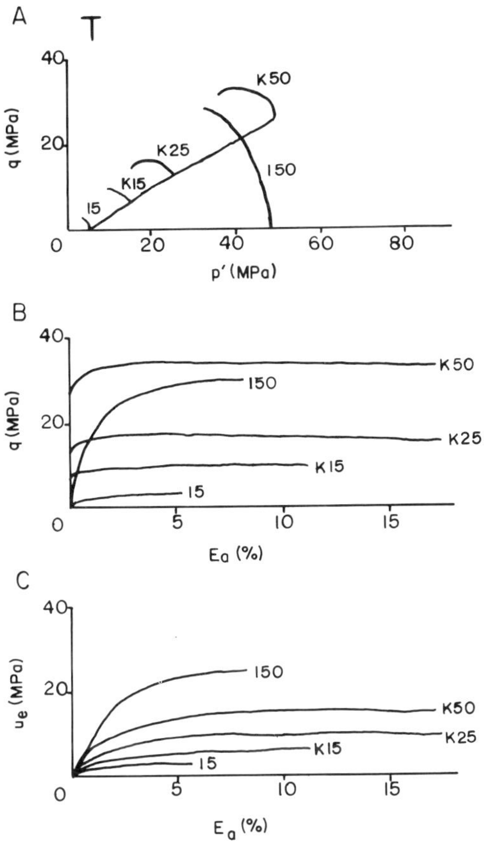

Fig. 2. Results of the Taiwan (T) test series. (**A**) Stress path in mean effective (p')–differential (q) stress space. (**B**) Stress (q)–axial strain (E_a) curves. (**C**) Excess pore pressure (u_e)–axial strain (E_a) curves. Porosities at the start of undrained shear: TI5 (23%), TK15 (20%), TK25 (16%), TI50 (14%).

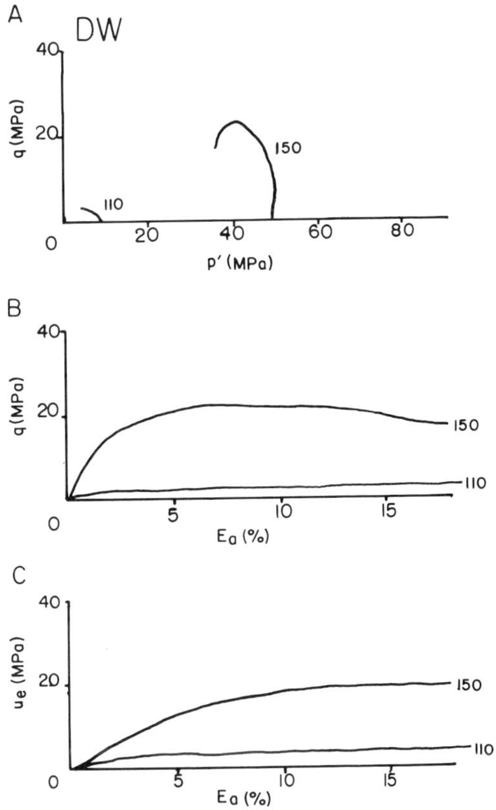

Fig. 3. Results of the Devil's Woodyard, Trinidad (DW) test series. (**A**) Stress path in mean effective (p')–differential (q) stress space. (**B**) Stress (q)–axial strain (E_a) curves. (**C**) Excess pore pressure (u_e)–axial strain (E_a) curves. Porosities at the start of undrained shear: DWI10 (35%), DWI50 (32%).

one did not show brittle failure in its stress path (Fig. 3A). The failure line remains more or less straight for all three materials, decreasing very slightly in slope at the higher stresses.

Discussion: some effects on undrained shear behaviour

Stress path

The orientation of the principal effective stresses has a direct influence on sediment fabric and, ultimately, its behaviour. If we compare two tests with different consolidation paths performed on the Taiwan mud volcano clay, TI50 and TK50, we see that they produce different results (Fig. 2A). If the structure of the clay had remained independent of consolidation path, the two undrained shear curves would have been parallel. Clearly, they are not, and the failure modes are also different, the anisotropically consolidated sample failing more 'contractantly' (i.e. decreasing effective stresses after failure). The more silty LB clay shows less sensitivity to consolidation path in its undrained shear behaviour, at least at the higher stresses. Tests LBI50 and LBK45 follow the same undrained shear path and both fail 'dilatantly'.

Dependence of behaviour on consolidation path has been observed in triaxial testing of soils at low and medium effective stresses (Skinner 1975; Gens 1982; Vaid 1985; Atkinson *et al.* 1987). Generally, the more anisotropic the stresses during consolidation, the greater the tendency for the sample structure to collapse during undrained shear (Gens 1982). This results in high pore pressure responses due to the changing shear stresses, and flatter, more contractant undrained shear curves. This is evident in the LB series at the lower stresses (Fig. 1A) and the Taiwan series (Fig. 2A).

Stress magnitude

Overpressured argillaceous sediments are subjected to relatively low effective stresses due to their abnormal pore fluid pressures. The test series, however, explores the changes in behaviour from the lower of the higher effective stresses to determine the possible effects of decreasing porosity and fabric changes on the behaviour.

All three test series show changes in behaviour with increasing effective stresses. The transition from critical state to dilatant failure in the LB series (between tests LBK25 and LBK45, Fig. 1A) is associated with grain crushing, as observed from SEM photography. This is thought to be the result of shear as opposed to consolidation because the cracks are vertical, even for the isotropically consolidated sample (LBI50), going through several grains at the highest stresses (Yassir 1989*a*). Dilatancy at failure is attributed to the large proportion of granular materials in the LB samples, which had the lowest clay content (10%). It is associated with dense sands at low effective stresses (Lee & Seed 1967). At higher stresses, it is thought that particle crushing becomes predominant in granular materials (Lee & Farhoomand 1967) and this crushing overcomes the tendency to dilate (Lee & Seed 1967; Skinner 1975). The two samples which did not dilate at failure, LBK15 and LBK25 (Fig. 1A), were not associated with obvious particle crushing. They both have flat undrained shear curves which suggest that the structure could not maintain the increasing shear stress. The load was therefore transferred to the pore fluid, which increased in pressure without further increase in shear stress (Fig. 1B&C). Dilatant conditions at failure at the higher stresses could be explained by the observed crack propagation (assuming that it occurred during undrained shear), causing a small decrease in pore fluid pressures (Johnston & Novello 1985).

The Taiwan clays also show some variations in the mode of failure, although the behaviour during undrained shear is more repeatable at different stresses than the LB material (Yassir 1989*a*). The contractant failure shown by the anisotropically consolidated samples at higher stresses (TK25 and TK50, Fig. 2A) could be explained by increasing particle orientation. The clay content of the material is still low (29%), so that the presence of coarser particles probably hinders more dramatic brittle failure. The DW clay, with a clay content of 62%, shows a change from critical state to brittle failure (Fig. 3A), despite the fact that both samples showed polished shear surfaces. The post-peak drop in strength of sample DWI50 is explained by slip along the shear plane, which is not observed in DWI10.

The change in behaviour with increasing stress is further reflected by the decreasing rate of pore pressure build-up in response to shear stress (Yassir 1989*b*).

There is therefore a strong indication that the consolidation stress magnitude, which controls the porosity, has a great effect on the manner in which the material behaves. Despite the low porosities at the start of undrained shear, especially for the LB and Taiwan material, distinct changes in behaviour were observed as the lowest porosities were approached. Behaviour leading to dramatic loss of strength is not seen to be associated with the higher stresses (above 25 MPa), particularly where the material dilates at failure. It is more likely to be associated with the lower stresses, where there is sufficient pore space to allow structural collapse; this is particularly applicable to natural, as opposed to remoulded samples.

Grain size

Grain size is, of course, a very important factor controlling the behaviour of a fine-grained sediment. The results show that the finer the particle size, the lower the strength. Whereas particle crushing was observed in the samples with the highest silt content (LB), with associated dilatant failure, particle orientation became more prominent with increasing clay content, often associated with contractant or brittle behaviour at failure.

The natural state of the sediment

It should be kept in mind that the sediment in nature will behave very differently from a laboratory tested material, especially if it is in the remoulded, slurry state. The natural relationships between porosity, effective stress and cementation, play a very important role in the behaviour of a sediment.

Skempton (1943) first recognized that not only are natural porosities higher than laboratory-determined porosities, but natural materials tested in the laboratory have higher porosities than remoulded materials for the same effective stress. The reason given for this is that, in nature, particle arrangement and cementation produce an open structure which is destroyed by remoulding and/or fast laboratory experiments. The open structure can survive considerable effective stresses (Leroueil & Vaughan 1990).

The destruction of clay cements at shallow depths is known to produce a dramatic collapse of structure and decrease in strength; the stresses initially sustained by the mineral skeleton are transferred to the pore water (Bjerrum 1954; Lambe & Whitman 1979). Ohtsuki *et al.* (1981) report a high porosity silty mudstone in Japan which shows dramtically contractant behaviour during undrained shear. High porosity chalks behave the same way (Jones & Leddra 1989). This dramatic loss of strength occurs during the breakdown of cements during shear, causing the behaviour of the material to change from elastic to plastic, with large associated pore pressure response.

It is reasonable to assume, therefore, that similar behaviour can be observed in fine-grained sediments, especially if they are underconsolidated.

Summary and conclusions

Some of the main factors affecting the behaviour of the fine-grained materials tested are given below.

(a) Consolidation history. This seems to be more the case for the finer grained, Taiwan material than for the LB material.

(b) Stress magnitude. Although the failure line remains straight for all three materials, stress magnitude had an effect on pore pressure generation, which decreased with decreasing porosity, and on post-failure mechanisms, causing particle crushing in the more silty materials and particle orientation in the more clay-rich material.

(c) Grain size. The strength decreases with increasing clay content. Also, the modes of failure change from critical state and dilatant for the lowest plasticity samples, to critical state and brittle for the finest samples.

In addition, as observed in low pressure soil mechanics, normally consolidated fine-grained sediments respond to undrained shear by significant pore pressure generation, leading to a decrease in effective stress. This is assumed to be particularly the case at the lower stress range, especially where abnormally high porosities have been preserved by underconsolidation and/or cementation. The work presented in this paper was aimed at achieving a better understanding of the effects of shear on overpressuring and deforming fine grained sediments. A great deal more work needs to be done in this field to enhance modelling of natural and induced sediment deformation.

The author would like to thank her colleagues at Sediment Deformation Research, University College London, UK, and especially M. E. Jones, for their assistance with her work.

References

ADDIS, M. A. 1987. *Mechanics of sedimentary compaction responsible for oilfield subsidence.* PhD thesis, University of London, UK.

ATKINSON, J., RICHARDSON, D. & ROBINSON, P. 1987. Compression and extension of Ko normally consolidated kaolin clay. *Journal of Geotechnical Engineering*, **113**, 1468–1480.

BARBER, A., TJOKROSAPOETRO, S. & CHARLTON, T. 1986. Mud volcanoes, shale diapirs and melanges in accretionary complexes, Eastern Indonesia. *American Association of Petroleum Geologists Bulletin*, **56**, 2068–2071.

BISHOP, A. W. & HENKEL, D. J. 1982. *The Measurement of Soil Properties in the Triaxial Test.* Arnold, London.

BJERRUM, L. 1954. Geotechnical properties of Norwegian marine clays. *Geotechnique*, **4**, 49.

GENS, A. 1982. *Stress-strain and strength characteristics of a low plasticity clay.* PhD thesis, University of London, U.K.

JOHNSTON, I. & NOVELLO, E. 1985. Cracking and critical state concepts for soft rocks. *Proceedings of the 11th International Conference on Soil Mechanics and Foundation Engineering, San Francisco*, 515–518.

JONES, M. E. & LEDDRA, M. J. 1989. Compaction and flow of porous sediments. *Proceedings of the International Symposium on Rock at Great Depth, Pau, August 1989*, **2**, 891–898.

LAMBE, T. W. & WHITMAN, R. V. 1979. *Soil Mechanics, SI version.* John Wiley and Sons Ltd.

LEE, K. & FARHOOMAND, I. 1967. Compressibility and crushing of granular soil in isotropic triaxial compression. *Canadian Geotechnical Journal*, **4**, 68–86.

LEE, K. & SEED, H. 1967. Drained strength characteristics of sands. *Proceedings of the American Society of Civil Engineers*, **SM6**, 117–141.

LEROUEIL, S. & VAUGHAN, P. 1990. The important and congruent effects of structure in natural soils and weak rocks. *Geotechnique*, (in press).

OHTSUKI, H., NISHI, K., OKAMOTO, T. & TANAKA, S. 1981. Time-dependent characteristics of strength and deformation of a mudstone. *Proceedings of the Symposium on Weak Rock, Tokyo*, **1**, 119–124.

SKINNER, A. E. 1975. *The effect of high pore water pressure on the mechanical behaviour of sediments.* PhD thesis, University of London, U.K.

SKEMPTON, A. 1943. Notes on the compressibility of

clays. *Quarterly Journal of the Geological Society of London*, **100**, 119–135.

VAID, V. P. 1985. Effect of consolidation and stress path on hyperbolic stress-strain relations. *Canadian Geotechnical Journal*, **22**, 172–176.

WESTBROOK, G. & SMITH, M. 1983. Long decollements and mud volcanoes: Evidence from the Barbados Ridge Complex for the role of high pore fluid pressure in the development of an accretionary complex. *Geology*, **11**, 279–283.

YASSIR, N. A. 1989*a*. *Mud volcanoes and the behaviour of overpressured clays and silts*. PhD thesis, University of London, U.K.

—— 1989*b*. Undrained shear characteristics of clay at high total stresses. *Proceedings of the International Symposium on Rock at Great Depth, Pau, August 1989*, **2**, 907–913.

Deformation in an accretionary melange, Alexander Island, Antarctica

P. A. R. NELL

British Antarctic Survey, NERC, High Cross, Madingley Road, Cambridge, CB3 0ET, UK

Present address: Department of Geology, The University of Manchester, M13 9PL, UK

Abstract: Alexander Island contains several belts of melange in a wide accretionary complex. One melange belt in the northwest of the island incorporates both oceanic and trench-fill material. It evolved by many different deformation mechanisms: dispersed independent particulate flow (IPF) and limited cataclasis at shallow levels; and diffusion mass transfer (DMT) and limited crystal plastic processes at deeper levels. Fluid pressures may have risen due to the subduction of young hot oceanic crust, which probably affected the structural evolution of the region by controlling the strength of the decollement and hence the taper of the accretionary prism.

The Deep Sea Drilling Project has revealed details of the tectonic processes operating at shallow levels in accretionary complexes, which are important regions of crustal growth. Complementary field-studies can reveal information about deeper-level mechanisms unavailable by other methods. This paper is the result of fieldwork in a melange belt from an ancient accretionary prism, and aims to show how changes in subduction conditions may have affected the taper of the prism.

The area examined is on northern Alexander Island which has several melange belts (Tranter 1987, 1988; Burn 1984), the largest is 250 km long, and trends N–S along the west of the island (Fig. 1). Only the northern part of this belt is described, here named the Debussy Heights melange belt (DHMB). In this paper the term 'melange' is adopted from Cowan (1985), and strongly-disrupted units comprising sandstone blocks in a mudstone matrix (in which the cleavage can be linked with stratal disruption) are described as melange.

Geological context of northern Alexander Island

The Antarctic Peninsula magmatic arc resulted from the subduction of palaeo-Pacific ocean-floor since at least 420 Ma ago (Milne & Millar 1989). Late Palaeozoic–Mesozoic accretionary prisms are exposed west of this magnetic arc (Storey & Garrett 1985). Alexander Island is located inside a major curve of the magmatic arc, and includes both a wide accretionary complex (the LeMay Group, Burn 1984) and a major fore-arc basin (the Fossil Bluff Group, Butterworth *et al.* 1988). From at least 145–80 Ma ago, the age of oceanic crust being subducted beneath Alexander Island decreased dramatically (Fig. 2a) as the southward-drifting Pacific/Phoenix ridge approached Antarctica. This ridge then split, and the progressive northward collision of the Phoenix/Antarctic spreading-ridge with the Antarctic Peninsula trench ended subduction beneath northern Alexander Island about 25 Ma ago (Barker 1982). The plate model of Engebretson *et al.* (1985) predicts that oceanic basement close to the ages of radiolarian cherts in the DHMB (Berriasian–mid Valanginian & Albian–earliest Turonian), was subducted between 95–90 Ma ago.

The ages of sedimentation and accretionary deformation in the LMG, summarized in Fig. 3, young progressively towards the Pacific, as expected in an accretionary complex constructed by the south-eastward subduction of palaeo-Pacific ocean-floor. Uplift caused by out-of-sequence thrusting in the Douglas Range (Fig. 3 column 4), was broadly coeval with the deposition of localized plant-bearing conglomerates, along the western margin of the FBG fore-arc basin (column 6). These conglomerates were derived from the accretionary complex to the west, and indicate that parts of the prism were emergent during the early Cretaceous. The prism was narrow, and probably steeply-tapered during this period. Accretion of the eastern parts of the DHMB (column 2) probably followed shortly afterwards.

From Knipe, R. J. & Rutter, E. H. (eds), 1990, *Deformation Mechanisms, Rheology and Tectonics*, Geological Society Special Publication No. 54, pp. 405–416.

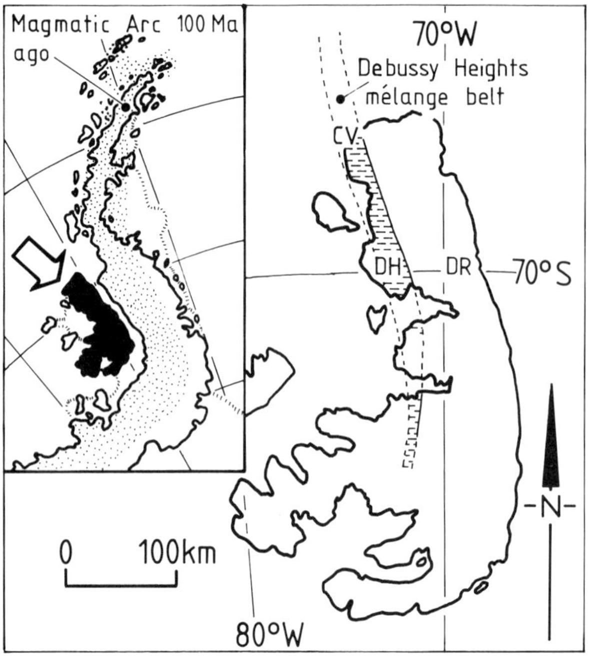

Fig. 1. Map of Alexander Island in relation to the Antarctic Peninsula magmatic arc 100 Ma ago (stippled). The dashed ornament indicates the location of the Debussy Heights melange belt. CV, Cape Vostok; DH, Debussy Heights, DR, Douglas Range.

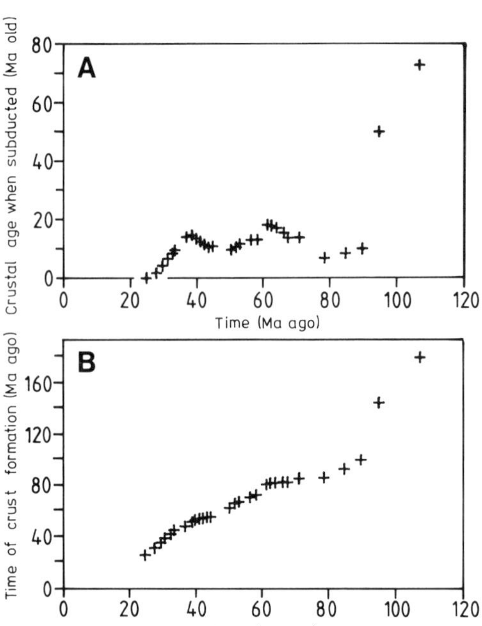

Fig. 2. Tentative subduction history age of oceanic basement beneath northern Alexander Island since 107 Ma ago. Estimates back to 85 Ma ago assume symmetric spreading at the Pacific/Phoenix and Phoenix/Antarctic ridges, and are based on marine magnetic anomalies in the SW Pacific (Halbouty *et al.* 1982). Estimates before 85 Ma ago use magnetic anomalies on the Pacific plate and the plate-motion model of Engebretson *et al.* (1985). (**A**) Plot of the age of the crust at the time of subduction, against time. (**B**) Plot of the time of formation of the oceanic slab being subducted, against time.

Blue amphibole and lawsonite occur locally in the older Lower Jurassic rocks of central Alexander Island, and indicate high-pressure metamorphism. In contrast, the north of the island is at zeolite prehnite–pumpellyite, pumpellyite–actinolite and greenschist facies (Burn 1984). This medium-pressure facies-series is comparable to the central Kii Peninsula, (Myashiro 1973). It is possible that the subduction of young, hot, oceanic-crust, may

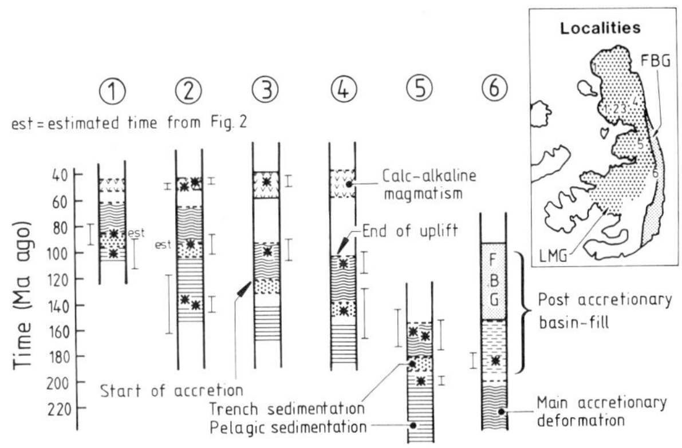

Fig. 3. Tectono-stratigraphic events on Alexander Island. Data, and estimated age of accretion, indicated by asterisks. Sources: palaeontological data of Thomson & Tranter (1986), Pimpirev & Dinnyovsky (1989), B. K. Holdsworth (pers comm.), zircon fission track cooling ages (I. B. Evans, pers comm.), and estimated ages of accretion (Fig. 2). Localities: 1 western DHMB, southern Lennon Glacier and western Debussy Heights; 2 eastern DHMB, western Sullivan Glacier; 3 central Sullivan Glacier; 4 Douglas Range; 5 Lully Foothills and LeMay Range; 6 western margin of FBG.

have led to higher geothermal gradients in the younger parts of the accretionary prism. However, Karig (1980) noted that blueschists were often associated with strike-slip faults, and rapid uplift on steeply-dipping transpressional faults could have caused localized blueschist preservation. Such faults exist on Alexander Island (Nell & Storey 1990), and this should not be ignored.

Parts of the DHMB are coherent and preserve a stratigraphy. Analyses of basalts from these regions show mid-ocean ridge basalts and ocean island affinities (Alabaster & Nell, unpublished data). There is a sedimentary transition from the pelagic and hemipelagic shales and cherts above these basalts to overlying turbidites. The stratigraphy is interpreted as the top 100–200 m of an oceanic crust associated with seamounts, overlain by pelagic and hemipelagic sediments, which were then covered by a trench-fill sequence as it approached the subduction zone. During accretion, this sequence was disrupted to form the DHMB. Serpentine is present as a late-stage alteration of some porphyritic basalts, but no exotic antigorite-serpentinites or high-grade metamorphic blocks exist. The coherent region, east of the DHMB, consists of thick beds of medium-grained greywacke-turbidite, interbedded with laminated units of siltstone and dark mudstone, with rare slump-folds and slump-sheets.

Figure 4 shows an interpretative E–W cross-section across most of northern Alexander Island based on a detailed, broad, structural traverse. The DHMB dips eastwards beneath an imbricated hangingwall of coherent, trench-fill turbidites. The western boundary is obscured beneath ice, but the belt is at least 18 km wide. The DHMB is a broad west-directed thrust zone at zeolite facies, although the coherent hangingwall is at prehnite–pumpellyite facies, or above. Both transected and steeply-plunging folds in the hangingwall of the DHMB suggest the early stages of subduction were oblique (Nell & Storey 1990). Low-angle extensional-faulting was found mainly in the melange belt, but minor late extensional duplexes exist in the lower 10 km of the hangingwall.

The melange fabric in the DHMB is inclined moderately to the east except where it is rotated by later folds. These folds formed with a set of west-directed out-of-sequence thrusts during horizontal shortening of the prism, and are later than the deformation forming the melange

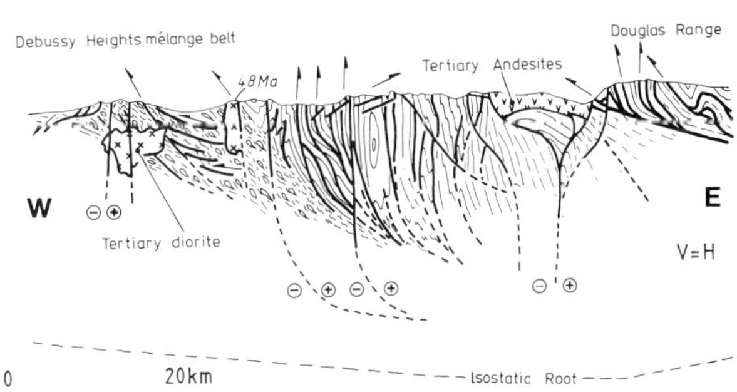

Fig. 4. Schematic E–W cross-section across northern Alexander Island at a latitude of 69° 50′S.

fabric. Cleavages associated with this deformation are affected by a later period of low-angle extensional faulting.

Minor calc-alkaline plutons and dykes (40–60 Ma old, Burn 1981) intrude the melanges of northern Alexander Island, cutting both the main melange-fabric and later out-of-sequence thrusts. Later N–S strike-slip faults cutting Tertiary igneous rocks also affect the forearc basin and are the latest structural event in the area (Storey & Nell 1988).

Shallow-level deformation in the Cape Vostok area

The Cape Vostok region, forming the northern part of the DHMB preserves shallow-level features, and is unaffected by the intense, deep-level deformation of the area to the south.

Field relationships

At 69°07′50″S, 71°59′30″W, on the eastern margin of the melange belt, there is a thin zone of melange along a fault, which is interpreted as an accretionary thrust. This locality consists of a westward-dipping turbidite sequence, with a melange-zone of arenite blocks in a mudstone matrix. The zone is probably a fault because the lower contact of the melange consistently cuts up-section westwards, at about 5°, for over 50 m. The local westerly dip at this locality is a result of late-stage refolding, although the regional sheet-dip is inclined towards the east. A traverse across the locality in the younging direction, east to west, revealed the following field relationships (Fig. 5).

(a) The extreme east of the exposure consists of greywackes and shales with only slight stratal-extension in the top 15 m. This sequence is a relatively undeformed footwall to the melange-zone.

(b) A 4 m thick melange-zone of siltstone and shale contains randomly-distributed blocks of sandstone and chert. The sandstone inclusions are up 10 cm long, and display folds, irregular rounded outlines, and flames of injected mudstone matrix. Spalling-off of sand grains produced diffuse clast margins and tails, but chert clasts are frequently angular, indicating instead that they were partially lithified before deformation. Parts of the melange exhibit a diffuse internal lamination parallel to the margins of the zone, due to variations in the relative proportions of fine sand, silt, and clay, and the top contains swirled mixtures of muddy sands. A spaced cleavage cuts the melange zone, and dips westwards more steeply than the internal lamination.

(c) The melange is overlain by a 3 m thick, structureless greywacke bed disrupted by healed listric, extensional faults which define several back-rotated fault-blocks. These listric faults curve into parallelism with the layering in the melange, cutting down-section to the west, and transporting higher-levels westwards.

(d) Further west in the hangingwall, a 200 m section of thin arenite beds displayed bedding-parallel extension on minor, 'welded' conjugate faults, resulting in strongly-extended but continuous layers, or trails of symmetrical lozenge-shaped blocks, separated by shales. These symmetrical conjugate arrays of extensional faults in the hangingwall contrast with the asymmetrical array of listric faults in (c) above.

Melange microstructures

Thin-sections of melange matrix exhibit 250 μm to 3 mm thick laminae of dark mud and buff

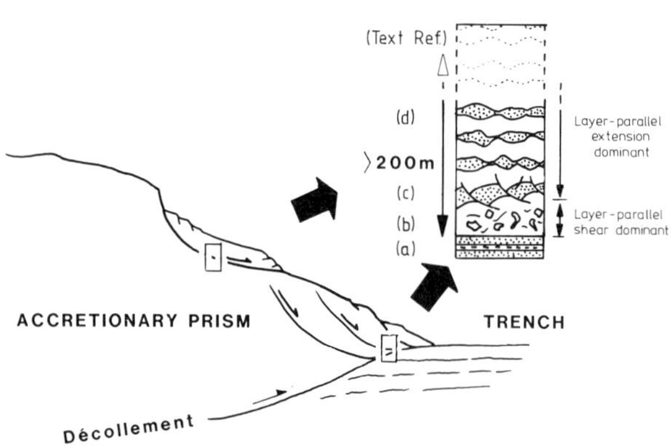

Fig. 5. Diagrammatic structural log through a zone of early stratal disruption near Cape Vostok, and alternative tectonic explanations. Faults in likely initial orientations before late-stage refolding. See text for details.

silt, extended and offset, up to a few millimetres, by a set of curved minor faults forming the mesoscopic cleavage. These microfaults (spaced about 3 mm) dominantly cut down-to-the-west across the layering. The net displacement on this microfault-system, shown by the offset layering, was a thinning of the zone perpendicular to the layering and an overall top-to-the-west shear. A series of layer-normal calcite veins further extend the lamination.

In the microlithons between microfaults, most grains, sand-sized or above, are intact, isolated, angular fragments. Some chert clasts are fractured, with trails of fragments rotated and dispersed along the lamination. The microfaults lack evidence of cataclasis, and are either discrete surfaces, or shear zones up to 200 μm wide which transpose the early lamination.

Almost all microfaults at less than 50° to the layering show pressure solution across the fracture due to the loss of calcite cement. However, microfaults at more than 60° are dilational, and contain calcite veins. Similarly, the layer-normal veins (−80° to +70°), show no shear displacement, and are tensile fractures filled with coarse granular calcite. Tensile fractures are present in both clasts and matrix, and at clast/matrix contacts forming pressure shadows. Some of the tensile fractures occur in the overlap zones of en-echelon microfaults. Most stop at shear fractures, and none are displaced by them, apparently because the displacement on the two fracture systems was linked.

Interpretation

The footwall (a, above) more than 15 m below the melange shows no evidence of stratal extension; but the hangingwall (d) shows layer-parallel extension, parallel to the melange-zone. The strain in these two regions is incompatible, and the melange (b), and the listric faults above the melange (c), appear to form an accommodation zone. Conjugate fractures in the hangingwall (d), suggest an irrotational strain. In this accommodation zone, both the melange microfaults, and the geometry of the listric faults, indicate that extension was associated with a major component of west-directed shear.

The melange-microstructures show that cataclastic deformation of chert clasts was followed by fragment dispersal by independent particulate flow (IPF) without intragranular deformation, during layer-parallel shearing. Compositional layering may have been preserved because the shear was layer-parallel. Lucas & Moore (1986) described cataclasis and sub-parallel layers of clay and silt, similar to those seen in the melange, from partially lithified rocks in a modern accretionary prism.

The deformation (Fig. 5) could be produced by down-slope mass movements in an extending glide block, or by gravity-driven extensional collapse of a tectonically-overstepped thrust hangingwall. Cataclastic deformation in the melange favours deeper-levels of deformation than a debris flow, or most modern slump-zones (Lucas & Moore 1986), although some submarine slides may exceed 1 km in thickness (Woodcock 1979). Initial intrusion of a weak muddy melange (Pickering *et al.* 1988) along a thrust, perhaps because of high fluid-pressures (Westbrook & Smith 1983), is also a possibility. Overprinting by later events would make this difficult to recognize.

Swirled, mixed muddy sands near the top of the melange suggest the sediments may have been liquidized. Dewatering during the initial deformation of the melange, or fluids released from deeper levels of the fault-zone, may have caused this. The expulsion of fluidized sand to form sand-volcanoes is common along the surface traces of active faults, and the process is not restricted to sedimentary deformation.

A change in deformation-response during microfaulting, probably occurred after some cementation, since calcite has been removed from most fault-zones, while some microfault-zones disaggregated the rock. Early low-angle fractures are cut by higher-angle fractures, culminating in the formation of tensile fractures perpendicular to the layering and extension direction. The earlier microfaults are modified by pressure-solution, perhaps while the tensile fractures formed, with their coarse, granular vein-fillings. These veins suggest fractures remained open normal to the minimum principal compressive stress direction (σ_3). Hobbs *et al.* (1976, p. 325) reported that tensile fractures are inconceivable below depths of a few hundred metres, unless fluid pressures were high enough to allow faulting at very low effective stresses. Such pressures need to overcome both the minimum principal compressive stress and the tensile strength of the rock. The relationships between the shear and tensile fractures in the rock may be explained by rising fluid pressures, causing a change from shear to tensile failure, as has been proposed for other melanges (Waldron *et al.* 1988).

The melange fabrics suggest that DMT processes and cementation followed cataclastic deformation, IPF, and sediment-liquidization, spanning lithification of the rocks. The long-lived progressive deformation, and the footwall-cutoff geometry, favours a thrust interpretation, rather than one of shear at the base of a major glide-block (or slide).

Deeper-level deformation in the Debussy Heights area

Stratal disruption in the Debussy Heights, further south, occurred under zeolite-facies metamorphism, at deeper structural levels than near Cape Vostok.

Field relationships

Domains of weakly-extended coherent strata preserving clear, small-scale sedimentary structures, up to 2 km wide, alternate with melange zones up to 800 m across. The 200 m thick transition zones between these domains consist of less-disrupted strata and alternate with more strongly disrupted regions, or zones of melange, on scales of a few metres. The position and nature of the main units of melange were not continuous along-strike. In cliff sections the smallest coherent domains are surrounded by and grade into melange, both along- and across-strike. The same lenticular geometry may be valid at a map-scale.

The main fabric in the melange matrix is a pervasive but weak, finely-spaced cleavage, which is deflected around, and encloses the inclusions. The cleavage anastomoses, and separates lenticular chips about 1 mm wide. Cleavage surfaces are typically dull, and show no lineations. The weak fabrics in inclusions are best described from thin-sections (see below). Inclusions range from individual grains to large rafts up to 100 m long. Clasts larger than 5 cm long comprise between 30–60% of most exposures, and thin-sections of the muddy melange matrix show a similar proportion of clasts larger than single grains.

In most exposures, clasts less than 2 m across are dominantly sub-equant and blocky, but elsewhere they have oblate–lenticular shapes aligned parallel to the matrix cleavage. The matrix cleavage, and clast-alignment, form the mesoscopic melange-fabric. In cliff sections, the larger rafts form irregular tabular bodies extended along-strike and down-dip by conjugate minor faults, pinch-and-swell structures, and boudinage. Smaller sandstone bodies also show similar oblate extension by numerous layer-normal veins, pinch-and-swell structures, and closely-spaced extensional microfaults with a few mm of displacement.

Cliff sections reveal minor compositional variations parallel to the melange fabric, such as the proportion of greywacke to mudstone, or trails of chert clasts: this emphasizes the melange fabric. Some apparently coherent domains displaying parallel lithological layers in cliff-sections contain many minor extensional faults. Coherent regions, and low-strain domains in the melanges, have conjugate sets of healed high-angle microfaults, indicating bedding-parallel extension. In contrast, one dominant set of extensional structures, which cut down-to-the-west across the fabric, occurs in the more strongly cleaved melanges. These structures are minor faults with displacements of up to 2 m, or an extensional crenulation cleavage of the melange fabric.

Folds contemporaneous with the melange fabric are absent from the Debussy Heights melanges, but folds of the cleavage and tabular clasts are present and due to later layer-parallel-shortening strains. This refolding distinguishes melange-forming strains from some of the later events, and is associated with a new closely-spaced axial-planar cleavage oblique to main melange fabric. The folds are produced by later thrust zones, up to 10 m wide, cutting across the melange fabric, which locally reorientate and intensify the melange fabric. In these fault-zones, polished slickensided surfaces are well developed, unlike the melange outside, and particularly in dilational sites at fault jogs (Sibson 1987) where chlorite slickenfibres are deposited. In these zones of type IV melange (Cowan 1985), lozenge-shaped inclusions are typical, and minor Riedel shears indicate the sense of movement.

Microstructures

In thin-section the matrix cleavage is an anastomosing web of narrow zones rich in dark minerals, which frequently displays strongly-aligned phyllosilicates under crossed polars in the particularly clay-rich domains. Partially dissolved Radiolaria at cleavage seams provide clear evidence of some modification by pressure-solution during cleavage formation.

Many layer-normal veins are present crossing enclaves in the melange. The ends of these veins are plugged by the melange matrix and have a drusy infilling of quartz, calcite, analcite, prehnite, epidote and chlorite, and separate boudins of sandstone. Some layer-normal fractures, separating boudins of sandstone are only filled by melange matrix, and it appears that the matrix moved into developing cracks on webs of cleavage surfaces, representing paired, conjugate shears.

Sandstone blocks contain many fractured and broken grains, and are crossed by regions of cataclastic grain-size reduction. Dispersal of the broken grains after fracturing was by independent particulate flow (IPF), with no evidence of crystal plastic deformation. The edges of many blocks are diffuse and irregular, with sand-grains partially engulfed by the matrix. Irregular lobes of sandstone are displaced, or detached, into the muddy melange matrix on minor cataclastic extensional microfaults. These microfaults can be traced, as shears, into the matrix cleavage.

All sections from coherent regions in the DHMB have a bedding-parallel fabric with aligned and kinked detrital micas. Solution at bedding-parallel grain boundaries produced a weak grain-shape fabric in greywackes, and a similar pitting of pebbles occurred in conglomerates. Most thin-sections contain numerous minor faults, but are less obvious when they become bedding-parallel in mud-rich laminae. The bedding-parallel cleavage in small-scale extensional duplexes is rotated by these faults at up to 70° to the sheet dip: locally indicating large amounts of extension. Where the faults cut greywackes, they formed zones of cataclastic grain-size reduction. Pressure solution removed some regions of microbreccia from fault segments, and caused insoluble grains to locally impinge into, or deflect, the fault zones. Most faults contain concentrations of opaque minerals and phyllosilicates, which are particularly marked in fault-segments at a low angle to bedding. Pressure-solution, sub-normal to bedding, both before and after faulting, is indicated. Fault-segments at a high angle to bedding, and dilational fault jogs, formed local sites for the deposition of granular quartz. Some vein textures indicate multiple stick-slip behaviour.

Interpretation

The rocks of the DHMB correspond to Cowan's (1985) Type I and II melange, which he attributed to coaxial, layer-parallel, extension in an olistostrome (Cowan 1982). This process produces strike-parallel extension, but it cannot explain all the features of the DHMB. No sharp contacts between melange and little deformed sedimentary successions exist in the melanges of the Debussy Heights. Such contacts provide positive evidence of an olistostromal or diapiric origin (Pickering *et al.* 1988), and neither mechanism alone could have produced the DHMB. Consequently, an alternative structural explanation for the stratal disruption must be sought.

These melanges retain some gross compositional layering, which is interpreted as transposed relict bedding. The extensional cleavage, and single fault-set observed in the melanges, indicate a component of top-to-the-west layer-parallel shear. The gross disposition of low- and high-strain domains is shown by the large-scale distribution of coherent regions and melange, and appears to be parallel to both bedding (in the coherent regions) and the melange fabric. This suggests a large component of layer-parallel shear, but does not preclude additional components of layer-parallel extension.

The sub-equant and blocky shapes of many of the inclusions cannot be explained as solely the result of structural slicing on P- or R-shears (Byrne 1984; Needham 1987). There is a difference between the simple-shear dominated melanges that Needham studied, and the Debussy Heights melange. The former show lozenge-shaped clasts, isolated by Riedel shear fractures; but in the Debussy Heights melange, blocky clasts bounded by high-angle fractures are common. Regions with phacoidal inclusions occur, but there are also indications of components of coaxial deformation. The typical structure of coherent, low-strain regions, is a weak bedding-parallel cleavage extended parallel to the bedding by conjugate microfaults. This cleavage is not oblique to bedding, as expected in the low-strain regions of a layer-parallel simple-shear zone. It is perhaps significant that cleavage-related folds are absent from the Debussy Heights melanges. It appears that neither layer-parallel shear alone, nor only layer parallel extension, can totally explain the structures of the DHMB.

The alternations of regions of conjugate extensional faults in the coherent domains, with some melanges exhibiting shear-band like faults and microfaults, suggest either strain partitioning or superimposition. The low-strain coherent domains and sandstone rafts in the melange contain more sandstone than the melange matrix, and their internal structures suggest a dominantly layer-parallel extensional strain history.

Two types of tectonic model could produce these relationships (Fig. 6). Karig (1986) noted that thrusting and resulting simple shear near the wedge toe and basal decollement, was replaced by a grossly irrotational flattening throughout the rest of the wedge later during the accretionary process. With steeply-dipping accretionary packets, late-stage flattening, perhaps accompanied by some flexural synthetic shearing during back-rotation, would result in a thickening of the accretionary complex. This is probably true of the steeply-dipping hangingwall of the DHMB, but not the relatively gently-

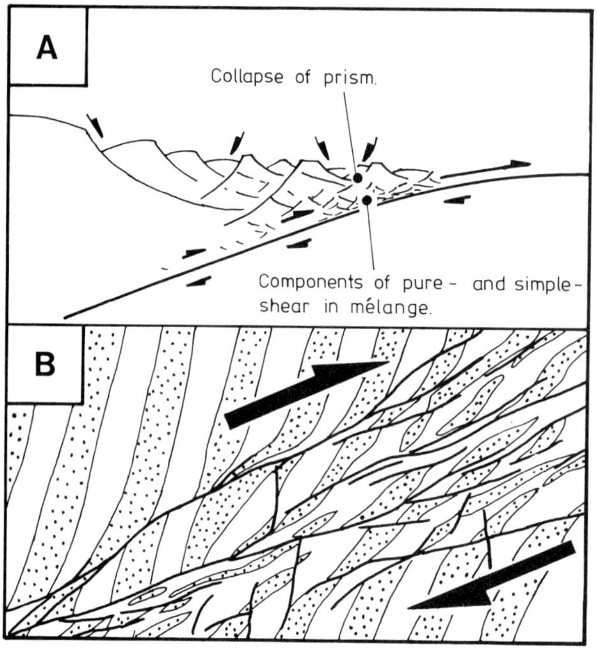

Fig. 6. Two models for producing extension in the Alexander Island accretionary complex: (**A**) On the scale of the entire 120 km wide accretionary wedge, 90 Ma ago; weakening of the decollement due to rising fluid pressures could have led to extensional collapse of, and complex strain-patterns in, the accretionary prism. (**B**) An out-of-sequence thrust, cutting down-section through a steeply-dipping, previously imbricated accretionary sequence could form late melange zones, 100 m to 10 km wide, exhibiting layer-parallel extension and block rotation.

dipping DHMB itself. The Debussy Heights area, in particular, retains a moderate–gentle overall dip despite later refolding. Strata extension of these rocks which probably never were steeply-dipping, implies a thinning of the accretionary complex, and would be equivalent to a spreading or collapse of the hangingwall. This model is capable of explaining all the relationships of the DHMB.

In the second type of model (Fig. 6b), an out-of-sequence shear-zone cutting down-section through suitably inclined, previously imbricated strata, could result in stratal extension. If thrusting was accommodated by rigid block rotation, extension of the competent units and shear in the incompetent units would result. The model is capable of producing stratal disruption at depth, but the antithetic shearing produced by a block-rotation model has not been found in the DHMB. Neither does bedding in coherent domains, nor the hangingwall of the DHMB, appear to be at an angle to the melange fabric.

The melange matrix fills the terminations of layer-normal veins by displacements along cleavage seams, and cleavage-parallel shear along the cleavage detaches fragments of sandstone from block margins. These features show that cleavage-parallel shear helped form the melange fabric. Cleavage seams often exhibit a strong preferred orientation of phyllosilicates. The cleavage does not perfectly fit the definition of scaly foliation of Moore *et al.* (1986), because of the relative rarity of polished striated surfaces, but they suggested that this fabric was produced by shear during dewatering, causing fabric-collapse. The DHMB melange fabric may have originated in this way, but was modified by pressure solution. The layer-normal veins formed perpendicular to the extension direction and represent tensile veins. They formed contemporaneously with at least the later stages of the melange fabric, under conditions of zeolite-facies metamorphism. Isolated, angular clasts of similar zeolite veins in the melange matrix, show that much disruption occurred during zeolite-facies metamorphism.

The layer-normal veins in the Debussy Heights appear analogous to the tensile veins in the Cape Vostok area. They suggest that high fluid pressures and low effective stresses were important during some of the development of the melanges. Some blocks separated on layer-normal fractures, and layer-normal fracturing caused stratal-disruption in the melanges during zeolite-facies metamorphism. Layer-normal veins are present in other melanges (eg. Kodiak Island, Byrne 1984; Franciscan, Cowan 1978), and may be a widespread indicator of periods of high fluid pressures and low effective stresses in such terrains. Waldron *et al.* (1988) found similar layer-normal vein arrays in a melange related to extension following ophiolite emplacement in Newfoundland, also interpreted as a result of high fluid pressures.

Coherent regions in the DHMB show much greater layer-parallel extensions than the very minor strains that have attributed to normal faulting during downbending of the subducting slab (Lundberg & Karig 1986). Fisher & Byrne (1988) interpreted similar layer-normal veins and microfaults from Kodiak Island as hydrofractures in flat-lying sediments, related to a vertical orientation of the principal stress during underthrusting. They considered the structures in coherent domains formed beneath the decollement, but is this interpretation correct? Until sediments are detached from the oceanic slab, any strains must be compatible with the subducting slab and irrotational layer-parallel extensions should be limited to those present in the ocean crust. Evidence from the Nankai Trough (Lundberg & Karig 1986), suggests that while the subducting slab does extend, such strains are probably minor, at least beneath the leading edge of the prism. While compaction in the tectonically loaded sediments beneath the decollement is likely, their extension is not, and the decollement, or the region above it, is a more favourable site for generating large extensional strains. If fluid pressures along the decollement were high, and its strength low, rocks lying beneath it should experience minimal shear stresses. The bedding-parallel fabrics in coherent blocks from the DHMB probably formed beneath the decollement by compaction, but not the extensional faults which cut the fabric. Extension in the shallowly-dipping parts of a prism could result from either partitioning of the shear in the decollement zone into regions of coaxial deformation during underthrusting, or changes in the length of the decollement shear zone during collapse and extension of the prism. The decollement zone would have a transpressive geometry if the prism extended above it, compatible with high-angle fracturing (Sanderson & Marchini 1986).

Dewatering during shear is expected in the toe of an accretionary complex, where most dewatering occurs (Bray & Karig 1985). If shearing in the decollement zone occurred during a compactional volume-loss (Moore *et al.* 1986), departures from simple-shear occur, and high-angle fractures could form (Ramsay & Huber 1983, p. 50). Coherent domains from Debussy Heights show that extension occurred after compaction, under low-grade metamorphic conditions, and was not due to this process.

Other mechanisms exist for producing extensional structures in an accretionary complex, apart from those already discussed. Sticking portions of thrust faults, 'footwall-plucking' (Platt & Leggett 1986), and stretching of hangingwalls over culminations (Butler 1982), can all produce localized minor extensional strains; but not the widespread, major extension of the DHMB.

Later structural events

Later out-of-sequence thrusting; and backthrusting postdates bedding-parallel extension and melange formation in the DHMB, and formed minor duplexes and ramp-related folds. The backthrusts clearly truncate earlier thrusts parallel to the melange fabric. Similar late thrusts cut earlier thrusts in the hangingwall of the melange belt, and probably served locally to re-thicken the prism at various times.

Thrust-related folds in coherent regions of the DHMB are associated with an axial-planar, finely-spaced (250–500 μm), discrete crenulation cleavage of the bedding-parallel fabric. Pressure solution concentrated opaque minerals and phyllosilicates in cleavage seams, and limited low temperature crystal-plasticity in the microlithons produced undulose extinction in bent, detrital, quartz grains. The thrust zones show textural evidence of cataclasis, which contrasts with the different fabrics produced by the lower strain rates associated with microfolding.

The crenulation cleavage axial-planar to folds (above) is truncated by later, low-angle extensional detachments which cut-down across bedding to the west. The extension direction, and σ_3, is constrained by flat-lying detachments and fibre-directions, to be essentially parallel to bedding. Fault jogs on these detachments confirm a top to the NW displacement. However, drusy fabrics in the zeolite–calcite–quartz mineralization along detachments requires fluid pressures to be maintained at high levels, as related tensile veins normal to bedding and σ_3 also imply.

This late extension occurred under zeolite-facies conditions, after melange formation and a later period of shortening by folding and thrusting. It apparently represents a thinning of the accretionary prism, above the decollement, under conditions of high fluid pressures and low effective stresses. These later stages of shortening and extension may reflect late-stage adjustments of the prism taper.

Conclusions

Microstructures in the DHMB indicate along-strike variations in the depth of deformation during stratal disruption and melange formation. Near Cape Vostok, deformation started before cementation, but continued to depths at which DMT became important in the melanges of the Debussy Heights area. The melange belt apparently records an early compaction during underthrusting, before stratal extension, in and above the decollement.

Stratal extension at Cape Vostok and Debussy Heights was probably associated with high fluid pressures which weakened the decollement. Deformation in the melanges probably involved components of both layer-parallel extension and layer-parallel shear, during collapse of the accretionary prism at both shallow and deep structural levels.

Discussion

Platt (1986) suggested that above 150°C, Coulomb (frictional) rheologies were inappropriate, making the model of Davis *et al.* (1983) invalid for the deeper levels of accretionary complexes. He proposed, instead, a viscous model; considering the approximately linear flow law for diffusional mass transfer. The likely rheology of the forearc needs consideration.

The rocks in the DHMB were once in the decollement zone at the base of the prism, and during subsequent stages of accretion, sampled conditions through the accretionary prism as they rose to the surface. Brittle faulting under low effective stresses dominated rock-deformation in the DHMB, and consequently, a Coulomb model, not a viscous model, is appropriate for the major deformation of these rocks. A reduction in subduction rates probably was not the direct cause of extension in the DHMB.

There are indications that the Alexander Island prism was steeply-tapered prior to accretion of the DHMB (above). Rising fluid pressures then, could have caused the accretionary prism to become supercritical and collapse (Davis *et al.* 1983). Structures from the DHMB indicate that fluid pressures were high during extensional periods, and perhaps the two were related.

Lallemand & LePichon (1987) investigated the effects of seamount subduction based on the critical-wedge model of Davis *et al.* (1983). Changes in the basal slope of the prism as a seamount was subducted, caused smaller changes in wedge topography; resulting in compression followed by extension in the prism. If fluid pressures were high, as they probably were in the DHMB, seamount subduction would have small effects on the taper of the prism, unless fluid pressures rose during the subduction of a seamount, when the effects would have been dramatic.

Norris & Henley (1976) demonstrated that water produced by metamorphic dehydration reactions may be retained. Hydraulic fracturing, induced by the thermal expansion of water is a common feature of rocks undergoing prograde metamorphism under linear geothermal gradients greater than 12° km^{-1}, and at depths greater than 5–10 km. But under high-pressure/low-temperature metamorphism, the pressure increase due to the expansion of water might not exceed the lithostatic load, and quantities of fluids would be retained in the pore spaces.

Oxburgh & Turcotte (1970) modelled the thermal effects of the approach of a spreading ridge towards a subduction zone, and showed that the subduction of younger, thinner and hotter oceanic crust resulted in higher geothermal gradients in the prism. This could trigger the release of previously-trapped water in the old parts of a prism which accreted under a low geothermal gradient. If this initiated spreading of the prism, thinning and unloading deep levels could release more fluids, and might lead to the fluid pressures exceeding the lithostatic load until the lower parts of the prism became dehydrated. The estimated age, since formation, of the oceanic crust being subducted beneath Alexander Island, decreased dramatically at the estimated time of DHMB accretion, due to the approach of the Pacific/Phoenix ridge (Fig. 2).

A critical wedge model is appropriate for the Alexander Island accretionary prism. The prism shows evidence of extension during melange accretion, when fluid pressures were high. This occurred when young, hot, oceanic crust was being subducted, which could have triggered the release of fluids, weakening the decollement, and leading to collapse of the prism. If so, the model predicts that basins with a high heat-flow may have formed offshore Alexander Island.

I am grateful for the essential support of the personnel at Rothera base, particularly J. Hall. The fieldwork would not have been possible without the assistance of D. Carroll and P. Marquis. I thank I. Goddard for making thin sections, and B. Storey, K. Pickering, and one anonymous referee, for helpful suggestions in improving an earlier draft of the manuscript.

References

Barker, P. F. 1982. The Cenozoic subduction history of the Pacific margin of the Antarctic Peninsula: ridge crest-trench interactions. *Journal of the*

Geological Society, London, **139**, 787–801.

BRAY, C. J. & KARIG, D. E. 1985. Porosity of Sediments in Accretionary Prisms and some implications for Dewatering Processes. *Journal of Geophysical Research*, **90**, 768–778.

BURN, R. W. 1981. Early Tertiary calc-alkaline activity on Alexander Island, *British Antarctic Survey Bulletin*, **53**, 175–193.

—— 1984. *The Geology of the LeMay Group, Alexander Island*. British Antarctic Survey Scientific Reports, No. 109.

BUTLER, R. W. H. 1982. Hangingwall strain: a function of duplex shape and topography. *Tectonophysics*, **88**, 235–246.

BUTTERWORTH, P. J., CRAME J. A., HOWLETT, P. J. & MACDONALD, D. I. M. 1988. Lithostratigraphy of Upper Jurassic–Lower Cretaceous strata of eastern Alexander Island, Antarctica. *Cretaceous Research*, **9**, 249–264.

BYRNE, T. 1984. Early deformation in melange terranes of the Ghost rocks formation, Kodiak Island, Alaska, *In*: RAYMOND, L. A. (ed.) *Melanges; Their nature, origin & significance*. Geological Society America Special Paper, **198**, 21–52.

COWAN, D. S. 1978. Origin of blueschist-bearing chaotic rocks of the Franciscan Complex, San Simeon, California. *Geological Society America Bulletin*, **89**, 1415–1423.

—— 1982. Deformation of partly dewatered and consolidated Franciscan sediments near Piedras Blancas Point, California. *In*: LEGGETT, J. K. (ed.) *Trench–Forearc Geology*. Geological Society London, Special Publication, **10**, 439–457.

—— 1985. Structural styles in Mesozoic and Cenozoic mélanges in the western Cordillera of North America. *Geological Society of America Bulletin*, **96**, 451–462.

DAVIS, D., SUPPE, J. & DAHLEN, F. A. 1983. Mechanics of fold-and-thrust belts and accretionary wedges. *Journal of Geophysical Research*, **88**, 1153–1172.

ENGEBRETSON, D. C., COX, A. & GORDON, R. D. 1985. *Relative Motions between Oceanic and Continental Plates in the Pacific Basin*. Geological Society America Special Paper, **206**.

FISHER, D. & BYRNE, T. 1988. Structural evolution of underthrusted sediments, Kodiak Island, Alaska. *Tectonics*, **6**, 755–793.

HALBOUTY, M. T. *et al.* 1982. *Plate tectonic maps of the circum-Pacific region*. American Association of Petroleum Geologists, Tulsa, USA.

HOBBS, B. E., MEANS, W. D. & WILLIAMS, P. F. 1976. *An outline of structural geology*. John Wiley & Sons, New York.

KARIG, D. E. 1980. Material transport within accretionary prisms and the "knocker" problem. *Journal of Geology*, **88**, 27–39.

—— 1986. Physical properties and mechanical state of accreted sediments in the Nankai Trough. *In*: MOORE, J. C. (ed.) *Structural fabrics in DSDP cores*. Geological Society America Memoir, **166**, 117–133.

LALLEMAND, S. & LE PICHON X. (1987). Coulomb wedge model applied to the subduction of seamounts in the Japan Trench. *Geology*, **15**, 1065–1069.

LUCAS, S. E. & MOORE, J. C. 1986. Cataclastic deformation in accretionary wedges: Deep Sea Drilling Leg 66, southern Mexico, and on land examples from Barbados and Kodiak Islands. *In*: MOORE, J. C. (ed.) *Structural fabrics in DSDP cores*. Geological Society of America Memoir, **166**, 89–103.

LUNDBERG, N. & KARIG, D. E. 1986. Structural features in cores from the Nankai Trough, Deep Sea Drilling Project Leg 87A. *In*: KAGAMI, H., KARIG, D. E. *et al.* (eds) *Initial Reports Deep Sea Drilling Project, Leg 87*. US Govt. Printing Office, Washington, 797–808.

MILNE, A. J. & MILLAR, I. L. 1989. The significance of mid-Palaeozoic basement in Graham Land, Antarctic Peninsula. *Journal of the Geological Society, London*, **146**, 207–210.

MOORE, J. C., ROESKE, S., LUNDBERG, N., SCHOONMAKER, J., COWAN, D. S., GONZALES, E. & LUCAS, S. E. 1986. Scaly fabrics from Deep Sea Drilling Project cores from forearcs. *In*: MOORE, J. C. (ed.) *Structural fabrics in DSDP cores*. Geological Society of America Memoir, **166**, 55–73.

MYASHIRO, A. 1973. *Metamorphism and Metamorphic Belts*. George Allen and Unwin, London.

NORRIS, R. J. & HENLEY R. W. 1976. Dewatering of a metamorphic pile. *Geology*, **4**, 333–336.

NEEDHAM, D. T. 1987. Asymmetric extensional structures and their implications for the generation of mélanges. *Geological Magazine*, **124**, 311–318.

NELL, P. A. R. & STOREY B. C. 1990. Strike slip tectonics within the Antarctic Peninsula forearc. *In*: THOMSON, M. R. A., CRAME, J. A. & THOMSON, J. W. (eds) *Geological evolution of Antarctica*. Cambridge University Press, Cambridge, (in press)

OXBURGH, E. R. & TURCOTTE, D. L. 1970. Thermal evolution of Island Arcs. *Geological Society of America Bulletin*, **81**, 1665–1668.

PICKERING, K. T., AGAR, S. M. & OGAWA, Y. 1988. Genesis & deformation of mud injections containing chaotic basalt–limestone–chert associations: Examples from the southwest Japan forearc. *Geology*, **16**, 881–885.

PIMPIREV, C. T. & DINNYOVSKY. D. S. 1989. Upper Jurassic–Lower Cretaceous calcareous nannofossils from the Le May Group, Alexander Island, Antarctica. *Comptes Rendus de L'Academie Bulgare des Sciences*, **T42**, 63–66.

PLATT, J. P. 1986. Dynamics of orogenic wedges and the uplift of high-pressure metamorphic rocks. *Geological Society America Bulletin*, **97**, 1037–1053.

—— & LEGGETT, J. K. 1986. Stratal extension in Thrust Footwalls, Makran Accretionary Prism: Implications for Thrust Tectonics. *American Association of Petroleum Geologists Bulletin*, **70**, 191–203.

RAMSAY, J. G. & HUBER, M. I. 1983. *The techniques*

of modern structural geology. Academic Press, London.

SANDERSON, D. J. & MARCHINI, W. R. D. 1986. Transpression. *Journal of Structural Geology*, **6** 449–458.

SIBSON, R. H. 1987. Earthquake rupturing as a mineralizing agent in hydrothermal systems. *Geology*, **15**, 701–704.

STOREY, B. C. & GARRETT, S. W. 1985. Crustal growth of the Antarctic Peninsula by accretion, magmatism and extension. *Geological Magazine*, **122**, 5–14.

—— & NELL, P. A. R. 1988. Role of strike-slip faulting in the tectonic evolution of the Antarctic Peninsula. *Journal of the Geological Society London*, **145**, 333–337.

THOMSON, M. R. A. & TRANTER, T. H. 1986. Early Jurassic fossils from central Alexander Island and their geological setting. *British Antarctic Survey Bulletin*, **70**, 23–39.

TRANTER, T. H. 1987. The structural history of the LeMay Group of Central Alexander Island, Antarctic Peninsula. *British Antarctic Survey Bulletin*, **77**, 61–80.

—— 1988. *The tectostratigraphic history of the LeMay Group of central Alexander Island, Antarctica*. PhD thesis, London, Council of National Academic Awards.

WALDRON, J. W. F., TURNER, D. & STEVENS, K. M. 1988. Stratal disruption and development of melange, Western Newfoundland: effect of high fluid pressure in an accretionary terrain during ophiolite emplacement. *Journal of Structural Geology*, **10**, 861–875.

WESTBROOK, G. K. & SMITH, M. J. 1983. Long decollement and mud volcanoes: Evidence from the Barbados Ridge Complex for the role of high pore-fluid pressure in the development of an accretionary complex. *Geology*, **11**, 279–283.

WOODCOCK, N. H. 1979. Sizes of submarine slides and their significance. *Journal of Structural Geology*, **1**, 137–142.

Vein structure and the role of pore fluids in early wet-sediment deformation, Late Miocene volcaniclastic rocks, Miura Group, SE Japan

KEVIN T. PICKERING[1], SUSAN M. AGAR[2] & DAVID J. PRIOR[2,3]

[1] *Department of Geology, University of Leicester, Leicester LE1 7RH, UK*

[2] *Department of Earth Sciences, University of Leeds, Leeds LS2 9JT, UK*

[3] *Present address: Department of Earth Sciences, University of Liverpool, PO Box 147, Liverpool L69 3BY, UK*

Abstract: The Misaki Formation, Miura peninsula, SE Honshu, Japan, comprises deep-shelf/slope volcaniclastic rocks deposited on a segment of the Izu-Bonin island arc prior to accretion about 2.5 Ma. Wet-sediment deformation structures occur as gravity-controlled slides, and sediment injections intruded into semi-lithified over-pressured sediments. Vein structures are locally abundant in the Misaki Formation, mainly as 1–10 cm wide layer-parallel zones, with individual veins about 0.1–3 cm apart. The veins are planar and up to 0.1–2.0 mm in width in their central parts. Vein structure occurs within blocks in chaotic horizons, interpreted as early sediment slides on the seafloor. The vein structures have a strike interpreted as forming parallel to the strike of the basin slope and are considered to arise from pervasive creep-related phenomena in unconsolidated sediments at shallow depths of burial. Back scattered electron (BSE) studies of the vein structure suggest high average atomic number contrasts between the infill of the vein structure and the surrounding sediments and, together with microprobe analyses, suggest that this is due to very fine disseminated pyrite. The pore-fluids involved in the vein evolution were derived from trapped sea water in which certain elements such as S, Fe and Ca were preferentially concentrated in the vein structure. Later generations of vein structure tend to have more complex geometry, are larger, show greater spacing, and appear to have a geometry that cannot be related to any regional trends.

A fundamental problem in the study of accretionary systems at active-convergent plate margins is the assessment of the role of pore fluids in progressively deforming, dewatering and lithifying sediments. A related problem concerns the recognition of early fabrics that may be associated with gravity-controlled slope failure, including sediment creep. The pore fluids, whether trapped and modified sea water or of deeper origin resulting from mineral dehydration reactions, exert a primary control on diagenesis and sediment rheology, and therefore deformation style. One of the early microfabrics to receive considerable attention in the last few years is 'vein structure' (Lundberg & Moore 1986).

Vein structure has been recognized in various forearc trench-slope sediments from the Deep Sea Drilling Project and described as: (a) dewatering veins on Legs 56 and 57 (Japan trench slope); (b) vein structure on Legs 67 and 84 (trench slope off Guatemala); (c) sigmoidal examples of spaced foliation on Leg 66. Lundberg & Moore (1986) have shown that these features, although variously named, are indistinguishable. Good detailed descriptions of vein structure from DSDP Leg 84 (Middle America trench slope) are given by Ogawa & Miyata (1985). Vein structure was also recognized on ODP Leg 110 in the Tiburon Rise sediments in the Atlantic abyssal plain off the Barbados accretionary complex (Moore *et al.* 1987, 1988). Recently, Kimura *et al.* (1989) have described vein structure from within decimetres of the seafloor in unconsolidated, volcaniclastic muds at the junction between the Mariana and Yap trenches. Detailed descriptions of vein structure can be found in Moore (1986).

There have been many interpretations of vein structure (cf. Knipe 1986; Lundberg & Moore 1986). Cowan (1982), who was the first to describe vein structure from the Middle America Trench off Guatemala (DSDP Leg 67), interpreted it as extensional fractures. Knipe (1986) and Leggett *et al.* (1987) interpreted vein structure as forming in response to gravity-induced downslope failure of sediment. Knipe (1986) also noted the possibility that veins may develop or be modified to a sigmoidal geometry as a result of shear parallel to bedding in unconsolidated slope sediments (cf. Kimura *et al.* 1989).

Vein structure is abundant in the Mio-Pliocene volcaniclastic successions of SE Central Japan, associated with early surfacial sediment slides. Since the chronology of deformation is decipherable in this area, and the regional context relatively easy to interpret, the area of SE Central Japan provides a useful field laboratory in which to understand the nature and origin of vein structure.

Regional context

The Miura–Boso peninsulas, SE Japan, are close to a triple junction, formed by the Japan, Sagami and Izu–Bonin (Ogasawara) trenches (Fig. 1). These peninsulas form the easternmost part of the Palaeogene–Neogene Shimanto Belt. Initial collision of the Izu–Bonin arc may have been as early as 11 Ma (Ogawa 1982; Ogawa & Naka 1984; Ogawa & Taniguchi 1988). At present, the Izu, Miura and Boso peninsulas are part of the forearc of the Honshu arc. In early to mid-Tertiary times, however, their basaltic basement formed part of the Izu–Bonin island arc that was accreted during the Upper Miocene–Pliocene through to the present day. During this collision, the Miura Group was accreted from the Izu–Bonin (Ogasawara) arc onto the Honshu arc, 3–2.5 Ma, as seen in the major regional Kurotaki unconformity of this age across Izu, Miura and Boso peninsulas, and the plate boundary incrementally jumped to the present position, the Sagami Trough south of the Miura Group (Ogawa & Horiuchi 1978; Ogawa & Taniguchi 1988). Post-collision deposition, unconformable above the deformed Miura Group, is the latest Pliocene–Pleistocene Kazusa Group on Miura and Boso peninsulas, and above the Shirahama Group on Izu peninsula.

Oblique collision between the Izu–Bonin and Honshu arcs has led to the post-accretion late Pliocene–Recent, dextral strike-slip between the Eurasian and Philippine Sea plates (Fig. 1). Deformation in the Miura and related blocks is dominated by neotectonic dextral oblique-slip and uplift. Steeply dipping normal and reverse faults cut the entire Miura Group and age equivalents, with fault-plane solutions on major neotectonic faults suggesting right-lateral displacements (Kaneko 1969). Recent dextral strike-slip movements have been identified along at least five major faults in Miura Peninsula (Kaneko 1969). Also, the Miura Group has been uplifted to its present outcrop. Palaeomagnetic studies suggest an approximately 30° clockwise rotation of the Miura peninsula (Miura block) after deposition of the Miura Group (Yoshida *et al.* 1984), since about 2.5 Ma. None of the Neogene sediments of Miura and Boso peninsulas show evidence of metamorphism and compaction has been weak.

The Middle and Late Miocene appears to be absent from many areas of the submarine Okinoyama Bank (to the west and southwest of Miura peninsula), suggesting that such areas were emergent during the deposition of the Miura Group. Parts of the Okinoyama Bank may have been volcanic eruptive centres supplying sediment to the Miura Group on Miura and Boso peninsulas. The absence, however, of any diagnostic gravity or magnetic anomalies above the Okinoyama Bank make this interpretation speculative.

This paper focuses on the early gravity-controlled wet-sediment deformation (cf. Pickering 1987 for terminology), in particular the microfabric of vein structure, that formed in a submarine forearc slope, in the late Miocene to earliest Pliocene Misaki Formation, SE Central Japan (Oda 1975; Nitsuma 1975; Ogawa & Horiuchi 1978; Horiuchi & Saito 1981, 1982; Horiuchi & Taniguchi 1985; Eto *et al.* 1987).

Sedimentology of the Misaki Formation

The Misaki Formation (10–3 Ma) is the oldest exposed part of the Miura Group; the base of the formation is unexposed. The thickness of the formation is estimated at 1200 m, but this is unlikely to represent the original thickness as abundant bedding-parallel thrusts have duplicated the stratigraphy. The Misaki and Hatsuse Formations, southern Miura peninsula, are correlated with the Awa Group exposed on Boso peninsula. The petrography, geochemistry and field mapping of the tuff horizons has produced a detailed correlation of key tuff marker beds (cf. Horiuchi & Taniguchi 1985).

Benthic foraminifera throughout the Misaki Formation suggest water depths from lower to middle bathyal depths, i.e. 800–2500 m (S. Hasegawa, University of Tokyo, pers. comm. 1989). Calcareous nannofossils show that the transition from the Misaki to Hatsuse Formation is in the late early Pliocene NN14-NN15 zone (J. Young, British Museum, UK, pers. comm. 1989), i.e. > 3.2 Ma.

The Misaki Formation comprises tuffaceous siltstones with scoria beds of pebble conglomerates and sandstones. This formation shows

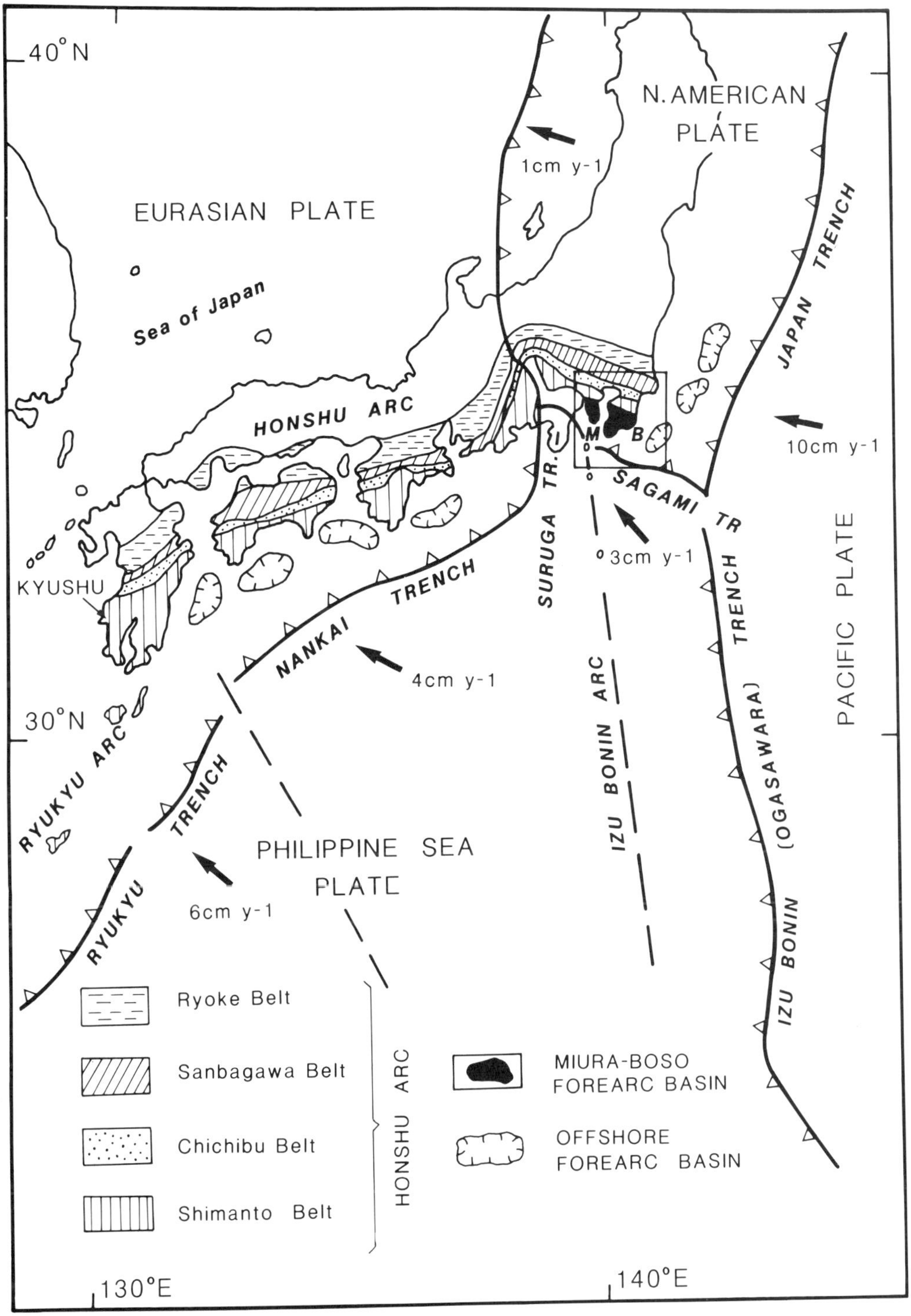

Fig. 1. Map to show the plate tectonic setting of the study area for this paper (boxed). Cretaceous, Palaeogene and Neogene arc-related rocks shown by tectono-stratigraphic belts. Arrows show plate vectors and rates of motion. I, Izu peninsula; M, Miura peninsula B, Boso peninsula.

lateral changes from western, relatively shallow, to eastern deeper, marine environments, and together with the palaeocurrents, suggests a regional palaeoslope that dipped ENE (Soh *et al*. 1989). The Misaki Formation, Miura peninsula, together with the coeval volcaniclastic successions of Boso peninsula, shows an overall fining towards the E/SE, suggesting that the volcaniclastic rocks were derived essentially from the west. The sites of the volcanic vents remain unknown, but were possibly in part the submerged Okinoyama Bank immediately to the W/SW of Miura peninsula. Alternatively, the volcanic vents may have been subducted or underthrust to lower crustal levels farther north. The Misaki Formation comprises the deposits of phreatomagmatic, basaltic, eruptions (Soh *et al*. 1989), hemipelagic and pelagic mudrocks, and locally shows considerable evidence of reworking by various bottom currents, including tides. Benthic foraminifera throughout the Misaki Formation suggest water depths of the order of 1000 m (S. Hasegawa pers. comm. 1989).

The Misaki Formation is locally eroded into by the 5–3 Ma Hatsuse Formation, about 800 m in thickness of coarse-grained, cross-stratified, coalescing submarine canyon deposits. The Hatsuse Formation appears to represent a shallowing upward from the Misaki Formation.

Early slope-related wet-sediment deformation

Normal faults

The earliest recognizable post-depositional structures in the Misaki Formation are local, small-scale, normal faults. Some of these faults are clearly synsedimentary. In other cases, the earliest recognizable faults show normal displacements and are cut by later bedding-parallel thrusts and wet-sediment injections. There is no evidence to indicate whether such faults formed at or very near to the seabed or at greater depths.

Multichannel seismic reflection profiles across the northern parts of the Izu-Bonin are demonstrate the importance of large-scale normal faulting (e.g. Fujioka *et al*. 1987, figs 8a,b,c). Many of these normal faults propagate close to the seafloor or intersect the seafloor and show differences in sediment thickness across the faults. The large-scale normal faults are clearly synsedimentary and, presumably, are associated with much smaller scale mainly normal faults.

Seafloor sediment slides and sediment injections

The Misaki Formation contains numerous zones of chaotic sediments that are typically brecciated and generally parallel or subparallel to bedding. These chaotic zones vary in thickness from centimetres to about 30–40 m thick, but generally are of the order of decimetres thick.

The thickest chaotic zone, containing boulder-size intraformational clasts showing vein structure, can be correlated across Miura peninsula (Fig. 2) and has an irregular upper surface. Immediately overlying the irregular surface of this chaotic deposit, and locally smoothing any residual uneven topography, there is a black scoriaceous, granule to coarse grain size, wavy laminated and parallel-laminated bed up to about 1 m thick. This bed is best seen on the foreshore at Misaki. Small pebbles of pumice occur as 'lag' like layers one pebble thick in the lower part of this scoriaceous bed and probably represent reworked material from the top of the chaotic unit. Other much thinner chaotic horizons also show uneven upper surfaces infilled by wavy lamination, parallel-lamination or current ripples.

It is possible, therefore, to demonstrate that many of the bedding-parallel chaotic horizons formed as gravity-driven sediment slides on the seafloor. Since they contain blocks/clasts with vein structure, the genesis of the vein structure must predate the failure of semi-lithified sediments close to the seafloor and probably reflect a process operating at very shallow depths of burial.

In addition to the surficial sediment slides on the seafloor, there are thin chaotic zones that intrude along fault zones, such as bedding-parallel thrusts, high-angle reverse faults and conjugate reverse faults. Such injections and chaotic zones may be cut locally by low-angle to bedding-parallel thrust faults. These latter structures post-date the generation of vein structure and the sediment slides and, therefore, are not considered further in this paper. All of the zones of chaotic sediment were interpreted as layer-parallel gravity-driven slides of semi-lithified marine sediments (cf. Kozima 1980, 1981). Although superficially similar, this study has shown that the origin of the chaotic zones is more complex, with some chaotic zones not being interpretable as submarine slides at the seabed.

Vein structure

The earliest regional fabric recognized within

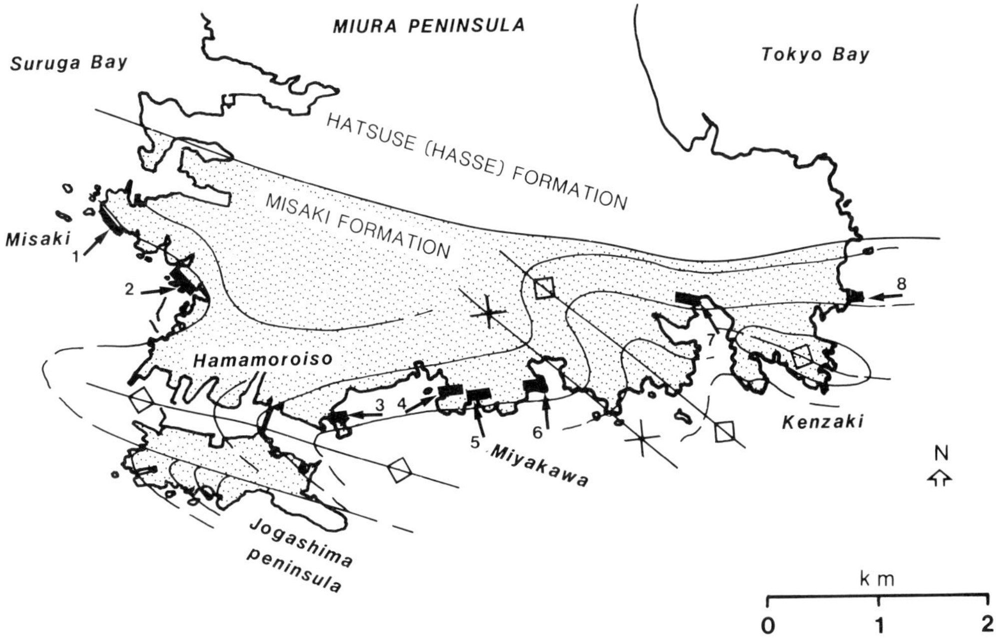

Fig. 2. Map to show lateral continuity of 40 m thick chaotic unit across southern Miura peninsula. Dashed lines show tuff bands correlated by Horiuchi & Taniguchi (1985). The chaotic unit occurs at approximately the same stratigraphic level, contains boulder-size intraformational clasts with vein structure, and is interpreted as a surficial, submarine, gravity slide deposit.

the Misaki Formation is preferentially developed within the fine-grained volcaniclastic siltstones, and is referred to as vein structure (Ogawa 1980). All other post-depositional fabrics, including the chaotic sediment zones transect the vein structure. There is more than one generation of vein structure, but only one set appears to have a systematic and regional orientation. Later vein structures are more locally distributed with a relatively complex geometrical relationship to other fabrics, including bedding. The non-systematic vein structure is not considered in detail in this paper.

The vein structure is generally restricted to layer-parallel zones 1–10 cm thick in which individual veins are approximately equidistant, ranging from 1–30 mm apart, planar and typically up to 0.1–2.0 mm in width in the central parts (Fig. 3). Vein terminations characteristically show a fine-scale bifurcation. Sigmoidal and en echelon vein arrays are rare, therefore, layer-parallel simple shear may not have been an important factor in the development of the vein structure.

Mapping of the vein structure by Ashi (ORI, Univ. Tokyo) demonstrated a preferred regional NNW–SSE strike orientation for individual veins on Boso and Miura peninsulas (Fig. 4). There are deviations from this strong NNW–SSE pattern, for example at Kenzaki, where the veins formed perpendicular to an anticlinal axis. The vein structure appears to have formed prior to tectonic tilting of the beds because the regional preferred orientation becomes evident after bedding is restored to horizontal. The anomalous, more east–west, vein directions occur where tectonic complications pose the greatest uncertainty for rotating bedding to horizontal. Alternatively, the anomalous readings may reflect local variations in the original dip direction of the seafloor where the original regional tilt was to the NNE–E (see below).

The veins are invariably darker in colour than the host rock. Thin-section microscopy suggests that the composition of the vein infills is identical to the host sediments, albeit much finer grained. There are, however, some subtle geochemical differences which are considered below. Back scattered electron (BSE) images of the vein structure show some alignment of the platy minerals parallel to the vein margins.

Delicate sponge spicules extend across the vein walls, suggesting that no grain fracture due to shearing occurred along the vein margins

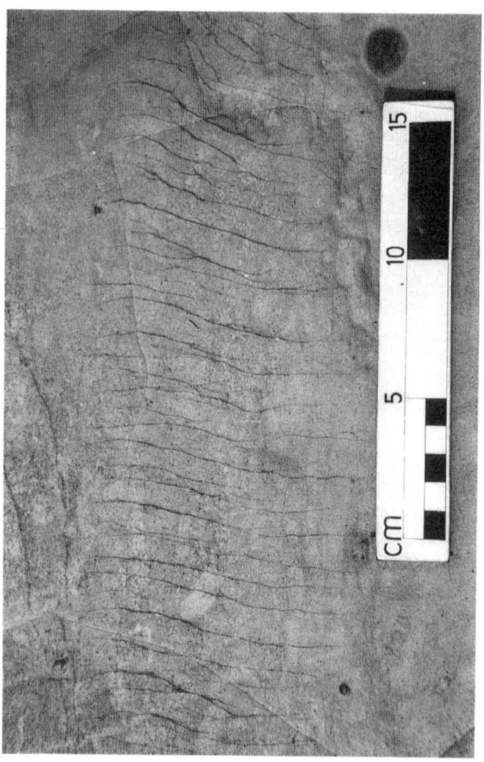

(a)

(b)

(c)

Fig. 3. (a & b) Transverse views of vein structure, Miyakawa. This structure is the earliest post-depositional, regional, fabric in the Miura Group and occurs in bands that tend to parallel bedding; individual veins are perpendicular to bedding and approximately equidistant. (**a**) Two generations of vein structure: 1, regionally systematic, earliest set; 2, non-systematic vein array; arrows show bedding trace. (**b**) close-up of vein structure in bioturbated silty mudstone to show bifurcating terminations and slightly sigmoidal shape. (**c**) Vein structure within clast in 40 m thick seafloor slide, Misaki.

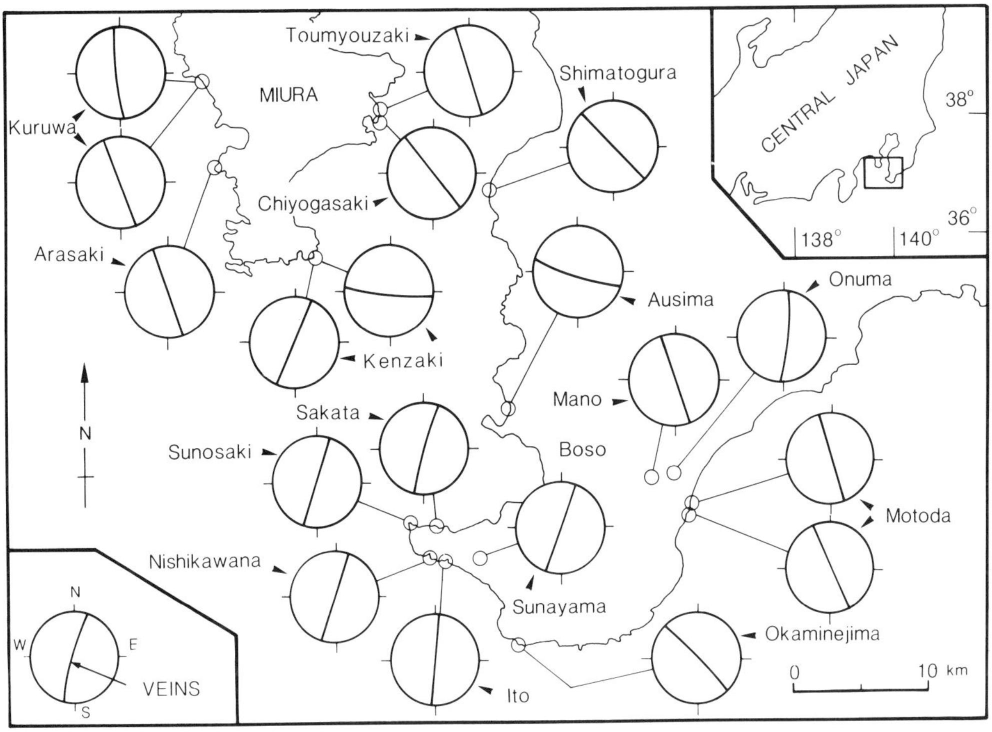

Fig. 4. Map to show the distribution and orientation of the regional vein structure across Miura and Boso peninsulas. Orientation of vein structure shown after correction for tectonic tilting and folding (with permission, J. Ashi, ORI, University of Tokyo).

during dilation. There is some microstructural evidence to suggest that this regional vein structure acted as conduits for continuous fluid flow, at least locally, throughout the sediments (see below). Later, larger scale, vein structure showing more irregular and branching geometries, appears to be more common in the vicinity of chaotic horizons and early faults, and may also have acted as pathways for fluid flow.

Fibrous calcite is identified in some veins. Calcite veins from the Hota Group on Boso peninsula were analysed for stable oxygen and carbon isotopes (W. Soh. ORI, University of Tokyo, pers. comm. 1989). The O isotope data indicate palaeotemperatures of about 30–40°C, and the C isotopes are light, suggesting input from organic carbon. The data is consistent with methanogenesis at shallow depths of burial. Similar calcite-filled veins have been documented from ODP Leg 110 in the Barbados accretionary prism (Behrmann *et al.* 1988). The timing of the calcite growth in relation to the genesis of the vein structure, however, remains unresolved. The palaeotemperatures of 30–40°C, together with the rarity of calcite infills in most of the vein structures, however, supports a later infill where pore fluids have preferentially precipitated calcite in such sites, at depths of burial of the order of 1 km.

Back scattered electron (BSE) images of vein structure. Back scattered electron (BSE) studies (conducted at Leeds University) reveal the vein structure as brighter zones of BSE signal (Fig. 5). These data indicate higher mean atomic number for the elements in the vein structure compared to the enclosing sediments. Pyrite framboids are particularly common in some of the vein structure, and very fine (sub-100 μm) pyrite may be responsible for the brighter BSE signal where individual pyrite grains cannot be resolved. The pyrite framboids show a significant zoning from core to rim. These zones may relate to both chemical and crystallographic variations, but are too fine to view by electron microprobe techniques (Goldstein *et al.* 1981) or electron channelling patterns (Lloyd 1987).

There are no obvious differences in microfabric between the infill of the vein structure

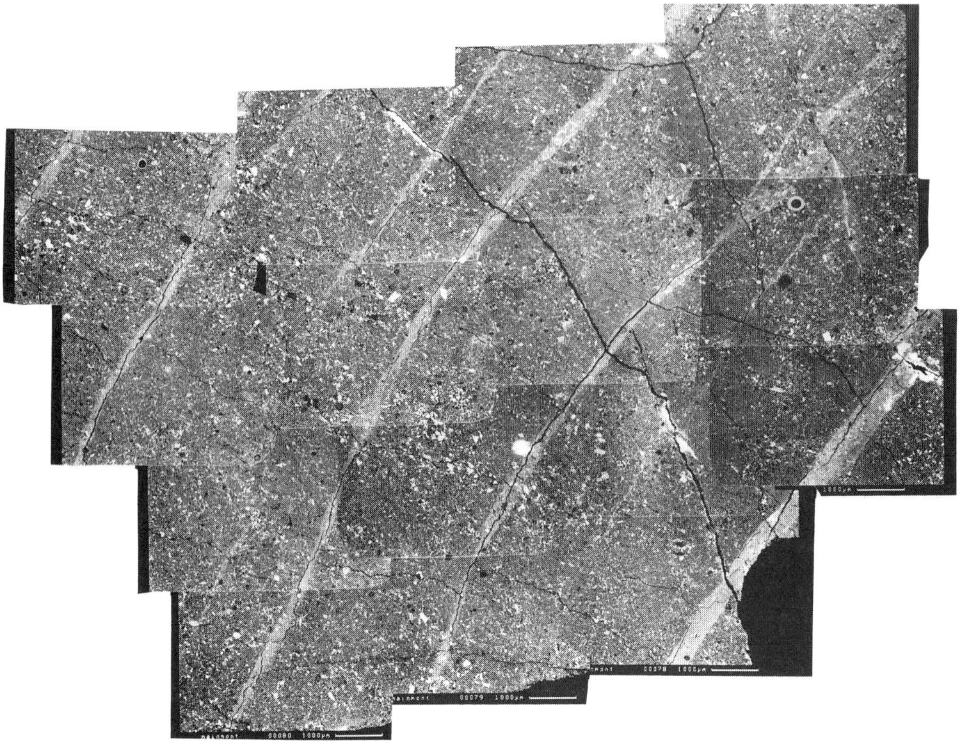

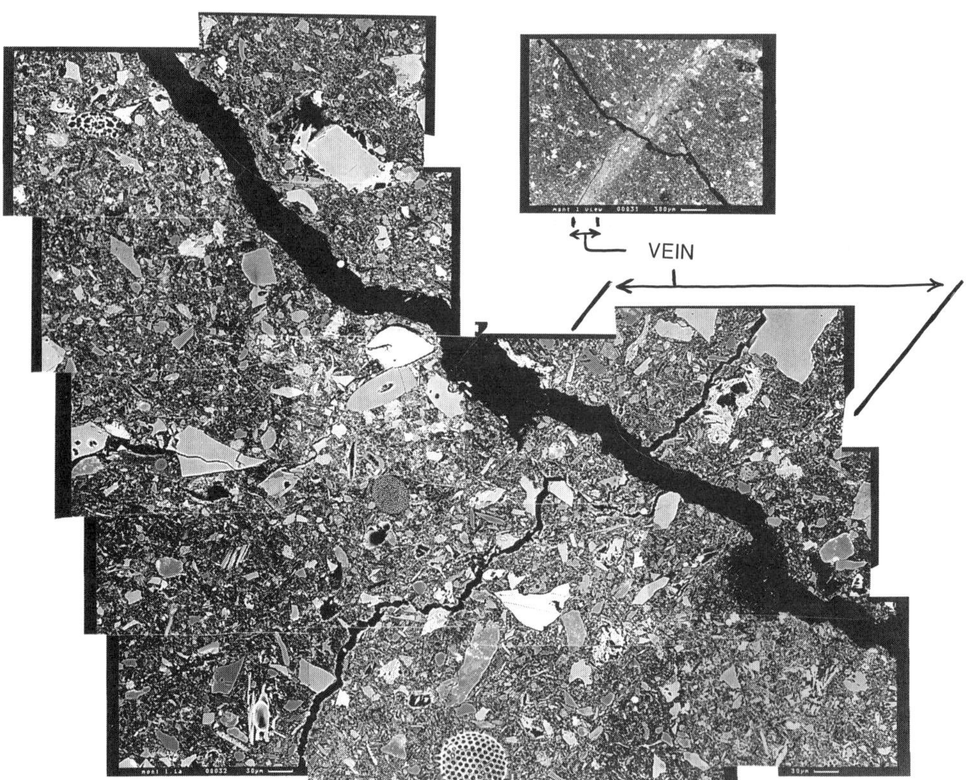
VEIN

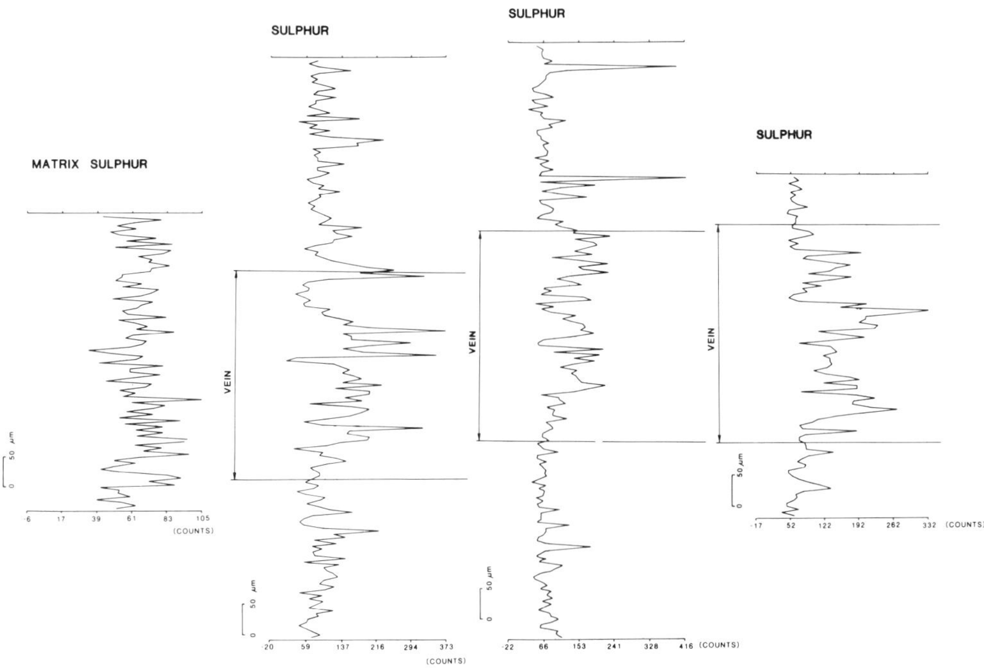

Fig. 6. Three representative probe analyses to show the enhanced S values in traverses across vein structure. Anomalously high S counts are pyrite framboids. 60 μm interval of low S counts towards top of vein in probe traverse second from left correlate with late open fracture infilled by resin.

and surrounding sediments. There is, however, a clear difference in bulk mean grain size such that the veins contain substantially greater amounts of clay-grade material.

Electron probe traverses across vein structure. Wavelength-dispersive electron microprobe analyses of the vein structure were undertaken (at Leicester University) to define the chemical signature of the vein structure compared to the surrounding sediments. Ten traverses with up to 196 measured points at 5 μm spacing, were made for various elements (Si, Al, K, Na, Ca, Mg, Mn, Fe, Ti, Ni, Cu, Co, V, S, P. Sr, Ce, Cs, Cd, Mo, La, Y, Cl, Br) across vein structure and matrix selected for the absence of both open cracks and large rock fragments. Four probe traverses were run through the resin used to impregnate the very friable sediment samples. The results are expressed as total counts per 10 second interval at each probe station. Figure 6 shows representative probe traverses for sulphur through selected vein structure and matrix.

The heterogeneous but essentially basaltic mineralogy of the volcaniclastic sediments and the chemistry of the resin used for impregnation tend to mask chemical differences between the vein structure and surrounding material apart from S and Fe. S, Fe, and Ca are the only elements that appear to be generally higher in the traverses of vein structure compared to the matrix (Table 1). S is the only major element to show a consistent increase in abundance in the veins (Fig. 6). The bulk high Ca content of the volcaniclastic (basaltic) sediments militates against such clear visual discrimination into the vein structure.

The enhanced S and Fe contents of the vein infills, together with the BSE study, suggests that the veins contain much fine-grained, disseminated, pyrite. Ca values also suggest some secondary carbonates (calcite) within the veins. S is the only element of the three which also occurs in significant quantities in the impreg-

Fig. 5. Montage of BSE micrographs showing vein structures. Inset shows location of montage and the average atomic number of infill of vein structure compared to the surrounding sediments. The BSE differences are masked on the montages because each micrograph was exposed individually. The montages illustrate the continuity of fabric from surrounding sediments into the vein.

Table 1. *S, Fe & Ca counts (10 s) at 5 μm spacing in vein structure, Misaki Formation, Miyakawa*

File	Comments	Mean S	Mean Fe	Mean Ca
14	Matrix out side veins $n = 97$	66	1091	7475
7	vein, $n = 69$	129	1396	8482
11	vein, $n = 70$	139	1382	8863
3	vein, $n = 55$*	118	1062	6976
6	vein, $n = 73$	137	1375	8647
10	vein, $n = 42$	159	1363	5380
13	vein, $n = 45$	70	1318	6988
4	vein, $n = 37$	103	1386	7166
5	vein, $n = 43$	135	1309	8116

* Fracture in vein from 80–135, μm

Note, veins contain greater amounts of S and Fe compared to surrounding, equivalent grain-size, sediments or 'matrix'. Ca is more variable, with probe traverses being susceptible to analysing a calcareous microfossil fragment both within and outside of veins. Traverses for about 20 elements completed but data not yet synthesised. n, number of stations.

nating resin. The much greater mean grain size and correspondingly high porosity of the matrix, and therefore resin content suggests that the S anomalies are not only real but probably underestimated in their relative abundance in the vein structure. The darker colour of the finer grained vein infills, together with the enhanced S and pyrite content, suggests that the major subtle geochemical difference between the veins and surrounding sediments was caused by reduction within the veins.

Vein structure and slope-related sediment creep. The presence of vein structure within clasts in chaotic horizons interpreted as sediment slides involving the failure of seafloor sediments (Fig. 3c) suggests that vein structure is a microfabric that forms at shallow depths of burial in unconsolidated or semilithified sediments. The consistent regional orientation of the systematic vein structure (approximately oriented north–south), parallel to the strike of the Miocene-Pliocene basin slope (inferred from palaeocurrents, unpublished data of Pickering), together with the timing of formation of the vein structure is interpreted to reflect creep processes on an eastward–dipping (ENE-dipping) basin slope. Subtle, local, variations in the orientation of vein structure, after rotating the data to remove tectonic tilting, are then explained as due to irregularities in the basin slope during deposition of the Misaki Formation.

Ogawa & Miyata (1985) suggested that the veins formed perpendicular to the minimum compressive stress. Ogawa (1980, 1982) and Ogawa & Miyata (1985) interpreted the vein structure as an early dewatering cleavage. The orientation of the vein structure shows a strong preferred N–S orientation, parallel with the inferred regional strike of the eastward-dipping slope of the Izu–Bonin arc/ridge.

There is no evidence to suggest that the veins formed by violent hydrofracture, but rather under pressure-dilation conditions (cf. Ritger 1985) with elevated pore-fluid pressures and a regional E–W extensional stress field. The restriction of the vein structure to the finer grained, silty, lithologies reflects the considerably lower permeability and, probably, locally elevated pore-fluid pressures of these sediments compared to the coarser grained siltstones and sandstones.

Faults associated with wet-sediment deformation during accretion

The Misaki and Hatsuse Formations are cut by numerous reverse and normal faults, many of which occur as conjugate pairs, similar to those described by Underhill & Woodcock (1987) from high-porosity sandstones. Abundant early reverse faults and thrusts, with offsets commonly up to a few metres, are common and intimately associated with thin wet-sediment injections along the fault surfaces. The presence of distinctive white tuffs and black to orange-black scoria beds locally shows the extent of the stratigraphic duplication caused by faulting. The low-angle conjugate sets of reverse faults appear similar to those produced experimentally in wet sediments by Carson & Berglund (1986), who show that the faults act as important conduits for dewatering.

Pre-existing normal faults at high angles to bedding were reactivated as reverse faults. A new set of low-angle reverse faults and thrusts formed (commonly as conjugate pairs). Synchronous with this faulting, pore fluids (and possibly gases) were expelled, often explosively, from below and within the exposed Miura Group sediments, to brecciate and disrupt the semi-lithified sediments. The disruption and injections occurred mainly along reverse faults, thrusts and as layer-parallel sill-like intrusions, locally cutting bedding (Fig. 7a). The escaping pore fluid/gas mixture hydraulically fractured the sediments and produced many of the faults.

The plate-scale N–S compression caused by arc–arc collision can explain the complex,

(a)

(b)

Fig. 7. Wet-sediment injections in the Misaki Formation, Miura peninsula. (**a**) Chaotic wet-sediment injection of scoriaceous and pumiceous material transecting cross-stratified scoria bed, Jogashima. (**b**) Thin injection at high angle to bedding along conjugate reverse faults (offsets shown by markers beds), Miyakawa; this injection is offset by early layer-parallel thrusts.

mainly conjugate, faulting associated with the wet-sediment injection structures. Minor faults occur both as low-angle and high-angle with respect to layering. In the former cases, the acute bisectrix is layer-parallel, whereas it is perpendicular to bedding for the high-angle reverse and normal conjugate faults. Both sets of faults are explicable within the same overall N–S compressional field (cf. Kakimi *et al.* 1966), where the maximum effective compressive stress remains layer-parallel, but the pore fluid pressure varies, perhaps cyclically.

Over-pressuring of the sediments probably led to the hydrostatic pressure exceeding the lithostatic, loading, pressure and also the maximum effective principal stress due to regional compression (with arc–arc collision). Brittle failure appears to have been by hydro-

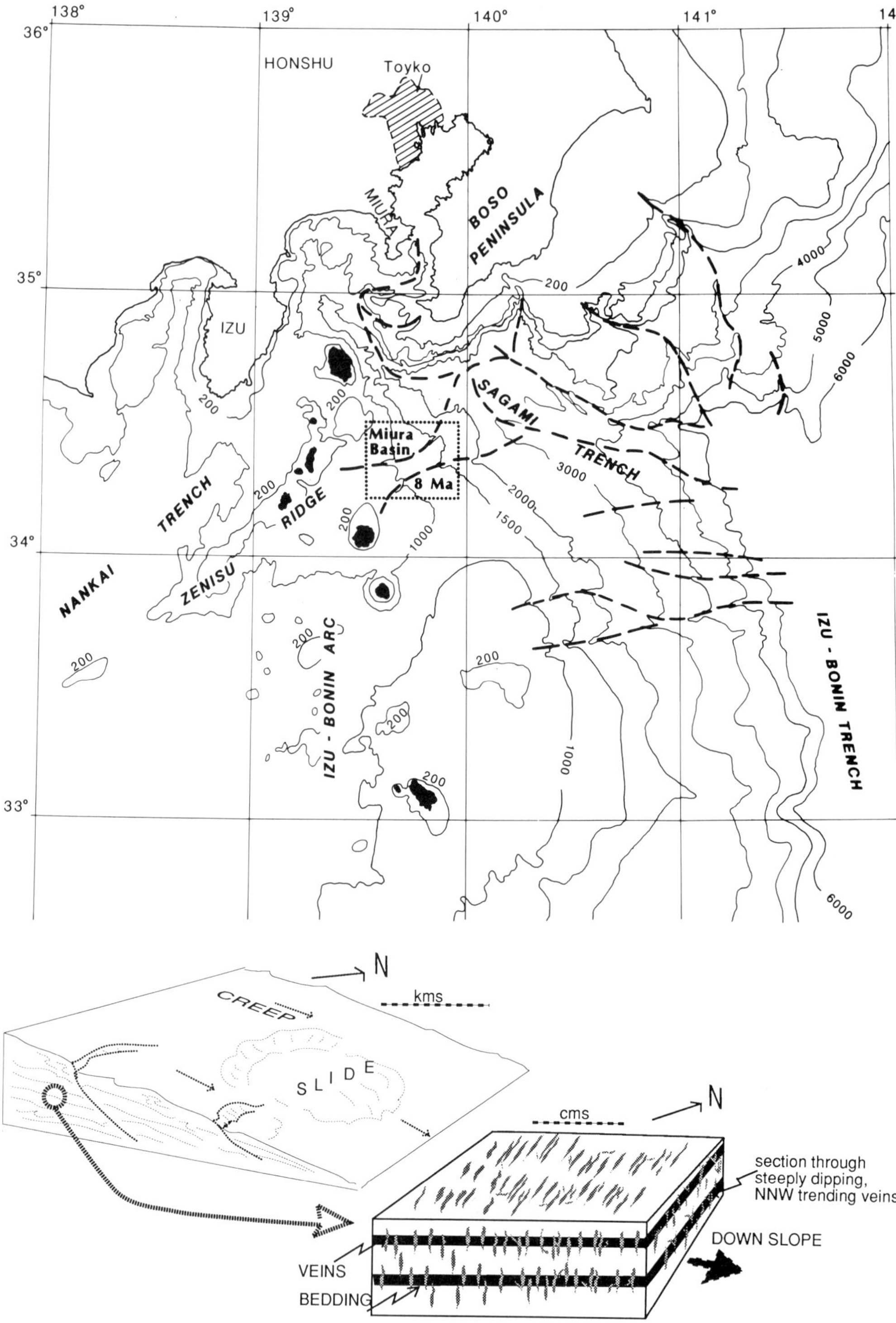

Fig. 8. (**a**) Map of the present-day Izu Collision Zone with a superimposed location of the Miura Group palaeoenvironments approx. 8 Ma, assuming present rates of plate motion. Dashed lines are axes of main canyons. (**b**) Sketch of slope environment about 8 Ma, in which the earliest vein structure with a regional, slope-contour parallel, orientation developed as a result of creep processes. Compare with vein orientation data in Fig. 4.

fracture with the acute bisectrix of the faults being perpendicular to bedding (Fig. 7b). During intervening periods when the pore-fluid pressures remained high but subcritical (with respect to hydrofracture), faulting occurred with the acute bisectrix being layer parallel. The association of wet-sediment injections with both fault sets, many of which show injections along the fault surfaces, tend to support the hypothesis for cyclically changing pore-fluid conditions under regional N–S compression. An alternative explanation for the two sets of faults is that the steeply-dipping (relative to layering) faults representing local areas of stress release (by hydrofracture) under a regional N–S compressive stress field that favoured the development of low-angle conjugate reverse faults, including thrusts that tend to show southward transport: these low-angle faults are now steeply dipping due to later tectonic, northward tilting.

Modern analogue in the Izu–Bonin arc

The northern parts of the present Izu–Bonin arc provide a perfect modern analogue for the plate-tectonic setting and sedimentary environments of the Misaki and Hatsuse Formations between 10–3 Ma.

Figure 8 is a reconstruction of the depositional location, superimposed on the present Izu–Bonin arc, of the Misaki Formation, Miura peninsula, (together with the correlative lithologies on Boso) about 10–5 Ma, assuming present plate convergence rates. The regional NE–ENE-dipping slope (Misaki Fm.) is cut by submarine canyons (Hatsuse Fm.) and subject to considerable sediment sliding and, presumably, creep processes. The earliest, regional, vein structure generated by gravity-controlled creep processes would strike parallel with the slope contours, perpendicular to bedding and show a NNW–SSE to NW–SE orientation (Fig. 8). Sediment sliding may incorporate semi-consolidated blocks of slope sediment and/or rotate packages of sediment so that later vein structure has a more complex geometry and orientation.

This research was funded by a NERC grant to KTP. J. Ashi, ORI, University of Tokyo, Japan, is thanked for the use of Fig. 4. Y. Ogawa, Department of Geology, Kyushu University, Japan, is thanked for introducing KTP to vein structure on Miura and Boso peninsulas and, together with W. Soh (ORI, University of Tokyo, Japan) and H. Taniguchi (Nihon University, Tokyo), are thanked for useful discussions in the field.

References

BEHRMANN, J. H., BROWN, K., MOORE, J. C., MASCLE, A. & TAYLOR, E. 1988. Evolution of structures and fabrics in the Barbados accretionary prism, insights from Leg 110 of the Ocean Drilling Program. *Journal of Structural Geology*, **10**, 577–591.

CARSON, B. & BERGLAND, P. L. 1986. Sediment deformation and dewatering under horizontal compression: experimental results, *In*: MOORE, J. C. (ed.) *Structural fabrics in Deep Sea Drilling cores from forearcs*. Geological Society of America Memoir, **166**, 135–150.

COWAN, D. S. 1982. Origin of vein structure in slope sediments on the inner slope of the Middle America trench off Guatemala. *In*: *Initial Reports Deep Sea Drilling Project 67*, US Government Printing Office Washington, DC, 645–649.

ETO, T., ODA, M., HASEGAWA, S., HONDA, N. & FUNAYAMA, M. 1987. Geologic age and paleo-environment based upon microfossils of the Cenozoic sequence in the middle and northern parts of the Miura Peninsula. *Science Report Yokohama National University, Series II*, **34**, 41–57 (in Japanese).

FUJIOKA, K., KINOSHITA, M., CHOI, J., FUSE, K., GAMO, T., HASUMOTO, K., ISHIBASHI, J., KOGA, K., MIYATA, H., NISHIYAMA, E., SAYANAGI, K., SHIMAMURA, K., SHITASHIMA, K., SUZUKI, K., TANAKA, Y., TOKUYAMA, H. & WATANABE, M. 1987. Preliminary report of the KT 86–10 cruise for the Mikura and Hachijo Basin. *Bulletin of Earthquake Research Institute (University of Tokyo)*, **62**, 61–132.

GOLDSTEIN, J. I., NEWBURY, D. E., ECHLIN, P., JOYCE, D. C., FIORI, C. & LIFSHIN, E. 1981. *Scanning electron microscopy and X-ray microanalysis*. Plenum, New York.

HORIUCHI, K. & SAITO, K. 1981. Tuff key beds and correlation of the Miura Group, Central Japan. *Proceedings Institute of Natural Sciences, Nihon University*, **20**, 125–136.

—— & —— 1982. Heavy mineral assemblages of tuff beds in the Miura Group, southern Miura Peninsula, Central Japan. *Proceedings Institute of Natural Sciences, Nihon University*, **17**, 47–58.

—— & TANIGUCHI, H. 1985. Study on the correlation of tuff key beds in the Miura Group, southern Miura Peninsula, Central Japan. *Proceedings Institute Natural Sciences College of Humanities & Sciences, Nihon University, Earth Sciences*, **20**, 11–31.

KAKIMA, T., HIRAYAMA, J. & KAGEYAMA, K. 1966. Tectonics stress-field deduced from the minor fault systems in the northern part of the Miura Peninsula. *Journal Geological Society of Japan*, **72**, 469–489.

KANEKO, S. 1969. Right-lateral faulting in the Miura Peninsula, south of Tokyo, Japan. *Journal of the Geological Society of Japan*, **75**, 199–208.

KIMURA, G., KOGA, K. & FUJIOKA, K. 1989. Deformed soft sediments at the junction between the

Mariana and Yap Trenches. *Journal of Structural Geology*, **11**, 463–472.

KNIPE, R. J. 1986. Microstructural evolution of vein arrays preserved in Deep Sea Drilling Project cores from the Japan Trench, Leg 57. *In*: MOORE, J. C. (ed.) *Structural fabrics in Deep Sea Drilling cores from forearcs*. Geological Society of America Memoir, **166**, 75–87.

KOZIMA, N. 1980. On the layers of disordered sedimentation in the Misaki Formation in the southwestern part of the Miura Peninsula, south Kwanto, Japan (Part 1). *Journal of the Geological Society of Japan* **86**, 313–326.

—— 1981. On the layers of disordered sedimentation in the Misaki Formation in the southwestern part of the Miura Peninsula, south Kwanto, Japan (Part 2). *Journal of the Geological Society of Japan*, **87**, 197–210.

LEGGETT, J. K., LUNDBERG, N., BRAY, C. J., CADET, J. P., KARIG, D. E., KNIPE, R. J. & VON HUENE, R. 1987. Extensional tectonics in the Honshu fore-arc, Japan: integrated results of DSDP Legs 57, 87 and reprocessed multichannel seismic reflection profiles. *In*: COWARD, M. P., DEWEY, J. F. & HANCOCK, P. L., (eds) *Continental Extensional Tectonics*. Geological Society, London, Special Publication, **28**, 593–609.

LLOYD, G. E. 1987. Atomic number and crystallographic contrast images with the SEM: a review of backscattered electron techniques. *Mineralogical Magazine*, **51**, 3–19.

LUNDBERG, N. & MOORE, J. C. 1986. Macroscopic structural features in Deep Sea Drilling Project cores from forearc regions. *In*: MOORE, J. C. (ed.) *Structural fabrics in Deep Sea Drilling Cores from forearcs*. Geological Society of America Memoir, **166**, 13–44.

MOORE, J. C., MASCLE, A & ODP LEG 110 SCIENTIFIC PARTY 1987. Expulsion of fluids from depth along a subduction-zone decollement horizon. *Nature*, **326**, 785–788.

——, —— & —— 1988. Tectonics and hydrogeology of the northern Barbados Ridge: results from Ocean Drilling Program Leg 110. *Geological Society of America Bulletin*, **100**, 1578–1593.

NITSUMA, N. 1976. Magnetic stratigraphy in the Boso Peninsula. *Journal of the Geological Society of Japan*, **82**, 163–181.

ODA, M. 1975. A chronological interpretation of the palaeomagnetic polarity records of the Upper Cenozoic of the Boso Peninsula based upon planktonic foraminifera. *Journal of the Geological Society of Japan*, **81**, 645–647.

OGAWA, Y. 1980. Beard-like veinlet structure as fracture cleavage in the Neogene siltstone in the Miura and Boso Peninsulas, central Japan. *Science Reports Department Geology Kyushu University*, **13**, 321–327.

—— 1982. Tectonics of some forearc fold belts in and around the arc-arc crossing area in central Japan. *In*: LEGGETT, J. K. (ed.) *Trench-Forearc Geology*. Geological Society, London, Special Publication, **10**, 49–61.

—— & HORIUCHI, K. 1978. Two types of accretionary fold belts in central Japan. *Journal of Physics of the Earth*, **26**, 5321–5336.

—— & MIYATA, Y. 1985. Vein structure and its deformational history in the sedimentary rocks of the Middle America Trench slope off Guatemala, Deep Sea Drilling Project Leg 84, 811–829. *In*: VON HUENE, R, AUBOUIN, J. & OTHERS (eds) *Initial Reports Deep Sea Drilling Project*, **84** US Government Printing Office Washington DC.

—— & NAKA, J. 1984. Emplacement of ophiolitic rocks in forearc areas: examples from central Japan and the Izu-Mariana-Yap island arc system. *In*: GASS, I. G., LIPPARD, S. J. & SHELTON, A. W. (eds) *Ophiolites and Oceanic Lithosphere*. Geological Society, London, Special Publication, **13**, 253–263. Oxford: Blackwell Scientific.

—— & TANIGUCHI, H. 1988. Geology and tectonics of the Miura–Boso Peninsulas and the adjacent area. *Modern Geology*, **12**, 147–168.

PICKERING, K. T. 1987. Wet-sediment deformation in the Upper Ordovician Point Leamington Formation: an active thrust-imbricate system during sedimentation, Notre Dame Bay, north-central Newfoundland. *In*: JONES, M. E. & PRESTON, R. M. F. (eds) *Deformation of Sediments and Sedimentary Rocks*. Geological Society, London, Special Publication, **29**, 213–239.

RITGER, S. D. 1985. Origin of vein structure in the slope deposits of modern accretionary prisms. *Geology*, **13**, 437–439.

SOH, W., TAIRA, A., OGAWA, Y., TANIGUCHI, H., PICKERING, K. T. & STOW, D. A. V. 1989. Submarine depositional processes for volcaniclastic sediments in the Mio-Pliocene Misaki Formation, Miura Group, central Japan. *In*: TAIRA, A. & MASUDA, F. (eds) *Sedimentary Facies in the Active Plate Margin*. TERRAPUB, Tokyo, 619–630.

UNDERHILL, J. R. & WOODCOCK, N. H. 1987. Faulting mechanisms in high-porosity sandstones; New Red Sandstone, Arran, Scotland. *In*: JONES, M. E. & PRESTON, R. M. F. (eds) *Deformation of Sediments and Sedimentary Rocks*. Geological Society, London, Special Publication, **29**, 91–105.

YOSHIDA, S., SHIBUYA, H., TORII, M. & SASAJIMA, S. 1984. Post-Miocene clockwise rotation of the Miura Peninsula and its adjacent area. *Journal Geomagnetism & Geoelectricity*, **36**, 579–584.

Centrifuge modelling of thrust faulting: strain partitioning and sequence of thrusting in duplex structures

SHUMIN LIU & JOHN M. DIXON

Experimental Tectonics Laboratory, Department of Geological Sciences, Queen's University, Kingston, Ontario K7L 3N6, Canada

Abstract: The kinematic evolution of fold-thrust duplex structures has been investigated by analog scale modelling using the centrifuge technique. Plasticine and silicone putty simulate rocks such as limestone and shale, respectively. Model stratigraphic units consist of microlaminations (55–250 μm thick) of these two materials, different ratios of thickness of the constituent laminae simulating different limestone/shale ratios (bulk competence). The models involve six stratigraphic units totalling 4 mm in thickness, with alternating bulk competence (low at the base, high at the surface). They were subjected to layer-parallel compression from one end, and kinematic evolution was monitored by cutting profile sections at six deformation stages.

Structural style varies with depth. The top competent unit develops upright buckle-folds which propagate from 'hinterland' to 'foreland'. The middle competent unit also buckles, but the folds have pronounced foreland vergence. The bottom competent unit first develops gentle folds that propagate forwards, but the foreland-dipping limbs of these folds are soon cut by thrust faults. As shortening continues, this unit is telescoped into a classic blind duplex structure: thrust faults root in the underlying incompetent unit, ramp through the competent unit, and merge into a roof thrust in the overlying incompetent unit. The 'wavelength' of the duplex structure is inherited from the initial buckling instability. The duplex develops by serial nucleation of faults from hinterland to foreland as in natural thrust belts. However, all the faults remain active while the stratigraphic pile deforms as a tapered wedge.

Palinspastic reconstruction by restoration of buckle folds and fault slip shows that layer-parallel shortening accounts for approximately half of the total shortening at all structural levels. Folding and faulting account for the remainder, with folding dominating at high levels and faulting at depth. The transfer is accommodated by decollement in the incompetent units.

The models bear striking resemblance to fold-thrust structures of the Appalachian Valley and Ridge Province of West Virginia and Virginia, USA.

Fold-thrust belts are a manifestation of horizontal compression of horizontally-layered successions of sedimentary rock at or near convergent tectonic plate margins. The stratified rocks are horizontally shortened and vertically thickened by three complementary mechanisms: the strata are homogeneously deformed by layer-parallel shortening and are buckled into fold structures, and the stratigraphic succession is imbricated by displacement on low-angle overthrust faults with stair-case trajectories. The relative importance of these three mechanisms varies with the nature of the stratigraphic succession: the first is favoured in sequences dominated by incompetent (ductile) units; the second is dependent on prominent bedding anisotropy and contrast in layer competency; and the third requires brittle behaviour by at least some components of the stratigraphic sequence. The three mechanisms can be variably partitioned in different parts of a fold-thrust belt, given that rheological behaviour is a function of rock type, temperature, confining and fluid pressure, differential stress and other variables.

The geometry of thrust systems has been described in detail by Bally *et al.* (1966), Dahlstrom (1969), Price & Mountjoy (1970), Boyer & Elliott (1982), and Mitra (1986), among others. Thrust faults in fold-thrust belts occur in an overlapping pattern. They generally have shallow dips and concave-upward profiles. The faults cut upward through the stratigraphic succession in the direction of hanging-wall transport, that is, generally towards the foreland. This overall pattern of cutting through the strata includes wide areas where the fault follows the layering ('bedding-glide zones' or 'flats') and narrow areas where the fault transects the layering ('ramps') (Rich 1934).

From Knipe, R. J. & Rutter, E. H. (eds), 1990, *Deformation Mechanisms, Rheology and Tectonics*, Geological Society Special Publication No. 54, pp. 431–444.

Thrusts flatten and merge into a master or sole fault at depth; they bifurcate upwards into splays which constitute either imbricate fans or duplexes. In imbricate fans the splays do not reconnect up-dip; in duplexes they merge up-dip into a roof thrust. Thrusts may die out into folds, both up-dip and along strike; indeed, folding and thrusting are inextricably linked (Rich 1934; Elliott 1976; Suppe 1983; Suppe & Medwedeff 1984; Jamison 1987). The displacement on one fault may be transferred to another through strained but unfaulted rock. While the general geometric properties of thrust systems are well understood, factors which control the nucleation of ramps (e.g. Mandl & Shippam 1981; Bombolakis 1986; Eisenstadt & DePaor 1987) and the formation of duplexes are less certain.

Folds and faults nucleate heterogeneously (serially) and propagate spatially (both across strike and along strike) in fold-thrust belts. The nucleation and propagation patterns are related to the vergence of compression: the structures in general propagate from hinterland to foreland, and the conventional view has been that early-formed structures lock up as younger ones develop at the toe of the fold-thrust belt. The common exception is 'out-of-sequence' thrusting, a process which involves re-activation of old faults or propagation of new faults through already-faulted hinterland portions of the belt. This phenomenon has been explained in the context of the 'critical Coulomb wedge' analysis of Davis *et al.* (1983). This analysis (and sand-box models based on it, e.g. Davis *et al.* 1983; Mulugeta & Koyi 1987; Mulugeta 1988) represents the deforming sedimentary succession as a homogeneous body that exhibits Coulomb rheological behaviour. In the accretionary wedge model, the wedge thickens by imbricate faulting until it attains a tapered profile that allows it to slide stably on its base. If new material accretes at the toe, the taper decreases and the wedge must then undergo internal failure throughout in order to build up the critical taper once again. This means that thrusts must slip simultaneously throughout the wedge if it is to maintain the critical taper. Bedding anisotropy and inter-layer ductility contrasts are omitted, and therefore the buckling mechanism of strain is precluded. The question therefore remains: does bedding affect the propagation and reactivation of faults in a deforming accretionary wedge?

In an on-going program of analogue modelling using the centrifuge technique (Ramberg 1981), we are investigating the three-dimensional geometry and the kinematic evolution of fold-thrust structures. In this paper we describe three models which shed light on some of the problems mentioned above. The models reveal a partitioning of strain among layer-parallel shortening, buckling and thrusting, and show that these processes are of varying significance in different parts of the fold-thrust belt at different times. They demonstrate that thrust ramps through competent units can be localized by an early buckling instability in the stratified model foreland. They confirm that while folds and faults nucleate and propagate serially from hinterland to foreland, early-formed structures in the hinterland portion of the fold-thrust belt continue to develop throughout the shortening. The models develop a classic duplex structure in their lower levels, rather than an imbricate fan; it is argued that the presence of prominent weak horizons in the stratigraphic succession provide a locus for the floor and roof thrusts of the duplex and allow decoupling of the duplexed horizon from the overlying strata.

The models, model materials and scaling

In this paper we describe the results of three fold-thrust experiments (Models TH-20, TH-22 and TH-24). The three models produced very similar results; only TH-24 is illustrated in detail.

The models are constructed from analogue materials: Plasticine (Harbutt's Gold Medal Brand) and silicone putty (Dow Corning dilatant compound 3179) have the appropriate rheological properties to represent, respectively, competent rocks such as limestone and sandstone, and incompetent rocks such as shale, under natural conditions typical of the upper few kilometres of the Earth's crust. The rheological similitude of these materials, as well as the method of construction of finely-laminated multilayers of the two materials, have been discussed in detail elsewhere (Dixon & Summers 1985; Dixon & Tirrul 1990). It will suffice to repeat here that an internally-laminated stratigraphic unit of a desired bulk competency can be prepared by repeated rolling and stacking of an initial couplet of layers of the two materials whose thickness ratio corresponds to the thickness ratio desired for the unit. Competent stratigraphic units have a higher proportion of plasticine, whereas less-competent units contain more silicone putty. Model ratios applicable to the experiments described here are given in Table 1. Note that 1 mm in the model represents 1 km in the prototype. Readers are referred to Ramberg (1981) for a detailed discussion of the theory of scale modelling and the centrifuge technique.

Table 1. *Model ratios applicable to Models TH-20, TH-22 and TH-24*

Quantity	Ratio	Equivalence
length	$l_r = 1.0 \times 10^{-6}$	4 mm in model = 4 km in prototype
specific gravity	$\rho_r = 0.6$	1.60 in model = 2.67 in prototype
acceleration	$a_r = 4.0 \times 10^3$	4000 g in model = 1 g in prototype
time	$t_r = 1.0 \times 10^{-10}$	1 hour in model = 1.15 Ma in prototype
stress	$\sigma_r = \rho_r l_r a_r = 2.4 \times 10^{-3}$	
viscosity	$\mu_r = \sigma_r t_r = 2.4 \times 10^{-13}$	

The dimensions and initial geometric configuration of the models are illustrated in Fig. 1. The three models all consist of a six-part stratigraphic sequence (Fig. 2), three competent and three incompetent units. For convenience these will be referred to as Units I–VI from top to bottom. Competent Units I, III and V each contain four laminae of plasticine of four different colours (yellow, red, black and blue from top to bottom). Incompetent Units II, IV and VI each contain four laminae, two of silicone putty and two of black plasticine in a 2:1 ratio of thickness. The competent units are each 1.0 mm thick, and the incompetent units measure 0.33 mm; the total thickness of the model stratigraphic pile is 4 mm.

The models contain stratigraphic units that simulate rock units such as limestone and shale with prototype thicknesses of 1000 m and 330 m, respectively. The units are internally laminated so that they simulate to some extent the bedding anisotropy of natural sedimentary units. By deforming the models in a centrifuge the natural vertical lithostatic gradient is simulated. This has a small effect on the rheological behaviour of the model materials:

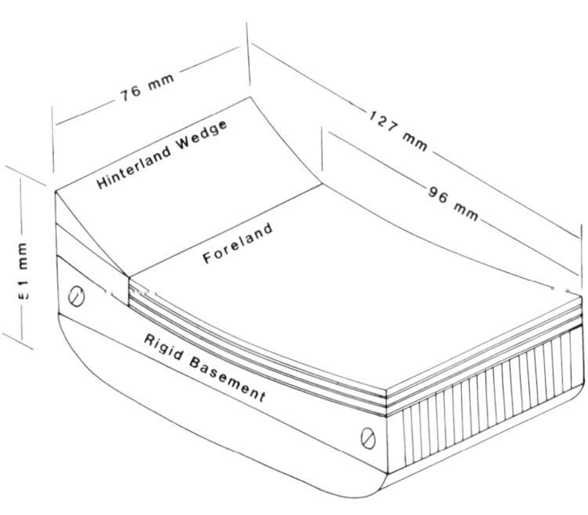

Fig. 1. Dimensions and initial internal configuration of models TH-20, TH-22 and TH-24.

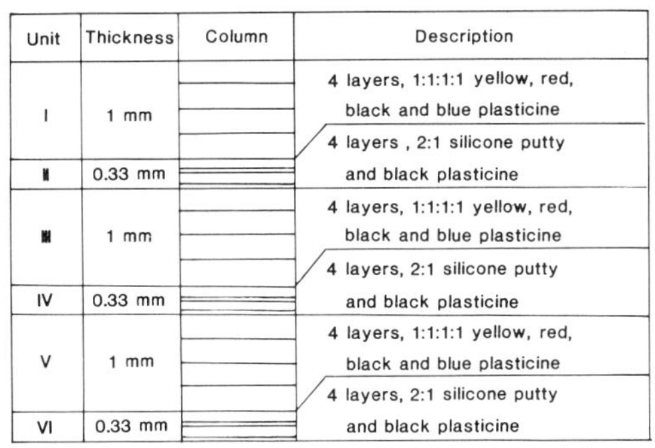

Fig. 2. Stratigraphic column of models TH-20, TH-22 and TH-24.

preliminary triaxial testing of plasticine at constant strain-rate indicates that its shear strength increases by a small amount (about 25% (from approximately 20–25 kPa) for Harbutts Gold Medal black and 50% (from approximately 12–17 kPa) for Harbutts Gold Medal blue) as the confining pressure increases from 0–400 kPa (for strain rates in the range $1.6 \times 10^{-5}\ s^{-1}$ and $2.5 \times 10^{-4}\ s^{-1}$, and stress measured at 10% strain). It is not known whether increased confining pressure promotes ductile flow or localized shear failure, although, as will be seen below, faulting is favoured over folding at deeper levels in the models. At 4000 g the confining pressure at the base of the 4 mm thick model stratigraphic pile is 250 kPa (specific gravity of plasticine is 1.70; of silicone putty, 1.12) so the pressure effect is probably negligible, given the other uncertainties and approximations involved in analogue scale modelling. In the model study described here we have made no attempt to simulate temperature or pore-pressure effects. An analysis of the magnitude of the stress applied to the edge of fold–thrust models similar to those discussed here is given in Dixon & Tirrul (1990).

The model stratigraphic pile represents a prototype which has six-fold stratigraphy with alternating competent and incompetent units.

An example of such a sequence is provided by the succession in the Appalachian Valley and Ridge Province in West Virginia, Virginia and Maryland (Perry 1978; Kulander & Dean 1986).

Deformation history

The models were deformed in the large centrifuge in the Experimental Tectonics Laboratory at Queen's University (Dixon & Summers 1985). They were subjected to horizontal compression from one end by gravitational collapse and spreading of a 'hinterland wedge' of plasticine (see Dixon & Tirrul 1990) for a detailed discussion of this mechanism of shortening). The wedge began to fail when the centripetal acceleration in the centrifuge reached about 2500 g. As the wedge collapses it loses gravitational potential; hence, the acceleration must be gradually increased to maintain the spreading. The experiments were run in stages so that the progressive development of structures could be monitored. The models were sectioned, photographed and then reassembled for a subsequent stage of deformation. For each new stage the driving wedge was rejuvenated by addition of a tapered slice of plasticine to its upper surface. Model TH-24, described in this paper, was run through six stages. The run time for each stage is shown in Table 2.

As mentioned above, the deformation at each stage commenced during the run-up phase (2500 g is reached at about 180 seconds of run-up). The deformation at each stage thus includes about 100 seconds of run-up in addition to the time at 4000 g as listed in Table 2. Thus, total deformation time for model TH-24 was about 1560 seconds, corresponding roughly to 0.5 Ma in the prototype. This may be about an order of magnitude faster than the evolution of natural fold-thrust belts. The high rate likely arises either because the basal decollement horizon (unit VI) is thicker and weaker than it should be in relation to the full stratigraphic pile, or because we have used a hinterland wedge that is steeper than it ought to be, given the thin (4 mm) stratigraphic pile. In view of the other uncertainties involved, this degree of scaling accuracy is deemed to be satisfactory for our present purposes (we are not attempting to predict the rate of evolution of thrust belts).

Previous models deformed by the mechanism used here (Dixon & Summers 1985, Dixon & Tirrul 1990) have experienced drag against the sides of the model chamber. This drag has a great effect on the deformation of the models especially when the shortening is large. It produces an arcuate (convex towards the foreland) shape of folds and faults, and causes extension parallel to the hinges of folds. In the present study we have eliminated the side drag by lubricating the sides of the models with petroleum jelly (vaseline) and wrapping the sides with polyethylene film. The same procedure was followed at each stage. This method successfully eliminated side drag. The front of the spreading wedge advanced uniformly, folds and thrust faults developed straight rather than convex traces (see below), and the deformation approximated plane-strain conditions with virtually no extension parallel to fold hinges. The wedge front advanced as a vertical, straight 'bulldozer blade', very like that in the conceptual model of Davis *et al.* (1983), although it tended to ride over a small part of the stratigraphic pile at advanced stages of deformation (see below, Fig. 4).

Surface deformation pattern

The nucleation and propagation of folds on the free surface of model TH-24 through six stages of deformation is shown in Fig. 3. Folding first nucleated at the hinterland end and the fold train propagated towards the foreland. The folds had straight hinges and were relatively cylindrical, although some branching and en-echelon patterns were evident. These patterns developed because folds nucleated at isolated points and propagated along their hinge lines. Folds

Table 2. *Run records for model TH-24*

Stage	Run-up seconds	Time at 4000 g, seconds	Run-down, seconds
I	280	320	400
II	280	320	400
III	280	80	400
IV	280	80	400
V	280	80	400
VI	280	80	400

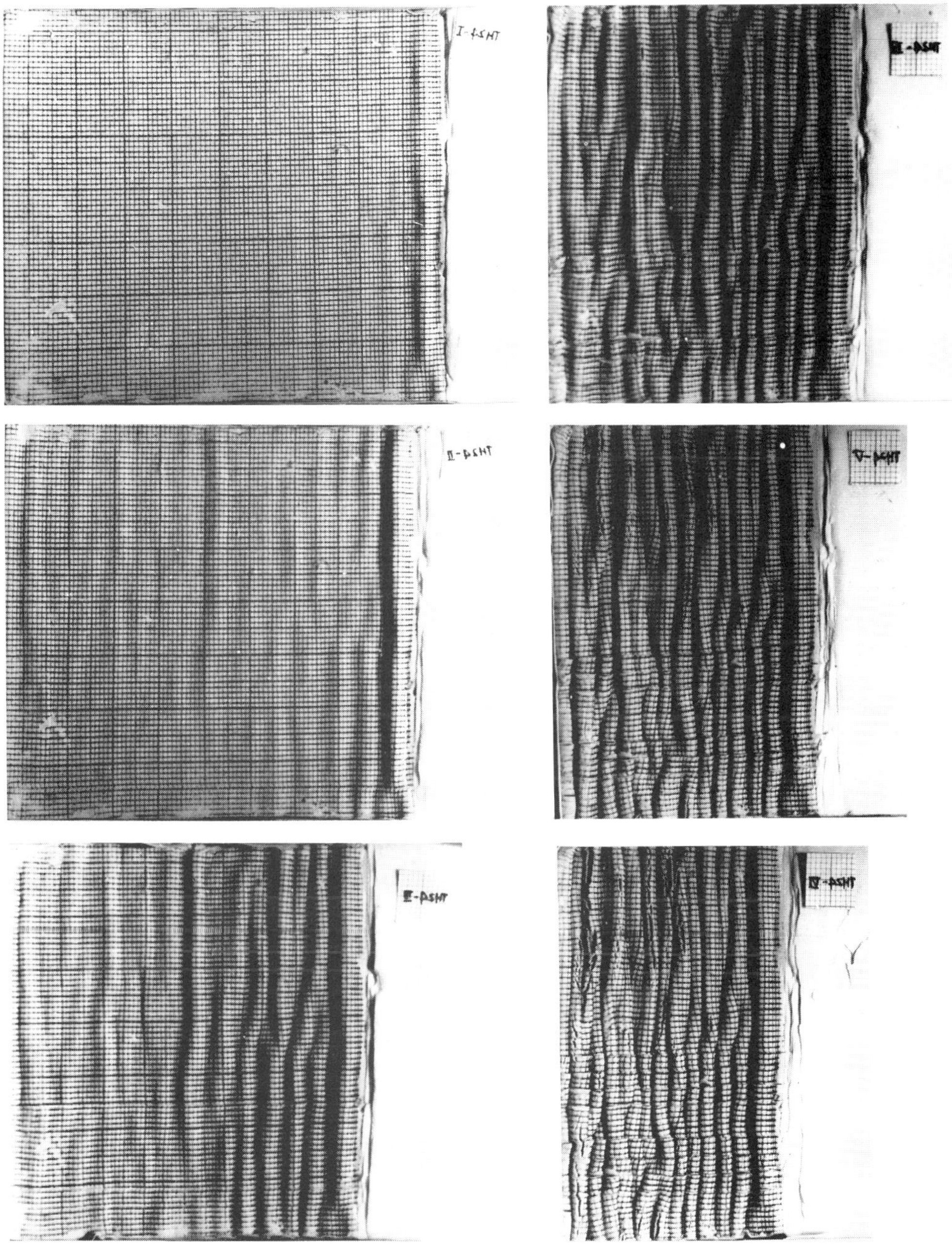

Fig. 3. Progressive deformation of the free surface of model TH-24 through stages I to VI. The model is compressed from the right-hand end by a gravitationally-spreading 'hinterland wedge' of plasticine (not shown in the photos. The model surface is illuminated from the right to enhance surface relief. The photos are printed backwards so that the models are shown in the same orientation (compression from the right) as in Fig. 4. The small labels measure 10 mm × 10 mm.

which propagated towards each other did not necessarily meet precisely. By stage IV, the folding had propagated across the full width of the model. Some fold crests were locally ruptured by extension fractures striking parallel to the fold hinges. As shortening proceeded (stages V and VI), the isolated fractures propagated along the fold hinges to form longer fractures.

Internal deformation in section view

The progressive evolution of model TH-24 is shown in profile view (vertical sections cut parallel to the shortening direction) in the photographs in Fig. 4. For each section the Roman numeral refers to the model stage, and the Arabic number indicates the position of the line of section relative to one edge of the model (distance in inches). The reproducibility of the model structures can be gauged by comparing sections of models TH-20 and TH-22 (Fig. 5) with sections of model TH-24 in Fig. 4. The sections of model TH-24 are reproduced as line drawings in Fig. 6 for clarity.

In the early stages of shortening (Figs 4 and 6, stages I-III) the deformation was characterized by harmonic buckle-fold trains in the competent units and localized groups of small folds in incompetent units. The small folds in the incompetent units were localized beneath the anticlines in the competent units (as in Dixon & Tirrul 1990). The folds nucleated serially, from hinterland towards foreland, and grew progressively in amplitude. During stage II, three thrusts ramped upward across the competent unit V; their displacements decreased from hinterland towards foreland, suggesting serial nucleation and progressive growth. At this stage, the localized groups of small folds in the incompetent units were evenly distributed through all the section, and the spacing of these fold groups was fairly constant in unit II and unit IV, but not in unit VI, which displayed an increasing spacing toward foreland (Fig. 7). This may be caused by somewhat heterogeneous slip along the rigid basement and the low amplitude of buckling in unit V which is not pronounced enough to localize buckling in unit VI with a regular group wavelength.

During stage III, the folds continued to increase in amplitude in unit I and unit III. New thrusts propagated across unit V, on the foreland side of those ramps formed at earlier stages, but the old thrusts continue to increase their displacement. This is even clearer if we compare the section of stage III with the one of stage IV (Figs 4 and 6A): the displacements of thrusts a3, b3, c3 and d3 (labelled in Fig. 6A) change from 1.4 mm, 0.4 mm, 0.2 mm and 0.2 mm at stage III to 3.6 mm, 1.4 mm, 0.2 mm and 0.8 mm at stage IV, respectively, even though four new hinterland-dipping thrusts have developed in front of these four thrusts at stage IV.

Figure 8 shows the accumulated displacements of thrust ramps a3 and b3 through the six stages. It is clear that these dislocations never ceased to be active; they developed continuously through all the stages. It is true that the thrusts developed successively, one by one, toward the foreland, but an older thrust does not die out as a new thrust forms in front of it. This is not in agreement with the simple conception that imbricate thrusts develop serially with early-formed thrusts becoming inactive as later-formed ones develop towards the foreland. On the other hand, the continued accumulation of displacement on early-formed faults within the older, hinterland end of the fold-thrust belt is in agreement with the critical Coulomb wedge model of Davis *et al.* (1983). No out-of-sequence foreland-verging thrusting could be detected in the lowest competent unit, although this did occur in unit III (compare stage IV with stage III in Fig. 6A).

Another remarkable character of the thrusts shown in the model is that the displacements of thrusts at stage VI did not decrease systematically from hinterland towards the foreland. Some thrusts in the far foreland side (e.g. ramp i3 in stage VI, Fig. 6A) showed larger displacement than other thrusts closer to the hinterland side. This variability can be attributed to heterogeneous buckling strain in the overlying units: unit-V thrusts with anomalously large displacement (ramps d3 and i3, Fig. 6A) are spatially associated with anomalously large anticlines in unit III. However, one could argue that this is a chicken vs. egg problem: is it instead the anomalously large fault slips which cause the large folds? Both possibilities can be defended. Ramp i3 nucleated in a large fold in unit V between stage IV and stage V (see Fig. 6A) at a site which at stage IV was overlain by a unit-III anticline of modest amplitude. By stage V this unit-III fold had grown dramatically. It would appear that the faulting allowed the overlying fold to grow. In contrast, ramp d3 nucleated at stage III beneath a unit-III fold of large size. Here it appears that the large fold accommodated anomalous slip on the underlying fault.

The incompetent unit IV played an important role as a decollement horizon (see Fig. 6B). It is a boundary between domains with contrasting structural style. Unit V is dominated by faulting, whereas that of the overlying units III and I are dominated by buckle-folding. It is also the roof

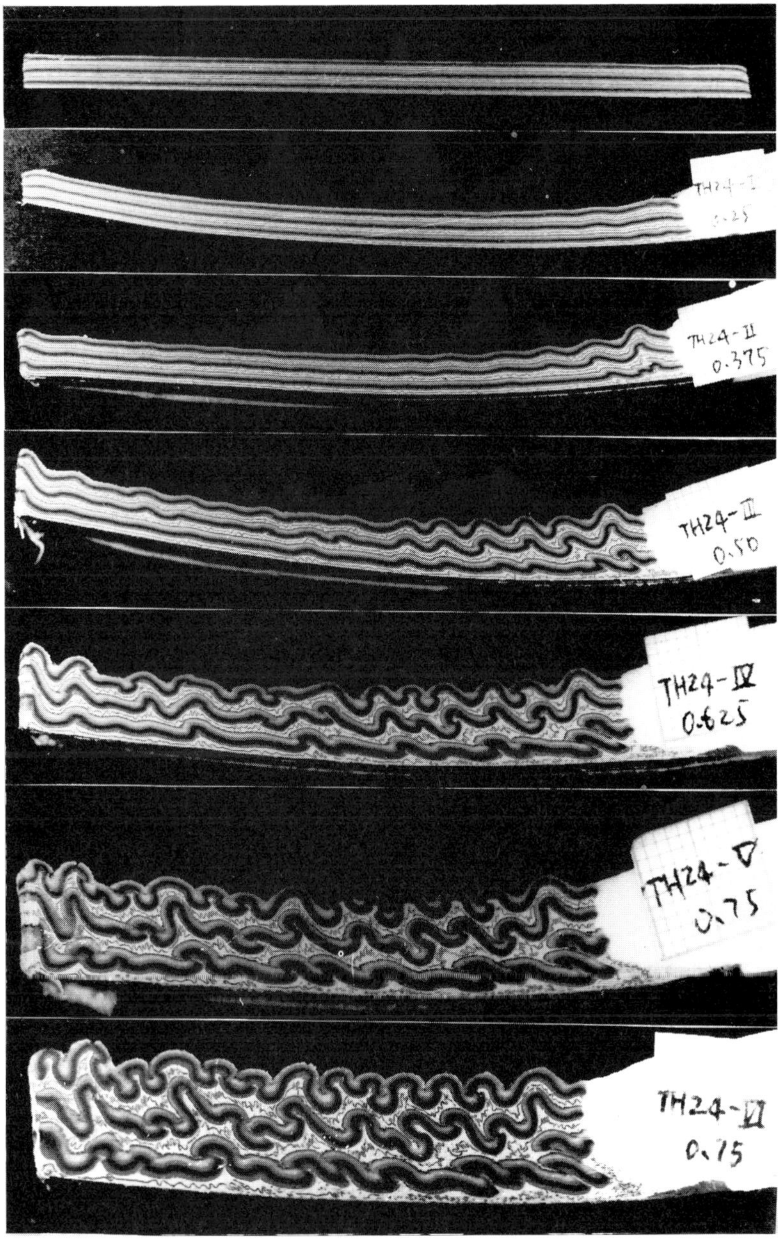

Fig. 4. Vertical sections through model TH-24 at stages I to VI. The positions of the sections relative to the lower edge of the model (as oriented in Fig. 3) are indicated in inches on the labels. The labels measure 10 mm × 10 mm.

of a duplex structure. The fault ramps that cut through unit V merged down-dip into a basal decollement in the lowest incompetent unit VI. In the up-dip (foreland) direction, these faults also merged into a roof thrust within incompetent unit IV. Thus, the faults which ramp through unit V (Fig. 6B) constitute a blind duplex structure rather than an imbricate fan. Although overlying competent units were cut by fault ramps, these faults were not linked

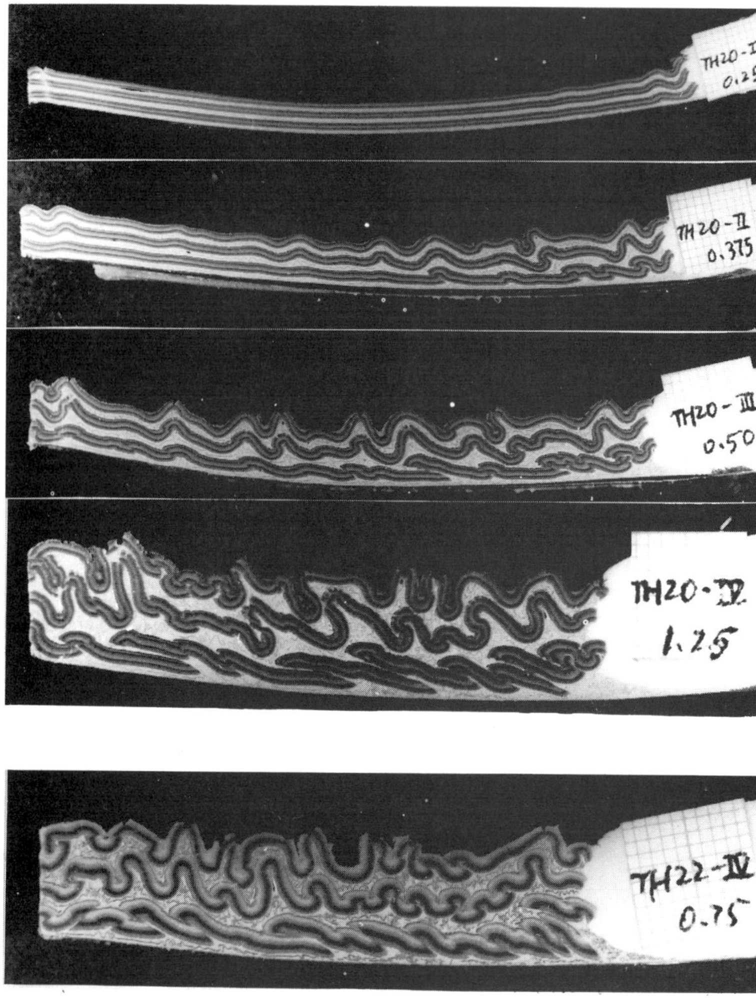

Fig. 5. Vertical sections through models TH-20 and TH-22, for comparison with model TH-24 (Fig. 4) to indicate the degree of reproductibility of the model structures. Stage IV in these models is equivalent to stage VI in TH-24.

directly with those in unit V (but see below). The presence of the prominent incompetent unit IV provided a decollement horizon for the roof thrust of the unit-V duplex, and apparently inhibited further upward propagation of the faults as separate imbricates.

The structural succession strongly resembles that of the Appalachian Valley and Ridge province of West Virginia where the Cambro-Ordovician carbonates are duplexed between a floor thrust in the Lower Cambrian Waynesboro (Rome) Formation and a roof thrust in the Middle Ordovician Martinsburg Formation (Kulander and Dean 1986; see Fig. 9). Geiser (1988) and Ferrill & Dunne (1989) have discussed accommodation of slip on duplex structures by fore- and back-thrusting, folding and layer-parallel shortening in the lower strata. These mechanisms were all evident in the model described here.

In the upper two competent units (I and III), nearly all the faults were clearly formed at a relatively late stage through the fore-limbs of folds which had previously grown to high amplitude. The faults resemble Heim's (1921) 'stretch thrusts' (see also Dixon & Tirrul 1990). A few faults also formed through the back limbs of folds. In unit V, faults were also related to buckle-folds, but the faults nucleated at a much earlier stage of the folding. After these faults

A)

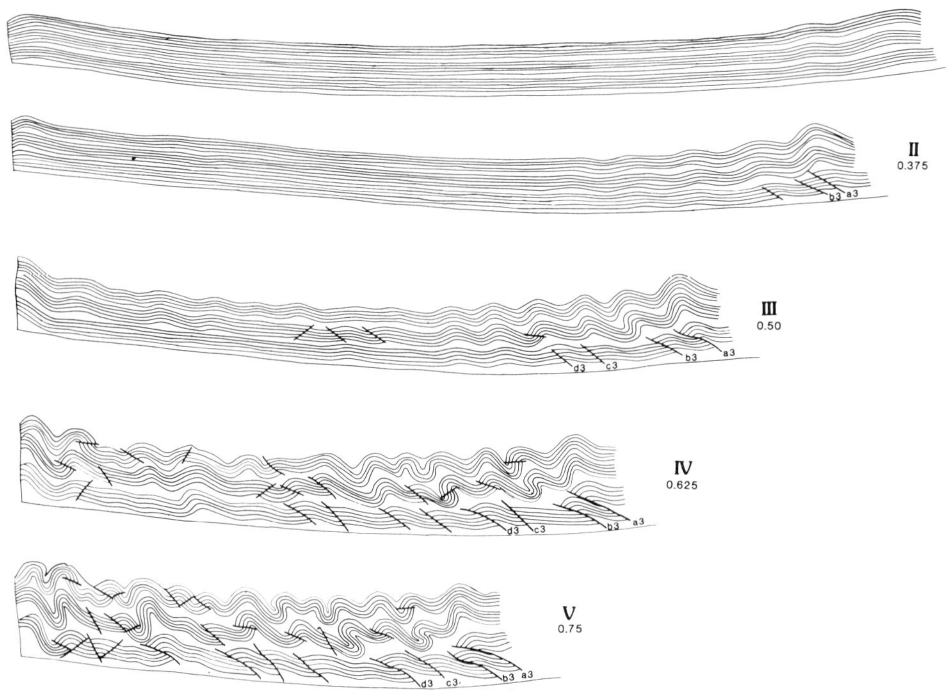

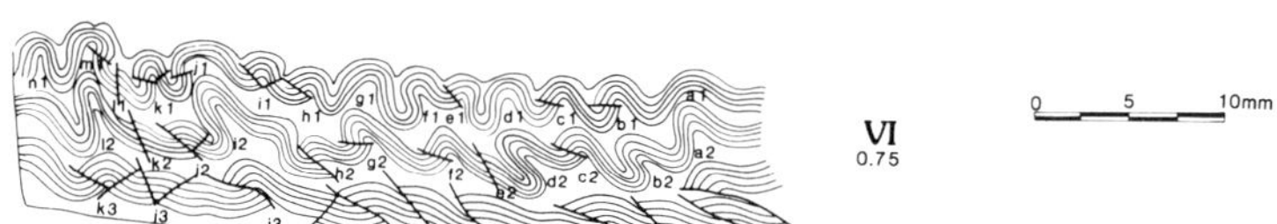

B)

Fig. 6. (**A**) Line drawings of the sections of model TH-24 shown in Fig. 4. Roman numerals indicate model stage (I to VI); Arabic numbers give positions (in inches) of the sections relative to one edge of the model (as in Fig. 4). Individual fault ramps are numbered a3, b3, c3 etc. (see text). (**B**) Section 0.75 of model TH-24, stage VI (as in part A), with heavy dashed lines added to indicate the trajectories of floor and roof thrusts associated with the unit-V duplex.

formed, the folds were modified in shape by fault-bend folding.

The basal decollement (unit VI) was not much weaker than the bulk of the stratigraphic pile, so the maximum stress trajectories sloped downward towards the foreland (Chapple 1978) in the lower parts of the model. This favours foreland-verging thrusting (Tirrul 1983) through foreland-dipping limbs of folds relative to back-thrusting through hinterland-dipping limbs. At higher levels and at the foreland end of the model, the axis of maximum compression is more nearly parallel to the layering and fore- and back-thrusting are more equally favoured.

Restored sections and strain partitioning

Natural fold-thrust belts are well-known to involve shortening by three different deformation mechanisms: buckling, faulting and layer-parallel shortening (see, e.g., Kulander & Dean 1986; Simon & Gray 1982). Mulugeta &

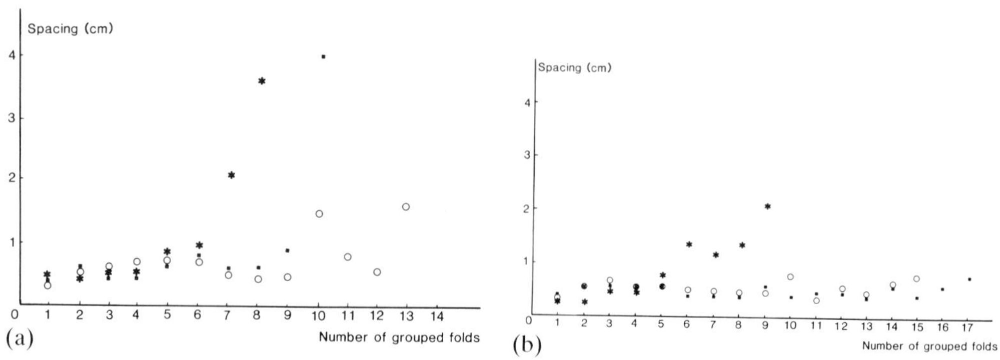

Fig. 7. Spacing of groups of small folds in the incompetent units II (squares), IV (circles) and VI (asterisks) at stage I (**a**) and stage II (**b**). The fold groups are numbered in order from the hinterland towards the foreland. Note that the spacing is relatively constant within units II and IV, but that it increases towards the foreland in unit VI.

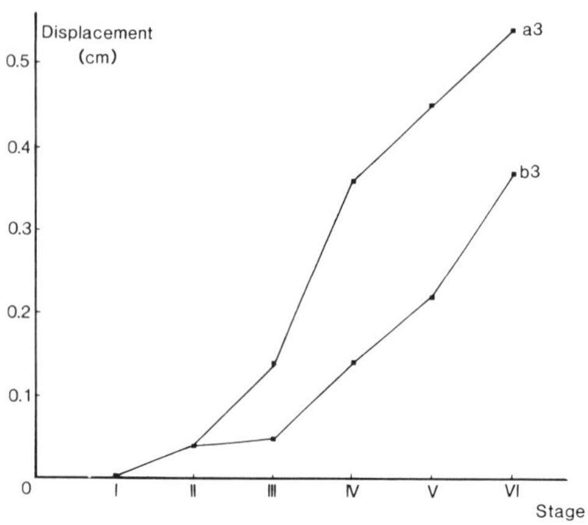

Fig. 8. Progressive increase in displacement for fault ramps a3 and b3 (see Fig. 6A for location) through stages I to VI.

Koyi (1987) have analysed the partitioning of shortening strain among these mechanisms in a sand model. Likewise, the shortening strain in model TH-24 involved these three deformation mechanisms, and we follow Mulugeta & Koyi's (1987) approach in analysing the partitioning. By measuring the amount of slip on faults and the length of beds around folds, we have constructed restored sections (Fig. 10) of the same model slices from stages I to VI that are shown in Fig. 6. The differences between the restored length and the known original length of each unit reflects layer-parallel shortening. Careful measurement of the thickness of the deformed units confirmed that constant-area balance was maintained (and we know that this must be, for the model deforms at constant volume, the model fold belt has a fixed length along strike, and the folds are straight--there is no hinge-parallel stretching). The amount of shortening accommodated by each of the three mechanisms is tabulated for each competent unit at stages I to VI in Table 3.

In the earlier stages of deformation, most of the strain was accommodated by layer-parallel shortening. After stage IV, this mechanism decline in importance, although it still accounts for a large percentage of total shortening. For example, for unit V, at stage VI the total shortening had reached 59%, comprising 34% by layer-parallel shortening and 25% by folding and faulting combined. In this unit, layer-parallel shortening had stopped by stage V, and subsequent shortening was entirely accomplished by folding and faulting. Thus, the different mechanisms vary in importance at a particular level as the deformation progresses. The mechanisms also vary in relative importance at different structural levels. For example, at stage VI, buckle shortening of units I, III and V was 27%, 28% and 13%, respectively, while shortening of these same units by faulting was 2.3%, 2.7% and 11.5%. Layer-parallel shortening was relatively uniform at 30%, 29% and 34%, respectively.

Faulting in unit V started when the total shortening was about 15% whereas it did not begin in units III and I until total shortening reached about 30% and 40%, respectively. This difference documents decollement within units II and IV.

The magnitude of total shortening in the model to the end of stage VI is very similar to the total shortening within the Central Appalachian Valley and Ridge Province of West Virginia and Virginia. Kulander & Dean (1986)

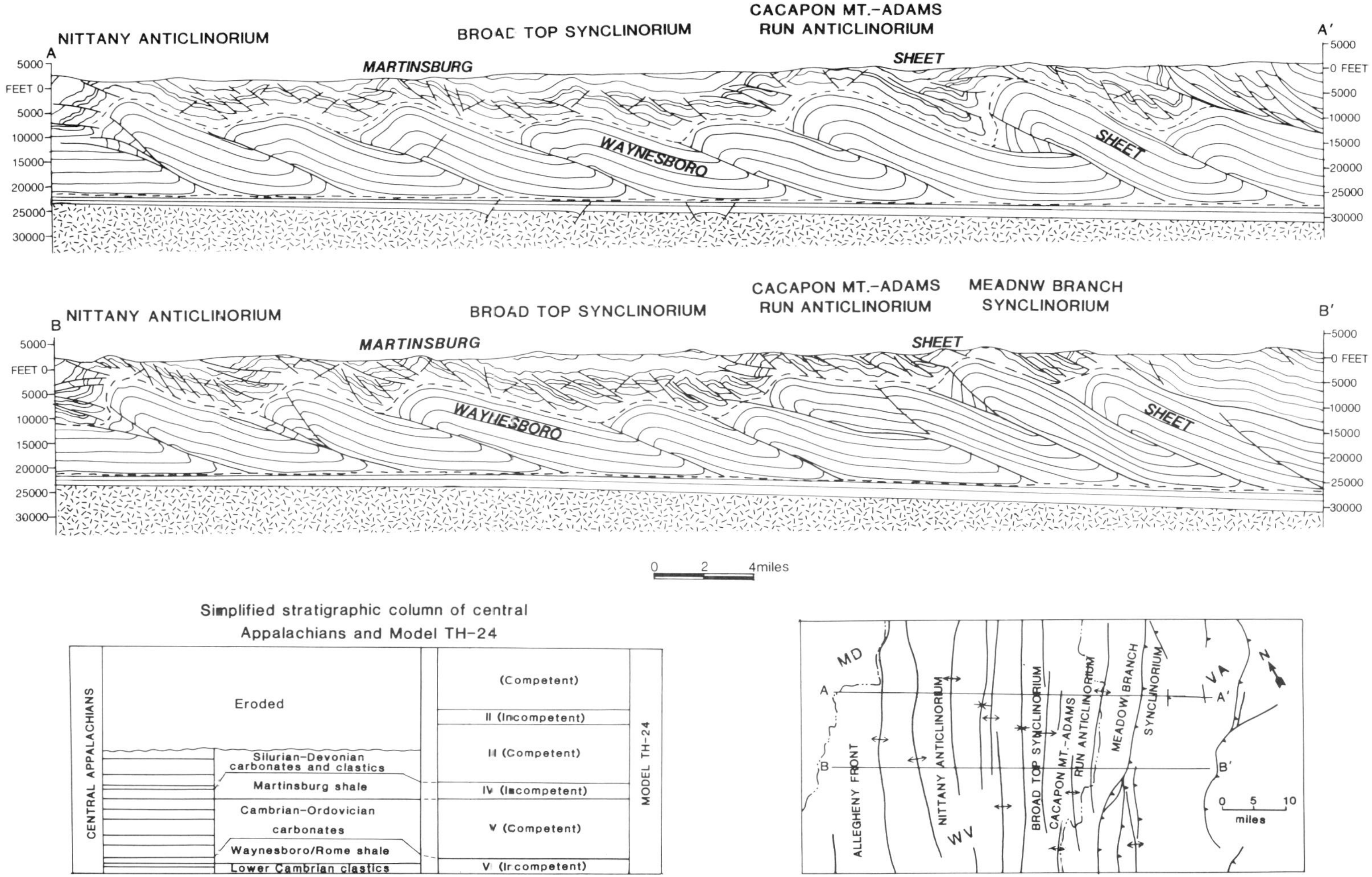

Fig. 9. Structural cross-sections through part of the Appalachian Valley and Ridge structural province, Virginia and West Virginia, USA (from fig. 6, Sections 2 and 3 of Kulander & Dean (1986)). Inset at lower right shows locations of cross-section lines A and B. The Waynesboro Sheet comprises Cambro-Ordovician carbonate strata and has a high bulk competency. The Martinsburg Sheet comprises Siluro-Devonian carbonates, also with a high competency. A basal decollement horizon (lower dashed line) follows the shale of the Lower Cambrian Waynesboro/Rome Formation. A second decollement horizon (upper dashed line) in the Upper Ordovician Martinsburg Formation separates the Waynesboro and Martinsburg sheets. The stratigraphic sequence in these sections is represented in models TH-20, TH-22 and TH-24 by units III (Martinsburg sheet), IV (Martinsburg Fm.), V (Waynesboro sheet), and VI (Waynesboro/Rome Fm), as shown in the inset at the lower left.

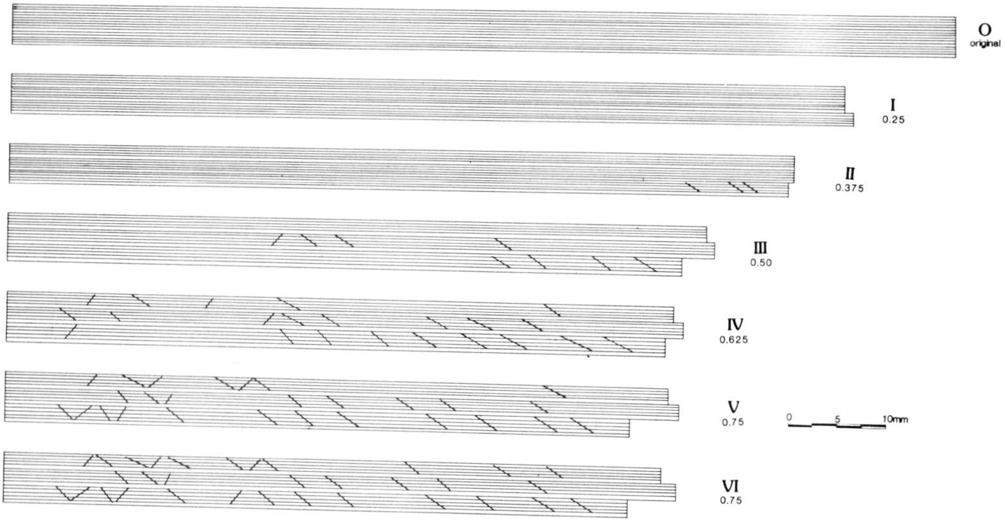

Fig. 10. Restored sections of model TH-24 (same sections as shown in Figs 4 and 6A). Labels as in Fig. 6. The short diagonal lines indicate positions of fault ramps in the competent units. The restoration takes account of fault slip and buckle-shortening. The difference in length between sections at stages I to VI and the undeformed state (section 0) is a manifestation of layer-parallel shortening.

Table 3. *Horizontal shortening and strain partitioning in Model TH-24*

	Unit I				Unit III				UnitV			
Stage	Total %	Fault %	Fold %	LPS %	Total %	Fault %	Fold %	LPS %	Total %	Fault %	Fold %	LPS %
I	12.9	0.0	1.2	11.7	12.9	0.0	1.2	11.7	12.5	0.0	1.4	11.1
II	19.6	0.0	2.9	16.7	19.6	0.0	2.9	16.7	18.3	0.7	1.5	16.1
III	33.1	0.0	7.1	26.0	32.7	0.3	7.4	25.0	31.5	2.0	1.8	27.7
IV	42.9	1.0	12.7	29.2	42.7	1.8	12.8	28.1	41.7	6.5	5.0	30.2
V	53.8	1.4	22.8	29.6	53.8	2.4	22.9	28.5	53.3	10.5	8.9	33.9
VI	59.4	2.3	26.9	30.2	59.4	2.7	28.0	28.7	58.8	11.5	13.4	33.9

estimated shortening of Cambro-Ordovician carbonates (modelled by unit V in model TH-24) by thrusting and folding in the range 24% to 32%, and not including layer-parallel shortening by plastic or pressure-solution deformation mechanisms. They state that layer-parallel shortening reached 30% at some locations.

The restored sections (Fig. 10) reveal a striking alignment of the fault ramps among the three competent units. The restored positions are aligned even though the faults did not all nucleate at the same stage of deformation, and they nucleated at positions that would have been shifted differentially by heterogeneous buckling, layer-parallel shortening and earlier faulting. As was discussed above, the fault ramps in one competent unit did not link to ramps in adjacent units; rather, they merged into decollement horizons in the incompetent units. Therefore, the apparent alignment of ramps between adjacent units must be due to localization of the ramps in the limbs of earlier-formed buckle folds which are harmonic through the full stratigraphic succession at their time of nucleation.

Conclusions

The analogue models described here contained a six-fold stratigraphic succession, with units that alternated in bulk competence (weak at the base and strong at the surface). Folds and faults nucleated and propagated from hinterland to foreland but early-formed faults at the hinterland end of the fold-thrust belt continued to grow as the deformation progressed. As the fold–thrust belt developed by propagation into undeformed foreland, the hinterland region continued to thicken by folding, faulting and layer-parallel shortening in order to maintain a critical taper. The structural style varied with depth: foreland-verging thrusts imbricated the bottom competent unit into a blind duplex which was separated from overlying units that were strongly buckled and only weakly faulted. Faults in the upper levels developed at late stages, after folds had grown to high amplitude. In contrast, the faults that bound horses in the duplex developed at an early stage, and were localized in the foreland-dipping limbs of low-amplitude buckle-folds. Bedding thus strongly influences the location of nucleation of faults.

Horizontal shortening strain in the models is partitioned among three deformation mechanisms: layer-parallel shortening, buckling and thrust faulting. The relative magnitudes of these mechanisms varied with depth at different stages of deformation, and they varied in time at a given structural level. Layer-parallel shortening was dominant at early stages but was gradually replaced by the other mechanisms. Buckling dominated over faulting at high levels where the compression was parallel to the layering; foreland-verging thrusting dominates over folding at depth where maximum principal stress trajectories slope towards the foreland due to basal drag.

The models developed a 'blind' duplex in the lowest competent unit. The ramp faults of the duplex merged into a roof thrust that was localized in a prominent weak horizon. This horizon was a boundary of differential shortening, accommodating different mechanisms above and below, and allowing previously-harmonic folds to become disharmonic. Because the folds became disharmonic, thrust ramps localized in forward-dipping fold limbs in the underlying competent unit could not readily propagate through aligned fold limbs in the overlying competent unit. Again, we see the influence of mechanical layering over the development of fold-thrust structures.

The models reproduced structural relationships exhibited by the Central Appalachian Valley and Ridge Province of West Virginia and Virginia, including the large-scale blind duplex of Cambrian–Ordovician carbonates beneath a roof thrust in the Martinsburg shale and the accommodation of duplex shortening by a combination of fore- and back-thrusting, buckling and layer-parallel shortening in the overlying cover strata. The models provide insight into the patterns of temporal evolution of these structures.

The work reported here constitutes part of the PhD research of S. Liu, who acknowledges with thanks an R. S. McLaughlin Fellowship from Queen's University. Our investigation of fold-thrust tectonics is supported by a grant from Exploration Research, ARCO Oil and Gas Company. The construction and operation of the Experimental Tectonics Laboratory at Queen's University has been supported by grants from the Natural Sciences and Engineering Research Council of Canada to J. M. Dixon. The manuscript has benefitted from constructive reviews by C. J. Talbot and an anonymous referee.

References

BALLY, A. W., GORDEY, P. L. & STEWART, G. A. 1966. Structure, seismic data and orogenic evolution of southern Canadian Rocky Mountains. *Bulletin of Canadian Petroleum Geology*, **14**, 337–381.

BOYER, S. E. & ELLIOTT, D. 1982. Thrust systems. *American Association of Petroleum Geologists Bulletin*, **66**, 1196–1230.

BOMBOLAKIS, E. G. 1986. Thrust-fault mechanics and the origin of a frontal ramp. *Journal of Structural Geology*, **8**, 281–290.

CHAPPLE, W. M. 1978. Mechanics of thin-skinned fold-and-thrust belts. *Geological Society of America Bulletin*, **89**, 1189–1198.

DAHLSTROM, C. D. A. 1969. Balanced cross sections. *Canadian Journal of Earth Sciences*, **6**, 743–757.

DAVIS, D., SUPPE, J. & DAHLEN, F. A. 1983. Mechanics of fold-and-thrust belts and accretionary wedges. *Journal of Geophysical Research*, **88**, 1153–1172.

DIXON, J. M. & SUMMERS, J. M. 1985. Recent developments in centrifuge modelling of tectonic processes: equipment, model construction techniques and rheology of model materials. *Journal of Structural Geology*, **7**, 83–102.

—— & TIRRUL, R. 1990. Centrifuge modelling of fold-thrust structures in a tripartite stratigraphic succession. *Journal of Structural Geology* (In press).

EISENSTADT, G. & DEPAOR, D. C. 1987. Alternative model of thrust-fault propagation. *Geology*, **15**, 630–633.

ELLIOTT, D. 1976. The energy balance and deformation mechanisms of thrust sheets. *Philosophical Transactions of the Royal Society of London*, **A283**, 289–312.

FERRILL, D. A. & DUNNE, W. M. 1989. Cover deformation above a blind duplex: an example from West Virginia, U.S.A. *Journal of Structural Geology*, **11**, 421–431.

GEISER, P. A. 1988. Mechanisms of thrust propagation: some examples and implications for the analysis of overthrust terranes. *Journal of Structural Geology*, **10**, 829–845.

HEIM, A. 1921. *Geologie der Schweiz, II. Die Schweizer Alpen*. Tauchnitz, Leipzig.

JAMISON, W. R. 1987. Geometric analysis of fold development in overthrust terranes. *Journal of Structural Geology*, **9**, 207–219.

KULANDER, B. R. & DEAN, S. L. 1986. Structure and tectonics of Central and Southern Appalachian Valley and Ridge and Plateau Provinces, West Virginia and Virginia. *American Association of Petroleum Geologists Bulletin*, **70**, 1674–1684.

MANDL, G. & SHIPPAM, G. K. 1981. Mechanical model of thrust sheet gliding and imbrication, *In*: MCCLAY, K. & PRICE, N. J. (eds) *Thrust and Nappe Tectonics*, Geological Society, London, Special Publication, **9**, 79–98.

MITRA, S. 1986. Duplex structures and imbricate thrust systems: geometry, structural position, and hydrocarbon potential. *American Association of Petroleum Geologists Bulletin*, **70**, 1087–1112.

MULUGETA, G. 1988. Modelling the geometry of Coulomb thrust wedges. *Journal of Structural Geology*, **10**, 847–859.

MULUGETA, G. & KOYI, H. 1987. Three-dimensional geometry and kinematics of experimental piggy-back thrusting. *Geology*, **15**, 1052–1056.

PERRY, W. J., JR. 1978. Sequential deformation in the central Appalachians. *American Journal of Science*, **278**, 518–542.

PRICE, R. A. & MOUNTJOY, E. W. 1970. Geologic structure of the Canadian Rocky Mountains between Bow and Athabasca Rivers--a progress report. *Geological Association of Canada Special Paper*, **6**, 7–25.

RAMBERG, H. 1981. *Gravity, Deformation and the Earth's Crust (Second Edition)*. London: Academic Press.

RICH, J. L. 1934. Mechanics of low-angle overthrust faulting as illustrated by the Cumberland thrust block. *American Association of Petroleum Geologists Bulletin*, **18, 1584–1596.**

SIMON, R. I. & GRAY, D. R. 1982. Interrelations of mesoscopic structures and strain across a small regional fold, Virginia Appalachians. *Journal of Structural Geology*, **4**, 271–289.

SUPPE, J. 1983. Geometry and kinematics of fault-bend folding. *American Journal of Science*, **283**, 684–721.

—— & MEDWEDEFF, D. A. 1984. Fault-propagation folding. *Geological Society of America Program with Abstracts*, **16**, 670.

TIRRUL, R. 1983. Structure cross-sections across Asiak foreland thrust and fold belt, Wopmay Orogen, District of Mackenzie. *Current Research, Part B, Geological Survey of Canada. Paper* **83–1B**, 253–260.

Deformation mechanics in analogue models of extensional fault systems

K. R. McCLAY

Department of Geology, Royal Holloway and Bedford New College, University of London, Egham, Surrey TW20 0EX, UK

Abstract: Scaled analogue models of fault structures are powerful tools for investigating the progressive development of extensional deformation. The results of a wide ranging programme of extensional modelling are briefly reviewed and are analysed in terms of the deformation mechanics and rheological behaviour of the modelling materials. Extensional models have been constructed using homogeneous dry quartz sand with isotropic behaviour; sand and dry china clay mixtures; with behaviour controlled by competency; and sand with dry vermiculite mica flakes to simulate anisotropic systems. Rheological tests show that these materials deform by Navier–Coulomb failure with friction angles between 30° and 35°. Extensional faults developed within the models are dilatant granular shear zones, whose widths are grain-size dependent, being widest in the sand models and narrowest in faults cutting the fine-grained clay layers. Initial fault–bedding cutoff angles are high, 60°–70°, but during progressive deformation shear strains may change this, depending upon the boundary conditions of the model. Measurements of angular shear strains within domino fault arrays show significant zones of shear either parallel to the basal detachment or parallel to bounding faults. Small bulk dilations are found in most models as a result of grain packing rearrangements during extension. The primary extensional architecture of the models is controlled by the underlying detachment configuration. Despite limitations to the models, it is believed that the rheologies, the modelling materials and the deformation mechanics within the models realistically simulate brittle deformation of sedimentary rocks in the upper crust.

Scaled analogue models have provided graphic new insights into the geometries and progressive evolution of brittle extensional fault systems (e.g. McClay & Ellis 1987 *a*, *b*; Vendeville *et al.* 1987; Vendeville & Cobbold 1988; Ellis & McClay 1988). Previous papers have largely focussed upon extensional fault geometries, fault sequences and progressive fault evolution. In this paper, attention is directed to the mechanics of deformation and rheologies of the model materials in order that the behaviour of the experimental models may be compared with that of sedimentary rocks in the upper crust.

The analogue modelling programme at RHBNC has, to date, largely focussed upon the development of brittle extensional fault systems in the upper 10 km of the Earth's crust. Under these conditions sedimentary rocks tend to have uniaxial compressive strengths in the order of 10–100 MPa and uniaxial tensile strengths typically 10 to 20 times lower (Paterson 1978). In order, therefore, to generate accurately scaled analogue models that simulate the extensional deformations of brittle sedimentary rocks, cohesion to near cohesionless materials must be used. (cf. Horsfield 1977; Naylor *et al.* 1986). Dry quartz sand is such a material and has been widely used to model the deformation of brittle sedimentary rock (e.g. Ellis & McClay 1988; Vendeville *et al.* 1987; Vendeville & Cobbold 1988; Horsfield 1977; Naylor *et al.* 1986; Sanford 1959). Dry quartz sand models are largely isotropic, whereas most rock sequences have degrees of anisotropy and competency. More recently, in addition to isotropic quartz sand, analogue model experiments have been constructed from sand together with dry china clay and dry vermiculite mica in order to simulate competency contrasts and anisotropy respectively (e.g. Ellis & McClay 1988). Here the rheological properties of sand, clay and mica are investigated and related to the deformation features of the extensional analogue models.

Rheology of modelling materials

The rheologies of sand, clay and vermiculite mica have been determined using a standard 60 mm × 60 mm soil mechanics simple shear apparatus operated at low values of normal stress, 12–30 kPa (Ellis 1988), in order to

From Knipe, R. J. & Rutter, E. H. (eds), 1990, *Deformation Mechanisms, Rheology and Tectonics*, Geological Society Special Publication No. 54, pp. 445–453.

measure the mechanical properties of the modelling materials at normal stress conditions approaching those found in the deformation experiments. The sand, clay and mica aggregates were tested in an uncompacted state, identical to the conditions in the extensional experiments.

Shear stress/shear displacement curves show that, for a normal stress of 12.2 kPa (the lowest practical value obtained in the tests), mica was the weakest material, displaying markedly non-linear behaviour, whereas clay also displayed non-linear behaviour but was noticeably stronger than both the sand and mica (Fig. 1a). Mohr envelopes are shown for all materials in Fig. 1b. In general all of these materials display characteristic linear Navier–Coulomb behaviour at moderate and high values of normal stress (tests also conducted by P. Buchanan pers. comm. 1989). Sand has an angle of friction of 31°, and an apparent cohesive strength of 1.05 kPa which partly reflects the effects of irregular grain size and surface roughness at low normal stresses.

Clay has an angle of friction of 35° and a low value of cohesive strength (approximately 1.7 kPa, Fig. 1a). Mica displays the lowest value of friction at 30° and deforms at lower shear stresses than either sand or clay (Fig. 1b). The data, though not comprehensive, do show that the modelling materials used in the extensional experiments exhibit near linear Navier–Coulomb behaviour.

Analogue model experiments: summary of results

The extensional analogue models carried out to date have largely been of two kinds: (a) those that involve extension above a basement that undergoes a stretching deformation; (b) those where the deformable hanging wall block is translated over a rigid, non deformable footwall, usually a simple listric or a ramp-flat listric geometry (e.g. McClay & Ellis 1987*a*; Ellis & McClay 1988). The experiments have been carried out in a specially designed glass-walled deformation rig shown schematically in Fig. 2. Initial model dimensions are typically 30 cm long, 20 cm wide and 10 cm deep. The models are constructed by carefully sieving layers of

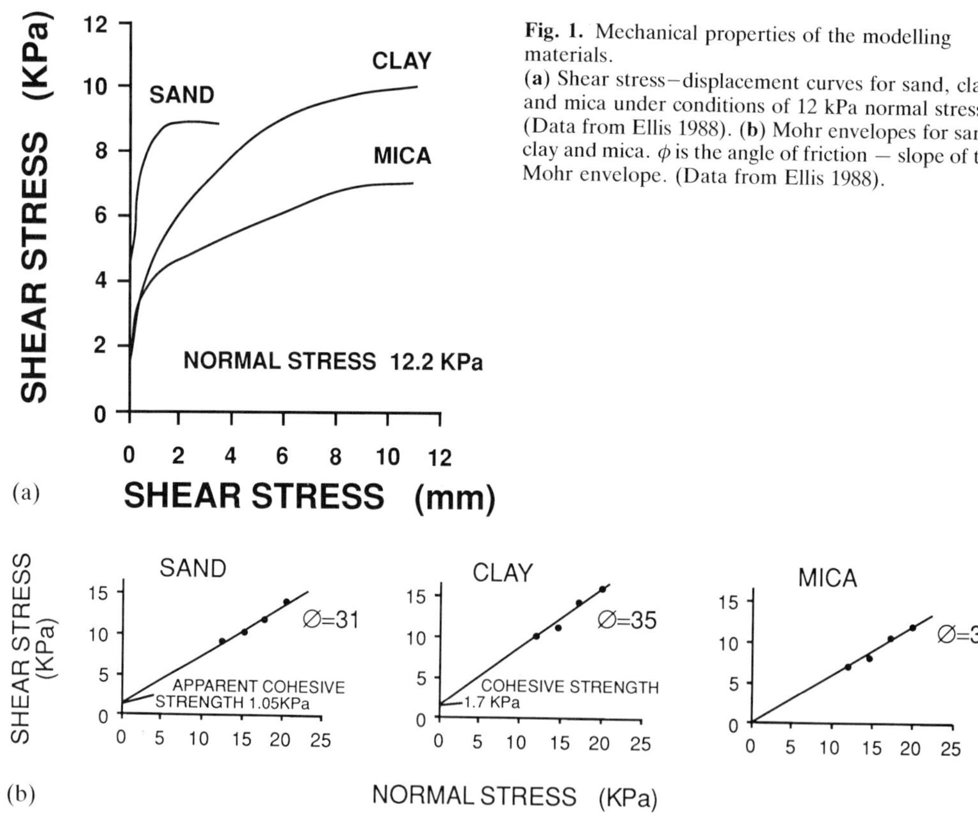

Fig. 1. Mechanical properties of the modelling materials.
(**a**) Shear stress–displacement curves for sand, clay and mica under conditions of 12 kPa normal stress (Data from Ellis 1988). (**b**) Mohr envelopes for sand, clay and mica. ϕ is the angle of friction — slope of the Mohr envelope. (Data from Ellis 1988).

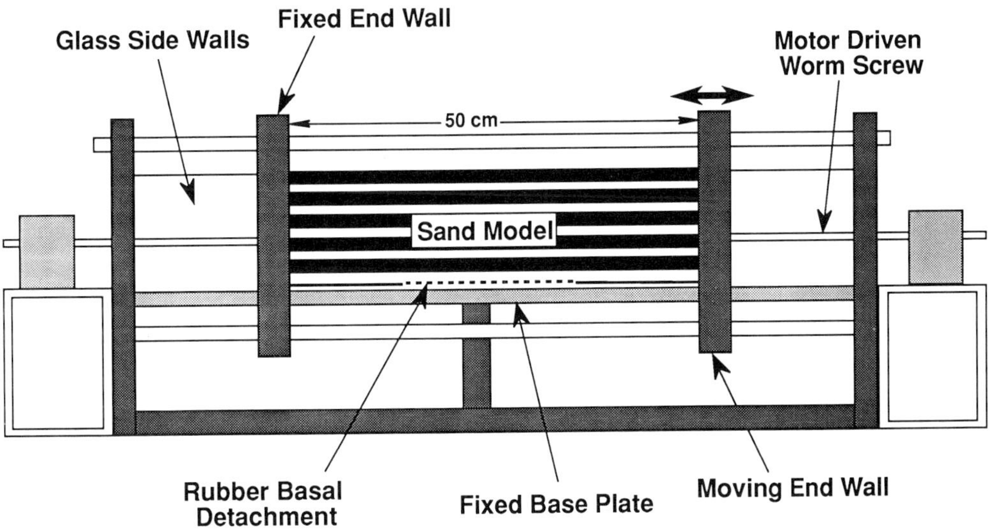

Fig. 2. Schematic diagram of the extensional experimental apparatus. Various detachment geometries may be introduced at the base of the sand model. The configuration as shown is for a stretching rubber base extending only over the central part of the model as indicated.

sand, clay or mica as appropriate, into the deformation rig and onto the predetermined detachment geometry. Extension is achieved by means of a motor driven worm screw which enables one or both of the moving walls to be extended at a constant displacement rate 4.2×10^{-3} cm s^{-1}. As extension proceeds, synrift sediments are added incrementally (usually at intervals of 1 cm of extension) maintaining a constant base level in order to record the evolution of the faults in the new synrift sediments and to prevent unstable surface slopes from forming. Scaling factors are 10^{-4} to 10^{-5} such that a 10 cm deep model will represent between 1 and 10 km in the crust. The models shown in Fig. 3 and described elsewhere in the literature are essentially two dimensional. This has been demonstrated by carefully impregnating the models and serially sectioning after deformation. The models are reproducible with most geometries having been repeated at least three times. The models are recorded using time lapse 35 mm photography and 16 mm cinematography.

Four examples are shown in Fig. 3. In the case of hangingwall deformation above rigid listric footwall detachments (Ellis & McClay 1988), extension is achieved using a thin plastic sheet which approximates to a decollement with zero friction. The resultant extensional geometries are characterised by roll-over anticlines into the concave upwards segments of the listric detachment together with geometrically-necessary crestal-collapse graben structures (Fig. 3a and b). The concave downwards segments of ramp/flat listric detachments produce zones of reverse faulting (McClay & Ellis 1987*a*; Ellis & McClay 1988) and hangingwall synclines (Fig. 3b). In the examples shown in Fig. 3a and b, internal deformation within the fault blocks appears to be limited to segments of the crestal-collapse structure and to the folds over the ramp in ramp/flat geometries (Fig. 3b). In contrast, extensional models developed above a stretching detachment (Fig. 3c and d) exhibit internal deformation with the development of domino-style fault arrays in which both faults and fault blocks change shape during progressive extension. In this paper attention is focused upon this style of deformation.

The fault geometries are controlled by the underlying detachment and are similar for each of the modelling materials used in this study. In detail, sand and sand mica mixtures produce broad fault–shear zones commonly with internal deformation within the fault blocks (Fig. 3b–d), where as clay and clay-sand mixtures produce discrete faults and fault blocks which have a greater rigidity and have less internal deformation (Fig. 3a). The addition of mica layers to a laminated sand model (Fig. 3b) promotes interbed slip and folding over ramps in ramp/flat detachments.

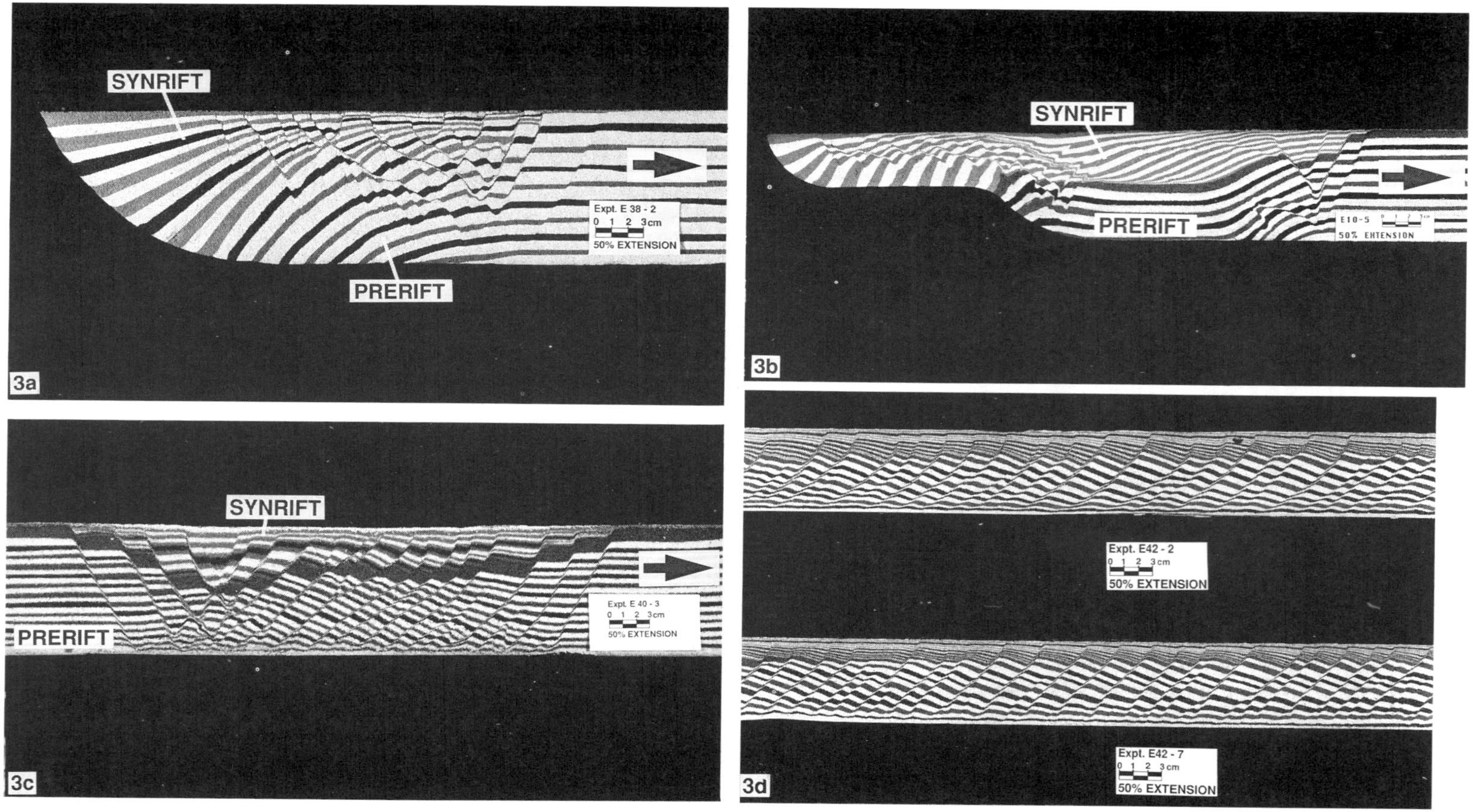

Fig. 3. Typical end results of two dimensional extension models. Each model has been impregnated and serially sectioned to show the internal deformation geometry away from any possible sidewall distortion. (**a**) Experiment E38. Simple listric fault with alternating layers of sand (coloured) and sand-clay mixture (light layers) in the pre-rift and syn-rift section. 50% extension. (**b**) Experiment E10. Ramp-flat listric extensional fault consisting of alternating layers of coloured and white sand, with thin layers of mica between them, in the pre-rift sequence and alternating layers of coloured and white sand only in the syn-rift. Note the folding and unfolding in the pre-rift sequence and small reverse faults in the syn-rift sequence. 50% Extension. (**c**) Experiment E40. Extension above a stretching horizontal basal detachment, the limits of which are indicated on the photograph. Extension was simultaneously both to the right and to the left. Pre-rift sequence consists of thin alternating layers of coloured and white sand only. The syn-rift sequence consists of thicker alternating layer of coloured and white sand. 50% Extension. (**d**) Experiment E42. Extension above a stretching horizontal basal detachment in which the rubber sheet extends beyond the limit of the sand pack. The ends of the model were unconfined allowing the sand to form a natural slope at the angle of rest, 31°. Both the pre-rift and syn-rift consist of alternating layers of coloured and white sand. The pre-rift sequence is the regularly spaced dark and white layers that occupy the basal two thirds of the model. Extension direction was to the right. 50% Extension.

Deformation mechanics in analogue models

The deformation within the analogue models is controlled by the model boundary conditions and the rheology of the modelling materials. These features determine the characteristics of the faults, the fault cut-off angles and the nature of the internal strain within the models.

Boundary conditions

In the two dimensional modelling programme carried out to date, two principal boundary conditions have been applied to the base of the models. These are that of extension above a horizontal (or tilted) basal detachment (rubber sheet) that undergoes stretching, thus simulating a type of pure shear deformation at the base of the sand pack and the second type of detachment where a fixed footwall detachment geometry (either simple listric or ramp-flat listric) is introduced with the hangingwall translated over this fixed footwall geometry (see Ellis & McClay 1988; McClay 1990 for details). In this case a weak decollement is simulated by using a plastic sheet to translate the hangingwall over the footwall. These basal boundary conditions determine the geometry of extension in the hangingwall. In the case of extension above a stretching rubber detachment, a dilatant boundary shear layer is formed at the base of the stretched section of the model (Fig. 3c & d), whereas for the listric and ramp-flat listric detachments the boundary shear layer occurs on the curved segment of the basal detachment (Fig. 3a & b). This boundary layer is generally less than 3 mm thick and consists of a frictional traction carpet of grains between the bulk of the model and the bottom of the deformation apparatus. For listric systems, the basal traction carpet does not appear to affect the geometry of the overlying hanging wall structures. In contrast, for the models above a stretching detachment, the basal traction-carpet is geometrically necessary to accommodate the domino-style extension and curved nature of the faults (Fig. 3c & d). Variations in the basal detachment geometry generate different fault curvatures in the lower part of the section (Fig. 3d.) and give rise to variations in the angular shear strains at the base of the models.

The frictional constraints imposed on the modelling materials sliding past the glass walls of the apparatus (Fig. 2) generate some curvature of the faults both in plan and in section. These effects generally disappear within 2 cm from the glass walls. The frictional drag along the side walls does not appear to significantly alter the overall geometries or the nucleation sequences of the internal faulting. The analysis described below was carried out on serial sections usually, a minimum of 4 cm away from these sidewall boundaries.

In most model configurations (Fig. 3a–c), the endwall boundaries are fixed. In a few experiments where the underlying basal detachment was stretched over the whole length of the model, the ends were not constrained and the sand was allowed to form a slope at the angle of rest (Fig. 3d). In such models there is less lateral constraint at these ends such that rotational deformations were not inhibited (cf. Fig. 3c & d).

Fault characteristics

Within the analogue models the faults are not discrete fracture surfaces but rather are granular shear zones of finite extent. The widths of the shear zones are directly proportional to the grain sizes of the modelling materials, from approximately 1.5 mm for sand with a grain of 300 μm, to less than 0.2 mm for faults in clay layers (Fig. 4a–c). There is no discernible difference in the width and fabric of the extensional faults in either the listric detachment or the models with stretched basal detachments. The widths of the faults–shear zones are normally approximately five times the average grain size of the modelling material used (Fig. 4a–c). Within the fault–shear zones considerable dilatancy occurs (cf. Mandl *et al.* 1977; Mandl 1988) and typical measurements indicate up to 12% volume increase in narrow zones as the packing of sand grains is altered from the central portions of the fault block to the interior of the fault–shear zone.

Fault cut-off angles

For the experimental configuration σ_1 is vertical and using a Navier–Coloumb rheology the dip of the faults at surface is given by

$$\theta = 90 - (45 - \phi/2)$$

where θ = dip of fault at surface and ϕ is the angle of friction of the model material (similar to Anderson's (1951) approach to extensional faulting in the upper crust). For homogeneous sand, ϕ = 31° and the expected extensional fault dip would be 60.5° (in areas away from frictional boundary constraints). Using carefully serially sectioned models, the initial fault dip (or cut-off angle relative to the horizontal pre-rift and syn-rift bedding at the top of the model) was measured from a range of sand models. A

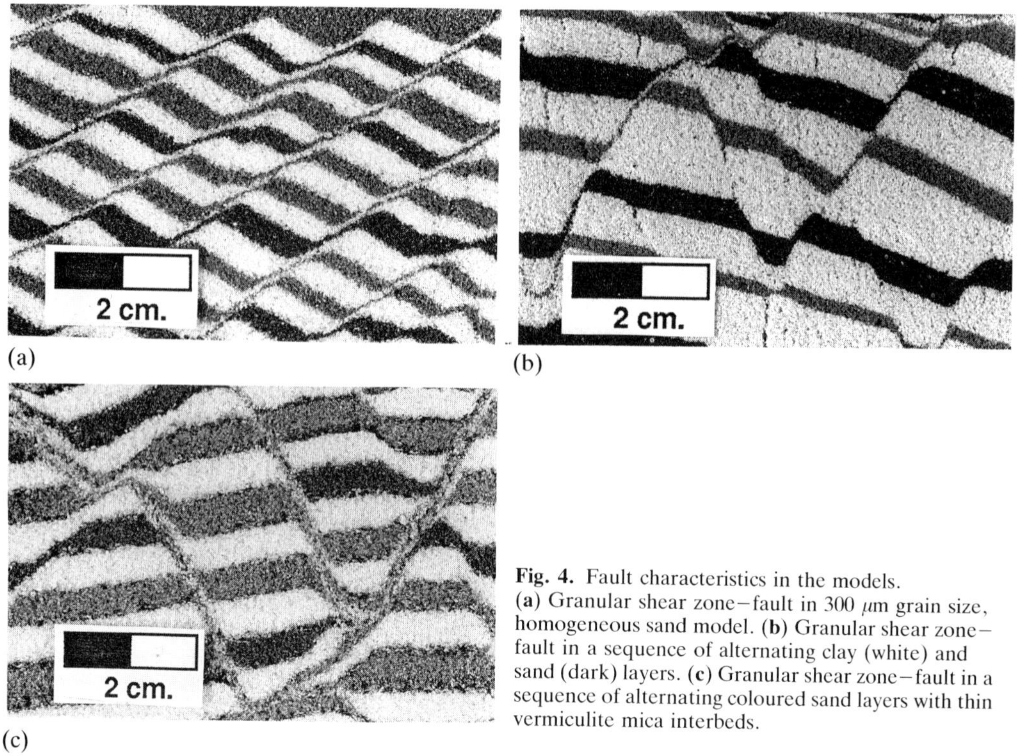

Fig. 4. Fault characteristics in the models. (**a**) Granular shear zone–fault in 300 μm grain size, homogeneous sand model. (**b**) Granular shear zone–fault in a sequence of alternating clay (white) and sand (dark) layers. (**c**) Granular shear zone–fault in a sequence of alternating coloured sand layers with thin vermiculite mica interbeds.

wide variation in initial fault (cut-off) angle was found from 55° through to 75° with a mean value around 65° (Fig. 5) significantly higher than the 60.5° predicted by the shearbox experiments. This, however, may be attributed to grain roughness and size variations, which would be important at low stresses in the upper part of the model. Similar departures from ideal behaviour have also been recorded for both clay and mica with increased fault cut-off angles. To date insufficient data has been collected in order to conduct a statistical evaluation of these other materials.

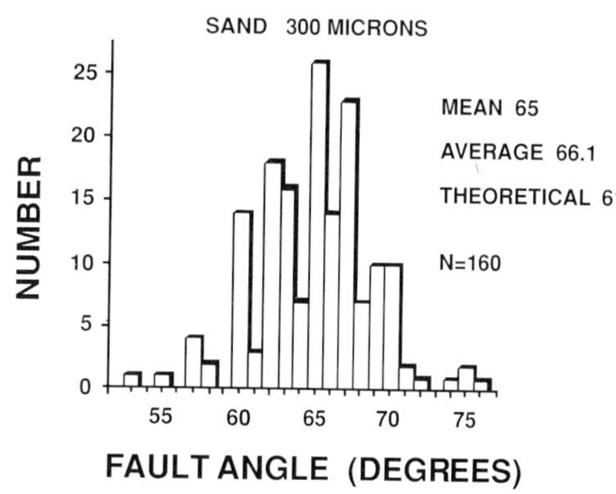

Fig. 5. Histogram of fault–bedding cut off angles in homogeneous sand models.

Strain in the extensional models

Internal strains within fault blocks are most obvious for those in which the basal detachment is stretched (Fig. 3c & d) and attention is therefore focussed upon this group. Both the faults/shear zones and the fault blocks have been subjected to rotation, internal shearing, layer thinning and extension together with area and volume increases due to dilation caused by changes in packing of the sand grains within the model. Careful measurements of the pre-rift sequence before and after extension indicate slight area (hence volume) increases of less than 5% (although this can be up to 11% in some cases). The accompanying layer thinning, most noticeable in the fault/shear zones (Fig. 4a–c), is difficult to assess quantitatively because of the lack of distinctive point markers in the models.

It is possible, however, to factorize the shear strain within the models using the assumption of a constant initial fault-bed cut off angle. For example, in homogeneous sand models the initial fault-bed cut off angle has a mean value of 65°. Using this angle and measuring the fault-bed angle within the extended model enables the angular shearing strain to be readily computed using the equation:

$$\cot \alpha^1 = \cot \alpha - \gamma \text{ (Ramsay \& Huber 1983)}$$

where α is the angle between bedding and the fault upon initiation and α^1 is the angle between the bedding and the fault after deformation.

Using this technique the angular shear strains have been computed for two horizontal stretching detachments, experiments similar to experiment E40 (Fig. 3c) and experiment E42 (Fig. 3d). The fault-bedding relationships are shown in Fig. 6a & 6c. The shear strain map for experiment E1 shows most of the shear strain concentrated towards the base (Fig. 6b) where the domino faults sole out into the basal detachment due to clockwise rotation of the faults relative to the bedding (negative shear strain). The only area of marked positive shear strain occurs close to the graben-bounding fault in the ieft hand side of the model (Fig. 6b), generated by a dominant sense of shearing parallel to that fault. In contrast, the section of model E42 shown in Fig. 6c was chosen to illustrate an example where the domino faults steepen and horse-tailed downwards. Here the shear strain map shows a zone of positive shear strain at the base of the model (i.e. where the faults steepen, Fig. 6d) and a band of strong negative shear in the centre of the model (Fig. 6d) where the faults accommodate the listric shape generated at higher levels. Although this form of analysis is in its infancy for the experimental programme, it does illustrate that the shear strain can be measured and might indicate that in nature fault-bedding angles may not be constant throughout deformation, a result also found by Dietrich & Ramsay for faults in Marble Canyon, Snake Range, Nevada (oral contribution TSG Annual Meeting, 1988).

Discussion and conclusions

Properties of the model materials

Testing of the model materials has shown that they deform by linear or near-linear Navier–Coulomb behaviour under the conditions expected in the extensional models. Clay has a small finite cohesive strength which gives it a degree of competency. This is attributed to surface adhesion effects and a small amount of fluid/moisture trapped by the very fine grain-size (< 10 μm). The properties of these materials make them suitable for simulating the brittle deformation of sedimentary rocks in the upper crust.

Extensional model results

The extensional models briefly illustrated in this paper and those described in the literature (Vendeville *et al.* 1987; Vendeville & Cobbold 1988; McClay & Ellis 1987a, b; Ellis & McClay 1988; McClay 1990) successfully simulate many extensional structures found in sedimentary basins. These models, however, cannot accurately incorporate compaction, thermal or isostatic effects which may be significant in some extensional settings (e.g. Wernicke & Axen 1988). They may however provide interpretative templates for the progressive evolution of extensional structures and for their geometrical and kinematic analysis.

Deformation mechanics

Analysis of the analogue extensional models indicates that the geometry and nature of the basal detachment surface is the fundamental parameter that controls the extensional deformation above it. The models are essentially two dimensional and cylindrical across the structures. End-wall constraints (cf. Fig. 3c & d) control the rotation within the model. A free end wall allow significant rotation and domino style faults to develop (Fig. 3d). Similar styles of domino faults have been generated by Vendeville *et al.* (1987). The faults developed within the extensional models are essentially grain-sliding shear zones in which the packing distribution of grains is altered and dilatancy occurs (Fig. 4). The widths of the shear zones are grain size dependent, being wider in sand and sand/mica mixtures and narrower in clay layers. The size of discernible faults within the models is also grain size dependant. Small faults within the models merge into grain boundary sliding and packing rearrangement which is thought to be responsible for layer thinning within the models. The net effect is 'hidden extension' whereby the total extension of the models cannot simply be accounted for by summation of the heaves on individual faults. In nature, pervasive small scale faulting which is not imaged on seismic sections, may account for extension discrepancies in a similar way (e.g. discussions by Ziegler, 1983; Wood & Barton, 1983; Kautz & Sclater, 1988).

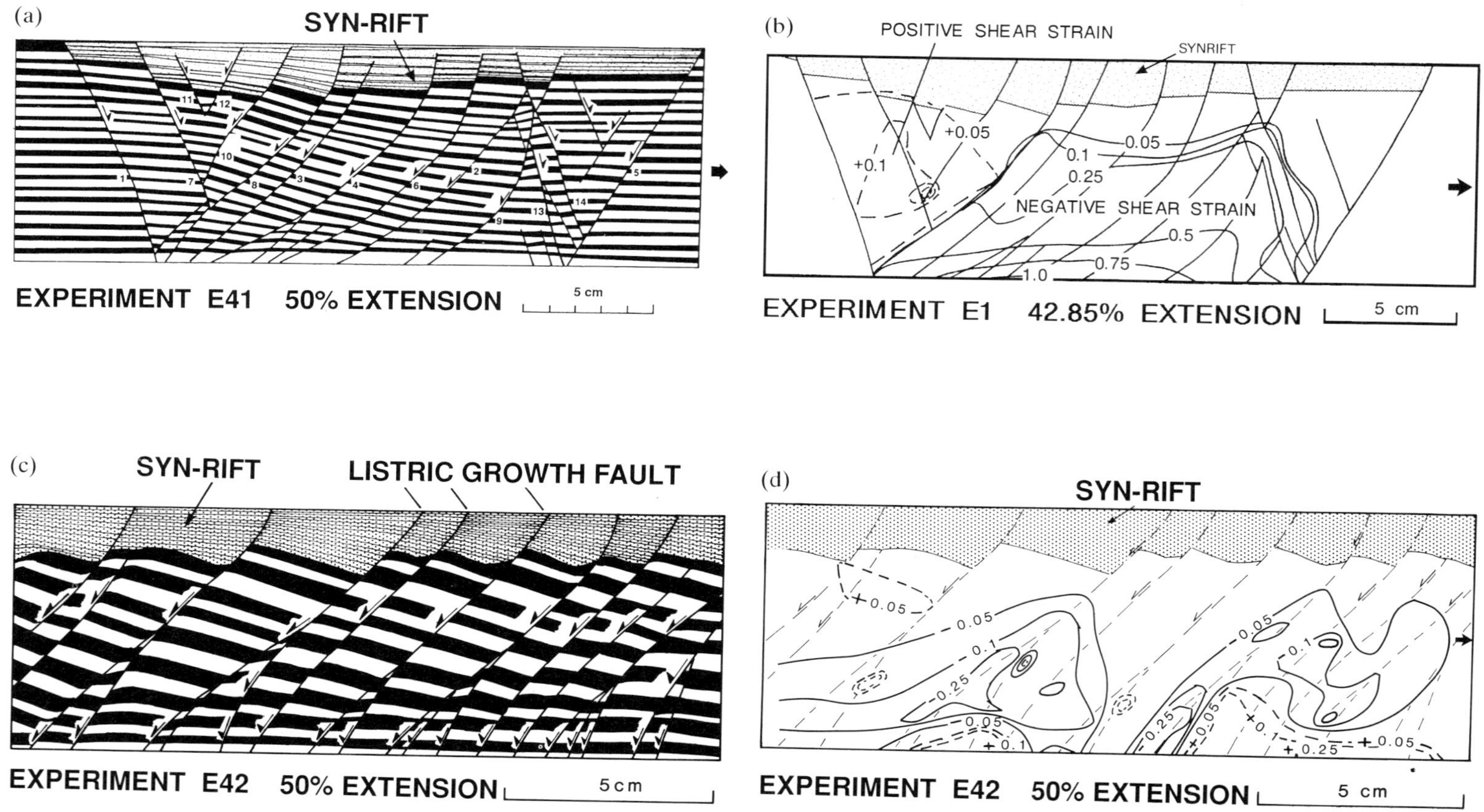

Fig. 6. Fault architecture and shear strain maps for domino-style fault models. (**a**) Fault architecture and fault sequence diagram for Experiment E1 (similar to E40, Fig. 3c above). Both the syn-rift and pre-rift sequences consist of alternating layers of coloured and white sand. The sequence is stippled. The limit of the stretching basal detachment is indicated at the base of the model. 42.85% Extension. (**b**) Shear strain distribution map experiment E1. Negative shear strains indicate clockwise rotation of the fault surface relative to the bedding surface. (**c**) Fault architecture diagram for a segment of Experiment E42 (Fig. 3d above) showing a domino-style array of faults which steepen and horsetail downwards towards the basal detachment. Syn-rift sediments are stippled. 50% Extension. (**d**) Shear strain distribution map for the segment of Experiment E42 shown in (c) above.

The fault cut-off angles in the models, although steeper than those predicted from the Navier–Coulomb properties of the modelling materials, are similar to the 60–70° dips of many extensional faults observed at the earth's surface (e.g. Walsh & Waterson, 1988). Shear strain analysis of deformation within the models indicates that for domino style faulting, the fault-bedding angle is not constant but changes in response to the deformational geometry imposed by the basal detachment and end-wall boundary conditions. In particular angular shear strains will occur in zones parallel to the basal detachment or parallel to graben bounding faults where the extension is over a stretched rubber sheet (Fig. 6b & d).

Limitations of analogue modelling

Although anologue modelling produces realistic geometric and kinematic models of extensional structures and the modelling materials simulate the deformation mechanics of brittle sedimentary rocks in the upper crust, there are, nevertheless, important limitations which must be born in mind when interpreting the results. The models cannot simulate isostatic, compactional, thermal, confining pressure or pore-pressure effects. The models of listric fault systems use plastic sheets to simulate a weak decollement thus constraining the displacement to that of constant slip. In addition, the models simulate basins that are filled with sediment. Both of these above conditions may not be found in natural listric fault systems. The models are grain size dependent and grain size effects are noticeable in the width of the shear-fault zones and the limits to which faults are discernible. However, bearing these provisos in mind, analogue models are powerful tools for understanding the mechanics and progressive deformation of extension in the brittle upper crust.

The research described in this paper was funded by a National Environmental Research Council Grant No. GR3/6658A. I am indebted to A. Scott for his invaluable assistance in running the experiments and in helping with the preparation of this paper. P. Elllis kindly allowed me to used the rheological data from his PhD thesis. A. Scott and P. Buchanan critically read the manuscript, S. Muir and P. Buchanan assisted with the drafting of the diagrams, and K. D'Souza is thanked for the photographic reproduction.

References

ANDERSON, E. M. 1951. *The dynamics of faulting*. Oliver and Boyd, Edinburgh.

ELLIS, P. G. 1988. *Analogue model studies of the kinematics of extensional faulting*. PhD thesis, University of London.

—— & MCCLAY K. R. 1988. Listric extensional fault systems-results of analogue model experiments. *Basin Research*, **1**, 55–70.

HORSFIELD, W. T. 1977. An experimental approach to basement controlled faulting. *Geologie en Mijnbouw*, **56**, 363–370.

KAUTZ, S. A. & SCLATER, J. G. 1988. Internal deformation in clay models of extensional block faulting. *Tectonics*, **7** 823–832.

MCCLAY, K. R. 1990. Physical models of structural styles during extension. *In*: TANKARD, A. J. & BALKWILL, H. (eds) *Tectonics and Stratigraphy of the North Atlantic Margins*. AAPG Memoir, **46**, 341–344.

—— & —— 1987*b*. Analogue models of extensional fault geometries. *In*: COWARD, M. P., DEWEY, J. F. & HANCOCK, P. L. (eds) *Continental Extension Tectonics*, Geological Society, London, Special Publication, **28**, 109–125.

MANDL, G. 1988. *Mechanics of Tectonic Faulting*. Elsever, Amsterdam.

——, DE JONG, L. N. J. & MALTA, A. 1977. Shear Zones in granular material. *Rock Mechanics*, **9**, 95–125.

NAYLOR, M. A., MANDL, G. & SIJPESTEIN, C. H. K. 1986. Fault geometries in basement induced wrench faulting under different initial stress states. *Journal of Structural Geology*, **8**, 737–752.

PATERSON, M. S. 1978. Experimental Rock Deformation – The Brittle Field Springer-Verlag, Berlin.

RAMSAY, J. G. & HUBER, M. I. 1983. *The Techniques of Modern Structural Geology. Volume 1: Strain Analysis*. Academic Press, London.

SANFORD, A. R. 1959. Analytical and experimental study of simple geologic structures. *Geological Society of American Bulletin*, **70**, 19–52.

VENDEVILLE, B & COBBOLD, P. R. 1988. How normal faulting and sedimentation interact to produce listric fault profiles and stratigraphic wedges. *Journal of Structural Geology*, **10**, 649–659.

VENDEVILLE, B. COBBOLD, P. R., DAVY, P., BRUN, J. P. & CHOUKROUNE, P. 1987. Physical models of extensional tectonics at various scales. *In*: COWARD, M. P., DEWEY, J. F. & HANCOCK, P. L. (eds) *Continental Extensional Tectonics. Geological Society, London, Special Publication*, **28**, 95–107

WALSH, J. J. & WATTERSON, J. 1988. Dips of normal faults in British Coal Measures and other sedimentary sequences. *Journal of the Geological Society, London*, **145**, No 5, 859–874.

WERNICKE, B. & AXEN, G. J. 1988. On the role of isostasy in the evolution of normal fault systems. *Geology*, **16**, 848–851.

WOOD, R. & BARTON, P. 1983. Crustal thinning and subsidence in the North Sea. *Nature*, **302**, 134–136.

ZIEGLER, P. 1983. Crustal thinning and subsidence in the North Sea. *Nature*, **304**, 561.

Slickenside lineations due to ductile processes

C. J. L. WILSON & THOMAS M. WILL

Department of Geology, The University of Melbourne, Parkville, Victoria 3052, Australia

Abstract: Faults with features characteristic of slickensides, but with plastically deformed wallrocks, were propagated in paraffin wax in a shear environment. Flow of the bulk of the sample produced a strong foliation with fabric strengthening in localized zones that preceded the initiation of the fault. The faults contain ridge-in-groove type slickenlines that have excess-lengths with respect to the measured slip displacement on the fault plane. With progressive movement on the fault surface, these surface markings are disrupted by both ductile shear bands and brittle fractures. The slickenside lineations observed on the fault surface are produced while the material behaves in a ductile manner which does not involve brittle processes.

The development of slickenside striations during fault nucleation and propagation in pyrophyllitic clay has been described by Will & Wilson (1989). Comparable results, including development of ridge-in-groove type lineations with excess-lengths with respect to the measured displacements have been developed in the paraffin wax experiments described in this paper. These wax experiments use a comparable starting material to those described by Means (1989) but differ in that a homogeneous wax block was deformed in a double-shear jig instead of in compression, and there was no pre-cut fault interface inserted in the sample. Means (1989) was able to produce a series of pseudo-faults, referred to as stretching faults, with wall rocks that lengthen or shorten. These faults involved ductile flow in the slip direction within the wall rocks adjacent to the fault. This is in contrast to the perception of a brittle fault where movement on a slickenside is confined to discrete narrow zones with little strain outside the zone. In experiments described in this paper, the development of slickenside lineations associated with fault nucleation, growth, and subsequent modification of the fault surface are discussed.

Experimental procedures

The samples used in this study were deformed at room temperature in a direct shear jig of the type described by Means (1987) and Will & Wilson (1989) and a slip rate of 3×10^{-3} m^{-1}. Paraffin wax (melting point 62° C) made black with candle dye, to improve the photographic record of the sample, was moulded into ingots in a space (20 × 12.7 × 10 mm) within warm sample holders. This produced a nearly uniform grain size and a random orientation of the wax crystallites across the sample. Owing to the large volume change during the crystallization of the wax (Mancktelow 1988) additional wax must be added to the cooling ingot, producing minor grain size differences adjacent to a sample margin. All microstructural observations recorded in this investigation were from the centre of the wax ingot.

Microstructural observations within the wax sample were obtained by preparing 5–10 μm thick microtome peels and observing these through a normal petrographic microscope. To minimise microtome damage and reduce deformation to the wax, all samples were embedded in a 3 mm wide jacket of ice. Both the protective ice and wax were sectioned. Comparison of variably oriented cuts suggests that insignificant microstructural change was induced by the microtome procedure, even though some samples were compressed by 30% as a consequence of the microtoming. The majority of samples were cut normal to the faults surface and parallel to the slickenside lineation.

Marker lines and inscribed circles on the external surface of the sample were used to measure the magnitude of total displacement, d_t, and the slip displacement, d_s, similar to the method described by Will & Wilson (1989). All experimentally produced fault surfaces were photographed using the techniques described by Will & Wilson (1989).

Localization of faulting

There is a heterogeneity of strain distribution parallel to the long edge of the sample (Fig. 1). Linear marker grids (Fig. 1a, b) inscribed at a high angle to potential fault surfaces develop a

From Knipe, R. J. & Rutter, E. H. (eds), 1990, *Deformation Mechanisms, Rheology and Tectonics*, Geological Society Special Publication No. 54, pp. 455–460.

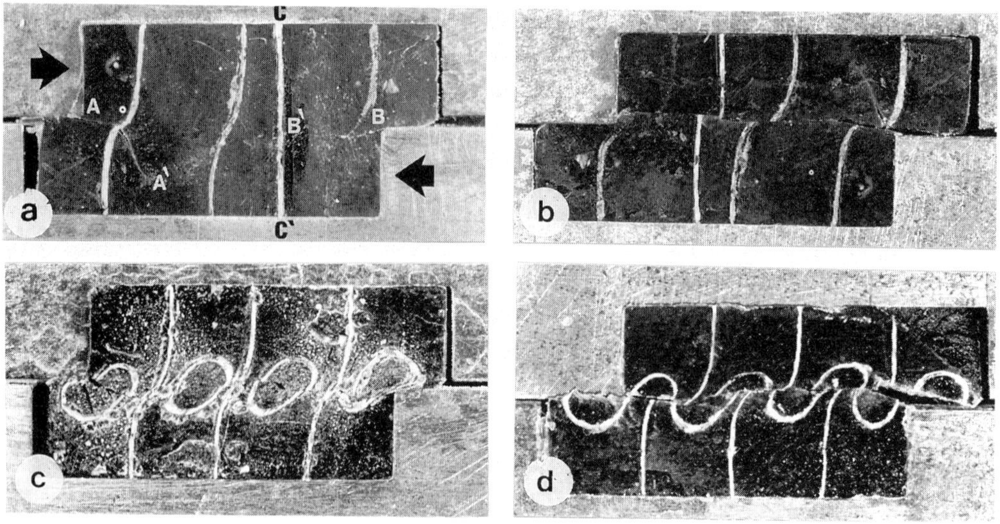

Fig. 1. Experimentally deformed samples with initially linear and circular strain markers. The long edge of the sample is 20 mm. (**a**) Sample WS59 deformed (d_t = 2.7 mm) and displaying curved markers and minor fracture surfaces at A–A′ and B–B′. There is no offset of markers and a new marker line C–C′ is inscribed on the surface. The two indenting surfaces and sense of movement of the apparatus are indicated by the arrows. (**b**) Further displacement (1 mm) in WS59 and the generation of a uniform slip displacement (d_s = 1.2 mm) on a fault surface that offsets all marker lines. (**c**) Experimentally deformed sample WS61a at the stage of fault initiation (d_t = 2.4 mm). Note that the stretching direction, as seen from ellipse elongation, is inclined at approx. 45° to the fault plane and strain distribution at the end of the sample is inhomogeneous. (**d**) Sample WS61b (d_t = 4.3 mm) with non-uniform offset of strain markers and strain ellipses.

sigmoidal shape in the centre of the specimen and 2-D circles produce 2-D average ellipses (Fig. 1c, d). The shape of the 2-D ellipses represents a uniform strain developed centrally and progressively propagated from the margin to the centre of the sample. The principal axes of each successive incremental strain ellipse are initially 45° to, and coincident along, the length of the sample, with the principal stretching direction lying in the *XZ* profile of the deformed sample (Fig. 2). However, with progressive deformation, the superposition of strain becomes more inhomogeneous (Fig. 1c) with non-coaxial incremental distortions beginning at the end of sample and progressively propagating to the centre. The initial ductile deformation is followed by local rupturing, again beginning at sample ends, followed by translation of separate halves of the ellipses past one another along a fault plane (Fig. 1d). Rotation of the passive marker lines is the same as the finite ellipses, and illustrates that the principal stretching direction is locally deflected into a central zone of high shear strain.

Fractures were initiated after the ductile bending of the markers, and comprise two types.

(1) *Extensional fractures* that were observed at the indenting end of the apparatus as it was pressing into the wax sample (Fig. 1a). The initial fracture front, as it propagated from the boundaries of the indenting end of the apparatus, became curved and convex towards the extension direction (A–A′ in Fig. 1a). The initial fracture may then dilate to produce a larger extensional opening, perpendicular to the extension direction. The initial fractures may either extend, be overprinted, or bifurcate to produce a central fault.

(2) *Central faults*. The earlier formed fractures were locally subparallel to the long edge of the sample and propagated as through-going faults. The path of the fault was not always planar but could have small lateral ramps and large flats, using the thrust terminology (Boyer & Elliot 1982), and on two dimensional surfaces had a curvilinear shape (Fig. 1). Once a fault was initiated through the centre of a sample, there appeared to be non-uniform displacement along the fault surface. This was seen by the apparent minor differences in offsets of the marker lines (Fig. 1b). For instance, marker lines in the centre of the sample showed a smaller offset than those closer to the margins. Marker lines added to the sample at the onset of brittle

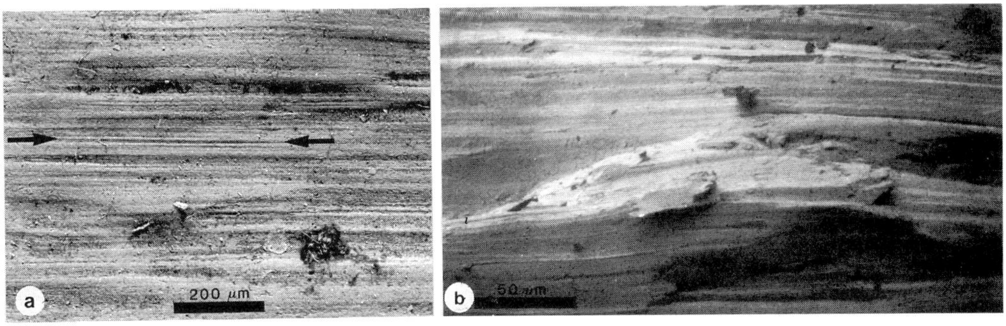

Fig. 2. SEM micrographs illustrating the footwall of fault surface in deformed wax samples dominated by ridge-in-groove lineations produced by sinistral movement on the fault. (**a**) Plan view of lineated surface produced immediately after the initiation of fault in sample WS41. Total displacement is 400 μm. The length of individual ridges and grooves seen on this surface exceeds d_t. (**b**) Oblique view of surface undulations on the striated fault surface with hummocks and depressions. Ridges and grooves, developed on overlapping inclined and interesecting fracture surfaces, appear to be superimposed on one another and hence restrict the measurement of lineation lengths.

faulting (between C–C′ on Fig. 1a) showed a smaller displacement than the early marker lines.

The shiny and reflective fracture surface is dominated by long and continuous ridge-in-groove type lineations (Means 1987) and is referred to as a slickenside. Comparison of the average slip displacement, d_s, on the fault surface and parallel to the slip vector, as measured from the offset markers and the maximum length of the grooves (between the back arrows in Fig. 2a), shows that the grooves can have an excess-length. Similar observations have been documented by Will & Wilson (1989). The excess-length is most prominent immediately after a fault surface has formed. With additional movement on the fault surface (e.g., Fig. 1d) the excess-length becomes difficult to measure as the ridges and grooves become bent by the development of cresent-shaped surface undulations and disappear under additional discontinuities (Fig. 3b).

Microstructures related to fault surface

Preceding the onset of faulting (Fig. 3a), a zone dominated by aligned wax crystallites progressively propagated from diametrically opposite ends of the sample, i.e. on the hanging-wall or foot-wall on the respective ends (Fig. 4), towards the centre of the sample. The development of this fabric occurred only in the area under compression, in marked contrast to the tensile side of the sample where there was no obvious grain alignment. During the progressive deformation, a region of fabric reorientation propagated laterally along the length of the sample, from the indenting end towards the centre of the sample, with an accompanying broadening of the zone of fabric reorientation or foliation development. At the interface of the tensile and compressional sides of the sample, a juxtaposition of randomly oriented wax with a narrow zone of highly oriented crystallites (Fig. 3a) has developed, that grades into a weaker fabric (Fig. 3b). The alignment of the wax in the area of weaker fabric development is conspicuously oblique (≤ 20°) to the interface. It is at this interface that the first fractures that ultimately produce the central faults have formed (A in Fig. 3c). In many specimens there are oblique concentrations of wax crystallites (B in Fig. 3c), and it is along these surfaces that secondary fractures nucleated. These secondary fractures truncate the central fault surface (Fig. 3b).

In all specimens where deformation continued after the formation of a central fault, there is significant modification of the earlier fabric. In samples displaying a uniform fabric and no well defined zonal distribution of wax crystallites, the pre-existing foliation was buckled to produce shear bands, C′, or extensional crenulation cleavages (Platt 1979) that correspond to the undulations seen on the surface of the fault (Fig. 4c). The sense of curvature suggests that stretching was taking place within the fault surface and is always consistent with the sense of fault movement. Continued deformation in samples that attained a zonal distribution of crystals produced open buckles (Fig. 3d). Marked discontinuities defined by fracture partings developed within and/or adjacent to areas of highly oriented fabric. These dis-

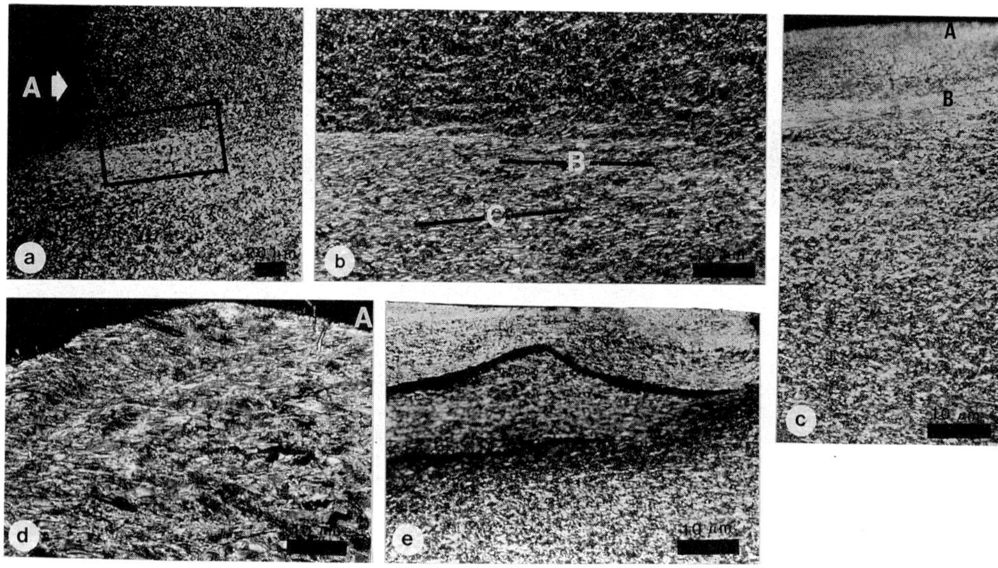

Fig. 3. Sections cut parallel to slip vector and perpendicular to fault surfaces in deformed samples of wax photographed with crossed nicols.
(**a**) Deformed but unfaulted sample (WS42) where the initial increment of deformation has produced an oriented farbic adjacent to the indenting end of the apparatus (A) that grades inwards into the deformed sample. The area on the tensile side of the specimen has little fabric development. (**b**) Enlargement of area shown in (a) showing oblique grain orientations parallel to B and C. (**c**) Faulted sample (WS51). The fault surface (A) is developed within a zone of oriented wax crystallites. Another zone (B), lying oblique to A, converges with A outside the area of the photomicrograph. There is a progressively weaker fabric development from the fault surface (A) to the margin of the sample (base of photomicrograph). (**d**) A weak secondary spaced cleavage rotates the principal foliation in the deformed wax and trends about 40° to the main foliation and fault surface at A (sample WS27). (**e**) Structures developed adjacent to the fault surface lying at the top of the micrograph in sample WS42. In this sample a zoned fabric similar to that seen in (c) was formed earlier.

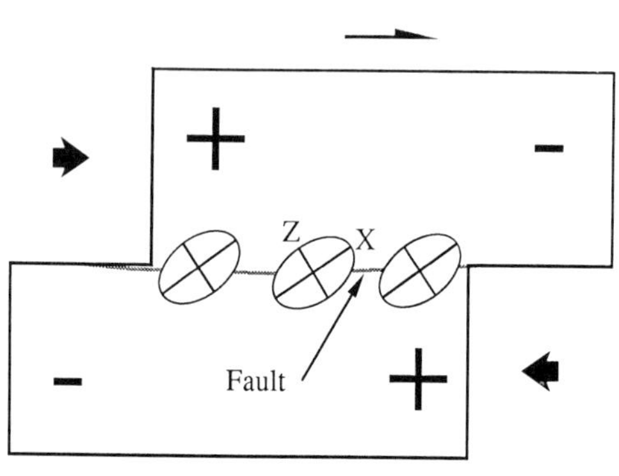

Fig. 4. The compressional (indicated with + sign and adjacent to indenting portion of apparatus) and tensile areas (−) in deformed sample are symmetrically opposed about a zone of high strain. Within this zone, there is a strong stretching component (parallel to *X*) and the initiation and localization of the central faults.

continuities propagated to the surface as irregular steps (Fig. 3e).

Comparison with natural samples

The experiments described here, and those by Will & Wilson (1989), have been performed in analogues that model argillaceous material without the presence of a free fluid. Clearly, naturally occurring fault planes and slickenside lineations will only rarely be formed under these conditions. Freely moving fluids and associated mineral transformations are the general case in nature and great care has to be exercised when experimentally deformed specimens are compared with naturally produced fault rocks. As we are not in a position in this study to document the incremental strain history experienced by fault plane rocks and fault plane markings, we are unable to compare the results of finite strain on naturally occurring rocks with our analogue material.

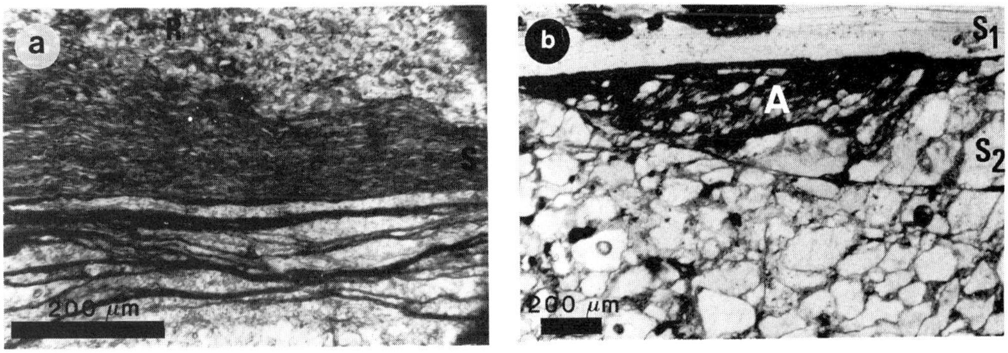

Fig. 5. Section parallel to the lineation and normal to ridge-in-groove type slickensides in faulted sediments. (**a**) Layer silicates subparallel to fault surface in a strain modified zone (S) in the hangingwall in contrast to the host rock (R) that has not experienced a change in fabric due to faulting. In the footwall (below S) there are multiple zones of layer silicates along which fractures with displacement have developed. Slate from Wattle Gully Mine, Victoria, Australia. (**b**) Exotic material precipitated from infiltrating fluids define the S-zone. The S-zone can be further divided into the S_1 zone, the zone of mainly exotic material, and S_2 zone, composed of deformed host rock material. Psammitic schist from Laurel Creek Reservoir, Pennsylvania, USA.

Layer silicates sub-parallel to the fault surface can be recognized in the hanging-wall shown in Fig. 5a, whereas the layer silicates in the foot-wall, which are intercalated with quartz, lie at a shallow angle to the slickenside. Anastomosing shear bands (at A) dipping at an angle of about 20° away from the fault plane, which is coated with a thin layer of fibrous layer silicates, can be seen in Fig. 5b. This orientation, together with the observation that the fault surface shown in Fig. 5b is part of a pair of nesting foot-wall and hanging-wall blocks (e.g., Means 1987) mirrors our experimentally derived results. These natural examples show that slickensides can be penetrative features, as indicated by the strain-modified sub-surface fabric.

Discussion

Deformation adjacent to the faults described here is more complex than one resulting from a homogeneous deformation alone, e.g. simple shear, or of the type described by Means (1989). In fact, most of the events of flow in these experiments and in natural ductile rocks result from small contributions of more or less complex heterogeneous deformations. In the deformed paraffin wax samples described here, the development of a fault is always preceded by the initiation of an oriented fabric. In this rheologically weak material (Mancktelow 1988), the heterogeneity created by the oriented crystals suggests that ductile deformation is necessary to create an anisotropy, along which a brittle fracture can propagate. This conclusion is also supported by some observations of natural samples, but is contrary to the situation described by Segall & Simpson (1986) who infer brittle fracturing as a precursor to the localization of ductile shear. The examples described in these wax experiments are another variation of the critical component described by Cox & Scholz (1988) as a *process zone*. In this case, it is the localization of ductile, rather than brittle deformation, that precedes the development of the shear rupture that produces the fault.

Because of the boundary conditions (Fig. 4) imposed on the sample, the natural fracture path required to develop a fault is dependent on the progressive propagation of the fabric through the sample. All fracture surfaces appear to evolve in areas that are characterised by a high degree, or local concentrations, of fabric development. The fabric is initiated in a zone of high strain concentration where there is evidence of extensive stretching. It is the local stretching component and especially the degree of fabric development that ultimately govern the excess-length property observed in the ridge-in-groove lineations.

The expansion of deformation along a fault surface involves both brittle fractures and/or shear bands. These produce the undulations and disruption of the fault surface and are propagated into the wall rocks as a ductile deformation in the form of extensional features. We suggest that the extent of this latter deformation is a function of the normal stress acting across the fault surface and the magnitude of displacement.

This work was supported by the National Science Foundation grant 860961 to W. D. Means. The experimental work was undertaken while C. J. L. W. spent a sabbatical leave at the SUNY Albany and T. M. W. was undertaking an MSc program. During this time, W. D. Means provided us with many inspired and stimulating discussions which contributed greatly to this work.

References

Boyer, S. E. & Elliott, D. 1982. Thrust systems. *American Association of Petroleum Geologists Bulletin*, **66**, 1196–1230.

Cox, S. J. D. & Scholz, C. H. 1988. On the formation and growth of faults: An experimental study. *Journal of Structural Geology*, **10**, 413–430.

Mancktelow, N. S. 1988. The rheology of paraffin wax and its usefulness as an analogue for rocks. *Bulletin Geological Institute University Uppsala New Series*, **14**, 181–193.

Means, W. D. 1987. A newly recognized type of slickenside striation. *Journal of Structural Geology*, **9**, 585–590.

—— 1989. Stretching faults. *Geology*, **17**, 893–896.

Platt, J. P. 1979. Extensional crenulation cleavage. *Journal of Structural Geology*, **1**, 95–96.

Segall, P. & Simpson, C. 1986. Nucleation of ductile shear zones on dilatant fractures. *Geology*, **14**, 56–59.

Will, T. M. & Wilson, C. J. L. 1989. Experimentally produced slickenside lineations in pyrophyllitic clay. *Journal of Structural Geology*, **11**, 657–667.

Transition between seismic and aseismic deformation in the upper crust

J. P. GRATIER & J. F. GAMOND

L.G.I.T., Observatoire de Grenoble, Universite Joseph Fourier, I.R.I.G.M., BP 53X, 38041 Grenoble, France

Abstract: Displacement on faults is often accommodated by a succession of two mechanical processes: aseismic sliding mass transfer and seismic cataclastic events. Pressure solution is attested both by dissolution markers of asperities which prevent sliding on and around the fault zone, and by the mechanism of growth of the mineral fibres by aseismic crack-seal in cavities opened by sliding. The cataclastic process is attested by observations of broken and kinked fibres, and by observations of euhedral crystals in the cavities opened by fault sliding. Natural examples are given to recognize and balance the mass transfers. Finally, a model is proposed to explain seismic/aseismic transitions on a given fault based on a gradual reduction in the cataclastic strength of the fault during sliding by successive breaking of asperities, and on the principle of minimum work consumed during the slip since the energy needed either for pressure solution sliding or for the cataclastic event may vary differently with the progressive sliding. Depending on the limiting processes for pressure solution slip, stable aseismic or unstable slip with seismic/aseismic transition is predicted.

The mechanical behaviour of the upper crust is commonly considered to be cataclastic. The creep strength of rocks is presumed to increase with depth, down to a transition zone at about 15–20 km, where dislocation creep mechanisms operate (Brace & Kohlstedt 1980; Kirby 1983; Carter & Tsenn 1987). This model is supported by two facts.

(i) The maximum frequency of earthquakes is located within the upper 15–20 km of the crust, as shown, for example, by depth distribution of earthquakes in the continental crust of the United States (Sibson 1982), and focal depth of intracontinental and intraplate earthquakes (Chen & Molnar 1983).

(ii) The experimental strength of rocks indicates a transition from frictional slip to plastic creep at temperatures and pressures appropriate for depths of 15–20 km (Paterson 1978). At relatively high strain rate (greater than 10^{-8} s^{-1}), a change from a pressure-sensitive friction law (Byerlee 1968), to a strongly temperature-dependent plastic law (Poirier 1985) is observed.

Other natural and experimental data are not, however, consistent with this model for distribution of deformation mechanisms in the crust. It is well known that within the upper crust not one, but two major deformation mechanisms operate: cataclastic deformation and pressure solution creep (see Knipe 1989; Cox & Etheridge 1989 for recent reviews), with transitions in the relative intensities of these two observed mechanisms (Hadizadeh & Rutter 1983). The aim of this paper is to show that these two mechanisms may accommodate the sliding on the same fault, with transitions between seismic (cataclastic) and aseismic (pressure solution) slip, during the sliding period.

Accommodation of sliding on faults by mass transfer

The question that must be answered is how to integrate earthquakes and pressure solution in an upper crust deformation model. The problem was already discussed, for example by De Bremaecker (1987), who proposed a clear distinction between two different processes.

An earthquake is a dynamic process, and at a given instant, only part of a long fault (the active part) is in motion. This active part sweeps the length of the fault at high velocity.

After the earthquake, the barriers and asperities that prevented sliding of the fault remain under stress and may be gradually (but very slowly) dissolved by pressure solution.

The aim of this paper is to illustrate, by natural observations, the succession of these two mechanisms on the same fault. Seismic/aseismic transition markers will be described on several faults. The observations suggest that the progressive dissolution of the asperities reduces their strength, breaking them one after the

From Knipe, R. J. & Rutter, E. H. (eds), 1990, *Deformation Mechanisms, Rheology and Tectonics*, Geological Society Special Publication No. 54, pp. 461–473.

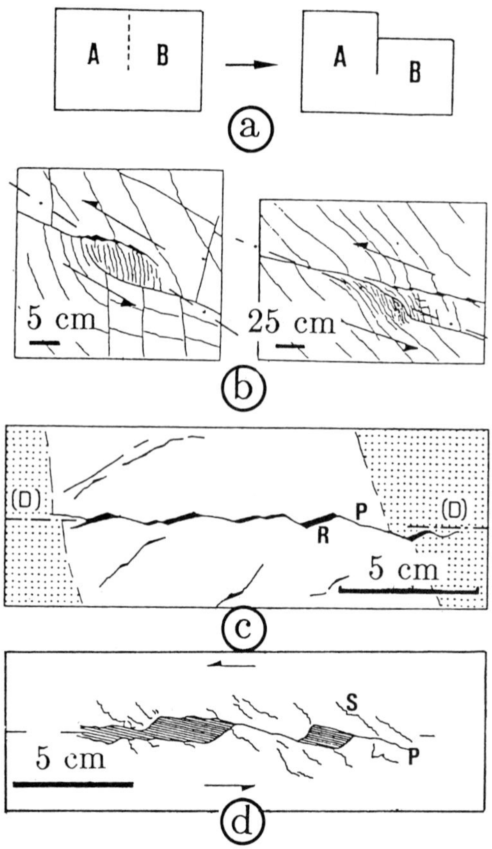

Fig. 1. Accommodation of sliding on faults by mass transfer; (**a**) internal deformation around fault (with volume change between A and B) either by change of density or by mass transfer; (**b**) compressive bridges between first generation en-echelon fractures in a fault zone (limestones in Languedoc); (**c**) pressure solution phenomena along second generation P fractures, associated with domino openings filled with quartz (black) (from gneiss in the Oisans Massif); (**d**) calcite domino opening associated with dissolution on P fractures and with diffuse dissolution around the fault (stylolites, S) (limestones in the Chaînes Subalpines).

other, leading to general fracturing of the whole fault zone i.e. earthquake. To model such behaviour the characteristics of the solution features on and around the faults have first to be discussed.

To accommodate a relatively large fault displacement, various types of internal deformation of the rocks are needed to accommodate volume changes near or along the fault (Fig. 1): such internal deformation is always needed near the termination of faults (Fig. 1a). If a fault consists of an array of en echelon fractures (Fig. 1b), first generation fractures define compressive or tensile bridges (Gamond & Giraud 1982), and internal deformation is also needed within these bridges (Segall & Pollard 1980, Rispoli 1981, Sibson 1989). Finally, if the fault surface is irregular (Fig. 1c), asperities preventing fault movement must also change shape (Elliot 1976), either by dissolution on the surface (Fig. 1c), or dissolution around the fault (Fig. 1d), to accommodate the displacement.

The knowledge of the mechanisms of such a volume change is of major interest for understanding the kinematics of sliding on faults. There are two mechanisms of volume change. As elastic strains probably never exceed 1% (Molnar 1983), a difference in internal deformation, with a volume change of more than 10% between parts A & B (Fig. 1a), and linked to the displacement on the fault, must be accommodated either by a change in density or by mass transfer. The change in density may be associated with cataclastic deformation and then with fast displacement during an earthquake, whereas mass transfer by pressure solution implies a very slow displacement which cannot be associated with the dynamic motion of an earthquake. The partitioning between these two mechanisms will be discussed on several natural examples.

Mass transfer around faults

For various natural examples of faults (from decimetre to kilometre in length) it is possible to show that the total displacement along a fault may be accommodated only (or at least mostly) by mass transfer, (e.g. dissolution of asperities or barriers).

Mass transfer may be demonstrated by a variation of chemical content and density of rocks across a heterogeneously deformed zone. When two portions in an initially homogeneous rock subjected to heterogeneous stress are deformed by pressure solution, they reveal a difference in chemical composition. For example, such chemical composition differences are observed (Fig. 2b) in layer composed of both soluble and insoluble species and cut by a fault (Fig. 2a). This behaviour may be explained as follows. Before faulting, the chemical composition and density of zones A & B were the same. During the slow displacement of the fault, soluble species were removed from the B zone leading to a concentration of insoluble species in this zone and thus to a difference in chemical composition between the two zones A & B. The sector with dissolution has been called the

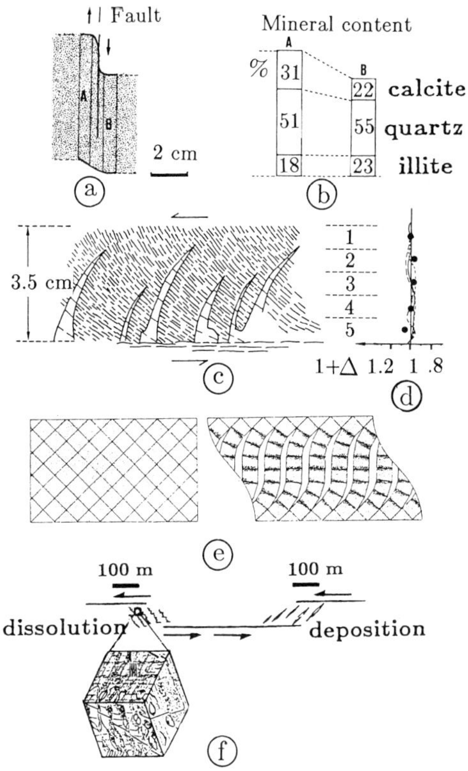

Fig. 2. Balance of mass transfers around faults. (**a**) Layer (slates) composed of both soluble and insoluble species, initially homogeneous, and cut by a fault. A difference in mineral content appears between parts A and B (**b**), giving a means of estimating the mass decrease from A to B (−22%). This value, which may be considered the same as the change in thickness (−23%, without change in density), indicates that the displacement on the fault occurred at a very low (aseismic) rate. (**c**) Shear zone (sandstones), with solution cleavage (s) and tension gashes (t). (**d**) estimated mass transfer between a reference zone (1) and some strips parallel to the displacement (2,3,4,5) by chemical (dots) and geometric (lines) analysis, indicating that each strip may be considered as a closed system. (**e**) Schematic shear zone model with mass transfer from dissolution (concentration of dots) to deposition (white veins), but in natural deformation the spacing of the zones of dissolution is smaller than those of the zones of deposition. (**f**) A 26% volume decrease (estimated by geometric analysis of various markers), within a 100 m wide compressive bridge between en-echelon faults, indicating a large open system for mass transfer (greater than 100 m), with deposition in a tensile bridge (from Morrocan marbles).

'exposed zone' and the zone either with deposition (closed system) or with the initial homogeneous composition (open system) has been called the 'protected zone' (Gratier 1983). The mass decrease of the exposed zone compared with the protected zone is defined as follows:

$$\Delta M/M_0 = (I_p/I_e) - 1,$$

where I_p and I_e are the total percentages of all insoluble minerals in the exposed zone (I_e), and in the protected zone (I_p), ΔM = change in mass, M_0 = initial mass.

This difference in composition is used to calculate the decrease in mass from A to B, and to compare this value with that for the change in thickness (Gratier 1983). In the example given here (Fig. 2a, b) these two values may be considered as the same: −22% (mass) and −23% (thickness), and this without any change in density between A and B. This demonstrates that the entire displacement along the fault was accommodated, at a very slow rate, only by mass transfer. If the volume of redeposited matter can be estimated in the protected zone, this method can also be used to establish the size of the closed system for mass transfer (size for which the dissolved mass is equal to the redeposited mass).

Another example is given in Fig. 2c, where the internal deformation around an irregular fault is accommodated by mass transfer from solution cleavage seams to tension gashes in a shear zone within a small closed system. In such a case, the balance of mass transfer may also be determined by comparative chemical analyses. For example, in Fig. 2d, the chemical analysis of strips of rock (sandstone) parallel to the shear zone displacement were compared with the mineral composition of a reference zone (this reference being away from the zone with curvature of the cleavage and tension gashes). Using the relation given above, the calculation of the volume change of each strip versus the reference zone (Fig. 2d), shows that no mass transfer occurs between these strips. The development of the tension gashes filled with quartz and calcite thus compensates for dissolution of these minerals along the solution cleavage surfaces between the veins, (see also Beach 1974). Such a closed system is confirmed by the geometric analysis of the angles between cleavage (S), veins (T) and the shear zone (Fig. 2d) suggested by Ramsay (1981). A schematic model is given for such behaviour (Fig. 2e), but in natural deformation the spacing of the zones of dissolution is always smaller than those of the zones of deposition (Gratier 1987).

These measurements clearly show that sliding on a fault may be accommodated by solution/deposition with closed mass transfer systems varying from few millimetres to several decimetres. In such a case, the mass transfer must take place by diffusion along paths with high fluid content since the temperature of deformation, deduced from fluid inclusion studies remains below 350°C (Gratier & Vialon 1980). This mass transfer mechanism is attested by other studies using stable isotope and fluid inclusion measurements (Kerrich *et al.* 1978). Various other examples confirm this mechanism for sliding (Carrio-Schaffhauser & Chenevas-Paule 1989), but the calculation of mass transfer using the difference in chemical composition is limited to samples of faults of several decimetres to one metre in size.

To investigate the volume change of rocks along fault systems with much larger closed systems, a method based on the geometric analysis of deformation markers (Ramsay & Huber 1983), is necessary. The geometric analysis of several deformation markers: stylolites, veins, deformed brachiopods in some Morrocan marble quarries (Fig. 2f) indicates a volume decrease of 26%, estimated to be a mean value for several hundred metres in size (Gratier 1976). On the contrary, rocks from other quarries, several hundred metres away show an increase in volume that is more deposition than dissolution. These two kinds of quarries are situated along a large strike alip fault, at the position of compressive and tensile bridge structures respectively. This means that mass transfer may also accommodate very slow displacement on large fault zones (dissolution and/or deposition within bridge structures several hundred metres wide). In this case, fluid displacement must take place by infiltration (Etheridge *et al.* 1984). Such a large displacement of fluids in tectonic or metamorphic regimes is attested by various types of measurements: fluid inclusion studies (Mullis 1983), carbon and oxygen isotope data (Dietrich *et al.* 1983; Rye & Bradbury 1988).

As a conclusion the observations of natural examples show two types of result.

Closed systems vary in size from few millimetres to several decimetres, with mass transfer occurring by diffusion in the shear zone or around a single fault, or, the open system is greater than 100 m in large regional fault zones with extensive area of volume change by dissolution or deposition, and with mass transfer occurring by infiltration.

It is now necessary to discover whether the displacement markers on the fault surfaces, e.g. the striae, also indicate such a very slow rate of sliding.

Mass transfer along the fault surface

Two completely different types of striae must be distinguished. One kind of striation is a mechanical in origin comparable to a mechanical scratch (Goguel 1948; Petit 1987), which may be associated with fast displacement during seismic slip. This kind of striation is well known on natural fault surfaces developed during earthquakes, and it can be reproduced easily in the laboratory when fracturing rocks at high stress level, high strain rate and low temperature (Paterson 1978). Other kinds of striae which are also very common on naturally formed faults (Durney & Ramsay 1973; Means 1987), are not scratches at all, but are rather crystal fibres, (Figs. 3 & 4). The development of such fibres may be explained as follows (Elliott 1976). If two rock parts are separated by an irregular fault surface, a slow displacement of one part of the rock may be accommodated by the dissolution of the asperities that prevent the displacement. At the same time, deposition may occur in the cavities created by the displacement on the fault. The crystals growing within these cavities sometimes have a typical fabric (fibre minerals), first described by Ramsay (1980) and explained by him as linked to a succession of microcrack openings followed by their immediate sealing. The redeposited crystalline material is derived by pressure solution from the rocks matrix (see Figs 1 & 2). On the natural example given, Fig. 3a, the mean width of each crack opening is about 10 to 50 μm and, as each crack is limited to one or two crystals, several thousand crack-events would be required to achieve a 1 cm displacement. The length of such crystal fibres may reach 20 cm on other examples (Fig. 4).

Another interesting feature is shown in Fig. 3a. Pressure solution–deposition slip on this fault is attested by crystal fibres of quartz and feldspars but some of these fibres are clearly kinked or broken and cut by cataclastic scratching faults. This means that cataclastic deformation may be found, on the same fault, associated with pressure solution markers. The presumed displacement rate on this fault has been plotted Fig. 3a with two successive processes: a very slow pressure solution slip then a fast cataclastic event.

A major problem has to be overcome. When considering microcracks within crystal fibres on a given fault, should these microcracks be thought of as part of a seismic rupture process

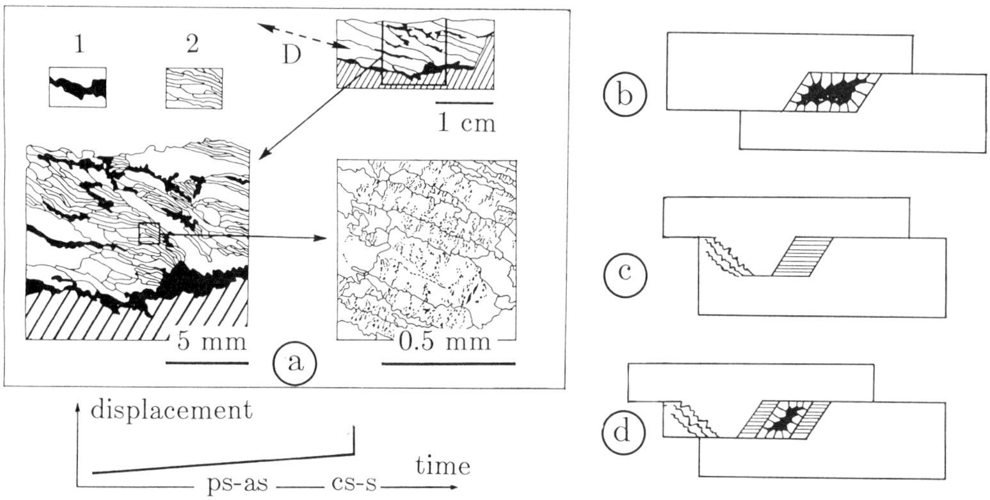

Fig. 3. Transition markers between seismic/aseismic sliding on the same fault; (**a**) Mineral fibres (aseismic slip), cut by small fractures (cataclastic event). Black area (1), cataclastic material; white area (2), crack-sealed quartz fibres; hatched area, gneiss, with general setting (top), and details of the fibres (bottom). The cavity opened by sliding is filled by mineral fibres indicating the displacement direction (D). The mechanism of such a sealing process is by successive microcracks opening (10 to 50 μm wide) immediately sealed (see Ramsay 1980; Cox & Etheridge 1983). Each microcrack is limited to only one or two crystals, each centimetre of displacement implies at least 10 000 microcrack events, during aseismic sliding with the rate limited by the solution–deposition rate (sample from gneiss in the Oisans massif). The diagram below the drawing gives the presumed sliding rate to time relation with successively pressure solution aseismic slip (ps-as) then cataclastic seismic slip (c-ss). (**b**) Euhedral radial crystals in a cavity opened by fault sliding indicate fast (wide) opening (seismic event). (**c**) Mineral fibres indicate slow (aseismic) slip. (**d**) Association of these two types of growth indicates aseismic to seismic transition.

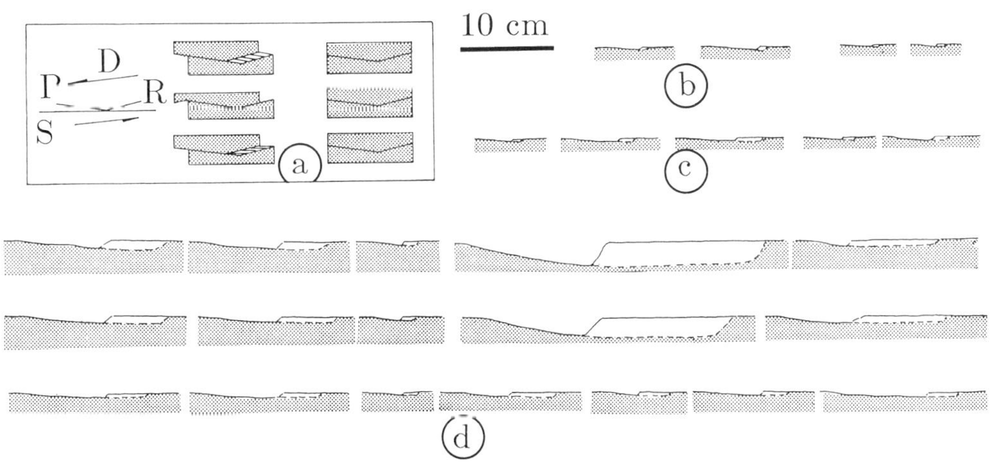

Fig. 4. (**a**) Various geometric relations between R and P fractures, displacement direction (D) and mean orientation of fault (S), illustrating various mass transfer systems: open systems with deposition (top) or with dissolution (middle) and closed systems (bottom). (**b,c,d**) Various natural geometric features in closed systems, measured on three different faults in the Chaînes Subalpines, on the same limestones (Upper Jurassic). White area, fibres; dotted area, limestones. On the faults (c) (5 fibres) and (d) (17 fibres) the length of the fibres which indicates the displacement accommodated by pressure solution, vary along the fault surface; (dotted part, limestones; white part, tectonic calcite fibres).

which occurred on this fault? According to the approach described here, the answer is no. The difference between seismic and aseismic deformation must be determined with reference to a ratio between displacement of an event and size of the asperities along the fault. If an event breaks several asperities, it is a cataclastic-seismic event for the fault. If an event only occurs within an asperity without significant associated displacement, it is not. This process of successive microcracks with immediate sealing, associated with dissolution, must then be considered as an aseismic slip. Of course some of these microcracks may be induced by the regional seismicity but the seismic/aseismic transition must be defined for a given fault.

Another indicator of such a transition between seismic and aseismic deformation is the observation of the change in the type of crystal growth within the cavities opened by sliding. If sliding occurs at very low rate, successive microcracks affecting only some grains may be immediately sealed with the typical fabric described above (Fig. 3c). On the contrary, if a relatively large displacement occurs, large cavities are opened and filled after their opening, by euhedral crystals with radial growth (Fig. 3b). An association of the two types of growth may be observed, attesting to a aseismic to seismic transition (Fig. 3d).

Volume transfer balance along faults may also be discussed through a simple geometrical plane-strain model in which volume loss is assumed to combine with sliding on the first and second-generation fractures (Gamond 1987). Left-lateral discontinuity (Fig. 4a) is considered to have two types of en echelon fractures, P & R, with respectively compressive and tensile behaviour. The volume transfer balance can be determined between the volume dissolved under transpression (on P fracture) and the volume deposited in the gaps (on R fracture) opened by sliding on the fault. On a section perpendicular to the fault surface, the change in area depends upon the geometry (P and R) of the asperities and their spacing, and upon the angle between the displacement (D indicated by the fibres) and the mean surface of the fault (F). Simple theoretical examples are given on Fig. 4a: open system with only dissolution (D parallel to R), or only deposition (D parallel to P), and closed system (D parallel to F).

In the case of closed systems (fibres parallel to the major discontinuity) various measurements were performed on natural faults, using a shape tracer in order to determine typical geometric features of the asperities. Measurements on a same fault surface in limestones, for example Fig. 4d, show that the length of the asperities (A) varies from 5–30 cm, as the length of the fibres (F) varies from 1.5–20 cm. The F/A ratio expresses the portion of the asperity dissolved during aseismic slip. Its mean value is 0.4, most values being between 0.5 and 0.35, with extreme values of 0.7 and 0.25. Measurements on other faults (Fig. 4b & c) in the same limestones (Upper Jurassic), but in different areas, show lower range of values for the asperity and fibre length, with lower F/A ratio (0.3). The spacing between the asperities is more difficult to estimate, but is probably of the same order of magnitude as the size of the asperities. The angle between dissolution surface and mean displacement on fault remains low, with a narrow range (from 5 to 15°). Conversely, the angle between the deposition surface and the fault varies more widely from 30 to 80°. On the given examples the change of length of the fibres along each fault surface is either random (Fig. 4c & 4d) or constant (Fig. 4b). Other measurements (Gamond 1983) indicated progressive decrease in the length of the fibres near the termination of a fault.

These features may be explained as follows. The variation in asperity size is linked to the geometry of the first and second-generation fractures along the fault zone and will be discussed later. A variation in fibre length may be due to different processes:

(i) some cavities may begin to form while sliding is still active, within bridge zones which were undergoing internal deformation (Fig. 5a)
(ii) some asperities may break during the sliding (Fig. 5b)
(iii) a decrease in displacement along the fault always appears near the end of the fault (Fig. 5c), associated with a heterogeneous volume change in the vicinity (see Fig. 1a).

From these observations, a model is derived which predicts why cataclasis versus pressure solution takes place, and the factors which affect or control these deformation processes can be determined.

Change in sliding mechanisms with time

The cataclastic shear fault strength is considered to be the yield stress needed to fracture a large part of the fault (earthquake). This cataclastic shear strength is proportional to the total length of the asperities along the fault. Progressive breaking of some asperities during aseismic slip (Fig. 5b), will lead to progressive decrease in this shear seismic strength.

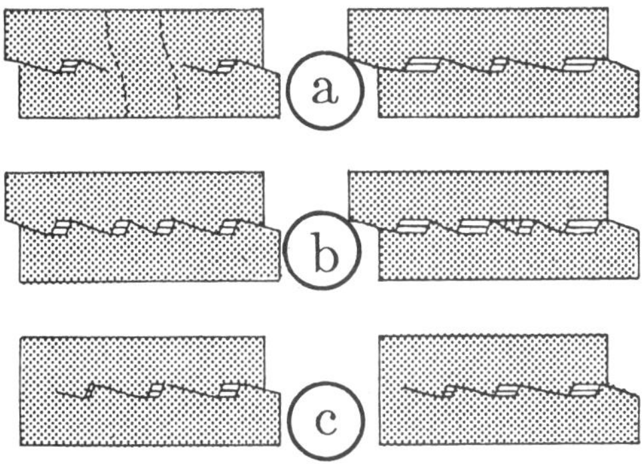

Fig. 5. Some explanations of the variation in the fibre length which can be observed on a given fault (see Fig. 4d). (**a**) Development of a new cavity in an internal deformation zone, while sliding is going on. (**b**) Breaking of an asperity during sliding. (**c**) Progressive decrease in displacement near the end of the fault (see Figs 1a and 2a).

Since the total length of the asperities array also decreases by progressive dissolution, this cataclastic shear strength value is also linked to the dependence of the two sliding mechanisms (cataclastic or pressure solution) on the asperity geometry. To study this effect, a simple model of sliding by pressure solution may be constructed for asperities of very simple shape. Pressure solution is a mechanism which implies successive processes: dissolution, transfer (by diffusion or infiltration), and deposition. A variety of thoeretical pressure solution creep laws have been established in which the overall rate is controlled by the slowest process (Raj & Ashby 1971; Rutter 1976 & 1983; Elliott 1976; Raj 1982; Etheridge *et al.* 1984; Gratier & Guiguet 1986; Gratier 1987). The same analysis was used to establish a theoretical pressure solution sliding law (Rutter & Mainprice 1979). With the different types of limiting processes, the sliding rates ($\dot{\gamma}$) are as follows:

when infiltration rate is the limiting process:

$$(I)\text{: } \dot{\gamma} \propto Kic$$

when kinetics of reaction is the limiting process:

$$(R)\text{: } \dot{\gamma} \propto k_c \sigma_n v / RT$$

when diffusion rate is the limiting process:

$$(D)\text{: } \dot{\gamma} \propto Dc\sigma_n w / RThd\alpha$$

where kc = kinetics of reaction; v = molar volume; R = gas constant; T = temperature (K); c = concentration of soluble component; σ_n = difference in normal stress between dissolution and deposition sites; K = permeability coefficient; i = pressure gradient; w = width of path transfer; D = diffusion coefficient; h = asperity height; d = asperity length, α = numerical coefficient.

Details of these laws will not be discussed here. For the purpose of the work, the most interesting relation is that between the sliding rate and the asperity length (d). In the first two relations (I and R), the sliding rate is not dependent on the asperity length. In the third one (D) the sliding rate depends upon the distance of mass transfer from dissolution to deposition, more or less linked to the asperity length. Variations in distance of mass transfer during the sliding also depends on the crystal growth mechanism and on the distribution of dissolution between the two parts of the fault (hatched upper part and shaded lower part, Fig. 6). If only one part of the fault is dissolved e.g. the upper part (Fig. 6), the distance of mass transfer decreases with the progressive sliding if each successive growth increment occurs at the fibres/upper part limit, and this distance remains constant if each growth increment occurs at the fibres/lower part limit (Fig. 6). In these cases, the numerical coefficient α varies from $1/d$ to 1, d being the length of the asperities on the upper part of the fault. The variation in the distance of mass transfer with the progressive sliding is more complex if dissolution occurs on both parts of the fault, and/or if fibres grow by the syntaxial mechanism (each new increment located in the median part of the fibres, see Durney & Ramsay 1969).

Assuming a linear relation between stress (or pressure gradient) and sliding rate (Rutter 1976; Urai *et al.* 1986), the stress values needed to drive a constant sliding rate may or may not change during the progressive sliding, depending on the sliding laws. As the strain energy consumed during pressure solution sliding depends on this stress value (Elliott 1976), at constant sliding rate, the change in rate of work with time differs depending on the laws considered. The energy needed to break the asperities, however, is always dependent upon the asperity length.

The energy values needed to accommodate either cataclastic or pressure solution sliding may be compared. When the two energy values vary in a similar manner during the progressive sliding, stable sliding is expected (e.g. pressure solution aseismic sliding Fig. 6), whereas, when the two energy values vary in a different manner, an unstable process is expected (Fig. 6), with successive mechanisms using minimum energy (aseismic/seismic slip during the progressive sliding).

The geometry of the asperities and fibres

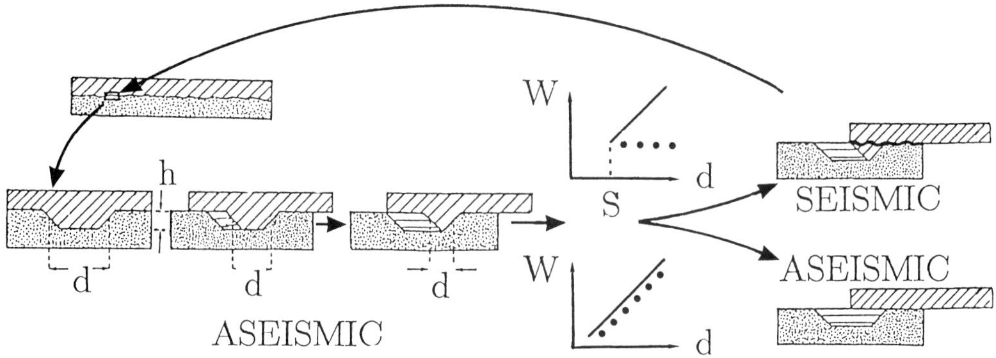

Fig. 6. Theoretical model of pressure solution sliding with two types of change. Stable aseismic sliding (total dissolution of the asperity), when the energy needed for pressure solution (dotted line) or breaking (solid line) of the asperity is the same during progressive sliding (bottom diagram), or unstable sliding (aseismic/seismic transition at S point), when the energy needed for pressure solution (dotted line) is not dependent upon asperity length and when the energy for breaking the asperity (solid line) becomes lower than the energy needed for pressure solution (top diagram).

measured on several faults may be used to estimate the ratio between displacement by pressure solution (length of fibres) and total (initial) length of asperities (0.3 to 0.4, Fig. 4).

Change in sliding mechanisms with depth

By way of example, we will examine the results of deformation in the upper crust through a large crustal thrusted region, such as the Alpine chain (Fig. 7), where significant internal deformation was associated with thrust motion. In this example, horizontal contraction of the Subalpine sedimentary cover varies from 5% to 30% when homogeneous deformation (folds and minor faults) of large elements (15 × 15 km of initial horizontal extent) is considered (Gratier *et al.* 1989). Cataclasis (zones where rocks are broken into lithic fragments) is fairly rare, and is generally located near fault zones. This mechanism alone cannot account for the whole internal deformation of the 15 × 15 km elements of the upper crust. Large diffuse deformation associated with folds, shear zones or faults is accommodated by pressure solution. The markers of this mass transfer deformation are pitted pebbles in molassic basins (McEwen 1981), stylolites in limestones, or solution cleavage in slates, metamorphic rocks and granites (Gratier & Vialon 1980). These dissolution markers are always associated with veins (deposition), and the study of the fluid inclusions trapped in these veins (Bernard *et al.* 1977; Jenatton 1981) indicates the depth of deformation from 1 to 10 km (Fig. 1).

Cataclasis and pressure solution are clearly associated in thrust sheet motion (e.g. limestones in the Chaînes Subalpines; Fig 8b), and in other orogenic belts (see for example Geiser 1988). Cataclasis is located near the ramp of the basal discontinuity whereas the deformation of the sheet moving over this ramp is accommodated by pressure solution (stylolites, solution cleavage and veins). As this internal deformation of the sheet is always necessary to accommodate its displacement this means that the sheet motion was slow enough and continuous to be associated with a diffuse creep process at depths lower than 5 km. In slates of the sedimentary cover of the Oisans massif the amount of strain accommodated by pressure solution over a large region (several km) may reach high values such as 0.4 (contraction) × 2.5 (extension), at depth from 5 to 10 km (deduced from fluid inclusion studies as for the preceding example, Gratier & Vialon 1980). Following the finite strain/time relations given by Pfiffner & Ramsay (1982), and taking into account the regional geological constraints, such deformation could take place over 3×10^6 years at a strain rate of $10^{-14}s^{-1}$. Other examples in the Alps indicate pressure solution creep at depths up to 15–20 km.

Using the theoretical sliding laws, and both natural observations and experimental results (see also Cox & Etheridge 1989), the order of magnitude of the pressure solution shear strength of rocks may be estimated through a typical section (sediments with high porosity, limestones, slates, metamorphic rocks (gneiss then schists), then granitic rocks), for a strain rate of $10^{-14}s^{-1}$. Classical values were used for transfer coefficients (Rutter 1976), with values of the size of the closed system for mass transfer

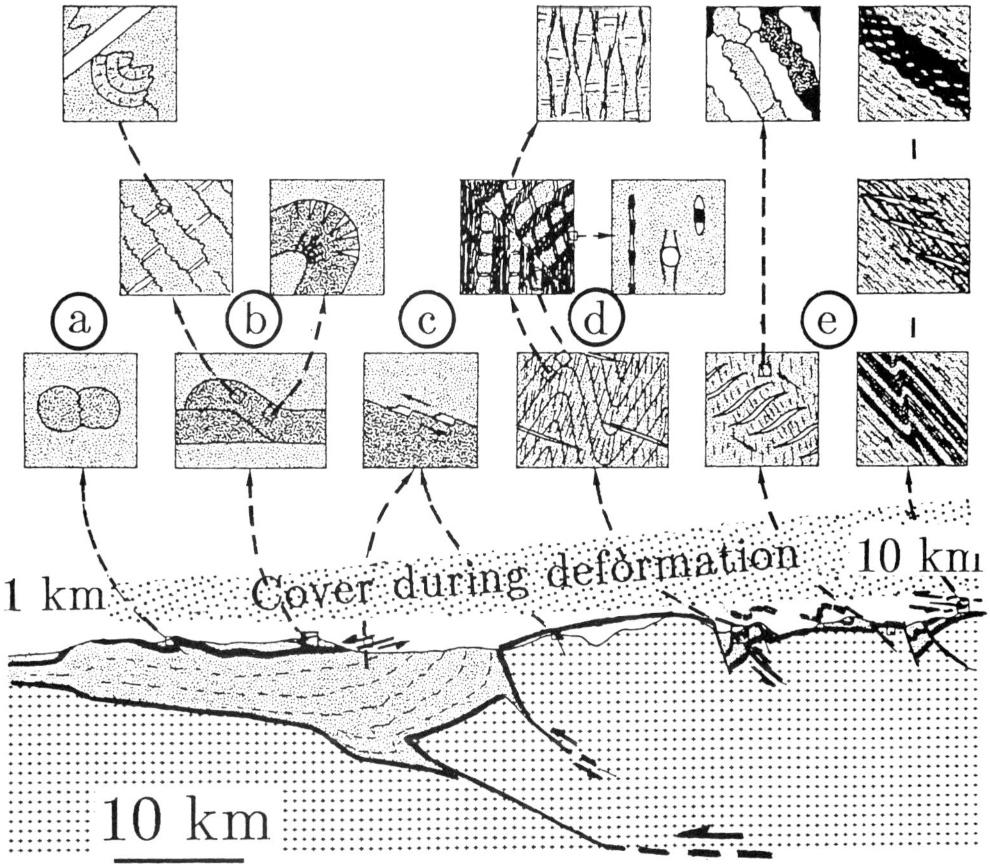

Fig. 7. Internal deformation of a thrust domain in the Alpine Chain; cross section from thrusting basement (dotted area, Oisans) to deformed sedimentary cover (shaded area, Chaînes Subalpines) near the passive margin. The diffuse deformation associated with folds (**a,b,d**), shear zones (**e**), and fault (**c**), is accommodated by pressure solution, attested by pitted pebbles (a, molassic basins), stylolites (b, limestones), solution cleavage (d,e, slates, metamorphic and granitic rocks), and deposition in tectonic veins. The process occurs at all scales from macro (bottom) to micro (top) scale. The thickness of the cover during the deformation was deduced from fluid inclusion studies (in tectonic veins).

from 0.1–10 mm (increasing values from granite-gneiss to sediments, shales, and limestones). A wide range of values was obtained, with zones of very low strength (which may be associated with flat shear zones either in extensive or compressive deformation), and with a mean value more than one order of magnitude below the cataclastic shear strength (Molnar & Tapponnier 1981; Sibson 1982). Following this approach, pressure solution creep is equivalent to a Newtonian fluid with orders of magnitude of viscosity from 10^{19} to 10^{21} Pa s, with at least one or two orders of magnitude of uncertainty linked to the meagre knowledge of the behaviour of a fluid phase trapped between two stressed solids. For example, from experimental results (Gratier & Guiguet 1986) a value of 10^{17} Pa s has been extrapolated for deformation of a fine-grained quartz aggregate in water at about 350°C. The two profiles (pressure solution and cataclastic shear strength) are drawn on Fig. 8. The strength of the upper crust is probably between these two profiles: this strength may change both with time and with space depending on the fluid content and on the mean strain rate of the upper crust.

Conclusions

A single fault surface often displays evidence of coupled seismic and aseismic slip processes.

For a given fault zone, the observations

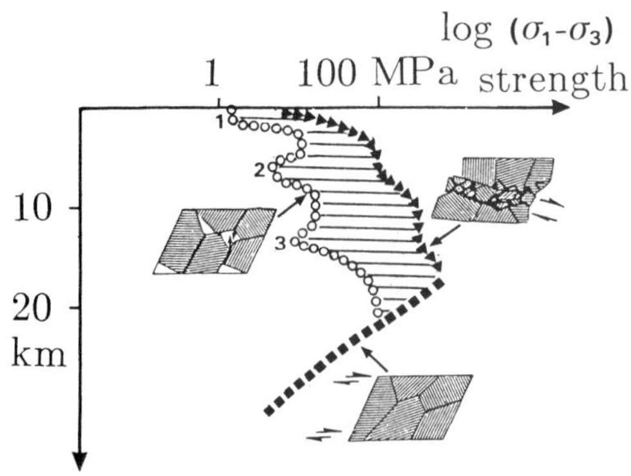

Fig. 8. Strength profile through the upper crust taking into account the two (observed) deformation mechanisms: cataclastic deformation (law derived from Mohr–Coulomb behaviour, triangles), and pressure solution creep (laws derived from theoretical, experimental and natural observations, circles), at geological strain-rate: 10^{-14} s^{-1}, (equivalent to a change in length of 3 mm a^{-1} for an element of 10 km size). Dislocation creep mechanisms (squares) operate within the lower crust. The model of the typical crustal section is composed of different rocks, (see Fig. 7): from top to bottom, sediments with high porosity (1), limestones, slates (2), metamorphic rocks (gneisses then schists (3)), granitic rocks. The strength values vary considerably with depth, depending on various parameters (possibility of mass transfer, solubility of solid in fluid, mean displacement rate at the boundaries of the system, etc.).

suggest the following sequence which integrates earthquakes and pressure solution mechanisms in an upper crust deformation model.

(a) The development of first generation fractures generally leads to arrays of *en-echelon* fractures, either P or R fractures, depending on the deformation conditions, (Fig. 9a).

(b) Second generation fractures appear in the bridges between the previous ones. These second generation fractures generally only need an incremental displacement to achieve the connection with the first ones (Fig. 9b).

(c) As the rate of displacement is generally imposed at the boundaries of the system, the behaviour of the zone depends on various parameters classified by order of importance: fluid content, solubility of certain minerals of the rocks in this fluid, geometry of mass transfer path and value of the coefficient of mass transfer (by diffusion or infiltration), size of the closed system, type of limiting process (reaction kinetics, mass transfer rate), stress level, temperature and pressure conditions. If dissolution–deposition mechanism can accommodate the imposed displacement rate, the motion on the fault is limited by the rate limiting solution-deposition process (Fig. 9c). If accommodation by mass transfer is not sufficient, the stress level increases and an earthquake is likely to occur (Fig. 9c′).

(d) During this very slow sliding, local fracturing of some asperities may occur without

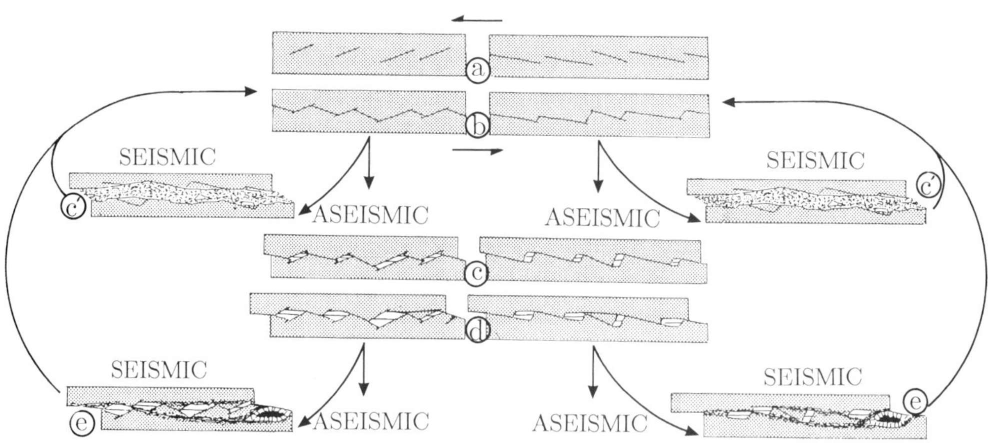

Fig. 9. Seismic/aseismic transition on two fault zones. (**a**) First generation fractures, either R (left) or P (right) fractures. (**b**) Second generation fractures (within compressive or tensile bridges), leading to an irregular fault surface. (**c**) Pressure solution sliding, with eventually local (aseismic) breaking of some asperities (**d**). (**e**) General fracturing of a large part of the fault (cataclastic displacement greater than asperity length, earthquake). If the two parts of the rock remain imbricated after seismic displacement, pressure solution sliding may be reactivated (b). An earthquake may occur (c′) at any time, if pressure solution sliding is not able to reduce the stress level linked to the constant displacement rate at the boundaries of the system.

general fracturing (Fig. 9d) but this local fracturing gradually lowers the cataclastic shear strength of the fault zone.

(e) The system behaviour depends on the variation in energy needed either to dissolve the asperities or to break them. As this energy change does not always have the same dependence on the geometry of the asperities (Fig. 6), the system may behave either as a stable system (dissolution of all the asperities) or as an unstable system (general breaking of a large part of the fault, earthquake, Fig. 9e). This cataclastic deformation may be followed by a new imbrication of the two parts of the fault, the process being reinitialised at the step corresponding to Fig. 9b.

The model implies alternately fast, localized processes (seismic-cataclasis) and large slow creep processes (aseismic-pressure solution). The same cataclastic/pressure solution coupled processes may occur to accommodate the deformation around the fault. Successive aseismic/seismic slip cycles probably lead to softening of the fault, since the size and strength of the asperities decrease. As the deformation of rocks by pressure solution is always associated with the development of a tectonic differentiation (Gratier 1987), softened and hardened bands are developed which may change the location of sliding deformation.

Finally, one of the major problems about earthquakes is to know whether such a phenomenon can be avoided. This study shows that natural sliding on faults is often accommodated by an aseismic mechanism. The question is: is it possible to increase this aseismic rate of sliding artificially in order to release the stress concentration? Following experimental results on pressure solution, the best strain rate activation is obtained when using a very good solvent, either water with salts (Rutter 1976; Raj 1982; Pharr & Ashby 1983) or NaOH and NH_4Cl solutions with quartz or calcite respectively (Gratier & Guiguet 1986). One method of preventing earthquakes could be to inject such good solvents of at least one of the minerals of the rocks, through a fault zone (at very low pressure of course to avoid hydraulic fracturing).

We thank D. Prior, A. McCaig, P, Molnar and an anonymous reviewer for their careful revision of this paper.

References

BEACH, A. 1974. A geochemical investigation of pressure solution and the formation of veins in a deformed Greywacke. *Contributions to Mineralogy and Petrology*, **46**, 61–68.

BERNARD, D., GRATIER, J. P. & PECHER, A. 1977. Application de la microthermométrie des inclusions fluides des cristaux syncinématiques à un problème tectonique. *Comptes-Rendus Sommaires de l'Académie des Sciences Paris*, **5**, 284–288.

BRACE, W. F. & KOHLSTEDT, D. L. 1980. Limits on lithospheric stress imposed by laboratory experiments. *Journal of Geophysical Research*, **85**, 6248–6252.

BYERLEE, J. D. 1968. Brittle-ductile transition in rocks. *Journal of Geophysical Research*, **73**, 4741–4750.

CARRIO-SCHAFFHAUSER, E. & CHENEVAS-PAULE, F. 1989. Quantification géométrique de la dissolution-cristallisation liée à une déformation cassante. *Bulletin de la Société Géologique de France*, **8**, 597–604.

CARTER, N. T. & TSENN, M. C. 1987. Flow properties of continental lithosphere. *Tectonophysics*, **136**, 27–63.

CHEN, W. P. & MOLNAR, P. 1983. Focal depths of intracontinental and intraplate earthquakes and their implications for the thermal and mechanical properties of the lithosphere. *Journal of Geophysical Research*, **88**, 4183–4214.

COX, S. F. & ETHERIDGE, M. A. 1983. Crack-seal fibre growth mechanisms and their significance in the development of oriented layer silicate microstructures. *Tectonophysics*, **92**, 147–170.

—— & —— 1989. Coupled grain-scale dilatancy and mass transfer during deformation at high fluid pressures: examples from Mount Lyell, Tasmania. *Journal of Structural Geology*, **11**, 147–162.

DE BREMAECKER, J. Cl. 1987. Thrust sheet motion and earthquake mechanisms. *Earth and Planetary Science Letters*, **83**, 159–166.

DIETRICH, D., MCKENZIE, J. A. & SONG, H. 1983. Origin of calcite in syntectonic veins as determined from carbon-isotope ratios. *Geology*, **11**, 547–551.

DURNEY, D. W. & RAMSAY, J. G. 1973. Incremental strains measured by syntectonic crystal growths. *In*: DE JONG, K. A. & SCHOLTEN, R. (eds) *Gravity and tectonics*. J. Wiley, New York, 67–96.

ELLIOTT, D. 1976. The energy balance and deformation mechanisms of thrust sheets. *Philosophical Transaction of the Royal Society*, **283**, 289–312.

ETHERIDGE, M. A., WALL, V. J., COX, S. F. & VERNON, R. H. 1984. High fluid pressure during regional metamorphism and deformation: implications for mass transport and deformation mechanisms. *Journal of Geophysical Research*, **89**, 4344–4358.

GAMOND, J. F. 1983. Displacement features associated with fault zones: a comparison between observed examples and experimental models. *Journal of Structural Geology*, **5**, 33–45.

—— 1987. Bridge structures as sense of displacement criteria in brittle fault zones. *Journal of Structural Geology*, **9**, 609–620.

—— & GIRAUD A. 1982. Identification des zones de failes à l'aide des associations de fractures de second ordre. *Bulletin de la Société Géologique de France*, **7**, 24, 755–762.

GOGUEL, J. 1948. *Introduction à l'étude mécanique des déformations de l'écorce terrestre*. Memoires du Service de la Carte Géologique France.

GEISER, P. A. 1988. Mechanisms of thrust propagation: some examples and implications for the analysis of overthrust terranes. *Journal of Structural Geology*, **10**, 829–845.

GRATIER, J. P. 1976. Deformation et changement de volume dans un marbre à stylolites de la ràgion de Rabat (Maroc). *Bulletin de la Société Géologique de France*, **7**, 18, 1461–1469.

—— 1983. Estimation of volume change by comparative chemical analyses in heterogeneously deformed rocks (folds with mass transfer). *Journal of Structural Geology*, **5**, 329–339.

—— 1987. Pressure solution-deposition creep and associated tectonic differentiation. *In*: JONES, M. E. & PRESTON, R. M. (eds) *Deformation of Sediments and Sedimentary Rocks*. Geological Society, London, Special Publication, **29**, 25–38.

—— & GUIGUET, R. 1986. Experimental pressure solution on quartz grains: the crucial effect of the nature of the fluid. *Journal of Structural Geology*, **8**, 845–856.

—— & VIALON, P. 1980. Deformation pattern in a heterogeneous material: folded and cleaved sedimentary cover immediately overlying a crystalline basement (Oisans, France). *Tectonophysics* **65**, 151–180.

——, MENARD, G. & ARPIN, R. 1989. Strain-displacement compatibility and restoration of the Chaines Subalpines of the Westerns Alps. *In*: COWARD, M. P., DIETRICH, D. & PARK, R. G. (eds) *Alpine Tectonics*. Geological Society, London, Special Publication, **45**, 65–81.

HADIZADEH, J. & RUTTER, E. H. 1983. The low temperature brittle-ductile transition in a quartzite and the occurrence of cataclastic flow in nature. *Geologische Rundschau*, **72**, 493–509.

JENATTON, L. 1981. *Microthermométrie des inclusions fluides des cristaux associés à l'ouverture des fentes alpines*. Thesis, University of Grenoble.

KERRICH, R., BECKINSALE, R. D. & SHACKLETON, N. J. 1978. The physical and hydrothermal regime of tectonic vein systems: evidence from stable isotope and fluid inclusion studies. *Neues Jahrbuch fur Mineralogie Abhandlungen*, **131**, 225–239.

KIRBY, S. H. 1983. Rheology of the lithosphere. *Reviews of Geophysics and Space Physics*, **21**, 1458–1487.

KNIPE, R. J. 1989. Deformation mechanisms: recognition from natural tectonites. *Journal of Structural Geology*, **11**, 127–143.

MCEWEN, T. J. 1981. Diffusional mass transfer processes in pitted pebble conglomerates. *Journal of Structural Geology*, **3**, 25–38.

MEANS, W. D. 1987. A newly recognized type of slickenside striation. *Journal of Structural Geology*, **9**, 585–590.

MOLNAR, P. 1983. Average regional strain due to slip on numerous faults of different orientations. *Journal of Geophysical Research*, **88**, 6430–6432.

—— & TAPPONIER, P. 1981. A possible dependence of tectonic strength on the age of the crust in Asia. *Earth and Planetary Sciences Letters*, **52**, 107–114.

MULLIS, J. 1983. Transported metamorphism of the external parts of the central Alps, shown by fluid inclusions. *Symposium of European Current Research on Fluid Inclusions, Orleans*.

PATERSON, M. S. 1978. *Experimental rock deformation: The brittle field*. Springer-Verlag, Berlin.

PETIT, J. P. 1987. Criteria for the sense of movement on fault surface in brittle rocks. *Journal of Structural Geology*, **9**, 597–608.

PFIFFNER, A. O. & RAMSAY, J. G. 1982. Constraints on geological strain rates: arguments from finite strain states of naturally deformed rocks. *Journal of Geophysical Research*, **87**, 311–321.

PHARR, G. M. & ASHBY, M. F. 1983. On creep enhanced by a liquid phase. *Acta metallurgica*, **31**, 129–138.

POIRIER, J. P. 1985. *Creep of crystals, high-temperature deformation processes in metals, ceramics and minerals*. Cambridge Univ. Press.

RAJ, R. 1982. Creep in polycrystalline aggregates by matter transport through a liquid phase. *Journal of Geophysical Research*, **87**, 4731–4739.

—— & ASHBY, M. F. 1971. On grain boundary sliding and diffusional creep. *Metallurgical Transactions*, **2**, 1113–1128.

RAMSAY, J. G. 1980. The crack-seal mechanism of rock deformation. *Nature*, **284**, 135–139.

—— 1981. Shear zone geometry: a review. *Journal of Structural Geology*, **2**, 83–99.

—— & HUBER, 1983. *Modern structural geology, vol. 2*. Academic Press.

RISPOLI, R. 1981. Stress fields about strike-slip faults inferred from stylolites and tension gashes. *Tectonophysics*, **75**, T29–36.

RUTTER, E. H. 1976. The kinetics of rock deformation by pressure solution. *Philosophical Transactions of the Royal Society*, **A283**, 43–54.

—— 1983. Pressure solution in nature, theory and experiment. *Journal of the Geological Society, London*, **140**, 725–740.

—— & MAINPRICE, D. H. 1979. On the possibility of slow fault slip controlled by diffusive mass transfer process. In: STILLER, H. (ed.) *Symposium on Physics of Earthquake sources*. Gerlands Beiter Geophysik, Leipzig, **88**, S154–162.

RYE, D. M. & BRADBURY, H. J. 1988. Fluid flow in the crust: an example from a Pyrenean thrust ramp. *American Journal of Science*, **288**, 197–235.

SEGALL, P. & POLLARD, D. D. 1980. Mechanics of discontinuous faults. *Journal of Geophysical Research*, **85**, 4337–4350.

SIBSON, R. H. 1982. Fault zone models, heat flow, and the depth distribution of earthquakes in the continental crust of the United States. *Bulletin of the Seismological Society of America*, **72**, 151–163.

—— 1989. Earthquake faulting as structural process. *Journal of Structural Geology*, **11**, 1–14.
URAI, J. L., SPIERS, C. J., ZWART, H. J. & LISTER, G. S. 1986. Weakening of rock salt by water during long-term creep. *Nature*, **324**, 554–557.

Vein distribution in a thrust zone: a case history from the Northern Apennines, Italy

MASSIMO COLI & FEDERICO SANI

Dipartimento Scienze della Terra, Università Firenze, 50121 Firenze, Italy

Abstract: In the Northern Apennines the Cervarola Thrust superposes the Cervarola Unit (Early–Middle Miocene sandstones and shales) on the Marnoso Arenacea Unit (Middle–Late Miocene marly sandstones) and has the geometry of a leading imbricate fan. Each thrust sheet consists of a thick turbiditic sandstones sequence (Mt Cervarola Fm.), which grades downward into shales (Scisti Varicolore Fm.). During thrusting, the Scisti Varicolore Fm. was intensely deformed and widely affected by shear veins, filled with calcite. A regional pattern has been detected in the vein distribution: veins are concentrated some ten metres above the thrust fault, in a band a few metres thick, and can be grouped into two systems. The principal one presents two sets of veins striking in the same direction as the thrust: one set dips at a high angle to the thrust fault (High Angle Set, HAS), whereas the other one lies sub-parallel to the thrust fault (Low Angle Set, LAS). Both of these sets of veins have dip-slip slickenfibres. The secondary system comprises two other sets of veins, which lie obliquely to the thrust fault. Slickenfibres are strike-slip and trend nearly parallel to the thrusting direction and to the intersection lineation between the two sets. Vein development can be explained in the light of the fluid-dynamic studies recently developed for present-day accretionary prisms.

In recent years a deep interest has been shown in the geometry of thrust systems and the role of fluids in the development of thrusts. A special effort had been put into the study of present-day accretionary prisms in order better to define, by drilling, coring and in situ measurement, the geometry and the fluid-dynamics of these structures (Moore 1986; Moore *et al.* 1988). Studies have been mainly developed in Mio-Plio-Quaternary clastic sediments, water rich and commonly poorly consolidated. The aim of this paper is to contribute to the understanding of those processes by the case history of the Cervarola Thrust, Northern Apennines, Italy. It is of Upper Tortonian age, when the tectonics, sedimentary situation and geometry were comparable to those of the present-day accretionary prisms, even though in a different geodynamic context.

Geological setting

This paper analyses the vein distribution in the central segment of the Cervarola Thrust (Baldacci *et al.* 1967; Boccaletti *et al.* 1986; Abbate & Bruni 1987), between the Giogo Pass and Mt Falterona (Fig. 1), in the Northern Apennines, where the Cervarola Unit overthrusts, with the interposition of the Castel Guerrino sub-unit, the Marnoso Arenacea Unit.

In the Giogo Pass – Mt Falterona area, the Cervarola Thrust has the geometry of a leading imbricate fan (*sensu* Boyer & Elliott, 1982) (Fig. 2) from which secondary thrusts splay off. Each thrust sheet comprises up to a few tens of metres of shales (Scisti Varicolore Fm.) and a thicker turbiditic sequence (Cervarola Fm.). The Castel Guerrino sub-unit represents a frontal-thrust sheet resulting from the deformation of the eastern sedimentary ramp of the Cervarola foredeep.

The tectonic deformation due to thrusting in the Cervarola and Castel Guerrino Fms resulted in jointing, faulting and flexural slip folding with associated pencil structures. In contrast, the Scisti Varicolore Fm. shows jointing, fissility, pencil structures, scaly fabric and veins. The Marnoso Arenacea Fm., which represents the underthrust sequence, shows flexural slip folding, jointing, pencil structure and web structure.

The Cervarola Unit

According to Ricci Lucchi (1986*a*, *b*) the Cervarola Unit constitutes a complete clastic foredeep sequence, which tectonically evolved as an orogenic wedge. From bottom to top it comprises the following formations (Bortolotti *et al.* 1970; Dallan Nardi & Nardi 1972; Abbate *et al.* 1982; Abbate & Bruni 1987).

From Knipe, R. J. & Rutter, E. H. (eds), 1990, *Deformation Mechanisms, Rheology and Tectonics*, Geological Society Special Publication No. 54, pp. 475–482.

Scisti Varicolore Fm. (Late Eocene–Aquitanian). At the bottom (i.e. at the lowermost outcropping level), this consists of red and pale grey coloured shales and mudstones, grading upward to brown coloured marlstones and siltstones.

Cervarola Fm. (Aquitanian–Langhian). At the base, this consists of thick beds of turbiditic sandstones, with thin interbeds of shale and silty shale and thin beds of calcarenite. Further up, the formation consists of regularly alternating thin fine-grained sandstone beds and silty marls.

The Castel Guerrino sub-unit

This sub-unit (Groscurth 1971; De Jager 1979; Ten Haaf & Van Wamel 1979), is represented by a regular alternation of grey marlstones and thick beds of medium-grained turbiditic middle-Miocene sandstones.

The Marnoso Arenacea Unit

This is a composite, migratory turbidite wedge (Ricci Lucchi 1986*b*), which consists of regularly alternating beds of medium- fine-grained turbiditic sandstones and marlstones, with rare calcarenites (Bortolotti *et al.* 1970; Ricci Lucchi 1975; Ten Haaf & Van Wamel, 1979).

Vein distribution

The survey of the vein distribution in the Scisti Varicolore Fm. which outcrops along the main thrust and the secondary splay-off thrusts, was carried out by mapping the whole area at the scale 1:10 000 and then by a series of cross sections each one going from the thrust fault up to the base of the Cervarola Fm. (Figs 1 & 2). During this stage, each vein was characterized in detail in terms of both directional data (azimuth, dip, pitch, slip) and geometric relationship with other veins and structural elements (bedding, fissility, scaly fabric, fold, pencil, thrust). In the plots of Fig. 3 each great circle trace represents at least five veins in the field.

In the Scisti Varicolore Fm., the intensity of deformation increases from top to bottom (i.e. towards the thrust faults) (fig. 4). The uppermost stratigraphic levels of the Scisti Varicolore Fm. show a deformation pattern in which jointing played the main role. Going downward in the sedimentary sequence, jointing grades into a spaced fissility with a spacing of a few centimetres. When the lithology becomes more shaly, jointing disappears, fissility increases and

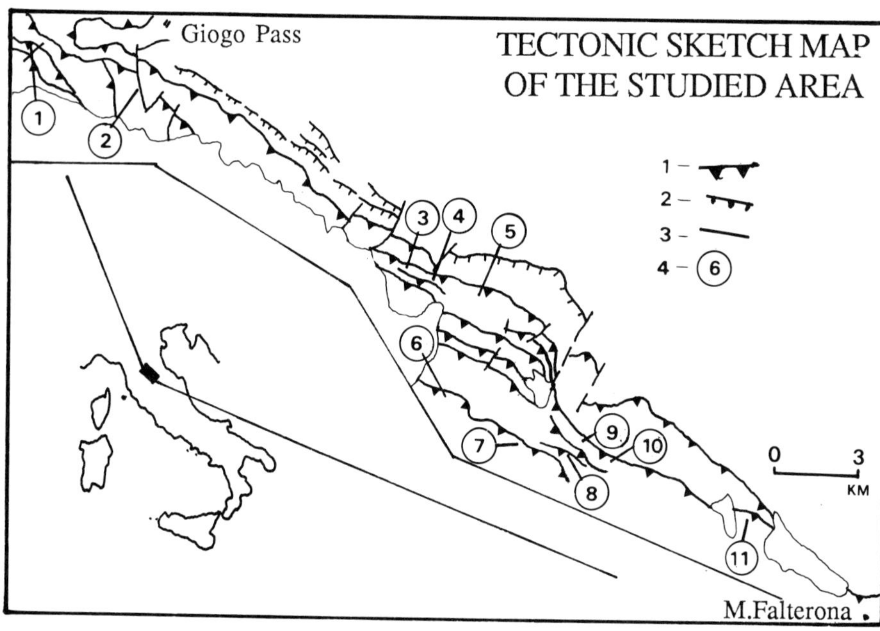

Fig. 1. Tectonic sketch of the Cervarola Thrust between the Giogo Pass and Mt Falterona. 1, thrust fault; 2, normal fault; 3, fault; 4, station reference number as in Fig. 3a.

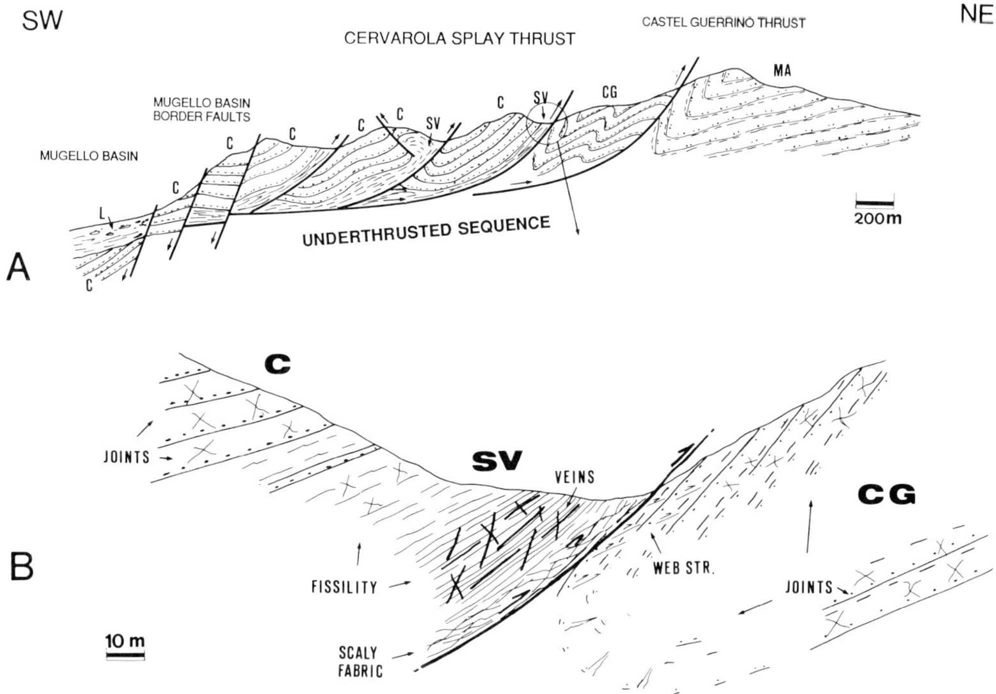

Fig. 2. (**A**) Conceptual cross-section of the Cervarola Thrust. It has the geometry of a leading imbricate fan thrust, where the Scisti Varicolore Fm. played as the main decollement level. Key: L, Mugello basin fluvial-lacustrine deposits; C, Cervarola Fm.; SV, Scisti Varicolore FM.; CG, Castel Guerrino Unit; MA, Marnoso Arenacea Fm. (**B**) Schematic cross section of the main thrust zone. The Castel Guerrino (CG) unit forms a footwall syncline affected by jointing and web structures, at the bottom of the overriding sequence the Scisti Varicolore (SV) shows scaly fabric, veins and fissility. These structural features grade upward to large spaced fissility and then to jointing in the Cervarola sandstones (C).

becomes more closely spaced (a few millimeters) and pencil structures and calcite veins appear. Veins reach a maximum concentration some ten metres above the thrust fault, in correspondence with the red and pale grey shales and mudstones. X-Ray data reveal the prevalence of smectitic group clays at this stratigraphic level in the Scisti Varicolore. Approaching the thrust fault, fissility grades into scaly fabric (*sensu* Moore *et al.* 1986) and vein occurrence rapidly decreases. In some localities, folded veins can be found at this lower structural level. They appear to have been deformed in a ductile fashion after their formation, probably when the surrounding rock was deformed by progressive simple-shear within the thrust-fault zone.

Veins are mineralized by calcite, and are of both shear and pure extensional type. The latter are rare, commonly sub-vertical, at about 90° to the thrust fault, with slickenfibres trending perpendicular to the thrusting direction and parallel to the pencil structures. Shear-veins are both of syntaxial and antitaxial type. No variations in the behaviour and structural pattern have been recognized among the two types. Many of them also present tracks of crack-seal origin. Shear-veins are lentiform, generally some metres long and up to 2 cm thick. Slickenfibres are sub-parallel to the rock-walls, and are up to about ten centimeters long. Within some shear-veins there are empty channels, which have the same dimension and orientation as the slickenfibres. These channels may represent the fluid-flow paths during the last stage of thrusting.

In the distribution of the shear-veins a regional pattern has been recognized, typified by two systems of veins (Fig. 3).

System 1 (Thrust strike parallel vein system, TSPVS)

This consists of two sets of veins, lying parallel to the strike of the thrust plane. At a regional scale, it is generally the predominant system.

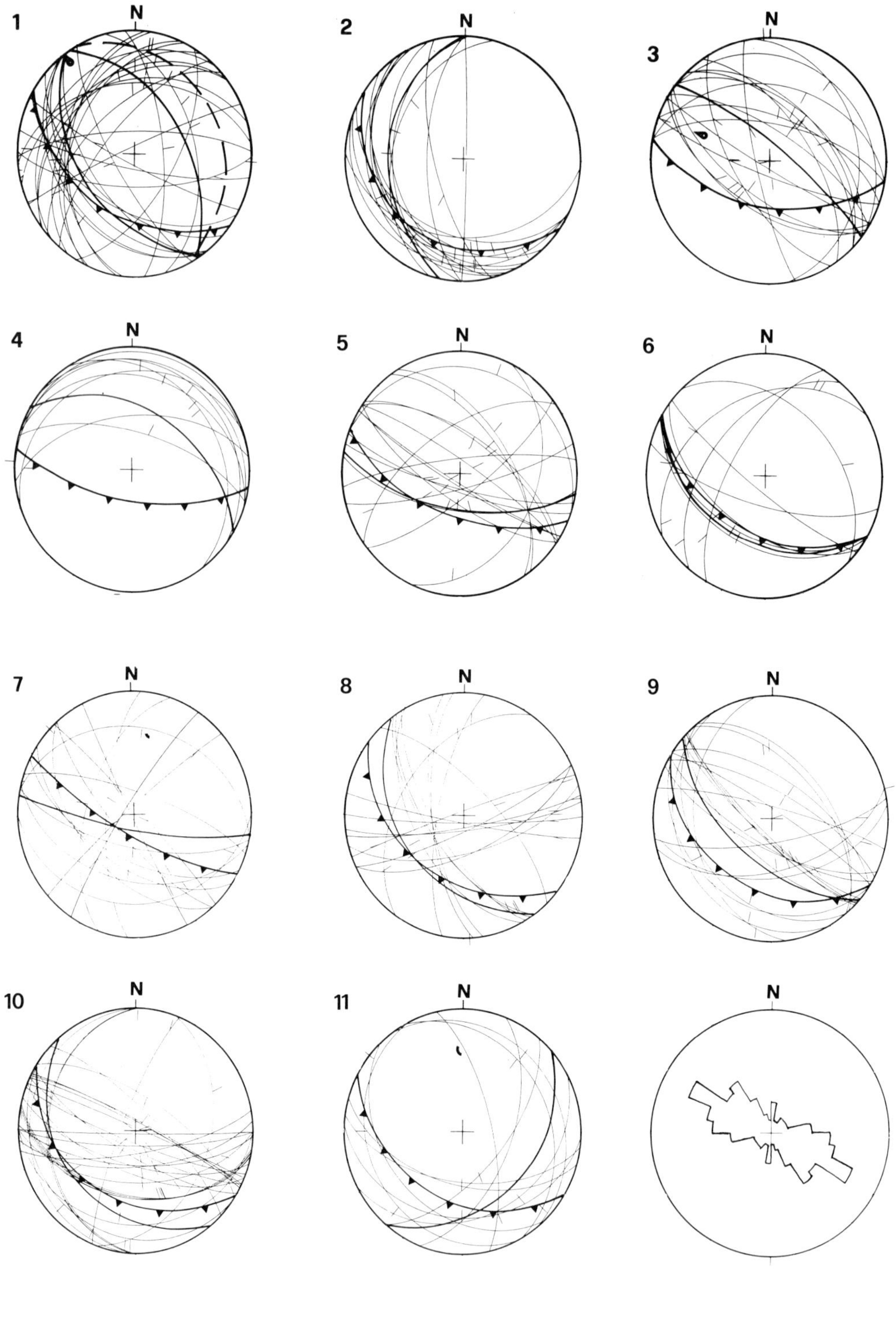
1
N
2
N
3
N
4
N
5
N
6
N
7
N
8
N
9
N
10
N
11
N
N

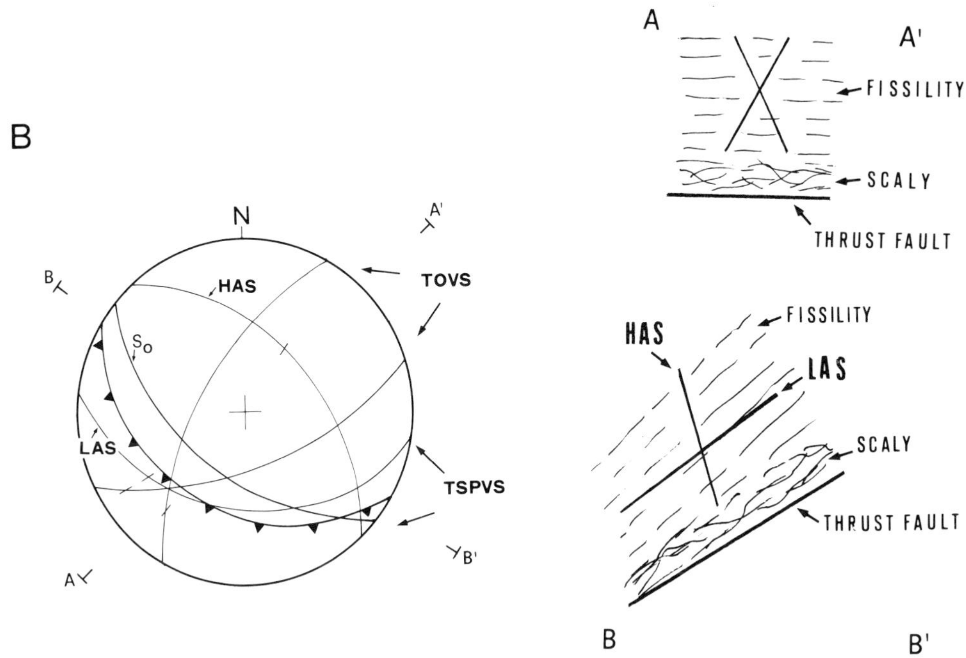

Fig. 3. (**A**) Stereo-projections (lower hemisphere, equal area net) of the shear-vein great circle traces. For each geographic area the stereoplot shows the mean orientation of the shear-veins, each trace being representative of at least five shear-veins surveyable in the field, small tick is the slickenfibre orientation. Station reference number as in Fig. 1. In the bottom right corner a general orientation rose of the vein strike direction is given. (**B**) Sketch of the shear-vein regional pattern recognized in the Cervarola thrust. Two systems of shear-veins had been detected, the principal one trends nearly-parallel to the main thrust (TSPVS) and dips according to (LAS) or against (HAS) the thrust plane. The secondary system (TOVS) trends at a high angle to the thrust plane. In both cases slickenfibres are parallel to the thrust slip line. S_0 = fissility.

One set of the system, (antithetic- or high angle-set, HAS) dips at a high angle to the thrust plane. Slickenfibres are dip-slip, with pitch values clustered at 70–90°. The sense of shear deduced from calcite slickenfibres can either be normal or reverse. The other set of veins (synthetic- or low angle-set, LAS) dips in the same direction as the thrust plane, at a low angle to it. Where bedding and fissility dip in the same direction as the thrust plane, the LAS is sub-parallel to them. The sense of shear on the LAS is usually reverse, synthetic with the thrusting direction. Slickenfibres are dip-slip, with pitch values ranging some 20° about the thrust slip vector.

System 2 (thrust oblique vein system, TOVS)

This system of shear veins is generally less well-developed compared to the TSPVS, although in some localities it appears as the predominant one due to the particular local structural setting. This system consists of two sets of veins, dipping some 50–80°, obliquely to the thrust fault. Slickenfibres are strike-slip, with pitch values lower than 40°. Their directions are consistent in the two sets and are parallel both to the thrust slip vector and the line of intersection of the two sets.

The slickenfibre orientations in the four vein sets are clustered around the thrust slip vector, although they indicate transcurrent, normal or reverse slip movements in particular veins.

The four vein sets are all present at most localities, but commonly one set predominates over the others in different exposures in the same area. Local situations with a particular lithological or structural setting can determine the absence of one or more of the four recognized regional sets, but in any case the kinematics of the sets indicate a movement picture related to thrusting along its main slip vector.

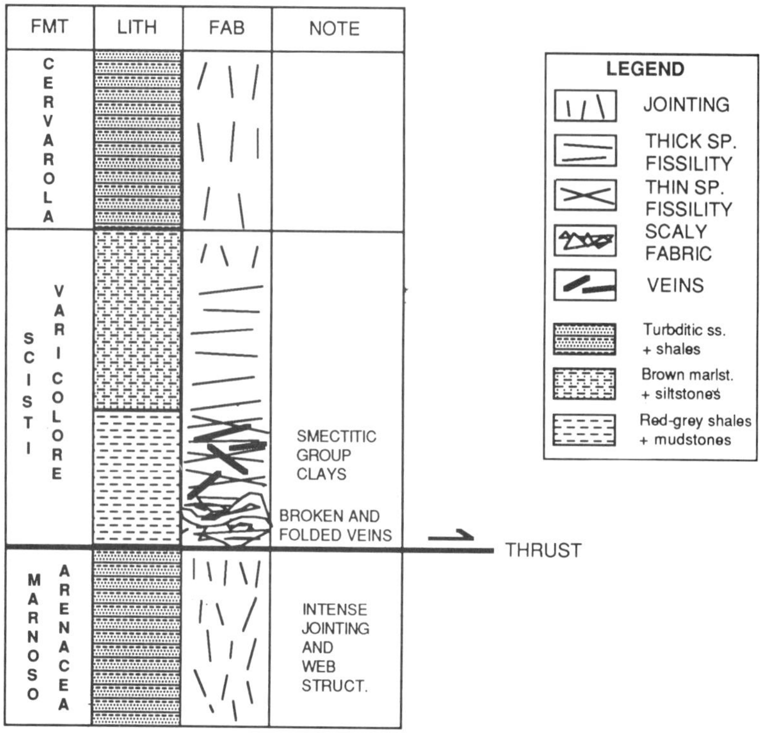

Fig. 4. Summary of the structural features and lithologies surveyed in the field along the Cervarola Thrust. Clay data are from unpublished X-Ray analysis.

A working hypothesis on vein development

The vein development forms a consistent regional pattern. Within different local structural and lithological assemblages it supports the inference of a genetic relationship between veins and thrusting.

A hypothesis on the vein development in the Cervarola Thrust can be built up matching the Cervarola Thrust field data with recent studies on vein occurrence and development in active accretionary prisms, such as Barbados and Kodiak Islands (Moore 1986; Moore *et al.* 1988), where the lithological and structural assemblages are comparable with those of the Cervarola Thrust complex.

During thrusting, the smectitic level in the Scisti Varicolore Fm., water rich but very fine grained (< 0.01 mm) and hence defining a level of low permeability, would have reached an overpressured state close to the lithostatic pressure, which favoured the location of a main decollement in this stratigraphic level. Rock lying just above the thrust fault was intensely affected by progressive simple-shear, with volume loss resulting from dewatering along the thrust fault, until a scaly fabric stage of deformation was reached. Overpressured fluids present in this structural level flowed towards the thrust front owing to dilatancy and secondary permeability resulting from diffuse faulting and shearing processes. This permeability would be higher than the primary permeability of the arenaceous formations above and below. The upper boundary of this level, which almost always contained fluid at near-lithostatic pressure, constituted a fluid-pressure interface, above which rock was still fluid saturated but, not yet having reached the scaly fabric stage of deformation, did not allow fluid (at a pore pressure lower than lithostatic) to migrate. Veins developed in this upper level when micro-failure resulting from the increase of fluid pressure above the lithostatic value occurred (Fig. 5). These failures resulted in sudden decrease of pore-pressure and in mineral deposition. The large size of the surveyed veins argue for a protracted stage of growth, which implies that fluid pressure was maintained at sufficiently high levels to keep fractures growing and that fluids were always supersaturated with

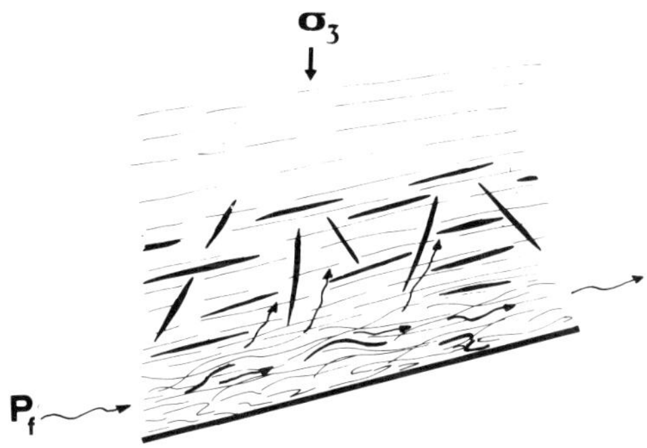

Fig. 5. Inferred fluid movements along the thrust zone. The incoming overpressured fluid P_f is channelled through the scaly fabric level where pore pressure overcomes the lithostatic pressure. Infiltration of overpressured fluid from this level increases pore pressure in the overlying level, causing fracture opening and veining resulting from fluid-phase deposition due to decrease of fluid pressure as the fracture dilates.

calcite. When fluid pressure decreased or crystal growth could match the fracture opening rate the vein was sealed (Cox *et al.* 1986). The large quantity of veins and the presence of many crack-seal tracks argue for a cyclic processes of vein development and hence for fluid pressure pumping effects.

This picture requires a fluid flow gradient from the sole thrust to the thrust front such as occurs in the currently active accretionary prisms (Moore *et al.* 1988), and drives fluid migration from depth towards the surface, channelling fluid flow in a narrow band just above the thrust fault.

Where deformation was more intense, rock volumes which were previously deformed only by extensional fracturing with vein development were affected by progressive simple-shear with development of a scaly fabric and at some point were included into the high strain level just above the thrust fault. Here, both early veins and the scaly fabric were broken and folded by the intense superimposed simple-shear processes.

Concluding remarks

We have described the occurrence of a geometric relationship between overthrusting and the development of mineralized syntectonic veins in the hanging wall. The formation of the veins can be explained in terms of the development of pore fluid overpressuring arising from the dewatering of smectitic clays at this stratigraphic level, leading in turn to the localization of the thrust plane at this level. The model is consistent with inferences of the same process in currently active accretionary complexes.

Support from MPI and M. Boccaletti is acknowledged.

References

ABBATE, E. & BRUNI, P. 1987. Modino-Cervarola o Modino e Cervarola? Torbiditi Oligo-Mioceniche ed evoluzione del margine nord-appenninico. *Memorie Societa' Geologica Italiana*, **39**, 19–33.

——, BOCCALETTI, M., BRAGA, G., COLI, M., DALLAN NARDI, L., MARCHETTI, G., NARDI, R., POCHINI, A. & PUCCINELLI, A. 1982. Le Unité Toscane. *In: Note illustrative della carta strutturale dell'Appennino Settentrionale*. C.N.R., Progetto Finalizato Geodinamica, pubbl. **429**, 46–65.

BALDACCI, F., ELTER, P., GIANNINI, G., GIGLIA, G., LAZZAROTTO, A., NARDI, R. & TONGIORGI, G. 1967. Nuove osservazioni sul problema della Falda Toscana e sulle interpretazioni dei flysch arenacei tipo "Macigno" dell'Appennino Settentrionale. *Memorie Societa' Geologica Italiana*, **6**, 213–244.

BOCCALETTI, M., CALAMITA, F., CENTAMORE, E., CHIOCCHINI, U., DEIANA, G., MICARELLI, A., MORATTI, G. & POTETTI, M. 1986. Evoluzione dell'Appennino tosco-umbro-marchigiano durante il Neogene. *Giornale di Geologia*, ser. 3, **48**, 227–233.

BORTOLOTTI, V., PASSERINI, P., SAGRI, M. & SESTINI, G. 1970. The Miogeosynclinal sequences. *In*: SESTINI, G. (ed.) *Development of the Northern Apennines Geosyncline*. Sedimentary Geology, Special Issue, **4**, 341–444.

BOYER, S. E. & ELLIOTT, D. 1982. Thrust Systems. *American Association of Petroleum Geologists Bulletin*, **66/9**, 1196–1230.

COX, S. F., ETHERIDGE, M. A. & WALL, V. J. 1986. The role of fluids in syntectonic mass transport, and the localization of metamorphic vein-type ore deposits. *Ore Geological Review*, **2**, 65–86.

DALLAN NARDI, L. & NARDI, R. 1972. Schema stratigrafico e strutturale dell'Appennino Settentrionale. *Memorie Accademia Lunigianese di Scienze "G. Cappellini"*, **42**, 1–212.

DE JAGER, J. 1979. The relation between tectonics and sedimentation along the "Sillaro lines" (Northern Apennines, Italy). *Geologica Ultrajectina*, **19**, 1–98. Utrecht.

GROSCURTH, J. 1971. *Zur geologie der Randgebiete des westlichen Teils des Mugello – Beckens ostlich der Prato-Sillaro "Linie" (N. Apennin, Prov. Florenz)*. Dissertation Freie Univ. Berlin.

MOORE, J. C. (ed.) 1986. Structural fabrics in Deep Sea Drilling Project cores from forearcs. *Geological Society of America Memoir*, **166**, 1–160.

——, ROESKE, S., LUNDBERG, N., SCHOONMAKER, J., COWAN, D. S., GONZALES, E. & LUCAS, S. E. 1986. Scaly fabric from Deep Sea Drilling Project cores from forearcs. *Geological Society of America Memoir*, **166**, 55–74.

——, MASCLE, A., TAYLOR, E., ANDREIEFF, P., ALVAREZ, F., BARNES, R., BECK, C., BEHRMANN, J., BLANC, G., BROWN, K., CLARK, M., DOLAN, J., FISCHER, A., GIESKES, J., HOUNSLOW, M., MCLELLAN, P., MORAN, K., OGAWA, Y., SAKAI, T., SCHOONMAKER, J., VROLIJK, P., WLKENS, R. & WILLIAMS, C. 1988. Tectonics and hydrogeology of the northern Barbados Ridge: Results from Ocean Drilling Program Leg 110. *Geological Society of America, Bulletin*, **100**, 1578–1593.

RICCI LUCCHI, F. 1975. Miocene paleogeography and basin analysis in the Periadriatic Apennines. *In* SQUYRES, C. (ed.) *Geology of Italy* (C. **2**, Castelfranco Veneto (Tripoli, 1977) 129–236.

—— 1986*a*. The foreland basin system of the Northern Apennines and related clastic wedges: a preliminary outline. *Giornale di Geologia, ser. 3*, **48**, 165–185.

—— 1986*b*. The Oligocene to Recent foreland basins of the Northern Apennines. *Special Publications, International Association of Sedimentology*, **8**, 105–139.

TEN HAAF, E. & VAN WAMEL, W. A. 1979. Nappes of the Alta Romagna. *Geologie en Mijnbouw*, **58**, 145–152.

Extensional veins and shear joint development in a thrust–fold zone (Northern Apennines, Italy)

FEDERICO SANI

Dipartimento di Scienze della Terra, University of Florence, 50121 Florence, Italy

Abstract: The Northern Apennines thrust system consists of clastic foredeep deposits that have been progressively involved in thrusting towards the northeast from the Oligocene to the Recent. The outcroping geological units are progressively younger towards the foreland implying that sedimentation and deformation developed together. The Umbrian Sequence is the most external unit and has been affected by Late Miocene to recent thrusting. The Marnoso-arenacea Formation, a marly arenaceous turbiditic sequence, belongs to this Unit. This formation provides good examples of pre-thrust and thrust-related structures. An extensional calcite vein system with two sub-vertical and orthogonal sets having constant orientation, developed before thrusting. A later conjugate shear joint system which cross-cuts the extensional veins, were developed in the early stages of thrusting. Folds related to the thrusting (i.e footwall synclines or hangingwall anticlines) spatially reoriented the previously formed extensional vein and shear-joint systems.

The northern Apennines orogenic chain is contains tectonic units accreted on to the Adriatic foreland. The sense of superposition of nappes is from SW to NE (Merla 1951; Baldacci *et al.* 1967; Boccaletti *et al.* 1980; Pieri & Groppi 1981; Castellarin *et al.* 1985; Ricci Lucchi 1986; Boccaletti *et al.* 1989).

The study area is the sector of the Apennines chain between P.so del Giogo and Mt Falterona where four geological units in tectonic contact (Fig. 1a) are exposed. These units, from SW to NE, are (1) Ligurian Units *s.l.* that generally occupy the uppermost position and were emplaced during more than one tectonic phase; (2) the Tuscan unit, containing four formations; Scisti Varicolori Formation — (oldest) variegated shales with thin sandstone beds; Falterona Sandstone formation — coarse thick bedded sandstones alternated with thin beds of marls (Fazzini 1964; Pellegrini 1965); Cervarola Sandstone formation — thin bedded sandstones alternated with marls; Vicchio marls — (youngest); (3) the Castel Guerrino Unit (Groscurth 1971; De Jager 1979) which overrides the Tuscan Unit and is a mostly marly-arenaceous unit. At Mt Falterona, this unit disappears under the Tuscan Unit (Fig. 1); (4) the Umbrian sequence which outcrops from the watershed down to the Adriatic Sea. This area is mostly occupied by the Marnoso-arenacea formation that belongs to the Umbrian Sequence and represents the turbiditic filling of an elongated (180 km) foreland basin formed as a consequence of the collision. The Marnoso-arenacea formation is a wide and thick (5300 m) clastic wedge sedimented from Langhian to Late Tortonian (Ricci Lucchi 1975, 1986). The Marnoso-arenacea formation is composed of an alternating succession of sandstones, siltstones, marls and clays beds. The sandstones are generally medium grained, calcareous and marly (Bortolotti *et al.* 1970).

Our structural analysis has assessed the mesostructures associated with the thrust system developed in the Marnoso-arenacea formation when the foredeep sediments were involved in thrusting because of the continental collision.

Structural analysis of Marnoso-arenacea formation

The most important structural feature of the Marnoso-arenacea formation is the very continuous and extensive thrust-faults striking NW–SE, from the Ligurids cover to the Mt Falterona–S. Benedetto area (Fig. 1). To the south these faults continue to Perugia and Lake Trasimeno. Each thrust-fault affecting the Marnoso-arenacea formation, can be considered as a 100 m – 1 km wide deformation band in which a progressive increase of the intensity of deformation towards the thrust occurs (i.e. towards NE). Deformation features within the thrusts include inclined or overturned strata, joints, veins (Fig. 2), faults and folds. However, open folds and joints also occur between thrust faults, because the whole area underwent a deformation during Late Miocene.

Folds are commonly related to thrust faults, such as hangingwall anticlines and footwall

From Knipe, R. J. & Rutter, E. H. (eds), 1990, *Deformation Mechanisms, Rheology and Tectonics*, Geological Society Special Publication No. 54, pp. 483–490.

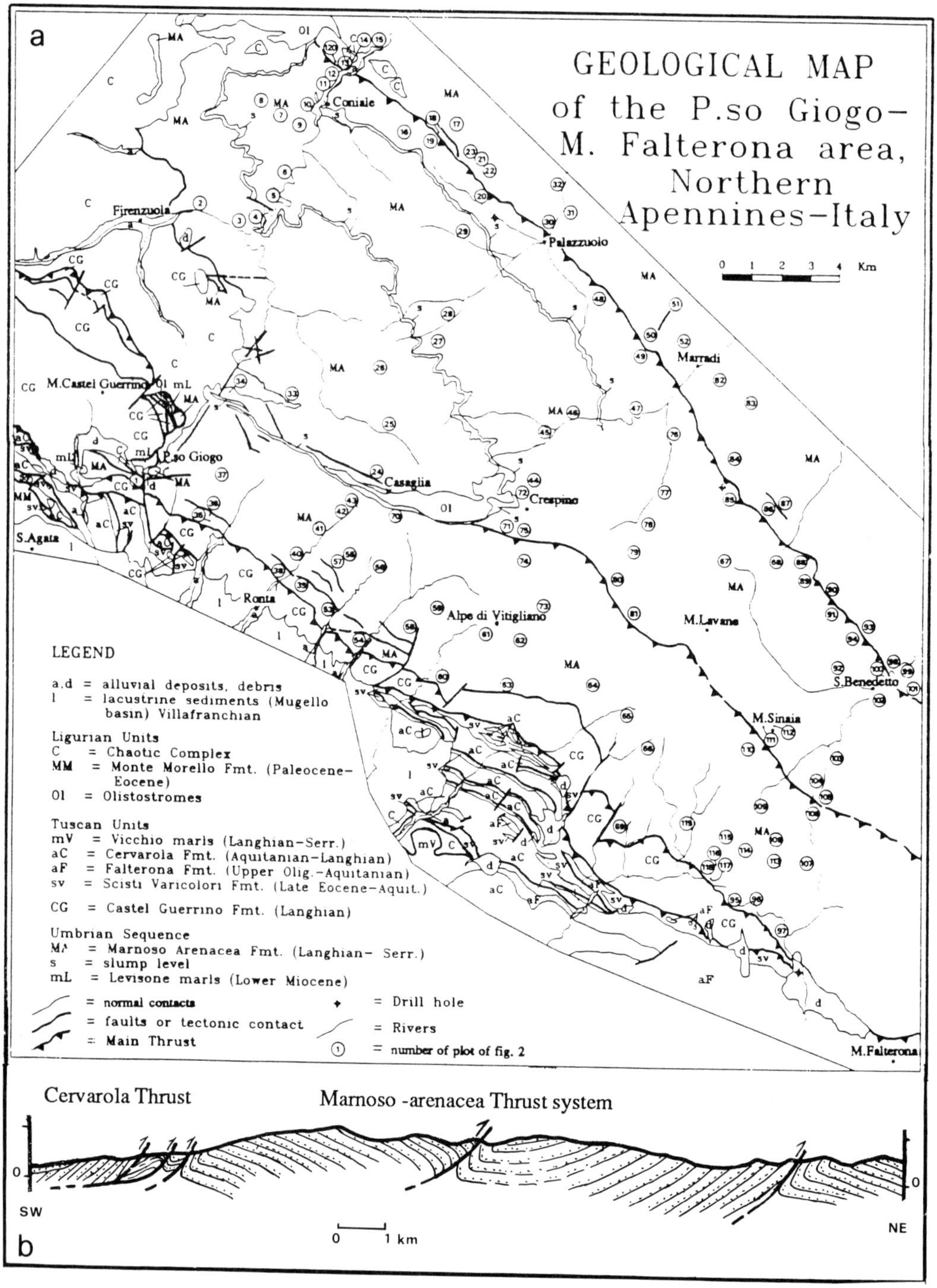

Fig. 1. (**a**) Simplified geological map of the studied area. (**b**) General section across the area. To the SW is the leading imbricate thrust fan of the Cervarola thrust. The Marnoso-arenacea thrust system has a more simple structure: only footwall synclines and/or hangingwall anticlines are present.

synclines (Fig. 3 a,b). In some cases a series of tight folds are present (Fig. 3c). The thrust-related folds are frequently cut by normal faults so that interpretation of the structure is often difficult.

Extensional veins and shear joints systems

This section describes the mesoscopic structures present within the Marnoso-arenacea Formation. Fracture systems defined by extensional veins and shear fractures are particularly well developed in sandstone beds and are dicussed in detail here.

Extensional veins. A very regular and constant pattern of extensional veins develops throughout the studied area (Figs 1 and 2). In the zones between thrust-faults (i.e. the area between the contact zone between Castel Guerrino Unit and Marnoso-arenacea formation and between the internal thrusts of Marnoso-arenacea formation, the Casaglia–Mt Lavane–Mt Sinaia Thrust and the Coniale–Palazzuolo–Marradi–S. Benedetto Thrust, see Fig. 1), where bedding dips are normally less than 30°, the extensional vein system is the most important mesostructural feature. It is composed of two sets; the first striking WNW–ESE and the second NNE–SSW (Fig. 2). The angle between the two sets is normally 80–90°, and an H or T pattern can be recognized (Hancock 1985; Pollard & Aydin 1988) (Fig. 4A). The two sets cross cut each other, which suggests a contemporaneous origin (Fig. 4A). Crystal fibres in the veins are perpendicular to the rock walls and the opening of fractures in this system normally ranges between 0.5–1.5 cm in 80% of cases and 1.5–5.0 in 20% and in most cases have a calcite filling; plume and fringe structures (Hodgson 1961; Price 1966; Hancock 1985) on fracture surfaces can be recognized so they can be considered connected with extensional failure.

A third set is locally present, whose nature and genesis are not clear because of the changes in orientation and frequency between locations (Fig. 2). Calcite fibres are normally absent from this 'local set' which appear to post-date the extensional vein system.

Towards the thrust zones where there is an increase in bedding dip and deformation, the two extensional vein sets have orientations which depend on the dip value (Figs 1 and 2). The first set, trending NNE–SSW does not undergo an important change in orientation because it strikes parallel to the sense of thrusting; the second set (WNW–ESE) is rotated according to the changes of bedding strike and dip. Therefore thrusting developed after the formation of the extensional vein system.

Shear joint system. In the thrust zones, another joint system developed; it is composed of two conjugate shear joint sets (type hOl according to Hancock 1985; Fig. 2, Fig. 4B). These sets form a dihedron angle (2θ) of about 60° and incongruous steps can be observed on joints surfaces. These features are characteristic of shear failure. Both sets are equally developed although a local prevalence is sometimes observed.

Relationships between the two systems (extensional veins and shear joints) are

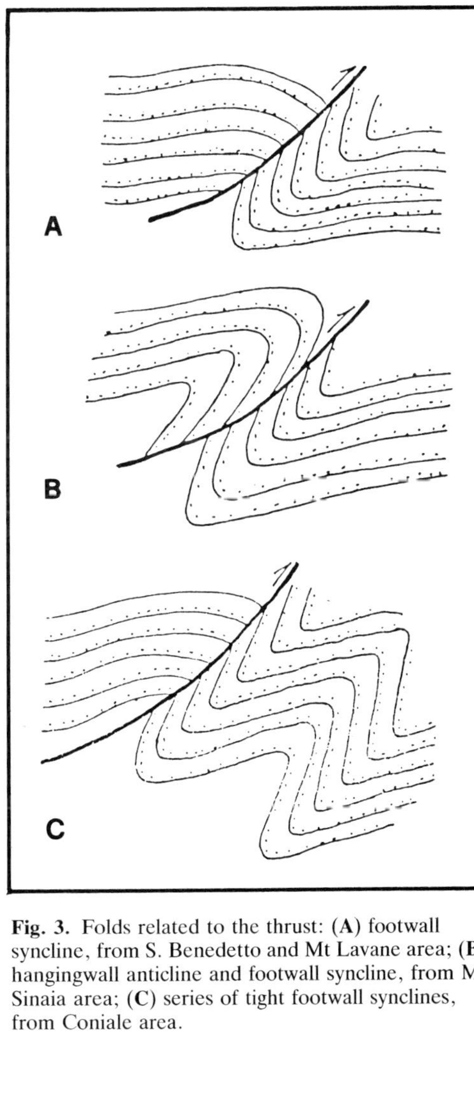

Fig. 3. Folds related to the thrust: (**A**) footwall syncline, from S. Benedetto and Mt Lavane area; (**B**) hangingwall anticline and footwall syncline, from Mt Sinaia area; (**C**) series of tight footwall synclines, from Coniale area.

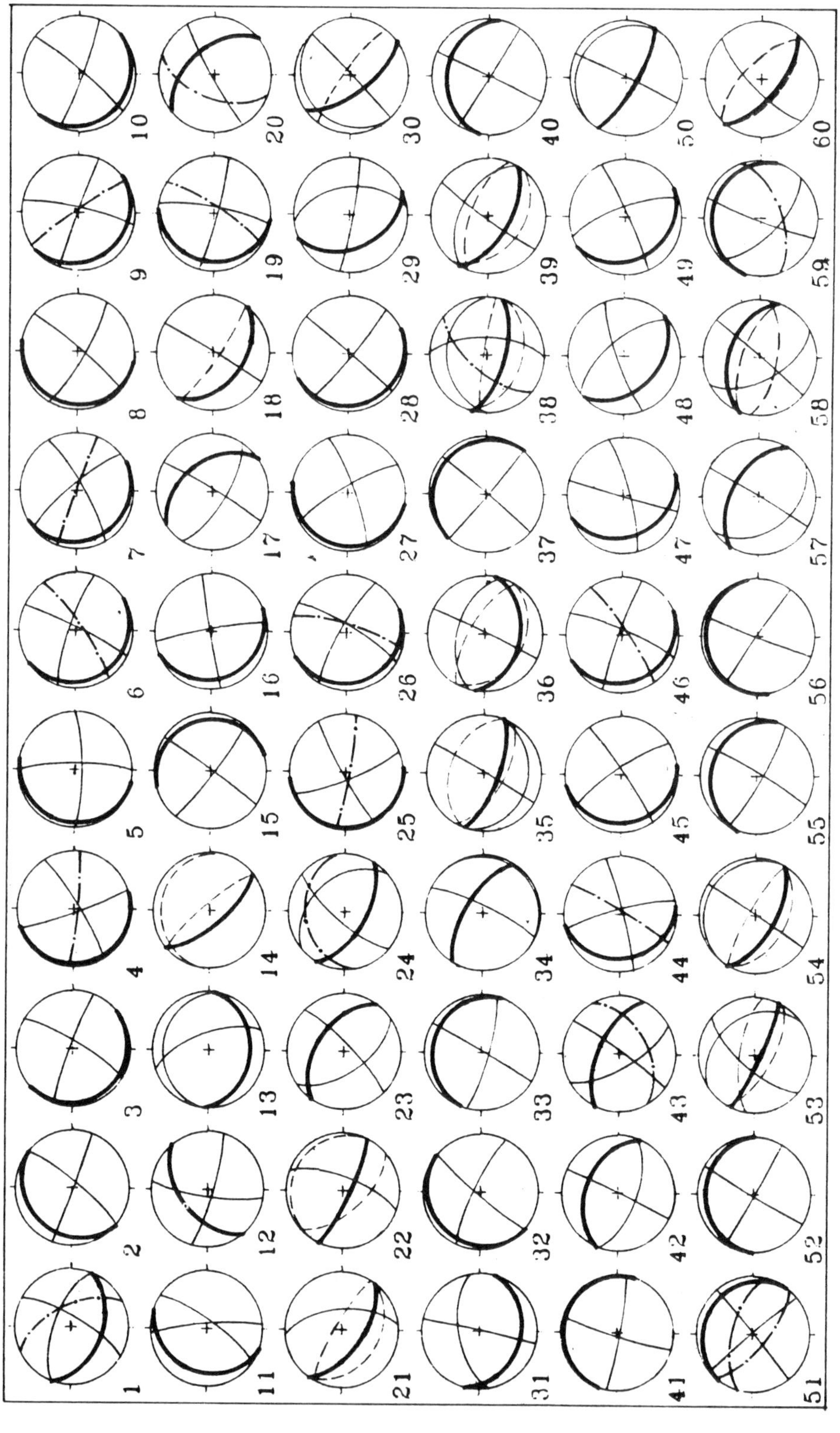
1
2
3
4
5
6
7
8
9
10
11
12
13
14
15
16
17
18
19
20
21
22
23
24
25
26
27
28
29
30
31
32
33
34
35
36
37
38
39
40
41
42
43
44
45
46
47
48
49
50
51
52
53
54
55
56
57
58
59
60

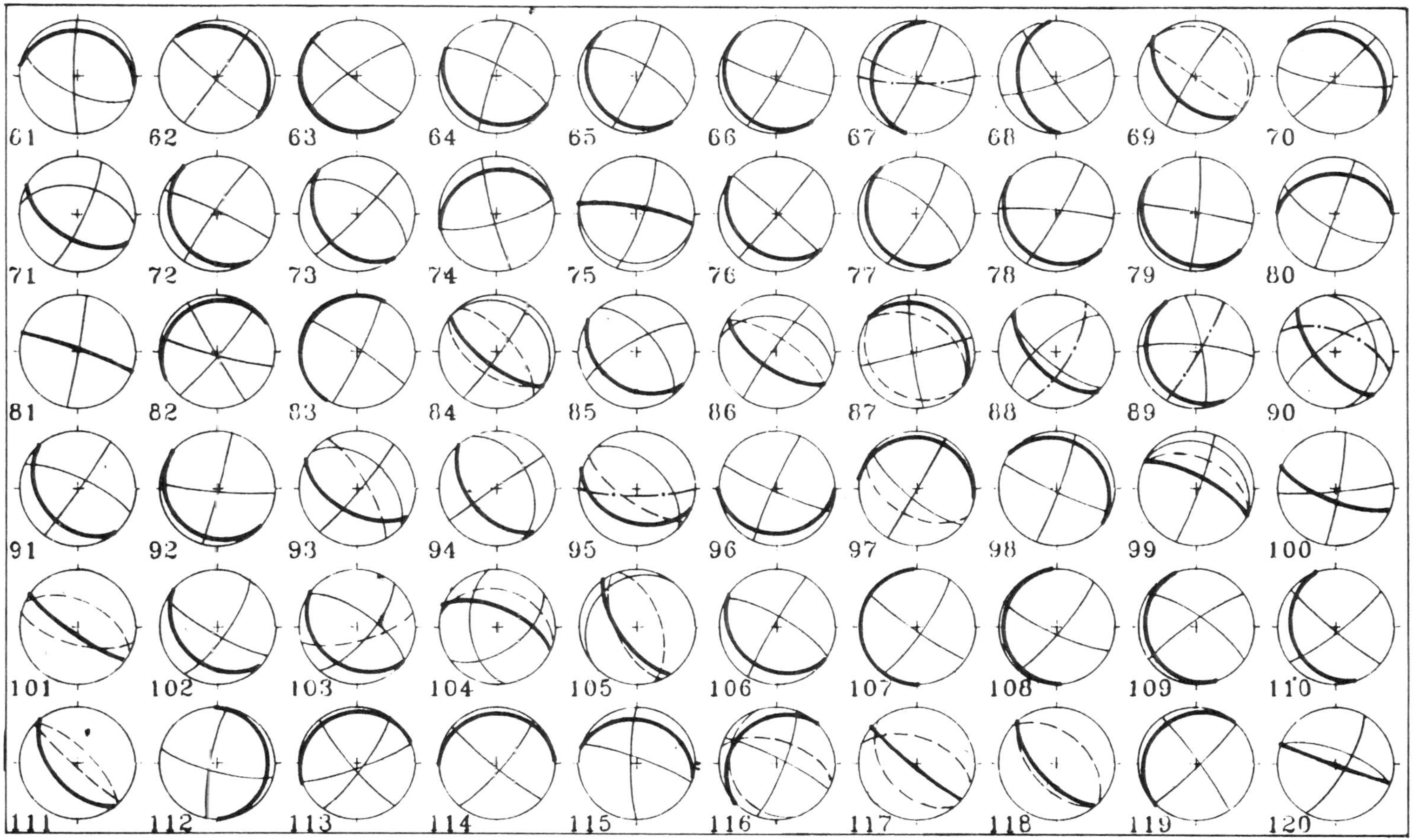

Fig. 2. Plots of extensicnal veins and shear joint systems as marked in Fig. 1a. Wulff net, lower hemisphere. Each great circle, representing a set, is an average of a large number of field data. Thick line, bedding (normal or overturned); Continuous line, extensional vein system; dash-dotted line, local set; dashed line, shear joint system.

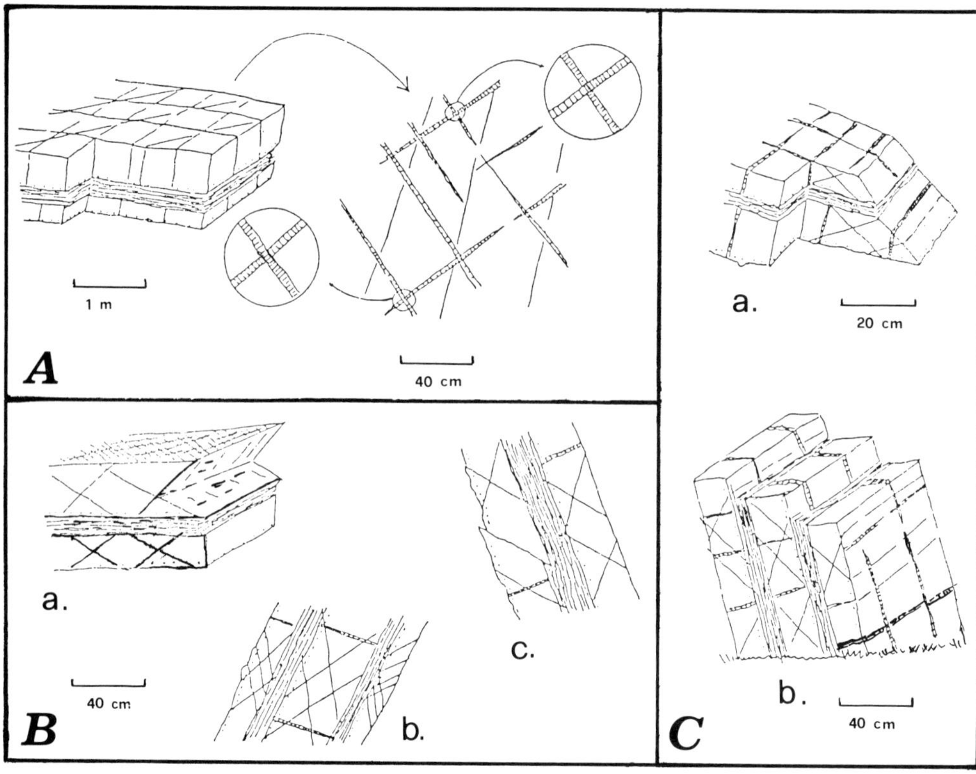

Fig. 4. (**A**) Extensional vein system (redrawn from photos). Loc.44 in Fig. 2. Local set is also present. The vein sets are well developed and cross cut each other. (**B**) Shear joint system (redrawn from photos): (**a**) loc.86 in normal sandstone beds; (**b**) loc.105 in overturned strata of the footwall syncline of Casaglia-Mt Sinaia thrust; (**c**) loc.58, little normal displacement in steeply dipping strata along shear joint surface. (**C**) Relationships between extensional vein and shear joint systems: (**a**) loc.116 a shear joint surface cross cut an extensional vein; (**b**) loc.84 same situation in overturned strata in the footwall syncline of the Casaglia–Mt Sinaia Thrust.

ubiquitous (Fig. 4C). Abutting and cross cutting criteria have been used to establish a relative chronology between the systems. Shear joints always abut against extensional veins and also cross cut them, thus their formation is later.

Both the systems of veins and joints have been spatially reoriented when involved in later folding (hangingwall anticlines and footwall synclines) associated with the thrusting.

Conclusions

The tectonic evolution can be assessed from the mesostructural analysis presented. Extensional veins developed during earlier burial while the last stages of foredeep filling (Late Tortonian) took place. The calcite vein filling is probably due to rapid uplift that brought calcite-rich fluids from marly levels of the formation undergoing compaction. The NNE–SSW set could have been formed by a maximum horizontal stress (σ_h) oriented in the same direction, which is also characteristic of the Apenninic orogenic stress field. The orthogonal set (WNW–ESE) is approximately parallel to open folds axes (like set II of Engelder & Geiser (1980) in the Appalachian Plateau) and formed during the same time (Fig. 5a). Another model to explain two orthogonal contemporaneous extensional sets supposes that σ_2 and σ_3 orientations alternate in time and have similar values (Hancock *et al.* 1987; Ramsay & Huber 1987). However in the Northern Apennines, the first model is preferred.

Later, the shear joint system resulted from regional compressive stress field developed (Fig. 5b), and when the thrust front reached Marnoso-arenacea formation clastic wedge or, at least, the inner side of its basin (Fig. 5c).

Extensional vein and shear joint systems were spatially reoriented and in a few cases acted as preferential shear planes for faulting. In many

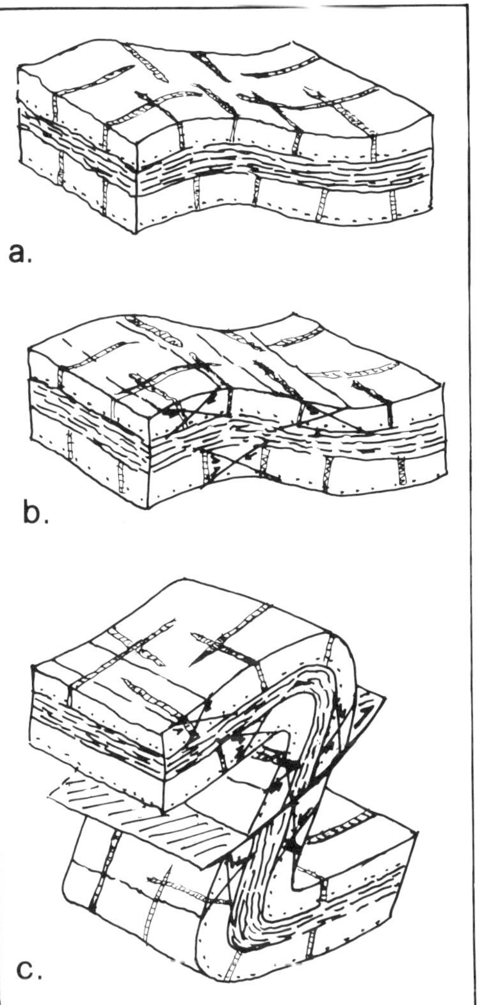

Fig. 5. Tectonic evolution of Marnoso-arenacea fm. (**a**) Extensional vein system development; (**b**) Shear joint system development; (**c**) Folding and thrusting that reorient previously formed structures.

cases the extensional veins have also been reactivated and generated calcite shear fibres. Small normal displacements (less than 5 cm) have been observed along some of the shear joints especially those in the vertical or overturned strata. These displacements appear to be due to later tectonic movements that took place after the formation of the shear joint system and induced tilting of the pre-fractured strata (Figs 4C and 5c).

I would like to thank M. Boccaletti for discussion and reading the manuscript, F. Burelli for digitizing the original map of Fig. 1 and for informatic advice; M. Morelli for help in drawing plots of Fig. 2. Special thanks to P. L. Hancock for kindness and very useful comments, and to S. Hippler for criticism.

References

BALDACCI, F., ELTER, P., GIANNINI, G., GIGLIA, G., LAZZAROTTO, A., NARDI, R. & TONGIORGI, G. 1967. Nuove osservazioni sul problema della Falda Toscana e sulle interpretazioni dei flysch arenacei tipo "Macigno" dell'Appennino Settentrionale. *Memorie Societa Geologica Italiana*, **6**, 213–244.

BOCCALETTI, M., CALAMITA, F., DEIANA G., GELATI R., MASSARI, F., MORATTI, G., RICCI LUCCHI, F. 1990. Migrating foredeep-thrust belt system in the Northern Apennines and Southern Alps. *Palaeogeography, Palaeoclimatology, Palaeoecology*, **77**, 3–14.

——, COLI M., DECANDIA F., GIANNINI, E., LAZZAROTTO, A. 1980. Evoluzione dell'Appennino Settentrionale secondo un nuovo modello strutturale. *Memorie Societa Geologica Italiana*, **21**, 359–373.

BORTOLOTTI, V., PASSERINI, P., SAGRI, M. & SESTINI, G. 1970. The Miogeosynclinal sequences. *In*: SESTINI, G. (ed.) *Development of the Northern Apennines Geosyncline*. Sedimentary Geology, Special Issue, 341–444.

CASTELLARIN, A., EVA C., GIGLIA G., VAI G. B. 1985. Analisi strutturale del Fronte Appenninico Padano. *Giornale di Geologia*, ser. 3, **47**, 47–75.

DE JAGER, J. 1979. The relation between tectonics and sedimentation along the "Sillaro line" (Northern Apennines, Italy). *Geologica Ultraiectina*, **19**, 1–98, Utrecht.

ENGELDER, T. & GEISER, P. 1980. On the use of regional joint sets as trajectories of paleostress fields during the development of the Appalachian Plateau, New York. *Journal of Geophysical Research*, **85**, 6319–6341.

FAZZINI, P. 1964. Geologia dell'Appennino Tosco-Emiliano tra il Passo dei Mandrioli e il Passo della Calla. *Bollettino della Societa Geologica Italiana*, **83**, 219–258.

GROSCURTH, J. 1971. *Zur geologie der Randgebiete des westlichen Teils des Mugello – Beckens ostlich der Prato-Sillaro "Linie" (N. Apennin, Prov. Florenz)*. Dissertation Freie Univ. Berlin.

HANCOCK, P. L. 1985. Brittle microtectonics: principles and practice. *Journal of Structural Geology*, **7**, 437–457.

——, AL-KADHI, A., BARKA, A. A. & BAVAN, T. G. 1987. Aspects of analysing brittle structures. *Annales Tectonicae*, **1**, 5–19.

HODGSON, R. A. 1961. Classification of structures on joint surfaces. *American Journal of Science*, **259**, 493–502.

MERLA, G. 1951. Geologia dell'Appennino Settentrionale. *Bollettino della Societa Geologica Italiana*, **70**, 95–382.

PELLEGRINI, M. 1965. Osservazioni geologiche nella zona del M. Falterona. *Bollettino della Societa Geologica Italiana*, **84**, 239–270.

PIERI, M. & GROPPI, G. 1981. *Subsurface geological structure of the Po Plain*. C.N.R. Progetto Finalizzato Geodinamica, pubbl. **14**.

POLLARD, D. D. & AYDIN, A. 1988. Progress in understanding jointing over the past century. *Geological Society of America Bulletin*, **100**, 1181–1204.

PRICE, N. J. 1966. *Fault and Joint Development in Brittle and Semi-Brittle rock*. Pergamon Press, Oxford.

RAMSAY, J. G. & HUBER, M. I. 1987. *The Techniques of Modern Structural Geology. Vol. 2: Folds and Fractures*. Academic Press.

RICCI LUCCHI, F. 1975. Miocene paleogeography and basin analysis in the Periadriatic Apennines. *In*: SQUYRES, C. (ed.) *Geology of Italy* **2**, Castelfranco Veneto (Tripoli, 1977), 129–236.

—— 1986. The Oligocene to Recent foreland basins of the Northern Apennines. *Special Publication of the International Association of Sedimentologists*, **8**, 105–139.

Convergence-related 'dynamic spreading' in a mid-crustal ductile thrust zone: a possible orogenic wedge model

ROBERT E. HOLDSWORTH[1] & COLIN J. GRANT[2]

[1] *Department of Geoscience, University of Reading, Whiteknights, Reading RG6 2AB, UK*

Present address: Department of Geological Sciences, University of Durham, Science Laboratories, South Road, Durham DH1 3LE, UK

[2] *Department of Earth Sciences, University of Liverpool, PO Box 147, Liverpool L69 3BX, UK*

Abstract: Collisional orogenic belts usually form as approximately wedge-shaped dynamic units irrespective of their bulk rheology. Quantitative models suggest that if the boundary conditions and mechanical properties of a wedge remain unchanged, it will undergo internal yielding (shortening or extension) until a stable, unchanging shape (or 'taper') is established. It follows that geological processes periodically leading to changes in wedge shape, or mechanical properties, will perturb the system, generating internal deformation as the wedge strives to re-establish a stable taper.

Geological applications of wedge models to upper-crustal features such as accretionary prisms and foreland thrust belts have highlighted the importance of internal, thrust-parallel shortening strains. The Caledonian, mid-crustal ductile thrust zone of Sutherland, N Scotland, displays many geometric and kinematic similarities to shallow-crustal thrust belts. In contrast to such belts, however, deformation fabric studies suggest a predominance of thrust-parallel *extension* strains which developed *synchronously* with large bulk shortening due to thrust telescoping. Using a viscous wedge model, it is suggested that the deformation occurs due to a process of *dynamic spreading*, possibly triggered by intermittent thrust-slice accretion, towards the foreland, at depth ('underplating') and/or strain softening in amphibolite-facies mylonite zones along the ductile thrusts. Similar deformation fabrics are recognized in many regions of ductile thrusting which suggests that a dynamic spreading model may be applicable to mid-crustal deformation regimes in collision zones worldwide.

In recent years, it has become apparent that convergent orogenic belts are analogous to accretionary prisms in that they form as roughly wedge-shaped units supported at their rear by a relatively undeforming 'rigid' buttress. The internal structure of orogens and accretionary prisms is complex, but, as a rule, they comprise linked thrust and fold systems which form in predictable and regular sequences (e.g. Elliott 1976; Boyer & Elliott 1982; Knipe & Needham 1986). This suggests that such wedges can be treated as mechanically continuous dynamic units on a *macroscopic* scale (Price 1973; Elliott 1976; Chapple 1978). Quantitative or semi-quantitative wedge models have now been formulated for several bulk rheologies, including plastic (Chapple 1978; Stockmal 1983), incohesive and cohesive Coulomb-type (Davis *et al.* 1983; Dahlen *et al.* 1984) and viscous (Platt 1986) materials. These models are distinct as a consequence of their rheologies, but from a qualitative standpoint they all share two common features:

(a) if the boundary conditions and mechanical properties of a wedge remain unchanged, it will deform by internal shortening or extension ('yielding') until a stable, non-deforming shape (or 'taper') is established;
(b) geological processes act upon the wedge periodically, causing changes in taper, and/or mechanical properties, intermittently perturbing the system and leading to internal deformation as the wedge attempts to re-establish a stable shape.

By selecting the most appropriate model rheology, it is possible to make certain qualitative deductions about orogenic wedge dynamics simply by studying deformational fabric patterns and structural histories at various scales. This helps geologists principally concerned with explaining the regional structural evolution to pinpoint likely controlling mechanisms. Most recent studies have concentrated on the applicability and deformational consequences of such wedge models in upper-crustal settings such as accretionary prisms (e.g. Davis *et al.* 1983), or foreland fold and thrust belts (e.g. Woodward 1987). The present work aims to extend the approach, in a speculative and qualitative manner, into what we regard as a typical example of an internal, ductile thrust zone formed at mid-crustal depths (amphibolite facies). Whilst there are clear geometric and kinematics similarities between such ductile thrust zones and their shallower crustal equivalents, the observation of significant differences in deformational fabric patterns implies important differences in static and dynamic controls; these can begin to be explored using the wedge-model concept.

The Moine rocks of Sutherland, Scotland

The Precambrian Moine rocks of the northern Scottish Highlands (Fig. 1) represent a sequence of sediments laid down unconformably upon continental basement gneisses thought to be equivalent to the Lewisian of the NW foreland (Johnstone *et al.* 1969). Both basement and cover were variably deformed and metamorphosed during Precambrian orogenesis, the so-called 'Ardgourian' event (*c.* 1000 Ma; Barr *et al.* 1986 and references therein). Subsequently, during the Caledonian orogeny (*c.* 470–430 Ma), NW–WNW-directed folding and ductile thrusting at mid-crustal levels (amphibolite facies) reworked and dismembered the Ardgourian orogen, so that the large-scale structure in the Moine is presently dominated by Caledonian structures.

In Sutherland (Fig. 1), three major metamorphic thrust sheets are recognized, each of which is bounded by Caledonian ductile thrusts of unknown, but probably large (e.g. ≫ 10 km), displacements (Fig. 1; Moorhouse & Moorhouse 1983; Barr *et al.* 1986; Moorhouse *et al.* 1988; Holdsworth 1989*a*). Each thrust sheet has distinct stratigraphic, structural and metamorphic characteristics (e.g. see Moorhouse *et al.* 1988) as each represents previously separated segments of Precambrian crust subsequently juxtaposed by thrusting. Syn-thrusting metamorphic grade decreases from kyanite grade in the east (Naver Thrust) to chlorite grade in the west (Moine Thrust),

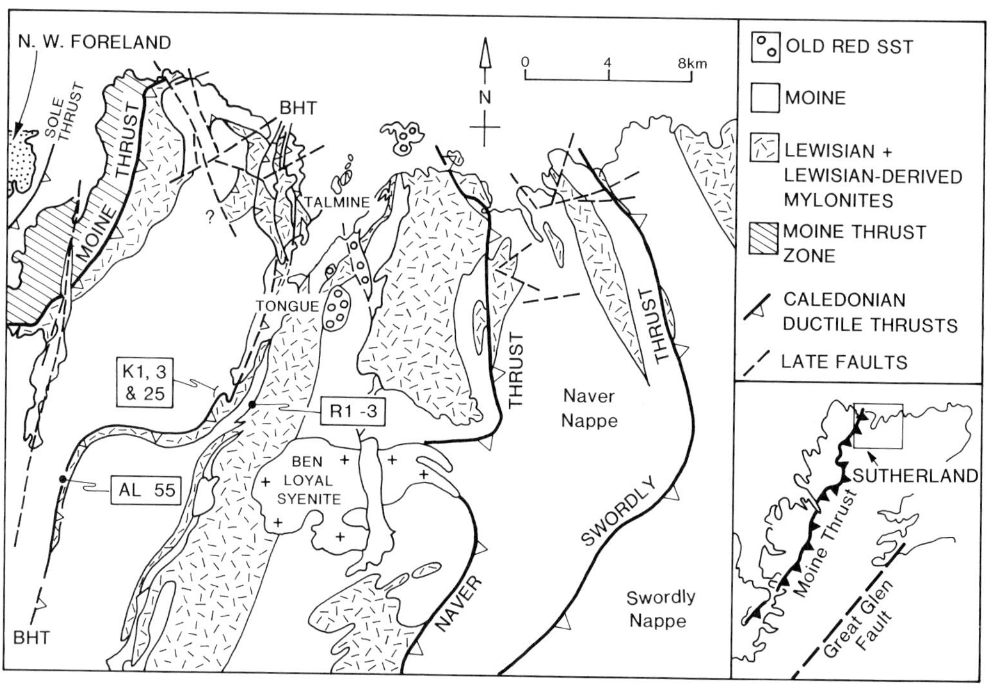

Fig. 1. Simplified geological map of the Sutherland Moine, showing location (see inset), major Caledonian thrusts and metamorphic nappes. BHT, Ben Hope Thrust. The location of specimens in which the quartz *c*-axis fabrics recorded in Fig. 6 is also shown.

strongly suggesting that the Caledonian thrust pile formed as a foreland-propagating system (Barr *et al.* 1986; Holdsworth 1989*a*). If erosion synchronous with uplift of the actively accreting thrust stack continually cut downwards, successively lower, later thrusts would then form at progressively shallower crustal levels and lower metamorphic grade. An apparent 'inversion' of brittle and ductile deformation styles is also observed in the underlying imbricates of the Moine Thrust Zone (e.g. Butler 1982), which suggests that foreland propagation continued right up until the final stages of thrust-related foreshortening in the Caledonian orogen.

Caledonian mesoscopic deformation patterns in the Moine Nappe

Figure 2 shows a schematic NW–SE cross-section through the Moine Nappe in Sutherland, which summarizes the findings of detailed field mapping in the region (e.g. Holdsworth 1989*a*). A series of low angle Caledonian ductile thrusts and related overfolds have brought about an intense interleaving of the Moine cover rocks with units of their underlying Lewisian basement which presently crop out as tectonic inliers throughout Sutherland (Figs 1 & 2). The ductile Moine Thrust is thought to form the basal thrust to the system, and is associated with a thick belt of mylonites (≤ 600 m) where it is exposed at the western edge of the Moine Nappe. The Moine Thrust is constrained to make a shallow cut through the crust on the basis of palinspastic restoration of the underlying imbricated Cambro-Ordovician shelf sequence (e.g. Butler & Coward 1984). A shallow trajectory would also explain why the Moine rocks in the hangingwall never exceed garnet-grades of regional metamorphism, despite the probable large (100 km+?) displacement across the Moine Thrust. The arrangement of folds and ductile thrusts in the Moine Nappe is analogous to structural configurations seen in shallower crustal thrust zones, but differences arise due to the intense, penetrative nature of the associated ductile deformation. Thrust ramp angles are very low (≪ 5°), suggesting very shallow detachment trajectories (Fig. 2), whilst intense shearing has profoundly modified the associated foreland-overturning buckle folds producing sheath-fold geometries on all scales (Holdsworth 1989*a*). As a result, the majority of Caledonian fold axes observed in the field

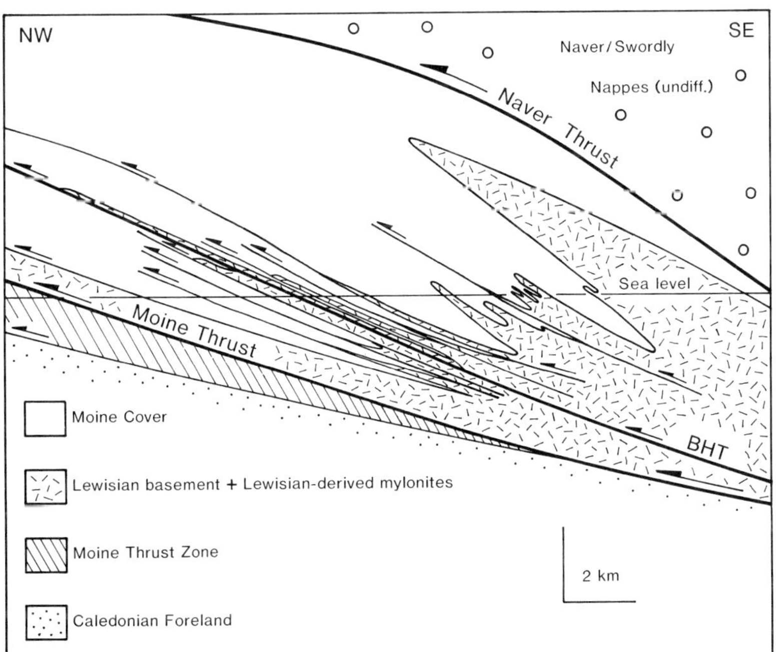

Fig. 2. Schematic NW–SE composite cross-section through the Moine Nappe showing Caledonian fold and ductile thrust structures. Upward and downward continuations of minor thrusts are for the sake of clarity not shown. Based on authors' field mapping and also after Barr *et al.* (1986).

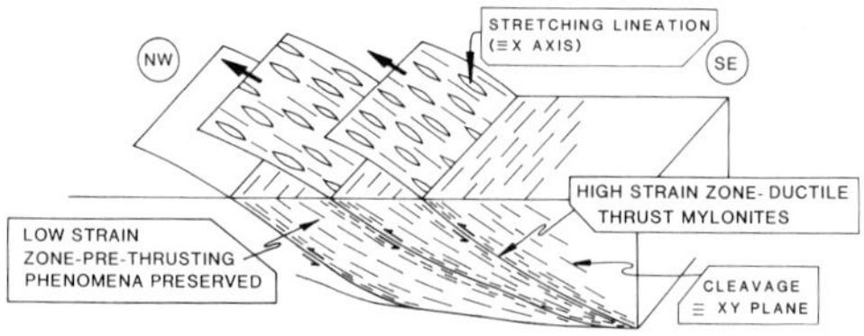

Fig. 3. Summary of Caledonian mesoscopic fabric patterns, Moine Nappe (after Holdsworth 1989*a*). Note that the cleavage and stretching lineations both lie close to parallelism with the ductile thrusts.

now lie at low angles to, or sub-parallel with a ubiquitous ESE–SE plunging mineral lineation (= thrust transport azimuth; Fig. 3). The Caledonian schistosity is axial planar to the folds and occurs throughout the Moine Nappe lying sub-parallel to the ductile thrusts (Fig. 3). Syn-thrusting strain intensity is heterogenous on all scales, with areas of lower deformation being identified on the basis that pre-Caledonian phenomena such as weak (?)Precambrian fabrics and sedimentary structures are preserved (Holdsworth 1989*a*). On approaching the ductile thrusts, the schistosity and lineation intensify to form thick belts of platey mylonite up to 80 m thick in which all traces of earlier features are completely obliterated. In some high strain zones associated with ductile thrusts (e.g. in the Talmine area, Fig. 1), local sequences of 'synmylonitization' folds have formed, probably due to perturbations in flow (see Holdsworth 1989*a*, 1990). Compass-clinometer measurements of schistosity and direct field observations unequivocally demonstrate that the Caledonian S-fabric does not significantly change its orientation between apparent high and low strain regions (Fig. 3). The apparent parallelism of the fabrics and ductile thrusts in regions of differing strain intensity is particularly significant and appears to be a characteristic feature of many deformational terrains formed at mid-crustal depths (e.g. Ramberg 1977, 1981; Sanderson *et al.* 1980; Sanderson 1982; Platt & Behrmann 1986; Mattauer 1986; Hamner 1988). A heterogeneous simple shear strain model cannot apply because marked changes in fabric orientation would occur between regions of different strain intensity (cf. Ramsay & Graham 1970). Whilst the initial formation of the major overfolds and ductile thrusts may be associated with an early phase of thrust-related shortening, the present sub-parallel fabric patterns strongly suggest that a *thrust-parallel extension* strain model applies to much of the deformation synchronous with thrust-related shortening (Holdsworth 1989*a*). In such a model, each thrust-bounded unit can be envisaged as having undergone an approximately homogenous component of coaxial flattening (*sensu lato*), together with a superimposed, foreland-directed component of heterogenous simple shear (Fig. 4; cf. Sanderson *et al.* 1980; Sanderson 1982). Within the bounds of internal strain compatibility, each thrust sheet can deform in-

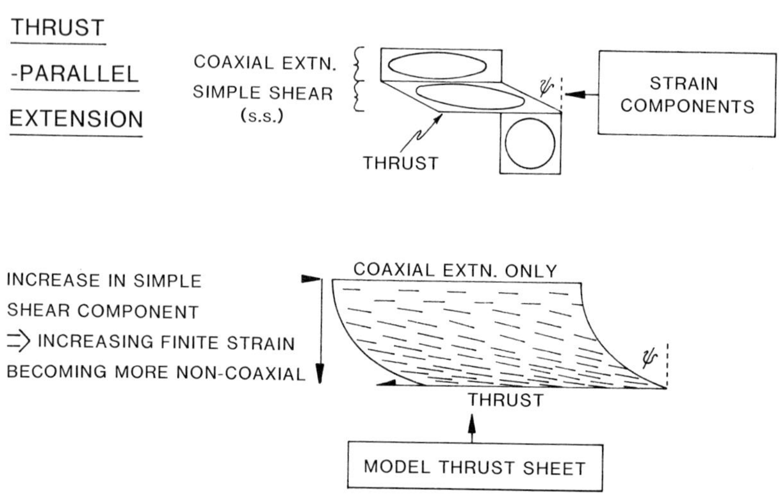

Fig. 4. Thrust-parallel extension strain model (assuming $k = 1$). The dashed lines in the thrust sheet represent the predicted cleavage trace (after Sanderson 1982). ($\gamma = \tan \psi$). Note that no discontinuity is present between domains of coaxial extension and simple shear, meaning that the former is of equal magnitude in both domains, i.e. strain compatibility is maintained.

dependently as the bounding ductile thrusts can act as strain discontinuities. It should be stressed that it is not possible to test quantitatively such a model, due to a lack of strain markers in the Sutherland Moine, but in general, a non-coaxial, 'extensional' pattern of flow is predicted. Non-coaxial deformation should be especially pronounced in the regions of highest finite strain, i.e. the ductile thrust zones (Fig. 4).

Caledonian microstructure in the Moine Nappe

The analysis of microstructures and quartz *c*-axis fabrics in deformed rocks can provide revealing insights into the kinematics of deformation and ductile flow (e.g. Lister & Williams 1979; Law *et al.* 1986). Existing studies of this type in thrust-related settings (including the Moine Thrust Zone in Sutherland; e.g. Law *et al.* 1986) have examined mylonites in which *dynamic* (i.e. unannealed) textures are developed. However, thin sections of deformed quartzo-feldspathic rocks in the Sutherland Moine consistently display an apparently annealed quartz-microstructure dominated by secondary recrystallization textures (Fig. 5; Evans & White ; Holdsworth 1989*a*). As part of a regional study, Grant (1989) has measured *c*-axis patterns from a number of deformed rocks spatially associated with a prominent Caledonian structure in the Sutherland area, the Ben Hope Thrust (Fig. 1). In summary, this study reveals the following.

(i) In most specimens, a diffuse girdle of *c*-axes is preserved lying at high angles to both the Caledonian foliation and lineation (Fig. 6a–c). Such patterns bear comparison with those found in unannealed mylonites (e.g. Law *et al.* 1986), strongly suggesting that preferred quartz *c*-axis orientations produced during plastic flow and dynamic recrystallization are preserved despite annealing.

(ii) Most quartz fabrics diplay a distinct asymmetry in the intensity distribution of *c*-axes which, in the majority of cases, are consistent with WNW-directed overthrusting (Fig. 6a & b).

(iii) Fabric skeletons are more variable in form, ranging from roughly symmetrical cross-girdle forms through to distinct asymmetric single girdles which are again mostly consistent with WNW-directed overthrusting (Fig. 6a & b).

(iv) In some regions (e.g. the results shown in Fig. 6b), as the strain rises, there is a tendency for the quartz *c*-axis patterns to become increasingly asymmetric both in terms of intensity distribution and skeletal outline. However, Grant (1989) has been unable consistently to demonstrate such a relationship in other regions.

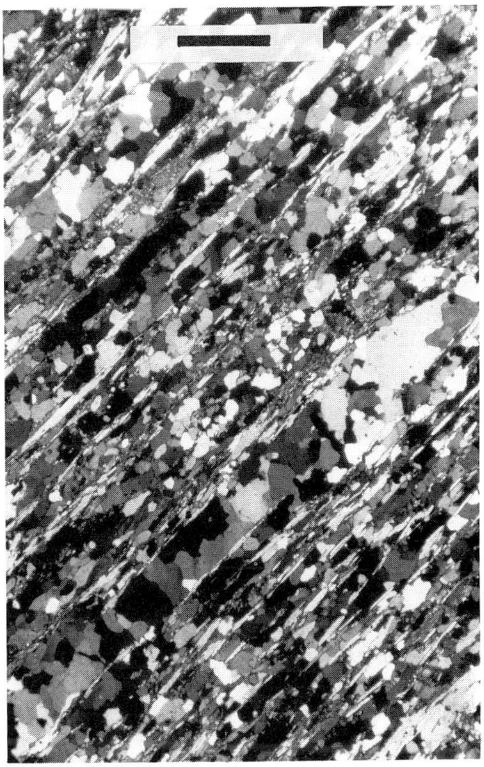

Fig. 5. Photomicrograph (crossed polars) of mylonitic Moine psammite, Ben Hope Thrust (NC 5499 5278). The section is orientated normal to the foliation and parallel to the lineation. The trails of aligned white micas and minerals other than quartz are inferred to preserve a type II S-C mylonite texture, in which the quartz has undergone almost complete secondary recrystallisation. Scale bar, 1 mm.

(v) Rarely, *c*-axis patterns measured from rocks lying close to certain lithological or tectonic contacts (e.g. Fig. 6c) show reversed senses of asymmetry consistent with *ESE-directed* shearing, down the dip of the regional foliation.

Although this study is at preliminary stage, the *c*-axis fabrics are generally consistent with the WNW-directed, non-coaxial 'extensional' flow pattern predicted by a thrust-parallel extensional model (cf. Law *et al.* 1986; Platt & Behrmann 1986). In addition, the form of the girdles seems to suggest plane strain (k = 1) deformation when compared to the results of

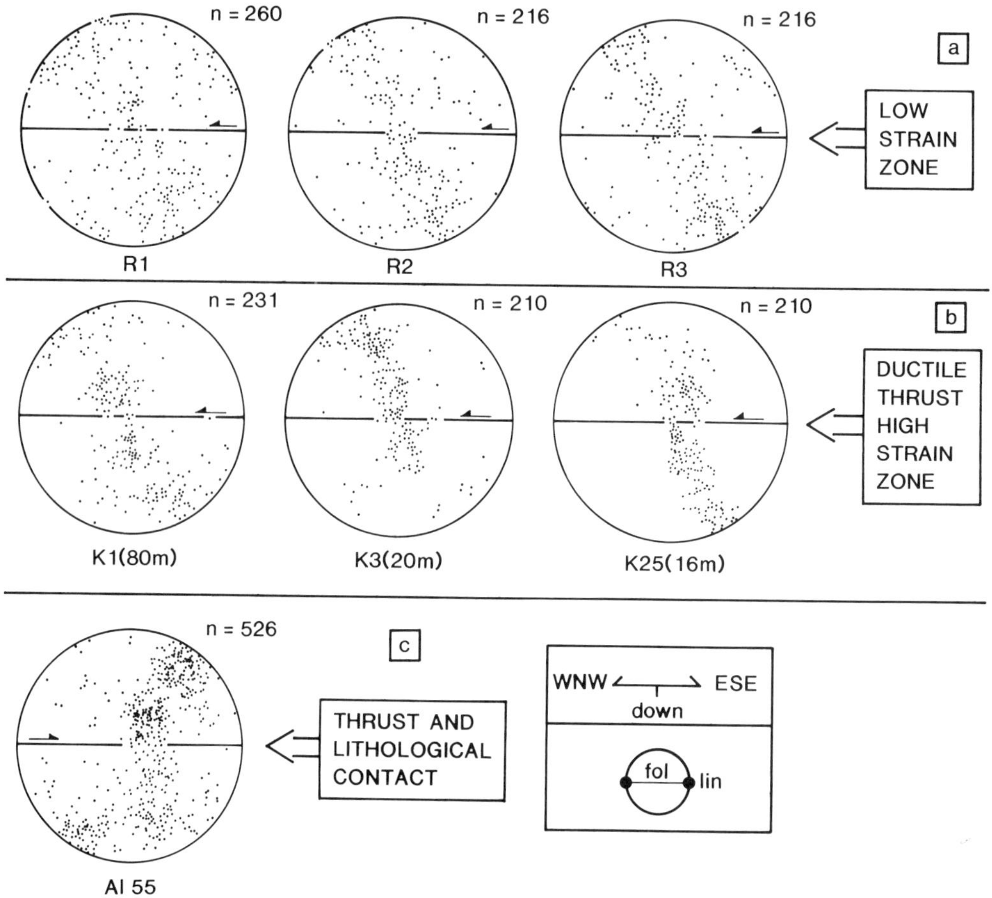

Fig. 6. Quartz *c*-axis fabrics (viewed toward the NNE) from (**a**) a low strain zone in Moine psammites preserving cross-lamination (all from NC 5618 5255); (**b**) the Ben Hope Thrust mylonites at Kinloch Broch (K1, NC 55100 5313; K3, NC 5499 5289; K25, NC 5492 5270; distance from thrust is shown in metres; and (**c**) from the Ben Hope Thrust *plane*, Leitir Mhuiseil (NC 4712 4822).

theoretical studies (e.g. Lister & Hobbs 1980). However, changes in the symmetry of *c*-axis fabrics can be interpreted in other ways (e.g. Jessell 1988), so that the *c*-axis data alone do not provide conclusive proof of the strain model proposed. Furthermore, quartz fabrics may be susceptible to 'resetting' as a consequence of annealing and/or late-stage coaxial or non-coaxial overprinting deformations during uplift (e.g. Lister & Williams 1979). This may explain the lack of a clearly defined correlation between degree of fabric asymmetry and intensity of finite strain (which is predicted by thrust-parallel extension model; see Fig. 4). It is possible that there is a late coaxial overprint, whilst the localized down-dip (ESE) movements along certain lithological contacts may have occurred still later in the Caledonian uplift history because Holdsworth (1989*b*) has recognized a late phase of ESE-directed 'back collapse' in the Sutherland Moine thrust wedge, based on studies of late fold and fault geometries.

Microstructures observed in thin sections for mineral species other than quartz provide further evidence for intense non-coaxial flow and high strain within the mylonites. For example, mylonitic Moine psammites are typically enriched in white mica which displays strong alignments and, together with minerals such as epidote and iron oxide, forms distinct layers or trails dispersed between elongate, strongly recrystallized quartzose domains (Fig. 5). The distribution and size of micas and other mineral species strongly controls the shape and grain size of recrystallized quartz due to grain boundary pinning. As a result, distinct planar domains of variable grain size and mica content have developed (Fig. 5), a grain-growth effect which has strongly enhanced the platey aspect of mylonites. The arrangement of micas and other minerals in distinct planar trails is reminiscent of Type II S–C mylonites in which the oblique, asymmetric quartz S-planes have been obliterated by secondary recrystallization (cf. Lister & Snoke 1984, Fig. 10d). The occasional preservation of white mica 'fish' and asymmetric wrapping textures in quartzo-feldspathic mylonites confirms their S–C character. In most cases, the sense of asymmetry is consistent with WNW-directed overthrusting parallel to the Caledonian mineral lineation. Intensely deformed micaceous lithologies, such as Moine pelite bands, also display good S–C fabrics (Types I and II, see Holdsworth 1989*a*). Such fabrics provide compelling evidence that intense non-coaxial deformation occurred within the ductile thrust mylonites (Lister & Snoke 1984). A lack of S–C fabrics and asymmetric microstructures in adjacent regions of low Caledonian strain may indicate a deformation closer to coaxial strain, as would be expected in the proposed thrust-parallel extension model (Fig. 4).

In summary, whilst the Caledonian microstructures are consistent with the extensional strain model proposed, they cannot be interpreted as unequivocal proof of its validity. It is not known how much of the total strain history is recorded by quartz *c*-axis fabric patterns, especially in mylonites of the type seen here where they are likely to have undergone a prolonged phase of 'annealing' at elevated temperatures during uplift of the orogen. This is a problem that needs to be addressed in future studies of mid-crustal deformation zones.

Discussion

Holdsworth (1989*a*) has proposed that the thrust-parallel extensional flow of the Moine Nappe may result from loading and foreland-directed 'extrusion' of thrust slices from beneath the overlying masses of the higher metamorphic nappes in Sutherland (Fig. 7). Such a gravitationally-induced spreading process must have occurred to some extent synchronously with the large foreshortening along major Caledonian bounding ductile thrusts (Moine, Naver, etc.), and also with movements along internal discontinuities such as the Ben Hope Thrust. It seems unlikely that the required large shortening displacements can also be accounted for by a simple spreading mechanism (e.g. see Chapple 1978).

It is illuminating, therefore, to view the problem in terms of dynamic orogenic wedge. The elevated temperatures and probable widespread occurrence of an active, water-rich metamorphic fluid phase at mid-crustal depths inevitably favours ductile deformation processes of the kind seen in the Sutherland Moine. Platt (1986) has argued convincingly that under such conditions, rocks are likely to deform under almost any deviatoric stress, albeit slowly at low stresses. By assuming an absence of long-term yield strength, a ductile thrust stack of the type seen in Sutherland can be qualitatively described in terms of bulk viscous rheology. According to Platt (1986, p. 1040), the basal traction (resistance to shear) exerted on the viscous wedge by the slab sliding down beneath it is balanced by components of gravitational and longitudinal stress principally related to the surface slope and taper of the wedge respectively. The counteraction of stresses is fundamental to the formation of a dynamic wedge shape: if the downgoing motion of the underlying slab (and hence the 'push') ceases, the wedge becomes unstable and it will eventually collapse and cease to exist. Platt argues further that internal deformation (shortening or extension) of the wedge arises due to, and hence relaxes, any perturbations in the longitudinal stresses. It is likely that the deformation will strive to achieve a 'condition of stability' in which the longitudinal stress components are reduced to zero. In such a condition, deformation in the wedge is restricted to sliding or shear flow parallel to the base, the basal traction being balanced merely by gravitational stresses induced by the surface slope. Geological processes leading to a change in wedge taper (and therefore surface slope) and/or basal traction will tend to upset the condition of stability, generating longitudinal stresses and hence internal deformation. In theory, any departure from the stable configuration will lead to internal deformation in a viscous rheology. If the taper/surface slope becomes too low, the wedge will deform by internal shortening, whilst if it is too high, internal extension will occur synchronous with wedge motion. The thrust-parallel extension model suggested for the thrust stack implies that the latter situation has occurred.

Two geological processes are likely to have occurred in the Sutherland Moine which could have perturbed the system and led to internal extension of the developing ductile thrust wedge (see Fig. 7).

(a) Progressive accretion of thrust sheets, at

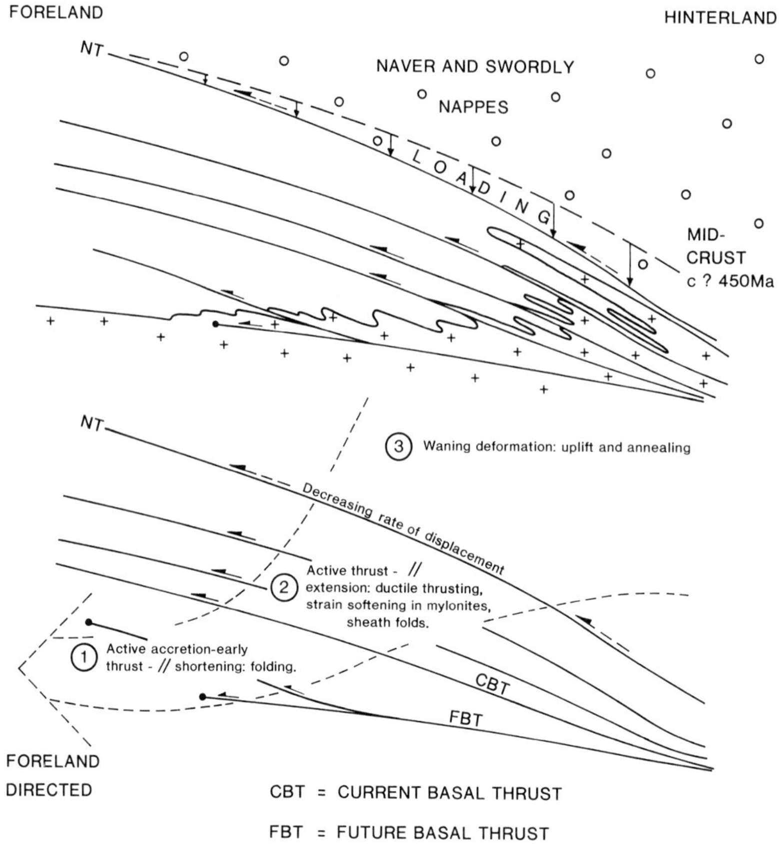

Fig. 7. Suggested model of Caledonian ductile thrusting and folding showing geological configuration during a stage of the thrust-sheet accretion process (upper diagram), and a summary of the proposed processes occurring at each level in the evolving thrust stack (lower diagram). Note that as a consequence of the foreland-propagation sequence, each unit of rock in the thrust wedge will pass through stages 1 to 3. Erosion synchronous with thrust-related uplift will also result in each successively lower set of structures forming at decreasing metamorphic grade.

depth, to the underside of the wedge ('underplating'; Platt 1986). This thickens the wedge and, via the process of isostatic compensation, will lead to an increase in surface slope. The actual process of accretion may involve an early phase of shortening and folding (transient work hardening), as required in Sutherland (Fig. 7; see Holdsworth 1989), but this will be progressively overprinted by thrust-parallel extension strains.

(b) Decreases in basal shear strength due to progressive strain softening in the mylonites associated with the ductile thrusts at any time forming the base of the accreting thrust stack (Fig. 7).

The first process is periodic and is an inevitable consequence of thrust-slice accretion in a foreland-propagating system forming in the middle crust. In Fig. 7, it is suggested that an early shortening (stage 1) is progressively superseded by the main extension phase (stage 2) which is, in turn followed by uplift, annealing and any associated late deformation (stage 3). The second process, strain softening, may result from several mechanisms occurring in the mylonites during stage 2 (cf. White *et al.* 1986 and references therein). In Sutherland, the geological evidence points especially to the probable importance of syn-metamorphic channelling of aqueous fluids into the ductile thrust zones (Holdsworth 1989*a*). For example, mylonitic quartzo-feldspathic lithologies are markedly enriched in white mica, whilst hornblendic rocks undergo a corresponding enrichment in biotite, at the expense of amphibole. Such changes are much less developed in areas outside of the high strain regions, and would lead to a marked reaction softening of the mylonites due to the enrichment in phyllosilicates (White *et al.* 1986). If such processes result from fluid channelling, the presence of water can also lead to profound weakening of quartz known as hydrolytic weakening (Griggs 1967).

The viscous wedge model is in certain respects *analogous* to the 'critical wedge' model developed for brittle (Coulomb-type) materials by Davis *et al.* (1983). As a consequence, Holdsworth (1989*a*) has a previously referred to the internal deformation patterns in the Sutherland Moine in terms of 'subcritical' or 'supercritical' states. This is probably unwise given the close connection made by most workers between such terms and a brittle rheology, which is clearly inappropriate in a region of ductile deformation. We wish to emphasize the importance of thrust-parallel extensional strains in ductile thrust zones, and would therefore suggest that the processes leading to such deformation should be termed *dynamic spreading*. This conveys the kinematic and underlying dynamic characteristics (a pushed, spreading wedge), and distinguishes the process from conventional 'static' gravity spreading previously discussed in a thrust belt context by Price (1973) and Elliott (1976).

Conclusions

Deformation fabrics formed during ductile thrusting at mid-crustal depths in the Caledonian Orogen in northern Scotland appear to result from a generally non-coaxial, thrust parallel extension strain. The evidence for extension is based largely on the sub-parallelism of the Caledonian schistosity and thrust (≡ shear) planes in regions of differing strain intensity. Whilst the quartz *c*-axes and other microstructural patterns are consistent with the strain model, they do not as yet offer conclusive proof. A pattern of internal extensional deformation contrasts strongly with the predominantly thrust-parallel *shortening* strains recognized in accretionary prisms and foreland thrust belts formed at shallower depths (e.g. Davis *et al.* 1983; Woodward 1987). The possible dynamic significance of such differences can be explored qualitatively using orogenic wedge models. Assuming a bulk viscous rheology, it is suggested that the internal extension of the thrust sheets that occurs synchronously with major thrust-related foreshortening may arise due to two processes (Fig. 7): (a) continual, episodic, thrust sheet accretion in the middle crust ('underplating'); (b) profound strain softening in the amphibolite-facies mylonite zones formed in association with the ductile thrusts.

The resulting response, here termed *dynamic spreading* may occur very widely in regions of ductile thrusting, and may be a characteristic feature of deformational regimes in collisional orogens at mid-crustal depths. Although it is often only possible to make qualitative deductions at the present time, field-based structural geologists can learn a great deal by attempting to assess the significance of their findings in terms of rheological models.

We are grateful to J. Platt and I. Main for comments on earlier manuscripts and to R. Law and M. Coward for helpful reviews.

References

Barr, D., Holdsworth, R. E. & Roberts, A. M. 1986. Caledonian ductile thrusting in a Precambrian metamorphic complex: the Moine of NW Scotland. *Geological Society of America Bulletin*, **97**, 754–764.

Boyer, S. E. & Elliott, D. 1982. Thrust systems. *American Association of Petroleum Geologist Bulletin*, **97**, 754–764.

Butler, R. W. H. 1982. A structural analysis of the Moine Thrust Zone between Loch Eriboll and Foinaven, N. W. Scotland. *Journal of Structural Geology*, **4**, 19–29.

—— & Coward, M. P. 1984. Geological constraints, structural evolution and deep geology of the northwest Scottish Caledonides. *Tectonics*, **3**, 347–365.

Chapple, W. M. 1978. Mechanics of thin-skinned fold-and-thrust belts. *Geological Society of America Bulletin*, **89**, 1189–1198.

Dahlen, F. A., Suppe, J. & Davis, D. 1984. Mechanics of fold-and-thrust belts and accretionary wedges: Cohesive Couloumb theory. *Journal of Geophysical Research*, **89**, 10,087–10,101.

Davis, D., Suppe, J. & Dahlen, F. A. 1983. Mechanics of fold-and-thrust belts and accretionary wedges. *Journal of Geophysical Research*, **88**, 1153–1172.

Elliott, D. 1976. The motion of thrust sheets. *Journal of Geophysical Research*, **81**, 949–963.

Evans, D. J. & White, S. H. 1984. Microstructural and fabric studies from the rocks of the Moine Nappe, Eriboll, NW Scotland. *Journal of Structural Geology*, **6**, 359–389.

Grant, C. J. 1989. *The kinematics and tectonic significance of ductile shear zones within the Northern Highland Moine*. PhD Thesis, University of Liverpool.

Griggs, D. 1967. Hydrolytic weakening of quartz and other silicates. *Geophysical Journal of the Royal Astronomical Society*, **14**, 19–31.

Hamner, S. 1988. Ductile thrusting at mid-crustal

level, southwestern Grenville Province. *Canadian Journal of Earth Sciences*, **25**, 1049–1059.

HOLDSWORTH, R. E. 1989*a*. The geology and structural evolution of a Caledonian fold and ductile thrust zone, Kyle of Tongue region, Sutherland, northern Scotland. *Journal of the Geological Society, London*, **146**, 809–823.

—— 1989*b*. Late, brittle deformation in a Caledonian ductile thrust wedge: new evidence for gravitational collapse in the Moine Thrust sheet, Sutherland, Scotland. *Tectonophysics*, **170**, (in press)

—— 1990. Progressive deformation structures associated with ductile thrusts in the Moine Nappe, Sutherland, N. Scotland. *Journal of Structural Geology*, **12**, (in press)

JESSELL, M. W. 1988. Simulation of fabric development in recrystallizing aggregates — II. Example model runs. *Journal of Structural Geology*, **10**, 779–794.

JOHNSTONE, G. S., SMITH, D. I. & HARRIS, A. L. 1969. The Moinian assemblage of Scotland. *In*: KAY, M. (ed.) *North Atlantic Geology and Continental Drift: a Symposium*. Memoir of the American Association of Petroleum Geologists, **12**, 159–180.

KNIPE, R. J. & NEEDHAM, D. T. 1986. Deformation processes in accretionary wedges — examples from the SW margin of the Southern Uplands, Scotland. *In*: COWARD, M. P. & RIES, A. (eds) *Collision Tectonics*. Geological Society, London, Special Publication, **19**, 51–65.

LAW, R. D., CASEY, M. & KNIPE, R. J. 1986. Kinematic and tectonic significance of microstructures and crystallographic fabrics within quartz mylonites from the Assynt and Eriboll regions of the Moine thrust zone, NW Scotland. *Transactions of the Royal Society of Edinburgh: Earth Sciences*, **77**, 99–125.

LISTER, G. S. & HOBBS, B. E. 1980. The simulation of fabric development during plastic deformation: the effect of deformation history. *Journal of Structural Geology*, **2**, 355–370.

—— & SNOKE, A. 1984. S-C Mylonites. *Journal of Structural Geology*, **6**, 617–638.

—— & WILLIAMS, P. F. 1979. Fabric development in shear zones: theoretical controls and observed phenomena. *Journal of Structural Geology*, **1**, 283–297.

MATTAUER, M. 1986. Intracontinental subduction, crust-mantle décollement and crustal-stacking wedges in the Himalayas and other collision belts. *In*: COWARD, M. P. & RIES, A. (eds) *Collision Tectonics*. Geological Society, London Special Publication, **19**, 37–50.

MOORHOUSE, S. J., MOORHOUSE, V. E. & HOLDSWORTH, R. E. 1988. Moine excursion 12: North Sutherland. *In*: ALLISON, I., MAY, F. & STRACHAN, R. A. (eds) *Excursion Guide to the Moine Geology of the Scottish Highlands*, Scottish Academic Press, 216–248.

MOORHOUSE, V. E. & MOORHOUSE, S. J. 1983. The geology and geochemistry of the Strathy Complex of the north-east Sutherland, Scotland, *Mineralogical Magazine*, **47**, 123–137.

PLATT, J. P. 1986. Dynamics of orogenic wedges and the uplift of high pressure metamorphic rocks. *Geological Society of America Bulletin*, **97**, 1037–1953.

—— & BEHRMANN, J. H. 1986. Structures and fabrics in a crustal-scale shear zone, Betic Cordillera, S. E. Spain. *Journal of Structural Geology*, **8**, 15–33.

PRICE, R. A. 1973. Large-scale gravitational flow of supracrustal rocks, southern Canadian Rockies. *In*: DE JONG, K. A. & SCHOLTEN, R. (eds) *Gravity and Tectonics*. Wiley, New York, 491–502.

RAMBERG, H. 1977. A tectonic model for the central part of the Scandarian Caledonides. *American Journal of Science*, **277**, 647–656.

—— 1981. *Gravity, Deformation and the Earths Crust*. Academic Press, London.

RAMSAY, J. G. & GRAHAM, R. H. 1970. Strain variation in shear belts. *Canadian Journal of Earth Sciences*, **7**, 786–813.

SANDERSON, D. J. 1982. Models of strain variation in nappes and thrust sheets: a review. *Tectonophysics*, **88**, 201–233.

——, ANDREWS, J. R., PHILLIPS, W. E. A. & HUTTON, D. H. W. 1980. Deformation studies in the Irish Caledonides. *Journal of the Geological Society, London*, **137**, 281–302.

STOCKMAL, G. S. 1983. Modelling of large-scale accretionary wedge deformation. *Journal of Geophysical Research*, **88**, 8271–8287.

WHITE, S. H. BRETAN, P. G. & RUTTER, E. H. 1986. Fault zone reactivation: kinematics and mechanisms. *Philosophical Transactions of the Royal Society of London*, **A317**, 81–97.

WOODWARD, N. B. 1987. Geological applicability of critical-wedge thrust belt models. *Geological Society of America Bulletin*, **99**, 827–832.

Structural implications of compactional strain caused by fault block rotation: evidence from two-dimensional numerical analogues

J. E. ILIFFE,[1] I. LERCHE[1] & K. NAKAYAMA[2]

[1] *University of South Carolina, Department of Geological Sciences, Columbia, South Carolina 29208, USA*

[2] *Japan Petroleum Exploration Co., Akasaka Twin Tower Bldg. East, 2–17–22, Akasaka Minatoku, Tokyo 107, Japan*

Abstract: The effects of compaction on tilted fault blocks are assessed using a two dimensional mathematical model which assumes homogeneous non-rigid blocks. Lithology is described by a depth-dependent porosity function.

Compaction is restricted to the upper parts of the hangingwalls of blocks and is greater in blocks bounded by sub-vertical faults rotated by 30° or more. The fault planes are unaffected.

Significant sub-vertical lines of maximum compaction are noted directly below block crests. Since secondary faults occur in similar orientations in rigid blocks, rotation in the Earth's gravitational field is proposed as a mechanism for secondary faulting.

Fluid migration from the compressed hangingwall to the undeformed footwall has implications for shale diapir formation.

By means of a two-dimensional forward mathematical model this paper assesses the amount and location of compactional strain in non-rigid fault blocks rotated in the Earth's gravitational field. The structural consequences of these results are also discussed. The model consists of three identical blocks of sediments bounded by planar faults (Fig. 1). The porosity and lithology of the blocks are given through an exponential porosity–depth function (Magara 1976). Changes in porosity, block thickness and length, overall block shape, and internal stress field caused by the rotation, are calculated for the different regions of the blocks.

Model description

The three unrotated identical blocks are given horizontal widths (W) and vertical heights (Th). Each of the blocks is bounded by faults with an angle (τ), and each has a number of layers (jx), of equal thickness ($z1$) (Fig. 1a). All these parameters may be varied in experiments.

The characteristics of blocks of different lithologies may be broadly recreated by varying the scaling constant (a) and surface porosity (ϕ_0), in the porosity depth function of Magara (1976) (Fig. 2).

$$\phi(z) = \phi_0 \exp(-z/a), \qquad (1)$$

where $\phi(z)$ is the calculated porosity at depth

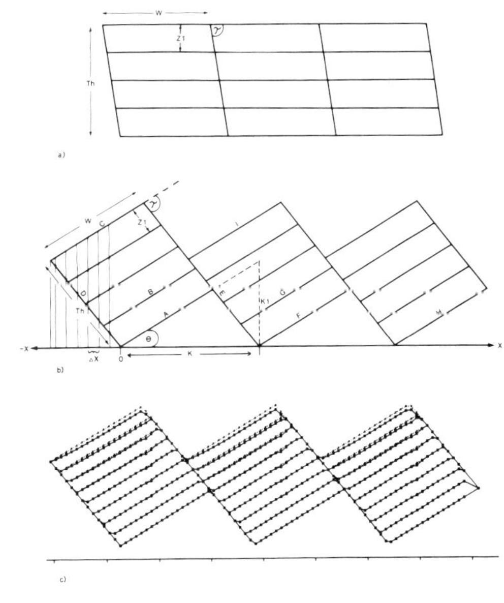

Fig. 1. (**a**) Cross-section showing parameters used to define the pre-rotated blocks. (**b**) Cross-section illustrating the method of post-rotation sampling of depth (z) in increments of Δx, and parameters used to describe the block in the model. (**c**) Model output section illustrating the results of post-rotation compaction on shale blocks tilted 30°.

From Knipe, R. J. & Rutter, E. H. (eds), 1990, *Deformation Mechanisms, Rheology and Tectonics*, Geological Society Special Publication No. 54, pp. 501–508.

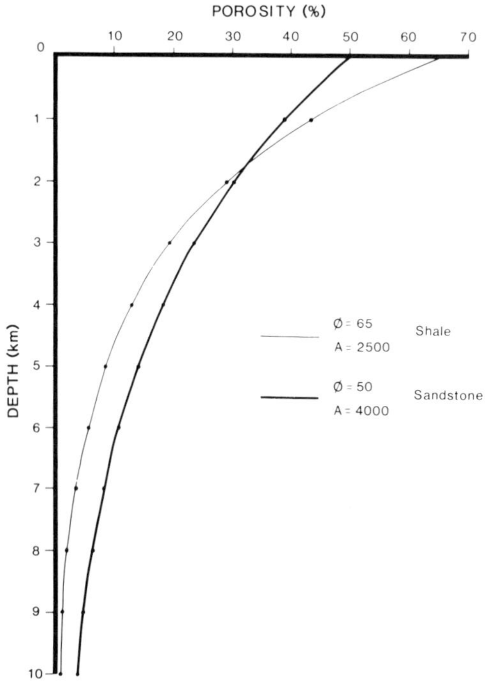

Fig. 2. Diagram comparing porosity depth curves of sandstone and shale using values of ϕ_0 and a from Sclater & Christie (1980) in equation 1.

(z) (measured from the top of the block downwards).

Rotation

The uncompacted fault blocks are rotated through an angle θ, and new equations of each stratal boundary are calculated, being careful to maintain the correct shape of each of the blocks. Before any other calculations are made a system of sampling stations is set up in the horizontal (x) direction, a distance Δx apart (Fig. 1b). The points of intersection of each stratal boundary with a line drawn vertically from each sampling station (x, z) are calculated. This information is then passed on to the section of the analysis which assesses the compactional effect.

Compaction

Equation 1 is used to calculate the porosity at any position in the block both before and after rotation, and hence determine the compaction effect. The formations have no internal strength and freely deform in the vertical plane.

There are four steps in this process.

(a) The initial (pre-rotation) porosity ϕ (z') along each stratal boundary is calculated from equation (1) at each sampling point in the x direction.

(b) Next, the pre-rotation total frame weight ($Wt(x', z')$) acting along each stratal boundary is calculated by summing the weight between the top of the block and the depth to the boundary.

(c) The new (post-rotation) total frame weight ($Wt(x, z)$) may now be calculated at each intersection point of a stratal boundary and a sampling point in the x direction. This calculation is done by summing the weights calculated for a set of depth values measured perpendicular to the strata over the new vertical depth. If the new weight ($Wt(x, z)$) is less than, or equal to, the old weight ($Wt(x' z')$, no compaction will occur at that point. If, however, $Wt(x, z)$ is greater than $Wt(x', z')$ then compaction occurs.

(d) The resulting amount of compaction (Δz) is calculated by solving the differential equations relating the porosity–depth function to the overburden calculation, in terms of the final porosity $\phi(z)$, original frame weight $Wt(x', z')$ and the resultant frame weight $Wt(x, z)$:

$$\Delta z = \frac{\phi(z')\,(Wt(x, z) - Wt(x', z'))}{(1 - \phi(z')^2)\,\rho g} \qquad (2)$$

where ρ is the rock matrix density and g is the acceleration due to gravity. The Δz values for each z (vertical) point at each x (horizontal) location are summed cumulatively from the base layer upwards. The resulting total displacement value for each z point is now subtracted from its pre-compaction vertical position to yield the compaction corrected section (Fig. 1c). This process is repeated for each x location. The new z values for each layer are joined to represent the deformation of the whole layer, which results in a two-dimensional compactional effect.

Results

A set of experimental conditions were designed to investigate the effects that variations in lithology, height to width ratio of the block, fault dip, and angle of rotation have on compaction and stress configurations. Changes in bed lengths, bed thicknesses, angles and overall shape of the output relative to the input are used to assess the size of various effects. In the tests two lithological types were considered, shale and homogeneous sandstone. The appropriate values of a and ϕ_0 were taken from Sclater & Christie (1980) (Fig. 2).

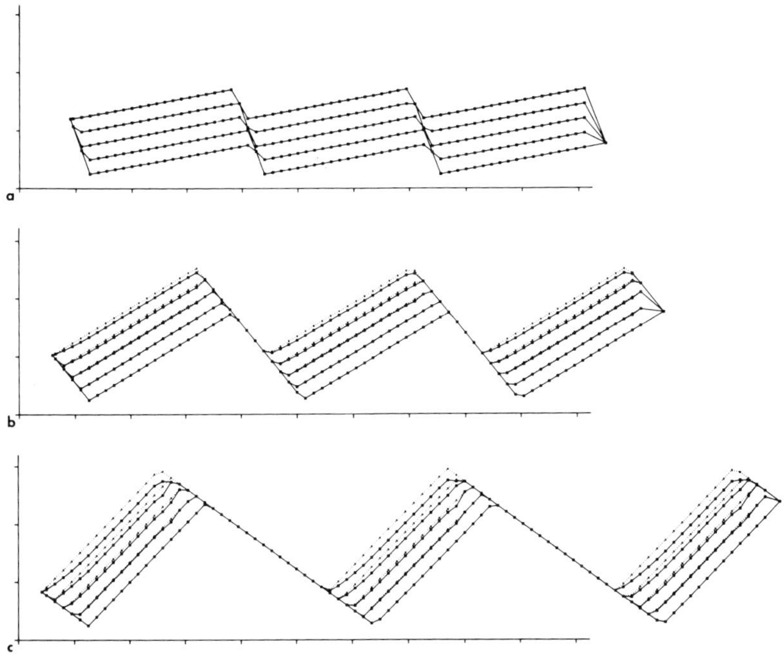

Fig. 3. (**a**) Fault block of shale, width (W) = 6 km, height (Th) = 2 km and angle of tilt (θ) = 10°; (**b**) θ = 30°; (**c**) θ = 45°.

The size effect of the blocks, specifically the ratio of the height of the block to its width, was tested using three basic block types. The long, flat, 'warehouse type' block (Fig. 3) was compared with a square block (Fig. 4), and with a tall, thin, 'skyscraper type' block (Fig. 5).

As the lithology and block dimensions were varied the fault angle and degree of rotation were also incrementally changed.

Figures 3, 4, and 5 show that the compaction effect is restricted to the downthrown hanging wall of the blocks, the footwall side of the block remains undeformed (Fig. 6) except for a few metres down the fault from the surface/fault intersection.

When a point on the top of a formation on the hangingwall side of the block is rotated the load acting is increased by the secant of the angle of rotation. A point on the top of the same formation towards the footwall side of the block has had its load decreased by the rotation and/or faulting. The sediment at this footwall side point will not compress.

The compaction effect is not only unevenly distributed laterally but also varies with depth: it is easier to squeeze a very porous surface mud compared to a fully lithified shale at 3 km depth.

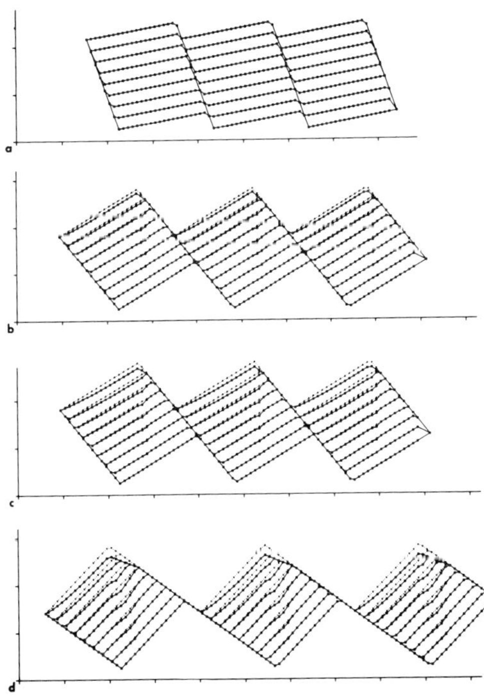

Fig. 4. (**a**) Fault block of shale, W = 4 km, Th = 4 km and θ = 10°; (**b**) θ = 30°; (**c**) fault block of sandstone, θ = 30°; (**d**) fault block of shale, θ = 45°.

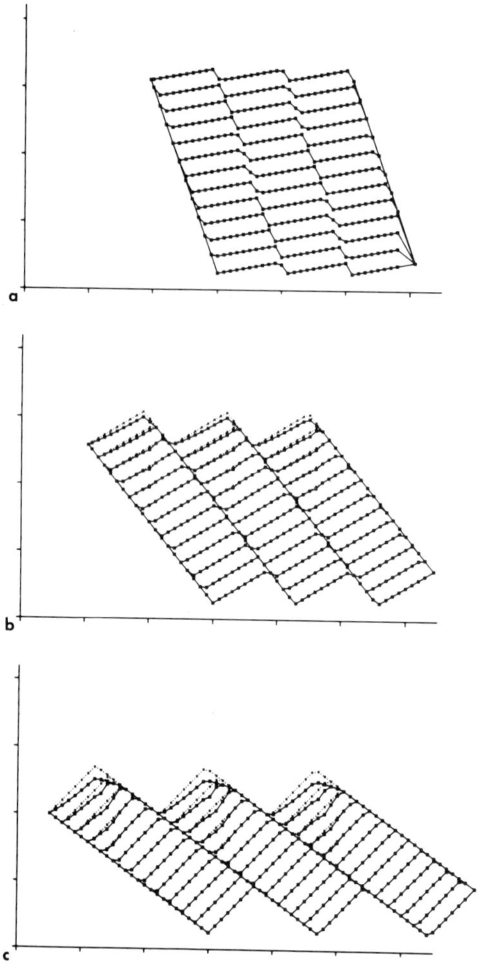

Fig. 5. (**a**) Fault block of shale, W_t = 2 km and Th = 6 km, θ = 10°; (**b**) θ = 30°; (**c**) θ = 45°.

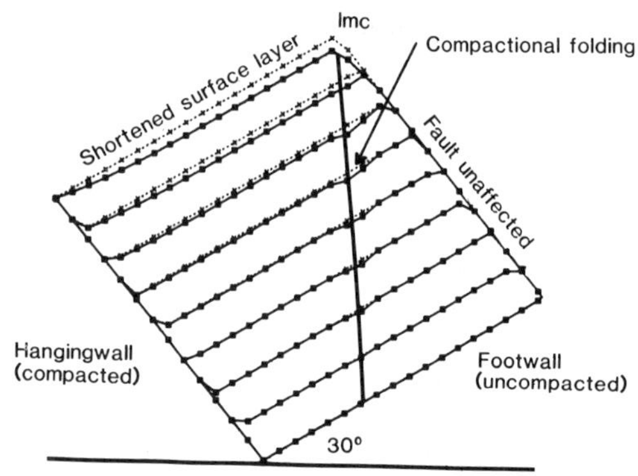

Fig. 6. One post-rotation compacted shale block tilted 30° with faults dipping 80°, exhibiting features of folds, line of maximum compaction, an uncompacted footwall and compacted hangingwall.

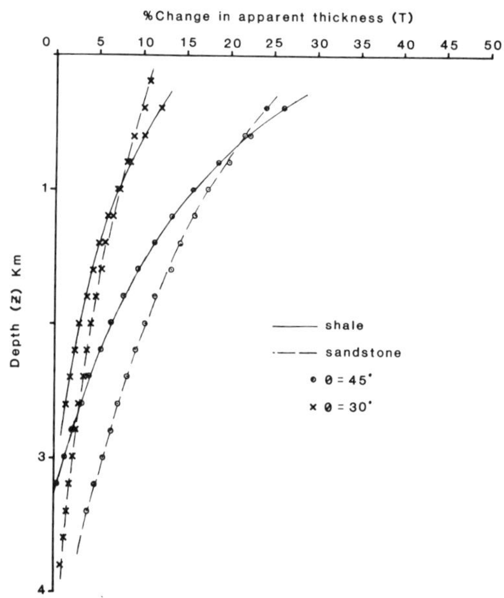

Fig. 7. Graphs to illustrate the percentage change in apparent thickness of layers 200 m thick, due to compaction following rotation of (**a**) 30° and (**b**) 45°, for a block of sandstone and shale.

Lithology

The effect of changes in lithology is exemplified in Figs 4b and c. The sandstone blocks compact more than shale blocks, because the sandstones compress to greater depths, more than compensating for the dramatic, near-surface, effect of the shale blocks. This compactional difference is measurable in the changes in apparent thicknesses of layers within the blocks (Fig. 7).

Height to width ratio and angle of rotation

Comparing warehouse and skyscraper-shaped blocks (Figs 3 and 5) it is obvious that block dimensions and angle of rotation are important parameters in compaction effects. In the skyscraper-type blocks, deformation in the upper formations is intense but the lower parts of the blocks are unaffected. The very wide warehouse-type blocks deform throughout their length and more intensely the more they are tilted (Fig. 3). As the angle of tilt is increased there is more block to be added to the load (because of its length) in the rough form of secθ times the original weight. The skyscraper-type blocks, because of their narrow width, have only a very small maximum potential increase in load (Fig. 5).

If the situation where most compaction occurs is considered (a 90° rotation) at the points where maximum change in weight occurs, (the bottom left hand corners of the blocks) the excess weight acting on these points can be crudely described as $\rho_{Rav}gW$. Since ρ_{Rav} (bulk density of rock) and g (gravity) are constants, the limiting factor is W, the width of the block. The depth to which compaction is evident is also controlled by the width of the block (compare Fig. 3b and Fig. 5b). The maximum depth, at which compaction occurs (z_{max}) may be written

$$z_{max} = W\sin\theta. \tag{3}$$

The compaction distribution of the blocks is not only related to the surface porosity ϕ_0 and scaling factor a i.e. lithological type, but also, in part, to the width of the block and angle of rotation.

When considering a block wider than the vertical scaling factor a, the rate of change of the compaction effect is roughly proportional to $\sin\theta$, which suggests that these effects are minimal until a tilt of 30° is reached. In such rotated blocks there is a reduction of stratigraphic thickness on the order of 6–10%. This effect is less in the deeper parts of the block.

Dip of fault

The distribution and maximum depth of compaction may be altered by changing fault angles. As the fault angle (τ) is decreased from 90°, the effective compactional area of the block is decreased by its physical rotation movement laterally out of the compacting region. In this case the maximum depth, at which compaction can occur is limited once again by the scaling factor a (equation 1) and width of the block, with

$$z_{max} = W\sin\theta + W\cos\theta\tan(\tau-\theta). \tag{4}$$

For a block 4 km wide, tilted at an angle of 45°, of infinite depth, and with the fault angle varied by 90°, 80° and 60°, z_{max} is 5.6, 4.75 and 3.6 km respectively. Fault angle therefore plays a significant role in how much compaction takes place and over what depth range.

Stress considerations

During the calculations to determine the change in position of a point or a formation due to compaction, two variables are calculated, $Wt(x', z')$ and $Wt(x, z)$, or original weight and final weight. If we begin with a block in equilibrium with regard to stress, then rotate it and compare the new weight acting on each point along formation boundaries with the original weight, it is possible to get a crude idea of the changes in magnitude of stress at each point. In blocks made up of compressible shale and sandstone, areas of stress increase are immediately recognized in the model from their visible compaction. The stress increase is relaxed by strain accommodated by alignment of shale grains and dewatering, the beds are thus acting as passive strain markers.

A phenomenon inherent in the rotated blocks are points along each formation at which compaction is greatest (Figs 3, 4 and 5). If the maximum compactional points are joined for each formation from the top of the block down, this line represents the region in the block where the stress has had the greatest change (line of maximum compaction or lmc) (Fig. 6). If the block were of brittle rock (such as granite) and the increase in stress overcomes the shear strength of the material, then a fault may appear. For granite transected by joints in various directions, the pre-existing joint which lies in the most favorable orientation (closest to that predicted here) will release the rotation-created stress by acting as a fault.

Tests to date suggest that the angle of intersection between this secondary fault and the main fault is a nearly constant 35° regardless of the degree of tilt for a fault angle of 80°. If the fault angle is 60°, the angle of the intersection becomes 45°.

This relationship has been independently observed in the field. Miller *et al.* (1983) working in the Snake Range, Nevada, USA, describe 'second generation' faults in which faults intersect older faults at an angle of 40°. They also describe the faults as beginning in a near vertical orientation, as predicted by this model.

Similar observations have been made by Angelier & Colletta (1983) in the Gulf of Suez, Egypt, Western Gulf of California, Mexico, and the Southern Basin and Range, USA. They describe the occurrence of near vertical (80–90° dip) tension gashes formed in early stages of extension, when blocks are tilted 5–10°, which later become active as faults as tilting progresses. The 80–90° dip of the tension gash, with a tilt of 20° in a fault with an original dip of 60°, yields an intersection of fault and tension gash at 40°.

Thorough testing of different block dimensions, fault angles, and angles of tilt has been carried out, providing firm corroborative support for the ruggedly stable nature of the effect.

Although the model was designed to investigate the compaction effect due to rotation of non-rigid, shallow sedimentary blocks such as those seen in the Niger Delta and Gulf Coast of the USA, the stress distributions outlined by the passive strain markers (i.e. the compacted beds) imply a gravity mechanism for secondary faulting. This observation can be extrapolated to blocks of more rigid material such as those in the Gulf of Suez and Basin and Range.

Discussion

Forward models such as the one described in this study have several advantages over field analyses. In the synthetic situation the designer knows exactly all the parameters of the system, and also has defined how the parameters will be manipulated. The problem with all synthetic models lies with their intrinsic assumptions. It is therefore important to question the limitations of assumptions set on experiments assessing the validity of a model such as the one described here.

In an effort to carry out such a task the limitations and validity of the assumption of a lithologically homogeneous block was investigated. Clearly, situations with interbedded varying lithologies would be more appropriate geologically.

Consider an interbedded sandstone and shale scenario. Figure 7 shows that above about 1000 m shale layers compact more than those of sandstone. Between 1000 m and 1600 m they compact the same amount, and their porosities are quite similar (Fig. 2). Below 1600 m sandstone layers compress more than shale.

Since the amount of compaction of layers is not much different between lithologies in the intermediate depth zone, the folding in the layers between 1000 and 1600 m depth in an interlayered situation would be just as prevalent as in the monolithologic blocks. The compaction effects in an interbedded setting above and below this zone will be slightly reduced.

Another assumption which requires validating is that of the boundary conditions of planar faults as opposed to listric faults. No compaction occurs on the faults themselves, although the rotation could just as easily be accommodated by listric faults. The simplicity of the mathematical functions required to rotate blocks on planar faults does not preclude extrapolation of results into listric faulted regions. Indeed the problem of the void beneath the rotated blocks would be solved by listric faults (Fig. 8).

The set of synthetic minor faults, visible in the listrically faulted block in Fig. 8, may be evidence of a time dependent sequential development of secondary faults in the position predicted by the model, not unlike the mode of development proposed by Angelier & Colletta (1983).

Block rotation compaction causes the surface of blocks to shorten, whereas compactional folding occurs at depth, which may introduce a small error in accurate calculations of extension. Although even using good field and seismic

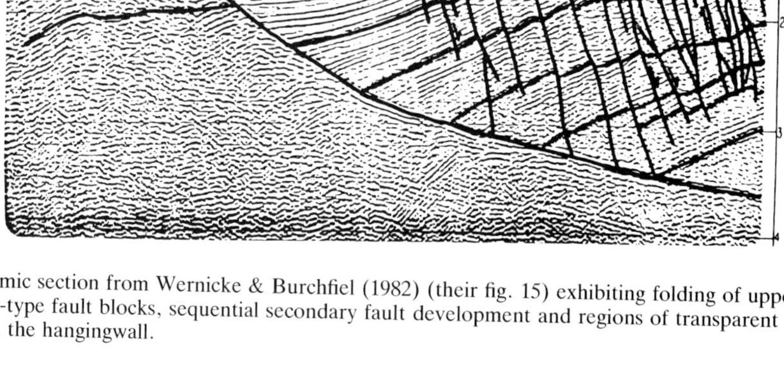

Fig. 8. Seismic section from Wernicke & Burchfiel (1982) (their fig. 15) exhibiting folding of upper layers of 'skyscraper'-type fault blocks, sequential secondary fault development and regions of transparent seismic reflectors in the hangingwall.

data it may not be possible to recognize this effect from other geological noise.

Compaction results in fluid (water and possibly hydrocarbons) loss from the layers, with the dewatering occurring after rotation so that flow is up through the easiest route, generally vertically. If the block were of interbedded sands and shales, this motion would be curtailed by the impermeable shale layers. Flow would then be directed along the permeable bed to the uncompacted sides of the block, and then upwards. This concentration of fluids in the footwall of the block might be a causative mechanism for mud diapirism. The effect is amplified by the dehydration of smectite to illite at around the depth at which the temperature attains 100°C (Bruce 1981).

Shale diapirs in the footwalls of faults in muddy deltaic settings, as seen in Tertiary of offshore Louisiana USA by Woodbury *et al.* (1973), and in the Upper Carboniferous of West Coast of Ireland (Rider, 1978), provide observational support for this suggestion.

Salt diapirism may also be triggered by fault block rotation. The variable changes in load in different regions of the block will encourage salt to flow into the uncompacted and unloaded footwall region of the block as suggested by the observations of Woodbury *et al.* (1973). Changes in porosity due to compaction caused by block rotation will also change other block properties such as thermal conductivity and seismic velocity. The authors are investigating the impact of these changes on block temperature and reflector character.

Conclusions

(1) The amount of compaction in a rotated fault block is governed by block lithology, width (W), angle of tilt (θ) and dip of faults (τ).

(2) Compaction is confined to the hanging-wall side of the blocks, is greater in the upper 1000 m, and is also greater for changes of tilt in excess of 30°.

(3) Steeper dipping faults have greater areas of compaction than do gentler dipping faults. The faults themselves are not affected by compaction since they are exhumed and unloaded, thus these effects are insensitive to fault geometry.

(4) A line drawn through the points along each formation which have the greatest compaction is a likely position for strain release in the form of a secondary fault. This result is consistent with the independent field observations.

(5) The secondary faulting would occur even in incompressible rocks provided the increase in stress was sufficient to overcome the shear strength of the material.

(6) Fluid migration may take place from the compacted side to the uncompacted side of the block and then vertically upwards, perhaps manifesting itself in nature as shale diapirs possibly associated with hydrocarbon accumulation sometimes observed in muddy deltaic regions.

(7) Although the model is designed to investigate the compaction effect due to rotation of non-rigid shallow sedimentary blocks such as those seen in the Niger Delta and Gulf coast of the USA, the stress distributions shown by the passive strain markers (i.e. the compacted beds) imply a gravity mechanism for secondary faulting. This observation can be extrapolated to blocks of more rigid material such as those in the Gulf of Suez and Basin and Range.

Thanks and appreciation go to the Industrial Associates of the University of South Carolina Basin Modelling Group for financial support. This paper benefitted from an astute review by A. Gibbs, for which the authors are grateful.

References

Angelier, T. & Colletta, B. 1983. Tension fractures and extensional tectonics. *Nature*, **301**, 49–51.

Bruce, C. H. 1981. Smectite Dehydration — Its Relation to Structural Development and Hydrocarbon Accumulation in Northern Gulf of Mexico Basin. *American Association of Petroleum Geologists Bulletin*, **68**, 673–683.

Magara, K. 1976. Water Expulsion from Elastic Sediments During Compaction — Direction and Volumes. *American Association of Petroleum Geologists Bulletin*, **60**, 543–553.

Miller, E. L., Grans, P. B. & Garing, T. 1983. The Snake Range décollement: An exhumed mid-Tertiary ductile-brittle transition. *Tectonics*, **2**, 234–263.

Rider, M. H. 1978. Growth Faults in the Carboniferous of Western Ireland. *American Association of Petroleum Geologists Bulletin*, **62**, 11, 2191–2213.

Sclater, J. G. & Christie, P. A. F. 1980. Continental stretching: an explanation of the post-mid-Cretaceous subsidence of the central North Sea basin. *Journal of Geophysical Research*, **85**, 3711–3739.

Wernicke, B. & Burchfiel, B. C. 1982. Modes of extensional tectonics. *Journal of Structural Geology*, **4**, 105–115.

Woodbury, H. O., Murray, I. B., Pickford, P. J. & Akers, W. H. 1973. Pliocene depocenters, outer continental shelf, Louisiana and Texas. *American Association of Petroleum Geologists Bulletin*, **57**, 2428–2439.

Alpine deformation on Naxos (Greece)

J. L. URAI[1,2], R. D. SCHUILING[1] & J. B. H. JANSEN[1]

[1] *Institute for Earth Sciences, PoBox 80021, 3508TA Utrecht, Netherlands*

[2] *Present address: Koninklijke/Shell Exploratie en Productie Laboratorium, PO Box 60, 2280 AB Rijswijk, Netherlands*

Abstract: A detailed structural study of Naxos (Attic–Cycladic massif, Greece) reveals two major deformation events. The first one is associated with large scale thrusting and high-pressure–low-temperature metamorphism during an early Alpine subduction episode. The second event occurred during continental extension and the associated development of localized thermal domes, where lower crustal rocks were brought into contact with upper crustal units along a major shallow dipping shear zone. We agree with a model of Naxos as a Cordilleran type Metamorphic Core Complex. However, our observations show that the sense of shear was 'upper plate moving North' during the second event, calling for a reinterpretation of existing tectonic models of the Cyclades.

Naxos is the largest island of the Cycladic archipelago in the Aegean sea, Greece. Geologically it belongs to the Attic–Cycladic massif (Trikkalinos 1947; Dürr *et al.* 1978; Robertson & Dixon 1984), which has undergone at least two Alpine regional tectono-metamorphic events involving a pre-Alpine basement, and marbles and schists of probably Mesozoic age (Henjes-Kunst & Kreuzer 1982; van der Maar & Jansen 1983; Andriessen *et al.* 1987; Dürr *et al.* 1978; Jansen & Schuiling 1976; Feenstra 1985).

The tectonic history of the Attic–Cycladic massif involves an early Alpine compressional phase involving subduction of continental margin material, generation of a nappe pile and associated high-pressure–low-temperature metamorphism (M_1) dated at approximately 50 Ma (Dürr *et al.* 1978; Andriessen *et al.* 1979; Ridley 1982, 1984; Smith & Woodcock 1982; Altherr *et al.* 1982; van der Maar & Jansen 1983; Wijbrans & McDougall 1986, 1988). This was followed by a phase of extensional tectonics, forming a back-arc basin in the Aegean (Horvath & Berckhemer 1982), with a thinned crust, high heat flow (Makris 1978) and rapid uplift of lower crustal rocks. The extensional event was associated with a regional greenchist facies metamorphic event (M_{2a}, 25 Ma), followed by more localized high-temperature–low pressure metamorphism (M_{2b}, 16 Ma), producing thermal domes (Jansen & Schuiling 1976; Andriessen *et al.* 1979; van der Maar & Jansen 1983; Wijbrans & McDougall 1988; Baker *et al.* 1989).

The present tectonic framework of the Aegean is that of a land-locked basin with movement along the subduction zone south of Crete, associated with roll-back of the subducted slab (McKenzie 1978; Le Pichon 1983; Spakman 1986; Genthon & Souriau 1987; Meulenkamp *et al.* 1987). The Neotectonic evolution of the region has been extensively studied; estimates of the maximum amount of (N–S) crustal extension from 15 Ma to present times are up to a factor of two (McKenzie 1978; Le Pichon & Angelier 1979; Mercier *et al.* 1989).

In a recent paper, Lister *et al.* (1984) proposed an interpretation of two islands in the Cyclades, Naxos and Ios, as Cordilleran type metamorphic core complexes, formed during crustal extension with upper crustal rocks superimposed upon deeper crustal units by movement along shallowly dipping detachment faults and shear zones (Crittenden *et al.* 1980; Lister *et al.* 1986). This model elegantly explains many of the features in the area, and proposes the existence of a major south-dipping shear zone-detachment fault system with 'upper plate moving south' sense of transport, active from approximately 20 Ma onwards. However, the small data set of kinematic indicators presented called for more work to test the proposed model.

Although the petrological and geochemical evolution on Naxos has been studied in great detail (Jansen & Schuiling 1976; Rye *et al.* 1976; Jansen 1977; Jansen *et al.* 1978; Kreulen 1980; Feenstra 1985; Baker *et al.* 1989), much less is known of the structural geology of the island. In particular there is confusion in the literature on the timing of the different deformation episodes and on the regional movement picture.

The purpose of the present work, then, was

From Knipe, R. J. & Rutter, E. H. (eds), 1990, *Deformation Mechanisms, Rheology and Tectonics*, Geological Society Special Publication No. 54, pp. 509–522.

to investigate in detail the Alpine deformation on Naxos and to determine the relationship between deformation and the different phases of metamorphism. Results are based on detailed mapping of a selected area along the NW-coast, detailed study of key locations on the rest of the island and thin section study of about 200 oriented samples.

Data

Because the geology of Naxos was described in detail by several workers (Jansen 1977; Feenstra 1985; Hecht 1979), in this description emphasis is given to the structural features only.

Briefly, the rocks on the island can be divided into four distinct units: (i) a migmatite complex, surrounded by (ii) a metamorphic complex of alternating marbles, pelitic rocks, metavolcanics and metabauxites; (iii) a granodiorite in the western part of the island, which intrudes the metamorphic complex, and (iv) a sequence of allochtonous, non-metamorphic rocks in tectonic contact with (ii) and (iii). A simplified geological map of Naxos is shown in Fig. 1, and Fig. 2 is a schematic illustration of the relationships between deformation and metamorphism discussed in this paper.

Other features important for the purpose of this study are: two mappable horizons of ultramafic rocks, and a series of M_{2b} isograds in the metamorphic complex. There are several synkinematic (M_{2b}) granitoids and a suite of (M_{2b}) pegmatites around the migmatite, these crosscut earlier foliations and are valuable structural markers (Figs 1 and 3).

The work of Jansen (1977), Bonneau *et al.* (1978) and Hecht (1979) has shown that the large-scale structure of the metamorphic complex consists of kilometre scale isoclinal folds with fold axes trending N–S. These folds are coaxially refolded by open, upright folds with N–S fold axes. The resulting structures are gently folded along an E–W axis, forming a structural dome with the regional foliation warping over the migmatite complex (cf. Fig. 4).

Structures in the metamorphic complex

On the outcrop scale, the prominent structural features in all but the most massive lithologies are a penetratively developed foliation parallel to the layering, and several types of lineation. This foliation is mostly shallowly dipping, except around the migmatite dome where dips over 45° are not uncommon. Lineations are invariably approximately N–S, shallowly plunging.

Overprinting relationships commonly show the presence of three generations (B_1–B_3; Hobbs *et al.* 1976) of folds. It should be noted here that these fold generations represent local age relationships only, and, as will be shown below, their formation overlapped considerably in time (cf. Fig. 2).

B_1 and B_2 structures. B_1 and B_2 folds are invariably tight to isoclinal and commonly rootless. In all suitable lithologies they have a well developed axial plane foliation. Lineations associated with B_1 and B_2 structures are stretching lineations formed by elongated mineral grains or aggregates. Fold axes are almost always parallel to the local stretching lineation. $B_1 \times B_2$ interference patterns are commonly of type 3 (Hobbs *et al.* 1976), with parallel B_1 and B_2 fold axes (Fig. 5a, c and d). In thin-bedded lithologies complicated interference patterns occur, and sheath folds can be found locally. In these localities fold axes show strong local variations and are non-parallel to the stretching lineation. Owing to the strong similarities in style, B_1 and B_2 structures can only be distinguished when overprinting relationships are found (Williams 1985); isoclinal folds without good overprinting relationships can be of either generation. In almost all observed cases, the pegmatites crosscut B_1 and B_2 structures. When pegmatites are deformed in B_1 or B_2 structures, the strains involved are relatively small, producing open folds or boudins indicating N–S extension (Fig. 5e,f).

Large scale B_1 and B_2 structures are similar in morphology to the ones seen in outcrop. Figure 3 is a map of an area in the NW part of the island, showing large isoclinal folds refolded in B_3 folds. These closely resemble the structures reported by Hecht (1979) in E-Naxos.

In the eastern part of the metamorphic complex the lithology consists predominantly of massive marble units, and small scale structures are relatively scarce. A metaconglomerate unit (Jansen 1977), intercalated in this series gives some indication of the high strains undergone by most marbles. Calcitic pebbles commmonly show aspect ratios of *c.* 1:6:30 with long axes parallel to the regional lineation.

B_3 structures. B_3 folds are open to tight, with a steep axial surface and N–S, gently plunging fold axes (Figs 3, 5g & 6). They refold earlier axial plane foliations, but only locally is an S_3 crenulation cleavage developed in the micaschists. A remarkable feature of B_3 folds is that,

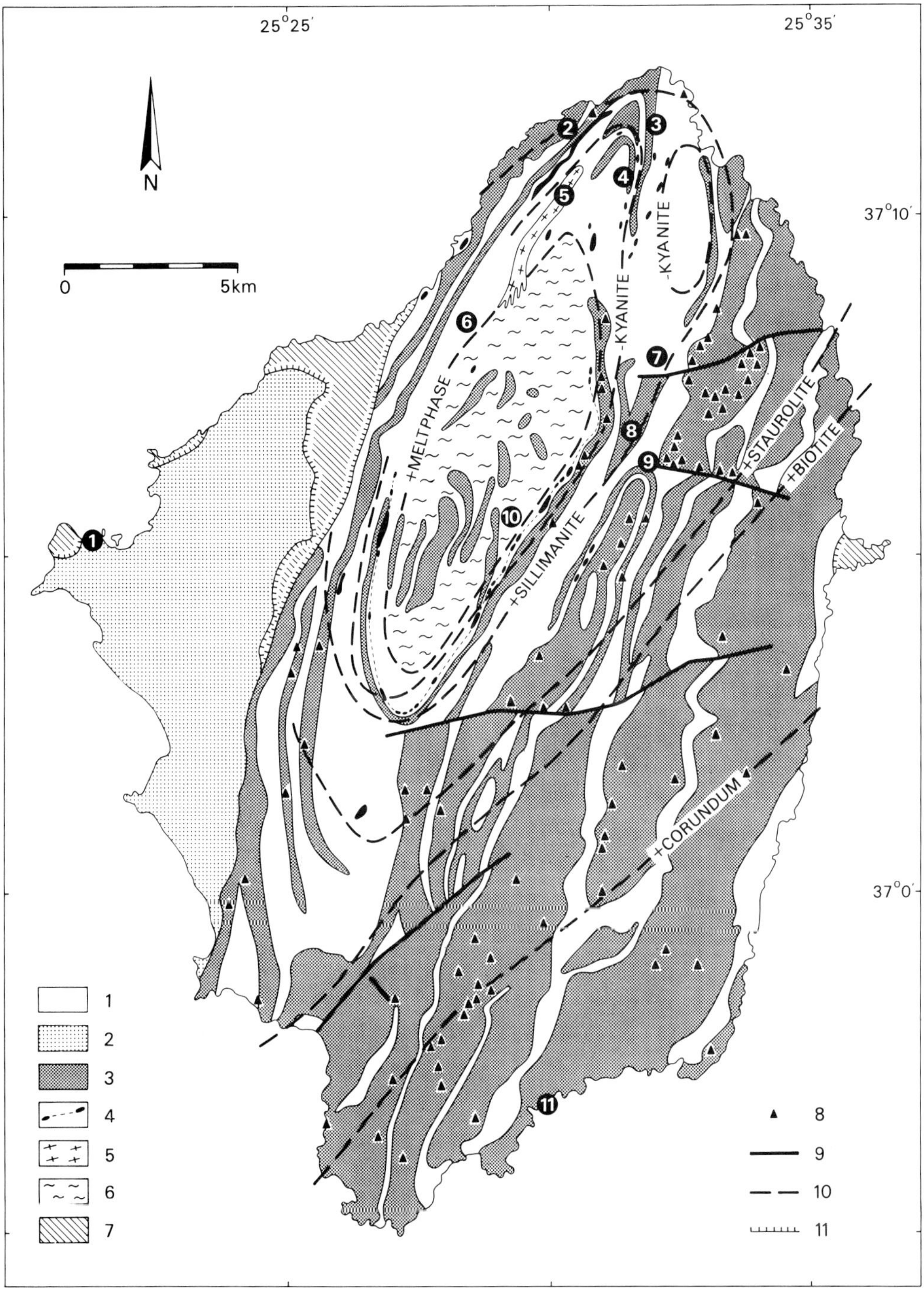

Fig. 1. Simplified geological map of Naxos (after Jansen 1977). Legend: 1, predominantly micaschist; 2, granodiorite; 3, predominantly marble; 4, ultrabasite; 5, synkinematic granite; 6, migmatite complex; 7, non-metamorphic nappe; 8, metabauxite; 9, fault; 10, isograds; 11, tectonic contact. Numbers in circles indicate location of samples described in Figs 5, 7 and 8. Detailed map of NW-tip of the island is shown in Fig. 3.

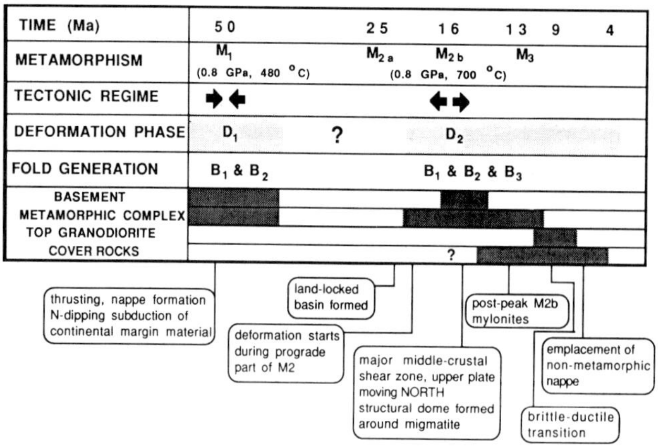

Fig. 2. Schematic diagram illustrating the relationships between deformation and metamorphism discussed in the text. Data on timing and conditions of metamorphism are mainly based on Jansen & Schuiling (1976) and Wijbrans & McDougall (1988). M3 is the contact metamorphism around the granodiorite, it is not discussed in the text. Note that time axis is not scaled. Data on tectonic evolution are from Robertson & Dixon (1984), Le Pichon & Angelier (1979), Lister *et al.* (1984) and Mercier *et al.* (1989). Note the two main deformation phases and their relation to the fold generations B1, B2 and B3. Lower part of the figure indicates periods of deformation of different units.

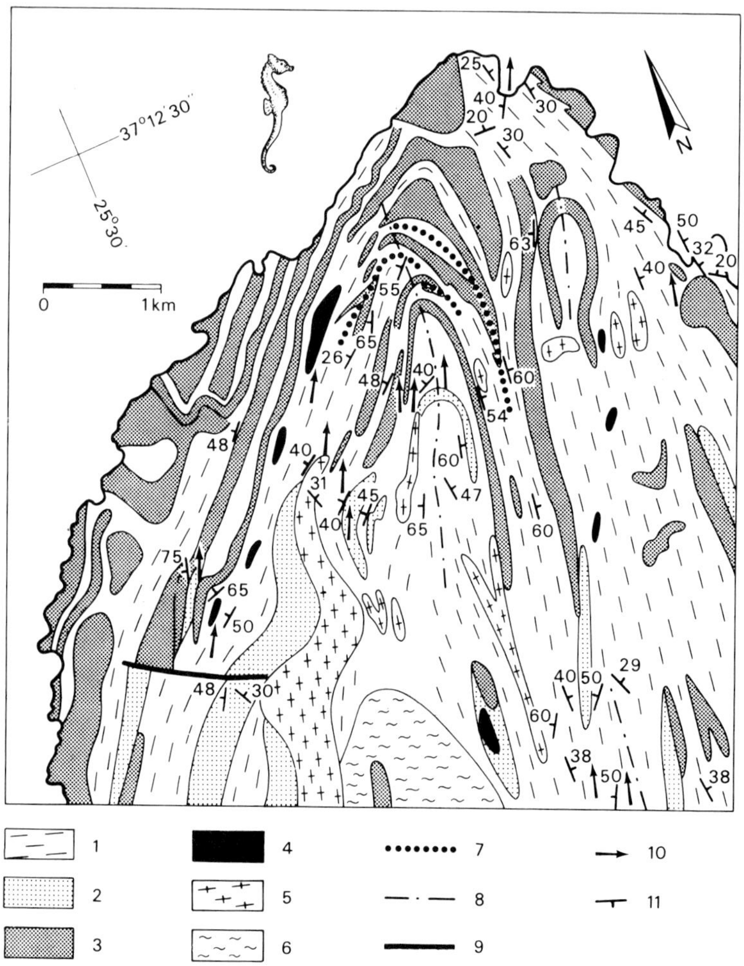

Fig. 3. Map of a part of NW Naxos, showing large scale B1 × B2 × B3 interference pattern. Legend: 1, micaschist; 2, amphibolite; 3, marble; 4, ultrabasite; 5, synkinematic granitoid; 6, migmatite; 7 B1 or B2 axial trace; 8, B3 axial trace; 9, fault; 10, locality with good sense of shear indicators for syn- and post-peak M2b movement, arrow indicates sense of motion of upper plate, azimuth shows trend of stretching lineation; 11, foliation.

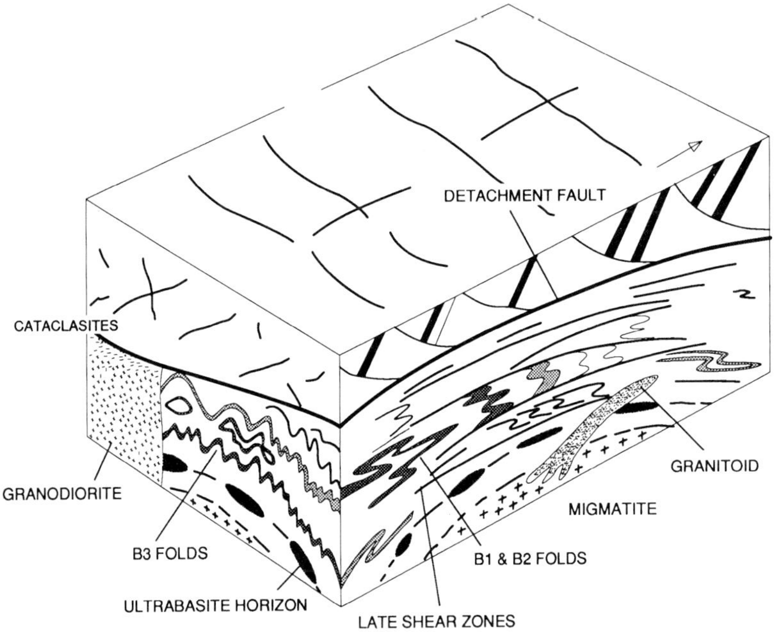

Fig. 4. Three dimensional drawing of the structures on the island, illustrating the structures discussed in this paper. Note that the non-metamorphic nappe is drawn much thicker than its present extent. Arrow in top right corner is pointing North.

in outcrop as well as on the regional scale, their fold axes are very nearly parallel to the stretching lineations associated with earlier fabrics (Fig. 6).

On a larger scale foliations and lineations show a slight warping over the migmatite dome, producing the elongated structural dome on the island.

Post-peak M_{2b} mylonites. Starting at the biotite-in isograd and increasing in intensity towards the migmatite, the effects of the M_{2b} metamorphism are clearly manifested in the field as a pronounced increase in grain-size in the schists and the marbles, together with the appearance of the high-temperature mineral assemblages described by Jansen (1977). Locally in this region the high grade fabric is overprinted in 10 cm to 1 m wide, localized zones of intense deformation. This is visible in the field as zones of extreme grainsize reduction in the marbles and pegmatites, associated with the development of a mylonitic foliation and lineation (Fig. 7). Mylonitic foliation is invariably parallel to the local orientation of bedding, or to the older, high grade schistosity, and is locally folded by B_3.

These mylonites are commonly found in a zone around the migmatite, and are somewhat more abundant in the west. Individual mylonites can rarely be followed along strike for more than 10 m.

It should be noted here that in SE Naxos, at lower grades of M_{2b} the rocks are all fine grained and strongly deformed. However, due to the lack of strong high grade M_{2b} overprint, in this region it is generally impossible to separate the effects of deformation events during M_1 and M_{2b}.

Structures outside the metamorphic complex

The main body of the *c*. 12 Ma granodiorite (Andriessen *et al.* 1979; Wijbrans & McDougall 1988) which intrudes the metamorphic complex in the W Naxos is undeformed. However, towards the contact with the non-metamorphic nappe it can be seen to be locally deformed and to develop a foliation. Ductile shear zones indicate the 'upper plate moving north' sense of shear, and microstructures indicate intermediate temperature deformation, with crystal plasticity in quartz, accompanied by incipient recrystallization, but brittle deformation in

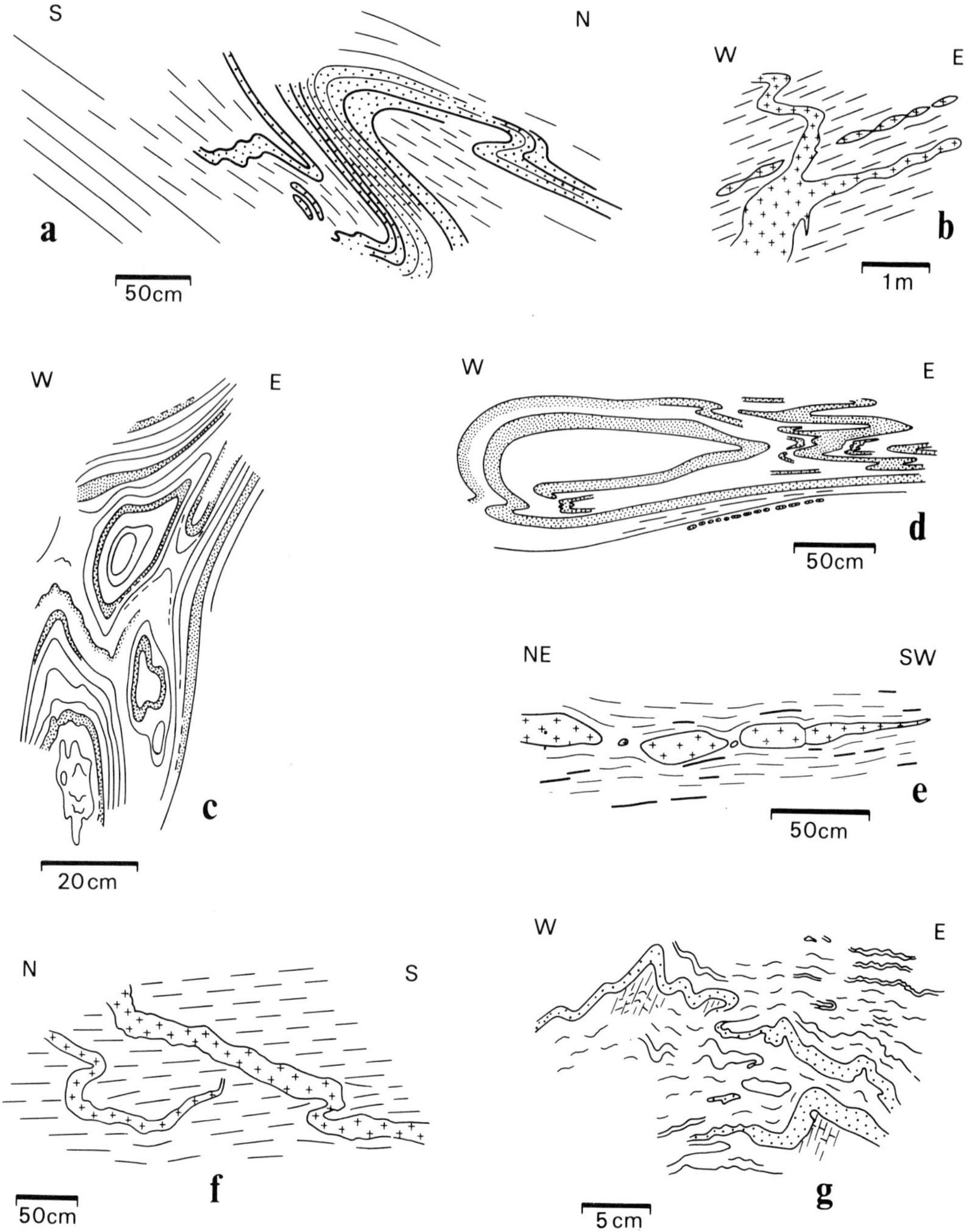

Fig. 5. Field drawings of small scale structures in the metamorphic complex. (**a**) B1 and B2 interference pattern in micaschist containing graphitic quartzite layers. Locality 7; (**b**) aplite (associated with the synkinematic granitoid) intruding an isoclinally folded sequence of micaschists and amphibolites, itself being deformed and boudinaged. Locality 6; (**c**) sheath folds in marble, locality 2; (**d**) complex B1 × B2 interference pattern in marble interbedded with micaschist, locality 3; (**e**) Boudinage in aplite intruding a micaschist sequence, locality as in (d); (**f**) gently folded (M2b) pegmatites in micaschist, Locality 3; (**g**) (B1 or B2) × B3 overprinting in micaschist containing quartzo-feldspathic layers, with local development of an S3 crenulation cleavage. E of locality 5.

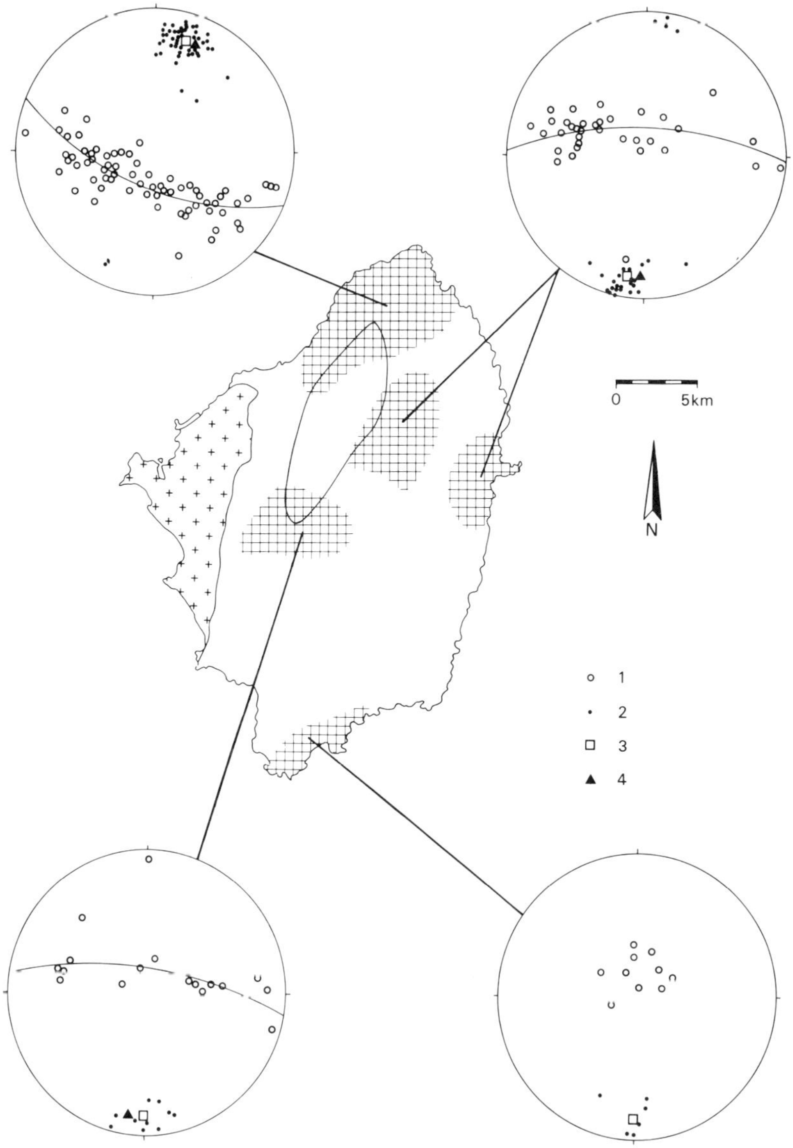

Fig. 6. Orientation data from selected areas (cross hatched) on the island, showing the parallelism of B3 folds axes with stretching lineations. Equal area projection. Legend: 1, pole to foliation plane; 2, lineation; 3, mean vector of lineations; 4, pole to best fit great circle to the foliations shown.

plagioclase, (Scholz 1988). These structures are in turn overprinted by brittle shear zones forming an irregular network of cataclasites and pseudotachylites (Fig. 7f). In these, field evidence (cf. Petit 1987) suggests northward transport of the upper plate.

The migmatite complex does not show signs of penetrative deformation after solidification. However in the marbles included in the migmatite, spectacular boudinage structures in thin amphibolite layers invariably imply a large component of subhorizontal N–S extension during M_{2b}. Also in these marbles, post-M_{2b} mylonites are not uncommon, indicating at least some late deformation inside the migmatite complex.

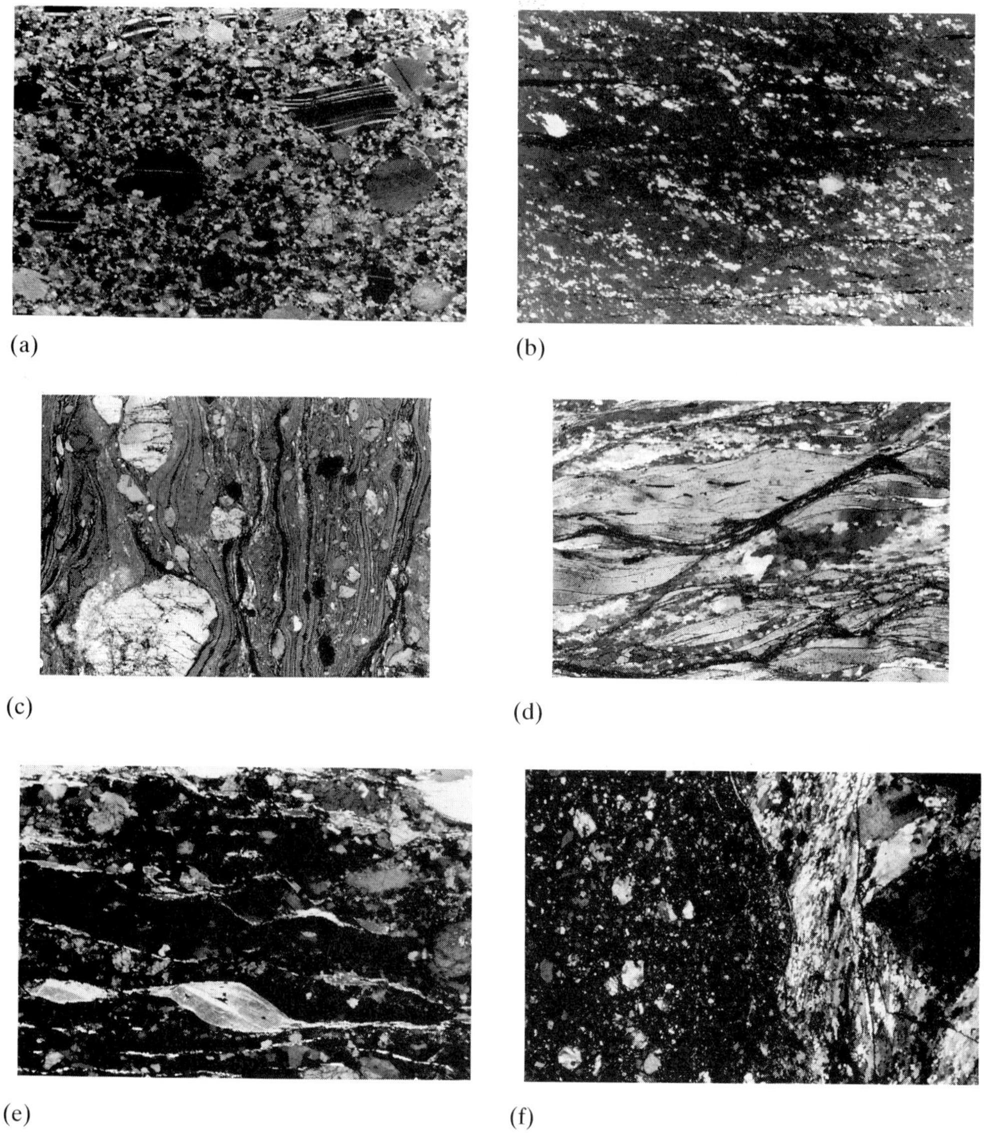

Fig. 7. Thin sections of late mylonites in the metamorphic complex and a cataclasite in the granodiorite. (**a**) Calcite mylonite, locality 3, long axis of photograph is 1.0 mm; (**b**) Mylonite in graphitic quartzite, with a well developed oblique foliation indicating left-lateral shear (Lister & Snoke 1984). Locality 7, long axis of photograph is 1.0 mm; (**c**) Extreme grainsize reduction in deformed M2b pegmatite, with brittle deformation in feldspar and tourmaline. Locality 6, long axis of photograph is 2.5 mm; (**d**) Shear bands and dynamic recrystallization in quartz–mica schist; indicating left-lateral shear. Locality 11, long axis of photograph is 2.5 mm; (**e**) Mica 'fish' structures (Lister & Snoke 1984), indicating left-lateral shear, locality 4, long axis of photograph is 4.5 mm; (**f**) Cataclasite from the granodiorite, showing plastically deformed quartz clasts embedded in the matrix. Locality 1, long axis of photograph is 4.5 mm.

Relationship between deformation and metamorphism

Field and microstructural investigations indicate a long and complex evolution of B_1 & B_2 structures.

(i) Most of the blueschists in the south of the island (Jansen & Schuiling 1976) show syntectonic microstructures, indicating deformation during M_1. On the other hand, some of the ultramafic bodies which were altered into incompetent, mica-rich aggregates during M_{2b}, (Jansen & Schuiling 1976) are located in an isoclinically folded sequence, but are not internally deformed after alteration, indicating that deformation (at least locally) was terminated before M_{2b}.

(ii) Metabauxites containing sedimentary pisoids (Feenstra 1985), show a drastic change in internal structure with increasing grade of M_{2b}. South of the corundum-in isograd the diasporites are intercalated in the surrounding, strongly deformed marbles as boudins and show only a weak internal deformation. North of this isograd, where the diasporites are transformed into emeries, the pisoids are strongly deformed and the emeries often develop a strong internal foliation and lineation.

We interpret this transition as evidence for deformation of the surrounding marbles during the prograde part of M_{2b}, when the diaspore-corundum transformation took place. The reason for this is as follows. During the diaspore-corundum transformation approximately 30% water-rich fluid is released, and this must have involved the development of locally lithostatic fluid pressures (Feenstra 1985). Although both diasporites and corundum-rich rocks can be safely assumed to be much stronger than marbles under the conditions involved (Cannon & Langdon 1983; Schmid *et al.* 1980), a fine-grained emery in which a lithostatic fluid pressure is maintained, can be considerably weaker than the marbles (Murrell 1985; Etheridge *et al.* 1984) and if the surrounding marbles are deforming, deformation can be temporarily repartitioned into the metabauxites. That the emeries have stopped deforming while temperatures were still high, is shown by the statically recrystallized microstructures in marble lenses completely enclosed in them.

(iii) Most of the high grade marbles (outside the emeries) have microstructures indicating high temperature dynamic recrystallization i.e. deformation during M_{2b}. These are characterized by lobate grain boundaries, large subgrains and orientation families; Urai *et al.*, (1986), see Fig. 7a. Our data are in disagreement with the interpretation of marble fabrics proposed by Covey-Crump & Rutter (1989) who interpreted the grainsize distributions on Naxos as a consequence of static annealing. However, none of the microstructures presented by these authors shows a true foam texture characteristic for static recrystallization, and most of their microstructures are overprinted by later deformation (our observations are in agreement with this) so that it is not possible to examine the original high temperature microstructure. Therefore we suggest that the grain size systematics observed by Covey-Crump & Rutter (1989) are more correctly interpreted as the product a dynamic grain growth episode.

(iv) Tremolite aggregates included in the marbles (Jansen & Schuiling 1976) are often undeformed, suggesting that deformation was terminated at these locations shortly after peak M_{2b} conditions (see also Fig. 8d).

(v) In NW Naxos a syntectonic (M_{2b}) granitoid body (Jansen & Schuiling 1976) is found associated with the migmatites. It intrudes isoclinally folded country rocks with a well developed foliation, and is itself deformed with a well developed foliation along the contact with the country rock. Microstructures indicate subsolidus deformation but still at relatively high temperature, involving crystal plasticity of both quartz and feldspar (Tullis & Yund 1980). Lineation and foliation in the granitoid are parallel to that in the surrounding rocks, and the foliation is folded by B_3. These structures in turn are cut by rare late ductile shear zones.

(vi) In thin sections of the post-peak-M_{2b} mylonites described above the peak-M_{2b} textures are clearly overprinted, with green or brown biotite commonly recrystallizing or growing at the expense of colourless mica in the micaschists, indicating temperatures in excess of 450°C.

(vii) Microstructures in B_3 folds indicate a complex history, with some B_3 folds being syn-M_{2b} with well equilibrated high grade microstructures (Fig. 8e), and others folding the late, lower grade mylonites discussed below. This implies that B_3 folds were forming during M_{2b} and the shape of the isograds can thus be expected to have been influenced by this. Unfortunately the accuracy of the isograd positions at present does not allow this to be unequivocally established.

Sense of shear syn- and post M_{2b}

On the outcrop scale, small-scale sense of shear indicators (Platt & Vissers 1980; Passchier 1986) are often found, indicating a large non-coaxial

(a)

(b)

(c)

(d)

(e)

Fig. 8. Micrographs of M2b — related structures in the high grade part of the metamorphic complex. Thin sections were cut parallel to the stretching lineation and perpendicular to the foliation. (**a**) Coarse-grained, dynamically recrystallized marble in the migmatite complex; locality 10, long axis of photograph is 4.5 mm; (**b**) Coarse-grained, dynamically recrystallized marble in the migmatite complex, with the microstructure overprinted by deformation and recrystallization after the peak of M2b; locality 2, long axis of photograph is 4.5 mm; (**c**) Graphitic quartzite with syntectonically grown ribbon grains. Although grain shape preferred orientation is lacking, quartz *c*-axis fabrics are clearly asymmetric, indicating 'upper plate moving North' sense of shear (see Krabbendam & Urai 1989). Locality 9, long axis of photograph is 4.5 mm; (**d**) Kyanite porphyroblast, being replaced by colourless mica, indicating that deformation has ended in this locality before the transformation. Locality 8, long axis of photograph is 4.5 mm; (**e**) B3 microfold in high grade biotite schist showing well equilibrated, polygonized microstructures typical of M2b in biotite. Thin section was cut perpendicular to the lineation and B3 fold axis, locality 8, long axis of photograph is 4.5 mm.

component of flow during the formation of B_1 and B_2 structures. Thin section studies yield similar results, with asymmetric quartz *c*-axis fabrics in quartzites and frequent shear band cleavages in micaschists. On both outcrop and thin section scale, *c*. 90% of these structures indicates an 'upper plate moving North' sense of shear, in both syn- M_{2b} and post-peak M_{2b} structures. Some of these observations are shown in Figs 6 and 7.

Discussion

The field and microstructural data presented above show how a complicated tectono-metamorphic history involving several major events can produce a relatively simple structural pattern, by reactivation and transposition of older structures. In the case of Naxos, it is only because of the rich lithological variation and steep metamorphic gradients during M_{2b}, that the deformation history can be deciphered, if only incompletely. Important points arising from this study are given below.

(1) Direct imprint of the M_1 high-pressure–low-temperature event has only been preserved in the south of Naxos. In other areas the rocks have suffered a sufficiently strong overprint to erase relict M_1 assemblages, if present. The M_1 event can be correlated with similar well documented structures on the island of Syros (Ridley 1982, 1984, 1986). The high grade metamorphosed bauxites, and the two ultrabasic horizons on the island, together with the pre-Alpine ages found in the migmatite all agree with a strong compressional event during M_1 (D_2 in Fig. 2), associated with subduction of continental margin material (Feenstra 1985; Jansen 1977; Wijbrans & McDougall 1988), and the formation of a nappe pile. Geological, (Robertson and Dixon 1984), and palaeomagnetic evidence (Kissel and Laj, 1988) support the interpretation of an approximately N-dipping subduction zone, with dominantly 'upper plate moving south' sense of transport (cf. Ridley 1986). At least part of the B_1 & B_2 structures on Naxos were certainly formed in this period, and small scale structures may not have been very different from those seen at present, except the asymmetry of sense of shear indicators.

The data presented in this study provide strong evidence for a second major deformation phase (D_2 in Fig. 2) which was already active during the prograde part of M_{2b}. This event has involved most units in the metamorphic complex inside the biotite isograd. In locations where this could be determined the strains involved where high, although orientations of B_1 and B_2 fold axes after D_1 are poorly known because of the strong later D_2 overprint. The complex relationships between deformation and metamorphism are best interpreted as the products of an essentially continuous movement during M_{2b}, but with frequent local changes in strain rate as a result of changes in mechanical properties during progressive metamorphism. The frequent indicators of noncoaxiality point to a large component of simple shear in most parts of the complex. Sense of shear is consistently 'upper plate moving north' during D_2.

With decreasing temperatures after the peak of M_{2b}, deformation demonstrably continued around the migmatite dome, with deformation being localized (Lister & Williams 1983) in narrow ductile shear zones between kinematically inactive lenses. These shear zones are always parallel to the local preexisting foliation and stretching lineations are exactly parallel to older ones. With the sense of shear also being the same, it is reasonable to interpret these structures as being the product of an essentially continuous movement. Fig 7 is a 3D sketch showing the large scale structure of the island.

(2) The strong deformation in the shear zones around the migmatite after the peak of M_{2b} has implications for the interpretation of geothermal gradients during M_{2b}. As the shear plane is more or less parallel to the isograd pattern (except the biotite-in isograd), there is a possibility of a narrowing of the isograds by tectonic telescoping. In addition, the E–W shortening responsible for the formation of B_3 structures has probably produced the present oval shape of the isograd pattern.

(3) The ('Cyclades') shear zone was active for a relatively long period (between *c*. 20 and 9 Ma), under conditions of declining temperature. We interpret its initiation to be correlated with the onset of crustal extension in the area, in agreement with results of Meulenkamp *et al.* (1987) based on tomographic estimates of the length of the subducted slab.

The last movements that can be documented on Naxos are the deformation of the boundaries of the granodiorite at brittle–ductile transition conditions, and the associated emplacement of the non-metamorphic nappe (*c*. 9.5 Ma). This was followed by normal faulting, and the arrival of Naxos at the surface at around 4 Ma.

(4) An interesting structural feature on Naxos is the parallelism of B_3 fold axes and B_1 fold axes and lineations. This is interpreted as a result of a deviation from plane strain deformation on the regional scale, due to an extra component of E–W horizontal shortening during the movement of the shear zone (C. W.

Passchier, pers. comm). Somewhat similar structures can also be formed in zones with a high component of wrench shear (cf. Ridley 1986), but in this case there is always a finite angle between the fold axes and stretching lineations. Our interpretation agrees well with B_3 folds being formed in a period overlapping with the formation of B_1 and B_2 structures, during D_2. Evidence for E–W shortening in the upper crust during the period from 13 Ma was given by Le Pichon & Angelier (1979).

(5) Our results agree with an interpretation of Naxos as a Cordilleran-type core complex (Lister *et al.* 1984). The major difference between our results is the sense of motion in the Naxos shear zone during D_2, which is opposite to that proposed by Lister *et al.* (1984). Note that our interpretations do not contradict as Lister *et al.* (1984) did not provide sense of shear data for Naxos. In a recent paper Faure & Bonneau (1988) demonstrated 'upper plate moving North' sense of shear on Mykonos, in agreement with results in this work. If this sense of motion can be shown to be present in other islands of the Cyclades, it has major implications for the large scale tectonic movement picture in the last 20 Ma.

The first author wishes to express his gratitude to professor H. J. Zwart for stimulating his interest in rock deformation and his guidance and support.

We thank R. Kreulen and J. Hecht for useful discussions on Naxos geology, and A. Garcia-Celma, C. Passchier and G. Lister for valuable improvements to the manuscript.

Financial support to J. L. U. by a C&C Huygens Fellowship of NWO is gratefully acknowledged.

References

ALTHERR, R., KREUZER, H., WENDT, I., LENZ, H., WAGNER, G. A., KELLER, J., HARRE, W. & HOHNDORF, A. 1982. A late Oligocene/early Miocene high temperature belt in the Attic-Cycladic crystalline complex (SE Pelagonian, Greece). *Geologische Jahrbuch*, **E 23**, 97–164.

ANDRIESSEN, P. A. M., BANGA, G. & HEBEDA, E. H., 1987. Isotopic age study of pre-Alpine rocks in the basal units of Naxos, Sikinos and Ios, Greek Cyclades. *Geologie en Mijnbouw*, **66**, 3–14,

—— BOELRIJK, N. A. I. M., HEBEDA, E. H., PRIEM, H. N. A., VERDURMEN, E. A. TH. & VERSCHURE, R. H. 1979. Dating the Events of Metamorphism and Granitic Magmatism in the Alpine Orogen of Naxos (Cyclades, Greece). *Contributions to Mineralogy and Petrology*, **69**, 215–225.

BAKER, J., BICKLE, M. J., BUICK, I. S., HOLLAND, T. J. B., & MATTHEWS, A. 1989. Isotopic and petrological evidence for the infiltration of water-rich fluids during the Miocene M2 metamorphism on Naxos, Greece. *Geochimica et Cosmochimica Acta*, **53**, 2037–2050.

BONNEAU, M. GEYSSANT, J., & LEPEVRIER, C., 1978. Tectonique Alpine dans le massif d'Attique-Cyclade (Grece): plis couches kilometriques dans l'ile de Naxos, consequences. Revue de Geographie Physique et de Geologie Dynamique, **10**, 109–122.,

CANNON, W. R. & LANGDON, T. G. 1983. Creep of Ceramics. *Journal of Materials Science*, **18**, 1–50.

COVEY-CRUMP, S. J. & RUTTER, E. H. 1989. Thermally induced grain growth of calcite marbles on Naxos Island, Greece, *Contributions to Mineralogy and Petrology*, **101**, 69–86.

CRITTENDEN, M. D., Jr., CONEY, P. J. & DAVIS, G. H. 1980. *Cordilleran metamorphic core complexes*. Geological Society of America Memoir, **153**.

DÜRR, S., ALTHERR, R., KELLER, J., OKURSCH, M. & SEIDEL, E. 1978. The Median Aegean Crystalline belt: Stratigraphy, Structure, Metamorphism, Magmatism. *In*: CLOSS, H., ROEDER, D. & SCHMIDT, K. (eds) *Alps, Appenines, Hellenides*. Inter-Union Commission on Geodynamics Scientific Report No. **38**, 455–477.

ETHERIDGE, M. A., WALL, V. J. & COX, S. F. 1984. High fluid pressures during regional metamorphism and deformation: implications for mass transport and deformation mechanisms. *Journal of Geophysical Research*, **89**, **B6**, 4344–4358.

FAURE, M. & BONNEAU, M. 1988. Données nouvelles sur l'extension néogéne de l'Égée: la déformation ductile du granite miocéne de Mykonos (Cyclades, Gréce). *Comptes Rendus de Seances de l'Academie des Sciences, Paris*, **307(11)**, 1553–1559.

FEENSTRA, A. 1985. Metamorphism of bauxites on Naxos, Greece. Geologica Ultraiectina, **39**, 1–206.

GENTHON, P. & SOURIAU, M., 1987. New constraints on Aegean subduction from seasat altimetry, *Journal of Geophysical Research*, **92**, **B1**, 619–632.

HECHT, J. 1979. *Geologische karte des Schmirgel führenden Gebietes zwischen Apiranthos und Koronos, Naxos-Griechenland; 1:10000*. Institute of Geology and Mineral Resources, Athens.

HENJES-KUNST, F. & KREUZER, H. 1982. Isotopic Dating of pre-Alpidic rocks from the island of Ios (Cyclades, Greece). *Contributions to Mineralogy and Petrology*, **80**, 245–253.

HOBBS, B. E., MEANS, W. D. & WILLIAMS, P. F. 1976. *An outline of structural geology*. Wiley.

HORVATH, F. & BERCKHEMER, H. 1982. Mediterranean backarc basins. *In*: BERCKHEMER, H. & HSÜ, K. (eds) *Alpine–Mediterranean geodynamics*. AGU Geodynamics Series, **7**, 141–173.

JANSEN, J. B. H. 1977. *Metamorphism on Naxos, Greece*. PhD Thesis, University of Utrecht.

—— & SCHUILING, R. D. 1976. Metamorphism on Naxos: Petrology and Geothermal gradients. *American Journal of Science*, **276**, 1225–1253.

——, VAN DE KRAATS, A. H., VAN DER RIJST, H. &

SCHUILING, R. D. 1978. Metamorphism of siliceous dolomites at Naxos, Greece. *Contributions to Mineralogy Petrology*, **67**, 279–288.

KISSEL, & LAJ, C. 1988. The Tertiary geodynamical evolution of the Aegean arc; a paleomagnetic reconstruction. *Tectonophysics*, **146**, 183–201.

KRABBENDAM, M. & URAI, J. L. 1989. Deformation mechanism switch due to grainsize stabilization by a second phase: an example from Naxos (Greece). *Terra Abstracts* **1(1)**, 379.

KREULEN, R 1980. CO_2-rich fluids during regional metamorphism on Naxos (Greece): Carbon isotopes and fluid inclusions. *American Journal of Science*, **280**, 745–771.

LE PICHON, X. 1983. Land-locked Oceanic Basins and Continental Collision: the Eastern Mediterranean as a Case Example. *In*: HSÜ, K. J. (ed.) *Mountain Building Processes*, Academic Press, 201–213.

—— & ANGELIER, J., 1979. The Hellenic arc and trench system: A key to the neotectonic evolution of the eastern Mediterranean area. *Tectonophysics*, **60**, 1–42.

LISTER, G. S., & SNOKE, A. W., 1984. S-C Mylonites. *Journal of Structural Geology*, **6**, 617–638.

—— & WILLIAMS, P. F. 1983. The partitioning of deformation in flowing rock masses, *Tectonophysics*, **92**, 1–33,

——, BANGA, G. & FEENSTRA, A. 1984. Metamorphic core complexes of Cordilleran type, in the Cyclades, Aegean sea, Greece, *Geology*, **12**, 221–225

——, ETHERIDGE, M. A. & SYMONDS, P. A., 1986. Detachment Faulting and the evolution of passive continental margins. *Geology*, **14**, 246–250.

MACKENZIE, D. 1978. Active tectonics of the Alpine-Himalayan belt: the Aegean sea and surrounding regions. *Geophysical Journal Royal of the Astronomical Society*, **55**, 217–254.

MAKRIS, J. 1978. Some geophysical constraints on the geodynamic situation in Greece. *Tectonophysics*, **46**, 251–268.

MERCIER, J. L., SOREL, D. & VERGELY, P. 1989. Extensional Tectonic regimes in the Aegean basins during the Cenozoic. *Basin Research*, **2**, 49–71.

MEULENKAMP, J. E., WORTEL, M. J. R., VAN WAMEL, W. A., SPAKMAN, W. & HOOGERDUYN. STRAATING, E. 1987. On the hellenic subduction zone and the geodynamic evolution of Crete since the late Miocene, *Tectonophysics*, **146**, 203–216.

MURRELL, S. A. F. 1985. Aspects of relationships between deformation and prograde metamorphism that causes evolution of water. *In*: THOMPSOM, A. B. & RUBIE, D. C. (eds) *Fluid–rock interactions during metamorphism*. Advances in Physical Geochemistry, **4**, 211–241.

PASSCHIER, C. W. 1986. Mylonites in the continental crust and their role as seismic reflectors. *Geologie en Mijnbouw*, **65**, 167–176.

PETIT, J. P. 1987. Criteria for the sense of movement on fault surfaces in brittle rocks. *Journal of Structural Geology*, **9**, 597–608.

PLATT, J. P. & VISSERS, R. L. M 1980. Extensional structures in anisotropic rocks. *Journal of Structural Geology*, **2**, 397–410.

RIDLEY, J. 1982. Arcuate lineation trends in a deep level ductile thrust belt, Syros, Greece. *Tectonophysics*, **83**, 347–360.

—— 1984. The significance of deformation associated with blueschist facies metamorphism on the Aegean island of Syros. *In*: DIXON, J. E. & ROBERTSON, A. H. F. (eds) *The Geological Evolution of the Eastern Mediterranean*. Geological Society, London, Special Publication, **17**, 545–550.

—— 1986. Parallel stretching lineations and fold axes oblique to a shear displacement direction-a model and observations, *Journal of Structural Geology*, **8**, 647–653.

ROBERTSON, A. H. F. & DIXON, J. E. 1984. Introduction: aspects of the geological evolution of the Eastern Mediterranean. *In:* DIXON, J. E. & ROBERTSON, A. H. F. (eds) *The Geological Evolution of the Eastern Mediterranean*. Geological Society, London, Special Publication, **17**, 1–74.

RYE, R. O., SCHUILING, R. D., RYE, D. M., & JANSEN, J. B. H., 1976. Carbon, Hydrogen and Oxygen Isotope studies of the regional metamorphic complex at Naxos, Greece. *Geochimica et Cosmochimica Acta*, **40**, 1031–1049.

SCHMID, S. M., PATERSON, M. S. & BOLAND, J. N. 1980. High temperature creep and dynamic recrystallization of Carrara Marble. *Tectonophysics*, **65**, 245–280.

SCHOLZ, C. 1987. The brittle-plastic transition and the depth of seismic faulting. *Geologische Rundschau*, **77**, 319–328.

SMITH, A. G. & WOODCOCK, N. H. 1982. Tectonic syntheses of the Alpine-Mediterranean region: a review. *In*: BERCKHEMER, H. & HSÜ, K. (eds) *Alpine–Mediterranean geodynamics*. AGU Geodynamics Series, **7**, 15–38.

SPAKMAN, W. 1986. Subduction beneath Eurasia in connection with the Mesosoic Tethys, *Geologie en Mijnbouw*, **65**, 145–153.

TRIKKALINOS, T. K. 1947. Beiträge zur Erforschung des Tektonischen Baues Griechenlands. II Über den Tektonischen Bau der Insel Naxos. *Annales Géologiques des Pays Helléniques*, **1**, 7–37.

TULLIS, J. & YUND, R. 1980. Hydrolitic weakening of experimentally deformed Westerly granite and Hale albite rock. *Journal of Structural Geology*, **2**, 439–451.

URAI, J. L., MEANS, W. D. & Lister, G. S. 1986. Dynamic recrystallization of minerals. *AGU Geophysical Monograph*, **36**, 161–199.

VAN DER MAAR, P. A & JANSEN, J. B. H. 1983. The geology of the polymetamorphic complex of Ios, Cyclades, Greece, and its significance for the Cycladic Massif. *Geologisches Rundschau*, **72**, 283–299.

WIJBRANS, J. R. & MCDOUGALL, I., 1986. $^{40}Ar/^{39}Ar$ dating of white micas from an alpine high pressure metamorphic belt on Naxos (Greece): the resetting of the argon isotopic system. *Contributions to Mineralogy and Petrology*, **93**, 187–194.

—— & ——, 1988. Metamorphic evolution of the Attic Cycladic Metamorphic Belt on Naxos (Cyclades, Greece) utilizing $^{40}Ar/^{39}Ar$ age spectrum measurements. *Journal of Metamorphic Geology*, **6**, 571–594.

WILLIAMS, P. F. 1985. Multiply deformed terrains — problems of correlation. *Journal of Structural Geology*, **7**. 269–280.

Index

References to tables appear in **bold**, those to figures in *italics*